Grundlehren der mathematischen Wissenschaften 115

A Series of Comprehensive Studies in Mathematics

Grundlehren der mathematischen Wissenschaften

A Series of Comprehensive Studies in Mathematics

2. Knopp: Theorie und Anwendung der unendlichen Reihe
3. Hurwitz: Vorlesungen über allgemeine Funktionentheorie und elliptische Funktionen
4. Madelung: Die mathematischen Hilfsmittel des Physikers
10. Schouten: Ricci-Calculus
14. Klein: Elementarmathematik vom höheren Standpunkte aus. 1. Band: Arithmetik, Algebra, Analysis
15. Klein: Elementarmathematik vom höheren Standpunkte aus. 2. Band: Geometrie
16. Klein: Elementarmathematik vom höheren Standpunkte aus. 3. Band: Präzisions- und Approximationsmathematik
20. Pólya/Szegö: Aufgaben und Lehrsätze aus der Analysis II: Funktionentheorie, Nullstellen, Polynome, Determinauten, Zahlentheorie
22. Klein: Vorlesungen über höhere Geometrie
26. Klein: Vorlesungen über nicht-euklidische Geometrie
27. Hilbert/Ackermann: Grundzüge der theoretischen Logik
30. Lichtenstein: Grundlagen der Hydromechanik
31. Kellogg: Foundations of Potential Theory
32. Reidemeister: Vorlesungen über Grundlagen der Geometrie
38. Neumann: Mathematische Grundlagen der Quantenmechanik
40. Hilbert/Bernays: Grundlagen der Mathematik I
43. Neugebauer: Vorlesungen über Geschichte der antiken mathematischen Wissenschaften. Band 1: Vorgriechische Mathematik
50. Hilbert/Bernays: Grundlagen der Mathematik II
52. Magnus/Oberhettinger/Soni: Formulas and Theorems for the Special Functions of Mathematical Physics
57. Hamel: Theoretische Mechanik
58. Blaschke/Reichardt: Einführung in die Differentialgeometrie
59. Hasse: Vorlesungen über Zahlentheorie
60. Collatz: The Numerical Treatment of Differential Equations
61. Maak: Fastperiodische Funktionen
62. Sauer: Anfangswertprobleme bei partiellen Differentialgleichungen
64. Nevanlinna: Uniformisierung
66. Bieberbach: Theorie der gewöhnlichen Differentialgleichungen
68. Aumann: Reelle Funktionen
69. Schmidt: Mathematische Gesetze der Logik I
71. Meixner/Schäfke: Mathieusche Funktionen und Sphäroidfunktionen mit Anwendungen auf physikalische und technische Probleme
73. Hermes: Einführung in die Verbandstheorie
74. Boerner: Darstellungen von Gruppen
75. Rado/Reichelderfer: Continuous Transformations in Analysis, with an Introduction to Algebraic Topology
76. Tricomi: Vorlesungen über Orthogonalreihen
77. Behnke/Sommer: Theorie der analytischen Funktionen einer komplexen Veränderlichen
78. Lorenzen: Einführung in die operative Logik und Mathematik
80. Pickert: Projektive Ebenen

continued after index

Edwin Hewitt Kenneth A. Ross

Abstract Harmonic Analysis

Volume I

Structure of Topological Groups

Integration Theory Group Representations

Second Edition

Springer-Verlag

New York Berlin Heidelberg London Paris

Tokyo Hong Kong Barcelona Budapest

Edwin Hewitt
Department of Mathematics GN-50
University of Washington
Seattle, WA 98195, USA

Kenneth A. Ross
Department of Mathematics
University of Oregon
Eugene, OR 97403, USA

Mathematics Subject Classification (1991): 43A70, 22B05, 22A10, 43A15

Library of Congress Cataloging in Publication Data. Hewitt, Edwin, 1920–. Abstract harmonic analysis. (Grundlehren der mathematischen Wissenschaften; 115). Bibliography: p. Includes indexes.
ISBN 0-387-94190-8
CONTENTS: v. 1. Structure of topological groups, integration theory, group representations. 1. Harmonic analysis. I. Ross, Kenneth A., joint author. II. Title. III. Series: Die Grundlehren der mathematischen Wissenschaften in Einzeldarstellungen; 115.
QA403.H4. 1979. 515′.2433. 79-13097

Printed on acid-free paper.

Production managed by Francine McNeill; manufacturing supervised by Vincent Scelta.
Typeset by Universitätsdruckerei H. Stürtz AG, Würzburg.
Printed and bound by Edwards Brothers, Inc., Ann Arbor, MI.
Printed in the United States of America.

9 8 7 6 5 4 3 2 1

ISBN 0-387-94190-8 Springer-Verlag New York Berlin Heidelberg (softcover)
ISBN 3-540-94190-8 Springer-Verlag Berlin Heidelberg New York (softcover)
ISBN 3-540-09434-2 Springer-Verlag Berlin Heidelberg New York (hardcover)
ISBN 0-387-09434-2 Springer-Verlag New York Berlin Heidelberg (hardcover)

Preface to the Second Edition

It has not been possible to rewrite the entire book for this Second Edition. It would have been gratifying to resurvey the theory of topological groups in the light of progress made in the period 1962—1978, to amplify some sections and curtail others, and in general to profit from our experience since the book was published. Market conditions and other commitments incurred by the authors have dictated otherwise. We have nonetheless taken advantage of the kindness of Springer-Verlag to make a number of improvements in the text and of course to correct misprints and mathematical blunders.

We are in debt to the readers who have written to us or spoken with us about the text, and we have tried to follow their suggestions. We are happy here to record our gratitude to Robert B. Burckel, W. Wistar Comfort, Robert E. Edwards, Robert E. Jamison, Jorge M. López, Theodore W. Palmer, Willard A. Parker, Karl R. Stromberg, and Fred Thoele, as well as to a host of others who have kindly made suggestions to us.

Our thanks are due as well to Springer-Verlag for their support of our work.

Seattle, Washington
Eugene, Oregon

Edwin Hewitt
Kenneth A. Ross

January 1979

Preface to the First Edition

When we accepted the kind invitation of Prof. Dr. F. K. SCHMIDT to write a monograph on abstract harmonic analysis for the *Grundlehren der Mathematischen Wissenschaften* series, we intended to write all that we could find out about the subject in a text of about 600 printed pages. We intended that our book should be accessible to beginners, and we hoped to make it useful to specialists as well. These aims proved to be mutually inconsistent. Hence the present volume comprises only half of the projected work. It gives all of the structure of topological groups needed for harmonic analysis as it is known to us; it treats integration on locally compact groups in detail; it contains an introduction to the theory of group representations. In the second volume we will treat harmonic analysis on compact groups and locally compact Abelian groups, in considerable detail.

The book is based on courses given by E. HEWITT at the University of Washington and the University of Uppsala, although naturally the material of these courses has been enormously expanded to meet the needs of a formal monograph. Like the other treatments of harmonic analysis that have appeared since 1940, the book is a lineal descendant of A. WEIL's fundamental treatise (WEIL [4])[1]. The debt of all workers in the field to WEIL's work is well known and enormous. We have also borrowed freely from LOOMIS's treatment of the subject (LOOMIS [2]), from NAĬMARK [1], and most especially from PONTRYAGIN [7]. In our exposition of the structure of locally compact Abelian groups and of the PONTRYAGIN-VAN KAMPEN duality theorem, we have been strongly influenced by PONTRYAGIN's treatment. We hope to have justified the writing of yet another treatise on abstract harmonic analysis by taking up recent work, by writing out the details of every important construction and theorem, and by including a large number of concrete examples and facts not available in other textbooks.

The book is intended to be readable by students who have had basic graduate courses in real analysis, set-theoretic topology, and algebra as given in United States universities at the present day. That is, we suppose that the reader knows elementary set theory, set-theoretic topology, measure theory, and algebra. Our ground rule is [although this is

---

[1] Here and throughout the book, numbers in square brackets designate works in the bibliography found at the end of the volume. These are arranged and numbered by authors.

not hard and fast] that facts and concepts from KELLEY [2], HALMOS [2], and VAN DER WAERDEN [1] may be used without explanation or proof.

In our effort to make the book useful for specialists, we have included a large corpus of material which a beginning reader may find it wise to omit. The following suggestions are for the beginning student who wishes to get to the root of the matter as quickly as possible. First read §§1—7, Subheads (8.1)—(8.7), (9.1)—(9.14), and all of §10. The reader who is already familiar with or wishes to take on faith the theory of integration on locally compact Hausdorff spaces may skip §§11—14. Section 15 is absolutely essential and should be read with careful attention. Sections 16—18 may be omitted. Sections 19 through 24 are vital, and should be read carefully. Sections 25 and 26 are rather special, but are so interesting that we hope all readers will work through them.

Sections 4—11 and 15—26 contain subsections entitled Miscellaneous Theorems and Examples. Some of these are worked out in detail; for others, proofs are sketched or omitted entirely. We refer *occasionally* in the main text to a result drawn from the Miscellaneous Theorems and Examples. All such results are easy and are supplied with proofs. The reader is counselled at least to read the statements of the Miscellaneous Theorems and Examples, and to use them as exercises *ad libitum.*

Many sections also contain historical notes. We have tried to trace the history of the principal theorems and concepts, but we obviously have not produced a complete history, and it would be foolish to claim that we have produced one correct in every detail. Also, while some of the results we give are new so far as we know, *failure to cite a reference for a given theorem should not be construed as a claim of originality on our part.*

For the reader's convenience, we have assembled in three appendices certain ancillary material not easily accessible elsewhere, which is essential for one part or another of the theory. These appendices may be read as the reader encounters references to them in the main text.

We have been assisted by many colleagues. W. WISTAR COMFORT, JOHN M. ERDMAN, ALBERT J. FRODERBERG, L. FUCHS, JAMES MICHELOW, RICHARD S. PIERCE, KARL R. STROMBERG, ELMAR THOMA, VEERAVALLI S. VARADARAJAN, and HERBERT S. ZUCKERMAN have all read various parts of the manuscript, made valuable improvements, and saved us on occasion from grievous error. Improvements have also been made by NORMAN J. BLOCH, FRANCIS X. CONNOLLY, GERALD L. ITZKOWITZ, and RICHARD T. SHANNON. For the help so generously given by our friends we are sincerely grateful. Our thanks are due to Mlles. LYNNE HARPER and JEANNE SLOPER for preparation of the typescript. Mr. ALBERT J. FRODERBERG has most generously assisted us in the task of proofreading.

The reading and research by both of us on which the book is based have been generously supported by the National Science Foundation, U.S.A., and by a fellowship of the John Simon Guggenheim Memorial Foundation granted to E. HEWITT. We extend our sincere thanks to the authorities of these foundations, without whose aid our work could not have been done.

Finally, we record our gratitude to Prof. Dr. F. K. SCHMIDT for his original suggestion that the book be written, and to the publishers for their dispatch, skill, and attention to our every wish in producing the book.

Seattle, Washington, EDWIN HEWITT
Rochester, New York KENNETH A. ROSS

October 1962

Table of Contents

Chapter One

Preliminaries

Abstract harmonic analysis has evolved over the past few decades on the basis of several theories. First we have the classical theory of Fourier series and integrals, set forth in many treatises, such as ZYGMUND [1] and BOCHNER [1]. Second, we have the algebraic theory of groups and their representations, which is also described in many standard texts (*e.g.* VAN DER WAERDEN [1]). Third, we have the theory of topological spaces, by now a fundamental tool of analysis, and also the subject of standard texts (*e.g.* KELLEY [2]). The latter two subjects were combined to form the notion of a *topological group*. This is an entity which is both a group and a topological space and in which the group operations and the topology are appropriately connected. The structure of topological groups was extensively studied in the years 1925—1940; and the subject is far from dead even today.

Using a fundamental construction published in 1933 by A. HAAR [3], A. WEIL [4] in 1940 showed that Fourier series and integrals are but special cases of a construct which can be produced on a very wide class of topological groups. Furthermore, several classical theorems about Fourier series and integrals [PARSEVAL's equality, PLANCHEREL's theorem, the HERGLOTZ-BOCHNER theorem, and the HAUSDORFF-YOUNG inequality, to name a few] could be meaningfully stated, and proved, in this general situation. The end of the development initiated by WEIL is nowhere in sight. Problems in one branch or another of harmonic analysis are occupying the attention of many mathematicians at the present day, and we cannot predict at all where future developments will lead.

At any rate, the foundations of abstract harmonic analysis now seem to be clear. The present volume is devoted to a study of these foundations. Our first task is to learn thoroughly what a topological group is. We naturally need an array of facts about groups and topological spaces *per se*, which are set down in the present chapter.

§ 1. Notation and terminology

In this section, we explain some of the notation and terminology used throughout the text. Standard concepts and notation are used without explanation.

The symbols $\subset$ and $\supset$ mean ordinary inclusion between sets; they do not exclude the possibility of equality. The void set is denoted by $\varnothing$. Frequently the sets we deal with are subsets of some universal set, say E. In this case we denote by A' the *complement* of the set A:

$$A' = \{x : x \in E \text{ and } x \notin A\}.$$

In every case it will be clear what the set E is. For sets A and B, the *symmetric difference* $A \triangle B$ is defined to be the set $(A \cap B') \cup (A' \cap B)$.

A family $\mathscr{A}$ of sets has the *finite intersection property* if $\bigcap\{A : A \in \mathscr{F}\} \neq \varnothing$ for all finite subfamilies $\mathscr{F}$ of $\mathscr{A}$. In particular, no set in $\mathscr{A}$ is void if $\mathscr{A}$ has the finite intersection property. A collection $\{A_\iota\}_{\iota \in I}$ of sets is said to *partition* a set X if $\bigcup_{\iota \in I} A_\iota = X$, each A_ι is nonvoid, and the sets A_ι are pairwise disjoint. A family $\{A_\iota\}_{\iota \in I}$ of sets is a *covering* or *cover* of X if $\bigcup_{\iota \in I} A_\iota \supset X$. Given a set X, $\mathscr{P}(X)$ denotes the family of all subsets of X.

Knowledge of elementary cardinal and ordinal arithmetic is assumed. The family of *countable* sets includes finite sets and the void set. Infinite countable sets have cardinal number $\aleph_0$ and the real line has cardinal number $2^{\aleph_0}$. We will write $\mathfrak{c}$ for $2^{\aleph_0}$. The cardinal number of an arbitrary set A is denoted by $\overline{\overline{A}}$.

A subset Y of a partially ordered set X is said to be *cofinal* in X if for every $x \in X$, there is a $y \in Y$ such that $x \leq y$.

The terms "mapping", "transformation", and "correspondence" are synonymous with "function". The term "operator" has a special meaning, which is explained in (B.2). A function f will often be defined by an expression

$$x \to f(x)$$

where x denotes a generic element of the domain of the function and $f(x)$ denotes its image under f. For a function f and a subset A of its domain, $f|A$ denotes the *restriction* of f to A.

If f is a function on X into Y and g is a function on Y into Z, then the *composition* of g by f is the function $g \circ f$ on X into Z defined by $g \circ f(x) = g(f(x))$ for $x \in X$.

Let X be a set and A any subset of X. 'The symbol ξ_A will denote the function defined on X such that

$$\xi_A(x) = \begin{cases} 1 & \text{for } x \in A, \\ 0 & \text{for } x \in A'. \end{cases}$$

The function ξ_A is called the *characteristic function* of A.

Let $\{X_\iota : \iota \in I\}$ be a nonvoid family of sets. We define $\mathsf{P}_{\iota \in I} X_\iota$ to be the set of all functions x from I into $\bigcup_{\iota \in I} X_\iota$ such that $x(\iota) \in X_\iota$ for all $\iota \in I$. This set is called the *Cartesian product* of the sets X_ι. We will almost invariably write the elements of $\underset{\iota\in I}{\mathsf{P}} X_\iota$ as (x_ι) where $x_\iota = x(\iota)$. Thus (x_ι) is an element of $\underset{\iota\in I}{\mathsf{P}} X_\iota$ and for each ι, x_ι is an element of X_ι. If the cardinal number of I is $\mathfrak{m}$ and if all of the sets X_ι are a fixed set X, we will frequently write $X^{\mathfrak{m}}$ for $\underset{\iota\in I}{\mathsf{P}} X_\iota$. If I is finite, say $I = \{1, 2, \ldots, m\}$, we sometimes write $\underset{\iota\in I}{\mathsf{P}} X_\iota$ as $\overset{m}{\underset{k=1}{\mathsf{P}}} X_k$ or $X_1 \times X_2 \times \cdots \times X_m$.

We reserve the symbols R for the set of all real numbers, K for the set of all complex numbers, Q for the set of all rational real numbers, and Z for the set of all integers. The symbol exp stands for the exponential function defined on K. The symbol T represents the set $\{\exp(2\pi i x): 0 \leqq x < 1\}$; it is a subset of K. The set of all numbers $\exp(2\pi i k/p^n)$, where p is a fixed prime, k runs through all integers, and n through all nonnegative integers, will be denoted by $Z(p^\infty)$. The function *signum* or sgn on K is defined by $\operatorname{sgn} z = \frac{z}{|z|}$ for $z \neq 0$ and $\operatorname{sgn} 0 = 0$.

For real numbers a and b, where $a \leqq b$, we define $[a, b] = \{x \in R: a \leqq x \leqq b\}$, $]a, b[= \{x \in R: a < x < b\}$, $]a, b] = \{x \in R: a < x \leqq b\}$, and $[a, b[= \{x \in R: a \leqq x < b\}$. We also define $]a, \infty[= \{x \in R: a < x\}$, with analogous definitions for $[a, \infty[$, $]-\infty, a[$, $]-\infty, a]$, and $]-\infty, \infty[$.

The sets K^n and R^n are *complex n-dimensional space* and *real n-dimensional space*, respectively $(n = 2, 3, 4, \ldots)$. For $\boldsymbol{a} = (a_1, \ldots, a_n)$ and $\boldsymbol{b} = (b_1, \ldots, b_n)$ in K^n, we define the *inner product* $\langle \boldsymbol{a}, \boldsymbol{b} \rangle$ as $\sum_{k=1}^{n} a_k \overline{b_k}$ and the *norm* $\|\boldsymbol{a}\|$ as $\langle \boldsymbol{a}, \boldsymbol{a} \rangle^{\frac{1}{2}}$. Let S_{n-1} denote the set of all $\boldsymbol{a} \in R^n$ such that $\|\boldsymbol{a}\| = 1$.

For a nonvoid set X and a positive real number p, we define $l_p(X)$ as the set of all complex-valued functions a on X such that $\sum_{x \in X} |a(x)|^p < \infty$. Obviously a function in $l_p(X)$ vanishes outside of a countable subset of X. The *norm* of $a \in l_p(X)$ is the number

$$\Big[\sum_{x \in X} |a(x)|^p\Big]^{1/p}.$$

See §12 for a complete discussion.

For an arbitrary nonvoid set X, let δ_{xy} denote the function on $X \times X$ such that $\delta_{xx} = 1$ for all $x \in X$ and $\delta_{xy} = 0$ if $x \neq y$ and $x, y \in X$. This function is called *Kronecker's delta function*.

The end of each proof is indicated by the symbol $\square$.

§2. Group theory

In this section, we establish the terminology and notation concerning groups that will be used throughout the book. We also prove a few theorems about groups, although most of these are quite standard[1]. We

[1] The only nonstandard theorem in §2 is (2.9).

do this to clarify analogous theorems appearing in subsequent sections, where topological considerations also play a rôle. For a very detailed account of the theory of groups, the reader is referred to KUROŠ [1]; for a shorter and more elementary version, see for example VAN DER WAERDEN [1].

For the detailed analysis given in §§ 9, 23—26, we require many refined theorems about Abelian groups. These are given in Appendix A, with complete proofs.

In dealing with abstract groups, we will adhere to multiplicative notation for the group operation, except in part of Appendix A, where it is convenient to write the group operation as addition. In dealing with R, Z, R^n, etc., which are obviously groups under addition, we will write the group operation as addition. In discussions involving a multiplicative group G, the symbol e will be reserved for the identity element of G.

We are principally concerned with groups in this book. For some purposes, however, it is useful to consider semigroups. A *semigroup* is a nonvoid set G and a mapping $(x,y) \rightarrow xy$ of $G \times G$ into G such that $x(yz) = (xy)z$ for all x, y, z in G. That is, a semigroup is any nonvoid set with an associative multiplication. We do *not* assume the existence of an identity or the validity of any cancellation law. Note too that semigroups may contain subgroups and that groups may contain subsemigroups that are not groups.

Let G be a group. For a fixed $a \in G$, the mappings $x \rightarrow ax$ and $x \rightarrow xa$ of G onto itself are called *left* and *right translation* by the element a, respectively. The mapping $x \rightarrow x^{-1}$ of G onto itself is called *inversion*.

Mappings $x \rightarrow axa^{-1}$ of G onto itself are called *inner automorphisms* of G. We will frequently write inner automorphisms as ϱ_a: $\varrho_a(x) = axa^{-1}$. The set of automorphisms of a group G is itself a group under the operation of composition. The set of all inner automorphisms of G forms a subgroup of the group of all automorphisms of G; and the mapping $a \rightarrow \varrho_a$ is a homomorphism of G into its automorphism group.

Let A and B be subsets of a group G. The symbol AB denotes the set $\{ab : a \in A, b \in B\}$, and A^{-1} denotes $\{a^{-1} : a \in A\}$. We write aB for $\{a\}B$ and Ba for $B\{a\}$. We write AA as A^2, AAA as A^3, etc. Also we write $A^{-1}A^{-1}$ as A^{-2}, etc. A subgroup H of G such that $H \neq G$ and $H \neq \{e\}$ is called a *proper* subgroup of G.

For a subgroup H of a group G, the symbol G/H will *always* denote the space of *left* cosets of H in G. Thus points of G/H are sets xH, and $G/H = \{xH : x \in G\}$. Let H be a normal subgroup of the group G. The mapping $x \rightarrow xH$ of G onto G/H is called the *natural mapping of G onto G/H*. It is obviously a homomorphism. The group G/H is called the *quotient group* of G by H. For an element a of G, let ${}_a\psi$ be the mapping of G/H onto itself defined by ${}_a\psi(xH)=(ax)H$, for all $xH\in G/H$. It is clear that ${}_a\psi$ is well defined, that ${}_a\psi$ is a one-to-one mapping of G/H onto itself, and that the set $\{{}_a\psi : a\in G\}$ forms a group under composition that is a homomorphic image of G under the mapping $a\rightarrow{}_a\psi$. Also, given cosets xH and yH, the mapping ${}_{yx^{-1}}\psi$ obviously carries xH onto yH.

Two elements a and b in G are said to be *conjugate* if some inner automorphism maps a onto b, *i.e.* if $a=xbx^{-1}$ for some $x\in G$. There exists a partition $\{A_\iota\}_{\iota\in I}$ of G such that two elements are conjugate if and only if they belong to the same A_ι; the sets A_ι are called the *conjugacy classes* of G.

As already noted, $R, Z, Q, R^n, K^n, \ldots$ are Abelian groups under addition. Observe also that T and $Z(p^\infty)$ are Abelian groups under multiplication. The groups R, T, Q, and $Z(p^\infty)$ are of special importance in the structure theory developed in §§ 9, 24—26. The group T in addition is of vital importance in harmonic analysis: this is explained in §§23 and 24.

The symbol $Z(m)$ denotes the finite cyclic group of m elements ($m=2, 3, \ldots$): we will frequently represent this group as the set of integers $\{0, 1, 2, \ldots, m-1\}$, with addition modulo m.

(2.1) First isomorphism theorem. *Let G be a group, H a normal subgroup of G, and A an arbitrary subgroup of G. Then $AH=HA$ is a subgroup of G, H is a normal subgroup of AH, and $H\cap A$ is a normal subgroup of A. The groups $(AH)/H$ and $A/(H\cap A)$ are isomorphic. In fact, the mapping τ defined by $aH\rightarrow(aH)\cap A=a(H\cap A)$, $a\in A$, is an isomorphism of $(AH)/H$ onto $A/(H\cap A)$.*

Proof. It is obvious that $AH=HA$, that HA is a subgroup of G, that H is a normal subgroup of HA, and that $H\cap A$ is a normal subgroup of A. Consider the mapping of A onto $(AH)/H$ defined by $a\rightarrow aH$. It is easy to see that this mapping is a homomorphism and that its kernel is $H\cap A$. By the fundamental homomorphism theorem for groups, AH/H and $A/(H\cap A)$ are isomorphic and the isomorphism can be given by $a(H\cap A)\rightarrow aH$; let τ be the inverse of this mapping. □

(2.2) Second isomorphism theorem. *Let G and $\widetilde{G}$ be groups with identity elements e and $\tilde{e}$, respectively, and let φ be a homomorphism of G onto $\widetilde{G}$. Let $\widetilde{H}$ be any normal subgroup of $\widetilde{G}$, $H=\varphi^{-1}(\widetilde{H})$, and $N=\varphi^{-1}(\tilde{e})$. Then the groups G/H, $\widetilde{G}/\widetilde{H}$, and $(G/N)/(H/N)$ are isomorphic.*

Proof. The mapping $x\rightarrow\varphi(x)\widetilde{H}$ is a homomorphism of G onto $\widetilde{G}/\widetilde{H}$ with kernel H. Hence G/H is isomorphic with $\widetilde{G}/\widetilde{H}$. Since $\widetilde{G}$ is isomorphic with G/N and $\widetilde{H}$ is isomorphic with H/N, $\widetilde{G}/\widetilde{H}$ is isomorphic with $(G/N)/(H/N)$. □

(2.3) Direct products. Let $\{G_\iota : \iota \in I\}$ be a nonvoid family of groups and let $\underset{\iota\in I}{\mathsf{P}} G_\iota$ be the Cartesian product of the sets G_ι. For (x_ι) and (y_ι) in $\underset{\iota\in I}{\mathsf{P}} G_\iota$, let $(x_\iota)(y_\iota)$ be the element $(x_\iota y_\iota)$ in $\underset{\iota\in I}{\mathsf{P}} G_\iota$. Under this multiplication, $\underset{\iota\in I}{\mathsf{P}} G_\iota$ is a group; it is called the *direct product* of the groups G_ι. The groups G_ι are called *factors*. The identity of $\underset{\iota\in I}{\mathsf{P}} G_\iota$ is (e_ι), where each e_ι is the identity of G_ι. Let $\underset{\iota\in I}{\mathsf{P}^*} G_\iota$ be the set of all $(x_\iota) \in \underset{\iota\in I}{\mathsf{P}} G_\iota$ such that $x_\iota = e_\iota$ for all but a finite set of indices [this set varying with (x_ι)]. Then $\underset{\iota\in I}{\mathsf{P}^*} G_\iota$ is a subgroup of $\underset{\iota\in I}{\mathsf{P}} G_\iota$; it is called the *weak direct product* of the groups G_ι. If the cardinal number of I is $\mathfrak{m}$ and if each $G_\iota = G$, then we will write $G^{\mathfrak{m}}$ for $\underset{\iota\in I}{\mathsf{P}} G_\iota$ and $G^{\mathfrak{m}*}$ for $\underset{\iota\in I}{\mathsf{P}^*} G_\iota$.

(2.4) Theorem. *Let G be a group with identity e, and let $N_1, N_2, \ldots, N_m$ be a collection of normal subgroups of G satisfying:*

(i) $N_1 N_2 \cdots N_m = G$;

(ii) $(N_1 N_2 \cdots N_k) \cap N_{k+1} = \{e\}$ *for* $k = 1, 2, \ldots, m-1$.

Then G is isomorphic with the direct product $\overset{m}{\underset{k=1}{\mathsf{P}}} N_k$.

Proof. Every element x of G can be written as $x_1 x_2 \cdots x_m$ where each x_k belongs to N_k. By virtue of (ii), this representation is unique. Hence for $x \in G$, we define $\tau(x) = (x_k) \in \overset{m}{\underset{k=1}{\mathsf{P}}} N_k$ where $x_1 x_2 \cdots x_m = x$. It is then simple to show that τ is an isomorphism of G onto $\overset{m}{\underset{k=1}{\mathsf{P}}} N_k$. $\square$

(2.5) Theorem. *Let G be a group with identity e, and let $\{N_\iota : \iota \in I\}$ be a nonvoid family of normal subgroups of G. For each $\iota \in I$, let M_ι be the smallest subgroup of G containing all N_λ for $\lambda \neq \iota$. Suppose that:*

(i) *the smallest subgroup of G containing all N_ι is G itself;*

(ii) *for every $\iota \in I$, the relation $M_\iota \cap N_\iota = \{e\}$ obtains.*

Then G is isomorphic with the weak direct product $\underset{\iota\in I}{\mathsf{P}^*} N_\iota$.

Proof. For $x_j \in N_{\iota_j}$ $(j = 1, 2)$, we have $x_1 x_2 x_1^{-1} x_2^{-1} = (x_1 x_2 x_1^{-1}) x_2^{-1} \in N_{\iota_2}$, and similarly $x_1 x_2 x_1^{-1} x_2^{-1} \in N_{\iota_1}$. Hence (ii) shows that $x_1 x_2 = x_2 x_1$. By (i) and the foregoing, every element of G can be written in the form $x_1 x_2 \cdots x_k$ where $x_j \in N_{\iota_j}$ $(j = 1, 2, \ldots, k)$. This representation is unique. For, if $x_1 x_2 \cdots x_k = y_1 y_2 \cdots y_k$ $(x_j, y_j \in N_{\iota_j})$, then $y_1^{-1} x_1 \cdots y_{k-1}^{-1} x_{k-1} = y_k x_k^{-1}$, and (ii) implies that $y_k = x_k$. It is now obvious that the mapping $x_1 x_2 \cdots x_k \to (y_\iota)$ [where $y_{\iota_j} = x_j$ for $j = 1, 2, \ldots, k$ and $y_\iota = e$ otherwise] is an isomorphism of G onto $\underset{\iota\in I}{\mathsf{P}^*} N_\iota$. $\square$

(2.6) Semidirect products. Let L be a group, and suppose that L contains a normal subgroup G and a subgroup H such that $GH = L$ and $G \cap H = \{e\}$. That is, suppose that one can select exactly one element h from each coset of G so that $\{h\}$ forms a subgroup, H. If H is also normal, then L is isomorphic with the direct product $G \times H$. If H fails to be normal, we can still reconstruct L if we know how the inner automorphisms ϱ_h behave on G. Namely, for $x_j \in G$ and $h_j \in H$ $(j=1, 2)$, we have

$$(x_1 h_1)(x_2 h_2) = x_1 h_1 x_2 h_1^{-1} h_1 h_2 = \big(x_1 \varrho_{h_1}(x_2)\big) h_1 h_2 .$$

The construction just given can be cast in abstract form. Let G and H be groups and suppose that there is a homomorphism $h \to \tau_h$ which carries H onto a group of automorphisms of G. That is, $\tau_h \circ \tau_{h'} = \tau_{hh'}$ for $h, h' \in H$. Let $G \circledS H$ denote the Cartesian product of G and H. For (x, h) and (x', h') in $G \circledS H$, define

$$(x, h)(x', h') = \big(x(\tau_h(x')), hh'\big).$$

Then $G \circledS H$ is a group; it is called a *semidirect product* of G and H. Its identity is (e_1, e_2), where e_1 and e_2 are the identities of G and H, respectively. The inverse of (x, h) is $\big(\tau_{h^{-1}}(x^{-1}), h^{-1}\big)$. Let $G_1 = \{(x, e_2) : x \in G\}$ and $H_1 = \{(e_1, h) : h \in H\}$. Then G_1 is a normal subgroup of $G \circledS H$, and H_1 is a subgroup. Since $(e_1, h) \cdot (x, e_2) \cdot (e_1, h)^{-1} = \big(\tau_h(x), e_2\big)$, the inner automorphism $\varrho_{(e_1, h)}$ for $(e_1, h) \in H_1$ reproduces the action of τ_h on G. Thus every semidirect product is obtained by the process described in the preceding paragraph.

(2.7) Linear groups. (a) Let $A = (a_{jk})_{j,k=1}^{n}$ be an $n \times n$ matrix, where the coefficients a_{jk} are complex numbers. The *transpose* ${}^tA$ of the matrix A is the matrix $(a_{kj})_{j,k=1}^{n}$ and the *conjugate* $\bar{A}$ of A is the matrix $(\bar{a}_{jk})_{j,k=1}^{n}$ where $\bar{a}_{jk}$ is the complex conjugate of the number a_{jk}. We define A^* to be the matrix ${}^t(\bar{A}) = (\overline{{}^tA})$. The *trace* of the matrix A is $\sum_{j=1}^{n} a_{jj}$ and is denoted by $\operatorname{tr} A$; the determinant of a matrix A is denoted by $\det A$.

(b) An $n \times n$ matrix A is said to be *orthogonal* if $A = \bar{A}$ and ${}^tA = A^{-1}$, *unitary* if $A^{-1} = A^*$, *symmetric* if $A = {}^tA$, *skew-symmetric* if $A = -{}^tA$, *Hermitian* if $A = A^*$, *skew-Hermitian* if $A = -A^*$, and *real* if $A = \bar{A}$.

Let F be any subfield of K, such as K or R. The set of all $n \times n$ matrices whose coefficients belong to F is denoted by $\mathfrak{M}(n, F)$. The set of all nonsingular matrices in $\mathfrak{M}(n, F)$ forms a group under multiplication, is called the *general linear group over* F, and is denoted by $\mathfrak{GL}(n, F)$. The subgroup of $\mathfrak{GL}(n, F)$ consisting of the matrices of determinant 1 is the *special linear group over* F and is denoted by $\mathfrak{SL}(n, F)$. The *unitary group* $\mathfrak{U}(n)$ and the *orthogonal group* $\mathfrak{O}(n)$ consist of all unitary matrices and all orthogonal matrices, respectively; they are subgroups of $\mathfrak{GL}(n, K)$. Finally, we define the *special unitary group* and the *special orthogonal group* as $\mathfrak{SU}(n) = \mathfrak{SL}(n, K) \cap \mathfrak{U}(n)$ and $\mathfrak{SO}(n) = \mathfrak{SL}(n, K) \cap \mathfrak{O}(n)$, respectively. The set $\mathfrak{M}(n, F)$, the group $\mathfrak{GL}(n, F)$, and all its subgroups can be regarded as subsets of F^{n^2} in an obvious way; we will frequently do so.

(2.8) Free groups and symmetric groups. (a) Let X be any nonvoid set. A *word* is either void (written $\boldsymbol{e}$) or a finite formal product $x_1^{\varepsilon_1} x_2^{\varepsilon_2} \cdots x_n^{\varepsilon_n}$ of elements of X where $\varepsilon_k = \pm 1$ [the elements x_k need not be distinct]. A word is *reduced* if it is void or if $\varepsilon_k = \varepsilon_{k+1}$ whenever $x_k = x_{k+1}$. The *length* of the reduced word $\boldsymbol{x} = x_1^{\varepsilon_1} x_2^{\varepsilon_2} \cdots x_n^{\varepsilon_n}$ is n and the *length* of $\boldsymbol{e}$ is 0. Let F denote the set of all reduced words of X. If $\boldsymbol{x} = x_1^{\varepsilon_1} x_2^{\varepsilon_2} \cdots x_n^{\varepsilon_n}$ and $\boldsymbol{y} = y_1^{\delta_1} y_2^{\delta_2} \cdots y_m^{\delta_m}$ belong to F, then the product $\boldsymbol{x}\boldsymbol{y}$ is defined as follows. Consider $x_1^{\varepsilon_1} x_2^{\varepsilon_2} \cdots x_n^{\varepsilon_n} y_1^{\delta_1} y_2^{\delta_2} \cdots y_m^{\delta_m}$. If this word is reduced, we define it to be $\boldsymbol{x}\boldsymbol{y}$. If it is not reduced, then $x_n = y_1$ and $\varepsilon_n = -\delta_1$. Then consider $x_1^{\varepsilon_1} x_2^{\varepsilon_2} \cdots x_{n-1}^{\varepsilon_{n-1}} y_2^{\delta_2} \cdots y_m^{\delta_m}$ and if this is reduced we define it to be $\boldsymbol{x}\boldsymbol{y}$. If it is not reduced, we continue in the indicated way until a reduced word is obtained and define it to be $\boldsymbol{x}\boldsymbol{y}$. With this multiplication, F is a group and is called the *free group generated by* X. The identity of F is $\boldsymbol{e}$ and the inverse of $x_1^{\varepsilon_1} x_2^{\varepsilon_2} \cdots x_n^{\varepsilon_n}$ is $x_n^{-\varepsilon_n} x_{n-1}^{-\varepsilon_{n-1}} \cdots x_1^{-\varepsilon_1}$. The proof that $\boldsymbol{x}(\boldsymbol{y}\boldsymbol{z}) = (\boldsymbol{x}\boldsymbol{y})\boldsymbol{z}$ for $\boldsymbol{x}, \boldsymbol{y}, \boldsymbol{z} \in F$ is a straightforward induction argument on the length of the word $\boldsymbol{y}$.

(b) Let X be any nonvoid set. For each $x \in X$, let F_x be the free group generated by x; *i.e.*, $F_x = \{e_x, x, x^{-1}, x^2, x^{-2}, \ldots\}$. Then the weak direct product $\underset{x \in X}{\mathsf{P}^*} F_x$ is the *free Abelian group generated by* X.

(c) If N is any nonvoid set, then the set of all one-to-one mappings of N onto N forms a group under the operation of composition. A set $\{f\}$ of these mappings is called *transitive* if for every $x, y \in N$, there is an $f_0 \in \{f\}$ such that $f_0(x) = y$.

If N is finite and has n elements, then the group of all one-to-one mappings of N onto N is called the *symmetric group* $\mathfrak{S}_n$ *on* n *letters*. The elements of $\mathfrak{S}_n$ are called *permutations* and any subgroup of $\mathfrak{S}_n$ is called a *permutation group on* n *letters*.

(2.9) Theorem. *Let* X *be any nonvoid set and let* $x_1^{\varepsilon_1} x_2^{\varepsilon_2} \cdots x_n^{\varepsilon_n}$ $(n \geqq 1)$ *be any reduced word of the free group generated by* X. *Let* ι *denote the identity permutation of* $\mathfrak{S}_{n+1}$ *or* $\mathfrak{S}_{n+2}$. *Suppose that* $x_1^{\varepsilon_1} x_2^{\varepsilon_2} \cdots x_n^{\varepsilon_n}$ *does not have the form* $y^{-1} y^{-1} \cdots y^{-1} = y^{-n}$ *and that* $n > 1$. *Then there is a mapping* $x \to P_x$ *of* X *into* $\mathfrak{S}_{n+1}$ *such that:*

(i) $P_x = \iota$ *if* x *is different from* $x_1, x_2, \ldots, x_n$;

(ii) $P_{x_1}^{\varepsilon_1} \circ P_{x_2}^{\varepsilon_2} \circ \cdots \circ P_{x_n}^{\varepsilon_n} \neq \iota$;

(iii) $P_{x_1}^{\varepsilon_1} \circ P_{x_2}^{\varepsilon_2} \circ \cdots \circ P_{x_n}^{\varepsilon_n} \neq P_{x_j}$ *for* $j = 1, 2, \ldots, n$.

If $x_1^{\varepsilon_1} x_2^{\varepsilon_2} \cdots x_n^{\varepsilon_n} = y^{-n}$, *there is a mapping of* X *into* $\mathfrak{S}_{n+2}$ *satisfying* (i), (ii), *and* (iii). *If* $n = 1$ *and* $\varepsilon_1 = 1$, *then there is a mapping of* X *into* $\mathfrak{S}_2$ *satisfying* (i) *and* (ii).

Proof. We here regard $\mathfrak{S}_{n+1}$ as the set of all one-to-one mappings of $\{1, 2, \ldots, n+1\}$ onto itself. For $x \notin \{x_1, x_2, \ldots, x_n\}$, define P_x as ι.

For $x \in X$ appearing in the sequence $\{x_1, x_2, \ldots, x_n\}$, suppose that $x = x_{j_1} = \cdots = x_{j_l}$. Then if $\varepsilon_{j_k} = 1$, let $P_x(j_k + 1) = j_k$; if $\varepsilon_{j_k} = -1$, let $P_x(j_k) = j_k + 1$. Make these definitions for $k = 1, 2, \ldots, l$. Since $\varepsilon_j = \varepsilon_{j+1}$ whenever $x_j = x_{j+1}$, there is no contradiction inherent in this definition. If $x \neq x_n$ and x appears in $\{x_1, x_2, \ldots, x_n\}$, let $P_x(n+1) = n+1$. If $\varepsilon_n = -1$, let t be the least integer such that $x_t = x_{t+1} = \cdots = x_n$. Then define $P_{x_n}(n+1) = t$. Extend the P_x's in any way to be permutations of $\{1, 2, \ldots, n+1\}$. It follows at once that $P_{x_1}^{\varepsilon_1} \circ P_{x_2}^{\varepsilon_2} \circ \cdots \circ P_{x_n}^{\varepsilon_n}(n+1) = 1$. This proves (ii). Since $P_{x_1}^{\varepsilon_1} \circ P_{x_2}^{\varepsilon_2} \circ \cdots \circ P_{x_n}^{\varepsilon_n}(n+1) = 1$ and $P_{x_j}(n+1) = n+1$ for $x_j \neq x_n$, (iii) is obvious for $x_j \neq x_n$. Since $P_{x_n}(n+1) = n$ if $\varepsilon_n = 1$ and $P_{x_n}(n+1) = t > 1$ if $\varepsilon_n = -1$, (iii) also holds for x_n.

The case $x_1^{\varepsilon_1} x_2^{\varepsilon_2} \cdots x_n^{\varepsilon_n} = y^{-n}$ is easily dealt with. For $x \neq y$, let $P_x = \iota$. Let $P_y(k) = k+1$ $(k = 1, 2, \ldots, n+1)$ and $P_y(n+2) = 1$. Then (i), (ii), and (iii) are immediate.

The case $n = 1$ and $\varepsilon_1 = 1$ is trivial: set $P_{x_1}(1) = 2$ and $P_{x_1}(2) = 1$. □

§ 3. Topology

In this section, we review briefly the parts of set-theoretic topology needed in the book. We give definitions only where terminology or notation is not universally agreed on, and we give proofs only where no correct proof as short as ours is available in standard texts. For a detailed account of set-theoretic topology, we refer the reader to Kelley [2].

In dealing with topological spaces, we require neighborhoods to be *open*. We frequently define a topology by its family of open sets, written as $\mathcal{O}$. We denote the interior of a set A in a topological space by A° and the closure of A by A^-. Let $\mathcal{O}_1$ and $\mathcal{O}_2$ be two topologies on a set X. If $\mathcal{O}_1 \subset \mathcal{O}_2$, then $\mathcal{O}_1$ is said to be *weaker* than $\mathcal{O}_2$ and $\mathcal{O}_2$ is said to be *stronger* than $\mathcal{O}_1$.

Let X be a topological space and suppose that for every pair of points $x, y \in X$, there is a homeomorphism f of X onto itself such that $f(x) = y$. Then X is called a *homogeneous* topological space.

In agreement with many writers on set-theoretic topology, although not with Kelley [2], we make T_0 separation part of the axioms of regularity and complete regularity. Thus in our terminology, regular and completely regular spaces satisfy Hausdorff's separation axiom.

For a topological space X, let $\mathfrak{w}(X)$ denote the least cardinal number of an open basis and let $\mathfrak{d}(X)$ denote the least cardinal number of a dense subset.

(3.1) Partitions of unity (Dieudonné [1]). *Let X be a normal space, F a closed subset of X, and $U_1, U_2, \ldots, U_n$ open sets such that $\bigcup_{k=1}^{n} U_k \supset F$.*

Then there exist continuous functions $h_1, \ldots, h_n$ on X with values in $[0,1]$ such that

(i) $\sum_{k=1}^{n} h_k(x) = 1$ *for all* $x \in F$;

(ii) $h_k(U_k') = 0$ *for* $k = 1, \ldots, n$.

Proof. (I) Suppose that $\bigcup_{k=1}^{n} U_k = X$; we will prove that there are closed sets $A_1, \ldots, A_n$ such that $\bigcup_{k=1}^{n} A_k = X$ and $A_k \subset U_k$ for $k=1, \ldots, n$. The proof is by induction and is obvious for $n=1$. Suppose that $U_1 \cup U_2 = X$. Then U_1' and U_2' are disjoint closed sets and there exist disjoint open sets V_1 and V_2 such that $U_1' \subset V_1$ and $U_2' \subset V_2$. Setting $A_1 = V_1'$ and $A_2 = V_2'$, we obtain the present assertion for $n=2$.

Suppose that the assertion is true for $n-1$ and that $\bigcup_{k=1}^{n} U_k = X$. Since $X = \left(\bigcup_{k=1}^{n-1} U_k\right) \cup U_n$, there are closed sets A and A_n such that $A \subset \bigcup_{k=1}^{n-1} U_k$, $A_n \subset U_n$, and $A \cup A_n = X$. For $k=1, \ldots, n-1$, define $V_k = U_k \cap A$. Then $\bigcup_{k=1}^{n-1} V_k = A$ and by the induction hypothesis applied to A, there are sets $A_1, A_2, \ldots, A_{n-1}$, closed in A, such that $\bigcup_{k=1}^{n-1} A_k = A$ and $A_k \subset V_k$ $(k=1, \ldots, n-1)$. Clearly the A_k's are also closed in X, each A_k is contained in U_k, and $\bigcup_{k=1}^{n} A_k = X$.

(II) We now prove the theorem in the case that $F = X$. Suppose that $\bigcup_{k=1}^{n} U_k = X$ and that $A_1, \ldots, A_n$ are determined as in (I). By URYSON's lemma, there are continuous functions f_k on X such that $f_k(X) \subset [0,1]$, $f_k(A_k) = 1$, and $f_k(U_k') = 0$ $(k=1, \ldots, n)$. For $k=1, \ldots, n$, let

$$h_k(x) = \frac{f_k(x)}{\sum_{j=1}^{n} f_j(x)} \quad \text{for} \quad x \in X.$$

Since $\sum_{j=1}^{n} f_j(x) \geqq 1$ for all x, it is clear that each h_k is continuous and (i) and (ii) are obvious.

(III) Finally, suppose that F is a closed subset of X and that $\bigcup_{k=1}^{n} U_k \supset F$. Define $U_0 = F'$ and note that $\bigcup_{k=0}^{n} U_k = X$. Thus by (II), there are continuous functions $h_0, h_1, \ldots, h_n$ on X with values in $[0,1]$ such that $\sum_{k=0}^{n} h_k(x) = 1$ for all $x \in X$ and $h_k(U_k') = 0$ $(k=0, 1, \ldots, n)$. Since $h_0(F) = 0$, we have $\sum_{k=1}^{n} h_k(x) = 1$ for $x \in F$. That is, $h_1, \ldots, h_n$ satisfy the conclusions of the theorem. □

(3.2) Compact and locally compact spaces. A topological space X is *compact* [*countably compact*] if every covering [every countable covering] of X by open sets admits a finite subcovering. A topological space X is *locally compact* [*locally countably compact*] if every point of X has a neighborhood U such that U^- is compact [countably compact] as a subspace of X. A topological space X is called *σ-compact* if it is a countable union of compact subspaces. A topological space X has the *Lindelöf property* if every covering of X by open sets admits a countable subcovering.

(3.3) Theorem. *Let X be a topological space having the property that for each $x \in X$ and neighborhood U of x, there is a neighborhood V of x such that $V^- \subset U$. Then the closure of a compact subset of X is again compact*[1]. *Also, if f is a continuous mapping of a topological space Y into X, and if A is a subset of Y with compact closure, then $f(A)$ also has compact closure.*

Proof. Let A be a compact subset of X. To show that A^- is compact, it suffices to show that every open cover of A also covers A^-. For then if $\mathscr{U}$ is an open cover of A^-, there is a finite subfamily of $\mathscr{U}$ covering A, and hence covering A^-.

Suppose that $\mathscr{U}$ is an open cover of A and let $W = \bigcup\{U : U \in \mathscr{U}\}$. For each $x \in A$, there is a neighborhood V_x of x such that $V_x^- \subset W$. Since A is compact, there exist $x_1, \ldots, x_n \in A$ such that $A \subset \bigcup_{k=1}^{n} V_{x_k}$. This implies that $A^- \subset \bigcup_{k=1}^{n} V_{x_k}^- \subset W$. That is, A^- is covered by $\mathscr{U}$.

Suppose now that f is a continuous mapping of Y into X and that the subset A of Y has compact closure. Then $f(A^-)$ is compact and hence $f(A^-)^-$ is also compact. This implies that $f(A)^-$ is compact. □

(3.4) Connectedness. A topological space X is *connected* if and only if it is not the disjoint union of two nonvoid sets that are both open and closed. A topological space X is *locally connected* if the connected open subsets of X form an open basis for the topology of X. A *component* of a topological space is a connected subset that is properly contained in no other connected subset. A topological space is *totally disconnected* if all of its components are points. A topological space X is *0-dimensional* if the family of all sets that are both open and closed is an open basis for the topology.

A topological space X is *arcwise connected* if for every pair of points x and y in X, there exists a continuous mapping f of $[0,1]$ into X such

[1] The closure of a compact set need not be compact. Consider $Z \times Z$ and let a subset F of $Z \times Z$ be closed if $F = Z \times Z$ or if the set $\{m \in Z : (m, n) \in F\}$ is finite for every $n \in Z$. Then $A = \{(m, 0) : m \in Z\}$ is compact and $A^- = Z \times Z$ is not.

that $f(0)=x$ and $f(1)=y$. A topological space X is *locally arcwise connected* if for every $x\in X$ and neighborhood U of x, there is a neighborhood V of x such that for every pair of points y and z in V, there exists a continuous mapping f of $[0,1]$ into U such that $f(0)=y$ and $f(1)=z$.

(3.5) Theorem. *Let X be locally compact, Hausdorff, and totally disconnected. Then X is 0-dimensional.*

Proof. Let a be any point of X and U any neighborhood of a such that U^- is compact. We need to show that there is an open and closed set V such that $a\in V\subset U$. We give the proof in three steps.

(I) Let F be a closed subset of U^- and suppose that every point x of F can be separated from a fixed point $b\in U^-$ by an open and closed subset of U^-. Then there is an open and closed subset C of U^- such that $b\in C'\cap U^-$ and $F\subset C$. To see this, let C_x be open and closed in U^-, $x\in C_x$, $b\notin C_x$, for each $x\in F$. A finite number $C_{x_1}, C_{x_2}, \ldots, C_{x_m}$ of these sets cover F. Hence if $C=C_{x_1}\cup C_{x_2}\cup\cdots\cup C_{x_m}$, C satisfies the conditions stated.

(II) Let M be the set of all points of U^- that cannot be separated from a by an open and closed subset of U^-. Then M is connected. To prove this, note first that M is closed in U^- and hence closed in X, and that $a\in M$. Assume that M is disconnected. Then there are nonvoid closed subsets E and F of M [and hence of X] that partition M. We may suppose that $a\in E$. Since U^- is compact and Hausdorff and hence normal, there is an open subset W of X such that $E\subset W$ and $W^-\cap F=\varnothing$. Thus we have $(W^-\cap W')\cap M=(W^-\cap W')\cap(E\cup F)=\varnothing$. By the definition of M, every point of $W^-\cap W'\cap U^-$ is separated from a by an open and closed subset of U^-. It follows from (I) that there is an open and closed subset C of U^- such that $a\notin C$ and $W^-\cap W'\cap U^-\subset C$. Since C is closed in U^-, it is closed in X, and hence $W\cap C'\cap U^-$ is an open subset of U^-. We also have $W\cap C'\cap U^-=W^-\cap C'\cap U^-$. Thus $W\cap C'\cap U^-$ is an open and closed subset of U^- containing a and disjoint from F. This contradicts the definition of M, and shows that M is connected.

(III) Since X is totally disconnected, the connected component of a is $\{a\}$. Hence (II) shows that for every $x\in U^-$, $x\neq a$, there is an open and closed subset C_x of U^- such that $a\in C_x$ and $x\notin C_x$. Applying (I) [with $F=U^-\cap U'$], we see that there is an open and closed subset C of U^- such that $a\in C$ and $C\cap U^-\cap U'=\varnothing$. Plainly C is an open and closed subset of X. ▯

(3.6) Theorem. *Let X be a completely regular space and let Y be an arcwise connected space. Let $x_1, x_2, \ldots, x_m$ be distinct points of X and $y_1, y_2, \ldots, y_m$ arbitrary points of Y. Then there is a continuous mapping ψ of X into Y such that $\psi(x_k)=y_k$ $(k=1, 2, \ldots, m)$.*

Proof. Let $U_1, U_2, \ldots, U_m$ be pairwise disjoint neighborhoods of $x_1, x_2, \ldots, x_m$, respectively, and let V_k be a neighborhood of x_k such that $V_k^- \subset U_k$ $(k=1, 2, \ldots, m)$. Let f_k be a continuous mapping of X into $[0,1]$ such that $f_k(x_k)=1$ and $f_k(V_k')=0$. Let y_0 be any point of Y and let τ_k be a continuous mapping of $[0,1]$ into Y such that $\tau_k(0)=y_0$ and $\tau_k(1)=y_k$ $(k=1, 2, \ldots, m)$. For $x\in V_k$, let $\psi(x)=\tau_k\circ f_k(x)$, $k=1, 2, \ldots, m$. For $x\in\left(\bigcup_{k=1}^{m} V_k\right)'$, let $\psi(x)=y_0$. It is easy to see that ψ has the properties claimed. □

(3.7) Let X be a metric space with metric d. For a positive real number α, an *α-mesh* in X is a finite set $\{x_1, \ldots, x_n\}\subset X$ such that for every $x\in X$, $\min[d(x_1,x), d(x_2,x), \ldots, d(x_n,x)]<\alpha$. A metric space X is said to be *totally bounded* if it admits an ε-mesh for every $\varepsilon>0$. The following facts are occasionally useful. A metric space X is compact if and only if it is complete and totally bounded. A subset A of a complete metric space has compact closure if and only if A is totally bounded. For proofs, see for example KELLEY [2], p. 198, Theorem 32.

A topological space X is called a *locally Euclidean space* if there is a positive integer n such that every $x\in X$ has a neighborhood U such that U is homeomorphic to the open unit ball of R^n: $\{(x_1, \ldots, x_n)\in R^n: \sum_{k=1}^{n}|x_k|^2<1\}$.

(3.8) Paracompact spaces. A cover $\mathscr{B}$ of a set X is a *refinement* of a cover $\mathscr{A}$ if each member of $\mathscr{B}$ is a subset of a member of $\mathscr{A}$. A family $\mathscr{A}$ of subsets of a topological space X is *locally finite* if each point of X has a neighborhood which intersects only finitely many members of $\mathscr{A}$. A family of subsets of a topological space is *σ-locally finite* if it is the countable union of locally finite subfamilies. A topological space is *paracompact* if it is regular and each open cover has an open locally finite refinement.

A regular topological space X is paracompact if and only if every open cover has an open σ-locally finite refinement. Every metrizable space is paracompact. A paracompact space is normal. [See KELLEY [2], pp. 156—160.]

(3.9) Theorem. *Let $\{X_\iota\}_{\iota\in I}$ be a family of topological spaces. The product space $\underset{\iota\in I}{\mathsf{P}} X_\iota$ is σ-compact if and only if all of the factors X_ι are σ-compact and all but a finite number of them are compact.*

Proof. Let X and Y be σ-compact topological spaces. Then $X=\bigcup_{n=1}^{\infty} X_n$ and $Y=\bigcup_{n=1}^{\infty} Y_n$ where X_n and Y_n are compact $(n=1, 2, \ldots)$. Then $X\times Y=\bigcup_{n=1}^{\infty}\bigcup_{m=1}^{\infty}(X_n\times Y_m)$ and hence $X\times Y$ is σ-compact. By finite induction we infer that $X_1\times X_2\times\cdots\times X_n$ is σ-compact whenever each X_k is σ-compact

$(k=1, \ldots, n)$. Now suppose that $\{X_\iota\}_{\iota\in I}$ is a family of σ-compact spaces and all but a finite number, say $X_1, \ldots, X_m$, are compact. Then $\underset{\iota\notin\{1,\ldots,m\}}{\mathsf{P}} X_\iota$ is compact by TIHONOV's theorem and hence

$$\underset{\iota\in I}{\mathsf{P}} X_\iota = X_1 \times X_2 \times \cdots \times X_m \times \underset{\iota\notin\{1,\ldots,m\}}{\mathsf{P}} X_\iota$$

is σ-compact.

Suppose that $\underset{\iota\in I}{\mathsf{P}} X_\iota = \bigcup_{n=1}^{\infty} A_n$, where each A_n is compact. Let π_ι denote the natural projection of $\underset{\iota\in I}{\mathsf{P}} X_\iota$ onto X_ι. Since $X_\iota = \bigcup_{n=1}^{\infty} \pi_\iota(A_n)$ and each $\pi_\iota(A_n)$ is compact, each X_ι is σ-compact. Assume that X_ι is noncompact for $\iota = \iota_1, \iota_2, \ldots$. Then $\pi_{\iota_n}(A_n) \neq X_{\iota_n}$ for $n=1, 2, \ldots$. Let (x_ι) be any point of $\underset{\iota\in I}{\mathsf{P}} X_\iota$ such that $x_{\iota_n} \notin \pi_{\iota_n}(A_n)$ for all $n=1, 2, \ldots$. Then $(x_\iota) \notin \bigcup_{n=1}^{\infty} A_n$; this contradiction shows that X_ι is compact except for a finite number of indices ι. □

(3.10) Nets. (a) Let D be a nonvoid set. A transitive partial ordering[1] $\geqq$ *directs* D if for every $\alpha, \beta \in D$, there is a $\gamma \in D$ such that $\gamma \geqq \alpha$ and $\gamma \geqq \beta$.

A *net* $x_\alpha = x(\alpha)$ is any function on a directed set. A net y_β, with domain E, is a *subnet* of a net x_α, with domain D, if there is a function N with domain E and range contained in D such that:

(i) $y_\beta = x_{N(\beta)}$ for each $\beta \in E$;

(ii) for each $\alpha \in D$ there is a $\beta \in E$ such that $N(\gamma) \geqq \alpha$ whenever $\gamma \geqq \beta$.

(b) Let x_α be a net with domain D and range contained in a topological space X. The net x_α is said to *converge* to an element $x \in X$ if for every neighborhood U of x there exists an element $\beta \in D$ such that if $\alpha \in D$ and $\alpha \geqq \beta$, then $x_\alpha \in U$. A point $x \in X$ is a *cluster point* of the net x_α if for every neighborhood U of x and every $\beta \in D$, there exists an $\alpha \in D$ such that $\alpha \geqq \beta$ and $x_\alpha \in U$.

A net x_α, with domain D, is *in* a set A if $x_\alpha \in A$ for all $\alpha \in D$.

(c) We now quote some useful facts about nets. Let X be a topological space. If A is a subset of X, then $x \in A^-$ if and only if there is a net in A converging to x. The space X is compact if and only if each net in X has a subnet converging to some point of X. The space X is Hausdorff if and only if each net in X converges to at most one point. A point $x \in X$ is a cluster point of a net x_α if and only if some subnet of x_α converges to x. Let f be a function on X to a topological space Y. Then f is continuous if and only if for each net x_α in X that converges to a point $x \in X$, the net $f(x_\alpha)$ in Y converges to $f(x)$. [See KELLEY [2], pp. 66, 136, 67, 71, and 86.]

---

[1] Here we use the notation $\geqq$ although we do *not* assume that $\geqq$ is reflexive.

(3.11) Dimension. There are several definitions of dimension of a topological space. We choose one adapted to our later needs (24.28).

Let X be a set and $\mathscr{A}$ a finite family of subsets of X. For $x \in X$, let $m(x)$ be the cardinal number of the subfamily $\{A \in \mathscr{A} : x \in A\}$. The *multiplicity* $m(\mathscr{A})$ is defined as $\max\{m(x) : x \in X\}$.

Let X be a compact Hausdorff space, and let n be a nonnegative integer. Then X is said *to have dimension* n if the two following conditions are satisfied:

(i) every finite open covering of X admits a finite closed refinement $\mathscr{F}$ for which $m(\mathscr{F}) \leqq n+1$;

(ii) there is some finite open covering $\mathscr{U}$ of X such that if $\mathscr{F}$ is a finite closed refinement of $\mathscr{U}$, then $m(\mathscr{F}) \geqq n+1$.

We write $\dim(X) = n$. If $\dim(X) = n$ for no nonnegative integer n, then X is said to have *infinite dimension*, and we write $\dim(X) = \infty$.

(3.12) Theorem. *For a compact Hausdorff space X, the definition of 0-dimensionality given in* (3.4) *agrees with the definition of dimension* 0 *given in* (3.11).

Proof. Suppose that open and closed sets form an open basis for X and that $\mathscr{U}$ is a finite open covering of X. Then $\mathscr{U}$ admits a finite refinement $\mathscr{F} = \{F_1, F_2, \ldots, F_n\}$ consisting of open and closed sets. Let $E_1 = F_1$ and $E_k = F_k \cap (F_1 \cup \cdots \cup F_{k-1})'$ $(k = 2, \ldots, n)$. Then $\mathscr{E} = \{E_1, E_2, \ldots, E_n\}$ is a refinement of $\mathscr{U}$ consisting of closed sets such that $m(\mathscr{E}) = 1$. Hence $\dim(X) = 0$.

Conversely, suppose that $\dim(X) = 0$. Let x be any point of X, U any neighborhood of x, and V a neighborhood of x such that $V^- \subset U$. The covering $\{U, V^{-\prime}\}$ of X admits a closed refinement $\mathscr{F} = \{F_1, F_2, \ldots, F_n\}$ for which $m(\mathscr{F}) = 1$. The F_j containing x is an open and closed set contained in U, and so the open and closed subsets of X form an open basis. □

Let X be a compact Hausdorff space and Y a closed subspace of X. Then $\dim(Y) \leqq \dim(X)$. This is easy to show, and we omit the proof.

Chapter Two

Elements of the theory of topological groups

In this chapter we initiate the program described at the beginning of Chapter One: namely, we define and study the structure of topological groups. Certain refined results in structure theory must be postponed to Chapter Five, where the powerful analytical tool of representation theory will have been made available. It seems appropriate to do at the outset what can be done with one's bare hands, so to say; and this we will now do. Section 10 contains the description of a few special topological groups which are important for later use; the remainder of the chapter deals with structural properties of one or another class of topological groups.

§ 4. Basic definitions and facts

A topological group is a set endowed with two structures: that of a group and that of a topological space. These structures are connected in such a way that algebraic properties of the group affect topological properties of the space, and *vice versa*. The present chapter is an introduction to the theory of the structure of topological groups. We begin with the exact definition of a topological group.

(4.1) Definition. Let G be a set that is a group and also a topological space. Suppose that:

(i) the mapping $(x, y) \to xy$ of $G \times G$ onto G is a continuous mapping of the Cartesian product $G \times G$ onto G;

(ii) the mapping $x \to x^{-1}$ of G onto G is continuous.

Then G is called a *topological group*[1].

In terms of open sets, condition (i) asserts that for every neighborhood U of xy, there are neighborhoods V and W of x and y, respectively, such that $VW \subset U$. Condition (ii) asserts that for every neighborhood U of x^{-1}, there is a neighborhood V of x such that $V^{-1} \subset U$.

We make the convention that whenever we write "connected group", "T_0 group", ..., we mean a *topological* group which as a topological space is connected, T_0, ...[2]

(4.2) Theorem. *Let G be a topological group. For $a \in G$, left and right translations by a are homeomorphisms of G. Inversion is also a homeomorphism of G.*

Proof. These facts are immediate consequences of (4.1). □

(4.3) Theorem. *Let G be a topological group, and let $\mathfrak{U}$ be an open basis at e. Then the families $\{xU\}$ and $\{Ux\}$, where x runs through all elements of G and U runs through all elements of $\mathfrak{U}$, are open bases for G.*

---

[1] Observe that we here assume *no separation axiom* for topological groups. Throughout §§ 4—8, the increased generality obtained by this convention seems worth while. However, beginning with § 9, the term "topological group" will mean "topological group that is a T_0 topological space", unless the contrary is specifically stated.

[2] Our sole violation of this excellent rule is in our use of "normal subgroup" to mean a subgroup of a group invariant under inner automorphisms, and not a normal topological space. "Normal" is simply too popular a term.

Proof. Let W be any nonvoid open subset of G, and a any element of W. The mapping $x\rightarrow a^{-1}x$ carries W onto the open set $a^{-1}W$, which contains e. Since $\mathcal{U}$ is an open basis at e, there is a $U\in\mathcal{U}$ such that $e\in U\subset a^{-1}W$. Hence $a\in aU\subset W$. Thus W is a union of sets aU; *i. e.*, $\{xU\}$ is an open basis for G. The proof for the family $\{Ux\}$ is similar. □

We next prove a theorem about product sets which will be frequently applied in the sequel.

(4.4) Theorem. *Let G be a topological group and A and B subsets of G. If A is open and B is arbitrary, then AB and BA are open. If A and B are compact, then AB is compact. If A is closed and B is compact, then AB and BA are closed. If A and B are closed, AB need not be closed.*

Proof. The first assertion is simple. We have $AB=\bigcup\{Ab:b\in B\}$; if A is open, so is Ab for all $b\in B$. Thus AB is the union of open sets and hence is open; similarly for the set BA.

Suppose now that A and B are compact. Then $A\times B$ is a compact subset of $G\times G$ [TIHONOV's theorem] and AB is the image of $A\times B$ under the continuous mapping $(x, y)\rightarrow xy$. Since a continuous image of a compact space is compact, AB is compact.

Next suppose that A is closed and B is compact. We will use nets to prove that AB is closed. The argument is typical of those used in applying nets and so, for the sake of the reader unfamiliar with nets, we will give a detailed argument; subsequent explanations will be briefer. Let D be a directed set and let $x_\alpha, \alpha\in D$, be a net in AB that converges to $x_0\in G$. To show that AB is closed, it suffices to prove that $x_0\in AB$ (3.10.c). For each $\alpha\in D$, we have $x_\alpha=y_\alpha z_\alpha$, where $y_\alpha\in A$ and $z_\alpha\in B$. Since B is compact, there are a $z_0\in B$ and a subnet z_β, $\beta\in E$, of z_α such that z_β converges to z_0 (3.10.c.). Clearly the net $x_\beta, \beta\in E$, converges to x_0. It is easy to see that (x_β, z_β) is a net in $G\times G$ that converges to (x_0, z_0). Thus the net $y_\beta=x_\beta z_\beta^{-1}$ converges to $x_0 z_0^{-1}$ since it is the composition of the net (x_β, y_β) with the continuous function $(x, y)\rightarrow xy^{-1}$ (3.10.c). Since A is closed and each y_β belongs to A, we have $x_0z_0^{-1}\in A$. Thus $x_0=(x_0z_0^{-1})z_0\in AB$. Similarly, BA is closed.

The last assertion of the theorem is easily proved by an example. In the additive group R, consider the closed sets Z and αZ, where α is any irrational number. The set $Z+\alpha Z$ consists of all numbers $m+n\alpha$, where m and n are integers. This is obviously a dense and nonclosed subset of R [it is actually a subgroup]. □

We now return to the analysis of the topology of a topological group. It can be completely described by properties of an open basis at e, as follows.

(4.5) Theorem. *Let G be a topological group, and $\mathfrak{U}$ an open basis at e. Then:*

(i) *for every $U \in \mathfrak{U}$, there is a $V \in \mathfrak{U}$ such that $V^2 \subset U$;*

(ii) *for every $U \in \mathfrak{U}$, there is a $V \in \mathfrak{U}$ such that $V^{-1} \subset U$;*

(iii) *for every $U \in \mathfrak{U}$ and every $x \in U$, there is a $V \in \mathfrak{U}$ such that $xV \subset U$;*

(iv) *for every $U \in \mathfrak{U}$ and $x \in G$, there is a $V \in \mathfrak{U}$ such that $xVx^{-1} \subset U$.*

Conversely, let G be a group, and let $\mathfrak{U}$ be a family of subsets of G with the finite intersection property for which (i), (ii), (iii), *and* (iv) *hold. The family of sets $\{xU\}$, where U runs through $\mathfrak{U}$ and x runs through G, is an open subbasis for a topology on G. With this topology, G is a topological group. [The family of sets $\{Ux\}$ is a subbasis for the same topology.] If the family $\mathfrak{U}$ also satisfies*

(v) *for $U, V \in \mathfrak{U}$, there is a $W \in \mathfrak{U}$ such that $W \subset U \cap V$,*

then $\{xU\}$ and $\{Ux\}$ are open bases for the topology of G.

Proof. If G is a topological group, then properties (i) and (ii) assert that the mappings $(x, y) \to xy$ and $x \to x^{-1}$ are continuous at e. Property (iii) asserts that U is open. Property (iv) follows from the fact that $x \to ax \to axa^{-1}$ is a homeomorphism of G (4.2).

To prove the converse, let $\mathfrak{U}$ satisfy conditions (i), (ii), (iii), and (iv) in the group G and have the finite intersection property. Then for $U \in \mathfrak{U}$, there are V and W in $\mathfrak{U}$ such that $V^2 \subset U$ and $W^{-1} \subset V$. Since $V \cap W \neq \emptyset$, we have $e \in VW^{-1} \subset V^2 \subset U$. Thus all elements of $\mathfrak{U}$ contain e. Let $\widetilde{\mathfrak{U}}$ be the family of all sets $\bigcap_{k=1}^{n} U_k$, where $U_1, \ldots, U_n \in \mathfrak{U}$. For $U_1, \ldots, U_n \in \mathfrak{U}$, there exist $V_1, \ldots, V_n \in \mathfrak{U}$ such that $V_k^2 \subset U_k$, $k = 1, \ldots, n$. Then we have $\left(\bigcap_{k=1}^{n} V_k\right)^2 \subset \bigcap_{k=1}^{n} V_k^2 \subset \bigcap_{k=1}^{n} U_k$. Thus property (i) holds for $\widetilde{\mathfrak{U}}$. Since $\left(\bigcap_{k=1}^{n} V_k\right)^{-1} = \bigcap_{k=1}^{n} V_k^{-1}$, property (ii) holds for $\widetilde{\mathfrak{U}}$. Since $x\left(\bigcap_{k=1}^{n} V_k\right) = \bigcap_{k=1}^{n} (xV_k)$ and $x\left(\bigcap_{k=1}^{n} V_k\right)x^{-1} = \bigcap_{k=1}^{n} (xV_kx^{-1})$, properties (iii) and (iv) hold for $\widetilde{\mathfrak{U}}$.

By definition of a subbasis, the nonvoid sets $\bigcap_{k=1}^{n} (x_k U_k)$, where $x_k \in G$ and $U_k \in \mathfrak{U}$, form an open basis for the topology of G. Let y be any element of $\bigcap_{k=1}^{n} (x_k U_k)$, and let $V_k \in \mathfrak{U}$ have the property that $x_k^{-1} y V_k \subset U_k$ $(k = 1, 2, \ldots, n)$ [see (iii)]. Then $y\left(\bigcap_{k=1}^{n} V_k\right) = \bigcap_{k=1}^{n} (yV_k) \subset \bigcap_{k=1}^{n} (x_k U_k)$. Thus the sets $y\widetilde{U}$, as $\widetilde{U}$ runs through $\widetilde{\mathfrak{U}}$, form an open basis at y, for each $y \in G$.

To prove that G is a topological group, let a and b be any elements of G and $\widetilde{U}$ any set in $\widetilde{\mathfrak{U}}$. By (i) and (iv) for $\widetilde{\mathfrak{U}}$, there are sets $\widetilde{V}$ and $\widetilde{W}$ in $\widetilde{\mathfrak{U}}$

such that $(b^{-1}\widetilde{W}b)\widetilde{V}\subset\widetilde{U}$. Thus $(a\widetilde{W})(b\widetilde{V})\subset ab\widetilde{U}$, so that (4.1.i) is verified. [Note that $\widetilde{V}$ and $\widetilde{W}$ depend only upon $\widetilde{U}$ and b, not upon a.] Property (4.1.ii) follows similarly from (ii) and (iv). The equivalence of the topologies generated by $\{xU\}$ and $\{Ux\}$ follows also from (iv).

The last statement of the theorem is now obvious. □

(4.6) Theorem. *Every topological group G has an open basis at e consisting of neighborhoods U such that $U=U^{-1}$ [sets having this property are called* symmetric*].*

Proof. For an arbitrary neighborhood U of e, let $V=U\cap U^{-1}$. Then plainly $V=V^{-1}$, V is a neighborhood of e, and $V\subset U$. □

(4.7) Corollary. *Let G be a topological group. For every neighborhood U of e, there is a neighborhood V of e such that $V^{-}\subset U$.*

Proof. Let V be a symmetric neighborhood of e such that $V^2\subset U$. Then if $x\in V^{-}$, we have $(xV)\cap V\neq\varnothing$. Hence $xv_1=v_2$ for some $v_1, v_2\in V$, and thus $x=v_2v_1^{-1}\in VV^{-1}=V^2\subset U$. □

(4.8) Theorem. *Let G be a T_0 topological group. Then G is regular and hence Hausdorff*[1].

Proof. By (4.7), G satisfies the axiom of regularity at e. By (4.2), G accordingly satisfies the axiom of regularity at every point. It is obvious that a regular space is Hausdorff. □

(4.9) Theorem. *Let G be a topological group, let U be any neighborhood of e, and let F be any compact subset of G. Then there is a neighborhood V of e such that $xVx^{-1}\subset U$ for all $x\in F$.*

Proof. Let W be a symmetric neighborhood of e such that $W^3\subset U$. Since $F\subset\bigcup_{x\in F}Wx$ and F is compact, there exist $x_1,\ldots,x_n\in F$ such that $F\subset\bigcup_{k=1}^{n}Wx_k$. Let $V=\bigcap_{k=1}^{n}x_k^{-1}Wx_k$. Clearly V is a neighborhood of e, and $x_kVx_k^{-1}\subset W$ for $k=1,\ldots,n$. If $x\in F$, then $x\in Wx_k$ for some k, $k=1,\ldots,n$. Thus $x=wx_k$ for some $w\in W$, and hence

$$xVx^{-1}=wx_kVx_k^{-1}w^{-1}\subset wWw^{-1}\subset W^3\subset U. \ \square$$

Theorem (4.9) shows in particular that property (4.5.iv) can be radically improved for compact groups: a single V can be found that works for *all* $x\in G$. This fact will be useful in several places in the sequel.

[1] We will show in (8.4) that a T_0 group is actually completely regular.

(4.10) Theorem. *Let G be a topological group with identity e, F a compact subset of G, and U an open subset of G such that $F \subset U$. Then there is a neighborhood V of e such that $(FV) \cup (VF) \subset U$. If G is locally compact, then V can be chosen so that $\big((FV) \cup (VF)\big)^-$ is compact.*

Proof. For each $x \in F$, there are a neighborhood W_x of e such that $xW_x \subset U$ and a neighborhood V_x of e such that $V_x^2 \subset W_x$. Since $F \subset \bigcup_{x \in F} xV_x$, there exist $x_1, \ldots, x_n \in F$ such that $F \subset \bigcup_{k=1}^{n} x_k V_{x_k}$. Let $V_1 = \bigcap_{k=1}^{n} V_{x_k}$. Then

$$FV_1 \subset \bigcup_{k=1}^{n} x_k V_{x_k} V_1 \subset \bigcup_{k=1}^{n} x_k V_{x_k}^2 \subset \bigcup_{k=1}^{n} x_k W_{x_k} \subset U.$$

Similarly, there is a neighborhood V_2 of e such that $V_2 F \subset U$. Letting $V = V_1 \cap V_2$, we obtain $\big((FV) \cup (VF)\big) \subset U$. If G is locally compact, then V can be chosen so that V^- is compact. It follows that $F(V^-)$ is closed and compact (4.4). Since $FV \subset F(V^-)$ and $F(V^-)$ is closed, we have $(FV)^- \subset F(V^-)$, and hence $(FV)^-$ is compact. Similarly, $(VF)^-$ is compact, so that $\big((FV) \cup (VF)\big)^-$ is compact. □

The fact that left and right translations are homeomorphisms of a topological group G makes it possible to introduce a notion of "uniform nearness" of two points in G, and also to define uniform continuity of real and complex functions on G, as well as other mappings. Consider the first notion. Given two points x and y in G, translate x [say on the left] by x^{-1}, and translate y by the same element: $x \to x^{-1}x = e$, $y \to x^{-1}y$. If $x^{-1}y$ is in a certain symmetric neighborhood U of e in G, then we may say that x and y are "U-near in the sense of left translation". Similarly if $yx^{-1} \in U$, we may say that x and y are "U-near in the sense of right translation". Both of these notions are *uniform* concepts: they may be applied to points x and y anywhere in the group G. If φ is a complex function on G, then we can say that φ is left [right] uniformly continuous if for every $\varepsilon > 0$, there is a neighborhood U of e in G such that $|\varphi(x) - \varphi(y)| < \varepsilon$ whenever x and y are U-near in the sense of left [right] translation. Thus for left [right] uniform continuity, we must have $|\varphi(x) - \varphi(xu)| < \varepsilon$ [$|\varphi(x) - \varphi(ux)| < \varepsilon$] for all $x \in G$ and all $u \in U$.

The notions of left and right uniform continuity of complex functions on G are natural generalizations of the notion of uniform continuity of a real or complex function defined, let us say, on R: for every $\varepsilon > 0$, there is a $\delta > 0$ such that $|\varphi(x) - \varphi(x+t)| < \varepsilon$ whenever $|t| < \delta$. Instead of a single $\delta > 0$ that works for all $x \in R$, we have a single neighborhood U of e that works for all $x \in G$. An important difference between the case of a general topological group and the case of the line R is that a general topological group may be noncommutative, so that the notions of left and right uniform continuity may be quite different.

It may not be inappropriate at this point to remark that topological groups, which are generalizations of and abstractions from the familiar elementary groups — R, T, $\mathfrak{G}\mathfrak{L}(n, K)$, and so on — admit many structures, mappings, etc., that are direct generalizations of corresponding notions that have been exhaustively investigated for the elementary classical groups. The similarities as well as the differences between general topological groups and the elementary models from which their theory springs are a source of much of the fascination of the theory of topological groups and of analysis on topological groups. Furthermore, a very fruitful source of theorems about topological groups is the observation of a fact for some particular group. Does a corresponding phenomenon then hold for a whole class of topological groups? We shall be concerned with this sort of question throughout the entire work.

We now give a precise definition of uniform structures on a topological group[1].

(4.11) Definition. Let G be a topological group. For every neighborhood U of e in G, let L_U be the set of all pairs $(x, y) \in G \times G$ such that $x^{-1}y \in U$, and let R_U be the set of all pairs $(x, y) \in G \times G$ such that $yx^{-1} \in U$. The family of all sets L_U [R_U], as U runs through all neighborhoods of e in G, is written as $\mathcal{S}_l(G)$ [$\mathcal{S}_r(G)$], and is called the *left* [*right*] *uniform structure on* G.

(4.12) Definition. Let G and H be topological groups, and let φ be a mapping of G into H. Let $\mathcal{U}$ and $\mathcal{V}$ be open bases at the identities of G and H, respectively. Suppose that for every $V \in \mathcal{V}$, there is a $U \in \mathcal{U}$ such that $(\varphi(x), \varphi(y)) \in L_V$ for all $(x, y) \in L_U$. Then φ is said to be a *uniformly continuous mapping* for the pair of uniform structures $(\mathcal{S}_l(G), \mathcal{S}_l(H))$. [It is evident that the uniform continuity of φ is independent of the choices of the open bases $\mathcal{U}$ and $\mathcal{V}$.] Uniform continuity for the pairs of uniform structures $(\mathcal{S}_l(G), \mathcal{S}_r(H))$, $(\mathcal{S}_r(G), \mathcal{S}_l(H))$, and $(\mathcal{S}_r(G), \mathcal{S}_r(H))$ is defined similarly.

(4.13) Definition. Let G be a topological group, and let ι be the identity mapping of G onto itself. If ι is uniformly continuous for the pairs $(\mathcal{S}_l(G), \mathcal{S}_r(G))$ and $(\mathcal{S}_r(G), \mathcal{S}_l(G))$, then the uniform structures $\mathcal{S}_l(G)$ and $\mathcal{S}_r(G)$ are said to be *equivalent*.

[1] There is an extensive theory of uniform spaces, quite independent of the theory of topological groups. We shall not reproduce any part of this general theory, confining ourselves merely to the description of and facts about uniform structures on groups, needed in the sequel. For the general theory of uniform spaces, see KELLEY [2], pp. 174 *ff*.

(4.14) We now list some facts about uniform continuity. The proofs are extremely simple, where they are not trivial, and are left to the reader.

(a) If the identity mapping of G onto itself is uniformly continuous for *one* of the pairs $(\mathcal{S}_l(G), \mathcal{S}_r(G))$ or $(\mathcal{S}_r(G), \mathcal{S}_l(G))$, then it is uniformly continuous for the other, and $\mathcal{S}_l(G)$ and $\mathcal{S}_r(G)$ are equivalent.

(b) If G is an Abelian topological group, the structures $\mathcal{S}_l(G)$ and $\mathcal{S}_r(G)$ are equivalent [in fact, they are identical].

(c) Every left or right translation of a topological group G is a uniformly continuous mapping of G onto itself for the pairs $(\mathcal{S}_l(G), \mathcal{S}_l(G))$ and $(\mathcal{S}_r(G), \mathcal{S}_r(G))$.

(d) Let a, b be any elements of G. The mapping $x \to a x b$ of G onto itself is uniformly continuous for the pairs $(\mathcal{S}_l(G), \mathcal{S}_l(G))$ and $(\mathcal{S}_r(G), \mathcal{S}_r(G))$.

(e) Inversion in G is uniformly continuous for the pairs $(\mathcal{S}_l(G), \mathcal{S}_r(G))$ and $(\mathcal{S}_r(G), \mathcal{S}_l(G))$.

(f) The structures $\mathcal{S}_l(G)$ and $\mathcal{S}_r(G)$ are equivalent if and only if inversion in G is uniformly continuous for the pair $(\mathcal{S}_l(G), \mathcal{S}_l(G))$ or for the pair $(\mathcal{S}_r(G), \mathcal{S}_r(G))$.

(g) The structures $\mathcal{S}_l(G)$ and $\mathcal{S}_r(G)$ are equivalent if and only if for every neighborhood U of e, there is a neighborhood V of e such that $x V x^{-1} \subset U$ for all $x \in G$.

Our main theorem about uniform continuity is the following.

(4.15) Theorem. *Let G and H be topological groups and let φ be a continuous mapping of G into H such that for every neighborhood W of the identity in H, there is a compact subset A_W of G for which $\varphi(A'_W) \subset W$. Then φ is uniformly continuous for each of the following pairs of uniform structures:*

$$(\mathcal{S}_l(G), \mathcal{S}_l(H)); \ (\mathcal{S}_l(G), \mathcal{S}_r(H)); \ (\mathcal{S}_r(G), \mathcal{S}_l(H)); \ (\mathcal{S}_r(G), \mathcal{S}_r(H)).$$

Proof. We will prove the theorem for the pairs of uniform structures $(\mathcal{S}_l(G), \mathcal{S}_l(H))$ and $(\mathcal{S}_l(G), \mathcal{S}_r(H))$; the pairs of uniform structures $(\mathcal{S}_r(G), \mathcal{S}_l(H))$ and $(\mathcal{S}_r(G), \mathcal{S}_r(H))$ present nothing new. All neighborhoods of the identity in G and in H used in the present proof will be taken as symmetric.

Let W be any neighborhood of the identity in H and let Y be another neighborhood of the identity in H such that $Y^2 \subset W$. Let x be an arbitrary point of G. By the continuity of φ, there is a neighborhood U_x of the identity in G such that $\varphi(x U_x) \subset (\varphi(x) Y) \cap (Y \varphi(x))$. For each $x \in G$, let V_x be a neighborhood of the identity in G such that $V_x^2 \subset U_x$.

Next let A_Y be a compact subset of G such that $\varphi(A'_Y) \subset Y$. The family of open sets $\{xV_x\}_{x \in A_Y}$ is an open covering of A_Y. Since A_Y is compact, there exist $x_1, \ldots, x_m \in A_Y$ such that $\bigcup_{k=1}^{m} x_k V_{x_k} \supset A_Y$. Let $V = \bigcap_{k=1}^{m} V_{x_k}$.

Now let x and y be elements of G such that $x^{-1}y \in V$. Suppose that $x \in A_Y$. Then $x \in x_k V_{x_k}$ for some k, $k=1, \ldots, m$, and so

$$y \in xV \subset x_k V_{x_k} V \subset x_k V_{x_k}^2 \subset x_k U_{x_k}.$$

It is obvious that $x \in x_k U_{x_k}$. Hence we have $\varphi(y) \in \varphi(x_k) Y$ and $\varphi(x) \in \varphi(x_k) Y$. That is, $\varphi(x_k)^{-1}\varphi(y) \in Y$ and $\varphi(x)^{-1}\varphi(x_k) \in Y^{-1} = Y$. Therefore $\varphi(x)^{-1}\varphi(y) = \varphi(x)^{-1}\varphi(x_k)\varphi(x_k)^{-1}\varphi(y) \in Y^2 \subset W$. If $y \in A_Y$, then the same argument shows that $\varphi(y)^{-1}\varphi(x) \in W$, and since W is symmetric, we have $\varphi(x)^{-1}\varphi(y) \in W$. Suppose finally that neither x nor y is in A_Y. Then $\varphi(x)$ and $\varphi(y)$ lie in Y, so that $\varphi(x)^{-1}\varphi(y) \in Y^2 \subset W$. This proves that φ is uniformly continuous for the pair of uniform structures $(\mathscr{S}_l(G), \mathscr{S}_l(H))$.

Again let $x, y \in G$ be such that $x^{-1}y \in V$. If $x \in A_Y$, then $x, y \in x_k U_{x_k}$ and hence $\varphi(y) \in Y\varphi(x_k)$ and $\varphi(x) \in Y\varphi(x_k)$. It follows that $\varphi(y)\varphi(x)^{-1} = \varphi(y)\varphi(x_k)^{-1}\varphi(x_k)\varphi(x)^{-1} \in Y^2 \subset W$. If $x, y \in A'_Y$, then $\varphi(y)\varphi(x)^{-1} \in Y^2 \subset W$. Hence φ is uniformly continuous for the pair of uniform structures $(\mathscr{S}_l(G), \mathscr{S}_r(H))$. □

(4.16) Corollary. *Every continuous mapping of a compact group G into a topological group H is uniformly continuous for all four pairs of uniform structures listed in* (4.15).

Proof. For every neighborhood W of e in H, let $A_W = G$, and use (4.15). □

(4.17) Corollary. *Let G be a compact group. Then the structures $\mathscr{S}_l(G)$ and $\mathscr{S}_r(G)$ are equivalent.*

Proof. The identity mapping ι of G onto itself is continuous and hence uniformly continuous for all pairs of uniform structures listed in (4.15). Thus by (4.13), the structures $\mathscr{S}_l(G)$ and $\mathscr{S}_r(G)$ are equivalent. The corollary also follows immediately from (4.14.g) and (4.9). □

Miscellaneous theorems and examples

We now list a number of examples of topological groups and give other illustrations of the definition of a topological group, uniform structures, etc. Some of the following assertions are proved only sketchily, and some are not proved at all. Verification in such cases is left to the reader.

(4.18) Elementary examples. (a) Let G be an arbitrary group, and let $\mathcal{O}$ be the family of all subsets of G [the discrete topology]. With this topology, G is a topological group. We shall often refer to such a G as a *discrete group*.

(b) Let G be an arbitrary group, and let $\mathcal{O}$ consist of $\varnothing$ and G alone. Then G is a topological group. We shall have little use for this topology [except in the trivial case in which $G=\{e\}$].

(c) Let G be an arbitrary infinite group and let $\mathcal{O}$ consist of G and the subsets of G having finite complements; this is the weakest possible T_1 topology on the set G. Then G is *not* a topological group. [In fact, G satisfies the T_1 separation axiom but not Hausdorff's separation axiom.]

(d) The additive group R of all real numbers with its usual topology is a locally compact, noncompact, Abelian group.

(e) The multiplicative group T with its topology as a subset of the complex number field K is a compact Abelian group.

(f) Let G be an arbitrary subgroup of the group $\mathfrak{GL}(n, K)$ with its topology as a subspace of n^2-dimensional complex Euclidean space. Then G is a topological group. [To see this, it is necessary only to note that the formula for multiplying two matrices and the formula for inverting a matrix employ only continuous functions of the entries of the matrices.]

(g) Let G be any group and let $\{\mathcal{O}_\iota\}_{\iota \in I}$ be any collection of topologies for G, each of which makes G into a topological group. Let $\mathcal{O}$ be the weakest topology stronger than all of the topologies $\mathcal{O}_\iota$. Then G is a topological group under the topology $\mathcal{O}$.

(h) Let H be any topological Abelian group and let G be an Abelian group containing H as a subgroup. Let $\mathcal{U}$ be an open basis at e in the group H. Then the same sets can be taken as an open basis at e in G. Properties (4.5.i)—(4.5.v) obviously hold, and G becomes a topological group with open sets as defined in (4.5). The subgroup H is open.

(i) Let E be a topological linear space as defined in (B.5). Then regarded as an additive group, E is clearly a topological group.

(4.19) Ordered groups. (a) Let G be a group with more than one element that is linearly ordered by a relation $<$ [write $x<y$ or $y>x$]. Suppose also that $x<y$ and $a \in G$ imply $ax<ay$ and $xa<ya$. For $a, b \in G$ such that $a<b$, let $]a, b[=\{x \in G: a<x<b\}$. Let the family of all sets $]a, b[$ be an open basis for a topology on G; note that G has no greatest or least element, so that $G=\bigcup\{]a, b[: a<b\}$. Then G is a normal T_0 group.

[Show that $a<b$ implies $b^{-1}<a^{-1}$. This proves that inversion is continuous. To prove multiplication continuous, it is sufficient to prove that if $a>e$ and there is some x for which $e<x<a$, then there is a $b>e$

such that $b^2 \leqq a$. If $x^2 \leqq a$, set $b = x$. If $x^2 > a$, set $b = a x^{-1}$. Then again $b^2 \leqq a$, because otherwise $b^2 = a x^{-1} a x^{-1} > a$ implies that $x^{-1} a x^{-1} > e$ and hence $a > x^2$, which is a contradiction.

We now outline a proof that G is normal; it can easily be modified to apply to any linearly ordered set with a topology having sets of the form $]a, b[$, $\{x \in G : x < a\}$, and $\{x \in G : x > b\}$ as an open basis. A subset C of G is *convex* if $x, y \in C$ and $x < z < y$ always imply $z \in C$. Let A and B be disjoint closed subsets of G and let $\{C_\lambda\}_{\lambda \in \Lambda}$ be the family of all nonvoid maximal convex subsets of $(A \cup B)'$. For each $\lambda \in \Lambda$, choose $x_\lambda \in C_\lambda$. Let $x \in A$. If $]y, x] \subset A$ for some $y < x$, let ${}_xU =]y, x]$. Otherwise, for some $\lambda_0 \in \Lambda$, we have $y \in B$, $y < x$, and $c \in C_{\lambda_0}$ imply $y < c$; that is, C_{λ_0} lies beween $\{y \in B : y < x\}$ and x. In this case, let ${}_xU =]x_{\lambda_0}, x]$. Define in an analogous manner sets U_x for $x \in A$ and sets ${}_xV$ and V_x for $x \in B$. Then $U = \bigcup\{{}_xU \cup U_x : x \in A\}$ and $V = \bigcup\{{}_xV \cup V_x : x \in B\}$ are disjoint open sets containing A and B, respectively.]

(b) (Dieudonné [2]). Consider a well-ordered set S. Let G be the set of all real-valued functions defined on S. For $f, g \in G$ and $\alpha \in S$, let $(f+g)(\alpha) = f(\alpha) + g(\alpha)$. Then G is plainly an [additive] Abelian group. We write $f > g$ if for some $\alpha_0 \in S$, we have $f(\alpha_0) > g(\alpha_0)$ and $f(\alpha) = g(\alpha)$ for all $\alpha < \alpha_0$. Then G is an ordered group. With the topology described in (a), G is a nondiscrete topological group. If S has a last element λ, then there is an open basis at $\mathbf{0}$ [the function identically 0] consisting of open intervals $]-g_t, g_t[$ where $g_t(\alpha) = 0$ if $\alpha < \lambda$ and $g_t(\lambda) = t$, $t > 0$. If S has no last element but has a countable cofinal subset, then there is a countable open basis at $\mathbf{0}$. If S has no cofinal countable subset, then there is no countable open basis at $\mathbf{0}$. In this case, the intersection of every countable family of open sets is open, every countable set is closed, and all compact subsets of G are finite.

(4.20) Independence of the axioms (4.1). (a) Properties (4.1.i) and (4.1.ii) are independent of each other, as this and the following example show. Let R have ordinary addition as its group operation, and let the sets $[a, b[$, for all $a, b \in R$ such that $a < b$, be a basis for open sets. The mapping $(x, y) \to x + y$ is continuous, but the mapping $x \to -x$ is not.

(b) Let R have ordinary addition as its group operation, and let the open sets be all sets of the form $A \cap C'$, where A is open in the usual topology of R and C is countable. Then the mapping $x \to -x$ is continuous, but the mapping $(x, y) \to x + y$ is not. [It is easy to see that $x \to -x$ is continuous. Since there are neighborhoods V of 0 such that $U^- \subset V$ for no neighborhood U of 0, addition cannot be continuous.]

(4.21) Topologies defined by subgroups. (a) Let G be any group and let $\mathscr{N} = \{N\}$ be a family of normal subgroups of G closed under the formation of finite intersections. Let the family of all sets of the form xN, as x runs through G and N runs through $\mathcal{N}$, be an open basis for G. Then G is a 0-dimensional topological group; in fact, every $N \in \mathcal{N}$ is both open and closed.

(b) In (a), let $\mathcal{N}$ consist of a single normal subgroup N. If $N \neq \{e\}$, G fails to satisfy the T_0 separation axiom. If A is any nonvoid finite subset of G that is not the union of cosets of N, then A is compact [like every finite space] but is nonclosed. Clearly the set N is the closure of $\{e\}$.

Every topology on a finite group making it into a topological group is obtained from a single normal subgroup N in this way.

(c) Let G be a non-Abelian group, let N be a nonnormal subgroup of G, and let $\mathcal{O}$ consist of $\varnothing$ and all unions of left cosets xN of N $(x \in G)$. Then G is *not* a topological group. [The mapping $x \to x^{-1}$ is continuous at $a \in G$ if and only if $aNa^{-1} \subset N$.]

(d) (Marshall Hall [1].) As another example of (a), consider any group G and the family $\mathcal{H} = \{H\}$ of subgroups of G having finite index $[G:H]$. First, let H_1 and H_2 be in $\mathcal{H}$. If $x, y \in H_1$ and $xH_2 = yH_2$, then $y^{-1}x \in H_1 \cap H_2$ and $x(H_1 \cap H_2) = y(H_1 \cap H_2)$. Hence if $x(H_1 \cap H_2)$ and $y(H_1 \cap H_2)$ are distinct cosets of $H_1 \cap H_2$ in H_1, xH_2 and yH_2 are distinct cosets of H_2 in G. That is, $[H_1 : H_1 \cap H_2] \leqq [G:H_2]$. It is easy to see that $[G:H_1 \cap H_2] = [G:H_1] \cdot [H_1 : H_1 \cap H_2]$, and so $H_1 \cap H_2$ has finite index in G.

Next, every $H \in \mathcal{H}$ contains a *normal* subgroup N in $\mathcal{H}$. For, consider the coset space $G/H = \{x_1H, x_2H, \ldots, x_nH\}$, where $[G:H] = n$ and $x_1 = e$. For $x \in G$, let $P(x)$ be the permutation of G/H defined by $P(x)\,(x_kH) = xx_kH$. It is clear that P is a homomorphism of G onto a [transitive] subgroup of the group of all permutations of G/H. It is also clear that $P(H)$ is exactly the subgroup of $P(G)$ leaving H fixed. Let N be $P^{-1}(\iota)$, where ι is the identity permutation of G/H. Then we have $N \subset H$, $[G:N] = \overline{\overline{(G/N)}} \leqq n!$, and N is normal. The family $\mathcal{H}$ thus satisfies conditions (4.5.i)—(4.5.v). Using the sets xH $(x \in G, H \in \mathcal{H})$ as a basis for open sets, we see that G is a topological group; G is a Hausdorff group if and only if for every $x \neq e$ in G, there is a subgroup $H \in \mathcal{H}$ not containing x.

(e) (Iwasawa [1]; proof adapted from Kuroš [1], pp. 236—237. See also von Neumann and Wigner [1].)

A free group F provides an example of the construction given in (d) resulting in a Hausdorff topology for F. For, if $g \in F$ and $g \neq e$, there is a normal subgroup of F of finite index not containing g. Let X be a set of free generators for F. Consider a reduced word $g = x_1^{\varepsilon_1} \cdots x_m^{\varepsilon_m}$, where ε_j is -1 or 1 and $x_j \in X$ $(j = 1, \ldots, m)$. By (2.9), there is a positive integer n and a mapping $x \to P_x$ of X into $\mathfrak{S}_{n+1}$ or $\mathfrak{S}_{n+2}$ such that $\psi = P_{x_1}^{\varepsilon_1} \circ \cdots \circ P_{x_m}^{\varepsilon_m}$ is not the identity permutation. Now extend the mapping $x \to P_x$ over all of F in the obvious way, sending products into products. We obtain a homomorphism of F into the finite group $\mathfrak{S}_{n+1}$ or $\mathfrak{S}_{n+2}$. The image of g

is ψ and therefore g is not in the kernel of the homomorphism just defined. This kernel is plainly a normal subgroup of F with finite index.

(4.22) Other topologies for $\boldsymbol{R}$ and $\boldsymbol{Z}$. (a) Consider once again the additive group R, and let H be any Hamel basis for R over the rational number field Q. Let J be a subset of H whose complement relative to H is countable. Let A_J be the set of all real numbers of the form $r_1h_1+r_2h_2+\cdots+r_mh_m$, where the r_j's are in Q and the h_j's are in J. Plainly A_J is a subgroup of R. Let the family of all subgroups A_J be used as the family $\mathcal{N}=\{N\}$ of (4.21.a). Then R is a topological group, and the smallest cardinal number of an open basis at 0 is $\mathfrak{c}$.

(b) Consider the weakest topology for R that is stronger than the topology defined in (a) and the usual topology. Then R is a topological group. There is a countable family of open sets whose intersection is 0, but the smallest cardinal number of an open basis at 0 is $\mathfrak{c}$. Furthermore, if $\{x_n\}_{n=1}^{\infty}$ is a sequence in R converging to 0, then x_n is equal to 0 for all sufficiently large n.

(c) Consider all of the functions $x\rightarrow\exp(i\alpha x)$ on R, where the parameter α runs through all real numbers. Let R have the weakest topology making all of these functions continuous. Then R is a topological group with topology strictly weaker than the usual topology. [Consider the sets $\{x\in R:|\exp(i\alpha x)-1|<\varepsilon\}$ for $\varepsilon>0$ and $\alpha\in R$ and use KRONECKER's theorem (26.19.b).]

(d) The additive group Z can be given a topology under which it is a topological group and for which there is no countable open basis at 0. Thus a countably infinite topological group need not have a countable open basis at the identity. [Let H be a Hamel basis for R over Q containing 1. For $\alpha_1, \alpha_2, \ldots, \alpha_m \in H\cap\{1\}'$ and $\varepsilon>0$, let $U(\alpha_1, \alpha_2, \ldots, \alpha_m; \varepsilon)=\{n\in Z: |\alpha_k n-p_k|<\varepsilon$ for some $p_k\in Z$, $k=1, 2, \ldots, m\}$. It is easy to verify that the family of sets $U(\alpha_1, \alpha_2, \ldots, \alpha_m; \varepsilon)$ satisfies conditions (4.5.i)—(4.5.v) and hence defines a topology on Z making Z into a topological group. KRONECKER's theorem (26.19.d) shows that every set $U(\alpha_1, \alpha_2, \ldots, \alpha_m; \varepsilon)$ is infinite. The cardinal number of every open basis at 0 is $\mathfrak{c}$.]

(4.23) Infinite Abelian groups. (a) Every infinite Abelian group G admits a nondiscrete Hausdorff topology under which it is a topological group. Since T is divisible, it follows from (A.7) that there are enough homomorphisms of G into T [*i.e.*, characters of G] to distinguish an arbitrary $x\neq e$ from e. The weakest topology for G under which all characters are continuous is then a nondiscrete Hausdorff topology for G under which G is a topological group. We will verify this in (26.14).

(b) (KERTÉSZ and SZELE [1].) Every infinite Abelian group G admits a nondiscrete Hausdorff topology under which there is a countable open basis at e. [First, if G has an element a of infinite order, let

A_k be the subgroup generated by a^{2^k} $(k=0, 1, 2, \ldots)$. Then we have $A_1 \supset A_2 \supset \cdots \supset A_k \supset \cdots$, and $\bigcap_{k=1}^{\infty} A_k = \{e\}$. Hence the A_k's can be taken as an open basis at e, in accordance with (4.21.a). Now suppose that every element of G has finite order. Then G is the weak direct product of p-primary groups G_p (A.3). If there are an infinite number of these G_p, say $G_{p_1}, G_{p_2}, \ldots, G_{p_n}, \ldots$, different from $\{e\}$, let B_k be the set of $x=(x_1, x_2, \ldots, x_n, \ldots) \in G$ for which $x_1, \ldots, x_k$ are the identity elements of $G_{p_1}, \ldots, G_{p_k}$, respectively $(k=1, 2, \ldots)$. Then $\{B_k\}_{k=1}^{\infty}$ is a decreasing sequence of subgroups of G whose intersection is $\{e\}$, and so the B_k's can be used as an open basis at e, as in (4.21.a). If there are only a finite number of p-primary subgroups G_p different from $\{e\}$, at least one of them, say G_q, is infinite. There is no loss of generality in supposing that G_q is countably infinite. Suppose that G_q contains a divisible subgroup different from $\{e\}$. Then G_q contains a subgroup A isomorphic to $Z(q^{\infty})$ (A.14). The subgroup A can be given the topology of $Z(q^{\infty})$ as a subspace of T. Let an open basis at e in G be neighborhoods of e in A obtained in this way. Then G is a topological group, with a countable basis at e. If G_q has no divisible subgroup different from $\{e\}$, form the transfinite sequence of subgroups $S_1=\{x^q : x \in G_q\}$, $S_2=\{x^{q^2} : x \in G_q\}, \ldots, S_n=\{x^{q^n} : x \in G_q\}$, $\ldots, S_\omega = \bigcap_{n=1}^{\infty} S_n$, $S_{\omega+1}=\{x^q : x \in S_\omega\}, \ldots$. The transfinite sequence $\{S_\alpha\}$ is obviously decreasing, and must be ultimately constant, $S_\alpha = S_{\alpha+1}$ for some *countable* ordinal α. Since G_q has no divisible subgroup except for $\{e\}$, we have $S_\alpha=\{e\}$. If α is finite, then every element of G_q has order $\leqq q^\alpha$, and so G_q is the weak direct product of an infinite number of finite cyclic groups (A.25). Thus $\{e\}$ is the intersection of a decreasing sequence of proper subgroups. If $S_n \underset{\neq}{\supset} S_{n+1}$ for all positive integers n, write the set $\{S_\beta : S_\beta \neq \{e\}\}$ in any way as a sequence $J_1, J_2, J_3, \ldots, J_n, \ldots$, and write $H_n = J_1 \cap \cdots \cap J_n$ $(n=1, 2, \ldots)$. Once again, $\{e\}$ is the intersection of a decreasing sequence of proper subgroups. As above, G then admits a nondiscrete topology with a countable open basis at e under which it is a topological group.]

(4.24) Examples on uniform structures. Let G be any T_0 group. If there are sequences $\{x_n\}_{n=1}^{\infty}$ and $\{y_n\}_{n=1}^{\infty}$ in G such that $\lim_{n\to\infty} x_n y_n = e$ and $\lim_{n\to\infty} y_n x_n = z \neq e$, then the left and right uniform structures of G are inequivalent. [Let U and W be disjoint neighborhoods of e and z, respectively. Let V be any neighborhood of e. If n is chosen so that $x_n y_n \in V$ and $y_n x_n \in W$, then $(x_n^{-1})^{-1} y_n \in V$ and $y_n (x_n^{-1})^{-1} \in W$; *i.e.*, $y_n (x_n^{-1})^{-1} \notin U$. Hence the uniform structures are inequivalent.]

(a) The groups $\mathfrak{GL}(n, F)$ and $\mathfrak{SL}(n, F)$, where $n \geqq 2$ and F is any subfield of K, are topological groups as subspaces of K^{n^2}. In none of them are right and left uniform structures equivalent. [To see this, let E be the identity matrix in $\mathfrak{GL}(n, F)$; for $\alpha, \beta \in F$ and $\alpha \neq 0$, let $A(\alpha, \beta)$ be the matrix $(a_{jk})_{j,k=1}^{n}$ in $\mathfrak{SL}(n, F)$ such that $a_{11}=\alpha$, $a_{nn}=\frac{1}{\alpha}$, $a_{jj}=1$ for $1<j<n$, $a_{1n}=\beta$, and $a_{jk}=0$ for all other pairs j, k. It is easy to see that $A(\alpha, \beta)\cdot A(\gamma, \delta)=A(\alpha\gamma, \alpha\delta+\beta\gamma^{-1})$. For $m=1, 2, \ldots$, let $X_m=A\left(\frac{1}{m}, \frac{1}{m}\right)$ and $Y_m=A\left(m, \frac{1}{m}\right)$. Then

$$\lim_{m\to\infty} X_m Y_m = \lim_{m\to\infty} A\left(1, \frac{1}{m^2}+\frac{1}{m^2}\right)=A(1, 0)=E,$$

and

$$\lim_{m\to\infty} Y_m X_m = \lim_{m\to\infty} A(1, 1+1)=A(1, 2).$$

Since $A(1, 2)\neq E$, the left and right uniform structures of $\mathfrak{GL}(n, F)$ and $\mathfrak{SL}(n, F)$ are inequivalent.]

(b) Consider the subgroup G of $\mathfrak{GL}(2, F)$ consisting of all matrices $\begin{pmatrix} x & y \\ 0 & 1 \end{pmatrix}$ where $x, y \in F$ and $x \neq 0$. Then G has inequivalent uniform structures.

$$\left[\text{We have } \lim_{m\to\infty} \begin{pmatrix} m & 1 \\ 0 & 1 \end{pmatrix}\begin{pmatrix} \frac{1}{m} & \frac{1}{m^2} \\ 0 & 1 \end{pmatrix} = \lim_{m\to\infty} \begin{pmatrix} 1 & \frac{1}{m}+1 \\ 0 & 1 \end{pmatrix} = \begin{pmatrix} 1 & 1 \\ 0 & 1 \end{pmatrix}\right.$$

and

$$\left.\lim_{m\to\infty} \begin{pmatrix} \frac{1}{m} & \frac{1}{m^2} \\ 0 & 1 \end{pmatrix}\begin{pmatrix} m & 1 \\ 0 & 1 \end{pmatrix} = \lim_{m\to\infty} \begin{pmatrix} 1 & \frac{1}{m}+\frac{1}{m^2} \\ 0 & 1 \end{pmatrix} = \begin{pmatrix} 1 & 0 \\ 0 & 1 \end{pmatrix}.\right]$$

(4.25) Linear groups. (a) The groups $\mathfrak{U}(n)$, $\mathfrak{O}(n)$, $\mathfrak{SU}(n)$, and $\mathfrak{SO}(n)$ are compact. [Since the mappings $A\to\bar{A}$, $A\to{}^tA$, $A\to A^{-1}$, $A\to A^*$, and $A\to\det A$ are continuous, the groups $\mathfrak{U}(n)$, $\mathfrak{O}(n)$, $\mathfrak{SU}(n)$, and $\mathfrak{SO}(n)$ are closed subsets of K^{n^2}. Since $\mathfrak{O}(n)$, $\mathfrak{SU}(n)$, and $\mathfrak{SO}(n)$ are closed subsets of $\mathfrak{U}(n)$, we need only show that $\mathfrak{U}(n)$ is compact. If $A=(a_{jk})_{j,k=1}^{n}$ is unitary, then ${}^tA\bar{A}=E$, the identity matrix, and hence $\sum_{m=1}^{n} a_{mj}\bar{a}_{mj}=1$ for $j=1, \ldots, n$. Thus $|a_{jk}|\leq 1$ for all j and k, so that $\mathfrak{U}(n)$ is closed and bounded. That is, $\mathfrak{U}(n)$ is compact.]

(b) The groups $\mathfrak{GL}(n, K)$, $\mathfrak{SL}(n, K)$, $\mathfrak{U}(n)$, $\mathfrak{SU}(n)$, $\mathfrak{GL}(n, R)$, $\mathfrak{SL}(n, R)$, $\mathfrak{O}(n)$, and $\mathfrak{SO}(n)$ are all locally Euclidean: they possess neighborhoods of the identity E that are homeomorphic to open balls in [real] Euclidean spaces of dimensions $2n^2$, $2n^2-2$, n^2, n^2-1, n^2, n^2-1, $\frac{n(n-1)}{2}$, and $\frac{n(n-1)}{2}$, respectively. [To prove this, we introduce the exponential of a matrix. For $A\in\mathfrak{M}(n, K)$, let $\exp(A)$ be the matrix $\sum_{l=0}^{\infty}\frac{1}{l!}A^l$, where $A^0=E$ and $A^l=AA\cdots A$ (l times). [If $\alpha=\max\{|a_{11}|,$

$|a_{12}|, \ldots, |a_{nn}|\}$, then it is easy to see that no entry of the matrix A^l has absolute value exceeding $(n\alpha)^l$. Hence each entry of the matrix $\sum_{l=0}^{m} \frac{1}{l!} A^l$ is the mth term of a convergent sequence, so that $\exp(A)$ is well defined; the convergence of $\sum_{l=0}^{m} \frac{1}{l!} A^l$ is uniform on compact subsets of $\mathfrak{M}(n, K)$.]

It is easy to show the following.

(1) If B is any matrix in $\mathfrak{GL}(n, K)$, then $\exp(B^{-1}AB) = B^{-1} \cdot \exp(A) \cdot B$.

(2) If $x_1, \ldots, x_n$ are the eigenvalues of A, then $\exp(x_1), \ldots, \exp(x_n)$ are the eigenvalues of $\exp(A)$. [Prove by induction on n.]

(3) The determinant of $\exp(A)$ is $\exp(\operatorname{tr} A)$.

(4) For every A, $\exp(A) \in \mathfrak{GL}(n, K)$.

(5) If $AB = BA$, then $\exp(A+B) = \exp(A) \cdot \exp(B)$.

(6) $\exp({}^tA) = {}^t(\exp(A))$.

(7) $\exp(\bar{A}) = \overline{\exp(A)}$.

We can now show that there is a neighborhood $\mathfrak{V}$ of 0 in $\mathfrak{M}(n, K)$ such that exp is a homeomorphism of $\mathfrak{V}$ onto a neighborhood of E in $\mathfrak{GL}(n, K)$. The entries of the matrix $\exp(A)$ are obviously entire functions of the entries $(a_{11}, a_{12}, \ldots, a_{nn})$ of A, say $\exp(A) = (F_{jk})_{j,k=1}^{n}$. Since $\exp(A) = E + A + \frac{A^2}{2!} + \cdots$, it is plain that $F_{jk}(a_{11}, a_{12}, \ldots, a_{nn}) = \delta_{jk} + a_{jk}$ plus terms of higher order in $a_{11}, \ldots, a_{nn}$. Now regard exp as a mapping of K^{n^2} into K^{n^2}. The Jacobian of exp at 0 is equal to 1, and hence, by the inverse function theorem (Apostol [1], p. 144), exp is one-to-one in some neighborhood $\mathfrak{V}$ of 0. As exp is continuous, if we take $\mathfrak{V}^-$ compact, it follows that exp is a homeomorphism on $\mathfrak{V}$, which carries $\mathfrak{V}$ onto a neighborhood of E in $\mathfrak{GL}(n, K)$.

We now prove in detail that $\mathfrak{U}(n)$ is locally Euclidean with dimension n^2. Let $\mathfrak{V}$ be a neighborhood of 0 in $\mathfrak{M}(n, K)$ such that: exp is a homeomorphism on $\mathfrak{V}$; $A \in \mathfrak{V}$ implies ${}^tA \in \mathfrak{V}$, $\bar{A} \in \mathfrak{V}$, and $-A \in \mathfrak{V}$. [Such a $\mathfrak{V}$ exists by the continuity of the mappings $A \to {}^tA$, $A \to \bar{A}$, and $A \to -A$.] Let $\mathfrak{W}$ consist of all the skew-Hermitian matrices in $\mathfrak{V}$, that is, matrices such that ${}^tA = -\bar{A}$. For $A = (a_{jk})_{j,k=1}^{n}$ to be skew-Hermitian, it is necessary and sufficient that a_{jj} be pure imaginary and that $a_{kj} = -\bar{a}_{jk}$ $(1 \leqq j < k \leqq n)$. Thus the set of all skew-Hermitian matrices is a [real] Euclidean space of dimension n^2, and $\mathfrak{W}$ can be taken as an open ball in R^{n^2} with center at 0. For A in $\mathfrak{W}$, we have $(\exp(A))^* = {}^t\overline{(\exp(A))} = \exp(\overline{{}^tA}) = \exp(-A) = (\exp(A))^{-1}$. Thus $\exp(A)$ is unitary. On the other hand, if $\exp(A)$ is unitary and lies in $\exp(\mathfrak{V})$, then A is skew-Hermitian. The mapping exp is therefore a homeomorphism of $\mathfrak{W}$ onto a neighborhood of E in $\mathfrak{U}(n)$.

Analogous mappings are obtained for the other linear groups mentioned above. We indicate them as follows.

For $\mathfrak{GL}(n, K)$, take $A \in \mathfrak{M}(n, K) \cap \mathfrak{V}$;
for $\mathfrak{SL}(n, K)$, take $A \in \mathfrak{V}$, where $\operatorname{tr} A = 0$;
for $\mathfrak{SU}(n, K)$, take $A \in \mathfrak{V}$, where $\operatorname{tr} A = 0$ and $a_{jk} = -\bar{a}_{kj}$;
for $\mathfrak{GL}(n, R)$, take $A \in \mathfrak{M}(n, R) \cap \mathfrak{V}$;
for $\mathfrak{SL}(n, R)$, take $A \in \mathfrak{M}(n, R) \cap \mathfrak{V}$, where $\operatorname{tr} A = 0$;
for $\mathfrak{O}(n)$, take $A \in \mathfrak{M}(n, R) \cap \mathfrak{V}$, where $a_{jk} = -a_{kj}$;
for $\mathfrak{SO}(n)$, take $A \in \mathfrak{M}(n, R) \cap \mathfrak{V}$, where $a_{jk} = -a_{kj}$.

For $\mathfrak{SL}(n, K)$ and $\mathfrak{SU}(n, K)$, the neighborhood $\mathfrak{V}$ of 0 in $\mathfrak{M}(n, K)$ must also have the property that $|\operatorname{tr} A| < 2\pi$ for all $A \in \mathfrak{V}$; this is plainly possible to secure.]

(4.26) Every nondiscrete, locally countably compact, T_0 group G has cardinal number greater than or equal to $\mathfrak{c}$. [In fact, every nonvoid open subset U of G has cardinal number greater than or equal to $\mathfrak{c}$. With no loss of generality, we may suppose that U^- is countably compact. Since G is nondiscrete and T_0 [hence Hausdorff], U is infinite. Let x_0 and x_1 be distinct points of U and let U_0 and U_1 be neighborhoods of x_0 and x_1, respectively, such that $U_0^- \cup U_1^- \subset U$ and $U_0^- \cap U_1^- = \varnothing$. Let x_{00} and x_{01} be distinct points of U_0, and U_{00} and U_{01} be neighborhoods of x_{00} and x_{01}, respectively, such that $U_{00}^- \cup U_{01}^- \subset U_0$ and $U_{00}^- \cap U_{01}^- = \varnothing$; define x_{10}, x_{11}, U_{10}, and U_{11} analogously. Continuing this process by finite induction, we construct nonvoid open sets $U_{j_1 \cdots j_m}$ for all finite sequences $j_1, j_2, \ldots, j_m$ of 0's and 1's, with the following properties:

$$U_{j_1 \cdots j_{m-1}0}^- \cap U_{j_1 \cdots j_{m-1}1}^- = \varnothing;$$
$$U_{j_1 \cdots j_{m-1}0}^- \cup U_{j_1 \cdots j_{m-1}1}^- \subset U_{j_1 \cdots j_{m-1}};$$
all sets $U_{j_1 \cdots j_m}$ are contained in U.

Now let $A_m = \bigcup \{U_{j_1 \cdots j_m}^- : j_k = 0 \text{ or } 1 \text{ for } k = 1, \ldots, m\}$ for each positive integer m. Let $D = \bigcap_{m=1}^{\infty} A_m$. The countable compactness of U^- implies that $\overline{\overline{D}} \geqq \mathfrak{c}$. In fact, for each sequence $\{j_m\}_{m=1}^{\infty}$ of 0's and 1's, the set $\bigcap_{m=1}^{\infty} U_{j_1 \cdots j_m}^- \subset D$ is nonvoid and the family of all such sets is pairwise disjoint.]

Notes

The notion of topological group in the form used today was apparently first introduced by F. Leja [1] in 1927; in 1925, O. Schreier [1] had given axioms for and studied groups that are Fréchet L-spaces in which the group operations are continuous. Another axiomatic treatment, inequivalent to Leja's, was published by R. Baer in 1929 [1].

Locally Euclidean groups were discussed *per se* by É. CARTAN in 1930 [1]. By the early 1930's topological groups were being treated with facility and precision by many writers. See for example D. VAN DANTZIG [1] and [3], A. HAAR [3], J. VON NEUMANN [3].

Some, although by no means all, of the early writers on topological groups restricted themselves to topological groups having countable open bases: this is the case for example in PONTRYAGIN [4] and [6], FREUDENTHAL [1], and HAAR [3]. The nugatory character of this restriction was stated emphatically by WEIL in [3] and [4]. WEIL's monograph [4] has thoroughly expunged countability restrictions from the theory of topological groups. It is interesting to note that the second edition of PONTRYAGIN's monograph [7] removes the countability restriction. Throughout the present book, we scrupulously avoid unneeded countability restrictions.

Many of the theorems of §§ 4—7 are found in one or more of the monographs BOURBAKI [1], LOOMIS [2], MONTGOMERY and ZIPPIN [1], PONTRYAGIN [7], WEIL [4]. It is pointless to trace every avatar of every theorem, and we will not attempt to do so. The following primary sources of results in § 4 may be of interest. Theorems (4.4), (4.7), and (4.8) are largely due to A. MARKOV [1]. Special cases of (4.6), (4.7), and (4.8) appear in D. VAN DANTZIG [3]. The earliest appearance of the axioms (4.5.i)—(4.5.v) known to us is in A. WEIL [3], p. 11. See also VAN DANTZIG [3], TG. 22. Theorem (4.9) must have been long known; the first reference we have found is a special case proved in MONTGOMERY and ZIPPIN [1], p. 55. Theorem (4.10) is due to VAN KAMPEN [1]. Uniform spaces are the creation of A. WEIL [3], who also described the uniform structures $\mathscr{S}_l(G)$ and $\mathscr{S}_r(G)$. Theorems (4.15)—(4.17) are essentially due to WEIL [*loc. cit.*]. See also VAN KAMPEN [1], p. 451.

§ 5. Subgroups and quotient groups

In this section, we take up some methods of forming new topological groups from a given topological group. We begin with a study of subgroups.

(5.1) Theorem. *Let G be a topological group and H a subgroup of G. With its relative topology as a subspace of G, H is a topological group.*

Proof. The mapping $(x, y) \to xy$ of $H \times H$ onto H and the mapping $x \to x^{-1}$ of H onto H are continuous, since they are restrictions of the corresponding mappings of $G \times G$ and G. □

(5.2) Theorem. *Let A and B be subsets of a topological group G. Then we have:*

(i) $(A^-)(B^-) \subset (AB)^-$;

(ii) $(A^-)^{-1} = (A^{-1})^-$;

(iii) $xA^-y = (xAy)^-$ *for all* $x, y \in G$.

If G *is a* T_0 *topological group, then we also have:*

(iv) *if* $ab = ba$ *for all* $a \in A$ *and* $b \in B$, *then* $ab = ba$ *for all* $a \in A^-$ *and* $b \in B^-$.

Proof. To prove (i), suppose that $x \in A^-$, $y \in B^-$, and that U is any neighborhood of e. Then there is a neighborhood V of e such that $(xV)(yV) \subset xyU$. Since $x \in A^-$ and $y \in B^-$, there are points $a \in A$ and $b \in B$ such that $a \in xV$ and $b \in yV$. Hence $ab \in (AB) \cap (xyU)$, so that $xy \in (AB)^-$.

Assertion (ii) follows from the fact that the mapping $z \to z^{-1}$ is a homeomorphism of G onto G, and (iii) follows from the fact that $z \to xzy$ is a homeomorphism of G onto G.

We now prove (iv). Suppose that $ab = ba$ for all $a \in A$, $b \in B$. The mapping $(a, b) \to aba^{-1}b^{-1}$ of $G \times G$ into G is continuous. Since $\{e\}$ is closed by (4.8), we infer that $H = \{(a, b) \in G \times G : aba^{-1}b^{-1} = e\}$ is closed. Clearly $A \times B \subset H$ and $(A \times B)^- = A^- \times B^-$. Thus, we have $A^- \times B^- \subset H$. That is, $ab = ba$ for all $a \in A^-$, $b \in B^-$. □

(5.3) Corollary. *If* H *is a subsemigroup, subgroup, or normal subgroup of a topological group* G, *then* H^- *is also a subsemigroup, subgroup, or normal subgroup, respectively, of* G. *If* G *is a* T_0 *topological group and* H *is an Abelian subsemigroup or subgroup of* G, *then* H^- *is also an Abelian subsemigroup or subgroup, respectively, of* G.

Proof. If H is a subsemigroup of G, then $H^2 \subset H$ and by (5.2.i), we have $(H^-)^2 \subset (H^2)^- \subset H^-$; that is, H^- is a subsemigroup of G. If H is a subgroup of G, then also $H^{-1} \subset H$. Hence $(H^-)^{-1} = (H^{-1})^- \subset H^-$, so that H^- is a subgroup of G. The remainder of the corollary is proved in like manner. □

(5.4) Theorem. *Let* G *be a topological group. The closure* $\{e\}^-$ *of* $\{e\}$ *in* G *is a closed normal subgroup of* G *and is the smallest closed subgroup of* G. *The closure of a point* $a \in G$ *is the coset* $a\{e\}^- = \{e\}^-a$.

Proof. The set $\{e\}^-$ is a closed normal subgroup of G by (5.3). It is clearly the smallest closed subgroup of G. Left translation is a homeomorphism, and therefore the closure of $\{a\}$ is $a\{e\}^-$ for every $a \in G$. Moreover, since $\{e\}^-$ is a normal subgroup, we have $a\{e\}^- = \{e\}^-a$. □

(5.5) Theorem. *A subgroup* H *of a topological group* G *is open if and only if its interior is nonvoid. Every open subgroup* H *of* G *is closed.*

Proof. Suppose that H has an interior point x. Then there is a neighborhood U of e in G such that $xU \subset H$. For every $y \in H$, we then have $yU = yx^{-1}xU \subset yx^{-1}H = H$. Hence H is open. If H is open, then by definition every point of H is an interior point.

If H is an open subgroup of G, then $H' = \bigcup\{xH : x \notin H\}$. Each set xH is open, and so H' is open; that is, H is closed. ▯

(5.6) Theorem. *Let $\mathscr{A}$ be a family of neighborhoods of e in a topological group G such that:*

(i) *for each $U \in \mathscr{A}$, there is a $V \in \mathscr{A}$ such that $V^2 \subset U$;*

(ii) *for each $U \in \mathscr{A}$, there is a $V \in \mathscr{A}$ such that $V^{-1} \subset U$;*

(iii) *for each $U, V \in \mathscr{A}$, there is a $W \in \mathscr{A}$ such that $W \subset U \cap V$.*

Let $H = \bigcap\{U : U \in \mathscr{A}\}$. Then H is a closed subgroup of G. If in addition,

(iv) *for every $U \in \mathscr{A}$ and $x \in G$, there is a $V \in \mathscr{A}$ such that $xVx^{-1} \subset U$, then H is a normal subgroup of G.*

Proof. Suppose that $x, y \in H$ and that $U \in \mathscr{A}$. Let $V \in \mathscr{A}$ be such that $V^2 \subset U$. Then $x, y \in V$, so that $xy \in V^2 \subset U$. Hence $xy \in H$. Similarly $x^{-1} \in H$ if $x \in H$. To see that H is closed, let a be any element of G that is not in H. Then $a \notin U$ for some $U \in \mathscr{A}$. Let $V_1, V_2, V \in \mathscr{A}$ be such that $V_1^2 \subset U$, $V_2^{-1} \subset V_1$, and $V \subset V_1 \cap V_2$; then $VV^{-1} \subset U$. Hence if $(aV) \cap V \neq \varnothing$, we have $a \in VV^{-1} \subset U$, a contradiction. Hence we have $a \in aV \subset H'$, and H' is accordingly open. That is, H is closed.

Suppose that (iv) holds and let $a \in H$ and $x \in G$. For $U \in \mathscr{A}$, let $V \in \mathscr{A}$ be such that $xVx^{-1} \subset U$. Plainly $xax^{-1} \in xVx^{-1} \subset U$, and since $U \in \mathscr{A}$ is arbitrary, we have $xax^{-1} \in H$. Hence H is a normal subgroup. ▯

As a sort of complement to (5.6), we show next how to generate open and closed subgroups from neighborhoods of e.

(5.7) Theorem. *Let U be any symmetric neighborhood of e in a topological group G. Then the set $L = \bigcup_{n=1}^{\infty} U^n$ is an open and closed subgroup of G.*

Proof. If $x \in U^k$ and $y \in U^l$, then $xy \in U^{k+l}$ and $x^{-1} \in (U^{-1})^k = U^k$. Hence L is a subgroup of G. By Theorem (5.5), L is open and closed. ▯

(5.8) Theorem. *A subgroup H of a topological group G is discrete if and only if it has an isolated point.*

Proof. Suppose that $x \in H$ and that x is isolated in the relative topology of $H \subset G$. That is, there is a neighborhood U of e in G such that $(xU) \cap H = \{x\}$. Then for arbitrary $y \in H$, we have $(yU) \cap H = (yU) \cap (yx^{-1}H) = yx^{-1}((xU) \cap H) = \{y\}$. Thus every point of H is isolated, so that H is indeed discrete. If H is discrete, then by definition *all* of its points are isolated. ▯

(5.9) Theorem. *Let G be a topological group and H a subgroup of G such that $U^- \cap H$ is closed in G for some neighborhood U of e in G. Then H is closed.*

Proof. Let U be a neighborhood of e in G such that $U^- \cap H$ is closed in G. Let V be a symmetric neighborhood of e in G such that $V^2 \subset U$. Now let x be any point in H^-, and let x_α, $\alpha \in D$, be a net in H such that x_α converges to x. Since $x^{-1} \in H^-$ (5.3), there is an element y in $Vx^{-1} \cap H$. There is an $\alpha_0 \in D$ such that $x_\alpha \in xV$ for all $\alpha \geqq \alpha_0$. Thus, if $\alpha \geqq \alpha_0$, we have $yx_\alpha \in (Vx^{-1})(xV) = V^2 \subset U$ and hence $yx_\alpha \in U^- \cap H$. Since the net yx_α, $\alpha \geqq \alpha_0$, converges to yx and $U^- \cap H$ is closed, we have $yx \in U^- \cap H$. Hence $x = y^{-1}yx \in H$ and therefore $H^- \subset H$, *i.e.*, H is closed. □

(5.10) Theorem. *Every discrete subgroup H of a T_0 group G is closed.*

Proof. Let U be a neighborhood of e in G such that $U \cap H = \{e\}$. By (4.7), there is a neighborhood V of e such that $V^- \subset U$. Then $V^- \cap H = \{e\}$, which is closed since G is Hausdorff (4.8). Theorem (5.9) now implies that H is closed. □

The hypothesis in (5.10) that G be T_0 is essential, as Example (5.37.g) shows. The following generalization of (5.10) also obtains.

(5.11) Theorem. *Let G be a T_0 group and H a subgroup of G that is locally compact in its relative topology. Then H is closed.*

Proof. Let U be a neighborhood of e in G such that $U^- \cap H$ is compact as a subset of H, and therefore as a subset of G. Since G is Hausdorff, $U^- \cap H$ is closed. Thus by (5.9), H is closed. □

(5.12) Definition. A topological group G is said to be *compactly generated* if it contains a compact subset F for which the subgroup generated by F is G: that is, $G = \{e\} \cup \bigcup_{n=1}^{\infty} (F \cup F^{-1})^n$.

(5.13) Theorem. *Let G be a locally compact topological group. Then the following assertions are equivalent:*

(i) *G is compactly generated;*

(ii) *there is an open subset U of G such that U^- is compact and U generates G;*

(iii) *there is a neighborhood U of e in G such that U^- is compact and U generates G.*

Proof. It is obvious that (iii) implies (ii) and that (ii) implies (i). Suppose then that G is compactly generated; we will prove that (iii) holds. Let F be a compact subset of G generating G. Then $F \cup \{e\}$ is compact, and by (4.10) there is an open set U containing $F \cup \{e\}$ such that U^- is compact. Clearly U generates G. □

(5.14) Theorem. *Let G be a locally compact group with identity e and let F be any compact subset of G. Then there is an open and closed compactly generated subgroup of G containing F.*

Proof. Since $F \cup \{e\}$ is compact, (4.10) implies that there is an open set U containing $F \cup \{e\}$ such that U^- is compact. Therefore $\bigcup_{n=1}^{\infty} (U \cup U^{-1})^n = \bigcup_{n=1}^{\infty} (U^- \cup (U^-)^{-1})^n$ is a compactly generated set, and it is an open and closed subgroup of G by (5.7). □

Just as in the purely algebraic theory of groups, subgroups of topological groups play an essential rôle in the formation of homomorphic images. Topological properties of the subgroups in question will also play an important part in our constructions. We now take up the consideration of continuous homomorphisms of topological groups and how they are obtained from normal subgroups. We begin by discussing a somewhat more general construction.

(5.15) Definition. Let G be a topological group, and let H be a subgroup of G. Let φ be the natural mapping $x \to xH$ of G onto G/H. Define a topology $\mathcal{O}(G/H)$ for G/H by the following rule: a subset $\{xH : x \in X\}$ of G/H is in $\mathcal{O}(G/H)$ if and only if $\varphi^{-1}(\{xH : x \in X\})$ is open in G. In other words, $\{xH : x \in X\}$ is open if and only if $\bigcup \{xH : x \in X\} = XH$ is open in G.

Since $\{xH : x \in X\} = \{xH : x \in XH\}$, it follows that every open set in G/H has the form $\{uH : u \in U\}$ where U is an open subset of G. Thus $\mathcal{O}(G/H)$ consists of all sets of the form $\{uH : u \in U\}$ where U is open. It is also easy to see that sets of the form $\{uH : u \in U\}$, where U is open in G and $a \in U$, form an open basis at each point aH of G/H.

(5.16) Theorem. *The family $\mathcal{O}(G/H)$ is a topology for G/H. Under this topology, the natural mapping φ of G onto G/H is continuous, and $\mathcal{O}(G/H)$ is the strongest topology on G/H under which the mapping φ is continuous.*

Proof. Let $\{uH : u \in U_\lambda\}_{\lambda \in \Lambda}$ be a family of open subsets of G/H where each U_λ is open in G. Then $\bigcup_{\lambda \in \Lambda} \{uH : u \in U_\lambda\} = \{uH : u \in \bigcup_{\lambda \in \Lambda} U_\lambda\}$ is open in G/H since $\bigcup_{\lambda \in \Lambda} U_\lambda$ is open in G. Similarly, the intersection of any finite number of sets in $\mathcal{O}(G/H)$ is again in $\mathcal{O}(G/H)$. Clearly $\varnothing$ and G/H are in $\mathcal{O}(G/H)$ and therefore $\mathcal{O}(G/H)$ is a topology on G/H. The remaining statements of the theorem are easy to verify: we omit the argument. □

The topological space G/H is called the *quotient space* of G by H. We now list a number of useful facts about G/H.

(5.17) Theorem. *The natural mapping φ of G onto G/H is an open mapping.*

Proof. If U is an open subset of G, then UH is open (4.4) and hence $\varphi(U)=\{uH : u\in U\}$ is open in G/H. ☐

It is easy to see that the natural mapping φ of G onto G/H need not be a *closed* mapping: $\varphi(A)$ may be nonclosed in G/H for closed subsets A of G. A simple example is provided by the additive group R and R/Z. Every coset $x+Z$ in R contains the number $x-[x]$ [$[x]$ is the integral part of x] and no other number in $[0,1[$. Thus $[0,1[$ can be taken as the space R/Z. It is not hard to see that the topology imposed on $[0,1[$ as a model of the space R/Z is the following. An open basis consists of all sets $]\alpha,\beta[$ where $0<\alpha<\beta<1$ and of all sets $[0,\alpha[\cup]\beta,1[$ where $0<\alpha<\beta<1$. Now let $A=\{\frac{3}{2},\frac{9}{4},\ldots,n+2^{-n},\ldots\}$. The set A is closed in R, but $\varphi(A)\subset[0,1[$ is the set $\{\frac{1}{2},\frac{1}{4},\ldots,\frac{1}{2^n},\ldots\}$, which is nonclosed.

In view of this example, the following theorem is of interest.

(5.18) Theorem. *If G is a topological group and H is a compact subgroup of G, then the natural mapping φ of G onto G/H is a closed mapping.*

Proof. Suppose that A is a closed subset of G; we will show that $\varphi(A)'$ is open in G/H. Consider x in G such that xH does not belong to $\varphi(A)$; then x does not belong to AH. Since AH is closed (4.4), there is a neighborhood U of x such that $U\cap(AH)=\varnothing$. Clearly $\{uH : u\in U\}$ is an open set in G/H containing xH, and $\{uH : u\in U\}$ is disjoint from $\varphi(A)$. Therefore $\varphi(A)'$ is open. ☐

(5.19) Theorem. *Let G be a topological group, H a subgroup of G, and U, V neighborhoods of e in G such that $V^{-1}V\subset U$. Let φ be the natural mapping of G onto G/H. Then $(\varphi(V))^-\subset\varphi(U)$.*

Proof. Let $xH\in G/H$ be in $(\varphi(V))^-$. Then $\{vxH : v\in V\}$ is a neighborhood of xH and hence contains some point of $\varphi(V)$. That is, there are points $v_1, v_2\in V$ such that $v_1xH=v_2H$, *i.e.*, $xH=v_1^{-1}v_2H\in\{wH : w\in V^{-1}V\}$ $\subset\{uH : u\in U\}=\varphi(U)$. ☐

(5.20) Theorem. *Let G be a topological group and H a subgroup of G. For $a\in G$, let ${}_a\psi$ be the mapping defined in* §2:

$${}_a\psi(xH)=(ax)H \quad \textit{for} \quad xH\in G/H.$$

Then ${}_a\psi$ is a homeomorphism of G/H. Thus G/H is a homogeneous space.

Proof. Since ${}_a\psi$ is a one-to-one mapping of G/H onto itself and $({}_a\psi)^{-1}={}_{a^{-1}}\psi$, we need only show that ${}_a\psi$ is an open mapping. Let $\{uH : u\in U\}$ be an open subset of G/H, where U is an open subset of G. Then

${}_a\psi(\{uH : u\in U\}) = \{auH : u\in U\} = \{vH : v\in aU\}$ is open, since aU is open in G. □

(5.21) Theorem. *Let G be a topological group and H a subgroup of G. Then G/H is a discrete space if and only if H is open in G. If H is closed, then G/H is a regular and hence Hausdorff space. If G/H is a T_0 space, then H is closed and G/H is a regular space.*

Proof. If H is open in G, then aH is open in G for all $a\in G$, and so $\varphi^{-1}(\{aH\}) = aH$ is open in G for every point $aH\in G/H$. That is, every point of G/H is an open set, and hence every subset of G/H is open. Conversely, if G/H is discrete, then the set $\{H\}$ is open in G/H, and thus $\varphi^{-1}(\{H\}) = H$ is open in G.

Now suppose that H is closed in G. Then aH is closed in G for all $a\in G$, and $(aH)' = \bigcup\{xH : xH \neq aH\}$ is open in G. This implies that the complement of each set $\{aH\}$ in G/H is open in G/H. Hence each point in G/H is a closed set. This property is equivalent to the T_1 separation axiom. This and (5.19) imply that G/H is regular.

Finally, suppose that G/H is a T_0 space. Then by (5.19), it is also a T_1 space, so that the set $\{xH : xH \neq H\}$ is open in G/H. This implies that $H = (\bigcup\{xH : xH \neq H\})'$ is closed in G. □

(5.22) Theorem. *Let G be a compact [locally compact] group, and let H be a subgroup of G. Then the space G/H is compact [locally compact].*

Proof. If G is compact, then G/H, being a continuous image of G (5.16), is also compact. Suppose that G is locally compact, and that U is a neighborhood of e in G such that U^- is compact. Then by (3.3) and (5.19), $\varphi(U)$ has compact closure in G/H. Since G/H is homogeneous (5.20), G/H is locally compact. □

(5.23) Lemma. *Let G be a topological group, H a subgroup of G, and φ the natural mapping of G onto G/H. Suppose that U is a symmetric neighborhood of e such that $((U^-)^3)^- \cap H$ is compact and that $\{xH : x\in X\}$ is a closed compact subset of G/H such that $\{xH : x\in X\} \subset \{uH : u\in U\}$. Then $U^- \cap (XH)$ is closed and compact in G.*

Proof. Evidently $XH = \varphi^{-1}(\{xH : x\in X\})$ is closed so that $U^- \cap (XH)$ is closed. Assume that $U^- \cap (XH)$ is not compact. Then there is a net x_α, $\alpha\in D$, in $U^- \cap (XH)$ such that no subnet converges to a point of $U^- \cap (XH)$. Since $U^- \cap (XH)$ is closed, no subnet of x_α converges to any point of G. Since each $x_\alpha H$ is in $\{xH : x\in X\}$, there is a subnet x_β, $\beta\in E$, of x_α and an $x_0\in X$ such that $x_\beta H$ converges to $x_0 H$ in the topology $\mathfrak{O}(G/H)$. By hypothesis, we have $x_0 H = u_0 H$ for some $u_0\in U$. Write $A = ((U^-)^3)^-$. Evidently x_β has no subnet converging to a point of $u_0(A\cap H)$; that is, no point of $u_0(A\cap H)$ is a cluster point of x_β (3.10.c). Hence for each $x\in u_0(A\cap H)$, there is a neighborhood U_x of e and a $\beta_x\in E$

such that

$$\beta \geqq \beta_x \ \text{implies} \ x_\beta \notin U_x x. \tag{1}$$

For each $x \in u_0(A \cap H)$, let V_x be a neighborhood of e such that $V_x^2 \subset U_x$. By hypothesis, $A \cap H$ is compact and therefore $u_0(A \cap H)$ is also compact. Hence there are points $x_1, \ldots, x_n$ in $u_0(A \cap H)$ such that

$$\bigcup_{k=1}^{n} V_{x_k} x_k \supset u_0(A \cap H). \tag{2}$$

Let V be any symmetric neighborhood of e such that $V \subset \left(\bigcap_{k=1}^{n} V_{x_k}\right) \cap U$. Clearly $\{v u_0 H : v \in V\}$ is a neighborhood of $u_0 H = x_0 H$ in G/H, and thus there is a $\beta_0 \in E$ such that $\beta \geqq \beta_0$ implies $x_\beta H \in \{v u_0 H : v \in V\}$. Let $\gamma \in E$ be such that $\gamma \geqq \beta_0$ and $\gamma \geqq \beta_{x_k}$ for $k = 1, \ldots, n$. Then $x_\gamma H = v u_0 H$ for some $v \in V$; that is, $x_\gamma = v u_0 h$ for some $h \in H$. Now

$$h = u_0^{-1} v^{-1} x_\gamma \in U V U^- \subset (U^-)^3 \subset A$$

and, consequently, $u_0 h \in u_0(A \cap H)$. Applying (2), we have $u_0 h \in V_{x_k} x_k$ for some k $(k = 1, \ldots, n)$, and hence

$$x_\gamma = v u_0 h \in V V_{x_k} x_k \subset V_{x_k} V_{x_k} x_k \subset U_{x_k} x_k.$$

Now (1) shows that this is impossible since $\gamma \geqq \beta_{x_k}$. It follows that $U^- \cap (XH)$ is compact. □

(5.24) Note. (a) In (5.23), suppose that H is compact. Then the hypotheses of (5.23) hold if we take $U = G$. In this case, Lemma (5.23) remains valid if the hypothesis that $\{xH : x \in X\}$ be closed in G/H is dropped. Thus we find: if H is a compact subgroup of G and if $\{xH : x \in X\}$ is a compact subset of G/H, then XH is compact in G. The details of the argument are very like those of (5.23), and are omitted.

(b) If G is locally compact, H is a subgroup of G, and $\{xH : x \in E\}$ is a closed compact subset of G/H, then there is a closed compact subset F of G such that $\{xH : x \in F\} = \{xH : x \in E\}$. In fact, if U is any neighborhood of e in G such that U^- is compact and φ denotes the natural mapping of G onto G/H, then $\{\varphi(sU) : s \in G\}$ is an open covering of G/H. A finite subfamily, say $\varphi(s_1 U), \varphi(s_2 U), \ldots, \varphi(s_m U)$, covers $\{xH : x \in E\}$. Then we can take $F = (s_1 U^- \cup s_2 U^- \cup \cdots \cup s_m U^-) \cap \varphi^{-1}(\{xH : x \in E\})$.

(5.25) Theorem. *Let G be a topological group and H a subgroup of G. If H and G/H are compact, then G itself is compact. If H and G/H are locally compact, then G is also locally compact.*

Proof. Suppose that H and G/H are compact. Then letting $U = X = G$ and applying Lemma (5.23), we see that $G = G^- \cap (GH)$ is compact[1].

[1] This can also be proved directly by using (5.18) and the fact that if a family $\mathscr{F}$ of closed subsets of G has the finite intersection property, then so does $\{\varphi(F) : F \in \mathscr{F}\}$.

Suppose that H and G/H are locally compact. Let V be a neighborhood of e in G such that $V^- \cap H$ is compact. Let U be a symmetric neighborhood of e such that $(U^-)^3 \subset V$ [use (4.5.i), (4.6), and 4.7)]. Then obviously $((U^-)^3)^- \cap H$ is compact. Let W be a neighborhood of e such that $\{xH : x \in W\}^-$ is compact in G/H. By (5.19), there is a neighborhood W_0 of e in G for which $\varphi(W_0)^- \subset \varphi(U)$, that is,

$$\{xH : x \in W_0\}^- \subset \{xH : x \in U\}.$$

Choose a subset X of G for which $\{xH : x \in X\} = \{xH : x \in W \cap W_0\}^-$. Then $\{xH : x \in X\}$ is clearly compact in G/H and

$$\{xH : x \in X\} \subset \{xH : x \in W_0\}^- \subset \{xH : x \in U\}.$$

By (5.23), $U^- \cap (XH)$ is closed and compact in G. Moreover, $e \in U \cap ((W \cap W_0)H) \subset U^- \cap (XH)$. Therefore e has a neighborhood $U \cap ((W \cap W_0)H)$ with compact closure; *i.e.*, G is locally compact. □

We now specialize the construction of and topology in G/H to the case in which H is a normal subgroup, so that G/H is a group.

(5.26) Theorem. *Let G be a topological group and let H be a normal subgroup of G. Let the quotient group G/H be given a topology as in* (5.15). *Then G/H is a topological group. Also the natural mapping φ of G onto G/H is an open, continuous homomorphism of G onto G/H. The group G/H is discrete if and only if H is open and is a T_0 [and hence regular] group if and only if H is closed.*

Proof. Let us show that G/H is a topological group by proving that the family of all neighborhoods of H [the identity!] in G/H satisfies conditions (4.5.i)—(4.5.v). Let $\{uH : u \in U\}$ be an arbitrary neighborhood of H in G/H, where U is a neighborhood of e. By (4.5.i), there is a neighborhood V of e in G such that $V^2 \subset U$. Clearly $\{vH : v \in V\}$ is open in G/H and, moreover, $\{vH : v \in V\}^2 = \{yH : y \in V^2\} \subset \{uH : u \in U\}$. Therefore (4.5.i) holds for the group G/H and the topology $\mathcal{O}(G/H)$. The verification of (4.5.ii) for G/H is similar.

We verify (4.5.iii) for G/H. Again let $\{uH : u \in U\}$ be an arbitrary neighborhood of H, where U is a neighborhood of e in G, and let $u_0 H$ be any element of $\{uH : u \in U\}$. If V is a neighborhood of e in G such that $u_0 V \subset U$ (4.5.iii), then $\{vH : v \in V\}$ is a neighborhood of H in G/H, and $u_0 H \cdot \{vH : v \in V\} = \{u_0 vH : v \in V\} \subset \{uH : u \in U\}$. This is just (4.5.iii) for G/H and the topology $\mathcal{O}(G/H)$.

To verify (4.5.iv) for G/H and the topology $\mathcal{O}(G/H)$, we need only note that $xH \cdot \{vH : v \in V\} \cdot (xH)^{-1} = \{xvx^{-1}H : v \in V\}$ and apply (4.5.iv) for the group G and its topology. Property (4.5.v) for G/H is virtually obvious: we omit the verification.

The other statements of the present theorem follow at once from (5.17), (5.16), (5.21), and (5.21) [combined with (4.8)], in that order. □

Theorem (5.26) admits a converse to the effect that all open, continuous homomorphic images of a topological group can be realized as quotient groups G/H with the topology $\mathcal{O}(G/H)$.

(5.27) Theorem. *Let G and $\widetilde{G}$ be topological groups with identities e and $\tilde{e}$, respectively, and let f be an open, continuous homomorphism of G onto $\widetilde{G}$. Then the kernel $f^{-1}(\tilde{e})$ of f is a normal subgroup H of G, and the sets $f^{-1}(\tilde{x}), \tilde{x} \in \widetilde{G}$, are exactly the distinct cosets of H in G. The mapping $\tilde{x} \to f^{-1}(\tilde{x}) = \Phi(\tilde{x})$ is a homeomorphism and an isomorphism of $\widetilde{G}$ onto the group G/H with the topology $\mathcal{O}(G/H)$.*

Proof. In view of (2.2), all that we need to do here is to show that the mapping Φ is a homeomorphism of $\widetilde{G}$ onto G/H. Let $\widetilde{U}$ be an open subset of $\widetilde{G}$. The set $\Phi(\widetilde{U})$ is the family of cosets $\{f^{-1}(\tilde{x}) : \tilde{x} \in \widetilde{U}\}$, which is open in G/H since $\bigcup\{f^{-1}(\tilde{x}) : \tilde{x} \in \widetilde{U}\} = f^{-1}(\widetilde{U})$ is open in G [recall that f is continuous]. Thus Φ is an open mapping, *i.e.*, Φ^{-1} is continuous. Let $\{uH : u \in U\}$ be an open subset of G/H, where U is open in G. Then we have $\Phi^{-1}(\{uH : u \in U\}) = \{\tilde{x} \in \widetilde{G} : f^{-1}(\tilde{x}) = uH$ for some $u \in U\} = \{f(u) : u \in U\} = f(U)$. Since f is by hypothesis an open mapping, the set $f(U)$ is open in $\widetilde{G}$. Thus Φ^{-1} and Φ are continuous, *i.e.*, Φ is a homeomorphism. □

We shall have frequent occasion throughout the book to discuss topological groups G and $\widetilde{G}$ for which there is a mapping f with domain G and range $\widetilde{G}$ which is both a group isomorphism and a homeomorphism. We permit ourselves the slight solecism of calling such groups *topologically isomorphic*. [Under a strict interpretation of the last phrase, two groups might be called topologically isomorphic if there were *two* mappings of G onto $\widetilde{G}$, one a group isomorphism and the other a homeomorphism.] The content of (5.26) and (5.27) is that G/H is topologically isomorphic to $\widetilde{G}$ if and only if the homomorphism f carrying G onto $\widetilde{G}$ is continuous and open $[H = f^{-1}(\tilde{e})]$. Thus $\widetilde{G}$ can be reconstructed not only as a group but also as a topological space from G and the kernel $f^{-1}(\tilde{e})$ provided that f is an open, continuous homomorphism. Simple examples show that these restrictions on f are needed. Consider the additive group R with its discrete topology $\mathscr{P}(R)$, and let f be the identity mapping ι of R onto itself. As we have already seen in (4.18.b), (4.18.d), and (4.22), R has many distinct topologies under which it is a topological group. The mapping ι is a group homomorphism [in fact an isomorphism] and is continuous as a mapping of R with the topology $\mathscr{P}(R)$ onto R with any other topology $\mathcal{O}$. It will be open, however, only if $\mathcal{O} = \mathscr{P}(R)$. Thus there are many continuous homomorphisms [in fact automorphisms] of R that cannot be reconstructed simply from the given topology on R and the kernel of the homomorphism.

Nevertheless, a reasonably large class of continuous homomorphisms are automatically open, as Theorem (5.29) *infra* shows. The following lemma is a trifling improvement on the classical BAIRE category theorem (B.19).

(5.28) Lemma. *A locally countably compact regular space X is not the union of a countable number of closed sets all having void interior.*

Proof. Suppose that $X = \bigcup_{n=1}^{\infty} A_n$, where each A_n is closed and $A_n^{\circ} = \varnothing$. For each n, let $D_n = A_n'$. Then each D_n is open and dense in X. We will show that $\bigcap_{n=1}^{\infty} D_n \neq \varnothing$, contradicting the equality $X = \bigcup_{n=1}^{\infty} A_n$.

Let U_0 be a nonvoid open subset of X such that U_0^- is countably compact. Using regularity and the fact that D_1 is open and dense, we choose a nonvoid open set U_1 such that $U_1^- \subset U_0 \cap D_1$. Having chosen a nonvoid open set U_n such that $U_n^- \subset U_{n-1} \cap D_n$, we choose a nonvoid open set U_{n+1} such that $U_{n+1}^- \subset U_n \cap D_{n+1}$. Since U_0^- is countably compact and each U_n^- is nonvoid, the intersection $\bigcap_{n=0}^{\infty} U_n^-$ is nonvoid. Points in this intersection must lie in $\bigcap_{n=1}^{\infty} D_n$.[1] □

(5.29) Theorem. *Let G be a locally compact group that is σ-compact: $G = \bigcup_{n=1}^{\infty} A_n$, where $A_1, A_2, \ldots$ are compact. Let f be a continuous homomorphism of G onto a locally countably compact T_0 group $\widetilde{G}$. Then f is an open mapping.*

Proof. Let $\mathfrak{U}$ be the family of all symmetric neighborhoods of the identity e in G and $\widetilde{\mathfrak{U}}$ the family of all neighborhoods of the identity $\tilde{e}$ in $\widetilde{G}$. Suppose that

$$\text{for every } U \in \mathfrak{U}, \text{ there is a } \widetilde{U} \in \widetilde{\mathfrak{U}} \text{ such that } \widetilde{U} \subset f(U). \tag{1}$$

Then consider an arbitrary open subset B of G. For each $x \in B$, there is a $U \in \mathfrak{U}$ such that $xU \subset B$. If $\widetilde{U}$ is as in (1), we have $f(x) \in f(x)\widetilde{U} \subset f(x)f(U) = f(xU) \subset f(B)$. Thus $f(B)$ is open in $\widetilde{G}$. To prove the present theorem, therefore, it suffices to establish (1).

To do this, let U be any set in $\mathfrak{U}$, and let $V \in \mathfrak{U}$ have the property that V^- is compact and

$$(V^-)^{-1}(V^-) \subset U. \tag{2}$$

[1] We have proved in fact that $\bigcap_{n=1}^{\infty} D_n$ is dense in X.

Such a V exists in view of (4.5.i), (4.7), and (5.2.ii). The family of sets $\{xV : x \in G\}$ is an open covering of G and hence of each A_n. Since A_n is compact, a finite number of the sets xV cover A_n, and therefore a countable number of the sets xV cover G itself: thus $G = \bigcup_{n=1}^{\infty} (x_n V) = \bigcup_{n=1}^{\infty} (x_n V^-)$, where each $x_n \in G$.[1] Plainly we have $\widetilde{G} = \bigcup_{n=1}^{\infty} f(x_n V^-) = \bigcup_{n=1}^{\infty} f(x_n) f(V^-)$.

We now assert that the set $f(V^-)$ has nonvoid interior in $\widetilde{G}$. The sets $x_n V^-$ are compact subsets of G, and so their continuous images $f(x_n) f(V^-)$ are compact subsets of $\widetilde{G}$. Since $\widetilde{G}$ is a T_0 group, it is also Hausdorff (4.8), and its compact subsets $f(x_n) f(V^-)$ are closed. Assume that $f(V^-)$ has void interior. Since left translation is a homeomorphism, all of the sets $f(x_n) f(V^-)$ then have void interior. Then $\widetilde{G} = \bigcup_{n=1}^{\infty} f(x_n) f(V^-)$ is the countable union of closed sets having void interior. This is impossible by (5.28), and accordingly $f(V^-)$ contains a nonvoid open subset $\widetilde{V}$ of $\widetilde{G}$.

To complete the proof, select any point $\tilde{x} \in \widetilde{V}$ and any point $x \in f^{-1}(\tilde{x}) \cap V^-$. Then by (2), we have $x^{-1} V^- \subset U$ and hence

$$f(U) \supset f(x^{-1} V^-) = f(x)^{-1} f(V^-) \supset \tilde{x}^{-1} \widetilde{V}.$$

The set $\tilde{x}^{-1} \widetilde{V}$ is a member of $\widetilde{\mathfrak{U}}$, and so the present proof is completed. □

(5.30) Note. Two special cases of topological groups satisfying the hypotheses on G in (5.29) are worth noting. First, if G is a locally compact group with a countable open basis, then it is obviously Lindelöf and therefore σ-compact. Second, if G is a compactly generated locally compact group, then G is clearly σ-compact.

(5.31) Theorem. *Let G be a topological group, H a normal subgroup of G, and L any subgroup of G. Let φ be the natural mapping of G onto G/H. Then the quotient group LH/H is topologically isomorphic with the subgroup $\varphi(L)$ of G/H.*

Proof. We have pointed out in (2.1) that LH is a subgroup of G. Plainly H is normal in LH, and so the quotient group LH/H exists and is a topological group with its topology $\mathcal{O}(LH/H)$. Considered simply as families of subsets of G, LH/H and $\varphi(L)$ are identical: the elements of LH/H are the cosets xH with $x \in LH$, and the elements of $\varphi(L)$ are the cosets xH with $x \in L$. These are obviously the same families of sets, and the identity mapping ι of $\{xH : x \in LH\}$ onto $\{xH : x \in L\}$ is an isomorphism of LH/H onto $\varphi(L)$.

---

[1] We actually use here the Lindelöf property for G. A locally compact space is σ-compact if and only if it has the Lindelöf property, so that we would not strengthen the present theorem by supposing that G has the Lindelöf property.

We will prove that ι is a homeomorphism. Let X be a subset of LH. Then $\{xH : x\in X\}$ is open in LH/H if and only if XH is relatively open in LH; that is, if and only if $XH = U\cap(LH)$, where U is open in G. The subset $\{xH : x\in X\}$ is open in $\varphi(L)$ if and only if it is the intersection with $\varphi(L)$ of an open subset of $\varphi(G)$; that is, if and only if there is an open subset V of G such that $\{xH : x\in X\} = \varphi(L)\cap\{vH : v\in V\}$.

Now if $\{xH : x\in X\}$ is open in LH/H, so that $XH = U\cap(LH)$ for an open subset U of G, we have

$$\begin{aligned}\{xH : x\in X\} &= \{xH : x\in U\cap(LH)\} = \{xH : x\in U\}\cap\{xH : x\in LH\}\\ &= \{xH : x\in U\}\cap\varphi(L).\end{aligned}$$

Hence $\{xH : x\in X\}$ is open in $\varphi(L)$. Conversely, suppose that $\{xH : x\in X\}$ is open in $\varphi(L)$. Then for some open subset V of G, we have $\{xH : x\in X\} = \varphi(L)\cap\{vH : v\in V\} = \{xH : x\in V\cap(LH)\}$ and hence $XH = \big(V\cap(LH)\big)H = (VH)\cap(LH.)$ Since VH is open, it follows that $\{xH : x\in X\}$ is open in LH/H. Thus the two topologies are identical, so that ι is a homeomorphism. □

We now take up the analogue for topological groups of the first isomorphism theorem (2.1). If G is a topological group, A is a subgroup of G, and H is a normal subgroup of G, then (2.1) shows that $(AH)/H$ and $A/(A\cap H)$ are isomorphic, the mapping τ defined by $\tau(aH) = a(A\cap H)$ being an isomorphism. The mapping τ does not need to be a homeomorphism, however, and in fact simple examples show that $(AH)/H$ and $A/(A\cap H)$ may be nonhomeomorphic. [See (5.39.d) below.] The isomorphism τ is always an open mapping, however, and under some circumstances it is a homeomorphism. We describe the situation precisely in the following two theorems.

(5.32) Theorem. *Let G be a topological group, A a subgroup of G, and H a normal subgroup of G. Let τ be the isomorphism $\tau(aH) = a(A\cap H)$ $(a\in A)$ described in the first isomorphism theorem, carrying $(AH)/H$ onto $A/(A\cap H)$. Then τ carries open subsets of $(AH)/H$ onto open subsets of $A/(A\cap H)$.*

Proof. An open subset of $(AH)/H$ is a subset $\{xH : x\in X\}$, where $X\subset A$, such that XH is open in AH regarded as a subspace of G. The set $\tau(\{xH : x\in X\})$ is the set $\{x(A\cap H) : x\in X\}$. Since $X(A\cap H) = (XH)\cap A$, it follows that $X(A\cap H)$ is an open subset of A in its relative topology as a subspace of G, and so, by the definition of the topology in $A/(A\cap H)$, $\{x(A\cap H) : x\in X\}$ is an open subset of $A/(A\cap H)$. □

(5.33) Theorem. *Notation is the same as in (5.32). Suppose that A is locally compact and σ-compact, that H is closed, and that AH is locally compact. Then τ is a homeomorphism, and so $(AH)/H$ and $A/(A\cap H)$ are topologically isomorphic.*

Proof. In view of (2.1) and (5.32), we need only show that τ^{-1} carries open subsets of $A/(A\cap H)$ onto open subsets of $(AH)/H$.

Consider first the natural mapping φ of G onto G/H. Restricted to the subgroup A, φ carries A *on*to the subgroup $(AH)/H$ of G/H. Since AH is locally compact and H is closed, the quotient group $(AH)/H$ is locally compact (5.22) and T_0 (5.21). Thus φ is a continuous homomorphism of the locally compact σ-compact group A onto the locally compact T_0 group $(AH)/H$. By (5.29), φ is an open mapping on A.

Now let $\{y(A\cap H):y\in Y\}, Y\subset A$, be an open subset of $A/(A\cap H)$; that is, $Y(A\cap H)$ is an open subset of A. Computing $\varphi(Y(A\cap H))$, we see at once that it is the subset $\{yH:y\in Y\}$ of $(AH)/H$. Since φ is an open mapping on A, the set $\{yH:y\in Y\}$ is open in $(AH)/H$. We also have, by the definition of τ, that $\{yH:y\in Y\}=\tau^{-1}\{y(A\cap H):y\in Y\}$. That is, τ^{-1} carries open subsets of $A/(A\cap H)$ onto open subsets of $(AH)/H$. □

The second isomorphism theorem (2.2) has a complete analogue for topological groups, as follows.

(5.34) Theorem. *Let G and $\widetilde{G}$ be topological groups with identity elements e and $\tilde{e}$, respectively; let f be an open continuous homomorphism of G onto $\widetilde{G}$. Let $\widetilde{H}$ be any normal subgroup of $\widetilde{G}$, $H=f^{-1}(\widetilde{H})$, and $N=f^{-1}(\tilde{e})$. Then the groups G/H, $\widetilde{G}/\widetilde{H}$, and $(G/N)/(H/N)$ are topologically isomorphic.*

Proof. Let ψ be the natural mapping of $\widetilde{G}$ onto $\widetilde{G}/H$. By (5.26), ψ is an open continuous homomorphism and hence $\psi\circ f$ is an open continuous homomorphism of G onto $\widetilde{G}/\widetilde{H}$ with kernel H. Hence $\widetilde{G}/\widetilde{H}$ is topologically isomorphic with G/H by (5.27). Since the mapping $\tilde{x}\to f^{-1}(\tilde{x})$ is a topological isomorphism of $\widetilde{G}$ onto G/N by (5.27) and the image of $\widetilde{H}$ under this mapping is H/N, $\widetilde{G}/\widetilde{H}$ is topologically isomorphic with $(G/N)/(H/N)$. □

Theorem (5.34) can be restated as follows.

(5.35) Theorem. *Let G be a topological group and H and N normal subgroups of G such that $N\subset H$. Then G/H is topologically isomorphic with $(G/N)/(H/N)$.*

We can now show to what extent the study of topological groups can be reduced to the study of Hausdorff groups.

(5.36) Theorem. *Let G be a topological group. Then $G/\{e\}^-$ is a Hausdorff group. If f is any continuous homomorphism of G onto a T_0 group $\widetilde{G}$ with identity $\tilde{e}$, then $f^{-1}(\tilde{e})\supset\{e\}^-$. If f is also open, then $\widetilde{G}$ is an open continuous homomorphic image of $G/\{e\}^-$.*

Proof. Since $\{e\}^-$ is a closed normal subgroup of G (5.4), $G/\{e\}^-$ is a T_1 group (5.21) and hence a Hausdorff group (4.8). Now $f^{-1}(\tilde{e})$ is a normal subgroup of G and is closed since f is continuous. Hence we have $f^{-1}(\tilde{e}) \supset \{e\}^-$. Finally, if f is an open continuous homomorphism, $\tilde{G}$ is topologically isomorphic with $G/f^{-1}(\tilde{e})$. By (5.35), $G/f^{-1}(\tilde{e})$ is topologically isomorphic with $(G/\{e\}^-)/(f^{-1}(\tilde{e})/\{e\}^-)$, which is an open continuous homomorphic image of $G/\{e\}^-$ (5.26). □

Miscellaneous theorems and examples

(5.37) Subgroups. (a) (Bourbaki [1], p. 10.) A subgroup H of a topological group G is closed if and only if there is an open subset U of G such that $U \cap H = U \cap H^-$ and $U \cap H \neq \varnothing$.

(b) If H is a nonclosed subgroup of a topological group, then $H^- \cap H'$ is dense in H^-.

(c) Let H be a dense subgroup of a topological group G, and let J be a normal subgroup of H. Then J^- is a normal subgroup of G.

(d) The intersection of all open subgroups and the intersection of all closed subgroups of a topological group G are closed normal subgroups of G.

(e) The center of a T_0 group G is a closed normal subgroup of G.

(f) Let G be a topological group such that $\bigcup_{n=1}^{\infty} V^n = G$ for every neighborhood V of e. Let H be a discrete normal subgroup of G. Then H is contained in the center of G. [There is a neighborhood U of e such that $(aU) \cap H = \{a\}$ for every $a \in H$. By continuity of multiplication, there is a neighborhood V of e such that $V^{-1}aV \subset aU$. Hence $x^{-1}ax = a$ for all $a \in H$ and $x \in V$.]

(g) Let G be any group and N a normal subgroup of G having at least two elements. Topologize G as indicated in (4.21.b). Then G is a topological group and $\{e\}$ is a discrete subgroup that is not closed; compare with (5.10).

(h) Let G be any topological group, with identity e. The subset $\{e\}^-$ of G has the weakest topology: $\varnothing$ and $\{e\}^-$ are its only open sets.

(i) The additive group R contains open and nonclosed subsemigroups, nondiscrete closed subsemigroups having isolated points, and discrete nonclosed subsemigroups. [An example of a discrete nonclosed subsemigroup of R is the semigroup generated by the set $\{1+r_n\}_{n=1}^{\infty}$ where $\{r_n\}_{n=1}^{\infty}$ is a sequence of positive real numbers converging to 0 for which the set $\{1, r_1, r_2, \ldots\}$ is rationally independent.]

(5.38) Quotient spaces. (a) If G is a compact group and H is a closed subgroup of G, then the natural mapping φ of G onto G/H is a closed mapping. [This follows from (5.18). To see that H needs to be closed, consider $G = T$ and $H = \{\exp(2\pi i r) : r \in Q\}$.]

(b) Let G be any topological group and H a subgroup of G. Let F be a closed subset of G that is the union of cosets xH. Then $\varphi(F)$ is closed in G/H.

(c) (Aleksandrov and Hopf [1], p. 66.) Let G be a topological group and H a subgroup of G. Introduce a topology $\mathcal{Q}$ into the set G/H of left cosets xH by the following rule. Given a coset aH, consider any open set U such that $aH \subset U \subset G$. Consider the family of all cosets xH such that $xH \subset U$. Let the collection of all such families of cosets be an open basis for the open sets of $\mathcal{Q}$. Then the topology $\mathcal{Q}$ is stronger than $\mathcal{O}(G/H)$, and $\mathcal{Q} = \mathcal{O}(G/H)$ if and only if for every coset aH and every open set U such that $aH \subset U$, there is an open set V such that $aH \subset V$ and such that $x \in G$ and $xH \cap V \neq \varnothing$ imply that $xH \subset U$. If H is compact, then $\mathcal{Q} = \mathcal{O}(G/H)$.

(d) Let G be a topological group and H a subgroup of G. Then $\mathcal{O}(G/H)$ is the weakest topology, $\{\varnothing, G/H\}$, if and only if $H^- = G$.

(e) (Adapted from Graev [4].) Let G be a topological group and H a subgroup of G. If H and G/H have countable open bases at each point, so does G. [Let $\{W_n\}_{n=1}^{\infty}$ be a sequence of symmetric neighborhoods of e in G such that $W_{n+1}^2 \subset W_n$ for each n and such that $\{W_n \cap H\}_{n=1}^{\infty}$ is an open basis at e for the group H. Let $\{\{uH : u \in U_n\}\}_{n=1}^{\infty}$ be an open basis at the point H of G/H, where the U_n are neighborhoods of e in G. For $n = 1, 2, \ldots,$ let

$$P_n = (W_{n+1} \cdot (H \cap W_n'))^{-\prime} \cap (U_n H)$$

and $Q_n = P_1 \cap P_2 \cap \cdots \cap P_n$. It is easy to see that every Q_n is a neighborhood of e. Also $\{Q_n\}_{n=1}^{\infty}$ is an open basis at e in G. To prove this, let Y be any neighborhood of e in G and V a neighborhood of e such that $V^2 \subset Y$. Let φ denote, as usual, the natural mapping of G onto G/H. There is an m such that $W_m \cap H \subset V \cap H$, and a $k > m$ such that $\varphi(U_k) \subset \varphi(V \cap W_{m+1})$. Then we have

$$\begin{aligned} Q_k &\subset P_k \cap P_m \subset (U_k H) \cap (W_{m+1} \cdot (H \cap W_m'))^{-\prime} \\ &\subset ((V \cap W_{m+1}) H) \cap ((V \cap W_{m+1}) \cdot (H \cap W_m'))' \subset (V \cap W_{m+1}) \cdot (H \cap W_m) \\ &\subset (V \cap W_{m+1}) \cdot (V \cap H) \subset V^2 \subset Y.] \end{aligned}$$

(f) Let G be a topological group and H a subgroup of G such that H and G/H contain dense subsets of cardinal numbers $\mathfrak{m}$ and $\mathfrak{n}$, respectively. Then G contains a dense subset of cardinal number less than or equal to $\mathfrak{m}\mathfrak{n}$. The cardinal numbers $\mathfrak{m}$ and $\mathfrak{n}$ may be finite or infinite. [Let A be dense in H and let $\{bH : b \in B\}$ be dense in G/H. If U is open in G and $U \neq \varnothing$, then there is a bH in $\{uH : u \in U\}$. Thus we have $(bH) \cap U \neq \varnothing$ and $H \cap (b^{-1}U) \neq \varnothing$. Hence $A \cap (b^{-1}U) \neq \varnothing$ and $(BA) \cap U \neq \varnothing$. Consequently, BA is dense in G.]

(g) An instructive example of a topological group, a closed subgroup, and the corresponding coset space is the following. Consider the group $\mathfrak{G}\mathfrak{L}(2, R)$, which consists of all 2×2 real matrices $\begin{pmatrix} x & y \\ z & w \end{pmatrix}$ for which $\Delta = xw - yz \neq 0$. Consider the subgroup $\mathfrak{H}$ consisting of all matrices $\begin{pmatrix} x & y \\ 0 & 1 \end{pmatrix}$ with $x \neq 0$. Consider the mapping T of $\mathfrak{G}\mathfrak{L}(2, R)$ onto $R^2 \cap \{(0,0)\}' = E$ defined by

$$T\begin{pmatrix} x & y \\ z & w \end{pmatrix} = \left(\frac{x}{\Delta}, \frac{z}{\Delta}\right).$$

Given matrices A and B in $\mathfrak{G}\mathfrak{L}(2, R)$, it is not hard to prove that $T(A) = T(B)$ if and only if $B^{-1}A \in \mathfrak{H}$. Hence, if we define $T^*(A\mathfrak{H}) = T(A)$ for $A\mathfrak{H} \in \mathfrak{G}\mathfrak{L}(2, R)/\mathfrak{H}$, then T^* is a one-to-one mapping of $\mathfrak{G}\mathfrak{L}(2, R)/\mathfrak{H}$ onto E. It is easy to see that T^* is continuous. To show that T^* is open, use the fact that for any $(u, v) \in E$, $T\begin{pmatrix} u & \frac{-v}{u^2+v^2} \\ v & \frac{u}{u^2+v^2} \end{pmatrix} = (u, v)$, and the fact that u, v, $\frac{-v}{u^2+v^2}$, and $\frac{u}{u^2+v^2}$ are continuous functions of u and v. Thus the topology of $\mathfrak{G}\mathfrak{L}(2, R)/\mathfrak{H}$ is the ordinary topology of E. For $A = \begin{pmatrix} a & b \\ c & d \end{pmatrix} \in \mathfrak{G}\mathfrak{L}(2, R)$, with $\Delta = ad - bc$, the homeomorphism ${}_A\psi$ of $\mathfrak{G}\mathfrak{L}(2, R)/\mathfrak{H}$ (5.20) corresponds to the homeomorphism η_A of E defined by

$$\eta_A(u, v) = \left(\frac{au+bv}{\Delta}, \frac{cu+dv}{\Delta}\right).$$

(5.39) Quotient groups. (a) The group R/Z is topologically isomorphic with the group T.

(b) Let G be a topological group and H a normal subgroup of G. The correspondence $L \to \varphi(L)$ is a one-to-one correspondence carrying the family of all subgroups of G containing H onto the family of all subgroups of G/H. Furthermore, a subgroup L is closed if and only if $\varphi(L)$ is closed.

(c) In (5.31), suppose that L is an arbitrary nonvoid subset of G, and let M be the smallest subgroup of G containing LH. Then $\varphi(L)$ is homeomorphic with the image of LH in M/H.

(d) Let θ be any irrational real number. Then $(\theta Z)/((\theta Z) \cap Z)$ is discrete and $((\theta Z)+Z)/Z$ is not. [Describe the group $((\theta Z)+Z)/Z$ as a subgroup of the compact group R/Z. Why does this example not contradict (5.33)?]

(e) The results obtained in (5.32) and (5.33) on the second isomorphism theorem can be extended. Notation is the same as in (5.32). In proving (5.33), we showed that if φ is open as a mapping of A onto AH/H, then τ is a homeomorphism. The converse is also true. [The condition that τ be a homeomorphism can be recast as follows: if Y is a subset of A such that $Y \cdot (A \cap H) = A \cap U$ for some open subset U of G, then $YH = V \cap (AH)$ for some other open subset V of G. Now if W is open in G, it is easy to see that $(A \cap W) \cdot (A \cap H) = A \cap (W(A \cap H))$. Thus if τ is a homeomorphism, we have $(A \cap W) \cdot H = V \cap (AH)$ for some open subset V of G. Hence φ is an open mapping of A onto AH/H.]

(f) (FREUDENTHAL [1].) Notation is as in (5.32). Suppose that τ is a homeomorphism, that A is locally compact, and that H is closed. Then AH is closed in G. [By (5.22), $A/(A \cap H)$ is locally compact. Hence AH/H is locally compact and therefore closed in G/H, by (5.11). This obviously implies that AH is closed in G.]

(g) Let G be a topological group and H a normal subgroup of G. Then G/H^- is topologically isomorphic with $(G/H)/\{H\}^-$, where $\{H\}^-$ is the closure in G/H of the identity element H. [Note that $\{H\}^-$ is H^-/H, and apply (5.35).]

(h) Let G be a topological group with a compact normal subgroup H such that G/H is compactly generated. Then G itself is compactly generated. [Let φ be the natural mapping of G onto G/H and suppose that $\{xH : x \in A\}$ is a compact subset of G/H that generates G/H. By (5.24.a), AH is compact. Therefore $(AH) \cup H$ is compact; plainly $(AH) \cup H$ generates G.]

(i) Let G be a locally compact group and H a normal subgroup of G. If both G/H and H are compactly generated, then G is also compactly generated. [Let $\{xH : x \in X\}$ be a closed compact subset of G/H that generates G/H. [Such a set exists by (3.3) and (4.7).] By (5.24.b), we may suppose that X is compact. If A is a compact subset of H that generates H, then $A \cup X$ is a compact subset of G that generates G.]

(j) Theorem (5.27) has the following analogue. Let f be a continuous homomorphism of a compact group G onto a Hausdorff group $\widetilde{G}$. Then f is an open mapping, and G/H and $\widetilde{G}$ are topologically isomorphic, where H denotes the kernel of f. [For $\tilde{x} \in \widetilde{G}$, let $\Phi(\tilde{x}) = f^{-1}(\tilde{x})$. Then Φ is an isomorphism of $\widetilde{G}$ onto G/H. As in the proof of (5.27), the continuity of f implies that Φ^{-1} is a continuous mapping of G/H onto $\widetilde{G}$. Since G/H is compact (5.22), Φ^{-1} is a homeomorphism. That is, $\widetilde{G}$ and G/H are topologically isomorphic; f is now an open mapping by (5.26) since $f(x) = \Phi^{-1}(xH)$ for $x \in G$.]

(5.40) Homomorphisms. (a) Let G and $\widetilde{G}$ be topological groups with identity elements e and $\tilde{e}$, respectively, and let f be a homomorphism of G into $\widetilde{G}$. If f is continuous at some point of G, then f is uniformly continuous for the pairs of uniform structures $(\mathscr{S}_l(G), \mathscr{S}_l(\widetilde{G}))$ and $(\mathscr{S}_r(G), \mathscr{S}_r(\widetilde{G}))$. [Suppose that f is continuous at $x_0 \in G$. Then if $\widetilde{U}$

is a neighborhood of $\tilde{e}$, there is a neighborhood U of e such that $f(x_0 U) \subset f(x_0)\tilde{U}$. Thus if $y^{-1}x \in U$, we have $f(x_0 y^{-1} x) \in f(x_0)\tilde{U}$. That is, $f(y)^{-1}f(x) \in \tilde{U}$. Consequently, f is uniformly continuous for the pair of uniform structures $(\mathscr{S}_l(G), \mathscr{S}_l(\tilde{G}))$; a similar argument applies to the pair of uniform structures $(\mathscr{S}_r(G), \mathscr{S}_r(\tilde{G}))$.]

(b) Let G and $\tilde{G}$ be topological groups, and let f be a homomorphism of G onto $\tilde{G}$. If $f(U)$ has nonvoid interior for all U in an open basis at e, then f is an open mapping.

(c) Let $G, \tilde{G}$, and $\tilde{\tilde{G}}$ be topological groups. Let f and $\tilde{f}$ be continuous homomorphisms of G onto $\tilde{G}$ and of $\tilde{G}$ onto $\tilde{\tilde{G}}$, respectively. If $\tilde{f} \circ f$ is a topological isomorphism, then f and $\tilde{f}$ are also topological isomorphisms.

(d) Let G be a locally compact group, $\tilde{G}$ a topological group, and φ a continuous homomorphism of G *into* $\tilde{G}$ such that φ is an open mapping of G *onto* $\varphi(G) \subset \tilde{G}$. Then $\varphi(G)$ is a locally compact subgroup of $\tilde{G}$; if $\tilde{G}$ is a T_0 group ,then $\varphi(G)$ is closed in $\tilde{G}$. [By (5.27), $\varphi(G)$ is topologically isomorphic with $G/\varphi^{-1}(\tilde{e})$. By (5.22), $G/\varphi^{-1}(\tilde{e})$ is locally compact. By (5.11), $\varphi(G)$ is closed if $\tilde{G}$ is a T_0 group.]

(e) The image of a closed subgroup under an open continuous homomorphism need not be closed. [Consider the mapping $t \to \exp[2\pi i t]$ of R onto T, and the image of the subgroup $\{n\sqrt{2} : n \in Z\}$.]

(f) The group R is an open continuous homomorphic image of a totally disconnected topological group. [Consider the group G of all Cauchy sequences $\boldsymbol{r} = (r_1, r_2, \ldots, r_n, \ldots)$ of rational numbers, and define addition coordinate-wise. For $t > 0$, let A_t consist of all $\boldsymbol{r} \in G$ such that $|r_n| \leq t$ for all n. For $\alpha > 0$, let $V_\alpha = \bigcup_{t<\alpha} A_t$. Then $\{V_\alpha + \boldsymbol{r} : \alpha > 0, \boldsymbol{r} \in G\}$ is an open basis for a topology on G under which G is a topological group. Let H be the subgroup of G consisting of all $\boldsymbol{r}$ such that $\lim_{n\to\infty} r_n = 0$. Then G/H with the topology $\mathcal{O}(G/H)$ is topologically isomorphic with R. Suppose that $\boldsymbol{s} \neq \boldsymbol{t}$, where $\boldsymbol{s}, \boldsymbol{t} \in G$. Then $s_{n_0} \neq t_{n_0}$ for some positive integer n_0; say $s_{n_0} < t_{n_0}$. Let $\theta \in]s_{n_0}, t_{n_0}[$ be irrational. Then $U = \{\boldsymbol{r} \in G : r_{n_0} < \theta\}$ is open and closed, $\boldsymbol{s} \in U$, and $\boldsymbol{t} \in U'$. Thus G is totally disconnected. The preceding is simply the usual construction of R from Q using sequences ,with the topology of R defined in a novel way.]

(g) (GRAEV [4].) Let G be a group and let f be a homomorphism of G onto a topological group $\tilde{G}$. Define a topology $\mathcal{O}$ on G by: $U \in \mathcal{O}$ if and only if $U = f^{-1}(\tilde{U})$, where $\tilde{U}$ is open in $\tilde{G}$. Then G with the topology $\mathcal{O}$ is a topological group.

(h) (GRAEV [4].) Let G be a topological group with topology $\mathcal{O}$, and let X be a subset of G. Let $\{\mathcal{O}_\lambda\}_{\lambda \in \Lambda}$ be the family of all topologies under which G is a topological group and which induce a topology on X weaker than the topology of X induced by $\mathcal{O}$. Let $\mathcal{O}_0$ be the weakest topology on G stronger than all of the $\mathcal{O}_\lambda$. Then if f is a homomorphism of G into a topological group $\widetilde{G}$ and f is continuous when restricted to the domain X, f is continuous on G in the topology $\mathcal{O}_0$.

(i) Let f be a homomorphism of the topological group H into the topological group G and let φ be the natural mapping of G onto $G/\{e\}^-$, where e is the identity in G. If $\varphi \circ f$ is continuous, then f is continuous. [If U is an open subset of G, then $U\{e\}^- = U$, and hence $f^{-1}(U) = (\varphi \circ f)^{-1}(\{x\{e\}^- : x \in U\})$.]

(j) Let f be a homomorphism of the topological group H onto the topological group G and let φ be the natural mapping of G onto $G/\{e\}^-$, where e is the identity in G. If $\varphi \circ f$ is an open mapping and $f^{-1}(\{e\}^-) = (f^{-1}(e))^-$, then f is an open mapping. [Let U be an open set in H. Clearly $\varphi^{-1} \circ \varphi \circ f(U)$ is open in G; we will show that $f(U) = \varphi^{-1} \circ \varphi \circ f(U)$. The inclusion $f(U) \subset \varphi^{-1} \circ \varphi \circ f(U)$ always holds. Let $y \in \varphi^{-1} \circ \varphi \circ f(U)$. Then for some $z \in U$, $\varphi(y) = \varphi(f(z))$ and clearly $y = f(x)$ for some $x \in H$. This implies that

$$f(x^{-1}z) = f(x)^{-1} f(z) = y^{-1} f(z) \in \{e\}^-,$$

and hence $x^{-1}z \in f^{-1}(\{e\}^-) = (f^{-1}(e))^-$. Now $x^{-1}U$ is a neighborhood of $x^{-1}z$ and hence intersects $f^{-1}(e)$. That is, for some $u \in U$ we have $f(x^{-1}u) = e$ and hence $y = f(x) = f(u) \in f(U)$.]

Notes

Like the contents of § 4, many of the results of § 5 go back to the early 1930's: D. VAN DANTZIG [2] and [3], H. FREUDENTHAL [1], D. VAN KAMPEN [1] and A. MARKOV [1] have all contributed to the theory of subgroups and quotient groups. The topology described in (5.15) was introduced by VAN DANTZIG [2], and independently [also in a more general framework] by R. BAER and F. LEVI [1]. MARKOV [1] proved (5.17), and also showed that not every continuous homomorphism is open. In view of (5.29), this phenomenon could not be recognized so long as attention was restricted to locally compact groups with countable open bases. VAN DANTZIG [3] did, however, state an incorrect result [TG. 20, p. 609]. FREUDENTHAL [1] also emphasized the distinction between continuous homomorphisms and open continuous homomorphisms. Theorem (5.22) appears in VAN KAMPEN [1], MARKOV [1], and FREUDENTHAL [1]. The first assertion in (5.25) is due to FREUDENTHAL [1]. The second appears in MONTGOMERY and ZIPPIN [1], p. 52; our proof is modelled on theirs. Theorems (5.26) and (5.27) are first explicitly stated in FREUDENTHAL [1], although MARKOV [1] comes fairly close. Theorem (5.29) is a modification of PONTRYAGIN [7],

Theorem 12. Theorem (5.31) is taken from BOURBAKI [1], p. 14. FREUDENTHAL [1] proved a special case of (5.33), as well as (5.34). Theorem (5.36) appears in BOURBAKI [1], p. 13.

N. YA. VILENKIN [14] has made a detailed study of topological groups *not* satisfying the T_0 separation axiom, and in particular of quotient groups G/H where H is a nonclosed normal subgroup of G.

§ 6. Product groups and projective limits

The principal method of constructing topological groups to be considered in this section is the formation of Cartesian products. This is an extremely important process, which frequently makes it possible to reduce problems concerning complicated topological groups to problems concerning simpler ones.

(6.1) Definition. Let $\{G_\iota : \iota \in I\}$ be a nonvoid family of topological groups, and let $\underset{\iota\in I}{\mathsf{P}} G_\iota$ be the direct product group (2.3). Then the group $\underset{\iota\in I}{\mathsf{P}} G_\iota$, with the Cartesian product topology, is called the *direct product* of the topological groups G_ι.

Whenever we refer to the direct product of topological groups G_ι, we will always mean the group $\underset{\iota\in I}{\mathsf{P}} G_\iota$ with the Cartesian product topology. The following theorem is just what one ought to conjecture about direct products of topological groups.

(6.2) Theorem. *The direct product* $\underset{\iota\in I}{\mathsf{P}} G_\iota$ *of the topological groups* G_ι *is a topological group, and* $\underset{\iota\in I}{\mathsf{P}^*} G_\iota$ *is a dense subgroup of* $\underset{\iota\in I}{\mathsf{P}} G_\iota$.

Proof. To show that $\underset{\iota\in I}{\mathsf{P}} G_\iota$ is a topological group, we need only to verify axioms (4.1.i) and (4.1.ii). Let us verify (4.1.i). Let (a_ι) and (b_ι) be points of $\underset{\iota\in I}{\mathsf{P}} G_\iota$ and let A be any neighborhood of $(a_\iota)(b_\iota)$. Then, by the definition of the Cartesian product topology, there are open subsets U_ι of G_ι such that $a_\iota b_\iota \in U_\iota$ for all $\iota \in I$, such that only a finite number of the U_ι are different from G_ι, and such that $\underset{\iota\in I}{\mathsf{P}} U_\iota \subset A$. Let V_ι and W_ι be neighborhoods of a_ι and b_ι, respectively, in G_ι such that $V_\iota W_\iota \subset U_\iota$, for all $\iota \in I$, and such that V_ι and W_ι are equal to G_ι if U_ι is equal to G_ι. Let $V = \underset{\iota\in I}{\mathsf{P}} V_\iota$ and $W = \underset{\iota\in I}{\mathsf{P}} W_\iota$. Then it is obvious that V and W are neighborhoods of (a_ι) and (b_ι), respectively, and that $VW \subset \underset{\iota\in I}{\mathsf{P}} U_\iota \subset A$. Axiom (4.1.ii) is verified similarly. Hence $\underset{\iota\in I}{\mathsf{P}} G_\iota$ is a topological group.

It is obvious that $\underset{\iota\in I}{\mathsf{P}^*} G_\iota$ is a subgroup of $\underset{\iota\in I}{\mathsf{P}} G_\iota$. To prove that it is dense, we must show that every nonvoid open subset of $\underset{\iota\in I}{\mathsf{P}} G_\iota$ contains a point of $\underset{\iota\in I}{\mathsf{P}^*} G_\iota$. Let $\underset{\iota\in I}{\mathsf{P}} U_\iota$, where each U_ι is open in G_ι and only a finite number of the U_ι are different from G_ι, be any nonvoid basic open set in $\underset{\iota\in I}{\mathsf{P}} G_\iota$. Let x_ι be any point of U_ι for all ι such that $U_\iota \neq G_\iota$, and let $x_\iota = e_\iota$ [the identity in G_ι] for all ι such that $U_\iota = G_\iota$. Then

$$(x_\iota) \in \left(\underset{\iota\in I}{\mathsf{P}} U_\iota\right) \cap \left(\underset{\iota\in I}{\mathsf{P}^*} G_\iota\right). \quad \square$$

(6.3) Theorem [Associativity of direct products]. *Let $\{G_\iota : \iota \in I\}$ be a nonvoid family of topological groups, and let $\{I_\lambda\}_{\lambda\in\Lambda}$ be a partition of I. Let A_λ be the topological group $\underset{\iota\in I_\lambda}{\mathsf{P}} G_\iota$, for each $\lambda \in \Lambda$. Then $\underset{\iota\in I}{\mathsf{P}} G_\iota$ is topologically isomorphic with $\underset{\lambda\in\Lambda}{\mathsf{P}} A_\lambda$.*

This theorem is obvious upon a little reflection, although to write out the proof in all of its details is somewhat cumbersome. We leave this task to the interested reader.

As is to be expected, special properties of the groups G_ι are reflected in special properties of the direct product $\underset{\iota\in I}{\mathsf{P}} G_\iota$. Throughout (6.4) to (6.9), to avoid tedious repetition of notation, we take $\{G_\iota : \iota \in I\}$ to be a nonvoid family of topological groups; we write e_ι for the identity in G_ι, and G for the direct product $\underset{\iota\in I}{\mathsf{P}} G_\iota$.

(6.4) Theorem. *The group G is a T_0 group if and only if each G_ι is a T_0 group. The group G is connected if and only if every group G_ι is connected, and G is compact if and only if each G_ι is compact. Finally, the group G is locally compact if and only if all of the groups G_ι are locally compact and all but a finite number of them are compact.*

Proof. These assertions follow immediately from the corresponding assertions for arbitrary topological spaces. $\square$

(6.5) Note. It is also well known that $G = \underset{\iota\in I}{\mathsf{P}} G_\iota$ is completely regular if and only if every G_ι is completely regular. The corresponding assertion for normal groups fails. We will show in (8.11) and (8.12) that there are many nonnormal products of normal groups.

(6.6) Theorem. *The group G is Abelian if and only if all of the groups G_ι are Abelian.*

Proof. Obvious. $\square$

We shall now take up subgroups and quotient groups of G, and their relation to the corresponding entities in the groups G_ι.

(6.7) Theorem. *Let A_ι be a subset of G_ι for all $\iota \in I$. Then if A_ι is a subsemigroup, subgroup, or normal subgroup of G_ι for all $\iota \in I$, it follows that $\underset{\iota\in I}{\mathsf{P}} A_\iota$ is a subsemigroup, subgroup, or normal subgroup, respectively, of G.*

Proof. Obvious. $\square$

(6.8) Theorem. *Let J be any nonvoid subclass of I. Let the mapping π_J of G onto $\underset{\iota\in J}{\mathsf{P}} G_\iota$ be defined by $\pi_J((x_\iota)_{\iota\in I}) = (x_\iota)_{\iota\in J}$. [The mapping π_J is called a* projection.] *Then π_J is an open continuous homomorphism of G onto $\underset{\iota\in J}{\mathsf{P}} G_\iota$. Its kernel consists of all $(x_\iota)\in G$ such that $x_\iota = e_\iota$ for all $\iota\in J$.*

Proof. This result too is obvious upon a little reflection, and we leave the verification to the reader. ☐

(6.9) Theorem. *For each $\iota\in I$, let H_ι be a subgroup of G_ι. Consider the two topological spaces $\underset{\iota\in I}{\mathsf{P}}(G_\iota/H_\iota)$ and $(\underset{\iota\in I}{\mathsf{P}} G_\iota)/(\underset{\iota\in I}{\mathsf{P}} H_\iota)$. Let Φ be the mapping of the first space onto the second defined by $\Phi[(x_\iota H_\iota)] = (x_\iota)\underset{\iota\in I}{\mathsf{P}} H_\iota = \underset{\iota\in I}{\mathsf{P}}[x_\iota H_\iota]$. Then Φ [the so-called* natural mapping *of the first space onto the second] is a homeomorphism. If each H_ι is a normal subgroup, so that both spaces are groups, Φ is an isomorphism. Thus, in this case, $\underset{\iota\in I}{\mathsf{P}}(G_\iota/H_\iota)$ and $(\underset{\iota\in I}{\mathsf{P}} G_\iota)/(\underset{\iota\in I}{\mathsf{P}} H_\iota)$ are topologically isomorphic.*

Proof. It is obvious that Φ is one-to-one and onto. Consider any open set in $\underset{\iota\in I}{\mathsf{P}}(G_\iota/H_\iota)$ of the form $\underset{\iota\in I}{\mathsf{P}}\{x_\iota H_\iota : x_\iota\in U_\iota\}$, where each U_ι is open in G_ι and only finitely many U_ι are different from G_ι. The image under Φ of this set is the set $\{\underset{\iota\in I}{\mathsf{P}}[x_\iota H_\iota] : (x_\iota)\in \underset{\iota\in I}{\mathsf{P}} U_\iota\}$. The union in $G = \underset{\iota\in I}{\mathsf{P}} G_\iota$ of the sets in this collection is the set $\underset{\iota\in I}{\mathsf{P}} U_\iota H_\iota$, which is obviously open in G. Thus Φ is an open mapping. The same argument shows that Φ^{-1} is open; *i.e.*, Φ is a homeomorphism.

Suppose next that H_ι is a normal subgroup of G_ι for all $\iota\in I$. Let $(x_\iota H_\iota)$ and $(y_\iota H_\iota)$ be any elements of $\underset{\iota\in I}{\mathsf{P}}(G_\iota/H_\iota)$. Then we have $\Phi[(x_\iota H_\iota)(y_\iota H_\iota)] = \Phi[(x_\iota y_\iota H_\iota)] = (x_\iota y_\iota)\underset{\iota\in I}{\mathsf{P}} H_\iota = (x_\iota)\underset{\iota\in I}{\mathsf{P}} H_\iota \cdot (y_\iota)\underset{\iota\in I}{\mathsf{P}} H_\iota = \Phi[(x_\iota H_\iota)]\,\Phi[(y_\iota H_\iota)]$. Thus Φ is a one-to-one homomorphism, and is therefore an isomorphism. ☐

(6.10) Suppose that we are given a topological group X and a collection of subgroups, $\{N_\iota\}_{\iota\in I}$, of X. Under what conditions can we assert that X is topologically isomorphic with the direct product of the subgroups N_ι? Let us first consider some obvious necessary conditions. Let $G = \underset{\iota\in I}{\mathsf{P}} G_\iota$, and for each $\iota_0\in I$, let $\widetilde{G}_{\iota_0} = G_{\iota_0} \times \underset{\iota\neq\iota_0}{\mathsf{P}}\{e_\iota\}$. Clearly each $\widetilde{G}_{\iota_0}$ is a normal subgroup of G, and is closed if and only if all of the G_ι $(\iota\neq\iota_0)$ are T_0 groups. If ι_0 and ι_1 are distinct indices in I, then every element of $\widetilde{G}_{\iota_0}$ commutes with every element of $\widetilde{G}_{\iota_1}$. Each product set $\widetilde{G}_{\iota_1}\widetilde{G}_{\iota_2}\cdots\widetilde{G}_{\iota_n}$ is a normal subgroup of G; $(\widetilde{G}_{\iota_1}\widetilde{G}_{\iota_2}\cdots\widetilde{G}_{\iota_n}) \cap \widetilde{G}_\iota = \{(e_\iota)\}$ if ι is different from $\iota_1, \ldots, \iota_n$; and the union of all of the subgroups $\widetilde{G}_{\iota_1}\widetilde{G}_{\iota_2}\cdots\widetilde{G}_{\iota_n}$ is dense in G. If I is finite, say $I = \{1, 2, \ldots, m\}$, then $G = \widetilde{G}_1\widetilde{G}_2\cdots\widetilde{G}_m$. Furthermore, if $I = \{1, 2, \ldots, m\}$, and if $\widetilde{U}_\iota$ is a neighborhood of (e_ι) in $\widetilde{G}_\iota$ [relative topology] for $\iota = 1, 2, \ldots, m$, then the product set $\widetilde{U}_1\widetilde{U}_2\cdots\widetilde{U}_m$ contains a neighborhood of (e_ι) in G.

We can now attempt to answer the question posed above. For the general case, where X has an infinite number of subgroups that satisfy the conditions observed for the groups G_ι, there are serious difficulties, and we know of no way to construct X as a *full* direct product. For the case of a finite number of subgroups a satisfactory answer to our question can be provided. We give it in the following two theorems.

(6.11) Theorem. *Let X be a topological group, and let $N_1, N_2, \ldots, N_m$ be normal subgroups of X with the following properties:*

(i) $N_1 N_2 \cdots N_m = X$;

(ii) $(N_1 N_2 \cdots N_k) \cap N_{k+1} = \{e\}$ $(k=1, 2, \ldots, m-1)$;

(iii) *if S_j is a neighborhood of e in N_j [relative topology] for $j=1, 2, \ldots, m$, then $S_1 S_2 \cdots S_m$ contains a neighborhood of e in X.*

Then X is topologically isomorphic with the direct product

$$N_1 \times N_2 \times \cdots \times N_m.$$

Proof. Conditions (i) and (ii) imply that X is isomorphic with the direct product $N_1 \times N_2 \times \cdots \times N_m$: every x in X can be written in precisely one way as a product $y_1 y_2 \cdots y_m$, where $y_j \in N_j$ $(j=1, 2, \ldots, m)$. Every element of N_j commutes with every element of N_k $(j \neq k; j=1, 2, \ldots, m; k=1, 2, \ldots, m)$. Thus $(y_1 y_2 \cdots y_m)(z_1 z_2 \cdots z_m) = y_1 z_1 y_2 z_2 \cdots y_m z_m$. It follows that the mapping ψ of $N_1 \times N_2 \times \cdots \times N_m$ onto X defined by $\psi(y_1, y_2, \ldots, y_m) = y_1 y_2 \cdots y_m$ is an isomorphism. The continuity of ψ is assured by the continuity of multiplication in X: if $y_1, y_2, \ldots, y_m$ are all sufficiently close to e, then $y_1 y_2 \cdots y_m$ is arbitrarily close to e. The continuity of ψ^{-1} is guaranteed by hypothesis (iii). □

(6.12) Theorem. *Let X be a locally countably compact T_0 group, and let $N_1, N_2, \ldots, N_m$ be locally compact, σ-compact normal subgroups of X satisfying conditions* (i) *and* (ii) *of* (6.11). *Then X is topologically isomorphic with $N_1 \times N_2 \times \cdots \times N_m$.*

Proof. The mapping ψ of (6.11) is a continuous isomorphism of $N_1 \times N_2 \times \cdots \times N_m$ onto X. The group $N_1 \times N_2 \times \cdots \times N_m$ is locally compact (6.4) and σ-compact (3.9). Theorem (5.29) implies that ψ is also an open mapping. □

(6.13) Definition. Let A be a set directed by a partial ordering $\leq$. [We will write $\alpha < \beta$ to indicate that $\alpha \leq \beta$ and $\alpha \neq \beta$.] For every $\alpha \in A$, let there be given a topological group G_α. Suppose that for every $\alpha, \beta \in A$ such that $\alpha < \beta$, there is an open continuous homomorphism $f_{\beta\alpha}$ of G_β into G_α. Suppose finally that if $\alpha < \beta < \gamma$, then $f_{\gamma\alpha} = f_{\beta\alpha} \circ f_{\gamma\beta}$. The object consisting of A, the groups G_α, and the mappings $f_{\beta\alpha}$ is called an *inverse mapping system*. Consider next the group $G = \underset{\alpha \in A}{\mathsf{P}} G_\alpha$ and the subset of G

consisting of all (x_α) such that if $\alpha<\beta$, then $x_\alpha=f_{\beta\alpha}(x_\beta)$. This subset is called the *projective limit* of the given inverse mapping system.

(6.14) Theorem. *Let there be given an inverse mapping system as in (6.13). The projective limit is a subgroup of the direct product G. If all of the groups G_α are T_0 groups, then the projective limit is a closed subgroup of the direct product G.*

Proof. It is easy to verify that the projective limit is a subgroup of G. If $(x_\gamma)\in G$ and (x_γ) is not in the projective limit, then for some α and β in A, we have $\alpha<\beta$ and $x_\alpha\neq f_{\beta\alpha}(x_\beta)$. If G_α is T_0, then there are disjoint open subsets U and V of G_α such that $x_\alpha\in U$ and $f_{\beta\alpha}(x_\beta)\in V$ [see (4.8)]. Then the set of all $(y_\gamma)\in G$ such that $y_\alpha\in U$ and $y_\beta\in f_{\beta\alpha}^{-1}(V)$ is an open set in G containing (x_γ) and disjoint from the projective limit. □

Miscellaneous theorems and examples

(6.15) Different topologies for product groups. (a) Let $\{G_\iota:\iota\in I\}$ be a nonvoid family of topological groups. Let W_ι be an open subset of G_ι for each $\iota\in I$, and let the family of all sets of the form $\underset{\iota\in I}{\mathsf{P}}W_\iota$ be a basis for open sets. With the usual operation $(x_\iota)(y_\iota)=(x_\iota y_\iota)$, and this topology, the Cartesian product $\underset{\iota\in I}{\mathsf{P}}G_\iota$ is a topological group.

(b) Let $G=\underset{\iota\in I}{\mathsf{P}}G_\iota$ be a direct product of topological groups, where e_ι is the identity of G_ι for all $\iota\in I$. Let I_1 and I_2 be complementary nonvoid subsets of I. We introduce another topology $\mathcal{Q}$ on the set G, as follows. Let an open basis at (e_ι) consist of all sets $\underset{\iota\in I_1}{\mathsf{P}}\{e_\iota\}\times\underset{\iota\in I_2}{\mathsf{P}}U_\iota$, where each U_ι is a neighborhood of e_ι in G_ι and only a finite number are distinct from G_ι. Then G, with the topology $\mathcal{Q}$, is a topological group. If all G_ι are compact for $\iota\in I_2$, then G is locally compact in the topology $\mathcal{Q}$, and is compact if and only if $\underset{\iota\in I_1}{\mathsf{P}}G_\iota$ is finite. The ordinary product topology of G agrees with the topology $\mathcal{Q}$ if and only if $\underset{\iota\in I_1}{\mathsf{P}}G_\iota$ is a discrete group in its Cartesian product topology.

(c) (BRACONNIER [1].) Let $\{G_\iota\}_{\iota\in I}$ be a nonvoid family of topological groups and let H_ι be a normal subgroup of G_ι for each $\iota\in I$. Let $H=\underset{\iota\in I}{\mathsf{P}}H_\iota$; plainly H is a normal subgroup of $G=\underset{\iota\in I}{\mathsf{P}}G_\iota$. Let Λ be a finite subset of I, and let U_ι be a neighborhood of e_ι in the subgroup H_ι for each $\iota\in\Lambda$; in other words, U_ι is the intersection of H_ι with an open subset of G_ι. Let the sets $\underset{\iota\in\Lambda}{\mathsf{P}}U_\iota\times\underset{\iota\notin\Lambda}{\mathsf{P}}H_\iota$ be taken as an open basis at $e=(e_\iota)$ in G. It follows from (4.5) that G is a topological group. The group H is an open subgroup of G.

(6.16) Local direct products (BRACONNIER [1])[1]. Let $\{G_\iota\}_{\iota\in I}$, $\{H_\iota\}_{\iota\in I}$, G, and H be as in (6.15.c). Suppose further that each H_ι is open. Let

[1] This construction is a fruitful source of useful examples. See in particular §§ 24 and 25.

G_0 be the subgroup of G consisting of all (x_ι) for which $x_\iota \in H_\iota$ for all but a finite number of indices ι. As a subspace of G topologized as in (6.15.c), G_0 is a topological group and is called the *local direct product of the G_ι's relative to the open normal subgroups H_ι*.

(a) The identity $\{e\}^- = \underset{\iota \in I}{\mathsf{P}} \{e_\iota\}^-$ obtains; thus G_0 is a T_0 group if and only if each G_ι is a T_0 group.

(b) The group G_0 is compact if and only if all G_ι are compact and $H_\iota = G_\iota$ for all but finitely many ι's. In this case, G_0 is the entire product $\underset{\iota \in I}{\mathsf{P}} G_\iota$. [If G_0 is compact, then each G_ι, as a continuous image of G_0, is also compact. The group G_0/H is compact and discrete and therefore is finite. It follows that $G_\iota = H_\iota$ except for finitely many ι's, and that $G_0 = \underset{\iota \in I}{\mathsf{P}} G_\iota$.]

(c) The group G_0 is locally compact if and only if all G_ι are locally compact and all but a finite number of the H_ι are compact. [This follows at once from (6.4).]

(d) The group G_0 is discrete if and only if each G_ι is discrete and $H_\iota = \{e_\iota\}$ for all but a finite number of indices ι. [If G_0 is discrete, then the direct product $\underset{\iota \in I}{\mathsf{P}} H_\iota$ must be discrete. This implies the present assertion.]

(6.17) Examples and counterexamples involving $\boldsymbol{R}$. (a) Let $\mathfrak{m}$ be any nonzero cardinal number. Then $R^{\mathfrak{m}}/Z^{\mathfrak{m}}$ is topologically isomorphic with $T^{\mathfrak{m}}$.

(b) The group R/Q is isomorphic with R and has the weakest topology $\{\varnothing, R/Q\}$. [The second statement follows from (5.38.d). To prove the first, consider a Hamel basis B for R over Q such that $1 \in B$. Plainly R is isomorphic with $\underset{b \in B}{\mathsf{P}^*} Q_{(b)}$, where each $Q_{(b)}$ is Q. The quotient group R/Q is isomorphic with $\underset{b \in B \cap \{1\}'}{\mathsf{P}^*} Q_{(b)}$, which is isomorphic with $\underset{b \in B}{\mathsf{P}^*} Q_{(b)}$. Both groups are in fact isomorphic with $Q^{\mathfrak{c}*}$.]

(c) Consider the group R^2, written as all pairs (x, y) of real numbers, with its usual topology. Let $H = \{(x, 0) \in R^2 : x \in Q\}$. Then R^2/H^- is topologically isomorphic with R and also with $(R^2/H)/(H^-/H)$. Thus R is an open continuous homomorphic image of a non T_0 group; the kernel is isomorphic with R and has the weakest topology. This construction can be generalized to any topological group G containing a nonclosed and nondense normal subgroup H.

(d) (Cartan and Dieudonné [1].) Let B be a Hamel basis for R over Q and let $B_1, B_2, \ldots, B_m$ be a partition of B, $m > 1$. For $j = 1, 2, \ldots, m$, let N_j be the set of all numbers $r_1 x_1 + r_2 x_2 + \cdots + r_s x_s$, where the r_k's are in Q and the x_k's are in B_j. Topologize each N_j with its relative topology as a subspace of R. Then the subgroups $N_1, N_2, \ldots, N_m$ of R satisfy

(6.11.i) and (6.11.ii). Thus R is isomorphic with $N = N_1 \times N_2 \times \cdots \times N_m$. However, R is not homeomorphic with N. Note also that N has a countable basis of open sets. [R is connected, while N has dimension 0.]

(e) Consider the group R^2 and the closed subgroup H consisting of all elements $(t, \alpha t)$ for $t \in R$ and some fixed $\alpha \in R$, $\alpha \neq 0$. Let φ be the natural mapping of R^2 onto R^2/Z^2. Then $\varphi(H)$ is a closed subgroup of R^2/Z^2 if and only if α is rational, and $\varphi(H)$ is a dense [proper] subgroup of R^2/Z^2 if α is irrational. Furthermore, φ is one-to-one on H if and only if α is irrational, and φ is an open mapping of H onto $\varphi(H)$ if and only if α is rational. Nevertheless, H, $\varphi(H)$, and φ do not contradict (5.29) for irrational α.

(6.18) Let $\mathfrak{m}$ be any cardinal number greater than $\aleph_0$. Consider the multiplicative group $\{-1, 1\}$ with the discrete topology and the direct product $G = \{-1, 1\}^{\mathfrak{m}}$. Let H be the subgroup of G consisting of all elements (x_ι) such that $x_\iota = 1$ except for a countable set of indices ι. Then H is a subgroup of G, H is countably compact, and H is not compact.

(6.19) (Kasuga [1].) Let G be any group and suppose that $\mathcal{O}_1$ and $\mathcal{O}_2$ are topologies for G that make G into a σ-compact locally compact T_0 group. If there is any Hausdorff topology $\mathcal{O}_3$ on G weaker than $\mathcal{O}_1$ and $\mathcal{O}_2$, then $\mathcal{O}_1 = \mathcal{O}_2$. Note that G need not be *a priori* a topological group under $\mathcal{O}_3$. [It is convenient to write $\mathcal{O} \times \mathcal{O}^*$ for the topology on $G \times G$ obtained when the first factor G has the topology $\mathcal{O}$ and the second factor G has the topology $\mathcal{O}^*$. Let $\Delta = \{(x, x) \in G \times G : x \in G\}$. Since $\mathcal{O}_3$ is a Hausdorff topology, Δ is a closed subset of $G \times G$ under the topology $\mathcal{O}_3 \times \mathcal{O}_3$. Since this topology is weaker than $\mathcal{O}_1 \times \mathcal{O}_2$, it follows that Δ is also closed under $\mathcal{O}_1 \times \mathcal{O}_2$. Give Δ the relative topology of $\mathcal{O}_1 \times \mathcal{O}_2$. Evidently $G \times G$, with the topology $\mathcal{O}_1 \times \mathcal{O}_2$, is σ-compact and locally compact, and hence the closed subset Δ is also σ-compact and locally compact. The mapping $(x, x) \to x$ of Δ onto G is a continuous isomorphism when G has the topology $\mathcal{O}_1$ and when G has the topology $\mathcal{O}_2$. By (5.29), the mapping $(x, x) \to x$ is also open when G has either of these topologies. Hence $(x, x) \to x$ is a topological isomorphism when G has the topology $\mathcal{O}_1$ or $\mathcal{O}_2$. It follows that the topologies $\mathcal{O}_1$ and $\mathcal{O}_2$ are the same.]

One can now easily deduce the following corollary. Let G be a T_0 group with topology $\mathcal{O}_1$ and H a subgroup of G. Suppose that H admits another topology $\mathcal{O}_2$ under which it is a σ-compact locally compact T_0 group and such that the identity mapping from H with the $\mathcal{O}_2$ topology onto H with the relative $\mathcal{O}_1$ topology is continuous. Then $\mathcal{O}_2$ is uniquely determined.

(6.20) Semidirect products. Let G and H be topological groups and let $G \circledS H$ be a semidirect product of G and H as in (2.6). Suppose also that the mapping $(x, h) \to \tau_h(x)$ is a continuous mapping of $G \times H$ onto G. In particular, each τ_h is a homeomorphism of G onto itself. Then the semidirect product $G \circledS H$ with the product topology is a topological group. This fact follows readily from the continuity of multiplication and inversion in G and H.

(6.21) The following result is related to (6.12). Let X be a locally compact T_0 group, and let $N_1, \ldots, N_m$ be closed normal subgroups of X satisfying (6.11.i) and (6.11.ii). If $N_1, \ldots, N_{m-1}$ are compactly generated, then X is topologically isomorphic with $N_1 \times N_2 \times \cdots \times N_m$. [An induction argument shows that it suffices to consider the case $m=2$. Let F denote a compact subset of N_1 that generates N_1. By (5.14), there is a compactly generated open subgroup H of G that contains F. It follows that $N_1 \subset H$. As in (6.11), there is an isomorphism ψ of $N_1 \times N_2$ onto G; it is defined by $\psi(x, y) = xy$ where $(x, y) \in N_1 \times N_2$. We next show that $\psi(N_1 \times (N_2 \cap H)) = H$. It is obvious that $\psi(N_1 \times (N_2 \cap H)) \subset H$. Moreover, if $h \in H$, then $h = xy$ where $x \in N_1$ and $y \in N_2$. Since x is also in H, we infer that $y \in H$, so that $h = xy \in N_1 \cdot (N_2 \cap H) = \psi(N_1 \times (N_2 \cap H))$. Since H is locally compact and N_1 and $N_2 \cap H$ are σ-compact and locally compact, (6.12) implies that ψ is a topological isomorphism of the open subgroup $N_1 \times (N_2 \cap H)$ of $N_1 \times N_2$ onto the open subgroup H of G. By (5.40.a), ψ is a topological isomorphism of $N_1 \times N_2$ onto G.]

(6.22) (WEIL [4], pp. 20 and 95.) (a) Let G be a T_0 group, H a normal subgroup of G, and suppose that there is a continuous homomorphism φ of G onto H such that $\varphi(x) = x$ for $x \in H$. Then H is closed, G is topologically isomorphic with $H \times \varphi^{-1}(e)$, and $\varphi^{-1}(e)$ is topologically isomorphic with G/H. [Let $L = \varphi^{-1}(e)$. Plainly L is a closed normal subgroup of G. For $x \in G$, we have $x = \varphi(x)\varphi(x)^{-1}x \in HL$ since $\varphi(\varphi(x^{-1})x) = \varphi(\varphi(x^{-1}))\varphi(x) = e$. Thus $G = HL = LH$. It is obvious that $H \cap L = \{e\}$, since φ is the identity mapping on H. Let U_1, U_2 be arbitrary neighborhoods of e in G, and let V and W be symmetric neighborhoods of e in G such that $V^2 \subset U_1 \cap U_2$, $W \subset V$, and $\varphi(W) \subset H \cap V$. Then it is easy to see that $W \subset (U_1 \cap H) \cdot (U_2 \cap L)$. Thus (6.11) implies that G is topologically isomorphic with $H \times L$, and so also L is topologically isomorphic with G/H.]

(b) Let G be a T_0 Abelian group, and let H be an open divisible (A.5) subgroup of G. Then G is topologically isomorphic with $H \times (G/H)$. [By (A.7), the identity mapping of H onto H admits an extension over G to a homomorphism φ. Since φ is the identity on the open subgroup H, φ is continuous. Now apply (a).]

Notes

Direct products of topological groups have been used since the inception of the theory of topological groups. As will appear in §§ 9, 24, and 25, direct products shed much light on the structure of locally compact Abelian groups. PONTRYAGIN [4] exploited countable direct products heavily, and in [6], §§ 20 and 21, treated special cases of finite and infinite direct products. The construction of arbitrary direct products appears in WEIL [4], § 4. Theorems (6.11) and (6.12) are from PONTRYAGIN [7], § 21. Projective limits are defined in WEIL [4], § 5, where one will also find remarks on the development of the concept. Example (6.17.d) corrects a mis-statement in FREUDENTHAL [1], 7.

Local direct products of topological groups play an important rôle in the constructions of BRACONNIER [1]. They have also been treated by GRAEV [1] and very extensively by VILENKIN [7]. For further references to VILENKIN's use of local direct products, see the notes to § 25.

§ 7. Properties of topological groups involving connectedness

In this section, we take up some elementary properties of topological groups that depend upon connectedness or disconnectedness of the group considered as a topological space. Since connectedness is a purely topological property, the results of this section have no analogue in the theory of groups without topology.

We begin with three theorems dealing with the component of the identity element.

(7.1) Theorem. *Let G be a topological group and let C be the component of the identity e. Then C is a closed normal subgroup of G.*

Proof. Since inversion is a homeomorphism of G, C^{-1} is a connected set containing e, and hence $C^{-1} \subset C$. If a is any point of C, then also $a^{-1} \in C$. Therefore aC is a connected set containing e, and so $aC \subset C$. Thus $C^2 \subset C$, so that C is a subgroup of G. If $a \in G$, then aCa^{-1} is a connected set containing e [$x \to axa^{-1}$ is a homeomorphism of G], so that $aCa^{-1} \subset C$. Thus C is a normal subgroup of G. Like all components in topological spaces, C is closed. □

(7.2) Theorem. *Let G be a topological group and let C be the component of the identity in G. Then for all $a \in G$, $aC = Ca$ is the component of a.*

Proof. The mapping $x \to ax$ is a homeomorphism of G, and C is a normal subgroup of G. □

(7.3) Theorem. *Let G be a topological group and let C be the component of the identity in G. Then G/C is a totally disconnected Hausdorff group.*

Proof. In view of (7.2) and (5.26), we need only show that the component of the identity C of G/C is C itself. Let $\{xC : x \in X\}$ be any subset of G/C properly containing $\{C\}$. We will show that $\{xC : x \in X\}$ is disconnected in G/C. Let φ be the natural mapping of G onto G/C, and let A be any subset of G. It is easy to verify that $\varphi(A \cap (XC)) = \varphi(A) \cap \{xC : x \in X\}$. The set XC properly contains C, and so is disconnected: $XC =$

$(U\cap(XC))\cup(V\cap(XC))$, where $(U\cap(XC))\cap(V\cap(XC))=\varnothing$, neither set is void, and U and V are open in G. Thus

$$\{xC: x\in X\}=(\varphi(U)\cap\{xC: x\in X\})\cup(\varphi(V)\cap\{xC: x\in X\}),$$

where $\varphi(U)$ and $\varphi(V)$ are open in G/C, as φ is an open mapping. For $x\in X$, we have $xC=(U\cap(xC))\cup(V\cap(xC))$. Thus since xC is connected, either $xC\subset U\cap(XC)$ or $xC\subset V\cap(XC)$. Consequently, $U\cap(XC)$ and $V\cap(XC)$ are unions of cosets of C, and so they have disjoint images under φ. Thus $(\varphi(U)\cap\{xC: x\in X\})\cap(\varphi(V)\cap\{xC: x\in X\})=\varnothing$, so that $\{xC: x\in X\}$ is disconnected. □

The following is a simple but frequently useful fact.

(7.4) Theorem. *Let G be a topological group, C the component of e, and U any neighborhood of e. Then $C\subset\bigcup_{n=1}^{\infty}U^n$; in particular, if G is connected, then $G=\bigcup_{n=1}^{\infty}U^n$.*

Proof. Let V be a symmetric neighborhood of e such that $V\subset U$. Then $\bigcup_{n=1}^{\infty}V^n$ is open and closed, by (5.7). Since C is connected, we have $C\subset\bigcup_{n=1}^{\infty}V^n\subset\bigcup_{n=1}^{\infty}U^n$. □

The two following theorems show how to construct compact, open subgroups in certain topological groups.

(7.5) Theorem. *Let G be a topological group and let U be a compact neighborhood*[1] *of the identity e. Then U contains a subgroup H of G that is compact, open, and closed.*

Proof. Since $U\subset U$, U is compact, and U is open, we may apply (4.10) and (4.6) to obtain a symmetric neighborhood V of e such that $UV\subset U$. We have $V=eV\subset UV\subset U$ and hence $V^2\subset UV\subset U$. By finite induction, we have $V^n=V^{n-1}V\subset UV\subset U$ for $n=3,4,\ldots$. We now set $H=\bigcup_{n=1}^{\infty}V^n$ and apply (5.7). □

(7.6) Theorem. *Let G be a compact group and let U be a closed neighborhood of the identity e. Then U contains an open and closed normal subgroup N of G. The group G/N is finite.*

Proof. By (7.5), U contains a compact, open, and closed subgroup H. Let $N=\bigcap_{x\in G}xHx^{-1}$. It is obvious that N is a closed normal subgroup of G contained in U. By (4.9), there is a neighborhood V of e such that $x^{-1}Vx\subset H$ for all $x\in G$. That is, $V\subset xHx^{-1}$ for all $x\in G$, so that $V\subset N$. This shows that N is also open (5.5). Since N is open, G/N is discrete

[1] Recall our convention that neighborhoods are open sets (page 9).

(5.21), and since G is compact, G/N is compact (5.22). Hence G/N is finite. □

Theorems (7.5) and (7.6) show that certain locally compact groups contain arbitrarily small compact and open subgroups[1].

(7.7) Theorem. *Let G be a totally disconnected or 0-dimensional group that is locally compact [compact]. Then every neighborhood of the identity e contains a compact open [compact open normal] subgroup.*

Proof. If G is totally disconnected, it is T_1 and hence Hausdorff (4.8). Therefore G is 0-dimensional, by (3.5). If U is any neighborhood of e, then U contains an open and closed neighborhood, which is obviously compact if U^- is compact. Now apply (7.5) and (7.6). □

Theorem (7.7) yields a new description of the component of the identity in locally compact groups.

(7.8) Theorem. *Let G be a locally compact group and let C be the component of the identity in G. Then C is the intersection of all open subgroups of G.*

Proof. Every open subgroup is closed. Since it also contains e, it must contain the connected component C of e. Thus C is contained in the intersection of all open subgroups of G. Now let x be any element of G not in C. Consider the quotient group G/C, which is totally disconnected (7.3) and locally compact (5.22). By (7.7), there is a compact open subgroup $\{uC : u \in U\}$ of G/C that does not contain the element xC of G/C. We may take U to be a neighborhood of e in G. Then UC is an open subgroup of G not containing x. □

(7.9) Corollary. *The following assertions about a locally compact group G are equivalent:*

(i) *G is connected;*

(ii) *G has no proper open subgroups;*

(iii) *for every neighborhood U of e, we have $\bigcup_{n=1}^{\infty} U^n = G$.*

Proof. This is obvious from (7.8) and (7.4). □

(7.10) Theorem. *Let G be a locally compact group and let C be the component of e in G. If G/C is compactly generated, then G is compactly generated.*

Proof. Since C is closed in G, C is locally compact. It follows immediately from (7.4) that any locally compact connected group is compactly generated. In particular, C is compactly generated; hence G is compactly generated by (5.39.i). □

[1] We say that a topological group contains arbitrarily small subsets of a given type if every neighborhood of the identity contains such a subset.

(7.11) Theorem. *Let G be a locally compact* 0*-dimensional group and let H be a closed subgroup of G. Then G/H is* 0*-dimensional.*

Proof. Let U be a neighborhood of e in G. By (7.7), U contains a compact open subgroup L of G. The set $\{xH : x \in L\}$ is open in G/H by (5.15) and closed in G/H because it is a compact subset of the Hausdorff space G/H (5.21). Thus H in G/H has arbitrarily small open and closed neighborhoods. Since G/H is homogeneous, the assertion follows. □

(7.12) Theorem. *Let G be a locally compact group and let C be the component of the identity e in G. Let f be an open, continuous homomorphism of G onto a T_0 group $\widetilde{G}$, and let $\widetilde{C}$ be the component of the identity $\tilde{e}$ in $\widetilde{G}$. Then $f(C)^- = \widetilde{C}$.*

Proof. Since $f(C)^-$ is connected, we have $f(C)^- \subset \widetilde{C}$. Clearly $f(C)^-$ is a closed normal subgroup of $\widetilde{G}$, and $H = f^{-1}(f(C)^-)$ is a closed normal subgroup of G containing C. By (5.34), $\widetilde{G}/f(C)^-$ is topologically isomorphic with G/H. By (5.35), G/H is also topologically isomorphic with $(G/C)/(H/C)$. Since G/C is totally disconnected (7.3), Hausdorff (7.3), and locally compact (5.22), G/C is 0-dimensional (3.5). It is easily verified that H/C is a closed normal subgroup of G/C. Hence by (7.11), $(G/C)/(H/C)$ is 0-dimensional and Hausdorff, hence totally disconnected. That is, $\widetilde{G}/f(C)^-$ is totally disconnected. As a continuous image of the connected set $\widetilde{C}$, $\{\tilde{x} f(C)^- : \tilde{x} \in \widetilde{C}\}$ is connected in $\widetilde{G}/f(C)^-$. Hence we have $\{\tilde{x} f(C)^- : \tilde{x} \in \widetilde{C}\} = \{f(C)^-\}$, and thus $\widetilde{C} \subset f(C)^-$. □

(7.13) Corollary. *Let G be a locally compact group and let C be the component of the identity e in G. Let f be an open, continuous homomorphism of G onto a T_0 group $\widetilde{G}$ and let $\widetilde{C}$ be the component of the identity $\tilde{e}$ in $\widetilde{G}$. If $f^{-1}(\tilde{e}) \subset C$ or $C \subset f^{-1}(\tilde{e})$, then $f(C) = \widetilde{C}$.*

Proof. If $f^{-1}(\tilde{e}) \subset C$ or $C \subset f^{-1}(\tilde{e})$, then $f^{-1}(\tilde{e}) \cdot C$ is equal to C or $f^{-1}(\tilde{e})$, respectively, and hence is a closed subgroup of G in either case. Since $f^{-1}(f(C)) = f^{-1}(\tilde{e}) \cdot C$, we have $f(C)' = f((f^{-1}(\tilde{e}) \cdot C)')$ so that $f(C)'$ is open in $\widetilde{G}$. Thus $f(C)$ is closed in $\widetilde{G}$, and (7.12) implies that $f(C) = \widetilde{C}$. □

In (10.14) we will give an example showing that the hypotheses of (7.13) are needed. We next give a result on connectedness reminiscent of (5.25).

(7.14) Theorem. *Let G be a topological group and let H be a subgroup of G. If H and G/H are connected, then G is connected.*

Proof. Assume that $G = U \cup V$ where U and V are disjoint nonvoid open sets. Since H is connected, each coset of H is either a subset of U or a subset of V. Thus the relation $G/H = \{xH : xH \subset U\} \cup \{xH : xH \subset V\} =$

$\{xH : x \in U\} \cup \{xH : x \in V\}$ expresses G/H as the union of disjoint nonvoid open sets. This contradicts the hypothesis that G/H is connected. □

The following fact about unitary groups $\mathfrak{U}(n)$ will be useful later on.

(7.15) Theorem. *For every positive integer n, the unitary group $\mathfrak{U}(n)$ is arcwise connected.*

Proof. As pointed out in (4.25), $\mathfrak{U}(n)$ is locally Euclidean and hence obviously locally arcwise connected. Since any connected, locally arcwise connected space is arcwise connected, we need only prove that $\mathfrak{U}(n)$ is connected. Let $\mathfrak{D}$ consist of all the diagonal matrices in $\mathfrak{U}(n)$; thus $(a_{jk})_{j,k=1}^n$ belongs to $\mathfrak{D}$ if and only if $|a_{jk}| = \delta_{jk}$ for all j and k. It is evident that $\mathfrak{D}$ is homeomorphic with T^n and so $\mathfrak{D}$ is connected. Thus we have $\mathfrak{D} \subset \mathfrak{C}$, where $\mathfrak{C}$ denotes the connected component of the identity in $\mathfrak{U}(n)$. It is well known that every unitary matrix A can be diagonalized by a unitary matrix P; see for example S. MacLane and G. Birkhoff: Algebra (New York: The Macmillan Co., 1967), page 413, Theorem 2C. That is, for any A in $\mathfrak{U}(n)$, there exists an element P in $\mathfrak{U}(n)$ such that $P^{-1}AP$ belongs to $\mathfrak{D}$, and hence to $\mathfrak{C}$. Since $\mathfrak{C}$ is a normal subgroup (7.1), it follows that A belongs to $\mathfrak{C}$. We have shown that $\mathfrak{U}(n) = \mathfrak{C}$ and therefore $\mathfrak{U}(n)$ is connected. □

Miscellaneous theorems and examples

(7.16) (a) Let G be a connected [locally connected] group and H a subgroup of G. Then G/H is connected [locally connected].

(b) (Schreier [1].) The component of the identity in a normal subgroup is itself a normal subgroup. [Let C be the component of the identity e in the normal subgroup H of G. For $x \in G$, xCx^{-1} is connected and $xCx^{-1} \subset xHx^{-1} \subset H$ since H is normal in G. Thus $xCx^{-1} \subset C$.]

(7.17) The center of a connected group. (a) Every discrete normal subgroup of a connected topological group G is contained in the center of G. [See (7.4) and (5.37.f).]

(b) (Hofmann [1].) Let G be a connected T_0 group and H a totally disconnected normal subgroup. Then H^- is contained in the center of G. [Fix an element $a \in H$. Then the mapping $x \to xax^{-1}$ carries G continuously into H. The image of G is connected and hence is a single point. Thus $a = xax^{-1}$.]

A related but much less trivial result along these lines appears in (26.10).

(7.18) Dimension zero and total disconnectedness. (a) (Erdős [1].) In the space l_2 with its usual metric topology, consider the set E of all points $(x_1, x_2, \ldots, x_n, \ldots)$ all of whose coordinates are real and rational. This is evidently a subgroup of the additive group l_2. The group E is obviously totally disconnected. [In fact, for every pair of distinct points $\boldsymbol{x}$ and $\boldsymbol{y}$ in E, there is an open and closed set U such that $\boldsymbol{x}\in U$ and $\boldsymbol{y}\in U'$.] However, E is not 0-dimensional. [Let V be any neighborhood of $\boldsymbol{0}=(0, 0, \ldots, 0, \ldots)$ in E such that $\|\boldsymbol{x}\|=\left(\sum_{n=1}^{\infty} x_n^2\right)^{\frac{1}{2}}<1$ for all $\boldsymbol{x}\in V$. For $\boldsymbol{y}\in l_2$ and a subset B of l_2, we write $d(\boldsymbol{y}, B)=\inf\{\|\boldsymbol{x}-\boldsymbol{y}\|: \boldsymbol{x}\in B\}$. Consider the set $A_1=\{x_1\in Q: (x_1, 0, \ldots, 0, \ldots)\in V\}$. Clearly A_1 is nonvoid, open, and bounded, and contains 0. There is therefore an $a_1\in A_1$ such that $\boldsymbol{p}^{(1)}=(a_1, 0, \ldots, 0, \ldots)\in V$ and $d(\boldsymbol{p}^{(1)}, V')<\varepsilon_1$, where ε_1 is a fixed but arbitrary positive number. Let $A_2=\{x_2\in Q: (a_1, x_2, 0, \ldots, 0, \ldots)\in V\}$. Clearly A_2 is nonvoid, open, and bounded, and contains 0. Thus there is an $a_2\in A_2$ such that $\boldsymbol{p}^{(2)}=(a_1, a_2, 0, \ldots, 0, \ldots)\in V$ and $d(\boldsymbol{p}^{(2)}, V')<\varepsilon_2$, ε_2 being a fixed but arbitrary positive number. Continuing by induction, we find a sequence $\{a_n\}_{n=1}^{\infty}$ of rational numbers such that $\boldsymbol{p}^{(n)}=(a_1, a_2, \ldots, a_n, 0, 0, \ldots)\in V$ for all n and $d(\boldsymbol{p}^{(n)}, V')<\varepsilon_n$. Now set $\boldsymbol{a}=(a_1, a_2, \ldots, a_n, \ldots)$. It is obvious that $\sum_{n=1}^{\infty} a_n^2\leq 1$, that $\boldsymbol{a}$ belongs to V^-, and that if $\varepsilon_n\to 0$, $\boldsymbol{a}$ belongs to $V'^-=V'$. Thus V is not closed.]

(b) For a topological group G with component of the identity C, G/C need not be 0-dimensional. [This is shown by the group E of (a).]

(c) A continuous isomorphic image of a non 0-dimensional group may be 0-dimensional. [The group E of (a) can be imbedded in $Q^{\aleph_0}$ with its direct product topology and addition. Plainly $Q^{\aleph_0}$ is 0-dimensional and hence $E\subset Q^{\aleph_0}$ is also 0-dimensional. The topology of E as a subspace of $Q^{\aleph_0}$ is obviously strictly weaker than its topology as a subspace of l_2.]

(d) A finite group with a nondiscrete topology under which it is a topological group is 0-dimensional and compact but not totally disconnected. There are compact groups of arbitrarily large cardinal number having the same property.

(7.19) Comments on theorem (7.7). (a) Let G be locally compact and totally disconnected, and suppose that the uniform structure $\mathscr{S}_l(G)$ is equivalent to the uniform structure $\mathscr{S}_r(G)$. Then every neighborhood of e contains a compact open normal subgroup. In view of (4.17), this result generalizes (7.7). [Let U be an arbitrary neighborhood of e and let H be an open compact subgroup of G contained in U. Let V be a neighborhood of e such that $x^{-1}Vx\subset H$ for all $x\in G$ (4.14.g). Then $V\subset\bigcap_{x\in G} xHx^{-1}$, so that $\bigcap_{x\in G} xHx^{-1}$ is a compact open normal subgroup of G contained in H.]

(b) Not all 0-dimensional locally compact groups have arbitrarily small open normal subgroups; such a group may indeed have *no* compact open normal subgroups. [The following example is adapted from MONTGOMERY and ZIPPIN [1], p. 57. Let G be any group containing an element a not equal to e, and consider the direct product $\underset{n\in Z}{\mathsf{P}} G_n$, where each G_n is G. Write the elements of this product as $\boldsymbol{x}=\boldsymbol{x}(n)$, where $\boldsymbol{x}$ is a function on Z with values in G. Let H be the subgroup of the product group consisting of all $\boldsymbol{x}$ such that $\boldsymbol{x}(n)=e$ for all $n\leqq n_0\in Z$ [n_0 depends upon $\boldsymbol{x}$]. Let T be the operator on H defined by $(T\boldsymbol{x})(n)=\boldsymbol{x}(n+1)$. Now let S be the set of all ordered pairs $(\boldsymbol{x}, m)$, where $\boldsymbol{x}\in H$ and $m\in Z$. For $(\boldsymbol{x}, m)$ and $(\boldsymbol{y}, n)$ in S, let $(\boldsymbol{x}, m)(\boldsymbol{y}, n)=(\boldsymbol{x}(T^m\boldsymbol{y}), m+n)$, where $T^0\boldsymbol{x}=\boldsymbol{x}$. It is easy to verify that S is a group under this operation. [Actually S is a semidirect product of H and Z, as defined in (2.6).] The identity of S is $(\boldsymbol{e}, 0)$, where $\boldsymbol{e}(n)=e$ for all $n\in Z$. For nonnegative integers m, let U_m be the set of all $(\boldsymbol{x}, 0)$ in S such that $\boldsymbol{x}(n)=e$ for $n\leqq m$. The family of all sets U_m satisfies (4.5.i)—(4.5.v) and hence defines a topology on S making S into a T_0 topological group.

Each U_m is topologically isomorphic with $\overset{\infty}{\underset{n=m+1}{\mathsf{P}}} G_n$, where each $G=G_n$ has the discrete topology. Therefore S is locally compact if G is finite. If G is 0-dimensional, then S is 0-dimensional. Suppose that N is an open normal subgroup of S. Then there is a neighborhood U_m contained in N. Denoting by $\boldsymbol{y}$ that point of H for which $\boldsymbol{y}(n)=e$ for $n\neq m+1$ and $\boldsymbol{y}(m+1)=a\neq e$, we see that $(\boldsymbol{y}, 0)\in U_m\subset N$. Since N is normal, all elements $(\boldsymbol{e}, k)(\boldsymbol{y}, 0)(\boldsymbol{e}, k)^{-1}=(T^k\boldsymbol{y}, 0)$ lie in N for $k\in Z$. For $k=m+2, m+3, \ldots$, the neighborhoods $(T^k\boldsymbol{y}, 0)\cdot U_0$ are pairwise disjoint. It follows that the net $\{(T^k\boldsymbol{y}, 0)\}_{k=m+2}^{\infty}$ has no cluster point. Thus N is noncompact.]

(c) Local compactness is necessary for the validity of (7.7). The group Q is 0-dimensional but contains no bounded open subgroups. The group Q also shows that (7.8) and (7.9) may fail in the absence of local compactness.

(7.20) (HARTMAN and MYCIELSKI [1].) Every T_0 group G is a closed subgroup of an arcwise connected and locally arcwise connected group $\widetilde{G}$. [Consider the set $\widetilde{G}$ of all functions f on $[0, 1[$ with values in G for which there is a sequence $0=a_0<a_1<a_2<\cdots<a_n=1$ such that f is constant on $[a_k, a_{k+1}[$ $(k=0, 1, \ldots, n-1)$. With $fg(t)=f(t)g(t)$ and $f^{-1}(t)=(f(t))^{-1}$ $(0\leqq t<1)$, $\widetilde{G}$ is obviously a group. The function identically equal to e is the identity of $\widetilde{G}$. For $\varepsilon>0$ and a neighborhood V of e in G, let $\widetilde{U}(V, \varepsilon)$ be all $f\in\widetilde{G}$ such that $\lambda(\{t\in[0, 1[: f(t)\notin V\})<\varepsilon$. Here λ is

Lebesgue measure on $[0, 1[$. Properties (4.5.i)—(4.5.v) are easy to verify. Hence $\widetilde{G}$ is a topological group. The constant functions form a closed subgroup of $\widetilde{G}$ topologically isomorphic with G. If f and g are elements of $\widetilde{G}$, and $0 \leqq \alpha \leqq 1$, let $h_\alpha(t) = f(t)$ for $0 \leqq t < \alpha$ and $h_\alpha(t) = g(t)$ for $\alpha \leqq t < 1$. Then $h_0 = g$, $h_1 = f$, and the mapping $\alpha \to h_\alpha$ is continuous. Moreover, if f and g lie in $\widetilde{U}\left(V, \frac{\varepsilon}{2}\right)$, then all of the h_α lie in $\widetilde{U}(V, \varepsilon)$.]

(7.21) (D. VAN DANTZIG [3].) Let G be a topological group and A a connected subset of G such that $e \in A$. Then the smallest subgroup of G containing A is also connected. [For each $a \in A$, Aa^{-1} is connected and contains e. Hence the set $AA^{-1} = \bigcup_{a \in A} Aa^{-1}$ is connected. Similarly the sets $(AA^{-1})^n$ and $\bigcup_{n=1}^{\infty} (AA^{-1})^n$ are connected.]

Notes

The special properties of connected groups were noticed by the first writers on topological groups: SCHREIER [1] proved (7.1) and LEJA [1] (7.4). VAN DANTZIG [3] also proved (7.1). The first statement of (7.7) appears to be in VAN DANTZIG [2]. It also is proved by VAN KAMPEN [1]. Our proofs of (7.14) and (7.15) are taken from CHEVALLEY [1], pp. 36—37.

§ 8. Invariant pseudo-metrics and separation axioms

In this section, we show that the topology of a topological group can be completely described by means of a family of left invariant pseudo-metrics and that every T_0 topological group is completely regular. We also explore other separation properties for topological groups.

(8.1) Definition. A metric or pseudo-metric d on a group G is said to be *left invariant* if $d(ax, ay) = d(x, y)$ for all $a, x, y \in G$. If $d(xa, ya) = d(x, y)$ for all $a, x, y \in G$, then d is said to be *right invariant*. If d is both left and right invariant, it is said to be *two-sided invariant*. Similarly, if H is any subgroup of G, then a metric or pseudo-metric d on G/H is said to be *left invariant* if $d(axH, ayH) = d(xH, yH)$ for all $a, x, y \in G$.

Our first theorem is rather technical, and its proof may not be obvious at first glance. It is, however, the key to the metrizability theorems for topological groups [(8.3), (8.5), and (8.6)], and it immediately implies complete regularity for T_0 groups (8.4).

(8.2) Theorem. *Let $\{U_k\}_{k=1}^{\infty}$ be a sequence of symmetric neighborhoods of e in a topological group G such that $U_{k+1}^2 \subset U_k$ for $k=1, 2, \ldots$. Let $H = \bigcap_{k=1}^{\infty} U_k$. Then there is a left invariant pseudo-metric σ on G such that:*

(i) *σ is uniformly continuous for the left uniform structure of $G \times G$;*

(ii) *$\sigma(x, y) = 0$ if and only if $y^{-1}x \in H$;*

(iii) *$\sigma(x, y) \leqq 2^{-k+2}$ whenever $y^{-1}x \in U_k$;*

(iv) *$2^{-k} \leqq \sigma(x, y)$ whenever $y^{-1}x \notin U_k$.*

If, in addition, $xU_kx^{-1} = U_k$ for all $x \in G$ and $k = 1, 2, \ldots$, then σ is also right invariant and

(v) *$\sigma(x^{-1}, y^{-1}) = \sigma(x, y)$ for $x, y \in G$.*

Proof. It is convenient to rename the sets U_k. Let $V_{2^{-k}} = U_k$ for $k=1, 2, \ldots$. We now define sets V_r for all dyadic rational numbers r, $0 < r < 1$, as follows. For

$$r = 2^{-l_1} + 2^{-l_2} + \cdots + 2^{-l_n},\ 0 < l_1 < \cdots < l_n, \tag{1}$$

let

$$V_r = V_{2^{-l_1}} V_{2^{-l_2}} \cdots V_{2^{-l_n}}. \tag{2}$$

We also define $V_r = G$ for all dyadic rational numbers $r \geqq 1$. We first show that

$$r < s \text{ implies } V_r \subset V_s. \tag{3}$$

We may suppose that $s < 1$, since the inclusion is obvious for $s \geqq 1$. Let r be given by (1) and s by $2^{-m_1} + \cdots + 2^{-m_p}$, $0 < m_1 < m_2 < \cdots < m_p$. There exists a unique integer k such that $l_j = m_j$ for $j < k$ and $l_k > m_k$. Letting $W = V_{2^{-l_1}} V_{2^{-l_2}} \cdots V_{2^{-l_{k-1}}}$, we then have

$$\begin{aligned}
V_r &= W V_{2^{-l_k}} V_{2^{-l_{k+1}}} V_{2^{-l_{k+2}}} \cdots V_{2^{-l_n}} \\
&\subset W V_{2^{-l_k}} V_{2^{-l_k-1}} V_{2^{-l_k-2}} \cdots V_{2^{-l_n+1}} V_{2^{-l_n}} V_{2^{-l_n}} \\
&\subset W V_{2^{-l_k}} V_{2^{-l_k-1}} V_{2^{-l_k-2}} \cdots V_{2^{-l_n+1}} V_{2^{-l_n+1}} \subset \cdots \\
&\subset W V_{2^{-l_k}} V_{2^{-l_k}} \subset W V_{2^{-l_k+1}} \subset W V_{2^{-m_k}} = V_{2^{-l_1}} V_{2^{-l_2}} \cdots V_{2^{-l_{k-1}}} V_{2^{-m_k}} \\
&\subset V_{2^{-m_1}} V_{2^{-m_2}} \cdots V_{2^{-m_{k-1}}} V_{2^{-m_k}} V_{2^{-m_{k+1}}} \cdots V_{2^{-m_p}} = V_s.
\end{aligned}$$

We next show that for every r of the form (1) and every positive integer l, we have

$$V_r V_{2^{-l}} \subset V_{r+2^{-l+2}}. \tag{4}$$

Since (4) is obvious if $r + 2^{-l+2} \geqq 1$, we suppose that $r + 2^{-l+2} < 1$. If $l > l_n$, then

$$V_r V_{2^{-l}} = V_{r+2^{-l}} \tag{5}$$

and (4) is again obvious. If $l \leqq l_n$, let k be the positive integer such that $l_{k-1} < l \leqq l_k$ [define $l_0 = 0$]. Let $r_1 = 2^{-l+1} - 2^{-l_k} - 2^{-l_{k+1}} - \cdots - 2^{-l_n}$ and $r_2 = r + r_1$. Then it is evident that $r < r_2 < r + 2^{-l+1}$. Applying (5) and (3) to r_2, we obtain

$$V_r V_{2^{-l}} \subset V_{r_2} V_{2^{-l}} = V_{r_2 + 2^{-l}} \subset V_{r + 2^{-l+1} + 2^{-l}} \subset V_{r + 2^{-l+2}}.$$

This proves (4).

For $x \in G$, let $\varphi(x) = \inf\{r : x \in V_r\}$. It is obvious that $\varphi(x) = 0$ if and only if $x \in H$. Finally, we define the function σ on $G \times G$ by the following rule:

$$\sigma(x, y) = \sup\{|\varphi(zx) - \varphi(zy)| : z \in G\}.$$

It is obvious that $\sigma(x, y) = \sigma(y, x)$ and that $\sigma(x, x) = 0$ for $x, y \in G$. It is easy to verify that $\sigma(x, z) \leqq \sigma(x, y) + \sigma(y, z)$ and that $\sigma(ax, ay) = \sigma(x, y)$ for $a, x, y, z \in G$. Therefore σ is a left invariant pseudo-metric on G.

To establish (iii), suppose that l is any positive integer, that $u \in V_{2^{-l}}$, and that $z \in G$. If $z \in V_r$, then by (4), we have $zu \in V_{r+2^{-l+2}}$. Hence $\varphi(zu) \leqq \varphi(z) + 2^{-l+2}$. Similarly, if $zu \in V_r$, then $z \in V_r V_{2^{-l}}^{-1} \subset V_{r+2^{-l+2}}$ and hence $\varphi(z) \leqq \varphi(zu) + 2^{-l+2}$. Thus $|\varphi(z) - \varphi(zu)| \leqq 2^{-l+2}$ for $u \in V_{2^{-l}}$ and $z \in G$. Consequently, we have $\sigma(u, e) \leqq 2^{-l+2}$ for $u \in V_{2^{-l}}$. Since σ is left invariant, we conclude that $\sigma(x, y) \leqq 2^{-l+2}$ whenever $y^{-1}x \in V_{2^{-l}} = U_l$. Thus (iii) holds.

We now prove (i). Consider (x, y) and (x_1, y_1) in $G \times G$. Using the left invariance of σ and the triangle inequality, we obtain

$$\begin{aligned}
|\sigma(x,y) - \sigma(x_1,y_1)| &= |\sigma(x_1^{-1}x, x_1^{-1}y) - \sigma(x_1^{-1}y, e) + \sigma(e, y^{-1}x_1) - \sigma(y^{-1}x_1, y^{-1}y_1)| \\
&\leqq |\sigma(x_1^{-1}x, x_1^{-1}y) - \sigma(x_1^{-1}y, e)| + |\sigma(e, y^{-1}x_1) - \sigma(y^{-1}x_1, y^{-1}y_1)| \\
&\leqq |\sigma(x_1^{-1}x, e)| + |\sigma(e, y^{-1}y_1)|.
\end{aligned}$$

This inequality and (iii) show that $|\sigma(x, y) - \sigma(x_1, y_1)| \leqq 2^{-k+3}$ whenever $x_1^{-1}x$ and $y_1^{-1}y$ are in U_k. This proves (i).

We next prove (iv). Suppose that $y^{-1}x \notin U_l = V_{2^{-l}}$. Then $\varphi(y^{-1}x) \geqq 2^{-l}$ and so $\sigma(x, y) = \sigma(y^{-1}x, e) \geqq |\varphi(ey^{-1}x) - \varphi(ee)| = \varphi(y^{-1}x) \geqq 2^{-l}$. This proves (iv). Assertion (ii) follows immediately from (iii) and (iv).

Finally, suppose that $xU_kx^{-1} = U_k$ for $x \in G$ and $k = 1, 2, \ldots$. Then $xV_rx^{-1} = V_r$ for all dyadic rational numbers $r > 0$, and hence $\varphi(xyx^{-1}) = \varphi(y)$ for all $x, y \in G$. Thus for $a, x, y \in G$, we have

$$\begin{aligned}
\sigma(xa, ya) &= \sup\{|\varphi(zxa) - \varphi(zya)| : z \in G\} \\
&= \sup\{|\varphi(azx) - \varphi(azy)| : z \in G\} \\
&= \sup\{|\varphi(zx) - \varphi(zy)| : z \in G\} = \sigma(x, y);
\end{aligned}$$

and so σ is right invariant. Also for $x, y \in G$, we have

$$\sigma(x^{-1}, y^{-1}) = \sigma(e, y^{-1}x) = \sigma(y, x) = \sigma(x, y). \ \square$$

(8.3) Theorem. *Let G be a T_0 topological group. Then G is metrizable if and only if there is a countable open basis at e. In this case, the metric can be taken to be left invariant.*

Proof. If G is metrizable, then plainly there is a countable open basis at e. Suppose now that $\{V_k\}_{k=1}^{\infty}$ is a countable open basis at e. Let $U_1 = V_1 \cap V_1^{-1}$. If $U_1, \ldots, U_{k-1}$ have been defined, let U_k be a symmetric neighborhood of e such that $U_k \subset U_1 \cap \cdots \cap U_{k-1} \cap V_k$ and $U_k^2 \subset U_{k-1}$. Such a set U_k exists by (4.5). The family $\{U_k\}_{k=1}^{\infty}$ satisfies the hypotheses of (8.2), and $H = \bigcap_{k=1}^{\infty} U_k = \{e\}$ by (4.8). Let σ be the left invariant pseudo-metric as in (8.2) for this family $\{U_k\}_{k=1}^{\infty}$. Clearly σ is a true metric; that is, $\sigma(x, y) = 0$ if and only if $x = y$. From (8.2.iii) and (8.2.iv), it follows that $\{x \in G : \sigma(x, e) < 2^{-k}\} \subset U_k \subset \{x \in G : \sigma(x, e) \leqq 2^{-k+2}\}$ for $k = 1, 2, \ldots$. This implies immediately that the topology defined by σ coincides with the given topology of G. $\square$

(8.4) Theorem. *Let G be a topological group, a an element of G, and F a closed subset of G not containing a. Then there is a continuous real function ψ on G such that $\psi(a) = 0$ and $\psi(x) = 1$ for each $x \in F$. Thus a T_0 group is completely regular.*

Proof. Let U_1 be a symmetric neighborhood of e such that $(aU_1) \cap F = \varnothing$. Choose neighborhoods $U_2, U_3, \ldots$ of e such that $\{U_k\}_{k=1}^{\infty}$ satisfies the hypotheses of (8.2). Such a sequence exists by (4.5). Let σ be a left invariant pseudo-metric constructed for $\{U_k\}_{k=1}^{\infty}$ as in (8.2). For $x \in G$, let $\psi(x) = \min\{1, 2\sigma(a, x)\}$. Then ψ is continuous by (8.2.i) and clearly $\psi(a) = 0$. If $x \in F$, we have $a^{-1}x \notin U_1$, and hence $\sigma(a, x) \geqq 2^{-1}$ by (8.2.iv). Consequently, $\psi(x) = 1$ for all $x \in F$. $\square$

(8.5) Theorem. *A locally countably compact group G admits a left invariant metric compatible with its topology if and only if $\{e\}$ is the intersection of a countable family of open sets.*

Proof. If G admits a metric compatible with its topology, then clearly $\{e\}$ is the intersection of a countable family of open sets. Suppose that $\{e\} = \bigcap_{n=1}^{\infty} U_n$, where each U_n is open in G. Let $V_1 = U_1$. By finite induction, we choose a sequence $\{V_n\}_{n=2}^{\infty}$ of neighborhoods of e such that $V_n^- \subset V_{n-1} \cap U_n$ and V_n^- is countably compact, $n = 2, 3, \ldots$. To show that $\{V_n\}_{n=2}^{\infty}$ is a basis at e, let W be an arbitrary neighborhood of e. If the inclusion $V_n \subset W$ fails for all n, then the sequence of countably compact sets $\{V_n^- \cap W'\}_{n=2}^{\infty}$ has the finite intersection property. Hence

$\bigcap_{n=2}^{\infty}(V_n^- \cap W')$ is nonvoid, contradicting the relation

$$\bigcap_{n=2}^{\infty}(V_n^- \cap W') = \left(\bigcap_{n=2}^{\infty} V_n^-\right) \cap W' \subset \left(\bigcap_{n=1}^{\infty} U_n\right) \cap W' = \{e\} \cap W' = \varnothing.$$

Consequently, G has a countable open basis at e and hence by (8.3) admits a left invariant metric compatible with its topology. □

(8.6) Theorem. *Let G be a compact group such that $\{e\}$ is the intersection of a countable family of open sets. Then G admits a two-sided invariant metric yielding the topology of G.*

Proof. By (8.5), G admits a left invariant metric σ compatible with its topology. For $x, y \in G$, let

$$\varrho(x, y) = \sup\{\sigma(xz, yz) : z \in G\}.$$

Since G is compact, σ is bounded and hence $\varrho(x, y)$ is finite for all $x, y \in G$. It is routine to verify that ϱ is a two-sided invariant metric on G. Let ε be an arbitrary positive number. By virtue of (4.9), there is a $\delta > 0$ such that

$$z^{-1} \cdot \{x \in G : \sigma(x, e) < \delta\} \cdot z \subset \{x \in G : \sigma(x, e) < \varepsilon\}$$

for all $z \in G$. Thus if $x \in G$ and $\sigma(x, e) < \delta$, then $\sigma(z^{-1}xz, e) = \sigma(xz, ez) < \varepsilon$ for all $z \in G$ and therefore $\varrho(x, e) \leqq \varepsilon$. That is, $\{x \in G : \sigma(x, e) < \delta\} \subset \{x \in G : \varrho(x, e) \leqq \varepsilon\}$. This and the obvious inclusion $\{x \in G : \varrho(x, e) < \varepsilon\} \subset \{x \in G : \sigma(x, e) < \varepsilon\}$ show that the topologies defined by σ and ϱ coincide. □

(8.7) Theorem. *Let G be a σ-compact locally compact group with identity e. Then for every countable family $\{U_n\}_{n=1}^{\infty}$ of neighborhoods of e, there is a compact normal subgroup N of G such that $N \subset \bigcap_{n=1}^{\infty} U_n$, and G/N is metrizable and has a countable basis for its open sets.*

Proof. We can write $G = \bigcup_{n=1}^{\infty} F_n$ where $\{F_n\}_{n=1}^{\infty}$ is an increasing sequence of compact subsets of G. Let V_0 be a neighborhood of e such that V_0^- is compact. Using (4.5) and (4.9), we construct a sequence $\{V_n\}_{n=1}^{\infty}$ of symmetric neighborhoods of e such that $V_n^2 \subset V_{n-1} \cap U_n$ and such that $xV_nx^{-1} \subset V_{n-1}$ for all $x \in F_n$ $(n = 1, 2, \ldots)$. As in the proof of (4.7), it follows that $V_n^- \subset V_{n-1}$ $(n = 1, 2, \ldots)$. Let $N = \bigcap_{n=1}^{\infty} V_n$. Obviously $N \subset \bigcap_{n=1}^{\infty} U_n$ and thus, by Theorem (5.6) applied to the family $\{V_n : n = 0, 1, 2, \ldots\}$, N is a closed normal subgroup of G. Since $N \subset V_0^-$, N is also compact.

Let φ denote the natural mapping of G onto G/N. We now show that $\{\varphi(V_n)\}_{n=1}^{\infty}$ is a basis at N in G/N. Let $\{wN : w \in W\}$ be an arbitrary neighborhood of N in G/N. For some n_0, we have $V_{n_0} \subset WN$. For in the contrary case, the family $\{V_n^- \cap (WN)'\}_{n=1}^{\infty}$ of compact sets has the finite intersection property and hence $\bigcap_{n=1}^{\infty} (V_n^- \cap (WN)')$ is nonvoid. This is impossible since $\bigcap_{n=1}^{\infty} (V_n^- \cap (WN)') = \left(\bigcap_{n=1}^{\infty} V_n^-\right) \cap (WN)' = \left(\bigcap_{n=1}^{\infty} V_n\right) \cap (WN)' = N \cap (WN)' = \varnothing$. Since $V_{n_0} \subset WN$, we have $\varphi(V_{n_0}) \subset \{wN : w \in W\}$.

Since G/N has a countable open basis at N, G/N is metrizable (8.3). Since $G/N = \bigcup_{n=1}^{\infty} \varphi(F_n)$, the group G/N is σ-compact, and hence Lindelöf. As a metric space satisfying the Lindelöf property, G/N has a countable basis for its open sets. □

While every T_0 group is completely regular (8.4), there are many nonnormal T_0 groups. We shall present two constructions of such groups. The first of them [(8.8) and (8.10)] is very general. The second, Theorem (8.12), is rather special, but shows that nonnormal groups exist in profusion. Our first construction, given in (8.8), bears a certain family resemblance to the well-known Stone-Čech compactification of a completely regular space.

(8.8) Theorem. *For every completely regular space X, there exists a topological group F with the following properties:*

(i) *X is a closed subspace of F;*

(ii) *algebraically, F is the free group generated by X;*

(iii) *for any continuous mapping φ of X into any topological group G, there exists a continuous homomorphism Φ of F into G such that*

$$\Phi(x) = \varphi(x) \text{ for all } x \in X.$$[1]

Proof. Let $\mathcal{G}$ be a family of topological groups such that $\overline{\overline{G}} \leq \max(\overline{\overline{X}}, \mathfrak{c})$ for $G \in \mathcal{G}$, distinct members of $\mathcal{G}$ are not topologically isomorphic, and every topological group H for which $\overline{\overline{H}} \leq \max(\overline{\overline{X}}, \mathfrak{c})$ is topologically isomorphic with some G in $\mathcal{G}$. Let $\{(G_\iota, \varphi_\iota)\}_{\iota \in I}$ consist of all pairs (G_ι, φ_ι) where $G_\iota \in \mathcal{G}$ and φ_ι is a continuous mapping of X into G_ι. If H is a topological group, $\overline{\overline{H}} \leq \max(\overline{\overline{X}}, \mathfrak{c})$, and φ is a continuous mapping of X into H, then there is an $\iota_0 \in I$ and a topological isomorphism τ of G_{ι_0} onto H such that $\tau \circ \varphi_{\iota_0} = \varphi$. In such a case, we will identify the pair (H, φ) with the pair $(G_{\iota_0}, \varphi_{\iota_0})$.

[1] The group F is called the *free topological group* generated by X.

Let $G_0 = \underset{\iota \in I}{\mathsf{P}} G_\iota$, and let e denote the identity of G_0. For $x \in X$, let $\nu(x) \in G_0$ be defined by $\nu(x)_\iota = \varphi_\iota(x)$. The function ν is easily shown to be a homeomorphism of X onto $\nu(X)$. [Note that X can be mapped homeomorphically into a product of real lines or apply KELLEY [2], p. 116, Lemma 5.] Henceforth we identify X with $\nu(X)$; that is, X is regarded as a subset of G_0. Now let F be the subgroup of G_0 generated by X.

Let l be any positive integer. For each permutation $P \in \mathfrak{S}_l$, let $U_P \in \mathfrak{U}(l)$ be the matrix $(u_{jk})_{j,k=1}^{l}$ such that $u_{jk}=1$ if $j=P(k)$ and $u_{jk}=0$ if $j \neq P(k)$. It is obvious that the mapping $P \to U_P$ is an isomorphism of $\mathfrak{S}_l$ into $\mathfrak{U}(l)$.

We next verify (ii). Let $x_1^{\varepsilon_1} x_2^{\varepsilon_2} \cdots x_n^{\varepsilon_n}$ be any reduced word formed from elements of X. By (2.9) and the preceding paragraph, there is a mapping A of the set [not sequence!] $\{x_1, \ldots, x_n\}$ into $\mathfrak{U}(l)$, where $l = n+1$ or $n+2$, such that $A(x_1)^{\varepsilon_1} A(x_2)^{\varepsilon_2} \cdots A(x_n)^{\varepsilon_n} \neq E$; here E is the $l \times l$ identity matrix. By (7.15) and (3.6), there is a continuous mapping φ of X into $\mathfrak{U}(l)$ such that $\varphi(x_k) = A(x_k)$ for $k=1, \ldots, n$. For some $\iota_0 \in I$, the pair $(\mathfrak{U}(l), \varphi)$ is $(G_{\iota_0}, \varphi_{\iota_0})$. Hence

$$(x_1^{\varepsilon_1} x_2^{\varepsilon_2} \cdots x_n^{\varepsilon_n})_{\iota_0} = A(x_1)^{\varepsilon_1} A(x_2)^{\varepsilon_2} \cdots A(x_n)^{\varepsilon_n} \neq E,$$

so that $x_1^{\varepsilon_1} x_2^{\varepsilon_2} \cdots x_n^{\varepsilon_n} \neq e$. That is, F is the free group generated by X.

A similar argument proves (i). Consider any element of $F \cap X'$. It can be written as $x_1^{\varepsilon_1} x_2^{\varepsilon_2} \cdots x_n^{\varepsilon_n}$, where $n > 1$ or $\varepsilon_1 = -1$, $\varepsilon_j = \pm 1$, and $\varepsilon_j = \varepsilon_{j+1}$ whenever $x_j = x_{j+1}$. Applying (2.9) [note in particular (2.9.iii)] and the isomorphism of $\mathfrak{S}_l$ into $\mathfrak{U}(l)$ described in the last paragraph but one [$l = n+1$ or $n+2$], we obtain matrices $A(x_1), \ldots, A(x_n) \in \mathfrak{U}(l)$ such that $B = A(x_1)^{\varepsilon_1} A(x_2)^{\varepsilon_2} \cdots A(x_n)^{\varepsilon_n}$ is distinct from each $A(x_k)$. Consider a neighborhood $\mathfrak{V}$ of B in $\mathfrak{U}(l)$ such that $\mathfrak{V}^- \cap \mathfrak{V}'$ is homeomorphic with a sphere S_{l^2-1} (4.25.b) and such that $A(x_1), \ldots, A(x_n)$ lie in $\mathfrak{V}^{-\prime}$. It is easy to see that $\mathfrak{U}(l) \cap \mathfrak{V}'$ is arcwise connected. Thus we may apply (3.6) to find a continuous mapping ψ of X into $\mathfrak{U}(l) \cap \mathfrak{V}'$ such that $\psi(x_k) = A(x_k)$ for $k=1, \ldots, n$. The pair $(\mathfrak{U}(l), \psi)$ is a pair $(G_{\iota_0}, \varphi_{\iota_0})$. The neighborhood in F of $x_1^{\varepsilon_1} x_2^{\varepsilon_2} \cdots x_n^{\varepsilon_n}$ consisting of all $(y_\iota) \in F$ for which $y_{\iota_0} \in \mathfrak{V}$ plainly excludes all points of X. Hence $F \cap X'$ is open, *i.e.*, X is closed in F.

Property (iii) is almost trivial. If φ is a continuous mapping of X into any topological group H, the image $\varphi(X)$ in H is contained in a subgroup J of H such that $\overline{\overline{J}} \leqq \max(\overline{\overline{X}}, \mathfrak{c})$. Thus (J, φ) is a pair $(G_{\iota_0}, \varphi_{\iota_0})$. Then φ is merely projection onto the ι_0-th axis in $F \subset \underset{\iota \in I}{\mathsf{P}} G_\iota$. As such, φ can obviously be extended to a continuous homomorphism [still a projection] of F into $J \subset H$. $\square$

The theory of free topological groups has been extensively developed [see for example GRAEV [4]]. We cannot set forth all of it here. We mention, however, the following fact.

(8.9) Theorem. *Let X be a completely regular space, F the topological group constructed in* (8.8), *and $\widetilde{F}$ any topological group such that:*

(i) *X is a subspace of $\widetilde{F}$;*

(ii) *the smallest closed subgroup of $\widetilde{F}$ that contains X is $\widetilde{F}$ itself;*

(iii) *for every continuous mapping φ of X into a topological group G, there is a continuous homomorphism Φ carrying $\widetilde{F}$ into G such that*

$$\Phi(x) = \varphi(x) \text{ for all } x \in X.$$

Then there is a topological isomorphism τ of F onto $\widetilde{F}$ such that $\tau(x) = x$ for all $x \in X$.

Proof. In view of (8.8.iii) and (iii), there are a continuous homomorphism Φ of F into $\widetilde{F}$ and a continuous homomorphism $\widetilde{\Phi}$ of $\widetilde{F}$ into F such that $\Phi(x) = \widetilde{\Phi}(x) = x$ for all $x \in X$. The mapping $\widetilde{\Phi} \circ \Phi$ is a continuous homomorphism of F into itself that is the identity mapping on X. In view of (8.8.ii), $\widetilde{\Phi} \circ \Phi$ is the identity mapping of F onto itself. Similarly, $\Phi \circ \widetilde{\Phi}$ is the identity mapping on the subgroup of $\widetilde{F}$ generated algebraically by X. Since this subgroup is dense in $\widetilde{F}$ and $\Phi \circ \widetilde{\Phi}$ is continuous, we see that $\Phi \circ \widetilde{\Phi}$ is the identity mapping on $\widetilde{F}$. Thus $\widetilde{\Phi} = \Phi^{-1}$ and we may take $\tau = \Phi$. ▯

(8.10) Theorem. *There exist nonnormal T_0 groups.*

Proof. Consider any nonnormal completely regular space X. As X is closed in its free topological group F, and every closed subspace of a normal space is normal, F cannot be a normal space. ▯

We will now show that some fairly simple groups are nonnormal.

(8.11) Theorem. *If $\mathfrak{m}$ is any uncountable cardinal number, then $Z^{\mathfrak{m}}$ is a nonnormal completely regular topological group.*

Proof. Since $Z^{\mathfrak{m}}$ is a T_0 group, it is completely regular (8.4). Write $Z^{\mathfrak{m}}$ as $\mathop{\mathsf{P}}_{\iota \in I} Z_\iota$, where $\overline{\overline{I}} = \mathfrak{m}$ and each Z_ι is Z. To show that $Z^{\mathfrak{m}}$ is not normal, let

$$A = \{(x_\iota) \in Z^{\mathfrak{m}}: \text{for } n \neq 0, \text{ there is at most one index } \iota \text{ for which } x_\iota = n\},$$

and

$$B = \{(x_\iota) \in Z^{\mathfrak{m}}: \text{for } n \neq 1, \text{ there is at most one index } \iota \text{ for which } x_\iota = n\}.$$

If $(x_\iota) \notin A$, then we have $x_{\iota_0} = x_{\iota_1} = n$ for some $n \in Z$, $n \neq 0$, and some $\iota_0, \iota_1 \in I$, $\iota_0 \neq \iota_1$. Hence $\{(y_\iota) \in Z^{\mathfrak{m}}: y_{\iota_0} = y_{\iota_1} = n\}$ is an open set containing

(x_ι) that is disjoint from A. Thus A is a closed set; similarly B is closed. Obviously $A \cap B = \emptyset$. Let U and V be open sets in $Z^{\mathfrak{m}}$ such that $U \supset A$ and $V \supset B$. To prove that $Z^{\mathfrak{m}}$ is nonnormal, it suffices to prove that $U \cap V \neq \emptyset$.

Let $(x_\iota^{(1)}) \in Z^{\mathfrak{m}}$ be defined by $x_\iota^{(1)} = 0$ for all $\iota \in I$. Since $(x_\iota^{(1)}) \in A \subset U$, there exist distinct $\iota_1, \ldots, \iota_{m_1} \in I$ such that

$$(x_\iota^{(1)}) \in \{(x_\iota) \in Z^{\mathfrak{m}} : x_{\iota_k} = 0 \text{ for } k = 1, \ldots, m_1\} \subset U.$$

Let $(x_\iota^{(2)}) \in Z^{\mathfrak{m}}$ be defined by $x_\iota^{(2)} = k$ if $\iota = \iota_k$ $(1 \leqq k \leqq m_1)$ and $x_\iota^{(2)} = 0$ otherwise. Since $(x_\iota^{(2)}) \in A \subset U$, there exist $\iota_{m_1+1}, \ldots, \iota_{m_2} \in I$, distinct from each other and from $\iota_1, \ldots, \iota_{m_1}$, such that

$$(x_\iota^{(2)}) \in \{(x_\iota) \in Z^{\mathfrak{m}} : x_{\iota_k} = k \text{ for } k = 1, \ldots, m_1 \text{ and}$$
$$x_{\iota_k} = 0 \text{ for } k = m_1 + 1, \ldots, m_2\} \subset U.$$

Continuing in this manner, we define by finite induction a sequence $\{(x_\iota^{(n)})\}_{n=1}^{\infty}$ of elements in $Z^{\mathfrak{m}}$, a sequence $\{\iota_k\}_{k=1}^{\infty}$ of indices from I, and a strictly increasing sequence of integers $\{m_n\}_{n=1}^{\infty}$ such that $x_\iota^{(n)} = k$ if $\iota = \iota_k$ $(1 \leqq k \leqq m_{n-1})$ and $x_\iota^{(n)} = 0$ otherwise and such that

$$(x_\iota^{(n)}) \in \{(x_\iota) \in Z^{\mathfrak{m}} : x_{\iota_k} = k \text{ for } k = 1, \ldots, m_{n-1} \text{ and}$$
$$x_{\iota_k} = 0 \text{ for } k = m_{n-1} + 1, \ldots, m_n\} \subset U.$$

We now define $(y_\iota) \in Z^{\mathfrak{m}}$ as follows. Let $y_\iota = k$ if $\iota = \iota_k$ $(k = 1, 2, \ldots)$ and $y_\iota = 1$ otherwise. Clearly $(y_\iota) \in B$. Hence for some finite subset J of I, we have

$$\{(x_\iota) \in Z^{\mathfrak{m}} : x_\iota = y_\iota \text{ for } \iota \in J\} \subset V.$$

Clearly there exists an n_0 such that $\iota_k \notin J$ whenever $k > m_{n_0}$. Finally, we define $(z_\iota) \in Z^{\mathfrak{m}}$ by the rule:

$$z_\iota = k \text{ if } \iota = \iota_k \text{ and } k \leqq m_{n_0};$$
$$z_\iota = 0 \text{ if } \iota = \iota_k \text{ and } m_{n_0} + 1 \leqq k \leqq m_{n_0+1};$$
$$z_\iota = 1 \text{ otherwise}.$$

Then $(z_\iota) \in \{(x_\iota) \in Z^{\mathfrak{m}} : x_\iota = y_\iota \text{ for } \iota \in J\} \subset V$ and

$$(z_\iota) \in \{(x_\iota) \in Z^{\mathfrak{m}} : x_{\iota_k} = k \text{ for } k = 1, \ldots, m_{n_0} \text{ and}$$
$$x_{\iota_k} = 0 \text{ for } k = m_n + 1, \ldots, m_{n_0+1}\} \subset U.$$

That is, $U \cap V \neq \emptyset$. □

(8.12) Theorem. *If the product $\underset{\iota \in I}{\mathsf{P}} G_\iota$ of T_0 topological groups G_ι is normal, then all but at most a countable number of the groups G_ι are countably compact.*

Proof. Suppose that G_ι is not countably compact for $\iota \in J \subset I$, where J is uncountable. Then for $\iota \in J$, there exists a countable closed discrete subset Z_ι of G_ι. [Choose a countable open covering $\{U_1, U_2, \ldots\}$ of G_ι with no finite subcovering, let $x_n \in (U_1 \cup \cdots \cup U_n)'$, and $Z_\iota = \{x_1, x_2, \ldots\}$.] Then $\underset{\iota\in J}{\mathsf{P}} Z_\iota \times \underset{\iota\notin J}{\mathsf{P}} \{e_\iota\}$ is a closed subset of $\underset{\iota\in I}{\mathsf{P}} G_\iota$ homeomorphic with $\underset{=}{Z}^J$. By (8.11), it is nonnormal. It follows that $\underset{\iota\in I}{\mathsf{P}} G_\iota$ is nonnormal. □

The reader has undoubtedly already observed that (8.11) and (8.12) are purely topological: no group structure is needed. [Theorem (8.12) is true for products of T_1 spaces.]

In view of Theorems (8.10) and (8.12), it is interesting to note that a large class of topological groups are normal.

(8.13) Theorem. *Every locally compact T_0 group is paracompact and hence normal.*

Proof. Let G be a locally compact T_0 group, and let U be a symmetric neighborhood of the identity in G such that U^- is compact. Let $L = \bigcup_{n=1}^{\infty} U^n$; then L is an open and closed subgroup of G, by (5.7). Since $U^- \subset U^2$, L is also σ-compact and is hence Lindelöf. Let $\mathscr{V}$ be any open covering of G. For each coset xL of L, there is a countable subfamily $\{V_{xL}^{(n)}\}_{n=1}^{\infty}$ of $\mathscr{V}$ such that $xL \subset \bigcup_{n=1}^{\infty} V_{xL}^{(n)}$. For $n=1, 2, \ldots$, let $\mathscr{W}_n = \{V_{xL}^{(n)} \cap (xL) : xL \in G/L\}$. Then $\mathscr{W} = \bigcup_{n=1}^{\infty} \mathscr{W}_n$ is clearly an open σ-locally finite refinement of the covering $\mathscr{V}$. Thus G is paracompact and hence normal (3.8). □

Miscellaneous theorems and examples

(8.14) Invariant pseudo-metrics for coset spaces. (a) Theorem (8.2) shows that the topology of a topological group can be described in terms of continuous left invariant pseudo-metrics: if Σ consists of all continuous left invariant pseudo-metrics on G, then the family of all sets $\{x \in G : \sigma(x, e) < \varepsilon\}$, where σ runs through Σ and ε through the positive real numbers, forms a basis at e for the topology of G.

If G is a topological group and H is any subgroup, then the topology of G/H can be described by continuous, but not necessarily invariant, pseudo-metrics. Our construction follows that of MONTGOMERY and ZIPPIN [1], p. 36. Let σ be a *right* invariant pseudo-metric on G. Define σ^* on G/H by

$$\sigma^*(xH, yH) = \inf\{\sigma(a, b) : a \in xH, b \in yH\}.$$

Let $\varepsilon > 0$ and choose $a \in xH$, $b, c \in yH$, and $d \in zH$ such that

$$\sigma(a, b) \leqq \sigma^*(xH, yH) + \varepsilon \quad \text{and} \quad \sigma(c, d) \leqq \sigma^*(yH, zH) + \varepsilon.$$

Since $dc^{-1}b \in zH$, we have

$$\sigma^*(xH, zH) \leqq \sigma(a, dc^{-1}b) \leqq \sigma(a, b) + \sigma(b, dc^{-1}b)$$
$$= \sigma(a, b) + \sigma(c, d) \leqq \sigma^*(xH, yH) + \sigma^*(yH, zH) + 2\varepsilon.$$

This proves the triangle inequality; the remaining axioms for a pseudo-metric are obvious. Let $\{xH : x \in Vy\}$ be a neighborhood of yH in G/H, where V is a neighborhood of e. If a continuous right invariant pseudo-metric σ and a positive real number ε are chosen such that $\{x \in G : \sigma(x, e) < \varepsilon\} \subset V$, then $\{xH : \sigma^*(xH, yH) < \varepsilon\} \subset \{xH : x \in Vy\}$. It is routine to verify that σ^* is continuous on $(G/H) \times (G/H)$. Thus the topology of G/H is describable by continuous pseudo-metrics. Also, as in (8.4), every T_0 coset space is completely regular.

(b) If G is a metric group and H is a *closed* subgroup of G, then G/H is metrizable. Let σ be a right invariant metric on G compatible with the topology of G (8.3), and let σ^* be as in (a). Then σ^* is a metric and the topology of G/H coincides with the σ^*-topology. In fact, for a fixed $y \in G$, we have

$$\{xH : \sigma^*(xH, yH) < \varepsilon\} = \{xH : \sigma(x, y) < \varepsilon\}$$

for all $\varepsilon > 0$.

(c) In (a) and (b), σ^* is not necessarily invariant. If H is a *compact* subgroup of G, then the topology of G/H can be described using only continuous left invariant pseudo-metrics. To see this, we first remark that if the hypothesis in (8.2) that U_k be symmetric is replaced by the condition that $U_{k+1}^{-1} \subset U_k$ for all k, then we obtain a left invariant pseudo-metric σ such that (8.2.i) and (8.2.iv) hold and such that $\sigma(x, y) \leqq 2^{-k+3}$ whenever $y^{-1}x \in U_k$. [Simply repeat the proof of (8.2) with slight modifications.]

Let $\{xH : x \in V_n\}$ be a family of neighborhoods of H, where V_n is a neighborhood of e in G, $n = 1, 2, \ldots$. Let U_1 and W_1 be neighborhoods of e such that $U_1^2 \subset V_1$, $W_1 \subset U_1$, W_1 is symmetric, and $hW_1h^{-1} \subset U_1$ for all $h \in H$ (4.9). By finite induction, we choose U_n such that $U_n^2 \subset W_{n-1}$ and find symmetric neighborhoods W_n of e such that $W_n \subset U_n \cap V_n$ and $hW_nh^{-1} \subset U_n$ for all $h \in H$. Then $(W_{n+1}H)^2 \subset W_nH$ and $(W_{n+1}H)^{-1} \subset W_nH$ for $n = 1, 2, \ldots$. Now let σ be the left invariant pseudo-metric which is guaranteed by the remark of the last paragraph [for U_k take W_kH, $k = 1, 2, \ldots$]. Define $\sigma^*(xH, yH) = \sigma(x, y)$ for $xH, yH \in G/H$. It is easy to verify that σ^* is a continuous left invariant pseudo-metric on G/H.

(d) Using (c), we can prove an analogue of (8.3) due to KRISTENSEN [1]. Let G be a topological group and let H be a compact closed subgroup of G. Then G/H is metrizable if and only if there is a countable open basis at H. Moreover, in this case the metric can be taken to be left invariant. To see this, let the sets $\{xH: x\in V_n\}$ in (c) be a basis at H. Note that the sets $\{xH: x\in W_n\}$ constructed in (c) also form a basis at H. The pseudo-metric σ^* of (c) is then a metric since $\bigcap_{n=1}^{\infty} W_n H = H$, and the topology of G/H coincides with the σ^*-topology, since

$$\{xH: \sigma^*(xH, yH) < 2^{-n}\} \subset \{xH: x\in yW_n\} \subset \{xH: \sigma^*(xH, yH) \leqq 2^{-n+3}\},$$

for $yH\in G/H$ and $n=1, 2, \ldots$.

(8.15) Examples concerning metrizability. (a) The construction of (8.2) applied to the additive real numbers R, with $U_n = V_{2^{-n}} =]-2^{-n}, 2^{-n}[$, gives $\sigma(x, y) = \min(|x-y|, 1)$ for all $x, y\in R$.

(b) (KAKUTANI [1], VAN DANTZIG [3].) For $X=(x_{jk})_{j,k=1}^{n}$ in the matrix group $\mathfrak{GL}(n, K)$, let $\|X\|$ be the norm of X regarded as a linear transformation of n-dimensional complex space onto itself:

$$\|X\| = \max\left\{\left(\sum_{j=1}^{n}\left|\sum_{k=1}^{n} x_{jk} v_k\right|^2\right)^{\frac{1}{2}} : |v_1|^2+|v_2|^2+\cdots+|v_n|^2=1\right\}.$$

Then it is easy to verify that $\|X+Y\| \leqq \|X\|+\|Y\|$, $\|XY\| \leqq \|X\|\|Y\|$, and $\|E\|=1$ [E is the identity $n\times n$ matrix]. The function $\sigma(X, Y) = \log\{1+\|X^{-1}Y-E\|+\|Y^{-1}X-E\|\}$ is a left invariant metric on $\mathfrak{GL}(n, K)$ compatible with the ordinary topology of $\mathfrak{GL}(n, K)$. However, the group $\mathfrak{GL}(n, K)$ admits no metric that is two-sided invariant. [This follows readily from (4.24); see (8.18) *infra*.]

(c) (ALEXANDER and COHEN [1].) Let G be any discrete group with identity e. Consider the group $G^{\aleph_0}$, realized as the group of all sequences $\boldsymbol{x}=(x_1, x_2, \ldots, x_n, \ldots)$ with $x_n\in G$. For $\boldsymbol{x}\in G^{\aleph_0}$, let $\delta(\boldsymbol{x})=0$ if $\boldsymbol{x}=\boldsymbol{e}=(e, e, \ldots, e, \ldots)$, and $\delta(\boldsymbol{x})=1/n$ if $x_1=\cdots=x_{n-1}=e$ and $x_n\neq e$, $n=1, 2, \ldots$. For $\boldsymbol{x}, \boldsymbol{y}\in G^{\aleph_0}$, let $\sigma(\boldsymbol{x}, \boldsymbol{y})=\delta(\boldsymbol{x}\boldsymbol{y}^{-1})$. Then σ is a two-sided invariant metric on $G^{\aleph_0}$ compatible with the topology of the product group $G^{\aleph_0}$. Let $\boldsymbol{x}_1, \boldsymbol{x}_2, \ldots, \boldsymbol{x}_n, \ldots$ be a fixed finite or infinite sequence of elements of $G^{\aleph_0}$ such that $\delta(\boldsymbol{x}_1) > \delta(\boldsymbol{x}_2) > \cdots > \delta(\boldsymbol{x}_n) > \cdots$. Then for every sequence $\{\alpha_n\}_{n=1}^{\infty}$ of integers, $\lim_{k\to\infty} \boldsymbol{x}_1^{\alpha_1}\boldsymbol{x}_2^{\alpha_2}\cdots\boldsymbol{x}_k^{\alpha_k}$ exists. The set of all such limits is a closed subgroup of $G^{\aleph_0}$. Every closed subgroup H of $Z^{\aleph_0}$ has this form. [To prove the last assertion, consider for every positive integer n the set of all $\boldsymbol{y}\in H$ such that $\delta(\boldsymbol{y})=1/n$. Let $\boldsymbol{y}_n$ be one of these such that $y_{n,n}$ is positive and is as small as possible. If there is no element $\boldsymbol{y}$ of H for which $\delta(\boldsymbol{y})=1/n$, let $\boldsymbol{y}_n=\boldsymbol{e}$. Then H is the set of all limits

$$\lim_{k\to\infty} \boldsymbol{y}_1^{\alpha_1}\boldsymbol{y}_2^{\alpha_2}\cdots\boldsymbol{y}_k^{\alpha_k}.$$

Removing the $\boldsymbol{y}_n$'s that are $\boldsymbol{e}$, we obtain the desired representation.]

(d) There exist nonmetrizable groups for which the identity is the intersection of a countable number of open sets. Thus the requirement of local countable compactness, or something like it, is needed in (8.5). [For example, see (4.22.b) and (4.22.d).]

(e) Every infinite Abelian group G admits a nondiscrete Hausdorff topology under which G is metrizable. [Combine (4.23.b) with (8.3).]

(8.16) (JONES [1] and HULANICKI [1].) Every locally compact T_0 group of cardinal number $\aleph_1$ is metrizable. [By (8.5), it suffices to prove that some point of G is the intersection of a countable family of open sets. Assume that this is not the case: no point of G is the intersection of a countable family of open sets. Well order G: $G=\{x_1, x_2, \ldots, x_\alpha, \ldots\}$, where α runs through all ordinal numbers less than Ω, the first uncountable ordinal. Let U_1 be a nonvoid open set such that U_1^- is compact and $x_1 \notin U_1^-$. Suppose that α is some ordinal less than Ω and that nonvoid open sets U_β have been defined for all $\beta<\alpha$ such that

$$\bigcap_{\beta\leqq\gamma} U_\beta \text{ is nonvoid for } \gamma<\alpha; \tag{1}$$

$$x_\beta \notin U_\beta^- \quad \text{for} \quad \beta<\alpha; \tag{2}$$

$$U_\beta^- \subset U_{\beta-1} \text{ whenever } \beta \text{ is a nonlimit ordinal and } 1<\beta<\alpha. \tag{3}$$

By assumption we have $\bigcap\limits_{\beta<\alpha} U_\beta \neq \{x_\alpha\}$. If α is not a limit ordinal, let U_α be a nonvoid open subset of G such that $U_\alpha^- \subset U_{\alpha-1}$, $U_\alpha \cap \big(\bigcap\limits_{\beta\leqq\alpha-1} U_\beta\big) \neq \emptyset$, and $x_\alpha \notin U_\alpha^-$. If α is a limit ordinal, then $\bigcap\limits_{\beta<\alpha} U_\beta = \bigcap\limits_{\beta<\alpha} U_\beta^- \neq \emptyset$, and we let U_α be a nonvoid open set such that $U_\alpha \cap \big(\bigcap\limits_{\beta<\alpha} U_\beta\big) \neq \emptyset$ and $x_\alpha \notin U_\alpha^-$. We then have $\bigcap\limits_{\alpha<\Omega} U_\alpha^- = \{x_1, x_2, \ldots, x_\alpha, \ldots\}' = \emptyset$, but also $\bigcap\limits_{\alpha<\Omega} U_\alpha^- \neq \emptyset$ by the compactness of U_1^-.]

(8.17) (GRAEV [4].) Let G be a T_0 group admitting an open basis at the identity e consisting of sets U such that $xUx^{-1}=U$ for all $x \in G$. Then G is topologically isomorphic with a subgroup of a direct product of topological groups each of which has a two-sided invariant metric. [The construction of σ in (8.2) can be carried out so that H is a normal subgroup, and so that the first neighborhood U_1 is contained in an arbitrarily preassigned neighborhood of e; recall the modification of (8.2) indicated in (8.14.c). Define a metric σ_H in the group G/H by $\sigma_H(aH, bH)=\sigma(a, b)$. Then σ_H is a metric on G/H making G/H into a metric group [although its topology may be different from the quotient group topology of G/H]. Forming the direct product of all of the groups G/H with their metric topologies, and mapping G into this direct product in the natural way, we obtain the result asserted.]

(8.18) Let G be any T_0 group having a countable open basis at the identity e. The following conditions on G are equivalent:

(i) G admits a two-sided invariant metric compatible with its topology;

(ii) the uniform structures $\mathscr{S}_l(G)$ and $\mathscr{S}_r(G)$ are equivalent;

(iii) there is a countable open basis $\{U_n\}_{n=1}^{\infty}$ at e such that $xU_nx^{-1}=U_n$ for $n=1, 2, \ldots$ and all $x\in G$;

(iv) if $\{x_n\}_{n=1}^{\infty}$ is a sequence of elements of G with limit e and $\{y_n\}_{n=1}^{\infty}$ is any sequence of elements of G, then $\{y_n x_n y_n^{-1}\}_{n=1}^{\infty}$ has limit e.

(8.19) Let G be a T_0 group with a subgroup H such that H and G/H have countable dense subsets and countable open bases at every point. Then G is metrizable and has a countable open basis for all open sets. [By (5.38.e), G has a countable open basis at e and is metrizable by (8.3). By (5.38.f), G has a countable dense subset. A metric space with a countable dense subset has a countable basis for all open sets.]

(8.20) (a) Let G be a compact T_0 group, and let $\{U_n\}_{n=1}^{\infty}$ be any countable family of neighborhoods of e in G. Then there is a closed normal subgroup N of G such that $N\subset\bigcap_{n=1}^{\infty}U_n$ and such that G/N admits a two-sided invariant metric compatible with its topology. [This is immediate from (8.7) and (8.6). However, we shall indicate a completely different and interesting proof due to L. GRABAR' (see PONTRYAGIN [7], § 20, Example 37, p. 126). For each n ($n=1, 2, \ldots$), let f_n be a real continuous function on G such that $f_n(e)=1$, $f_n(U_n')=0$, and $f_n(G)\subset[0, 1]$. Let $f=\sum_{n=1}^{\infty}2^{-n}f_n$; f is a continuous function on G into $[0, 1]$ such that $f(e)=1$ and $f\left(\bigcap_{n=1}^{\infty}U_n'\right)=0$. For $a, b\in G$, let

$$\varrho(a, b)=\sup\{|f(xay)-f(xby)| : x, y\in G\}.$$

Then ϱ is a two-sided invariant pseudo-metric on G. Let $N=\{a\in G:\varrho(a, e)=0\}$. Then N is a closed normal subgroup of G contained in $\bigcap_{n=1}^{\infty}U_n$. If for aN and bN in G/N, we define $\varrho^*(aN, bN)=\varrho(a, b)$, we obtain a two-sided invariant metric for G/N.

The foregoing construction can be carried out in any T_0 group. However, in general the metric ϱ^* on G/N will *not* yield the quotient group topology of G/N. If G is compact, the metric ϱ^* does yield the topology of G/N. Let ε be any positive number. Since f is uniformly continuous on G (4.16), there is a neighborhood V of e such that $|f(u)-f(v)|<\varepsilon$ whenever $u^{-1}v\in V$. Let W be a neighborhood of e such that $y^{-1}Wy\subset V$ for all $y\in G$ (4.9). Then $a\in W$ implies that $\varrho(a, e)=\varrho^*(aN, N)\leqq\varepsilon$. Hence the ϱ^*-topology on G/N is weaker than the quotient space topology of G/N. Since G/N is compact in the quotient space topology and Hausdorff in the ϱ^*-topology, the two topologies coincide.]

(b) The function ϱ of (a) may be a metric on G incompatible with the topology of G. [For example, let $G = \mathfrak{G}\mathfrak{L}(n, K)$, and choose f so that $f(E) = 1$, $f(X) < 1$ for $X \neq E$, and $f(\mathfrak{G}\mathfrak{L}(n, K)) \subset [0, 1]$. Then the ϱ constructed from this function f as in (a) cannot be compatible with the topology of $\mathfrak{G}\mathfrak{L}(n, K)$, since $\mathfrak{G}\mathfrak{L}(n, K)$ admits no two-sided invariant metric.]

(8.21) (H. H. Corson, oral communication.) If X is a completely regular space, then there is a complex topological linear space (B.5) such that X is a closed subset of L. [Let $\mathfrak{C}(X)$ denote the family of all continuous complex-valued functions on X, define addition and scalar multiplication pointwise, and endow $\mathfrak{C}(X)$ with the topology of pointwise convergence; then a net f_α of elements in $\mathfrak{C}(X)$ converges to $f \in \mathfrak{C}(X)$ if and only if $f_\alpha(x) \to f(x)$ for each $x \in X$. Define $\mathfrak{C}(\mathfrak{C}(X))$ in the same manner and let $L = \mathfrak{C}(\mathfrak{C}(X))$. For $x \in X$, let $\nu(x) \in L$ be defined by $\nu(x)(f) = f(x)$. Then ν is a homeomorphism of X onto $\nu(X)$. To show that $\nu(X)$ is closed, let $\varphi \in \nu(X)^- \cap (\nu(X)')$. Then there is a net x_α, $\alpha \in D$, such that $\nu(x_\alpha) \to \varphi$. For each $\alpha \in D$, let $N_\alpha = \{x_\beta \in X : \beta \geqq \alpha\}$. For each finite set $F \subset X$ such that $F \cap N_\alpha^- = \varnothing$, choose $f_{\alpha,F} \in \mathfrak{C}(X)$ such that $f_{\alpha,F}(F) = 0$ and $f_{\alpha,F}(N_\alpha^-) = 1$. Now $f_{\alpha,F}$ is a net whose domain is $\{(\alpha, F) : \alpha \in D \text{ and } F \cap N_\alpha^- = \varnothing\}$ and where $(\alpha, F) \geqq (\alpha_0, F_0)$ if and only if $\alpha \geqq \alpha_0$ and $F \supset F_0$. It is easy to check that $f_{\alpha,F} \to 0$, the function identically zero. We also have for each α and F,

$$\varphi(f_{\alpha,F}) = \lim_\beta \nu(x_\beta)(f_{\alpha,F}) = \lim_\beta f_{\alpha,F}(x_\beta) = 1.$$

Hence

$$0 = \lim_\alpha \nu(x_\alpha)(0) = \varphi(0) = \lim_{(\alpha, F)} \varphi(f_{\alpha,F}) = 1.$$

This contradiction shows that $\nu(X)^- \cap (\nu(X)') = \varnothing$; that is, $\nu(X)$ is closed.]

(8.22) (a) Theorem (8.8) can be modified as follows. If property (iii) is required to hold only for T_0 groups G, then the free topological group F can also be taken to be T_0. [Repeat the proof of (8.8) *verbatim*, replacing the term "topological group" by "T_0 group".]

(b) (Kakutani [5].) Many imbeddings similar to that described in (8.8) can be produced. We cite two examples. Let X be any completely regular space. Then there is a topological group A such that:

(i) X is a closed subspace of A;

(ii) algebraically, A is the free Abelian group generated by X;

(iii) every continuous mapping of X into a topological Abelian group G can be extended to a continuous homomorphism of A into G.

There is also a topological group E such that the left and right uniform structures of E are equivalent and such that:

(i′) X is a closed subspace of E;

(ii′) algebraically, E is the free group generated by X;

(iii′) every continuous mapping of X into a topological group G whose structures are equivalent can be extended to a continuous homomorphism of E into G.

[The proofs follow that of (8.8). In the Abelian case, only continuous mappings of X into T need be used to prove (i) and (ii). In the case of E, one needs only to note that all of the groups $\mathfrak{U}(n)$ have equivalent left and right uniform structures.]

(8.23) (GRAEV [4].) (a) Let X be any nonvoid completely regular space and let ψ be an open continuous mapping of X into a topological group G. Then the extension Ψ of ψ to a homomorphism carrying the free topological group F of X into G is an open mapping. [Let V be a neighborhood of the identity in F and x a point of X. Then $(xV)\cap X$ is open in X, and so $\psi((xV)\cap X)=\Psi((xV)\cap X)$ is open in G. Hence $\Psi(x^{-1})\Psi((xV)\cap X)=\Psi(V\cap(x^{-1}X))$ is a neighborhood of the identity in G, and $\Psi(V)\supset\Psi(V\cap(x^{-1}X))$. This implies that Ψ is an open mapping.]

(b) Every T_0 group G is the quotient group F/H of a free topological group F by a closed normal subgroup H. [Consider the identity mapping of G onto itself and apply (a).]

(8.24) (VILENKIN [2], [6], and GRAEV [4].) Every non T_0 group G is a quotient group of a certain T_0 group with respect to a nonclosed normal subgroup. [Let e be the identity of G, and let e^- denote the set $\{e\}^-$. Let A be a completely regular space with a dense subset D such that there exists a one-to-one correspondence ω of $A\cap D'$ onto e^-. Form the Cartesian product *space* $X=(G/e^-)\times A$. In each coset xe^- in G, select a point, denoted by $\psi(xe^-)$, and let $\psi(e^-)=e$. Let φ be the natural homomorphism of G onto G/e^-. Then obviously $\varphi\circ\psi(xe^-)=xe^-$ for all $xe^-\in G/e^-$. Let f be the mapping of X *onto* G defined by:

$$f(xe^-,a)=\begin{cases}\psi(xe^-)\,\omega(a) & \text{if } a\in A\cap D',\\ \psi(xe^-) & \text{if } a\in D.\end{cases}$$

Let F be the free T_0 group generated by X; see (8.22.a). Since F is the free group generated by X, f can be extended to‘ a homomorphism also written as f, of F onto G. Since $\varphi\circ f(xe^-,a)=xe^-$ for $xe^-\in G/e^-$, $\varphi\circ f$ is a continuous open mapping of X onto G/e^-. By (5.40.i), f is also continuous. Moreover, (5.40.j) implies that f is an open mapping provided that $f^{-1}(e^-)=(f^{-1}(e))^-$. Then (5.27) will show that G is the quotient group of the T_0 group F by the nonclosed normal subgroup $f^{-1}(e)$. The continuity of f yields at once the inclusion $f^{-1}(e^-) \supset (f^{-1}(e))^-$. For $a \in D$, we have $f(e^-, a) = e$ so that $\{(e^-, a) : a \in D\} \subset f^{-1}(e)$. Since D is dense in A, it follows that

$$f^{-1}(e^-) = \{(e^-, a) : a \in A\} = \{(e^-, a) : a \in D\}^- \subset (f^{-1}(e))^- .]$$

Notes

Invariant metrics on groups were introduced by VAN DANTZIG [3]. Theorems (8.2) and (8.3) are due to KAKUTANI [1] and G. BIRKHOFF [1]; BIRKHOFF did not point out that his metric is left invariant. A closely related construction, which also proves (8.4), is attributed to PONTRYAGIN by WEIL [3], p. 13. Theorem (8.7) is due to KAKUTANI and KODAIRA [1]; for our version see also MONTGOMERY and ZIPPIN [1], p. 58.

Free topological groups have been studied by many writers. The concept was invented by A. MARKOV [2]; a complete exposition appeared in MARKOV [4]. MARKOV proved (8.8) by very complicated means; simpler constructions were given by NAKAYAMA [1], KAKUTANI [5], and GRAEV [3]. Our treatment follows KAKUTANI'S. MARKOV [3] used his method to construct a connected group of arbitrary cardinal $\geqq \mathfrak{c}$ all of whose elements have order 2. See GRAEV [4], Theorem 28, for a simplified proof.

Theorems (8.11) and (8.12) are due to A. H. STONE [1]. The normality of locally compact T_0 groups was pointed out to us by E. A. MICHAEL.

Throughout the remainder of this book, ALL TOPOLOGICAL GROUPS WILL BE TAKEN TO BE $\boldsymbol{T_0}$—and hence completely regular—unless the contrary is explicitly specified.

§ 9. Structure theory for compact and locally compact Abelian groups

The principal result of this section (9.8) is that every locally compact, compactly generated Abelian group is topologically isomorphic with a product of a compact Abelian group and a finite number of copies of R and Z. This famous theorem is the continuous analogue of the structure theorem for finitely generated Abelian groups. A *discrete* Abelian group is compactly generated if and only if it is finitely generated; and every such group has the form $Z^a \times F$, where a is a nonnegative integer and F is a finite group (A.27).

Theorem (9.1) is of independent interest; it implies, for example, that if x is any element in a locally compact group, then the smallest closed group containing x is either compact or topologically isomorphic with Z.

In Theorem (9.15) we show that all infinite 0-dimensional compact groups are homeomorphic with products of two-element spaces. We close the section with a brief introduction to compact topological semigroups.

(9.1) Theorem. *Let H be any subgroup of the additive group R, with its relative topology formed from the usual topology of R. Let f be a continuous homomorphism of H into a locally compact group G. Then either f is a topological isomorphism of H onto $f(H)$, or $(f(H))^-$ is a compact Abelian subgroup of G. If f is not a topological isomorphism, then for every neighborhood U of e in G, there is a positive number $\mu \in H$ such that for every $h \in H$, the set $f([h, h+\mu[\cap H) \cap U$ is nonvoid.*

Proof. In steps (I), (II), and (III), we suppose that $H=R$ or $H=Z$.

(I) Suppose first that there is a neighborhood U of e in G such that $\{t \in H : f(t) \in U\}$ is a bounded subset of H:

$$f(t) \in U \quad \text{implies} \quad |t| \leqq \alpha .$$

Then f is one-to-one, since otherwise the set $f^{-1}(e)$ would be a subgroup of H different from $\{0\}$; such a subgroup is necessarily unbounded. If V is a neighborhood of e in G such that $V^- \subset U$ and V^- is compact, then we have $V^- \cap f(H) = V^- \cap f([-\alpha, \alpha] \cap H)$. Since H is R or Z, $[-\alpha, \alpha] \cap H$ is a compact subset of H; therefore $V^- \cap f([-\alpha, \alpha] \cap H)$ is compact. Hence the closure of $V \cap f(H)$ in the relative topology of $f(H)$ is compact. That is, $f(H)$ is locally compact. By (5.29), f is a topological isomorphism.

(II) Next suppose that f is not a topological isomorphism. With no loss of generality we may suppose that $G=(f(H))^-$. Then G is Abelian, by (5.3). Consider any a in H and let A be any nonvoid open subset of G. Then there are an element $t \in H$ and a symmetric neighborhood U of e in G such that $f(t)\, U \subset A$. By (I), there is an element $v \in H$ such that $v > |a| + |t|$ and $f(v) \in U$. Thus we have $f(v+t) = f(t) f(v) \in f(t)\, U \subset A$, so that the set $f(]a, \infty[\cap H)$ is dense in G.

Next let V be any symmetric neighborhood of e in G such that V^- is compact. Let a be any nonnegative element in H. We have just proved that $G = f(]a, \infty[\cap H)\, V$. Since V^- is compact, there are elements $t_1, t_2, \ldots, t_m \in]a, \infty[\cap H$ such that $V^- \subset \bigcup_{j=1}^{m} f(t_j)\, V$. Let $\mu = \max(t_1, t_2, \ldots, t_m)$. Let x be any element of G, and let $\tau_x = \inf\left(f^{-1}(x V^-) \cap [0, \infty[\right)$. Since $x V^-$ is closed, we have $\tau_x \in H$ and

$f(\tau_x) \in x V^-$, and hence $f(\tau_x) \in x f(t_j) V$ for some $j = 1, \ldots, m$. Thus $f(\tau_x - t_j) \in x V \subset x V^-$. Since t_j is positive, we find that $\tau_x - t_j < 0$. Thus $\tau_x < \mu$, and consequently $x \in f([0, \mu[\cap H) V^- \subset f([0, \mu] \cap H) V^-$. That is,

$$G = f([0, \mu] \cap H) V^-.$$

Now $[0, \mu] \cap H$ is compact since H is R or Z, and so $f([0, \mu] \cap H)$ is compact. By (4.4), $f([0, \mu] \cap H) V^-$ is compact. We have thus proved that G is compact.

(III) The last assertion of the theorem is now easy to verify if $H = R$ or $H = Z$. For a given neighborhood U of e in G, choose a symmetric neighborhood V of e in G such that $V^- \subset U$ and V^- is compact. By (II) if $h \in H$ there is a $\tau \in [0, \mu[\cap H$ such that $f(-h) \in f(\tau) V^-$. Thus $f(h + \tau) \in (V^-)^{-1} = V^- \subset U$.

(IV) Now suppose that H is any subgroup of R. If $H = \{0\}$, the theorem is trivial. If H contains a smallest positive element, H is topologically isomorphic with Z and so comes under (I)—(III). If H is not $\{0\}$ and contains no least positive element, then H is obviously dense in R. If we can extend the continuous homomorphism f to be a continuous homomorphism $\tilde{f}$ of R into G, then we can apply (I) and (II) to infer that f is a topological isomorphism of H or $(f(H))^-$ is compact. The last assertion is also obvious in this case: since $\tilde{f}$ is continuous, we have $\tilde{f}([h, h + \mu]) \subset \tilde{f}([h, h + \mu[\cap H)^-$, so that if an open set U intersects $\tilde{f}([h, h + \mu[)$, then it intersects $f([h, h + \mu[\cap H)$.

To extend f continuously over R, we recall that f is uniformly continuous (5.40.a). Let x be any element of $R \cap H'$ and let $\{h_n\}_{n=1}^{\infty}$ be any sequence of elements of H such that $\lim_{n \to \infty} h_n = x$. Then for every neighborhood U of e in G, there is a positive integer m_0 such that $f(h_m) f(h_n)^{-1} \in U$ if $m, n \geqq m_0$. If U^- is compact, then we have $f(h_m) \in U f(h_{m_0})$ for all $m > m_0$, and there is a cluster point in $U^- f(h_{m_0})$ of the net $\{f(h_m)\}_{m \geqq m_0}$. Let this cluster point be $\tilde{f}(x)$. It is easy to see that $\tilde{f}(x)$ is unique, both for the sequence $\{h_n\}_{n=1}^{\infty}$ and also for any other sequence $\{h'_n\}_{n=1}^{\infty}$ with limit x. Furthermore, $\tilde{f}$ is a continuous homomorphism of R into G agreeing with f on H. We omit the details. □

(9.2) Definition. Let G be a topological group. If G contains a dense cyclic subgroup, then G is said to be *monothetic*. Let τ be a continuous homomorphism of R into G. Then $\tau(R)$ is called a *one-parameter subgroup* of G. If G contains a dense one-parameter subgroup, then G is said to be *solenoidal*.

Theorem (9.1) shows that a locally compact monothetic [solenoidal] group is topologically isomorphic with Z [R] or is compact. A detailed analysis of the structure of compact monothetic and compact solenoidal groups appears in § 25.

We now establish the structure of locally compact, compactly generated Abelian groups. In (9.3)—(9.7), we give preliminaries to the main theorem, (9.8).

(9.3) Theorem. *Let G be a locally compact, compactly generated Abelian group. Then G contains a discrete subgroup N with a finite number of linearly independent generators such that G/N is a compact group.*

Proof. Let U be a symmetric neighborhood of e in G such that U^- is compact and $\bigcup_{n=1}^{\infty} U^n = G$ (5.13). Since $(U^-)^2$ is compact (4.4) and $GU = G$, there is a finite subset $\{a_1, a_2, \ldots, a_m\}$ of G such that $U^2 \subset (U^-)^2 \subset \bigcup_{j=1}^{m} a_j U$. Let A be the smallest subgroup of G containing $\{a_1, a_2, \ldots, a_m\}$. It is plain that $U^2 \subset AU$ and hence $AU^2 \subset A^2 U = AU$. Then we have $U^3 = U^2 U \subset AU^2 \subset AU$, $U^4 = U^3 U \subset AUU \subset AU$, and so on. Hence $G = \bigcup_{n=1}^{\infty} U^n = AU$.

Let C_j be the smallest subgroup of G containing a_j $(j = 1, 2, \ldots, m)$. We have $G = C_1^- C_2^- \cdots C_m^- U^-$, so that if $C_1^-, C_2^-, \ldots, C_m^-$ are compact, then G is compact, and we can take $N = \{e\}$. If some C_j^- is noncompact, then by (9.1), the group C_j is an infinite cyclic discrete subgroup of G.

Let b_1 be any one of the a_j for which $\{b_1^n : n \in Z\}$ is a discrete infinite subgroup of G. We define elements $b_2, b_3, \ldots$ by induction. Suppose that $b_1, b_2, \ldots, b_k$ have been selected from among the elements $a_1, a_2, \ldots, a_m$ in such a way that $b_1, b_2, \ldots, b_k$ generate a discrete subgroup N_k of G and $b_1, b_2, \ldots, b_k$ are linearly independent: if $b_1^{\alpha_1} b_2^{\alpha_2} \cdots b_k^{\alpha_k} = e$, where the α_j's are any integers, then $\alpha_1 = \alpha_2 = \cdots = \alpha_k = 0$. Consider the quotient group G/N_k. If it is compact, then our process terminates and we take $N = N_k$. If G/N_k is noncompact, then, writing φ for the natural homomorphism of G onto G/N_k, we have $G/N_k = \varphi(A)\,\varphi(U^-)$, and since $\varphi(U^-)$ is compact, $\varphi(A)^-$ cannot be compact. Clearly we have $\varphi(A) = \varphi(C_1) \cdots \varphi(C_m)$. If each $\varphi(C_j)^-$ is compact, then $\varphi(C_1)^- \cdots \varphi(C_m)^-$ is compact and closed so that $\varphi(A)^- \subset \varphi(C_1)^- \cdots \varphi(C_m)^-$. But $\varphi(A)^-$ is not compact, and thus there must be some j for which $\varphi(C_j)^-$ is not compact. Since G/N_k is locally compact (5.22), we infer from (9.1) that $\varphi(C_j)$ is a discrete, infinite cyclic subgroup of G/N_k. Let $b_{k+1} = a_j$. It is clear that no power b_{k+1}^n $(n \neq 0)$ can be equal to any $b_1^{\alpha_1} \cdots b_k^{\alpha_k}$, because all of the cosets $b_{k+1}^n N_k$ are distinct. Thus $b_1, b_2, \ldots, b_{k+1}$ are linearly independent. Let N_{k+1} be the subgroup of G generated by $b_1, b_2, \ldots, b_{k+1}$. Let V be a neighborhood of e in G such that $V \cap N_k = \{e\}$ and such that the open subset $\{vN_k : v \in V\}$ of G/N_k intersects $\{b_{k+1}^n N_k : n \in Z\}$ only in N_k. Then $V \cap N_{k+1} = \{e\}$, so that N_{k+1} is a discrete subgroup of G.

As the set $\{a_1, a_2, \ldots, a_m\}$ is finite, our process must terminate with a group N_k for which G/N_k is compact. □

(9.4) Theorem. *Let G be a locally compact Abelian group having a discrete subgroup N with a finite number of generators such that G/N is topologically isomorphic with $T^r \times F_0$, where r is a nonnegative integer and F_0 is a finite [Abelian] group. Then G is topologically isomorphic with $T^a \times R^b \times Z^c \times F_1$, where a, b, c are nonnegative integers and F_1 is a finite Abelian group.*

Proof. (I) Suppose first that G is connected. Then G/N is also connected, as the continuous image of a connected space, and so G/N has the form T^p. Let φ be the natural mapping of G onto T^p. We may also consider T^p as R^p/Z^p; let ψ be the natural mapping of R^p onto T^p. Since Z^p and N are discrete subgroups of R^p and G, respectively, there are a neighborhood U of $\mathbf{0}$ in R^p and a neighborhood V of the identity in G such that ψ is one-to-one on U, φ is one-to-one on V, and $\varphi(V) = \psi(U)$. [In fact, choose a symmetric neighborhood U_0 of $\mathbf{0}$ in R^p such that $U_0^2 \cap Z^p = \{\mathbf{0}\}$. Then choose a symmetric neighborhood V_0 of e in G such that $V_0^2 \cap N = \{e\}$ and $\varphi(V_0) \subset \psi(U_0)$. Finally, choose $U \subset U_0 \cap \psi^{-1}(\varphi(V_0))$ and let $V = V_0 \cap \varphi^{-1}(\psi(U))$.] With no loss of generality, we may suppose that $U = \{\boldsymbol{x} \in R^p : \|\boldsymbol{x}\| < \alpha\}$ for some positive number α [$\|\ \|$ denotes the norm described in § 1]. For $\boldsymbol{x} \in U$, let $\Phi(\boldsymbol{x})$ be the unique element in $V \subset G$ such that $\varphi(\Phi(\boldsymbol{x})) = \psi(\boldsymbol{x})$. It is obvious that if $\boldsymbol{x}$, $\boldsymbol{y}$, and $\boldsymbol{x}+\boldsymbol{y}$ are in U, then $\Phi(\boldsymbol{x}+\boldsymbol{y}) = \Phi(\boldsymbol{x})\,\Phi(\boldsymbol{y})$, and that $\Phi(-\boldsymbol{x}) = \Phi(\boldsymbol{x})^{-1}$.

We extend Φ to be a mapping, also denoted by Φ, of R^p into G by the following rule. For every $\boldsymbol{x} \in R^p$, there are an integer n and an element $\boldsymbol{y} \in U$ such that $n\boldsymbol{y} = \boldsymbol{x}$. Then we set $\Phi(\boldsymbol{x}) = (\Phi(\boldsymbol{y}))^n$. To see that Φ is still single-valued, suppose that n_1, n_2 are integers, that $\boldsymbol{y}_1, \boldsymbol{y}_2 \in U$, and $n_1 \boldsymbol{y}_1 = n_2 \boldsymbol{y}_2$. If n_1 or n_2 is 0, then plainly $\Phi(n_1 \boldsymbol{y}_1) = \Phi(n_2 \boldsymbol{y}_2) = e$. If neither n_1 nor n_2 is 0, then $\frac{k\boldsymbol{y}_1}{n_2} \in U$ for $k = 0, \pm 1, \pm 2, \ldots, \pm n_2$, and we have $\Phi(\boldsymbol{y}_1) = \left(\Phi\left(\frac{\boldsymbol{y}_1}{n_2}\right)\right)^{n_2}$; similarly we have $\Phi(\boldsymbol{y}_2) = \left(\Phi\left(\frac{\boldsymbol{y}_2}{n_1}\right)\right)^{n_1}$. Thus $(\Phi(\boldsymbol{y}_1))^{n_1} = \left(\Phi\left(\frac{\boldsymbol{y}_1}{n_2}\right)\right)^{n_1 n_2} = \left(\Phi\left(\frac{\boldsymbol{y}_2}{n_1}\right)\right)^{n_1 n_2} = (\Phi(\boldsymbol{y}_2))^{n_2}$. It is likewise easy to see that Φ is an open continuous homomorphism of R^p into G. Since $\Phi(R^p)$ contains a neighborhood of the identity element of G and G is connected, it also follows that $\Phi(R^p) = G$. [See (7.4).]

Since $\varphi \circ \Phi(\boldsymbol{x}) = \psi(\boldsymbol{x})$ for $\boldsymbol{x} \in U$ and U generates the group R^p, we have $\varphi \circ \Phi = \psi$. It follows that the kernel, say H, of Φ is contained in the kernel Z^p of ψ. If $H = \{\mathbf{0}\}$, G is topologically isomorphic with R^p. Suppose that $H \neq \{\mathbf{0}\}$. By Theorem (A.26), we can choose a basis

$\boldsymbol{e}_1, \boldsymbol{e}_2, \ldots, \boldsymbol{e}_p$ for Z^p and positive integers $d_1, d_2, \ldots, d_k$ $(1 \leqq k \leqq p)$ such that $d_1\boldsymbol{e}_1, d_2\boldsymbol{e}_2, \ldots, d_k\boldsymbol{e}_k$ are independent generators for H. Since $\boldsymbol{e}_1, \boldsymbol{e}_2, \ldots, \boldsymbol{e}_p$ are independent generators for Z^p, the elements $\boldsymbol{e}_1, \boldsymbol{e}_2, \ldots, \boldsymbol{e}_p$ are a basis for R^p as a vector space over R. Thus every coset $\boldsymbol{y}+H$ contains an element $x_1\boldsymbol{e}_1 + x_2\boldsymbol{e}_2 + \cdots + x_p\boldsymbol{e}_p$, where $0 \leqq x_1 < d_1, \ldots, 0 \leqq x_k < d_k$, and $x_{k+1}, \ldots, x_p$ are arbitrary real numbers. Furthermore, distinct elements of this type lie in distinct cosets of R^p with respect to H. It follows at once that G is topologically isomorphic with $T^k \times R^{p-k}$.[1]

(II) Now let G be any group satisfying the hypotheses of the theorem. With no loss of generality, we may take G/N equal to $T^r \times F_0$, where F_0 is finite. Let φ be the natural homomorphism of G onto G/N and let π be the projection of $T^r \times F_0$ onto T^r. Then $\pi \circ \varphi$ is an open continuous homomorphism of G onto T^r. The kernel, say M, of this homomorphism is a finite union of distinct cosets: $M = N \cup x_1N \cup x_2N \cup \cdots \cup x_lN$. Since N is closed (5.10), each x_jN is also closed and does not contain the identity of G. Hence M is a discrete subgroup of G with a finite number of generators.

Consider then the group $G/M = T^r$. If $r=0$, *i.e.*, $M=G$, then G is a discrete group with a finite number of generators, and by (A.27) G has the form $Z^c \times F_1$, where c is a nonnegative integer and F_1 is a finite Abelian group.

Suppose finally that $G/M = T^r$ with $r > 0$. Since M is discrete, $\pi \circ \varphi$ is one-to-one on some neighborhood U of the identity in G. Thus $\pi \circ \varphi$ is a homeomorphism on U, since it is continuous and open. As T^r is locally connected, we may suppose that U is connected and symmetric. The set U^n is a continuous image of the connected product $U \times U \times \cdots \times U$ [n times] for $n = 2, 3, \ldots$. Hence $\bigcup_{n=1}^{\infty} U^n$ is also connected and is an open subgroup of G (5.7). Since $\bigcup_{n=1}^{\infty} U^n$ is also closed, it is the connected component of the identity in G; we write $C = \bigcup_{n=1}^{\infty} U^n$. Since T^r is a connected group, it is generated by the neighborhood $\pi \circ \varphi(U)$. Thus we have $T^r = \pi \circ \varphi(C) = \pi \circ \varphi(G)$, and $T^r = C/(C \cap M)$ (5.32). Since C is connected, we may apply Step (I) of the present proof to infer that C is topologically isomorphic with $T^k \times R^{r-k}$, where k is a nonnegative integer.

Since $\pi \circ \varphi(C) = \pi \circ \varphi(G)$, we clearly have $G = CM$. By the purely algebraic form of the first isomorphism theorem (2.1), we know that $G/C = CM/C$ is algebraically isomorphic with $M/M \cap C$. [Since CM/C and $M/M \cap C$ are discrete, this algebraic isomorphism is automatically

---

[1] Observe that step (I) of the present proof holds for *any* discrete subgroup N of G: N need not have a finite number of generators.

a homeomorphism.] It follows that G/C has a finite number of generators. By (A.27), G/C is topologically isomorphic with $Z^c \times F_1$, where c is a nonnegative integer and F_1 is a finite Abelian group. Since C is open and divisible, (6.22.b) shows that G is topologically isomorphic with $C \times (G/C)$ and hence with $Z^c \times F_1 \times T^k \times R^{r-k}$. □

In order to complete the description of the structure of compactly generated, locally compact Abelian groups, we must borrow a theorem that is proved in Chapter Six: see (24.7). The proof of (24.7) is of course independent of the present section.

(9.5) Theorem. *Let G be a compact Abelian group and let U be a neighborhood of e in G. There is a closed subgroup H of G such that $H \subset U$ and G/H is topologically isomorphic with $T^a \times F$, where a is a nonnegative integer and F is a finite Abelian group.*

This is Theorem (24.7).

(9.6) Theorem. *Let G be a locally compact, compactly generated Abelian group. Every neighborhood U of e in G contains a compact subgroup H such that G/H is topologically isomorphic with $T^a \times R^b \times Z^c \times F_1$, where a, b, and c are nonnegative integers and F_1 is a finite Abelian group.*

Proof. By (9.3), there is a discrete subgroup N of G with finitely many generators such that G/N is compact. Let φ denote the natural mapping of G onto G/N. Let W be a symmetric neighborhood of e in G with compact closure such that $W \subset U$ and $W^3 \cap N = \{e\}$. By (5.17), $\varphi(W)$ is open in G/N. Thus (9.5) implies that there is a closed subgroup $\widetilde{H} \subset \varphi(W)$ of the group G/N such that $(G/N)/\widetilde{H}$ is topologically isomorphic with $T^a \times F$: here a is a nonnegative integer and F is a finite Abelian group. Let $H_0 = \varphi^{-1}(\widetilde{H})$ and $H = H_0 \cap W$. It is obvious that $\varphi(H_0) = \widetilde{H}$; we now prove that $\varphi(H) = \widetilde{H}$. Clearly we have $\varphi(H) \subset \widetilde{H}$; if $\tilde{x} \in \widetilde{H}$, then $\tilde{x} = \varphi(w)$ for some $w \in W$. Since $\varphi(w) \in \widetilde{H}$, we have $w \in H_0$ and therefore $\tilde{x} = \varphi(w) \in \varphi(H_0 \cap W) = \varphi(H)$. Since φ is one-to-one on W^- and W^- is compact, φ is a homeomorphism of W^- onto $\varphi(W^-)$. In particular, φ is a homeomorphism of H onto $\varphi(H) = \widetilde{H}$. Since $\widetilde{H}$ is compact, it follows that H is compact. We now show that H is a subgroup of G. If $x, y \in H$, then $xy^{-1} \in H_0$ and for some $h \in H$, we have $\varphi(h) = \varphi(xy^{-1})$. Hence $xy^{-1}h^{-1} \in N$ and plainly $xy^{-1}h^{-1} \in HH^{-1}H^{-1} \subset W^3$, so that $xy^{-1}h^{-1} = e$. That is, $xy^{-1} = h \in H$.

Let ψ denote the natural mapping of G onto G/H. To prove the present theorem it suffices by (9.4) to show that $\psi(N)$ is a discrete subgroup of G/H and that $(G/H)/\psi(N)$ is topologically isomorphic with $T^a \times F$; *i.e.*, with $(G/N)/\widetilde{H}$. [Note that $\psi(N)$ is finitely generated since N is.] We next show that $H_0 = HN$. If $h \in H$ and $x \in N$, then $\varphi(xh) = \varphi(h) \in \varphi(H) = \widetilde{H}$

and so $xh\in H_0$. Conversely, if $x\in H_0$, then there is an h in H such that $\varphi(x)=\varphi(h)$; thus we have $x=xh^{-1}h\in NH$. By (5.34), we see that $(G/H)/\psi(N)$ is topologically isomorphic with $G/\psi^{-1}(\psi(N))$ and likewise $(G/N)/\widetilde{H}$ is topologically isomorphic with $G/\varphi^{-1}(\widetilde{H})=G/H_0$. Since $\psi^{-1}(\psi(N))=HN=H_0$, $(G/H)/\psi(N)$ is topologically isomorphic with $(G/N)/\widetilde{H}$.

Finally, $\psi(N)$ is discrete in G/H. Note first that $N\cap H\subset N\cap W\subset N\cap W^3=\{e\}$. Clearly $N'\cup\{e\}$ is an open set containing H so that by (4.10), there is a neighborhood V of e in G such that $HV\subset N'\cup\{e\}$. That is, we have

$$(HV)\cap N=\{e\}. \tag{1}$$

This implies that $\psi(V)\cap\psi(N)=\{H\}$. In fact, if $vH=xH$, where $v\in V$ and $x\in N$, then $vh=x$ for some $h\in H$ and so by (1), we have $vh=x=e$. □

Before proceeding to the main theorem, we set down an obvious lemma.

(9.7) Lemma. *Suppose that G is a topological Abelian group, that H is a subgroup of G, and that $G/H=R^a\times Z^b\times F$, where a and b are nonnegative integers and F is a compact Abelian group. If φ denotes the natural mapping of G onto G/H, then every compact subgroup of G is contained in $\varphi^{-1}(\{0\}\times\{0\}\times F)$.*

Proof. If E is a compact subgroup of G, then $\varphi(E)$ is a compact subgroup of G/H. It follows that $\varphi(E)\subset\{0\}\times\{0\}\times F$, so that $E\subset\varphi^{-1}(\varphi(E))\subset\varphi^{-1}(\{0\}\times\{0\}\times F)$. □

(9.8) Theorem. *Every locally compact, compactly generated Abelian group G is topologically isomorphic with $R^a\times Z^b\times F$ for some nonnegative integers a and b and some compact Abelian group F.*

Proof. We will apply ZORN's lemma in conjunction with the preceding theorems.

By (9.6), there is a compact subgroup $H\subset G$ such that G/H is topologically isomorphic with $R^a\times Z^b\times F$, where F is a compact Abelian group. Here, and in similar situations later on, we will regard G/H as equal to $R^a\times Z^b\times F$; we will also identify $R^a\times Z^b$ with the subgroup $R^a\times Z^b\times\{e'\}$, and F with the subgroup $\{0\}\times\{0\}\times F$ [e' is the identity of F]. Let φ denote the natural mapping of G onto G/H. Let $M=\varphi^{-1}(F)$; by (5.24.a), M is compact. In view of (9.7), M is the largest compact subgroup of G. By (5.34), G/M is topologically isomorphic with $(R^a\times Z^b\times F)/F$ and hence with $R^a\times Z^b$.

Let $\mathscr{L}$ denote the family of all closed subgroups L of G such that $G=LM$. Partially order $\mathscr{L}$ by the rule:

$$L_1\leqq L_2 \text{ whenever } L_1\supset L_2.$$

Obviously G belongs to $\mathcal{L}$, so that $\mathcal{L}$ is nonvoid. Suppose next that $\{L_\alpha\}_{\alpha\in A}$ is a linearly ordered subset of $\mathcal{L}$; we will show that $\{L_\alpha\}_{\alpha\in A}$ has an upper bound in $\mathcal{L}$. Let $L=\bigcap_\alpha L_\alpha$. Clearly L is a closed subgroup of G and $L\subset L_\alpha$ for all $\alpha\in A$. To show that $L\in\mathcal{L}$, consider an arbitrary $x\in G$. We direct the set A by writing $\alpha\geqq\alpha'$ whenever $L_\alpha\geqq L_{\alpha'}$. For each $\alpha\in A$, we have $G=L_\alpha M$ and therefore $x=y_\alpha m_\alpha$ for some $y_\alpha\in L_\alpha$ and $m_\alpha\in M$. Now $\{m_\alpha:\alpha\in A\}$ is a net in M and has a convergent subnet q_γ, $\gamma\in E$; let m_0 denote the limit of q_γ. Let N denote the function from E into A that relates the net with its subnet [see (3.10)]. That is, $q_\gamma=m_{N(\gamma)}$ for $\gamma\in E$. Consider any α_0 in A. There exists a $\gamma_0\in E$ such that $N(\gamma)\geqq\alpha_0$ whenever $\gamma\geqq\gamma_0$. Thus for $\gamma\geqq\gamma_0$, we have

$$xq_\gamma^{-1}=xm_{N(\gamma)}^{-1}=y_{N(\gamma)}\in L_{N(\gamma)}\subset L_{\alpha_0}.$$

It follows that the limit xm_0^{-1} of the net xq_γ^{-1}, $\gamma\in E$, belongs to L_{α_0}. Since α_0 is arbitrary, we have $xm_0^{-1}\in\bigcap_\alpha L_\alpha=L$. Consequently, $x=xm_0^{-1}m_0\in LM$. Thus $G=LM$ and L is an upper bound for $\{L_\alpha\}_{\alpha\in A}$.

By ZORN's lemma, there is a maximal element L_0 in $\mathcal{L}$; obviously we have $G=L_0M$. Next we show that $L_0\cap M=\{e\}$. Assume that there is an element z in $L_0\cap M$ such that $z\neq e$. Let U be a neighborhood of e such that $z\notin U$. Being compactly generated, G is σ-compact, and its closed subgroup L_0 is also σ-compact. By (5.33), $L_0/(L_0\cap M)$ is topologically isomorphic with $L_0M/M=G/M$ and hence with $R^a\times Z^b$. In particular, $L_0/(L_0\cap M)$ is compactly generated. Since $L_0\cap M$ is compact, (5.39.h) shows that L_0 itself is compactly generated. By (9.6), there is a compact subgroup H_0 of L_0 such that $H_0\subset L_0\cap U$ and L_0/H_0 is topologically isomorphic with $R^{a_0}\times Z^{b_0}\times F_0$, where F_0 is compact. Let φ_0 denote the natural mapping of L_0 onto L_0/H_0; let $L_1=\varphi_0^{-1}(R^{a_0}\times Z^{b_0})$ and $M_0=\varphi_0^{-1}(F_0)$. It is trivial that $L_1\subset L_0$, $L_1\cap M_0=H_0$, and $L_0=L_1M_0$. By (9.7), every compact subgroup of L_0 is contained in M_0. In particular, we have

$$L_1\cap M=L_1\cap(L_1\cap M)\subset L_1\cap M_0=H_0\subset U.$$

Thus z does not belong to $L_1\cap M$ and since $z\in M$, we see that $z\notin L_1$. In other words, we have $L_1\subsetneqq L_0$. By (5.24.a), M_0 is compact and so M_0M is a compact subgroup of G. Therefore $M_0M=M$ since M contains all compact subgroups of G. Using this, we see that

$$G=L_0M=L_1M_0M=L_1M.$$

Thus L_1 belongs to $\mathcal{L}$; and $L_1\gneqq L_0$ since $L_1\subsetneqq L_0$. This contradiction shows that $L_0\cap M=\{e\}$.

By (5.33), we see that L_0 is topologically isomorphic with G/M and hence with $R^a\times Z^b$. Finally, by (6.12), G is topologically isomorphic with $L_0\times M$ and hence with $R^a\times Z^b\times M$. ☐

We now give some applications of (9.8).

(9.9) Definition. Let G be a topological group and a an element of G. The element a is said to be *compact* if the smallest closed subgroup of G containing a is compact.

(9.10) Theorem. *Let G be a locally compact Abelian group. The set B of all compact elements of G is a closed subgroup of G.*

Proof. Let $M_a = \{e, a, a^{-1}, a^2, a^{-2}, \ldots, a^n, a^{-n}, \ldots\}^-$, for all $a \in G$. Since $M_{ab} \subset M_a \cdot M_b$ and $M_{a^{-1}} = M_a$, (4.4) shows that B is a subgroup of G. Now suppose that $x \in B^-$. There is an open compactly generated subgroup H of G that contains x (5.14). The group H is topologically isomorphic with $R^a \times Z^b \times F$ by (9.8), where a and b are nonnegative integers and F is a compact Abelian group. We may regard $R^a \times Z^b \times F$ as an open subgroup of G. If the R^a-coordinate of x or the Z^b-coordinate of x were different from zero, there would be an entire neighborhood of x in H, and hence in G, disjoint from B, since R^a and Z^b contain no nonzero compact elements. It follows that x is in F, and so x is in B.[1] □

The following technical result is useful in proving some consequences of (9.8). It is also needed in the proof of Theorem (24.36).

(9.11) Theorem. *Let H be a closed subgroup of R^c different from $\{\mathbf{0}\}$. There are vectors $\boldsymbol{x}_1, \ldots, \boldsymbol{x}_a, \boldsymbol{y}_1, \ldots, \boldsymbol{y}_b$ in H that are linearly independent in R^c, such that H is exactly the set of vectors that can be written in the form*

$$\text{(i)} \qquad \alpha_1 \boldsymbol{x}_1 + \cdots + \alpha_a \boldsymbol{x}_a + m_1 \boldsymbol{y}_1 + \cdots + m_b \boldsymbol{y}_b,$$

where $\alpha_1, \ldots, \alpha_a \in R$ and $m_1, \ldots, m_b \in Z$. Thus H is topologically isomorphic with $R^a \times Z^b$ and is also compactly generated.

Proof. The proof is by induction on the number c. For $c = 1$, the theorem is quite elementary. We may thus suppose that $c \geqq 1$, that the theorem has been proved for R^c, and that $H \subset R^{c+1}$. There are two cases.

Case I. The subgroup H is discrete. Let $\boldsymbol{y}_0$ be a vector in H such that $t = \|\boldsymbol{y}_0\| = \min\{\|\boldsymbol{u}\| : \boldsymbol{u} \in H, \boldsymbol{u} \neq \mathbf{0}\}$. Plainly t is positive. Let $L = \{\alpha \boldsymbol{y}_0 : \alpha \in R\}$. Let $\boldsymbol{u} \in H \cap L'$, let α be any real number, and let m be an integer such that $|\alpha - m| \leqq \frac{1}{2}$. Then we have

$$t \leqq \|\boldsymbol{u} - m\boldsymbol{y}_0\| \leqq \|\boldsymbol{u} - \alpha \boldsymbol{y}_0\| + \|\alpha \boldsymbol{y}_0 - m\boldsymbol{y}_0\| \leqq \|\boldsymbol{u} - \alpha \boldsymbol{y}_0\| + \frac{t}{2},$$

so that

$$\|\boldsymbol{u} - \alpha \boldsymbol{y}_0\| \geqq \frac{t}{2}. \tag{1}$$

Consider the quotient group R^{c+1}/L, and the natural mapping φ of R^{c+1} onto R^{c+1}/L. It is obvious that R^{c+1}/L is topologically isomorphic to R^c

[1] In non-Abelian locally compact groups, the set B of all compact elements may fail to be a group or even a closed set; see (9.26.c) and (9.26.d).

and that φ is a linear mapping. In view of (1), $\varphi(H)$ is a discrete and hence closed subgroup of R^{c+1}/L (5.10).

By our inductive hypothesis, there are vectors $\varphi(\boldsymbol{y}_1), \ldots, \varphi(\boldsymbol{y}_b)$ in $\varphi(H)$ that are linearly independent in R^{c+1}/L and such that $\varphi(H)$ consists precisely of the vectors

$$m_1\varphi(\boldsymbol{y}_1) + \cdots + m_b\varphi(\boldsymbol{y}_b) \qquad (m_k \in Z). \tag{2}$$

Plainly we can choose $\boldsymbol{y}_k \in H$. Now for $\boldsymbol{u} \in H$, we can find $m_1, \ldots, m_b \in Z$ such that $\varphi(\boldsymbol{u}) = \varphi(m_1\boldsymbol{y}_1 + \cdots + m_b\boldsymbol{y}_b)$. This implies that $\boldsymbol{u} - m_1\boldsymbol{y}_1 - \cdots - m_b\boldsymbol{y}_b = m_0\boldsymbol{y}_0$ for some $m_0 \in Z$, or

$$\boldsymbol{u} = m_0\boldsymbol{y}_0 + \cdots + m_b\boldsymbol{y}_b.$$

If $\nu_0, \nu_1, \ldots, \nu_b \in R$ and $\sum_{k=0}^{b} \nu_k\boldsymbol{y}_k = \boldsymbol{0}$, then also $\boldsymbol{0} = \varphi\left(\sum_{k=0}^{b} \nu_k\boldsymbol{y}_k\right) = \sum_{k=0}^{b} \nu_k\varphi(\boldsymbol{y}_k) = \sum_{k=1}^{b} \nu_k\varphi(\boldsymbol{y}_k)$. Hence $\nu_1 = \nu_2 = \cdots = \nu_b = 0$, and finally $\nu_0 = 0$. Thus $\boldsymbol{y}_0, \boldsymbol{y}_1, \ldots, \boldsymbol{y}_b$ are linearly independent in R^{c+1}, and the induction is complete for discrete subgroups.

Case II. The subgroup H is nondiscrete. In this case the construction of L must be carried out differently. We will show that H contains a 1-dimensional linear subspace of R^{c+1}. Let $\{\boldsymbol{u}_n\}_{n=1}^{\infty}$ be a sequence of vectors in R^{c+1} and $\{\lambda_n\}_{n=1}^{\infty}$ a sequence of positive real numbers such that $\|\boldsymbol{u}_n\| = 1$ and $\lambda_n\boldsymbol{u}_n \in H$ for all n and such that $\lim_{n\to\infty} \lambda_n = 0$. Let $\boldsymbol{x}_0$ be any vector in R^{c+1} such that a subsequence of $\{\boldsymbol{u}_n\}_{n=1}^{\infty}$ converges to $\boldsymbol{x}_0$. Let α be any real number and ε any positive number. Choose a positive integer m such that $\lambda_m \leqq \varepsilon$ and $\|\boldsymbol{x}_0 - \boldsymbol{u}_m\| < \frac{\varepsilon}{|\alpha| + \varepsilon}$. Let p be an integer such that $|p\lambda_m - \alpha| < \lambda_m \leqq \varepsilon$. Then

$$\begin{aligned}\|\alpha\boldsymbol{x}_0 - p\lambda_m\boldsymbol{u}_m\| &\leqq \|\alpha\boldsymbol{x}_0 - p\lambda_m\boldsymbol{x}_0\| + \|p\lambda_m\boldsymbol{x}_0 - p\lambda_m\boldsymbol{u}_m\| \\ &= |\alpha - p\lambda_m| + |p\lambda_m|\cdot\|\boldsymbol{x}_0 - \boldsymbol{u}_m\| < \varepsilon + (|\alpha| + \varepsilon)\frac{\varepsilon}{|\alpha| + \varepsilon} = 2\varepsilon.\end{aligned}$$

Since $p\lambda_m\boldsymbol{u}_m \in H$ and H is closed, it follows that $\alpha\boldsymbol{x}_0 \in H$ and hence that $L = \{\alpha\boldsymbol{x}_0 : \alpha \in R\} \subset H$.

Consider now the quotient group R^{c+1}/L and the natural mapping φ of R^{c+1} onto R^{c+1}/L; clearly φ is linear and R^{c+1}/L is topologically isomorphic with R^c. By (5.31), $\varphi(H)$ is topologically isomorphic with H/L and is hence locally compact (5.22) and closed in R^{c+1}/L (5.11). By our inductive hypothesis, there are linearly independent vectors $\varphi(\boldsymbol{x}_1), \ldots, \varphi(\boldsymbol{x}_a), \varphi(\boldsymbol{y}_1), \ldots, \varphi(\boldsymbol{y}_b)$ in R^{c+1}/L such that $\varphi(H)$ is just the set of all vectors

$$\alpha_1\varphi(\boldsymbol{x}_1) + \cdots + \alpha_a\varphi(\boldsymbol{x}_a) + m_1\varphi(\boldsymbol{y}_1) + \cdots + m_b\varphi(\boldsymbol{y}_b)$$

where $\alpha_j \in R$ and $m_k \in Z$. We may obviously suppose that the vectors $\boldsymbol{x}_j$ and $\boldsymbol{y}_k$ lie in H.

All vectors $\sum_{j=1}^{a} \alpha_j \boldsymbol{x}_j$ $(\alpha_j \in R)$ lie in H. For, if $\boldsymbol{v} \in H$ and $\varphi(\boldsymbol{v}) = \sum_{j=1}^{a} \alpha_j \varphi(\boldsymbol{x}_j)$, then $\boldsymbol{v} - \sum_{j=1}^{a} \alpha_j \boldsymbol{x}_j = \alpha_0 \boldsymbol{x}_0$ for some $\alpha_0 \in R$, and so $\sum_{j=1}^{a} \alpha_j \boldsymbol{x}_j = \boldsymbol{v} - \alpha_0 \boldsymbol{x}_0 \in H$. The proofs that H consists exactly of the vectors in R^{c+1} of the form $\sum_{j=0}^{a} \alpha_j \boldsymbol{x}_j + \sum_{k=1}^{b} m_k \boldsymbol{y}_k$ $(\alpha_j \in R,\ m_k \in Z)$ and that $\boldsymbol{x}_0, \boldsymbol{x}_1, \ldots, \boldsymbol{x}_a, \boldsymbol{y}_1, \ldots, \boldsymbol{y}_b$ are linearly independent in R^{c+1} are like the corresponding proofs for Case I. □

(9.12) Theorem. *Let τ be a topological isomorphism of $R^a \times Z^b \times F$ into $R^c \times Z^d \times E$, where a, b, c, d are nonnegative integers and F and E are compact groups [not necessarily Abelian]. Then $a \leq c$ and $a + b \leq c + d$.*[1]

Proof. Let η be a topological isomorphism of $R^a \times Z^b$ onto $R^a \times Z^b \times \{e\}$, where e is the identity of F. Then $\tau \circ \eta$ is a topological isomorphism of $R^a \times Z^b$ into $R^c \times Z^d \times E$. Hence in this proof, we may disregard the factor F and suppose that the domain of τ is $R^a \times Z^b$.

(I) Assume that $a > c$. Since $\tau(R^a \times \{\mathbf{0}\})$ is a connected subgroup of $R^c \times Z^d \times E$, it follows that

$$\tau(R^a \times \{\mathbf{0}\}) \subset R^c \times \{\mathbf{0}\} \times E.$$

Let η be a topological isomorphism of R^a onto $R^a \times \{\mathbf{0}\}$, and let θ be the projection of $R^c \times Z^d \times E$ onto $R^c \times E$. Then $\tau_0 = \theta \circ \tau \circ \eta$ is a topological isomorphism of R^a into $R^c \times E$.

Let π be the projection of $R^c \times E$ onto R^c. Then the mapping $\psi = \pi \circ \tau_0$ is a continuous homomorphism of R^a into R^c. It is easy to see that ψ is actually a *linear* mapping of R^a into R^c, and since $a > c$, the linear subspace $S = \{\boldsymbol{x} \in R^a : \psi(\boldsymbol{x}) = \mathbf{0}\}$ is different from $\{\mathbf{0}\}$. For $\boldsymbol{x} \in R^a$, write $\tau_0(\boldsymbol{x}) = (\boldsymbol{y}, z)$, with $\boldsymbol{y} \in R^c$ and $z \in E$. It is obvious that $\tau_0(\boldsymbol{x}) = (\mathbf{0}, z)$ for every $\boldsymbol{x} \in S$; that is, the inclusion $\tau_0(S) \subset \{\mathbf{0}\} \times E$ obtains. Since τ_0 is a topological isomorphism, $\tau_0(S)$ is closed in $\{\mathbf{0}\} \times E$ (5.11). Since E is compact, S is also compact. This is a contradiction, since S is a Euclidean space of positive dimension. Therefore the inequality $a \leq c$ obtains.

(II) We return to the mapping τ of $R^a \times Z^b$ into $R^c \times Z^d \times E$. Let φ be the projection of $R^c \times Z^d \times E$ onto $R^c \times Z^d$. Plainly $\varphi \circ \tau$ is a continuous homomorphism of $R^a \times Z^b$ into $R^c \times Z^d$. Suppose that $\varphi \circ \tau(\boldsymbol{x}) = \mathbf{0}$, where $\boldsymbol{x} \in R^a \times Z^b$. Then $\tau(\boldsymbol{x})$ belongs to $\{\mathbf{0}\} \times E \subset R^c \times Z^d \times E$, and hence $\tau(\boldsymbol{x})$ is a compact element of $R^c \times Z^d \times E$. This means that $\boldsymbol{x}$ itself is a compact element in $R^a \times Z^b$, and so $\boldsymbol{x} = \mathbf{0}$. Thus $\varphi \circ \tau$ is a one-to-one mapping and φ is one-to-one on $\tau(R^a \times Z^b)$. Since $R^a \times Z^b$ is locally

[1] We are indebted to Roger W. Richardson for the inequality $a \leq c$ and its proof.

compact, $\tau(R^a \times Z^b)$ is closed in $R^c \times Z^d \times E$ (5.11). Now (5.18) implies that $\varphi \circ \tau(R^a \times Z^b)$ is closed in $R^c \times Z^d$. It follows that $\varphi \circ \tau(R^a \times Z^b)$ is a locally compact group. By (5.29), $\varphi \circ \tau$ is a topological isomorphism. We may obviously regard $\varphi \circ \tau(R^a \times Z^b)$ as a closed subgroup of R^{c+d}. By (9.11), $\varphi \circ \tau(R^a \times Z^b)$ contains $a+b$ vectors that are linearly independent in R^{c+d}. That is, $a+b \leqq c+d$. ▯

(9.13) Corollary. *Suppose that $R^a \times Z^b \times F$ and $R^c \times Z^d \times E$ are topologically isomorphic, where a, b, c, d are nonnegative integers and F and E are compact groups. Then $a=c$, $b=d$, and F and E are topologically isomorphic.*

Proof. The isomorphism τ must carry compact elements onto compact elements; therefore $\tau(\{0\} \times \{0\} \times F) = \{0\} \times \{0\} \times E$. By (9.12), we have $a=c$ and $b=d$. ▯

We now present a useful consequence of (9.8).

(9.14) Theorem. *Let G be a locally compact Abelian group, and let C be the component of e in G. Then C is topologically isomorphic with $R^n \times E$, where n is a nonnegative integer and E is a compact connected Abelian group. The number n is the largest possible dimension of a subgroup of G topologically isomorphic with R^a for a nonnegative integer a. If H is a compactly generated open subgroup of G, then H is topologically isomorphic with $R^n \times Z^b \times F$, where b is a nonnegative integer and F is a compact Abelian group.*

Proof. Let H be a compactly generated subgroup of G. By (9.8), H is topologically isomorphic with $R^a \times Z^b \times F$ for nonnegative integers a and b and a compact Abelian group F. The subgroup C is compactly generated (7.4) and connected, so that C is topologically isomorphic with $R^n \times E$, where n is a nonnegative integer and E is a compact connected Abelian group. [No factor Z^b appears because C is connected and Z is not.] The set C is defined as the union of all connected subsets of G containing e, and so there is a topological isomorphism carrying R^a into $R^n \times E$. From (9.12) we infer that $a \leqq n$.

Let H now be any *open* compactly generated subgroup of G: let H be topologically isomorphic with $R^c \times Z^d \times P$, where c and d are nonnegative integers and P is a compact Abelian group. The group C is the intersection of all open subgroups of G (7.8). Hence C is contained in a topological isomorph of $R^c \times P$ [which is plainly open in H; hence its isomorph is open in G]. Thus there is a topological isomorphism carrying R^n into $R^c \times P$, so that (9.12) implies that $n \leqq c$. Since we can take $a=c$ in the preceding paragraph, it follows that $c=n$. ▯

(9.15) Theorem. *Let G be a 0-dimensional, infinite, compact group, and let $\mathfrak{m}$ be the least cardinal number of an open basis at the identity e of G.*

Then G, regarded only as a topological space, is homeomorphic with the space $\{0, 1\}^{\mathfrak{m}}$, where $\{0, 1\}$ is discrete[1].

Proof. We give the proof in a number of steps.

(I) Let $\{U_\iota\}_{\iota\in I}$ be an open basis at e having cardinal number $\mathfrak{m}$. Using (7.7), choose a compact open normal subgroup V_ι of G such that $V_\iota\subset U_\iota$, for each $\iota\in I$. Now well order the family $\{V_\iota\}_{\iota\in I}$, and rewrite it as $\{V_1, V_2, \ldots, V_\alpha, \ldots\}$, where α runs through all ordinals less than, say, the first ordinal μ with cardinal number $\mathfrak{m}$. [Note that $\mathfrak{m}$ must be infinite.] With no loss of generality, we may suppose that $V_1=G$. For every ordinal β, $1<\beta<\mu$, let $N_\beta=\bigcap_{\alpha<\beta} V_\alpha$, and let $N_1=G$. It is clear that each N_β is a compact normal subgroup of G. Fix an ordinal $\beta<\mu$, and let X be any subset of G that is the intersection of sets of the form $a_1 V_{\beta_1}\cup a_2 V_{\beta_2}\cup\cdots\cup a_s V_{\beta_s}$, where $a_j\in G$ and $\beta_j<\beta$ $(j=1, 2, \ldots, s)$. It is obvious that $N_\beta X=X$.

(II) We next define certain open and closed subsets of G. Let B be the set of ordinals $\beta<\mu$ for which $V_\beta \subsetneqq N_\beta V_\beta$. Both V_β and $N_\beta V_\beta$ are open and closed normal subgroups of G. The quotient groups $N_\beta V_\beta/V_\beta$ and $G/N_\beta V_\beta$ are compact (5.22) and discrete (5.21) and hence finite. Let $b_\beta^{(1)} N_\beta V_\beta, \ldots, b_\beta^{(m_\beta)} N_\beta V_\beta$ be the distinct cosets of $N_\beta V_\beta$ in G, and let $a_\beta^{(1)} V_\beta, \ldots, a_\beta^{(k_\beta)} V_\beta$ be the distinct cosets of V_β in $N_\beta V_\beta$. For each integer j, $j=1, 2, \ldots, k_\beta$, let

$$A_\beta^{(j)}=b_\beta^{(1)} a_\beta^{(j)} V_\beta\cup b_\beta^{(2)} a_\beta^{(j)} V_\beta\cup\cdots\cup b_\beta^{(m_\beta)} a_\beta^{(j)} V_\beta.$$

Thus each $A_\beta^{(j)}$ is a finite union of cosets of V_β, and each $A_\beta^{(j)}$ intersects every coset of $N_\beta V_\beta$ in a coset of V_β. It is obvious that $A_\beta^{(j)}\cap A_\beta^{(l)}=\varnothing$ if $j\neq l$ and that $\bigcup_{j=1}^{k_\beta} A_\beta^{(j)}=G$.

We next show that if X is a nonvoid subset of G and $j=1, 2, \ldots, k_\beta$, then $(N_\beta X)\cap A_\beta^{(j)}\neq\varnothing$. Since $\bigcup_{l=1}^{m_\beta} b_\beta^{(l)} N_\beta V_\beta=G$, there is some $b_\beta^{(l)}$ for which $(N_\beta X)\cap(b_\beta^{(l)} N_\beta V_\beta)\neq\varnothing$. Since $a_\beta^{(j)}$ is in $N_\beta V_\beta$, we have $b_\beta^{(l)} N_\beta V_\beta= b_\beta^{(l)} a_\beta^{(j)} N_\beta V_\beta=N_\beta b_\beta^{(l)} a_\beta^{(j)} V_\beta$. Thus there are elements $a_1, a_2\in N_\beta$, $x\in X$, and $y\in V_\beta$ such that $a_1 x=a_2 b_\beta^{(l)} a_\beta^{(j)} y$. Hence $a_2^{-1} a_1 x=b_\beta^{(l)} a_\beta^{(j)} y$. Since $a_2^{-1} a_1 x\in N_\beta X$ and $b_\beta^{(l)} a_\beta^{(j)} y\in A_\beta^{(j)}$, we have $(N_\beta X)\cap A_\beta^{(j)}\neq\varnothing$.

Now let $\beta_1, \beta_2, \ldots, \beta_s$ be any distinct elements of B; we may suppose that $\beta_1<\beta_2<\cdots<\beta_s$. We will show that

$$A_{\beta_1}^{(j_1)}\cap A_{\beta_2}^{(j_2)}\cap\cdots\cap A_{\beta_s}^{(j_s)}\neq\varnothing, \tag{1}$$

if $1\leq j_l\leq k_{\beta_l}$ $(l=1, 2, \ldots, s)$. It follows from (I) that $A_{\beta_1}^{(j_1)}=N_{\beta_2} A_{\beta_1}^{(j_1)}$. By the preceding paragraph, we have $A_{\beta_1}^{(j_1)}\cap A_{\beta_2}^{(j_2)}=(N_{\beta_2} A_{\beta_1}^{(j_1)})\cap A_{\beta_2}^{(j_2)}\neq\varnothing$.

[1] For a related result, see (25.35) *infra*.

Applying (I) and the preceding paragraph again, we get $A_{\beta_1}^{(j_1)} \cap A_{\beta_2}^{(j_2)} \cap A_{\beta_3}^{(j_3)} = (N_{\beta_3}(A_{\beta_1}^{(j_1)} \cap A_{\beta_2}^{(j_2)})) \cap A_{\beta_3}^{(j_3)} \neq \emptyset$. Finite induction now shows that (1) holds.

To complete our construction, we need one more fact, *viz.*: if x, y are in G and $x \neq y$, then there is a $\beta \in B$ and a j, $1 \leqq j \leqq k_\beta$, such that $A_\beta^{(j)}$ contains one and only one of the points x and y. Since $x^{-1}y \neq e$, there is an $\alpha < \mu$ such that $x^{-1}y \notin V_\alpha$. Let β be the least of all these α's. Then $x^{-1}y \in N_\beta$, so that $x^{-1}y \in N_\beta V_\beta$ and $x^{-1}y \notin V_\beta$. Hence we have $V_\beta \subsetneqq N_\beta V_\beta$, so that $\beta \in B$. In the notation used above to define the sets $A_\beta^{(j)}$, we have $x, y \in b_\beta^{(t)} N_\beta V_\beta$ for some t, $1 \leqq t \leqq m_\beta$. Since $x^{-1}y \notin V_\beta$, we have $x \in b_\beta^{(t)} a_\beta^{(j)} V_\beta$ and $y \in b_\beta^{(t)} a_\beta^{(l)} V_\beta$, where $j \neq l$. Hence $x \in A_\beta^{(j)}$ and $y \in A_\beta^{(l)}$, so that $y \notin A_\beta^{(j)}$.

(III) To finish the proof, we form the Cartesian product $Y = \mathop{\mathsf{P}}\limits_{\beta \in B} \{1, 2, \ldots, k_\beta\}$, each finite space $\{1, 2, \ldots, k_\beta\}$ being given the discrete topology. Define a mapping Φ of G into Y by $\Phi(x) = (j_\beta)_{\beta \in B}$, where j_β is the integer such that $x \in A_\beta^{(j)}$, for each $\beta \in B$. Since all of the sets $A_\beta^{(j)}$ are open and closed, Φ is a continuous mapping. Since the sets $A_\beta^{(j)}$ separate points of G, Φ is one-to-one. Hence Φ is a homeomorphism, and $\Phi(G)$ is a compact and hence closed subspace of Y. In view of (1), $\Phi(G)$ is dense in Y, and thus $\Phi(G) = Y$.

The constructions in (I) and (II) above show that $\overline{\overline{B}} \leqq \mathfrak{m}$. Since G and Y are homeomorphic, we must have $\overline{\overline{B}} = \mathfrak{m}$, inasmuch as $\overline{\overline{B}}$ is the smallest cardinal number of an open basis at an arbitrary point of Y. To show that Y is homeomorphic to $\{0, 1\}^{\mathfrak{m}}$, write B as the union of a family of pairwise disjoint countably infinite sets.

The proof will be complete when we have shown that a countably infinite Cartesian product X of finite discrete spaces is homeomorphic with $\{0, 1\}^{\aleph_0}$. Write this product X as the set of all sequences $\boldsymbol{x} = (x_0, x_1, x_2, \ldots)$, where $x_n \in \{0, 1, \ldots, a_n - 1\}$ and the a_n's are integers greater than 1 ($n = 0, 1, 2, \ldots$). For $\boldsymbol{x} \in X$, let $\tau(\boldsymbol{x}) = A \cdot \sum\limits_{j=0}^{\infty} \frac{x_j}{(a_0+1)(a_1+1)\cdots(a_j+1)}$, where A^{-1} is the number $\sum\limits_{j=0}^{\infty} \frac{a_j - 1}{(a_0+1)(a_1+1)\cdots(a_j+1)}$. It is easy to see that τ is continuous: if $\boldsymbol{x}, \boldsymbol{y} \in X$ and $x_j = y_j$ for $j = 0, 1, \ldots, m$, then $|\tau(\boldsymbol{x}) - \tau(\boldsymbol{y})| \leqq A \cdot \sum\limits_{j=m+1}^{\infty} \frac{a_j - 1}{(a_0+1)(a_1+1)\cdots(a_j+1)}$. The mapping τ is also one-to-one. To see this, consider $\boldsymbol{x}, \boldsymbol{y} \in X$ and suppose that $x_j = y_j$ for $j = 0, \ldots, m-1$, and that $x_m > y_m$. Then we have

$$\tau(\boldsymbol{x}) - \tau(\boldsymbol{y}) \geqq \frac{A}{(a_0+1)\cdots(a_m+1)}$$
$$\times \left[1 - \frac{a_{m+1}-1}{a_{m+1}+1} - \frac{a_{m+2}-1}{(a_{m+1}+1)(a_{m+2}+1)} - \cdots\right] > \frac{A}{(a_0+1)\cdots(a_m+1)}$$
$$\times \left[1 - \frac{a_{m+1}-1}{a_{m+1}+1} - \frac{1}{a_{m+1}+1} - \frac{1}{(a_{m+1}+1)(a_{m+2}+1)} - \cdots\right]$$
$$\geqq \frac{A}{(a_0+1)\cdots(a_m+1)} \left[\frac{1}{a_{m+1}+1}\left(1 - \frac{1}{3} - \frac{1}{9} - \frac{1}{27} - \cdots\right)\right] > 0.$$

Thus τ is a one-to-one continuous mapping of X into $[0, 1]$ and accordingly is a homeomorphism. Hence $\tau(X)$ is a compact, 0-dimensional subspace of $[0, 1]$ having no isolated points. In particular, $\tau(X)$ contains no interval, and there is a point $t \in]\frac{1}{3}, \frac{2}{3}[$ such that $t \notin \tau(X)$. Let $A_0 = \{\boldsymbol{x} \in X : \tau(\boldsymbol{x}) < t\}$ and $A_1 = \{\boldsymbol{x} \in X : \tau(\boldsymbol{x}) > t\}$. Then A_0 and A_1 are complementary nonvoid open and closed subsets of X, and $|\tau(\boldsymbol{x}) - \tau(\boldsymbol{x}')| \leqq \frac{2}{3}$ if $\boldsymbol{x}, \boldsymbol{x}'$ lie in A_0 or if $\boldsymbol{x}, \boldsymbol{x}'$ lie in A_1. In like fashion we partition A_0 and A_1: $A_0 = A_{00} \cup A_{01}$, $A_1 = A_{10} \cup A_{11}$, where the A_{ij} are open and closed in X and $|\tau(\boldsymbol{x}) - \tau(\boldsymbol{x}')| \leqq (\frac{2}{3})^2$ if $\boldsymbol{x}$ and $\boldsymbol{x}'$ lie in the same A_{ij}. Proceeding by finite induction, suppose that sets $A_{i_1 i_2 \cdots i_l}$ have already been defined for $l = 1, \ldots, k$ and $\{i_1, i_2, \ldots, i_k\}$ any sequence of 0's and 1's. Suppose too that $\{A_{i_1 i_2 \cdots i_{l-1} 0}, A_{i_1 i_2 \cdots i_{l-1} 1}\}$ is a partition of $A_{i_1 i_2 \cdots i_{l-1}}$ into open and closed sets and that $|\tau(\boldsymbol{x}) - \tau(\boldsymbol{x}')| \leqq (\frac{2}{3})^l$ if $\boldsymbol{x}$ and $\boldsymbol{x}'$ lie in the same set $A_{i_1 i_2 \cdots i_l}$, for all relevant choices of i's and l's. Then $\tau(A_{i_1 i_2 \cdots i_k})$ is contained in an interval of length $(\frac{2}{3})^k$, and the middle third of this interval contains a point $t' \notin \tau(X)$. Define $A_{i_1 i_2 \cdots i_k 0}$ as $\{\boldsymbol{x} \in A_{i_1 i_2 \cdots i_k} : \tau(\boldsymbol{x}) < t'\}$ and $A_{i_1 i_2 \cdots i_k 1}$ as $\{\boldsymbol{x} \in A_{i_1 i_2 \cdots i_k} : \tau(\boldsymbol{x}) > t'\}$. This defines the sets $A_{i_1 i_2 \cdots i_k}$ for all positive integers k and sequences $\{i_1, i_2, \ldots, i_k\}$ of 0's and 1's.

Now for an arbitrary infinite sequence of 0's and 1's, say $\boldsymbol{i} = (i_1, i_2, i_3, \ldots, i_n, \ldots)$, define $\Phi(\boldsymbol{i}) = A_{i_1} \cap A_{i_1 i_2} \cap A_{i_1 i_2 i_3} \cap \cdots$. For every $\boldsymbol{i}$, the set $\Phi(\boldsymbol{i})$ is a single point because $|\tau(\boldsymbol{x}) - \tau(\boldsymbol{x}')| \leqq (\frac{2}{3})^k$ whenever $\boldsymbol{x}$ and $\boldsymbol{x}'$ lie in a single set $A_{i_1 i_2 \cdots i_k}$. The set of all points $\Phi(\boldsymbol{i})$ is X itself, because for each positive integer k, the sets $A_{i_1 i_2 \cdots i_k}$ partition X. For the same reason, Φ is one-to-one. Finally, to see that Φ is continuous, let $\boldsymbol{x} = \Phi(\boldsymbol{i})$ be any point of X and let U be any neighborhood of $\boldsymbol{x}$. Then the compactness of X guarantees that there is a positive integer k such that $A_{i_1 i_2 \cdots i_k} \subset U$. Hence if $\boldsymbol{i}'$ is a sequence of 0's and 1's such that $i_1 = i_1', i_2 = i_2', \ldots, i_k = i_k'$, we have $\Phi(\boldsymbol{i}') \in A_{i_1 i_2 \cdots i_k} \subset U$. Thus Φ is a one-to-one continuous mapping of $\{0, 1\}^{\aleph_0}$ onto X and is thus a homeomorphism[1]. □

We now make some observations about compact topological semigroups. A *topological semigroup* S is a semigroup endowed with a Hausdorff topology for which the mapping $(x, y) \to xy$ of $S \times S$ into S is continuous. If $zx = zy$ always implies that $x = y$, then S is said to

[1] With only a little more effort we can prove that every totally disconnected compact metric space Y with no isolated points is homeomorphic with $\{0,1\}^{\aleph_0}$. It is easy to see that Y can be partitioned into a finite number of open and closed sets $B_1, \ldots, B_{a_1}$ each of diameter not exceeding 1. Each B_i can be partitioned into a finite number of open and closed sets $B_{i1}, B_{i2}, \ldots, B_{ia_2}$ [a_2 is the same positive integer for each i], where each B_{ij} has diameter not exceeding $\frac{1}{2}$, and so on. This leads by an argument like that involving Φ to the fact that Y is homeomorphic with the product of a countably infinite number of finite discrete spaces. Then using the result proved in the theorem, we find that Y is homeomorphic with $\{0, 1\}^{\aleph_0}$.

satisfy the *left cancellation law*; if $xz = yz$ implies that $x = y$, then S satisfies the *right cancellation law*. An element f in a semigroup such that $f^2 = f$ is called an *idempotent*.

There are many differences between topological groups and arbitrary topological semigroups. One striking difference is that topological semigroups need not be homogeneous. For example, let $S = \{0, 1, \frac{1}{2}, \frac{1}{3}, \ldots\}$. Define $xy = \max(x, y)$ in S and give S the topology it inherits from R. Then S is a nonhomogeneous compact topological semigroup.

(9.16) Theorem. *A compact semigroup S satisfying the left and right cancellation laws is a topological group.*

Proof. Suppose first that S also satisfies

$$aS = Sa = S \quad \text{for all} \quad a \in S. \tag{1}$$

Fix a_0 in S and choose $e \in S$ such that $ea_0 = a_0$. Then for any $a \in S$, we have $a_0 x = a$ for some $x \in S$ and hence $ea = ea_0 x = a_0 x = a$. Thus e is a left identity for S. The existence of a left inverse is immediate. As is well known, S is therefore a group. To show that S is a topological group, we need only to verify that inversion is continuous. Let U be an open subset of S; we have to show that $V = U^{-1}$ is open. Let x_α, $\alpha \in D$, be a net of elements in V' converging to $x_0 \in S$. Each x_α^{-1} is in U' and thus by the compactness of U', there is a subnet x_β^{-1}, $\beta \in E$, of x_α^{-1}, and an element $y \in U'$ such that $\lim_\beta x_\beta^{-1} = y$. By the continuity of multiplication, $x_0 y = \lim_\beta x_\beta x_\beta^{-1} = e$. That is, $x_0 = y^{-1} \in V'$. Therefore V' is closed and V is open.

The previous paragraph shows that the theorem will be proved if we prove (1). Let $a \in S$ be fixed and let A be any nonvoid closed subset of S such that $aA \subset A$. [For example, S itself is such a set.] Evidently $A \supset aA \supset a^2 A \supset \cdots$ and each $a^k A$ is a closed set. The sequence $\{a^k\}_{k=1}^\infty$ is a net in S and therefore there is a subnet a^{k_α}, $\alpha \in D$, and a $y \in S$ such that $\lim_\alpha a^{k_\alpha} = y$. For $x \in A$, we have $yx = \lim_\alpha a^{k_\alpha} x$. For each positive integer k, there is an $\alpha_0 \in D$ such that $\alpha \geqq \alpha_0$ implies $k_\alpha \geqq k$. That is, $a^{k_\alpha} x \in a^k A$ for $\alpha \geqq \alpha_0$. Consequently, $yx \in a^k A$ for all k; *i.e.*, $yx \in \bigcap_{k=1}^\infty a^k A$. We thus have

$$yA \subset \bigcap_{k=1}^\infty a^k A. \tag{2}$$

Now consider $x \in \bigcap_{k=1}^\infty a^k A$. Then for each integer $k \geqq 1$, there is an $x_k \in A$ for which $x = a^k x_k$. There is a subnet x_{k_β}, $\beta \in E$, of x_{k_α} and an element $z \in A$ such that $\lim_\beta x_{k_\beta} = z$. Then

$$x = \lim_\beta (a^{k_\beta} x_{k_\beta}) = (\lim_\beta a^{k_\beta})(\lim_\beta x_{k_\beta}) = yz \in yA.$$

This and (2) yield

$$yA = \bigcap_{k=1}^{\infty} a^k A. \tag{3}$$

Setting A equal to S and then to aS in (3), we obtain $yS = \bigcap_{k=1}^{\infty} a^k S$ and $yaS = \bigcap_{k=1}^{\infty} a^k aS$. It follows that $yS = yaS$. Since S satisfies the left cancellation law, we obtain $S = aS$. The equality $S = Sa$ is proved in like manner. □

(9.17) Lemma. *Let S be a compact semigroup and let a be any element of S. Then $\{a, a^2, a^3, \ldots\}^-$ contains a closed Abelian group G, namely, $G = \bigcap_{k=1}^{\infty} \{a^k, a^{k+1}, \ldots\}^-$.*

Proof. For $k = 1, 2, \ldots$, let $A_k = \{a^k, a^{k+1}, \ldots\}$. These sets have the finite intersection property and therefore $G = \bigcap_{k=1}^{\infty} A_k^-$ is nonvoid. It is easy to verify that G is a closed Abelian subsemigroup of S. Therefore, as shown in the proof of (9.16), it suffices to establish the equality

$$yG = G, \text{ for all } y \in G. \tag{1}$$

Assume that $yG \subsetneqq G$ for some $y \in G$. Let $z \in G \cap (yG)'$; then $yx \neq z$ for all $x \in G$. Therefore, for each $x \in G$, there exist neighborhoods U_x, V_x, and W_x of y, x, and z, respectively, in S, such that $W_x \cap (U_x V_x) = \varnothing$. Since $\bigcup_{x \in G} V_x \supset G$ and G is compact, there exist $x_1, \ldots, x_n \in G$ such that $\bigcup_{k=1}^{n} V_{x_k} \supset G$. Let $U = \bigcap_{k=1}^{n} U_{x_k}$, $V = \bigcup_{k=1}^{n} V_{x_k}$, and $W = \bigcap_{k=1}^{n} W_{x_k}$. Clearly U, V, and W are open sets, $y \in U$, $z \in W$, $V \supset G$, and

$$W \cap (UV) = \varnothing. \tag{2}$$

Since $y \in G$, there is an integer $q \geqq 1$ such that $a^q \in U$, and since $z \in G$, there are positive integers $r_1, r_2, r_3, \ldots$ such that $q < r_1 < r_2 < \cdots$ and $a^{r_j} \in W$ $(j = 1, 2, 3, \ldots)$. The net $\{a^{r_j - q}\}_{j=1}^{\infty}$ has a cluster point u in S which is necessarily also in G. For some k, $k = 1, \ldots, n$, we have $u \in V_{x_k}$ and thus for some m, $m = 1, 2, \ldots$, we must have $a^{r_m - q} \in V_{x_k}$. Now

$$a^{r_m} = a^q a^{r_m - q} \in U V_{x_k} \subset UV.$$

Since a^{r_m} is also in W, this contradicts (2); therefore (1) holds. □

(9.18) Theorem. *Every compact semigroup contains at least one idempotent.*

Proof. This is immediate from (9.17). □

(9.19) Definition. A nonvoid subset M of a semigroup S is called a *left ideal* [*right ideal*] if $SM \subset M$ [$MS \subset M$]. If M is a left ideal and a right ideal, M is called a *two-sided ideal*. A semigroup S with no ideals different from S itself is called *simple*. A semigroup S is called *completely simple* if it is simple and satisfies:

(i) if e and f are idempotents in S such that $ef=f=fe$, then $e=f$;

(ii) for every $x\in S$, there are idempotents e and f in S such that $ex=x=xf$.

(9.20) Lemma. *A compact simple semigroup S is completely simple.*

Proof. We first verify (9.19.ii). Let $x\in S$. Since S is simple, we have $S=SxS$. Thus $x=axb$ for some $a, b\in S$, and evidently $x=a^k x b^k$ for $k=2, 3, \ldots$. By (9.17), there is an idempotent e in the set $\{a, a^2, \ldots\}^-$. Let a^{k_α}, $\alpha\in D$, be a net in $\{a, a^2, \ldots\}$ converging to e. There is a subnet b^{k_β}, $\beta\in E$, of b^{k_α}, and a point $b_0\in\{b, b^2, \ldots\}^-$ such that $\lim_\beta b^{k_\beta}=b_0$. Then $exb_0=\lim_\beta a^{k_\beta}xb^{k_\beta}=x$. Therefore $ex=e(exb_0)=exb_0=x$. Similarly $xf=x$ for some idempotent $f\in\{b, b^2, \ldots\}^-$.

Suppose now that e and f are idempotents in S such that $ef=f=fe$. Since S is simple, there are elements a and b of S such that $e=afb$. Let $y=eaf$ and $z=fb$. Then $yfz=e$ and $yez=e$, and therefore $y^kfz^k=e$ for $k=2, 3, \ldots$. There is an idempotent $g\in\{y, y^2, \ldots\}^-$ and a net y^{k_α}, $\alpha\in D$, converging to it. Further, there is a subnet z^{k_β}, $\beta\in E$, of z^{k_α} converging to a point $z_0\in\{z, z^2, \ldots\}^-$. Therefore $gfz_0=e$. Since $y=eaf=eafe\in eSe$ and eSe is a closed subsemigroup of S, we have $g\in eSe$. Therefore $ge=g$ so that $g=ge=ggfz_0=gfz_0=e$; *i.e.*, $e=efz_0=fz_0$. Consequently, $f=fe=ffz_0=fz_0=e$. ☐

(9.21) Theorem. *A compact semigroup S contains a smallest two-sided ideal K. The ideal K is closed and completely simple.*

Proof. Let E be the set of idempotents in S; by (9.18) E is nonvoid. Let $K=\bigcap_{e\in E} SeS$. If $e_1, \ldots, e_m\in E$, then each set Se_kS, $k=1, \ldots, m$, contains the element $e_1e_2\cdots e_m=e_1(e_1e_2\cdots e_{k-1})\,e_k\,(e_{k+1}\cdots e_m)\,e_m$ and hence the family $\{SeS: e\in E\}$ of closed sets has the finite intersection property. Therefore K is nonvoid. It is easy to see that K is a two-sided ideal of S.

Let M be any two-sided ideal of S and let $x\in M$. Then SxS is a closed ideal and $SxS\subset M$. By (9.18), SxS contains an idempotent e_0. It follows that $K\subset Se_0S\subset M$. Therefore K is the intersection of all two-sided ideals of S, and is therefore the smallest two-sided ideal of S.

If $x\in K$, then KxK is a closed ideal of S and $KxK\subset K$. Hence $KxK=K$. It follows that K is simple. By (9.20), K is completely simple. ☐

The smallest two-sided ideal of a compact semigroup S, whose existence was established in (9.21), is called the *kernel* [sometimes the *Suškevič kernel*] of S.

(9.22) Lemma. *A completely simple semigroup S with identity e is a group.*

Proof. Let x be any element of S. Since S is simple, there are elements a and b in S such that $axb=e$. Then bax and xba are idempotents. Thus by (9.19.i), we have $bax=e$ and $xba=e$. That is, ba is an inverse for x. Therefore S is a group. □

(9.23) Theorem. *A simple compact semigroup S is a pairwise disjoint union of closed subgroups.*

Proof. Let E be the set of all idempotents in S. We first prove that if e and f are in E, then eSf is a group. By (9.16), it suffices to verify the cancellation laws in eSf. The semigroup eSe is simple; indeed if eye, $exe \in eSe$, then $y=aexeb$ for some $a, b \in S$ and hence $eye = (eae)(exe)(ebe)$. This implies that eSe is simple. By (9.20), eSe is completely simple. Plainly e is an identity for eSe and thus by (9.22), eSe is a group. Suppose now that

$$(ezf)(exf) = (ezf)(eyf),$$

where $x, y, z \in S$. Let $(ezfe)^{-1}$ be the inverse of $ezfe$ in eSe. Then $exf = (ezfe)^{-1}(ezfe)xf = (ezfe)^{-1}(ezfe)yf = eyf$. The right cancellation law for eSf is proved in like manner.

Since S is completely simple (9.20), we have

$$S = \bigcup\{eSf : e, f \in E\};$$

see (9.19.ii). We now show that the family of sets $\{eSf : e, f \in E\}$ is pairwise disjoint. Consider eSf and let e_0 be its identity. Clearly we have $eSf \subset e_0 S e_0$. Thus ef is in the group $e_0 S e_0$ so that $e_0 S e_0 = ef(e_0 S e_0)ef = e(fe_0 S e_0 e)f \subset eSf$. Hence

$$e_0 S e_0 = eSf. \tag{1}$$

Suppose now that G and H are in $\{eSf : e, f \in E\}$ and have a point a in common. Let e_g and e_h be the identities of G and H, respectively. Then $e_g a = a = e_h a$. Let $b \in G$ be such that $ab = e_g$. Then $e_g = e_g ab = e_h ab = e_h e_g$. Similarly $e_h = e_h e_g$ and so $e_h = e_g$. By (1), we have $G = e_g S e_g = e_h S e_h = H$. □

(9.24) Corollary. *Let S be a compact semigroup, with kernel K. Then K is the union of pairwise disjoint closed subgroups.*

Miscellaneous theorems and examples

(9.25). Let L be a subgroup of R with a topology strictly stronger than its topology as a subspace of R, under which it is a topological group. Suppose that H_0 is a topological group and that there is a continuous homomorphism of H_0 onto L. Then Theorem (9.1) does not hold if H is replaced by H_0. Let H_1 be any continuous homomorphic image of a subgroup of R with its usual topology. Then Theorem (9.1) holds for H_1. [The second assertion uses (5.40.c).]

(9.26) Concerning compact elements. (a) (PONTRYAGIN [7], § 40, Example 74.) Let G be a locally compact Abelian group, let C be the component of e in G, and let B be the subgroup of all compact elements of G. Then BC is an open subgroup of G. [Since B and C are subgroups [(9.10) and (7.1)], BC is clearly a subgroup of G. Let H be any compactly generated open subgroup of G; H is topologically isomorphic with $R^n \times Z^b \times F$. Let τ be a topological isomorphism carrying $R^n \times Z^b \times F$ onto H. Then $\tau(R^n) \cdot \tau(F)$ is an open subgroup of G, and $\tau(F) = B \cap H$ is the largest compact subgroup of H. Thus $\tau(R^n) \cdot \tau(F) \cdot CB$ is open in G. Since $\tau(R^n)$ is a connected subgroup of G, we have $\tau(R^n) \subset C$, and so $\tau(R^n) \cdot \tau(F) \cdot CB = CB$.]

(b) If G is a locally compact, compactly generated Abelian group, then the subgroup B of all compact elements is itself compact; it is the largest compact subgroup of G. For an arbitrary locally compact Abelian group, the subgroup B of all compact elements need not be compact. [The first assertion is immediate from (9.8). The second assertion follows from the fact that there are infinite torsion groups. Simply endow such a group with the discrete topology. As a less trivial example, consider any group $\Omega_{\mathbf{a}}$ [(10.5) *infra*].]

(c) (E. THOMA, letter to the writers.) Consider the discrete group G generated by elements a and b and satisfying the relations $a^2 = b^2 = e$. Then a and b are compact elements, but ab is not a compact element. Thus the compact elements of G do not form a group.

(d) Consider the general linear group $\mathfrak{GL}(2, K)$. For $n = 2, 3, \ldots$, let $z_n = \exp[2\pi i(1/n)]$. An easy induction argument shows that

$$\begin{pmatrix} z_n & 1 \\ 0 & 1 \end{pmatrix}^k = \begin{pmatrix} z_n^k & z_n^{k-1} + z_n^{k-2} + \cdots + z_n + 1 \\ 0 & 1 \end{pmatrix}.$$

In particular, we see that

$$\begin{pmatrix} z_n & 1 \\ 0 & 1 \end{pmatrix}^n = \begin{pmatrix} z_n^n & z_n^{n-1} + z_n^{n-2} + \cdots + z_n + 1 \\ 0 & 1 \end{pmatrix} = \begin{pmatrix} 1 & 0 \\ 0 & 1 \end{pmatrix},$$

since the sum of all n-th roots of unity is zero for $n = 2, 3, \ldots$. Therefore $\begin{pmatrix} z_n & 1 \\ 0 & 1 \end{pmatrix}$ generates a finite subgroup of $\mathfrak{GL}(2, K)$ and is a compact element. To show that the set of all compact elements of $\mathfrak{GL}(2, K)$ is not closed, we need only observe that $\begin{pmatrix} 1 & 1 \\ 0 & 1 \end{pmatrix} = \lim_{n\to\infty} \begin{pmatrix} z_n & 1 \\ 0 & 1 \end{pmatrix}$ is not a compact element. This is obvious since $\begin{pmatrix} 1 & 1 \\ 0 & 1 \end{pmatrix}^k = \begin{pmatrix} 1 & k \\ 0 & 1 \end{pmatrix}$ for $k = 1, 2, \ldots$.

(e) (MARKOV [1].) A topological Abelian group G is topologically isomorphic with R^n for some positive integer n if and only if: (1) $G \neq \{e\}$; (2) G is locally compact; (3) G is connected; and (4) G contains no compact subgroup except $\{e\}$. [This follows readily from (9.14).]

(f) (VAN KAMPEN [1], p. 460.) Let G be a locally compact Abelian group and C the component of the identity. If G/C is compact, then G is topologically isomorphic with $R^n \times F$ for a nonnegative integer n and a compact Abelian group F. [By (7.4), C is compactly generated. By (5.39.h), G is compactly generated and hence Theorem (9.8) tells us that G is topologically isomorphic with $R^n \times Z^m \times F$, where F is a compact Abelian group. If C_0 is the component of the identity e in F, then C is topologically isomorphic with $R^n \times C_0$. Since $R^n \times \{0\} \times C_0 \subset R^n \times \{0\} \times F \subset R^n \times Z^m \times F$, Theorem (5.35) implies that $(R^n \times Z^m \times F)/(R^n \times \{0\} \times F)$ is topologically isomorphic with $(G/C)/(F/C_0)$. That is, Z^m is topologically isomorphic with a continuous image of the compact group G/C. Consequently, Z^m is compact and $m = 0$.]

(9.27) Topological semigroups. (a) The group G given in (9.17) is the largest subgroup of $\{a, a^2, \ldots\}^-$. [Use the fact that $\{a, a^2, \ldots\}^- = \{a, a^2, \ldots\} \cup G$.]

(b) The partition of a simple compact semigroup into subgroups is unique; see (9.23).

(c) (GELBAUM, KALISCH, and OLMSTED [1].) Unlike groups, semigroups need not be homogeneous. In particular, the following implication may fail:

$$U \text{ open in } S \text{ and } x \text{ in } S \text{ imply } xU \text{ open in } S. \qquad (1)$$

However, given any commutative topological semigroup S with identity e, we can find a stronger topology under which S is a topological semigroup satisfying (1) and such that the neighborhoods at e are the same under these two topologies. [A topological semigroup not satisfying (1) is $[0, 1]$ with the usual multiplication and topology, since $0 \cdot [0, 1] = \{0\}$. To prove the second assertion, let $\mathcal{O}$ denote the given topology of S and let $\mathcal{U}$ be a basis of open sets at e. We define the new topology $\mathcal{O}_1$ on S by requiring that $\{xU : U \in \mathcal{U}\}$ be a neighborhood basis at x, for each $x \in S$. Denote this open basis by $\mathcal{B}_1$; we must verify that $\mathcal{B}_1$ is really a basis for a topology. This will be the case if given $aU, bV \in \mathcal{B}_1$ where $U, V \in \mathcal{U}$ and $c \in aU \cap bV$, there exists a $W \in \mathcal{U}$ such that $cW \subset aU \cap bV$. We have $c = au = bv$ where $u \in U$ and $v \in V$. By continuity of the mapping $x \to ux$ at e, there is a $U_1 \in \mathcal{U}$ such that $uU_1 \subset U$.

Similarly, $vV_1 \subset V$ for some $V_1 \in \mathscr{U}$. Choosing W in $\mathscr{U}$ such that $W \subset U_1 \cap V_1$, we find that

$$cW \subset (cU_1) \cap (cV_1) = (auU_1) \cap (bvV_1) \subset (aU) \cap (bV).$$

To show that multiplication is continuous in the $\mathcal{O}_1$-topology, let $a, b \in S$ and $U \in \mathscr{U}$. If $V \in \mathscr{U}$ and $V^2 \subset U$, then $aVbV = abV^2 \subset abU$. Obviously the $\mathcal{O}_1$-topology of S satisfies (1). Finally, we see that $\mathcal{O}_1$ is stronger than $\mathcal{O}$, because if $W \in \mathcal{O}$ and $a \in W$, then the continuity of $x \to ax$ at e implies that $aU \subset W$ for some $U \in \mathscr{U}$.]

(9.28) Monothetic semigroups. A topological semigroup S is said to be *monothetic* if it contains an element a such that $\{a, a^2, \ldots\}$ is dense in S. In this context only, a will be called a *generator* of S.

(a) (HEWITT [4].) Let S be a compact monothetic semigroup, and let a denote a generator of S. Let G denote the closed subgroup $\bigcap_{k=1}^{\infty} \{a^k, a^{k+1}, \ldots\}^-$, and let e be the identity of G. One of the following situations must hold:

(i) $S = G$, in which case S is a compact monothetic group;

(ii) there is a positive integer m such that $S \cap G' = \{a, a^2, \ldots, a^m\}$ and all of the elements $a, a^2, \ldots, a^m$ are isolated points of S;

(iii) $S \cap G' = \{a, a^2, \ldots\}$ and all of the elements $a, a^2, \ldots$ are isolated points of S.

[By (9.17), G is a closed subset of S that is a group; thus by (9.16), G is a topological group. We next prove that if $a^n \in G$ and $p > n$, then $a^p \in G$. Let k be an arbitrary positive integer and U a neighborhood of a^p. By continuity of multiplication, there is a neighborhood V of a^n such that $Va^{p-n} \subset U$. For some integer $l \geqq k$, a^l is in V since $a^n \in \{a^k, a^{k+1}, \ldots\}^-$. Then $a^{l+p-n} \in U$ and $l+p-n > k$, so that U intersects $\{a^k, a^{k+1}, \ldots\}$. Thus $a^p \in \{a^k, a^{k+1}, \ldots\}^-$; as k is arbitrary, a^p is in G.

The last paragraph shows that one of the following equalities holds:

$$G \cap \{a, a^2, \ldots\} = \{a, a^2, \ldots\}, \tag{1}$$

$$G \cap \{a, a^2, \ldots\} = \{a^{m+1}, a^{m+2}, \ldots\} \quad \text{where} \quad m \geqq 1, \tag{2}$$

$$G \cap \{a, a^2, \ldots\} = \varnothing. \tag{3}$$

Any element of S that is not an isolated point must lie in every set $\{a^k, a^{k+1}, \ldots\}^-$ and hence in G. In particular, we have $S \cap \{a, a^2, \ldots\}' \subset G$ and $S \cap G' = \{a, a^2, \ldots\} \cap G'$. Also, all points in $S \cap G'$ are isolated. Using these remarks, we see that (1) implies (i), (2) implies (ii), and (3) implies (iii).]

(b) A compact monothetic semigroup S with two distinct generator. a and b is a topological group. [Assume that S is not a topological groups

Then a and b are isolated points by (a), and hence $a \in \{b, b^2, \ldots\}$ and $b \in \{a, a^2, \ldots\}$. Therefore $a = b^m$ and $b = a^n$ for some integers m and n greater than 1. Thus $a = a^{mn}$ and a is in $G = \bigcap_{k=1}^{\infty} \{a^k, a^{k+1}, \ldots\}^-$. Consequently, $S = G$ and S is a topological group.]

Notes

At the present time, it is difficult to assign exact priorities to all of the results (9.1)—(9.14). Theorem (9.3) for *connected* locally compact Abelian groups was proved by PONTRYAGIN [3] and [4], VAN KAMPEN [1], and MARKOV [1]. The earliest statement known to us of (9.1) [in a trivially more special form] is in WEIL [4], p. 96. The fact that (9.3) holds for compactly generated groups and not merely connected groups was apparently first noticed by WEIL [4], p. 97. Theorem (9.5) is due to VAN KAMPEN [1], p. 458. The first explicit statement of (9.8) known to us is in WEIL [4], p. 99, where a proof attributed to VAN KAMPEN [1] is given. A second proof of (9.8) based on the PONTRYAGIN-VAN KAMPEN duality theorem appears in WEIL [4], p. 110. Theorem (9.14) for groups with countable open bases is due to PONTRYAGIN [2] and [4]. A generalization of (9.14) appears in VAN KAMPEN [1], p. 460.

Theorem (9.15) is due to IVANOVSKIĬ [1] and KUZ'MINOV [1]. For a commutative G, see VILENKIN [17]. The proof given here is based on HULANICKI [4]. KUZ'MINOV proves a different result: *every* compact group is a continuous image of some space $\{0, 1\}^{\mathfrak{m}}$. In this connection, see (25.35).

In (9.1)—(9.15), we merely scratch the surface of what is known about the structure of locally compact groups. For non-Abelian groups, in particular Lie groups, we refer the reader to the monographs of MONTGOMERY and ZIPPIN [1], CHEVALLEY [1], and PONTRYAGIN [7], Chaps. VII—XI. Locally compact Abelian groups are examined in some detail in §§ 24—26 *infra*. Even for this case there is an enormous literature: for a partial bibliography, see the notes to § 25.

Theorems (9.16)—(9.22) are due to NUMAKURA [1]. Again, we have given only a glimpse of the known structure of topological semigroups. For a useful survey of the field, see A. D. WALLACE [1].

§ 10. Some special locally compact Abelian groups

We now describe a class of locally compact Abelian groups which are of importance in the structure theory of locally compact Abelian groups and are also interesting on their own account. To define these groups, we begin with some quite elementary facts about integers.

(10.1) Let $\boldsymbol{a}=(a_0, a_1, a_2, \ldots, a_n, \ldots)$ be any sequence of integers each of which is greater than 1. If M is any positive integer, we obtain by successive divisions:

$$\left.\begin{aligned}
&M=x_0+M_0 a_0, \quad \text{where} \quad x_0\in\{0, 1, \ldots, a_0-1\}\\
&\text{and } M_0 \text{ is a nonnegative integer;}\\
&M_0=x_1+M_1 a_1, \quad \text{where} \quad x_1\in\{0, 1, \ldots, a_1-1\}\\
&\text{and } M_1 \text{ is a nonnegative integer;}\\
&M_1=x_2+M_2 a_2, \quad \text{where} \quad x_2\in\{0, 1, \ldots, a_2-1\}\\
&\text{and } M_2 \text{ is a nonnegative integer;}\\
&\cdots\cdots\\
&M_{k-1}=x_k+M_k a_k, \quad \text{where} \quad x_k\in\{0, 1, \ldots, a_k-1\}\\
&\text{and } M_k \text{ is a nonnegative integer;}\\
&\cdots\cdots.
\end{aligned}\right\} \tag{1}$$

The M_k's decrease strictly as long as they are positive, and so there is a least integer m such that $M_m=0$; defining M_{-1} as x_0, we have in all cases $0\neq M_{m-1}=x_m$, and

$$M=x_0+x_1 a_0+x_2 a_0 a_1+\cdots+x_m a_0 a_1 \cdots a_{m-1}. \tag{2}$$

It is also easy to see that every positive integer can be written in exactly one way as a sum of the form (2). Thus we have a one-to-one correspondence between the set of all nonnegative integers and the set of all sequences $(x_0, x_1, x_2, \ldots, x_l, \ldots)$ such that $x_l\in\{0, 1, \ldots, a_l-1\}$ for $l=0, 1, 2, \ldots$, and only a finite number of the x_l's are different from 0.

For a positive integer N, let us carry out the divisions (1), obtaining

$$N=y_0+N_0 a_0, \quad N_0=y_1+N_1 a_1, \quad N_1=y_2+N_2 a_2, \ldots, N_{n-1}=y_n, \tag{3}$$

where the N_j's and y_j's satisfy conditions as in (1). Then we get

$$N=y_0+y_1 a_0+y_2 a_0 a_1+\cdots+y_n a_0 a_1 \cdots a_{n-1}. \tag{4}$$

Similarly, we write

$$M+N=z_0+P_0 a_0, \quad P_0=z_1+P_1 a_1, \quad P_1=z_2+P_2 a_2, \ldots, P_{p-1}=z_p, \tag{5}$$

and

$$M+N=z_0+z_1 a_0+z_2 a_0 a_1+\cdots+z_p a_0 a_1 \cdots a_{p-1}. \tag{6}$$

Obviously the z's are determined from the x's and y's. To make this computation directly, we proceed as follows. From (1), (3), and (5), we have $M+N=z_0+P_0 a_0=x_0+y_0+(M_0+N_0) a_0$, so that $x_0+y_0=$

$(P_0 - M_0 - N_0)\, a_0 + z_0 = t_0\, a_0 + z_0$. Plainly z_0 is the least nonnegative residue of $x_0 + y_0$ modulo a_0, and this also determines $t_0 = P_0 - M_0 - N_0$ directly from x_0 and y_0. Using (1), (3), and (5) again, we have $P_0 = z_1 + P_1\, a_1 = t_0 + M_0 + N_0 = t_0 + x_1 + y_1 + (M_1 + N_1)\, a_1$. That is, $x_1 + y_1 + t_0 = z_1 + (P_1 - M_1 - N_1)\, a_1 = z_1 + t_1\, a_1$. Plainly z_1 is the least nonnegative residue of $x_1 + y_1 + t_0$ modulo a_1, and this fact also defines $t_1 = P_1 - M_1 - N_1$ in terms of x_0, x_1, y_0, and y_1.

To complete the construction by finite induction, suppose that we have: $x_0 + y_0 = t_0\, a_0 + z_0$, $x_1 + y_1 + t_0 = t_1\, a_1 + z_1, \ldots,$ $x_k + y_k + t_{k-1} = t_k\, a_k + z_k$, where z_j is the least nonnegative residue of $x_j + y_j + t_{j-1}$ modulo a_j $(j = 0, 1, \ldots, k;\ t_{-1} = 0)$; and $t_j = P_j - M_j - N_j$ $(j = 0, 1, \ldots, k)$. Then we have $z_{k+1} + P_{k+1}\, a_{k+1} = P_k = t_k + M_k + N_k = t_k + x_{k+1} + y_{k+1} + (M_{k+1} + N_{k+1})\, a_{k+1}$; thus $x_{k+1} + y_{k+1} + t_k = z_{k+1} + (P_{k+1} - M_{k+1} - N_{k+1})\, a_{k+1} = z_{k+1} + t_{k+1}\, a_{k+1}$. Since $z_{k+1} \in \{0, 1, \ldots, a_{k+1} - 1\}$, the induction is complete.

Thus addition in the semigroup of nonnegative integers can be transferred to the set of sequences $(x_0, x_1, x_2, \ldots)$ $[0 \leq x_n < a_n$ for all n, and $x_n = 0$ for n sufficiently large$]$ without reference to the correspondence indicated in (2). If $(x_0, x_1, x_2, \ldots, x_m, 0, 0, \ldots)$ and $(y_0, y_1, y_2, \ldots, y_n, 0, 0, \ldots)$ are two such sequences, we define their sum by the following rule. Write $x_0 + y_0 = t_0\, a_0 + z_0$, where $z_0 \in \{0, 1, \ldots, a_0 - 1\}$ and t_0 is an integer. Write $x_1 + y_1 + t_0 = t_1\, a_1 + z_1$, where $z_1 \in \{0, 1, \ldots, a_1 - 1\}$ and t_1 is an integer. When $z_0, z_1, \ldots, z_{k-1}$ and $t_0, t_1, \ldots, t_{k-1}$ have been defined, write $x_k + y_k + t_{k-1} = t_k\, a_k + z_k$, where $z_k \in \{0, 1, \ldots, a_k - 1\}$ and t_k is an integer. This defines sequences $(t_0, t_1, t_2, \ldots)$ and $(z_0, z_1, z_2, \ldots)$; note that each t_k is either 0 or 1. It is clear that the largest integer p such that $z_p \neq 0$ is $\max(m, n)$ or $\max(m, n) + 1$. We define $(x_0, x_1, x_2, \ldots, x_m, 0, 0, \ldots) + (y_0, y_1, y_2, \ldots, y_m, 0, 0, \ldots)$ to be $(z_0, z_1, z_2, \ldots, z_p, 0, 0, \ldots)$.

A moment's reflection shows that in the definition of addition just given, there is no need for the added sequences to have only finitely many nonzero entries. Another moment's reflection shows that there is no need for the sequences to begin with x_0: sequences $(\ldots, x_{-n}, x_{1-n}, \ldots, x_{-2}, x_{-1}, x_0, x_1, x_2, \ldots, x_n, \ldots)$ can be added, *provided only that* x_k *is* 0 *for* $k < n_0$, n_0 *being some integer depending on the sequence.* We proceed to a formal definition.

(10.2) Definition. Let $\boldsymbol{a}$ be a fixed but arbitrary doubly infinite sequence of positive integers: $\boldsymbol{a} = (\ldots, a_{-n}, a_{1-n}, \ldots, a_{-2}, a_{-1}, a_0, a_1, a_2, \ldots, a_n, \ldots)$, where each a_n is greater than 1. Consider the Cartesian product $\mathsf{P}_{n \in Z} \{0, 1, \ldots, a_n - 1\}$. Let $\Omega_{\boldsymbol{a}}$ be the set of all $\boldsymbol{x} = (x_n)$ in this product space such that x_n is 0 for all $n < n_0$, where n_0 is an integer that depends upon $\boldsymbol{x}$. For $\boldsymbol{x}$ and $\boldsymbol{y} = (y_n)$ in $\Omega_{\boldsymbol{a}}$, let $\boldsymbol{z} = (z_n)$ be defined as follows. Suppose that $x_{m_0} \neq 0$ and $x_n = 0$ for $n < m_0$, and that $y_{n_0} \neq 0$ and $y_n = 0$ for $n < n_0$. Then let $z_n = 0$ for $n < p_0 = \min(m_0, n_0)$. Write

$x_{p_0} + y_{p_0} = t_{p_0} a_{p_0} + z_{p_0}$, where $z_{p_0} \in \{0, 1, \ldots, a_{p_0} - 1\}$ and t_{p_0} is an integer. Suppose that $z_{p_0}, z_{p_0+1}, \ldots, z_k$ and $t_{p_0}, t_{p_0+1}, \ldots, t_k$ have been defined. Then write $x_{k+1} + y_{k+1} + t_k = t_{k+1} a_{k+1} + z_{k+1}$, where $z_{k+1} \in \{0, 1, \ldots, a_{k+1} - 1\}$ and t_{k+1} is an integer. This defines by induction a sequence $\boldsymbol{z} = (z_n) \in \Omega_{\boldsymbol{a}}$. We define the sum $\boldsymbol{x} + \boldsymbol{y}$ to be the sequence $\boldsymbol{z}$. To complete the definition of addition in $\Omega_{\boldsymbol{a}}$, we define $\boldsymbol{0} + \boldsymbol{x} = \boldsymbol{x} + \boldsymbol{0} = \boldsymbol{x}$ for all $\boldsymbol{x} \in \Omega_{\boldsymbol{a}}$, where $\boldsymbol{0}$ is the sequence in $\Omega_{\boldsymbol{a}}$ that is identically zero. The subset of $\Omega_{\boldsymbol{a}}$ consisting of all $\boldsymbol{x}$ for which $x_n = 0$ for $n < 0$ will be denoted by $\Delta_{\boldsymbol{a}}$.

The set $\Omega_{\boldsymbol{a}}$ with the addition defined above will be called the *$\boldsymbol{a}$-adic numbers*. The set $\Delta_{\boldsymbol{a}}$ with the addition defined above will be called the *$\boldsymbol{a}$-adic integers*. If all of the integers a_n are equal to some fixed integer $r > 1$, we write Ω_r and Δ_r for $\Omega_{\boldsymbol{a}}$ and $\Delta_{\boldsymbol{a}}$, and call these objects the *r-adic numbers* and *r-adic integers* respectively. When $\Delta_{\boldsymbol{a}}$ is considered by itself, there is obviously no need to consider negative indices n, and we will take the liberty of writing elements of $\Delta_{\boldsymbol{a}}$ as $\boldsymbol{x} = (x_0, x_1, x_2, \ldots)$ and of ignoring a_n for $n < 0$.

(10.3) Theorem. *With the operation + defined in* (10.2), *the set $\Omega_{\boldsymbol{a}}$ is an Abelian group, containing $\Delta_{\boldsymbol{a}}$ as a subgroup. The group $\Delta_{\boldsymbol{a}}$ contains a subgroup isomorphic to Z.*

Proof. It is obvious that $\boldsymbol{0} + \boldsymbol{x} = \boldsymbol{x}$ for all $\boldsymbol{x} \in \Omega_{\boldsymbol{a}}$ and $\boldsymbol{x} + \boldsymbol{y} = \boldsymbol{y} + \boldsymbol{x}$ for all $\boldsymbol{x}, \boldsymbol{y} \in \Omega_{\boldsymbol{a}}$. We now find the inverse of an element $\boldsymbol{x} \in \Omega_{\boldsymbol{a}}$ different from $\boldsymbol{0}$. Suppose that x_m is different from 0 and $x_n = 0$ for $n < m$. Define the element $\boldsymbol{y}$ in $\Omega_{\boldsymbol{a}}$ by the following rule: $y_n = 0$ for $n < m$; $y_m = a_m - x_m$; $y_n = a_n - x_n - 1$ for $n > m$. It is plain that $\boldsymbol{x} + \boldsymbol{y} = \boldsymbol{0}$.

To complete the proof that $\Omega_{\boldsymbol{a}}$ is an Abelian group, we need only to show that the associative law holds. Let $\boldsymbol{x}, \boldsymbol{y}, \boldsymbol{z}$ be any elements of $\Omega_{\boldsymbol{a}}$ and let m be the least integer for which x_m or y_m or z_m is different from 0. It is then trivial that the values of $(\boldsymbol{x} + \boldsymbol{y}) + \boldsymbol{z}$ and $\boldsymbol{x} + (\boldsymbol{y} + \boldsymbol{z})$ at the index n are equal, and equal to 0, for all $n < m$. Consider now the set Σ_m of all elements $\boldsymbol{u} \in \Omega_{\boldsymbol{a}}$ such that $u_n = 0$ for $n < m$ and only a finite number of the values u_n are different from 0. It follows from (10.1) that the mapping $\boldsymbol{u} \to u_m + u_{m+1} a_m + u_{m+2} a_m a_{m+1} + \cdots$ is an isomorphism of Σ_m onto the additive semigroup of nonnegative integers. Consequently the associative law holds in Σ_m, since addition of integers is associative. Return now to our arbitrary $\boldsymbol{x}, \boldsymbol{y}$, and $\boldsymbol{z}$. Consider any fixed index $p \geqq m$ and the "truncated" elements $\boldsymbol{x}', \boldsymbol{y}', \boldsymbol{z}'$ such that $x'_n = x_n$ for $n \leqq p$ and $x'_n = 0$ for $n > p$, with $\boldsymbol{y}'$ and $\boldsymbol{z}'$ defined similarly. Then $(\boldsymbol{x}' + \boldsymbol{y}') + \boldsymbol{z}' = \boldsymbol{x}' + (\boldsymbol{y}' + \boldsymbol{z}')$ since $\boldsymbol{x}', \boldsymbol{y}'$, and $\boldsymbol{z}'$ are in Σ_m. The value of $(\boldsymbol{x}' + \boldsymbol{y}') + \boldsymbol{z}'$ at the index p is obviously the same as the value of $(\boldsymbol{x} + \boldsymbol{y}) + \boldsymbol{z}$ at the index p; a similar remark applies to $\boldsymbol{x}' + (\boldsymbol{y}' + \boldsymbol{z}')$ and $\boldsymbol{x} + (\boldsymbol{y} + \boldsymbol{z})$. This proves the associative law in $\Omega_{\boldsymbol{a}}$ and completes the proof that $\Omega_{\boldsymbol{a}}$ is an Abelian group.

It is obvious from the foregoing that $\Delta_{\boldsymbol{a}}$ is a subgroup of $\Omega_{\boldsymbol{a}}$. If $\boldsymbol{x}$ and $\boldsymbol{y}$ are in $\Omega_{\boldsymbol{a}}$ and $x_n = y_n = 0$ for $n < 0$, then $\boldsymbol{x} + \boldsymbol{y}$ has the value 0 at all indices $n < 0$, and $-\boldsymbol{x}$ also has the value 0 at all indices $n < 0$. Let $\boldsymbol{u}$ be the element of $\Delta_{\boldsymbol{a}}$ such that $u_n = 0$ for $n < 0$, $u_0 = 1$, and $u_n = 0$ for $n > 0$. Then $\{\boldsymbol{0}, \boldsymbol{u}, 2\boldsymbol{u}, \ldots, k\boldsymbol{u}, \ldots\}$ is a subsemigroup of $\Delta_{\boldsymbol{a}}$ isomorphic to the additive nonnegative integers, and so $\Delta_{\boldsymbol{a}}$ contains the subgroup $\{\boldsymbol{0}, \pm\boldsymbol{u}, \pm 2\boldsymbol{u}, \ldots, \pm k\boldsymbol{u}, \ldots\}$, which is isomorphic to Z. [Of course every element of $\Omega_{\boldsymbol{a}}$ of infinite order generates a subgroup isomorphic with Z.] □

As is to be expected, both $\Omega_{\boldsymbol{a}}$ and $\Delta_{\boldsymbol{a}}$ are topological groups under a suitable topology.

(10.4) Definition. For each integer k, let Λ_k be the set of all $\boldsymbol{x} \in \Omega_{\boldsymbol{a}}$ such that $x_n = 0$ for all $n < k$. For distinct elements $\boldsymbol{x}, \boldsymbol{y} \in \Omega_{\boldsymbol{a}}$, let $\sigma(\boldsymbol{x}, \boldsymbol{y})$ be the number 2^{-m}, where m is the least integer for which $x_m \neq y_m$. For all $\boldsymbol{x} \in \Omega_{\boldsymbol{a}}$, let $\sigma(\boldsymbol{x}, \boldsymbol{x}) = 0$.

(10.5) Theorem. *The sets* $\ldots, \Lambda_{-k}, \ldots, \Lambda_{-2}, \Lambda_{-1}, \Lambda_0, \Lambda_1, \Lambda_2, \Lambda_3, \ldots, \Lambda_k, \ldots$ *satisfy conditions* (4.5.i)—(4.5.v). *Hence they define a topology on* $\Omega_{\boldsymbol{a}}$ *as in* (4.5) *under which* $\Omega_{\boldsymbol{a}}$ *is a topological group. Under this topology,* $\Omega_{\boldsymbol{a}}$ *is Hausdorff, locally compact, σ-compact, and* 0*-dimensional. The sets* Λ_k *are compact subgroups of* $\Omega_{\boldsymbol{a}}$. *[Note that* $\Lambda_0 = \Delta_{\boldsymbol{a}}$.*] The function* σ *is an invariant metric on* $\Omega_{\boldsymbol{a}}$ *compatible with the topology of* $\Omega_{\boldsymbol{a}}$.

Proof. It is evident that the sets Λ_k are subgroups of $\Omega_{\boldsymbol{a}}$, so that (4.5.i), (4.5.ii), and (4.5.iii) are trivially satisfied. Since $\Omega_{\boldsymbol{a}}$ is Abelian, (4.5.iv) is also trivially satisfied, and (4.5.v) is obvious. Plainly $\boldsymbol{0}$ is in all Λ_k, so that (4.5) may be applied, and $\Omega_{\boldsymbol{a}}$ is a topological group in which the sets $\boldsymbol{x} + \Lambda_k$ are a basis for open sets. [Note that this is a special case of (4.21.a).] Since $\bigcap_{k=0}^{\infty} \Lambda_k = \{\boldsymbol{0}\}$, $\Omega_{\boldsymbol{a}}$ is a Hausdorff group (4.8).

Considered as a set, Λ_k may be regarded as the complete Cartesian product $\mathsf{P}_{n=k}^{\infty} \{0, 1, \ldots, a_n - 1\}$. For $\boldsymbol{x} \in \Lambda_k$ and $l > k$, the set $\boldsymbol{x} + \Lambda_l$ is the set of all elements $(\ldots, 0, \ldots, 0_{(k-1)}, x_k, x_{k+1}, \ldots, x_{l-1}, z_l, z_{l+1}, \ldots)$ of $\Omega_{\boldsymbol{a}}$, where z_j is an arbitrary number in $\{0, 1, \ldots, a_j - 1\}$ for $j = l, l+1, \ldots$. This shows that the relative topology of Λ_k as a subspace of $\Omega_{\boldsymbol{a}}$ is the Cartesian product topology of $\mathsf{P}_{n=k}^{\infty} \{0, 1, \ldots, a_n - 1\}$. This is a compact topology, and so the open subgroups Λ_k are compact and closed. Therefore $\Omega_{\boldsymbol{a}}$ is locally compact and 0-dimensional.

The set A of all $\boldsymbol{x} \in \Omega_{\boldsymbol{a}}$ such that $x_n = 0$ for $n \geqq 0$ is plainly a countably infinite set. Since $\Omega_{\boldsymbol{a}} = A + \Lambda_0 = \bigcup_{\boldsymbol{x} \in A} (\boldsymbol{x} + \Lambda_0)$, $\Omega_{\boldsymbol{a}}$ is σ-compact.

It remains to verify our assertions about the function σ. It is obvious that $\sigma(\boldsymbol{x}, \boldsymbol{y}) = \sigma(\boldsymbol{y}, \boldsymbol{x})$ and that $\sigma(\boldsymbol{x}, \boldsymbol{y})$ is positive unless $\boldsymbol{x} = \boldsymbol{y}$. To verify the identity

$$\sigma(\boldsymbol{x} + \boldsymbol{z}, \boldsymbol{y} + \boldsymbol{z}) = \sigma(\boldsymbol{x}, \boldsymbol{y}), \tag{1}$$

it suffices to consider the case in which $x_m \neq y_m$ and $x_n = y_n$ for all $n < m$. Applying (10.2), we write $x_m + z_m + t_{m-1} = t_m a_m + v_m$ and $y_m + z_m + t_{m-1} = t'_m a_m + v'_m$, where $v_m, v'_m \in \{0, 1, \ldots, a_m - 1\}$ and t_{m-1}, t_m, t'_m are integers. If $v_m = v'_m$, we would have $x_m = y_m \pmod{a_m}$, which is impossible. It follows that σ is invariant.

The function σ satisfies an inequality stronger than the usual triangle inequality, *viz.*:

$$\sigma(\boldsymbol{x} + \boldsymbol{y}, \boldsymbol{0}) \leqq \max\{\sigma(\boldsymbol{x}, \boldsymbol{0}), \sigma(\boldsymbol{y}, \boldsymbol{0})\}. \tag{2}$$

To prove (2), let $\boldsymbol{x}$ and $\boldsymbol{y}$ be different from $\boldsymbol{0}$ [if either is $\boldsymbol{0}$, (2) is trivial], and let p and q be the least indices for which we have $x_p \neq 0$ and $y_q \neq 0$, respectively. If $p \neq q$, then the least index at which $\boldsymbol{x} + \boldsymbol{y}$ is different from 0 is $\min(p, q)$, so that in this case $\sigma(\boldsymbol{x} + \boldsymbol{y}, \boldsymbol{0}) = 2^{-\min(p,q)} = \max(2^{-p}, 2^{-q})$. If $p = q$, then the least index r at which $\boldsymbol{x} + \boldsymbol{y}$ is different from 0 may not exist or may be any integer greater than or equal to p. Thus we have $\sigma(\boldsymbol{x} + \boldsymbol{y}, \boldsymbol{0}) = 0$ or $\sigma(\boldsymbol{x} + \boldsymbol{y}, \boldsymbol{0}) = 2^{-r} \leqq 2^{-p}$. Thus (2) is verified. The relations (1) and (2) obviously imply that the triangle inequality holds for σ.

Finally, to see that σ is compatible with the topology of $\Omega_{\boldsymbol{a}}$, we need only note that $\Lambda_k = \{\boldsymbol{x} \in \Omega_{\boldsymbol{a}} : \sigma(\boldsymbol{x}, \boldsymbol{0}) < 2^{-k+1}\}$ for all $k \in Z$. □

(10.6) Note. All of the groups Λ_k are compact and monothetic. Let $\boldsymbol{u}$ be the element of Λ_k such that $u_n = 0$ if $n \neq k$ and $u_k = 1$. Then it is clear that the set $\{l\boldsymbol{u}\}_{l=1}^{\infty}$ is dense in Λ_k. The set $\{l\boldsymbol{u}\}_{l=-\infty}^{\infty}$ is *a fortiori* dense in Λ_k.

(10.7) Our success in introducing addition in $\Omega_{\boldsymbol{a}}$ in such a way as to make it an Abelian group leads at once to the thought that perhaps a natural multiplication can also be introduced in $\Omega_{\boldsymbol{a}}$, which can be interpreted as ordinary multiplication of rational numbers for elements of $\Omega_{\boldsymbol{a}}$ having only a finite number of nonzero values.

It is simple to describe this multiplication in $\Delta_{\boldsymbol{a}}$ for an arbitrary sequence $\boldsymbol{a} = (a_0, a_1, a_2, \ldots)$. In this paragraph, we write the elements of $\Delta_{\boldsymbol{a}}$ as sequences $\boldsymbol{x} = (x_0, x_1, x_2, \ldots)$. Let $\boldsymbol{u}$ denote the sequence $(1, 0, 0, \ldots) \in \Delta_{\boldsymbol{a}}$. For $l, l' \in Z$, let $(l\boldsymbol{u})(l'\boldsymbol{u})$ be defined as $(ll')\boldsymbol{u}$. It is plain that in this way we reproduce multiplication of integers in the subset $\{l\boldsymbol{u} : l \in Z\}$ of $\Delta_{\boldsymbol{a}}$. Thus the mapping $l \to l\boldsymbol{u}$ is an isomorphism under both addition and multiplication. To define $\boldsymbol{x}\boldsymbol{y}$ for arbitrary $\boldsymbol{x}, \boldsymbol{y} \in \Delta_{\boldsymbol{a}}$, we "truncate" $\boldsymbol{x}$ and $\boldsymbol{y}$. For every nonnegative integer m, let $\boldsymbol{x}^{(m)}$ be the element $(x_n^{(m)})$ of $\Delta_{\boldsymbol{a}}$ such that $x_n^{(m)} = x_n$ if $n \leqq m$ and $x_n^{(m)} = 0$ if $n > m$. Define $\boldsymbol{y}^{(m)}$ similarly. Then $\boldsymbol{x}^{(m)}\boldsymbol{y}^{(m)}$ is already defined for all $m = 0, 1, 2, \ldots$. We define $\boldsymbol{x}\boldsymbol{y}$ as $\lim\limits_{m \to \infty} \boldsymbol{x}^{(m)}\boldsymbol{y}^{(m)}$. One can prove that this limit exists and defines an associative, commutative, and distributive product in $\Delta_{\boldsymbol{a}}$. Furthermore, the mapping $(\boldsymbol{x}, \boldsymbol{y}) \rightarrow \boldsymbol{x}\boldsymbol{y}$ of $\Delta_{\boldsymbol{a}} \times \Delta_{\boldsymbol{a}}$ onto $\Delta_{\boldsymbol{a}}$ is continuous. Thus $\Delta_{\boldsymbol{a}}$ is a compact topological ring[1].

(10.8) We now take up the problem of defining a reasonable multiplication in the larger groups $\Omega_{\boldsymbol{a}}$. Here it is convenient to restrict ourselves to the case in which all a_n are equal: that is, we consider only the groups Ω_r for an arbitrary integer $r > 1$. As in the case of addition in $\Omega_{\boldsymbol{a}}$, we define multiplication in Ω_r as an extension of ordinary multiplication. Consider any positive rational number x whose denominator is a power of r:

$$x = x_l r^l + x_{l+1} r^{l+1} + \cdots + x_m r^m, \tag{1}$$

where $l \leqq m$, l and m are integers, and the x_j's are integers in the set $\{0, 1, \ldots, r-1\}$. If we require further that $x_l \neq 0$ and $x_m \neq 0$, then this representation is unique. Given another rational number of the same sort,

$$y = y_s r^s + y_{s+1} r^{s+1} + \cdots + y_t r^t,$$

their product xy has the form $z_{l+s} r^{l+s} + z_{l+s+1} r^{l+s+1} + \cdots + z_u r^u$. [Note that z_{l+s} may be 0.] As in (10.1), we can compute the numbers z_j directly from the x_j's and y_j's by the following rule. Write $x_l y_s = v_{l+s} r + z_{l+s}$, where $z_{l+s} \in \{0, 1, \ldots, r-1\}$ and v_{l+s} is an integer. Next write $x_{l+1} y_s + x_l y_{s+1} + v_{l+s} = v_{l+s+1} r + z_{l+s+1}$, where $z_{l+s+1} \in \{0, 1, \ldots, r-1\}$ and v_{l+s+1} is an integer. Proceeding in an obvious way by induction, we see that the product of two rational numbers of the form (1) can be computed directly from the x_j's and y_j's. This leads to a definition of multiplication in Ω_r.

(10.9) Definition. Let $\boldsymbol{x}$ and $\boldsymbol{y}$ be any elements of Ω_r. If $\boldsymbol{x} = \boldsymbol{0}$ or $\boldsymbol{y} = \boldsymbol{0}$, let $\boldsymbol{x}\boldsymbol{y}$ be defined as $\boldsymbol{0}$. If neither $\boldsymbol{x}$ nor $\boldsymbol{y}$ is $\boldsymbol{0}$, suppose that $x_l \neq 0$ and $x_n = 0$ for $n < l$ and $y_s \neq 0$ and $y_n = 0$ for $n < s$. Define an element $\boldsymbol{z} \in \Omega_r$ by the following rules. For $n < l+s$, set $z_n = 0$. Write $x_l y_s = v_{l+s} r + z_{l+s}$, where $z_{l+s} \in \{0, 1, \ldots, r-1\}$ and v_{l+s} is an integer. Suppose that the numbers $v_{l+s}, v_{l+s+1}, \ldots, v_{l+s+k-1}$ and $z_{l+s}, z_{l+s+1}, \ldots, z_{l+s+k-1}$ have been defined. Then we write $x_{l+k} y_s + x_{l+k-1} y_{s+1} + x_{l+k-2} y_{s+2} + \cdots + x_l y_{s+k} + v_{l+s+k-1} = v_{l+s+k} r + z_{l+s+k}$, where $z_{l+s+k} \in \{0, 1, \ldots, r-1\}$ and v_{l+s+k} is an integer. This defines the numbers z_n by induction for all $n \geqq l+s$. Let the product $\boldsymbol{x}\boldsymbol{y}$ be the element $\boldsymbol{z}$.

(10.10) Theorem. *With the addition of* (10.2) *and the multiplication of* (10.9), Ω_r *is a commutative ring with unit. The ring* Ω_r *has divisors of* $\boldsymbol{0}$ *if r is not a prime power. It is a field*[2] *if and only if r is a prime power.*

---

[1] A *topological ring* A is a ring endowed with a topology such that, as an additive group, A is a topological group and such that the mapping $(x, y) \rightarrow xy$ is continuous. A *topological field* F is a field that is a topological ring for which the mapping $x \rightarrow x^{-1}$ is a homeomorphism on $F \cap \{0\}'$.

[2] The field Ω_p is usually called the *p-adic number field*.

Proof. In view of (10.3), where we proved *inter alia* that Ω_r is an Abelian group, Ω_r will be proved to be a commutative ring as soon as the identities $\boldsymbol{x}\boldsymbol{y}=\boldsymbol{y}\boldsymbol{x}$, $\boldsymbol{x}(\boldsymbol{y}\boldsymbol{z})=(\boldsymbol{x}\boldsymbol{y})\boldsymbol{z}$, and $\boldsymbol{x}(\boldsymbol{y}+\boldsymbol{z})=\boldsymbol{x}\boldsymbol{y}+\boldsymbol{x}\boldsymbol{z}$ ($\boldsymbol{x}, \boldsymbol{y}, \boldsymbol{z}\in\Omega_r$) are established. These are obvious for $\boldsymbol{x}, \boldsymbol{y}, \boldsymbol{z}$ having only a finite number of nonzero entries, since in this case our multiplication is nothing but multiplication of nonnegative rational numbers. These identities are extended to arbitrary elements of Ω_r by the argument used in the proof of (10.3) to prove the associative law for addition. We omit the details.

The element $\boldsymbol{u}\in\Omega_r$ such that $u_n=\delta_{0n}$ [Kronecker's delta symbol] is plainly a multiplicative unit for Ω_r.

Let r be a positive integer that is not a prime power; then r has the form ab where a and b are integers greater than 1 and $(a, b)=1$.[1] The ring Ω_r abounds in divisors of $\mathbf{0}$. We give one example. Let $x_n=0$ for $n<0$, $x_0=a$, and $y_n=0$ for $n<0$, $y_0=b$. Let x_1, y_1, and u_1 be chosen so that $0\leqq x_1<r$, $0\leqq y_1<r$, u_1 is an integer, and $x_1 b+y_1 a+1=u_1 r$. Such x_1, y_1, and u_1 can be found, since $(a, b)=1$. When $x_1, \ldots, x_n, y_1, \ldots, y_n$, and $u_1, \ldots, u_n$ have been chosen, let x_{n+1}, y_{n+1}, and u_{n+1} be determined so that $x_{n+1} b+y_{n+1} a+x_n y_1+\cdots+x_1 y_n+u_n=u_{n+1} r$. This defines $\boldsymbol{x}$ and $\boldsymbol{y}$ by induction, and it is obvious that $\boldsymbol{x}\boldsymbol{y}=\mathbf{0}$.

Now let p be a prime and α a positive integer. To prove that Ω_{p^α} is a field, we need only to show that every nonzero element $\boldsymbol{x}$ of Ω_{p^α} has a multiplicative inverse. Suppose that $x_l\neq 0$ and that $x_n=0$ for $n<l$. If $(x_l, p)=1$, we define an element $\boldsymbol{y}\in\Omega_{p^\alpha}$ by the following rules. For $n<-l$, let $y_n=0$. Let y_{-l} be the [unique] integer such that $y_{-l}x_l=1+v_0 p^\alpha$, where $y_{-l}\in\{1, 2, \ldots, p^\alpha-1\}$ and v_0 is a nonnegative integer. Suppose that $y_{-l}, y_{-l+1}, \ldots, y_{-l+k-1}$ and $v_0, v_1, \ldots, v_{k-1}$ have been defined. Let y_{-l+k} be the unique integer such that $y_{-l+k} x_l+y_{-l+k-1} x_{l+1}+\cdots+y_{-l} x_{l+k}+v_{k-1}=v_k p^\alpha$, where $y_{-l+k}\in\{0, 1, 2, \ldots, p^\alpha-1\}$ and v_k is a nonnegative integer. This description defines y_n by induction for all $n\geqq -l$. It is obvious from (10.9) that $\boldsymbol{x}\boldsymbol{y}=\boldsymbol{u}$. Suppose next that $(x_l, p)>1$, so that $x_l=p^\beta x_l'$, where $(x_l', p)=1$ and $1\leqq\beta<\alpha$. Let $\boldsymbol{w}\in\Omega_{p^\alpha}$ be defined by $w_n=p^{\alpha-\beta}\delta_{0n}$ ($n\in Z$) and let $\boldsymbol{z}=\boldsymbol{w}\boldsymbol{x}$. Then plainly $z_n=0$ for $n<l$. We also have $p^{\alpha-\beta}x_l=p^\alpha x_l'=t_l p^\alpha+z_l$, so that $z_l=0$ and $t_l=x_l'$. Furthermore, to compute z_{l+1}, we have $p^{\alpha-\beta}x_{l+1}+t_l=p^{\alpha-\beta}x_{l+1}+x_l'=t_{l+1}p^\alpha+z_{l+1}$. This implies that $z_{l+1}\neq 0$ and that $(z_{l+1}, p)=1$. Hence the element $\boldsymbol{z}\in\Omega_{p^\alpha}$ satisfies the conditions of the previous case and so admits a multiplicative inverse $\boldsymbol{z}^{-1}$. That is, we have $\boldsymbol{u}=\boldsymbol{z}\boldsymbol{z}^{-1}=\boldsymbol{x}(\boldsymbol{w}\boldsymbol{z}^{-1})$, so that $\boldsymbol{x}$ too has a multiplicative inverse.[2] □

---

[1] The symbol (a, b) as usual means the greatest common divisor of a and b.

[2] The last statement of this theorem for $\alpha>1$, and its proof, were kindly pointed out to us by JAMES MICHELOW.

(10.11) Theorem. *The mapping* $(\boldsymbol{x}, \boldsymbol{y}) \to \boldsymbol{x}\boldsymbol{y}$ *is a continuous mapping of* $\Omega_r \times \Omega_r$ *onto* Ω_r. *Thus* Ω_r *is a locally compact ring. If* r *is a prime power* p^α, *the mapping* $\boldsymbol{x} \to \boldsymbol{x}^{-1}$ *is a homeomorphism of* $\Omega_{p^\alpha} \cap \{\boldsymbol{0}\}'$ *onto itself. Thus* Ω_{p^α} *is a locally compact field.*

Proof. Let Λ_k be any neighborhood of $\boldsymbol{0}$, as defined in (10.4). Let $\boldsymbol{c}$ and $\boldsymbol{d}$ be arbitrary elements of Ω_r. If $\boldsymbol{c} \neq \boldsymbol{0}$, let s be the least integer such that $c_s \neq 0$; if $\boldsymbol{c} = \boldsymbol{0}$, let $s = 0$. If $\boldsymbol{d} \neq \boldsymbol{0}$, let t be the least integer such that $d_t \neq 0$; if $\boldsymbol{d} = \boldsymbol{0}$, let $t = 0$. Let $l = \max(k-s, k-t, s)$. Then if $\boldsymbol{x} - \boldsymbol{c} \in \Lambda_l$ and $\boldsymbol{y} - \boldsymbol{d} \in \Lambda_l$, we have $\boldsymbol{x}\boldsymbol{y} - \boldsymbol{c}\boldsymbol{d} = \boldsymbol{x}(\boldsymbol{y} - \boldsymbol{d}) + (\boldsymbol{x} - \boldsymbol{c})\,\boldsymbol{d} \in \Lambda_k$. Thus the mapping $(\boldsymbol{x}, \boldsymbol{y}) \to \boldsymbol{x}\boldsymbol{y}$ is continuous.

Suppose now that r is a prime power, say p^α. Let $\boldsymbol{c}$ be a nonzero element of Ω_{p^α}, and let s be the least integer such that $c_s \neq 0$. Let Λ_k be any neighborhood of $\boldsymbol{0}$ as in (10.4) such that $k > |s|$. If $(c_s, p) = 1$, the inverse $\boldsymbol{d} = \boldsymbol{c}^{-1}$ has the form $(\ldots, 0, \ldots, 0, d_{-s}, d_{-s+1}, \ldots)$, where $d_{-s} \neq 0$, as was shown in the proof of (10.10). If $\boldsymbol{x} - \boldsymbol{c} \in \Lambda_{k+2s}$, the rule given in the proof of (10.10) for computing inverses shows that $\boldsymbol{x}^{-1} - \boldsymbol{c}^{-1} \in \Lambda_k$. If $(c_s, p^\alpha) = p^\beta$ with $1 \leq \beta < \alpha$, then as in the proof of (10.10), we define $\boldsymbol{w}$ by $w_n = p^{\alpha-\beta}\,\delta_{0n}$, and write $\boldsymbol{z} = \boldsymbol{w}\boldsymbol{c}$. The element $\boldsymbol{z}$ has the form $(\ldots, 0, \ldots, 0, z_{s+1}, z_{s+2}, \ldots)$, where $z_{s+1} \neq 0$ and $(z_{s+1}, p) = 1$. Let k be an integer greater than $|s+1|$. Suppose that $\boldsymbol{x} - \boldsymbol{c} \in \Lambda_{k+2(s+1)}$. Then we have $\boldsymbol{w}\boldsymbol{x} - \boldsymbol{w}\boldsymbol{c} \in \boldsymbol{w} \cdot \Lambda_{k+2(s+1)} \subset \Lambda_{k+2(s+1)}$. The previous case shows that $(\boldsymbol{w}\boldsymbol{x})^{-1} - (\boldsymbol{w}\boldsymbol{c})^{-1} \in \Lambda_k$. Hence we have $\boldsymbol{x}^{-1} - \boldsymbol{c}^{-1} = \boldsymbol{w}[(\boldsymbol{w}\boldsymbol{x})^{-1} - (\boldsymbol{w}\boldsymbol{c})^{-1}] \in \boldsymbol{w} \cdot \Lambda_k \subset \Lambda_k$. This proves that inversion is continuous. Since $(\boldsymbol{x}^{-1})^{-1} = \boldsymbol{x}$, the mapping $\boldsymbol{x}^{-1} \to \boldsymbol{x}$ is also continuous. □

The groups $\Delta_{\boldsymbol{a}}$ are useful in constructing certain examples of compact connected groups.

(10.12) Definition. Consider the locally compact Abelian group $R \times \Delta_{\boldsymbol{a}}$, written additively, where $\boldsymbol{a} = (a_0, a_1, a_2, \ldots)$ is any sequence of integers all greater than 1. Let $\boldsymbol{u} = (u_n)$ be $(1, 0, 0, \ldots, 0, \ldots)$. Let B be the subgroup $\{(n, n\boldsymbol{u})\}_{n=-\infty}^{\infty}$ of $R \times \Delta_{\boldsymbol{a}}$. Let $\Sigma_{\boldsymbol{a}}$ be the group $(R \times \Delta_{\boldsymbol{a}})/B$. We call $\Sigma_{\boldsymbol{a}}$ the *$\boldsymbol{a}$-adic solenoid.*

(10.13) Theorem. *The group* $\Sigma_{\boldsymbol{a}}$ *is a compact, connected Abelian group containing a continuous homomorph of* R *as a dense subgroup.*

Proof. It is easy to see that B is an infinite cyclic discrete subgroup of $R \times \Delta_{\boldsymbol{a}}$ and is hence closed (5.10). Thus $\Sigma_{\boldsymbol{a}}$ is a T_0 group (5.21) and is locally compact. Since $R \times \Delta_{\boldsymbol{a}}$ is Abelian, so is $\Sigma_{\boldsymbol{a}}$.

Let φ denote the natural homomorphism of $R \times \Delta_{\boldsymbol{a}}$ onto $\Sigma_{\boldsymbol{a}}$. Then we have $\varphi(R \times \{\boldsymbol{0}\}) = \{(\xi, \boldsymbol{0}) + B : \xi \in R\}$, which is a subgroup of $\Sigma_{\boldsymbol{a}}$. If $\eta \in R$ and m is any integer, then $(\eta, m\boldsymbol{u}) + B = (\eta - m, \boldsymbol{0}) + B$ is in $\varphi(R \times \{\boldsymbol{0}\})$. Thus $\{(\eta, m\boldsymbol{u}) + B : \eta \in R, m \in Z\} = \varphi(R \times \{\boldsymbol{0}\})$; this equality and the fact that $\{m\boldsymbol{u} : m \in Z\}$ is dense in $\Delta_{\boldsymbol{a}}$ show that $\varphi(R \times \{\boldsymbol{0}\})$ is dense in $\Sigma_{\boldsymbol{a}}$. If $\boldsymbol{x}=(1, 0, 1, 0, 1, 0, \ldots)$, then $(0, \boldsymbol{x})+B$ is not in $\varphi(R\times\{\boldsymbol{0}\})$, so that $\varphi(R\times\{\boldsymbol{0}\})$ is a proper subgroup of $\Sigma_{\boldsymbol{a}}$. Since $R\times\{\boldsymbol{0}\}$ is homeomorphic with R, its continuous image $\varphi(R\times\{\boldsymbol{0}\})$ is connected and its closure $\Sigma_{\boldsymbol{a}}$ is also connected.

To show that $\Sigma_{\boldsymbol{a}}$ is compact, we will apply Theorem (9.1). Assume that the continuous homomorphism established above of $R\times\{\boldsymbol{0}\}$ into $\Sigma_{\boldsymbol{a}}$ is a topological isomorphism. Then by (5.11), $\varphi(R\times\{\boldsymbol{0}\})$ is a closed subgroup of $\Sigma_{\boldsymbol{a}}$. This contradicts the fact that $\varphi(R\times\{\boldsymbol{0}\})$ is a proper dense subset of $\Sigma_{\boldsymbol{a}}$. Hence (9.1) shows that $\varphi(R\times\{\boldsymbol{0}\})^{-}=\Sigma_{\boldsymbol{a}}$ is compact. □

(10.14) The groups $R\times\Delta_{\boldsymbol{a}}$ and $\Sigma_{\boldsymbol{a}}$ show that the hypotheses of Corollary (7.13) are essential. Since $\Sigma_{\boldsymbol{a}}$ is connected, the component of the identity in $\Sigma_{\boldsymbol{a}}$ is $\Sigma_{\boldsymbol{a}}$. The component of the identity in $R\times\Delta_{\boldsymbol{a}}$ is obviously $R\times\{\boldsymbol{0}\}$. Since $\varphi(R\times\{\boldsymbol{0}\})$ is *not* the entire group $\Sigma_{\boldsymbol{a}}$, the conclusion of (7.13) is false for $R\times\Delta_{\boldsymbol{a}}$, $\Sigma_{\boldsymbol{a}}$, and the mapping φ. Note that the kernel of φ is B, which is not contained in $R\times\{\boldsymbol{0}\}$, nor is $R\times\{\boldsymbol{0}\}$ contained in B. Thus the hypotheses of (7.13) are violated in the present example. Note too that (7.12) holds in the present example.

The group $\Sigma_{\boldsymbol{a}}$ can be realized in a quite concrete way.

(10.15) Theorem. *Consider the set* $[0, 1[\times\Delta_{\boldsymbol{a}}$, *with the following binary operation* $+$:

(i) $(\xi, \boldsymbol{x})+(\eta, \boldsymbol{y})=(\xi+\eta-[\xi+\eta], \boldsymbol{x}+\boldsymbol{y}-[\xi+\eta]\,\boldsymbol{u})$;

$\boldsymbol{u}$ *is as defined in* (10.12). *For* $n=1, 2, \ldots$, *let*

$$\text{(ii)}\quad U_n=\left\{(\xi, \boldsymbol{x})\in[0, 1[\times\Delta_{\boldsymbol{a}}: 0\leq\xi<\frac{1}{n}\ \text{and}\ x_0=\cdots=x_{n-1}=0\right\}$$
$$\cup\left\{(\xi, \boldsymbol{x})\in[0, 1[\times\Delta_{\boldsymbol{a}}: 1-\frac{1}{n}<\xi<1\ \text{and}\right.$$
$$\left. x_0=1,\, x_1=\cdots=x_{n-1}=0\right\}=U_n^{(1)}\cup U_n^{(2)}.$$

With the operation $+$ *and the neighborhoods* U_n *taken as an open basis at* $(0, \boldsymbol{0})$, $[0, 1[\times\Delta_{\boldsymbol{a}}$ *is a topological group topologically isomorphic with* $\Sigma_{\boldsymbol{a}}$.

Proof. Every coset $(\xi, \boldsymbol{x})+B\in\Sigma_{\boldsymbol{a}}$ contains the element $(\xi-[\xi], \boldsymbol{x}-[\xi]\,\boldsymbol{u})$. In other words, $(\xi, \boldsymbol{x})+B$ contains an element of $[0, 1[\times\Delta_{\boldsymbol{a}}$. Suppose that the elements $(\xi, \boldsymbol{x})$ and $(\eta, \boldsymbol{y})$ of $[0, 1[\times\Delta_{\boldsymbol{a}}$ belong to the same coset of B. Then $(\xi-\eta, \boldsymbol{x}-\boldsymbol{y})=(n, n\boldsymbol{u})$ for some integer n. This implies that $n=0$, so that $(\xi, \boldsymbol{x})=(\eta, \boldsymbol{y})$. Thus each coset $(\xi, \boldsymbol{x})+B$ contains exactly one element of $[0, 1[\times\Delta_{\boldsymbol{a}}$, and we have an obvious one-to-one mapping of $\Sigma_{\boldsymbol{a}}$ onto $[0, 1[\times\Delta_{\boldsymbol{a}}$. The definition (i) of addition in $[0, 1[\times\Delta_{\boldsymbol{a}}$ is obviously the same as for $\big((\xi, \boldsymbol{x})+B\big)+\big((\eta, \boldsymbol{y})+B\big)=(\xi+\eta, \boldsymbol{x}+\boldsymbol{y})+B$.

An open basis at the identity B in $\Sigma_{\boldsymbol{a}}$ is obtained as $\{\varphi(]-\varepsilon, \varepsilon[\times\Lambda_k)\}$ for $0<\varepsilon<1$, and $k=1, 2, \ldots$: this is pointed out in (5.15). It follows that $\left\{\varphi\left(\left]-\frac{1}{n}, \frac{1}{n}\right[\times \Lambda_n\right)\right\}_{n=1}^{\infty}$ is an open basis at B. The set $\left[0, \frac{1}{n}\right[\times \Lambda_n$ maps onto $U_n^{(1)}$ and the set $\left]-\frac{1}{n}, 0\right[\times \Lambda_n$ onto $U_n^{(2)}$. Thus $[0, 1[\times \Lambda_{\boldsymbol{a}}$ with the structure defined by (i) and (ii) is a replica of $\Sigma_{\boldsymbol{a}}$. □

Miscellaneous theorems and examples

(10.16) (a) The only proper closed subgroups of Ω_p, where p is a prime, are the subgroups Λ_k defined in (10.4). Let $\boldsymbol{a}=(\ldots, a_{-1}, a_0, a_1, \ldots)$ be a doubly infinite sequence of integers that are not all equal to the same prime number. Then $\Omega_{\boldsymbol{a}}$ contains a compact open subgroup H distinct from all the subgroups Λ_k.

[Let H denote a proper closed subgroup of Ω_p. We first prove that if H contains an element $\boldsymbol{x}$ such that $x_n=0$ for $n<k$ and $x_k \neq 0$, then $\Lambda_k \subset H$. It suffices to prove that $\boldsymbol{x}$ generates a dense subgroup of Λ_k. Consider $\boldsymbol{y} \in \Lambda_k$ and a neighborhood $\boldsymbol{y}+\Lambda_l$ of $\boldsymbol{y}$ where $l>k$. The group Λ_k/Λ_l is a finite discrete group: it is isomorphic to the cyclic group $Z(p^{l-k})$ under the mapping $\boldsymbol{x}+\Lambda_l \rightarrow x_k + x_{k+1}p + \cdots + x_{l-1}p^{l-k-1}$. From this we see that $\boldsymbol{x}+\Lambda_l$ generates the entire group Λ_k/Λ_l. In particular, $m\boldsymbol{x}+\Lambda_l=\boldsymbol{y}+\Lambda_l$ for some integer m, and hence $m\boldsymbol{x} \in \boldsymbol{y}+\Lambda_l$. Therefore $\{m\boldsymbol{x}\}_{m=-\infty}^{\infty}$ is dense in Λ_k and $\Lambda_k \subset H$. Since $H \neq \Omega_p$, there is a least integer j such that $\Lambda_j \subset H$. The foregoing implies at once that $H \subset \Lambda_j$.

Now consider $\Omega_{\boldsymbol{a}}$, where $\boldsymbol{a} \neq (\ldots, p, p, p, \ldots)$ for all primes p. Suppose that a_k is a composite number for some integer k: $a_k=rs$ for positive integers r and s greater than 1. Consider the element $\boldsymbol{x}$ such that $x_k=r$ and $x_n=0$ for $n \neq k$, and the smallest closed subgroup H of $\Omega_{\boldsymbol{a}}$ containing $\boldsymbol{x}$. Then we have $H=\bigcup_{j=0}^{s-1}(j\boldsymbol{x}+\Lambda_{k+1})$, so that $\Lambda_{k+1} \subsetneqq H \subsetneqq \Lambda_k$. Suppose finally that a_n is prime for all $n \in Z$. Choose k so that $a_k \neq a_{k+1}$. The numbers a_k and a_{k+1} are relatively prime. Hence there is an integer m such that $0 \leqq m < a_{k+1}$ and $a_k m + 1 = 0 \pmod{a_{k+1}}$. Let $\boldsymbol{x}$ be the element such that $x_k=1$, $x_{k+1}=m$, and $x_n=0$ for all other n in Z. It is easy to see that $a_k\boldsymbol{x}$ belongs to Λ_{k+2}. It follows that the group Γ in Λ_k/Λ_{k+2} generated by $\boldsymbol{x}+\Lambda_{k+2}$ has at most a_k elements. Plainly Λ_k/Λ_{k+2} itself has $a_k a_{k+1}$ elements, and therefore Γ is a proper subgroup of Λ_k/Λ_{k+2}. It follows that $H=\bigcup_{j=0}^{a_k-1}(j\boldsymbol{x}+\Lambda_{k+2})$ is a compact open subgroup of $\Omega_{\boldsymbol{a}}$, and it is a *proper* subgroup of Λ_k. Since $\boldsymbol{x}$ belongs to H and not to Λ_{k+1}, we infer that H is equal to no Λ_n.]

(b) Some groups $\Omega_{\boldsymbol{a}}$ contain proper closed subgroups that are not open. [For example, let $\boldsymbol{a}$ be defined by $a_n=2$ for $n \leqq 1$ and $a_n=3$ for $n \geqq 2$. Let $\boldsymbol{x}$ be defined by $x_n=0$ for $n \leqq 0$ and $x_n=1$ for $n \geqq 1$. Then $\{\boldsymbol{0}, \boldsymbol{x}\}$ is a closed subgroup of $\Omega_{\boldsymbol{a}}$ since $\boldsymbol{x}+\boldsymbol{x}=\boldsymbol{0}$.]

(c) No group $\Omega_{\boldsymbol{a}}$ is compactly generated, although every element of every $\Omega_{\boldsymbol{a}}$ is a compact element; see Definition (9.9).

Notes

The construction of $\Omega_{\boldsymbol{a}}$ and $\Delta_{\boldsymbol{a}}$ goes back in all essentials to K. HENSEL [1], who explicitly constructed $\Omega_{\boldsymbol{r}}$ and $\Delta_{\boldsymbol{r}}$. A construction more general than that of $\Omega_{\boldsymbol{a}}$ and $\Delta_{\boldsymbol{a}}$ was given by H. PRÜFER [1], in a form different from that found in the text. J. VON NEUMANN [1] recast PRÜFER's construction in a form that closely resembles our special case. D. VAN DANTZIG [4] has explored generalizations of the construction of $\Delta_{\boldsymbol{a}}$, and in [5] has given a wealth of details about $\Delta_{\boldsymbol{a}}$ for $\boldsymbol{a}=(2, 3, 4, \ldots)$. VAN DANTZIG [1] contains the definition of $\Sigma_{\boldsymbol{a}}$ for $\boldsymbol{a}=(n, n, n, \ldots)$, and the construction of $\Sigma_{\boldsymbol{a}}$ in general is given in VAN DANTZIG [6], p. 419.

Chapter Three

Integration on locally compact spaces

Harmonic analysis on a locally compact group is the study of certain spaces and algebras of functions and measures defined on the group. The function spaces in question are defined by measurability and integrability conditions; manipulation of the relevant measures requires delicate techniques from measure and integration theory. Also, the fundamental theorems of the theory of representations of locally compact groups [see §22] depend wholly upon the study of a certain space of square integrable functions. Thus the theories of measure and integration play a vital rôle in the present book.

There are many treatments of integration theory in the extant literature. None of those available to us is precisely what we need. Integration theory *per se* is not our main theme; but we must be sure of the validity of the theorems from integration theory on which our subsequent analysis rests, as applied in the very general situation presented by a locally compact group. Therefore we have set down the constructions and proved the theorems from integration theory that are needed in the sequel. We also entertain a hope that the present chapter may prove useful to students who wish a rapid introduction to integration theory at our level.

We shall assume as known the rudiments of abstract measure and integration theory. See for example SAKS [1], Chapters I—III, or HALMOS [2], Chapters I—V.

§ 11. Extension of a linear functional and construction of a measure

For the sake of completeness, we first repeat a number of definitions and review a few technicalities.

(11.1) Notation and terminology. Let Y denote an arbitrary nonvoid set. A nonvoid family $\mathcal{S}$ of subsets of Y is called a *ring* [or *ring of sets*] if $A \cup B \in \mathcal{S}$ and $A \cap B' \in \mathcal{S}$ whenever $A, B \in \mathcal{S}$. If the set Y is in the ring $\mathcal{S}$, then $\mathcal{S}$ is called an *algebra* [*of sets*]. A *σ-ring* [*σ-algebra*] *of sets* is a ring [algebra] such that $\bigcup_{k=1}^{\infty} A_k \in \mathcal{S}$ whenever $\{A_1, A_2, \ldots, A_n, \ldots\} \subset \mathcal{S}$.

If Y is a topological space, and $\mathcal{O}$ is the family of all open subsets of Y, then the family of *Borel sets* in Y is defined as the smallest σ-algebra of sets containing $\mathcal{O}$. Let $\mathcal{P}$ be the family of all subsets of Y of the form $\{y \in Y : f(y) > 0\}$ for some real-valued continuous function f on Y. The smallest σ-algebra of sets containing $\mathcal{P}$ is called the family of *Baire sets* in Y. If Y is a normal space, then $\mathcal{P}$ is exactly the family of open subsets of Y that are unions of countably many closed sets[1].

A *finitely additive measure* μ is a nonnegative extended real-valued set function defined on a ring $\mathcal{S}$ of subsets of a nonvoid set Y such that:

(i) $\mu(\varnothing) = 0$;

(ii) $\mu\left(\bigcup_{k=1}^{n} A_k\right) = \sum_{k=1}^{n} \mu(A_k)$ if each $A_k \in \mathcal{S}$ and the A_k's are pairwise disjoint.

If property (ii) can be replaced by the stronger property

(iii) $\mu\left(\bigcup_{k=1}^{\infty} A_k\right) = \sum_{k=1}^{\infty} \mu(A_k)$ if each $A_k \in \mathcal{S}$, $\bigcup_{k=1}^{\infty} A_k \in \mathcal{S}$, and the A_k's are pairwise disjoint,

then μ is called a *countably additive measure* or simply a *measure*. A complex-valued set function μ satisfying (i) and (iii) [in (iii), the series is required to be absolutely convergent] is called a *complex measure*. Given a nonnegative set function μ on a ring $\mathcal{S}$, a subset A of Y is said to be *σ-finite* if it can be written as $\bigcup_{k=1}^{\infty} A_k$ where each $A_k \in \mathcal{S}$ and $\mu(A_k) < \infty$.

Let $\mathcal{S}$ be a ring of subsets of Y. An extended real-valued function f on Y is *$\mathcal{S}$-measurable* if the set $\{x \in Y : f(x) > \alpha\}$ belongs to $\mathcal{S}$ for every $\alpha \in R$. A complex-valued function f on Y is *$\mathcal{S}$-measurable* if its real and imaginary parts are $\mathcal{S}$-measurable. If Y is a topological space and $\mathcal{S}$ is the σ-algebra of Borel [Baire] sets of Y, an $\mathcal{S}$-measurable function is called *Borel* [*Baire*] *measurable*.

---

[1] Observe that our definitions of Borel and Baire sets differ from the definitions in HALMOS [2], pp. 219, 220. Our definitions are appropriate to our construction of measures, which differs in various points from HALMOS's.

Let μ be a measure on $\mathscr{S}$. Let f be *any* function on Y with values in $[0, \infty]$. Then the *Lebesgue integral* $\int_Y f(x)\,d\mu(x)$ [or simply $\int_Y f\,d\mu$] is defined as

$$\sup\left\{\sum_{k=1}^{n}[\inf(f(x): x\in A_k)]\,\mu(A_k):\{A_1,\ldots,A_n\}\right.$$
$$\left.\text{is a partition of } Y \text{ with each } A_k\in\mathscr{S}\right\}.$$

If f is any extended real-valued function, if $f^+=\max(f, 0)$ and $f^-=-\min(f, 0)$, and if either $\int_Y f^+\,d\mu$ or $\int_Y f^-\,d\mu$ is finite, then $\int_Y f\,d\mu$ is defined to be $\int_Y f^+\,d\mu-\int_Y f^-\,d\mu$.[1] If $f=f_1+if_2$ is a complex-valued function on Y, where f_1 and f_2 are real-valued, and if $\int_Y f_j\,d\mu$ is defined and finite $(j=1, 2)$, then $\int_Y f\,d\mu$ is defined to be $\int_Y f_1\,d\mu+i\int_Y f_2\,d\mu$.

(11.2) Definition. Let X be any topological space. Let $\mathfrak{C}(X)$ denote the set of all bounded complex-valued continuous functions on X. Let $\mathfrak{C}_0(X)$ denote the set of all complex-valued continuous functions f on X such that for every $\varepsilon>0$, there exists a compact subset F of X [depending upon f and ε] such that $|f(x)|<\varepsilon$ for all $x\in F'$. Let $\mathfrak{C}_{00}(X)$ denote the set of all complex-valued continuous functions f on X such that there exists a compact subset F of X [depending upon f] such that $f(x)=0$ for all $x\in F'$. For $f\in\mathfrak{C}(X)$, let $\|f\|_u=\sup\{|f(x)|:x\in X\}$.

(11.3) Remarks. It is obvious that $\mathfrak{C}(X)$ is a complex normed linear space, with pointwise linear operations: $(f+g)(x)=f(x)+g(x)$, $(\alpha f)(x)=\alpha(f(x))$. It is also obvious that $\mathfrak{C}_0(X)$ is a linear subspace of $\mathfrak{C}(X)$ and that $\mathfrak{C}_{00}(X)$ is a linear subspace of $\mathfrak{C}_0(X)$. It is elementary to show that $\mathfrak{C}(X)$ and $\mathfrak{C}_0(X)$ are complete in the metric induced by the norm $\|\ \|_u$; that is, $\mathfrak{C}(X)$ and $\mathfrak{C}_0(X)$ are complex Banach spaces. Also $\mathfrak{C}_{00}(X)$ is a dense linear subspace of $\mathfrak{C}_0(X)$. Finally, we note that $\mathfrak{C}_0(X)$ may consist of 0 alone, even for very common spaces. However, if X is a nonvoid *locally compact* Hausdorff space, then there are always nonzero functions in $\mathfrak{C}_0(X)$. If X is a compact topological space, then $\mathfrak{C}_0(X)$ coincides with the space $\mathfrak{C}(X)$. If X is a locally compact, noncompact, Hausdorff space, then we have $\mathfrak{C}_{00}(X)\subset\mathfrak{C}_0(X)\subsetneqq\mathfrak{C}(X)$; the equality $\mathfrak{C}_{00}(X)=\mathfrak{C}_0(X)$ may obtain even for locally compact, noncompact, normal spaces. However, if G is a locally compact, noncompact *group*, then we have $\mathfrak{C}_{00}(G)\subsetneqq\mathfrak{C}_0(G)$. [See (11.43.e).]

[1] We adopt the usual conventions, here and throughout this book, in dealing with the extended real numbers ∞ [sometimes written $+\infty$] and $-\infty$. That is: $t<\infty$ if $t\in R$ or $t=-\infty$; $-\infty<t$ if $t\in R$ or $t=\infty$; $t+\infty=\infty+t=\infty$ if $t\in R$ or $t=\infty$; $(-\infty)+\infty$ and $\infty+(-\infty)$ are undefined; $t-\infty=-\infty+t=-\infty$ if $t\in R$ or $t=-\infty$; $t\cdot\infty=\infty\cdot t=\infty$ if $t\in R$ and $t>0$; $0\cdot\infty=\infty\cdot 0=0\cdot(-\infty)=(-\infty)\cdot 0=0$; $t\cdot\infty=\infty\cdot t=-\infty$ if $t\in R$ and $t<0$; similar formulas for $t\cdot(-\infty)$; $(\infty)(\infty)=(-\infty)(-\infty)=\infty$ and $(\infty)(-\infty)=(-\infty)(\infty)=-\infty$.

(11.4) Definition. Let Y be any nonvoid set, and let $\mathfrak{F}(Y)$ be any linear space of complex-valued functions defined on Y. The symbol $\mathfrak{F}^r(Y)$ will denote the set of all real-valued functions in $\mathfrak{F}(Y)$; and the symbol $\mathfrak{F}^+(Y)$ will denote the set of all nonnegative functions in $\mathfrak{F}^r(Y)$. For example, if X is a topological space, $\mathfrak{C}_{00}^+(X)$ denotes the set of all nonnegative continuous functions on X each of which vanishes outside of a compact subset of X. Where no confusion can arise, we will write $\mathfrak{F}$, $\mathfrak{F}^r$, and $\mathfrak{F}^+$ for $\mathfrak{F}(Y)$, $\mathfrak{F}^r(Y)$, and $\mathfrak{F}^+(Y)$, respectively.

In our discussion of integration theory, it is advantageous to consider an arbitrary nonvoid locally compact Hausdorff space X, the linear space $\mathfrak{C}_{00}(X)$, and an arbitrary complex linear functional I defined on $\mathfrak{C}_{00}(X)$ that is nonnegative: $I(f) \geqq 0$ if $f \in \mathfrak{C}_{00}^+(X)$. To avoid an occasional solecism, we will also suppose that I is not the zero functional. **These notations will be fixed throughout Chapter Three.**

(11.5) Theorem. *For $f \in \mathfrak{C}_{00}$, we have $|I(f)| \leqq I(|f|)$.*

Proof. Write $I(f) = \varrho \exp(i\theta)$, $0 \leqq \varrho < \infty$, $0 \leqq \theta < 2\pi$, and write $\exp(-i\theta) f = g_1 + i g_2$ with $g_1, g_2 \in \mathfrak{C}_{00}^r$. Then $\varrho = \exp(-i\theta) I(f) = I(\exp(-i\theta) f) = I(g_1) + i I(g_2) = I(g_1)$, since I is obviously real for real-valued functions in $\mathfrak{C}_{00}$. We have $g_1 \leqq |g_1| \leqq |g_1 + i g_2| = |f|$,[1] and thus $|I(f)| = \varrho = I(g_1) \leqq I(|f|)$. □

(11.6) Theorem. *Let A be any compact subset of X. There is a nonnegative number β depending only upon A such that $|I(f)| \leqq \beta \|f\|_u$ for all $f \in \mathfrak{C}_{00}$ such that $f(A') = 0$.*

Proof. Consider any $\omega \in \mathfrak{C}_{00}^+$ such that $\omega(A) = 1$ and apply (11.5). □

The next result expresses a sort of continuity property enjoyed by I.

(11.7) Theorem. *Let $\mathfrak{D}$ be any nonvoid subset of $\mathfrak{C}_{00}^+$ such that for $f_1, f_2 \in \mathfrak{D}$, there is an $f_3 \in \mathfrak{D}$ for which $f_3 \leqq \min(f_1, f_2)$, and such that $\inf\{f(x) : f \in \mathfrak{D}\} = 0$ for every $x \in X$. Then for every $\varepsilon > 0$, there is an $f_\varepsilon \in \mathfrak{D}$ such that $f_\varepsilon(x) < \varepsilon$ for all $x \in X$; furthermore we have $\inf\{I(f) : f \in \mathfrak{D}\} = 0$.*

Proof. Fix $\varepsilon > 0$, and let $A_f = \{x \in X : f(x) \geqq \varepsilon\}$ for each $f \in \mathfrak{D}$. All of the sets A_f are compact, and their intersection is void. Hence $A_{f_1} \cap A_{f_2} \cap \cdots \cap A_{f_n} = \varnothing$ for some $f_1, f_2, \ldots, f_n \in \mathfrak{D}$. If $f_\varepsilon \leqq \min(f_1, f_2, \ldots, f_n)$, then $f_\varepsilon(x) < \varepsilon$ for all $x \in X$. To show that $\inf\{I(f) : f \in \mathfrak{D}\} = 0$, consider any fixed $g \in \mathfrak{D}$, any function $h \in \mathfrak{D}$ such that $h \leqq \min(g, f_\varepsilon)$, and apply (11.6) with the set $\{x \in X : g(x) > 0\}^-$ as A. □

For extending the functional I, two classes of functions are useful. We now define them.

---

[1] For real-valued functions f and g defined on an arbitrary set Y, the expression $f \leqq g$ means $f(x) \leqq g(x)$ for all $x \in Y$. Also the expression $f \neq 0$ means that f is not equal to the function identically zero: that is, f assumes somewhere on Y a value different from 0.

(11.8) Definition. A function f on X with values in $[0, \infty]$ [that is, ∞ is a possible value for f] is said to be *lower semicontinuous* if the following conditions hold at every point x_0 in X. If $f(x_0) < \infty$, then for every $\varepsilon > 0$, there is a neighborhood U of x_0 in X such that $f(x) > f(x_0) - \varepsilon$ for all $x \in U$. If $f(x_0) = \infty$, then for every positive number A, there is a neighborhood U of x_0 such that $f(x) > A$ for all $x \in U$. This set of functions is denoted by $\mathfrak{M}^+$.

(11.9) Definition. A function f on X with values in $[0, \infty]$ is said to be *upper semicontinuous* if for every $x_0 \in X$ and every $\varepsilon > 0$, there is a neighborhood U of x_0 such that $f(x) \leqq f(x_0) + \varepsilon$ for all $x \in U$. The set of all such functions is denoted by $\mathfrak{N}^+$.

We next state without proof some simple properties of functions in $\mathfrak{M}_+$.

(11.10) Theorem. (i) *If $f \in \mathfrak{M}^+$ and α is a nonnegative real number, then $\alpha f \in \mathfrak{M}^+$.*

(ii) *If $f_1, f_2, \ldots, f_m \in \mathfrak{M}^+$, then* $\min(f_1, f_2, \ldots, f_m) \in \mathfrak{M}^+$.

(iii) *If $\mathfrak{D} \subset \mathfrak{M}^+$ and $\mathfrak{D}$ is nonvoid, then* $\sup\{f : f \in \mathfrak{D}\} \in \mathfrak{M}^+$.

(iv) *If $f_1, f_2, \ldots, f_m \in \mathfrak{M}^+$, then $\sum_{j=1}^{m} f_j \in \mathfrak{M}^+$.*

(v) *If $f \in \mathfrak{M}^+$, then* $f = \sup\{g : g \in \mathfrak{C}_{00}^+, g \leqq f\}$.

We now make our first extension of the functional I.

(11.11) Definition. For $f \in \mathfrak{M}^+$, let $\bar{I}(f) = \sup\{I(g) : g \in \mathfrak{C}_{00}^+, g \leqq f\}$. [Note that $\bar{I}(f)$ may be $+\infty$.]

The following assertions are obvious.

(11.12) Theorem. (i) *If $f \in \mathfrak{C}_{00}^+$, then $f \in \mathfrak{M}^+$ and $I(f) = \bar{I}(f)$.*

(ii) *If $f_1 \leqq f_2$ and $f_1, f_2 \in \mathfrak{M}^+$, then $\bar{I}(f_1) \leqq \bar{I}(f_2)$.*

(iii) *If $f \in \mathfrak{M}^+$ and α is a nonnegative real number, then $\bar{I}(\alpha f) = \alpha \bar{I}(f)$.*

We next state and prove a continuity property of $\bar{I}$.

(11.13) Theorem. *Let $\mathfrak{D}$ be a nonvoid subset of $\mathfrak{M}^+$ such that for $f_1, f_2 \in \mathfrak{D}$, there is an $f \in \mathfrak{D}$ such that $f \geqq \max(f_1, f_2)$. Then $\bar{I}(\sup\{f : f \in \mathfrak{D}\}) = \sup\{\bar{I}(f) : f \in \mathfrak{D}\}$. In particular, if $f_1, f_2, \ldots, f_n, \ldots \in \mathfrak{M}^+$ and $f_1 \leqq f_2 \leqq \cdots \leqq f_n \leqq \cdots$, then $\bar{I}(\lim_{n\to\infty} f_n) = \lim_{n\to\infty} \bar{I}(f_n)$.*

Proof. Let $f_0 = \sup\{f : f \in \mathfrak{D}\}$. If $\{f_0\} \cup \mathfrak{D} \subset \mathfrak{C}_{00}^+$, the present theorem follows from (11.7), since $\{f_0 - f : f \in \mathfrak{D}\}$ satisfies the hypotheses of (11.7), and $0 = \inf\{I(f_0 - f) : f \in \mathfrak{D}\} = \inf\{I(f_0) - I(f) : f \in \mathfrak{D}\} = I(f_0) - \sup\{I(f) : f \in \mathfrak{D}\}$.

In the general case, it obviously suffices to prove that $\bar{I}(f_0) \leqq \sup\{\bar{I}(f) : f \in \mathfrak{D}\}$. Let $\mathfrak{E} = \{g \in \mathfrak{C}_{00}^+ : g \leqq f$ for some $f \in \mathfrak{D}\}$. Let g_0 be any function in $\mathfrak{C}_{00}^+$ such that $g_0 \leqq f_0$. By (11.10.v), we have

$$f_0 = \sup\{\sup[g \in \mathfrak{C}_{00}^+ : g \leqq f] : f \in \mathfrak{D}\} = \sup\{g : g \in \mathfrak{E}\},$$

and accordingly

$$g_0 = \min(g_0, f_0) = \sup\{\min(g_0, g) : g \in \mathfrak{E}\}.$$

By the previous case, we have

$$I(g_0) = \sup\{I(\min(g_0, g)) : g \in \mathfrak{E}\} \leq \sup\{I(g) : g \in \mathfrak{E}\} = \sup\{\bar{I}(f) : f \in \mathfrak{D}\}. \quad \square$$

(11.14) Corollary. *If $f_1, f_2 \in \mathfrak{M}^+$, then $\bar{I}(f_1 + f_2) = \bar{I}(f_1) + \bar{I}(f_2)$.*

Proof. Note that $f_1 + f_2 = \sup\{g_1 + g_2 : g_j \in \mathfrak{C}_{00}^+,\ g_j \leq f_j \text{ for } j = 1, 2\}$; then apply (11.13) and the additivity of I. $\square$

(11.15) Corollary. *For $\mathfrak{D} \subset \mathfrak{M}^+$, we have $\bar{I}(\sum_{f \in \mathfrak{D}} f) = \sum_{f \in \mathfrak{D}} \bar{I}(f)$.*[1]

Proof. For finite $\mathfrak{D}$, this follows from (11.14); for infinite $\mathfrak{D}$, use (11.13). $\square$

We now extend $\bar{I}$ to *all* nonnegative functions defined on X.

(11.16) Definition. Let h be any function defined on X and assuming values in $[0, \infty]$. [In the present discussion, such functions will be called nonnegative, and the set of all such functions will be denoted by $\mathfrak{F}$.] Then the [perhaps infinite] number $\bar{\bar{I}}(h)$ is defined as $\inf\{\bar{I}(g) : g \in \mathfrak{M}^+, g \geq h\}$.

(11.17) Theorem. (i) *If $g \in \mathfrak{M}^+$, then $\bar{I}(g) = \bar{\bar{I}}(g)$.*

(ii) *If $h_1, h_2 \in \mathfrak{F}$ and $h_1 \leq h_2$, then $\bar{\bar{I}}(h_1) \leq \bar{\bar{I}}(h_2)$.*

(iii) *If $h \in \mathfrak{F}$ and $0 \leq \alpha < \infty$, then $\bar{\bar{I}}(\alpha h) = \alpha \bar{\bar{I}}(h)$.*

(iv) *If $h_1, h_2 \in \mathfrak{F}$, then $\bar{\bar{I}}(h_1 + h_2) \leq \bar{\bar{I}}(h_1) + \bar{\bar{I}}(h_2)$.*[2]

We omit the proofs of these simple assertions.

(11.18) Theorem. *If $\{h_n\}_{n=1}^{\infty} \subset \mathfrak{F}$ and $h_1 \leq h_2 \leq \cdots \leq h_n \leq \cdots$, then $\bar{\bar{I}}(\lim_{n\to\infty} h_n) = \lim_{n\to\infty} \bar{\bar{I}}(h_n)$.*

Proof. It suffices to prove that $\bar{\bar{I}}(\lim_{n\to\infty} h_n) \leq \lim_{n\to\infty} \bar{\bar{I}}(h_n)$, and we may suppose that $\lim_{n\to\infty} \bar{\bar{I}}(h_n) < \infty$. For an arbitrary positive number ε, let $g_n \geq h_n$ be a function in $\mathfrak{M}^+$ such that

$$\bar{I}(g_n) - \varepsilon \cdot 2^{-n} < \bar{\bar{I}}(h_n) \quad (n = 1, 2, 3, \ldots). \tag{1}$$

---

[1] If Y is a nonvoid set and α is a function defined on Y with values in $[0, \infty]$, then $\sum_{y \in Y} \alpha(y)$ is defined as the supremum of all sums $\sum_{y \in Y_1} \alpha(y)$, as Y_1 runs through all finite subsets of Y.

[2] The functional $\bar{\bar{I}}$ is not in general additive. For example, let $X = [0, 1]$ and $I(f) = \int_0^1 f(t)\,dt$. If h_1 is the characteristic function of a set $A \subset [0, 1]$ that is not Lebesgue measurable and h_2 is the characteristic function of $[0, 1] \cap A'$, then $\bar{\bar{I}}(h_1 + h_2) = 1 < \bar{\bar{I}}(h_1) + \bar{\bar{I}}(h_2)$. In some cases, of course, $\bar{\bar{I}}$ is additive on all of $\mathfrak{F}$.

Let $g'_n = \max(g_1, g_2, \ldots, g_n)$ $(n=1, 2, 3, \ldots)$. Then we clearly have

$$g'_{n+1} + \min(g'_n, g_{n+1}) = g'_n + g_{n+1},$$

and from (11.14), (11.12.ii), and (1) we obtain

$$\left.\begin{aligned} \bar{I}(g'_{n+1}) &= \bar{I}(g'_n) + \bar{I}(g_{n+1}) - \bar{I}(\min(g'_n, g_{n+1})) \\ &\leqq \bar{I}(g'_n) + \bar{I}(g_{n+1}) - \bar{\bar{I}}(h_n) < \bar{I}(g'_n) + \bar{\bar{I}}(h_{n+1}) - \bar{\bar{I}}(h_n) + \varepsilon \cdot 2^{-n-1}. \end{aligned}\right\} \quad (2)$$

Adding (2) over $n=1, 2, \ldots, m-1$, and taking account of (1), we have

$$\bar{I}(g'_m) < \bar{\bar{I}}(h_m) + \varepsilon \qquad (m=1, 2, 3, \ldots). \tag{3}$$

The present theorem follows from (3) and (11.13). □

(11.19) Corollary. *If $h_n \in \mathfrak{F}$ for $n=1, 2, 3, \ldots$, then $\bar{\bar{I}}\left(\sum_{n=1}^{\infty} h_n\right) \leqq \sum_{n=1}^{\infty} \bar{\bar{I}}(h_n)$.*

This follows at once from (11.18).

(11.20) Definition. For each subset A of X, let $\iota(A) = \bar{\bar{I}}(\xi_A)$.

(11.21) Theorem. *The set function ι is a Carathéodory outer measure on $\mathscr{P}(X)$. That is:*

(i) $0 \leqq \iota(A) \leqq \infty$ *for all* $A \subset X$;

(ii) $\iota(A) \leqq \iota(B)$ *if* $A \subset B \subset X$;

(iii) $\iota(\varnothing) = 0$;

(iv) $\iota\left(\bigcup_{n=1}^{\infty} A_n\right) \leqq \sum_{n=1}^{\infty} \iota(A_n)$ *if* $A_1, A_2, \ldots, A_n, \ldots \in \mathscr{P}(X)$.

Proof. Assertions (i) and (iii) are obvious; (ii) is a special case of (11.17.ii) and (iv) of (11.19). □

(11.22) Theorem. *For every $A \subset X$, $\iota(A) = \inf\{\iota(U) : U$ is open and $U \supset A\}$.*

Proof. The assertion is obvious if $\iota(A) = \infty$. If $\iota(A) < \infty$ and $0 < \varepsilon < 1$, there is a function $g \in \mathfrak{M}^+$ such that $g \geqq \xi_A$ and $\bar{I}(g) - \varepsilon < \iota(A)$. Let $U = \{x \in X : g(x) > 1 - \varepsilon\}$. Since g is lower semicontinuous, U is open; and we have $\frac{1}{(1-\varepsilon)} g \geqq \xi_U$. Thus $\iota(U) = \bar{\bar{I}}(\xi_U) \leqq \bar{I}\left(\frac{1}{(1-\varepsilon)} g\right) = \frac{1}{(1-\varepsilon)} \bar{I}(g)$ $< \frac{1}{(1-\varepsilon)} [\iota(A) + \varepsilon]$. Since ε is arbitrary, the proof is complete. □

(11.23) Theorem. *Let U be an open subset of X. Then $\iota(U) = \sup\{\iota(F) : F$ is compact and $F \subset U\} = \sup\{\iota(V) : V$ is open, V^- is compact, and $V^- \subset U\}$.*

Proof. Suppose that $\beta < \iota(U)$. Let $f \in \mathfrak{C}_{00}^+$ be such that $f \leqq \xi_U$ and $I(f) > \beta$. For $0 < \alpha < 1$, let $F_\alpha = \{x \in X : f(x) \geqq \alpha\}$ and $V_\alpha = \{x \in X : f(x) > \alpha\}$. Let $V = \{x \in X : f(x) > 0\}$. Plainly we have $V \subset U$, $\iota(V) \leqq \iota(U)$, and $I(f) \leqq \iota(V)$. By (11.20) and (11.18), we have $\beta < I(f) \leqq \iota(V) = \bar{\bar{I}}(\lim_{\alpha \downarrow 0} \xi_{F_\alpha}) =$

$\bar{\bar{I}}(\lim\limits_{\alpha\downarrow 0} \xi_{V_\alpha}) = \lim\limits_{\alpha\downarrow 0} \iota(F_\alpha) = \lim\limits_{\alpha\downarrow 0} \iota(V_\alpha)$ so that $\iota(F_\alpha) > \beta$ and $\iota(V_\alpha) > \beta$ for α sufficiently small. □

(11.24) Theorem. *If $A \subset X$ and A is contained in a compact set F, then $\iota(A) < \infty$.*

Proof. There is a function $f \in \mathfrak{C}_{00}^+ \subset \mathfrak{M}^+$ such that $f \geqq \xi_F$. Thus $\iota(A) \leqq \iota(F) \leqq \bar{I}(f) = I(f) < \infty$. □

(11.25) Theorem. *Let $\mathcal{U}$ be any family of pairwise disjoint open sets. Then $\iota(\cup\{U : U \in \mathcal{U}\}) = \sum\limits_{U \in \mathcal{U}} \iota(U)$. Also, there is a unique closed subset E of X such that $\iota(E') = 0$ and such that $\iota(E \cap U) > 0$ if U is open in X and $U \cap E \neq \varnothing$. [The set E is called the* support *of ι and is denoted by the symbol $S(\iota)$.]*

Proof. The first assertion follows at once from (11.15), since ξ_A is lower semicontinuous if A is open. To prove the second, let W be the union of all open sets of ι-measure 0. Corollary (11.15) implies that $\iota(W) = 0$. Let $E = W'$. Then if U is open and $U \cap E \neq \varnothing$, we have $\iota(U) > 0$, and $\iota(U) \leqq \iota(E \cap U) + \iota(W \cap U) = \iota(E \cap U)$. The uniqueness of E is easy to show; we omit the argument. □

(11.26) Definition. If $A \subset X$ and $\iota(A) = 0$, we call A an *ι-null* set. If $A \cap F$ is ι-null for every compact subset F of X, then A is said to be a *locally ι-null* set. The family of locally ι-null subsets of X will be denoted by $\mathcal{N}_\iota$.[1] A property that holds on X except for an ι-null set is said to hold *ι-almost everywhere.* A property that holds on X except for a locally ι-null set is said to hold *locally ι-almost everywhere.* A complex- or extended real-valued function f on X that vanishes ι-almost everywhere [locally ι-almost everywhere] is called an *ι-null* [*locally ι-null*] function. In using these terms we shall often drop the prefix ι- where no confusion can occur.

(11.27) Theorem. *Let h be a nonnegative extended real-valued function on X. Then $\bar{\bar{I}}(h) = 0$ if and only if the set $A = \{x \in X : h(x) > 0\}$ is a null set. If $\bar{\bar{I}}(h) < \infty$, then $B = \{x \in X : h(x) = \infty\}$ is a null set. Finally, if $\{h_n\}_{n=1}^\infty$ is a sequence of nonnegative functions on X such that $\lim\limits_{n\to\infty} \bar{\bar{I}}(h_n) = 0$, then there is a subsequence $\{h_{n_k}\}_{k=1}^\infty$ of $\{h_n\}_{n=1}^\infty$ such that $\lim\limits_{k\to\infty} h_{n_k}(x) = 0$ almost everywhere on X.*

Proof. Suppose that $\bar{\bar{I}}(h) = 0$. Then we have $\xi_A \leqq \lim\limits_{n\to\infty} nh$ and so by (11.18),

$$\iota(A) = \bar{\bar{I}}(\xi_A) \leqq \bar{\bar{I}}(\lim_{n\to\infty} nh) = \lim_{n\to\infty} \bar{\bar{I}}(nh) = \lim_{n\to\infty} n\bar{\bar{I}}(h) = 0.$$

If $\iota(A) = 0$, then $h \leqq \lim\limits_{n\to\infty} n\xi_A$ and $\bar{\bar{I}}(h) = 0$ by the same computation.

[1] Observe that $\mathcal{N}_\iota$ is closed under the formation of countable unions and arbitrary subsets.

Suppose now that $\bar{\bar{I}}(h) < \infty$. Then for $\varepsilon > 0$, we have $\xi_B \leqq \varepsilon h$ so that $\iota(B) = \bar{\bar{I}}(\xi_B) \leqq \varepsilon \bar{\bar{I}}(h)$. That is, $\iota(B) = 0$.

Finally, suppose that $\lim\limits_{n\to\infty} \bar{\bar{I}}(h_n) = 0$. Choose $n_1 < n_2 < \cdots$ such that $\sum\limits_{k=1}^{\infty} \bar{\bar{I}}(h_{n_k}) < \infty$. Then by (11.19), we have $\bar{\bar{I}}\left(\sum\limits_{k=1}^{\infty} h_{n_k}\right) < \infty$. The previous paragraph shows that $\sum\limits_{k=1}^{\infty} h_{n_k}(x) < \infty$ almost everywhere and hence $\lim\limits_{k\to\infty} h_{n_k}(x) = 0$ almost everywhere. □

We adopt the classical definition of measurability for subsets of X, due to CARATHÉODORY.

(11.28) Definition. A subset A of X is said to be *ι-measurable* if, for every subset S of X, the inequality

(i) $\iota(S) \geqq \iota(S \cap A) + \iota(S \cap A')$

obtains. The family of all ι-measurable subsets of X is denoted by $\mathcal{M}_\iota$. A function on X is said to be *ι-measurable* if it is $\mathcal{M}_\iota$-measurable. In using these terms we shall often drop the prefix ι- where no confusion can arise.

(11.29) Theorem. *The family $\mathcal{M}_\iota$ is a σ-algebra of subsets of X, and the set function ι is countably additive on $\mathcal{M}_\iota$.*

All statements of this theorem hold for every Carathéodory outer measure. Proofs may be found, for example, in HALMOS [2], pp. 44–47, and in SAKS [1], pp. 43–46. We shall not repeat these proofs.

(11.30) Theorem. *Every Borel subset of X is measurable, and every locally null set is measurable.*

Proof. (I) We first show that a subset B of X is measurable provided that

$$\iota(U) \geqq \iota(U \cap B) + \iota(U \cap B') \tag{1}$$

for all *open* subsets U of X such that $\iota(U) < \infty$. Let S be any subset of X. If $\iota(S) = \infty$, then clearly $\iota(S) \geqq \iota(S \cap B) + \iota(S \cap B')$. If $\iota(S) < \infty$, then for open sets $U \supset S$ such that $\iota(U) < \infty$, we have $\iota(U) \geqq \iota(U \cap B) + \iota(U \cap B') \geqq \iota(S \cap B) + \iota(S \cap B')$. Hence by (11.22), we have

$$\iota(S) = \inf\{\iota(U) : U \text{ is open and } U \supset S\} \geqq \iota(S \cap B) + \iota(S \cap B'),$$

and B is measurable.

(II) Let V be an open set. To show that V is measurable, let U be an arbitrary open set such that $\iota(U) < \infty$. Let ε be an arbitrary positive number. Use (11.22) to choose an open subset H of X such that $H \supset U \cap V'$ and

$$\iota(H) < \iota(U \cap V') + \frac{\varepsilon}{2}\,. \tag{2}$$

Applying (11.23), choose an open set W such that $W^- \subset U \cap V$ and

$$\iota(W) + \frac{\varepsilon}{2} > \iota(U \cap V). \qquad (3)$$

Let $W_0 = U \cap H \cap (W^-)'$; then W and W_0 are disjoint open sets. Since $U \cap V' \subset W_0 \subset H$, (2) shows that $|\iota(W_0) - \iota(U \cap V')| < \frac{\varepsilon}{2}$. Combining this with (3), we obtain $|\iota(W) + \iota(W_0) - (\iota(U \cap V) + \iota(U \cap V'))| < \varepsilon$. Using this and (11.25), we have $\iota(U) \geqq \iota(W \cup W_0) = \iota(W) + \iota(W_0) \geqq \iota(U \cap V) + \iota(U \cap V') - \varepsilon$. As ε is arbitrary, (1) holds with $B = V$, so that V is measurable.

The foregoing and (11.29) show that all Borel sets are measurable.

(III) Finally, let A be a locally null set and U any open set for which $\iota(U) < \infty$. For $\varepsilon > 0$, let F be a compact subset of X such that $F \subset U$ and $\iota(F) + \varepsilon > \iota(U)$ (11.23). Then $\iota(U) = \iota(F) + \iota(U \cap F')$ [U and F are measurable], so that $\iota(U \cap F') < \varepsilon$. Thus $\iota(U \cap A) \leqq \iota(F \cap A) + \iota(U \cap F' \cap A) \leqq 0 + \iota(U \cap F') < \varepsilon$. Hence $\iota(U \cap A) = 0$, so that $\iota(U \cap A) + \iota(U \cap A') = \iota(U \cap A') \leqq \iota(U)$. □

(11.31) Theorem. *Let A be a subset of X. The following statements are equivalent:*

(i) *A is measurable;*

(ii) *$\iota(U) \geqq \iota(U \cap A) + \iota(U \cap A')$ for all open subsets U of X such that $\iota(U) < \infty$;*

(iii) *$A \cap U$ is measurable for all open subsets U of X such that $\iota(U) < \infty$;*

(iv) *$A \cap F$ is measurable for all compact subsets F of X.*

Proof. Clearly (i) implies (iv). Suppose that (iv) holds and let U be an open subset of X such that $\iota(U) < \infty$. By (11.23), we see that $U = \left(\bigcup_{n=1}^{\infty} F_n\right) \cup N$ where each F_n is compact and $\iota(N) = 0$. It follows that $A \cap U = \left(\bigcup_{n=1}^{\infty} (F_n \cap A)\right) \cup (N \cap A)$ is measurable. Thus (iii) holds.

It is trivial that (iii) implies (ii). The fact that (ii) implies (i) is proved in Part (I) of (11.30). □

(11.32) Theorem. *Let A be a measurable set of finite measure. Then*

(i) *$\iota(A) = \sup\{\iota(F) : F$ is compact and $F \subset A\}$.*

Proof. For $\varepsilon > 0$, choose first an open set $V \supset A$ such that $\iota(V \cap A') < \frac{\varepsilon}{4}$. Then apply (11.23) to select a compact subset E of V such that $\iota(V \cap E') < \frac{\varepsilon}{2}$. Finally choose an open set W such that $V \supset W \supset V \cap A'$ and $\iota(W) < \frac{\varepsilon}{2}$. Then define $F = E \cap W'$. Obviously F is compact and $F \subset A$. Also we have $\iota(A \cap F' =) \iota((A \cap E') \cup (A \cap W)) \leqq \iota(V \cap E') + \iota(W) < \frac{\varepsilon}{2} + \frac{\varepsilon}{2}$, and hence $\iota(F) > \iota(A) - \varepsilon$. □

(11.33) Note. Theorem (11.32) holds if $A \in \mathcal{M}_\iota$ and A is contained in the union of a countable family of sets of finite measure [apply (11.18) and (11.32)]. However, (11.32.i) does not universally hold for measurable sets. Consider for example the space R^2 with the topology in which sets $\{(x_0, y): \alpha < y < \beta\}$ [x_0 fixed] are an open basis. Plainly this is the topology of $R_d \times R$, where R_d denotes the *discrete* reals[1]. Thus $R_d \times R$ is a locally compact Abelian group. For $f \in \mathfrak{C}_{00}(R_d \times R)$, let $\{x_1, \ldots, x_m\}$ be the set of numbers x such that $f(\{x\} \times R) \neq 0$. Let $I(f) = \sum_{j=1}^{m} \int_{-\infty}^{\infty} f(x_j, y)\, dy$, where $\int_{-\infty}^{\infty} \cdots dy$ denotes ordinary Riemann integration. The closed and discrete subgroup $R_d \times \{0\}$ has curious measure-theoretic structure. It is easy to see that for $A \subset R_d \times \{0\}$, $\iota(A) = 0$ if A is countable and $\iota(A) = \infty$ if A is uncountable. The set $R_d \times \{0\}$ is measurable since it is closed. Since the only compact subsets of $R_d \times \{0\}$ are finite, (11.32) fails for $R_d \times \{0\}$. Note also that $R_d \times \{0\}$ is locally null but not null.

This $0-\infty$ phenomenon persists for *all* locally null sets and arbitrary X and I. That is, every locally null set $A \subset X$ has measure 0 or ∞. The set A is measurable (11.30). If $\iota(A) < \infty$, then $\iota(A) = \sup\{\iota(F): F$ is compact and $F \subset A\}$ (11.32). Hence $\iota(A) = 0$. Note that a measurable set A is locally null if and only if $\iota(F) = 0$ for all compact sets $F \subset A$.

(11.34) Definition. Let $\mathcal{S}$ be a σ-algebra of subsets of X containing all open sets and let μ be a finitely additive measure defined on $\mathcal{S}$. Suppose that: for every open set V, we have $\mu(V) = \sup\{\mu(F): F$ is compact and $F \subset V\}$; for all $A \in \mathcal{S}$, we have $\mu(A) = \inf\{\mu(V): V$ is open and $V \supset A\}$. Then μ is said to be *regular*.

By repeating *verbatim* the proof of (11.32) with ι replaced by μ, we see that if $A \in \mathcal{S}$ and $\mu(A) < \infty$, then $\mu(A) = \sup\{\mu(F): F$ is compact and $F \subset A\}$. Theorems (11.29), (11.22), (11.23), and (11.30) can be combined in the statement that ι is a countably additive, regular measure on a σ-algebra of subsets of X containing all open sets and all sets of ι-measure 0.

Our principal reason for constructing ι from I is to represent I as the integral $\int_X f\, d\iota$. We now show that this can be done.

(11.35) Lemma. *Let A and B be disjoint measurable subsets of X and α and β nonnegative real numbers. Then*

$$\bar{\bar{I}}(\alpha \xi_A + \beta \xi_B) = \alpha \bar{\bar{I}}(\xi_A) + \beta \bar{\bar{I}}(\xi_B).$$

Proof. In view of (11.17.iv) and (11.17.iii), it suffices to show that

$$\bar{\bar{I}}(\alpha \xi_A + \beta \xi_B) \geqq \alpha \bar{\bar{I}}(\xi_A) + \beta \bar{\bar{I}}(\xi_B). \tag{1}$$

---

[1] Given a topological group G, G_d will always denote this group with the discrete topology.

The cases $\iota(A)=\infty$, $\iota(A)=0$, $\iota(B)=\infty$, $\iota(B)=0$, $\alpha=0$ and $\beta=0$ are trivial. Thus we may suppose that $0<\iota(A)<\infty$, $0<\iota(B)<\infty$, $\alpha>0$, and $\beta>0$.

Suppose that A and B are *compact*. Then A and B are contained in disjoint open sets U_0 and V_0, respectively, which we may obviously take to be of finite measure. It is also clear that $\bar{\bar{I}}(\alpha\xi_A+\beta\xi_B)$ is finite. Let ε be an arbitrary positive number. Then there is an $f\in\mathfrak{M}^+$ such that $f\geqq\alpha\xi_A+\beta\xi_B$ and

$$\bar{I}(f)-\frac{\varepsilon}{3}<\bar{\bar{I}}(\alpha\xi_A+\beta\xi_B). \tag{2}$$

Let δ be a number such that $0<\delta<\min\left\{\frac{\varepsilon}{3\iota(U_0)}, \frac{\varepsilon}{3\iota(V_0)}, \alpha, \beta\right\}$. Since f is lower semicontinuous, there are open sets U and V such that $A\subset U\subset U_0$, $B\subset V\subset V_0$, $f(x)>\alpha-\delta$ for $x\in U$, and $f(x)>\beta-\delta$ for $x\in V$. Thus we have $f\geqq(\alpha-\delta)\,\xi_U+(\beta-\delta)\,\xi_V$ [note that $U\cap V=\varnothing$!], and the functions $(\alpha-\delta)\,\xi_U$ and $(\beta-\delta)\,\xi_V$ are in $\mathfrak{M}^+$. Applying (11.14) and other properties already established, we obtain

$$\left.\begin{aligned}\bar{I}(f)&\geqq\bar{I}\big((\alpha-\delta)\,\xi_U+(\beta-\delta)\,\xi_V\big)=(\alpha-\delta)\bar{I}(\xi_U)+(\beta-\delta)\bar{I}(\xi_V)\\&=(\alpha-\delta)\,\iota(U)+(\beta-\delta)\,\iota(V)\geqq\alpha\,\iota(A)+\beta\,\iota(B)\\&\quad-\big(\delta\,\iota(U_0)+\delta\,\iota(V_0)\big)>\alpha\bar{\bar{I}}(\xi_A)+\beta\bar{\bar{I}}(\xi_B)-\frac{2\varepsilon}{3}.\end{aligned}\right\} \tag{3}$$

Combining (2) and (3) and recalling that ε is arbitrary, we obtain (1).

Consider now arbitrary A and B and $\varepsilon>0$. By (11.32), there are compact sets E and F such that $E\subset A$, $F\subset B$, $\alpha\,\iota(E)>\alpha\,\iota(A)-\frac{\varepsilon}{2}$, and $\beta\,\iota(F)>\beta\,\iota(B)-\frac{\varepsilon}{2}$. Making use of (1) for compact sets, we have

$$\bar{\bar{I}}(\alpha\xi_A+\beta\xi_B)\geqq\bar{\bar{I}}(\alpha\xi_E+\beta\xi_F)=\alpha\bar{\bar{I}}(\xi_E)+\beta\bar{\bar{I}}(\xi_F)>\alpha\bar{\bar{I}}(\xi_A)+\beta\bar{\bar{I}}(\xi_B)-\varepsilon. \quad \square$$

(11.36) Theorem. *Let f be any nonnegative measurable function on X. Then $\bar{\bar{I}}(f)=\int_X f\,d\iota$.*

Proof. Let $A_\infty=\{x\in X: f(x)=\infty\}$. For every positive integer n, and for $k=1, 2, \ldots, n\cdot 2^n-1$, let $A_{k,n}=\left\{x\in X: \frac{k}{2^n}\leqq f(x)<\frac{k+1}{2^n}\right\}$. Define the function h_n by $h_n=\sum_{k=1}^{n2^n-1}\frac{k}{2^n}\,\xi_{A_{k,n}}+n\,\xi_{A_\infty}$. Then we have $h_1\leqq h_2\leqq\cdots\leqq h_n\leqq\cdots$ and $\lim_{n\to\infty}h_n(x)=f(x)$ for all $x\in X$; also $\int_X f\,d\iota=\lim_{n\to\infty}\int_X h_n\,d\iota$ by the theorem on integrating monotone increasing sequences, and $\bar{\bar{I}}(f)=\lim_{n\to\infty}\bar{\bar{I}}(h_n)$ by (11.18). Finally, by (11.35), we have

$$\bar{\bar{I}}(h_n)=\sum_{k=1}^{n2^n-1}\frac{k}{2^n}\,\iota(A_{k,n})+n\,\iota(A_\infty)=\int_X h_n\,d\iota. \quad \square$$

(11.37) Corollary [F. RIESZ'S representation theorem]. *For every $f \in \mathfrak{C}_{00}^{+}$, we have $I(f) = \int_X f\,d\iota$.*

(11.38) Note. Corollary (11.37) asserts nothing about the *uniqueness* of the measure ι that gives the integral representation for I. There are pathological cases in which two different measures yield the same integral for all functions in $\mathfrak{C}_{00}$.[1] However, we have the following.

Let μ and ν be measures defined on σ-algebras $\mathscr{M}_\mu$ and $\mathscr{M}_\nu$, respectively, of subsets of X. Suppose that μ and ν are regular (11.34), and that $\mu(F)$ and $\nu(F)$ are finite for compact F. [Note that $\mathscr{M}_\mu$ and $\mathscr{M}_\nu$ both contain all Borel sets.] If $\int_X \varphi\,d\mu = \int_X \varphi\,d\nu$ for all $\varphi \in \mathfrak{C}_{00}^{+}$, then $\mu(A) = \nu(A)$ for all $A \in \mathscr{M}_\mu \cap \mathscr{M}_\nu$. To see this, suppose first that F is compact. Let $\{U_n\}_{n=1}^{\infty}$ be a decreasing sequence of open sets such that $F \subset \bigcap_{n=1}^{\infty} U_n$, $\lim_{n\to\infty} \mu(U_n) = \mu(F)$, and $\lim_{n\to\infty} \nu(U_n) = \nu(F)$. Let ψ_n be a function in $\mathfrak{C}_{00}^{+}$ such that $\psi_n(F) = 1$, $\psi_n(U_n') = 0$, and $\psi_n(X) \subset [0, 1]$. Let $\varphi_n = \min(\psi_1, \ldots, \psi_n)$. Then we have $\lim_{n\to\infty} \varphi_n = \xi_F$ μ-almost everywhere and ν-almost everywhere. Hence

$$\mu(F) = \int_X \xi_F\,d\mu = \int_X \lim_{n\to\infty} \varphi_n\,d\mu = \lim_{n\to\infty} \int_X \varphi_n\,d\mu = \lim_{n\to\infty} \int_X \varphi_n\,d\nu = \int_X \xi_F\,d\nu = \nu(F).$$

Now for any open set U, we have

$$\begin{aligned} \mu(U) &= \sup\{\mu(F) : F \text{ is compact and } F \subset U\} \\ &= \sup\{\nu(F) : F \text{ is compact and } F \subset U\} = \nu(U), \end{aligned}$$

and likewise for $A \in \mathscr{M}_\mu \cap \mathscr{M}_\nu$, we have

$$\begin{aligned} \mu(A) &= \inf\{\mu(U) : U \text{ is open and } U \supset A\} \\ &= \inf\{\nu(U) : U \text{ is open and } U \supset A\} = \nu(A). \end{aligned}$$

We next establish a technical but useful property of ι.

(11.39) Theorem. *There exists a family $\mathscr{F}$ of nonvoid compact subsets of X such that:*

(i) *the sets in $\mathscr{F}$ are pairwise disjoint;*

(ii) *$\iota(F \cap U) > 0$ if U is open, $F \in \mathscr{F}$, and $F \cap U \neq \varnothing$;*

(iii) *every open set of finite positive measure has nonvoid intersection with only countably many $F \in \mathscr{F}$, and every point of X has a neighborhood meeting only countably many $F \in \mathscr{F}$;*

(iv) *the set $N = (\bigcup\{F \in \mathscr{F}\})'$ is locally null.*

Proof. The void family of compact sets trivially satisfies (i) and (ii). A simple well-ordering argument shows that there is a maximal family $\mathscr{F}$

[1] See for example HALMOS [2], p. 231, Example (10).

of nonvoid compact subsets of X satisfying (i) and (ii)[1]. Property (iii) is an immediate consequence of (ii).

If U is an open set with finite measure and $F_1, F_2, \ldots, F_n, \ldots$ are the members of $\mathscr{F}$ having nonvoid intersection with U, then $\iota(U \cap N') + \iota(U \cap N) = \iota(U \cap (F_1 \cup F_2 \cup \cdots \cup F_n \cup \cdots)) + \iota(U \cap (F_1 \cup F_2 \cup \cdots \cup F_n \cup \cdots)') = \iota(U)$, since U and $(F_1 \cup F_2 \cup \cdots \cup F_n \cup \cdots)$ are measurable. This implies that N and N' are measurable.

Finally, assume that N is not locally null. Since N is measurable, there is a compact set $E \subset N$ such that $\iota(E) > 0$; this is pointed out in the last sentence of (11.33). Consider the family $\mathscr{U}$ of all open sets U in X such that $\iota(E \cap U) = 0$. Let $B = \bigcup\{E \cap U : U \in \mathscr{U}\} = E \cap (\bigcup_{U \in \mathscr{U}} U)$. Then $\iota(B) = 0$, since in the contrary case there would be a compact set $D \subset B$ such that $\iota(D) > 0$, and D would be contained in the union of a finite number of sets $D \cap U$, each of measure 0. Let $S_E = E \cap B'$. Then we have $\iota(S_E) = \iota(E) > 0$, and if V is open in X and $S_E \cap V = S_E \cap (E \cap V) \neq \varnothing$, then $\iota(S_E \cap V) = \iota(E \cap V) > 0$. Hence we could add S_E to the family $\mathscr{F}$ and obtain a strictly larger family of compact sets satisfying (i) and (ii). This proves (iv). $\square$

The following result will also be useful in the sequel.

(11.40) Theorem. *Let f be a nonnegative, measurable function on X such that $\int_X f\,d\iota < \infty$. Then there is a function f' defined on X such that for every $\alpha \geqq 0$, the set $\{x \in X : f'(x) > \alpha\}$ is σ-compact, $0 \leqq f' \leqq f$, and $\int_X f'\,d\iota = \int_X f\,d\iota$.*

Proof. For every positive integer n, and for $k = 1, 2, \ldots$, let $A_{k,n} = \left\{x \in X : \frac{k}{2^n} \leqq f(x) < \frac{k+1}{2^n}\right\}$. It is plain that $A_{k,n}$ is measurable and that $\iota(A_{k,n}) < \infty$. By (11.32), there is a σ-compact set $B_{k,n}$ such that $B_{k,n} \subset A_{k,n}$ and $\iota(A_{k,n} \cap B'_{k,n}) = 0$. Let $f'_n = \max\left\{\sum_{k=1}^{\infty} \frac{k}{2^l} \xi_{B_{k,l}} : l = 1, 2, \ldots, n\right\}$. It is plain that

$$\sum_{k=1}^{\infty} \frac{k}{2^n} \iota(A_{k,n}) \leqq \int_X f'_n\,d\iota \leqq \int_X f\,d\iota. \tag{1}$$

Now let $f' = \lim_{n\to\infty} f'_n$. Since $\int_X f\,d\iota = \lim_{n\to\infty} \sum_{k=1}^{\infty} \frac{k}{2^n} \iota(A_{k,n})$, inequality (1) and the theorem on integration of monotone increasing sequences imply that $\int_X f'\,d\iota = \int_X f\,d\iota$. It is also easy to see that for $\alpha \geqq 0$, the set $\{x \in X : f'(x) > \alpha\}$ is equal to $\bigcup\left\{B_{k,n} : \frac{k}{2^n} > \alpha\right\}$ and hence is σ-compact. $\square$

---

[1] Recall that the union of a void family is $\varnothing$. Thus if $\mathscr{F}$ were void, we would define N to be X.

(11.41) Corollary. *If f is any ι-measurable function such that $\int_X |f|\, d\iota < \infty$, then there is a Borel measurable function f' equal to f ι-almost everywhere.*

We now state and prove an elementary theorem about measurable functions which is analogous to (11.31).

(11.42) Theorem. *Let f be a real- or complex-valued function on X. Then the following statements are equivalent:*

(i) *f is measurable;*

(ii) *$f\xi_U$ is measurable for all open subsets of X such that $\iota(U) < \infty$;*

(iii) *$f\xi_F$ is measurable for all compact subsets F of X;*

(iv) *$f\varphi$ is measurable for all functions φ in $\mathfrak{C}_{00}^+$.*

Proof. Clearly (i) implies (iv). Suppose that (iv) holds and let F be a compact subset of X. Then, as in (11.38), there is a sequence $\{\varphi_n\}_{n=1}^{\infty}$ of functions in $\mathfrak{C}_{00}^+$ such that $\xi_F = \lim_{n\to\infty} \varphi_n$ almost everywhere. Therefore $f\xi_F = \lim_{n\to\infty} f\varphi_n$ almost everywhere, so that $f\xi_F$ is measurable. This shows that (iii) holds. Statement (iii) implies (ii) in the same way $[f\xi_U = \lim_{n\to\infty} f\xi_{F_n}$ almost everywhere for a suitably chosen sequence $\{F_n\}_{n=1}^{\infty}$ of compact sets].

Suppose that (ii) holds. To show (i) it suffices to consider nonnegative functions f. Let α be a positive number. Then for open sets U such that $\iota(U) < \infty$, we have

$$\{x \in X : f(x) > \alpha\} \cap U = \{x \in X : f(x)\, \xi_U(x) > \alpha\};$$

by (ii), this set is measurable. By (11.31), the set $\{x \in X : f(x) > \alpha\}$ is measurable. Hence f is measurable. $\square$

Miscellaneous theorems and examples

(11.43) Examples concerning $\mathfrak{C}_0(X)$ and $\mathfrak{C}_{00}(X)$. (a) Let X be any Hausdorff space. Then $f \in \mathfrak{C}(X)$ is in $\mathfrak{C}_0(X)$ if and only if $\{x \in X: |f(x)| > \alpha\}^-$ is compact for every $\alpha > 0$, and $f \in \mathfrak{C}_{00}(X)$ if and only if $\{x \in X: |f(x)| > 0\}^-$ is compact.

(b) Consider the space Q of rational numbers with its usual topology as a subspace of R. Then $\mathfrak{C}(Q)$ is a large space, but $\mathfrak{C}_0(Q)$ consists of 0 alone. [If $f \in \mathfrak{C}(Q)$ and $f(a) \neq 0$, then the set $\{x \in Q: |f(x)| > \frac{1}{2}|f(a)|\}^-$ is noncompact, so that $f \notin \mathfrak{C}_0(Q)$.]

(c) If X is a locally compact Hausdorff space, select any point $a \in X$ and any neighborhood U of a such that U^- is compact. By complete regularity, there is a continuous function f with domain X and range contained in $[0, 1]$ such that $f(a) = 1$ and $f(U') = 0$. Thus $f \neq 0$ and $f \in \mathfrak{C}_{00}(X)$.

(d) Let Ω be the smallest uncountable ordinal, and let X be a well-ordered set of order type Ω. With the order topology [see (4.19.a)], X is a countably compact, locally compact, noncompact, normal space, on which every continuous complex-valued function is bounded. [Assume that f is continuous and unbounded on X. For each positive integer n, there is an $\alpha_n \in X$ such that $|f(\alpha_n)| \geqq n$. Since Ω is the least uncountable ordinal, there is an element $\nu \in X$ such that $\nu \geqq \alpha_n$ for all $n = 1, 2, \ldots$. Thus f is unbounded on the compact set $\{\beta \in X : \beta \leqq \nu\}$, which is impossible.] Every compact subset F of X is bounded: there is an $\alpha \in X$ such that $\beta \leqq \alpha$ for all $\beta \in F$. Also, we have $\mathfrak{C}_0(X) = \mathfrak{C}_{00}(X)$. [Suppose that $f \in \mathfrak{C}_0(X)$. Then for every positive integer n, there is an $\alpha_n \in X$ such that $|f(\beta)| < \frac{1}{n}$ for all $\beta \geqq \alpha_n$. As before, there is an element $\nu \in X$ such that $\nu \geqq \alpha_n$ for all $n = 1, 2, 3, \ldots$. If $\beta \geqq \nu$, then we have $|f(\beta)| < \frac{1}{n}$ for all positive integers n, and so $f(\beta) = 0$. Since the set $\{\beta \in X : \beta \leqq \nu\}$ is compact, it follows that $f \in \mathfrak{C}_{00}(X)$.]

(e) Let G be a locally compact, noncompact group. Then we have $\mathfrak{C}_{00}(G) \subsetneqq \mathfrak{C}_0(G)$. [Let U be a neighborhood of e in G such that U^- is compact. Then the family of sets $\{xU : x \in G\}$ is an open covering of G that admits no finite subcovering. If it did, G would be the union of a finite number of compact sets and hence would be compact. Thus there is an infinite subset $\{x_1, x_2, x_3, \ldots, x_n, \ldots\}$ of G such that $x_n \notin \bigcup_{k=1}^{n-1} x_k U$ for $n = 2, 3, \ldots$. Now let V be a symmetric neighborhood of e such that $V^2 \subset U$. Then it is plain that the sets $\{x_n V\}_{n=1}^{\infty}$ are pairwise disjoint. Let f be a function in $\mathfrak{C}(G)$ such that $f(G) \subset [0, 1]$, $f(e) = 1$, and $f(V') = 0$. Let g be the function such that $g(x) = \sum_{n=1}^{\infty} 2^{-n} f(x_n^{-1} x)$ for $x \in G$. It is then easy to see that $g \in \mathfrak{C}_0(G)$ and $g \notin \mathfrak{C}_{00}(G)$.]

(11.44) The values of a continuous measure. (a) [R. E. JAMISON, oral communication.] Let X be a locally compact Hausdorff space, and suppose that I is a nonnegative linear functional on $\mathfrak{C}_{00}(X)$ whose corresponding measure ι is continuous, *i.e.*, $\iota(\{x\}) = 0$ for all $x \in X$. Suppose that B is an ι-measurable subset of X such that

$$\iota(B) = \sup\{\iota(F) : F \text{ compact and } F \subset B\},$$

and that β is any real number such that $0 < \beta < \iota(B)$. Then there is a compact subset E_0 of B such that $\iota(E_0) = \beta$.

[Let F be a compact subset of B such that $\iota(F) > \beta$, and let $\mathscr{F}$ be the family of all compact subsets E of F such that $\iota(E) \geqq \beta$. Let $\{E_\gamma\}_{\gamma \in \Gamma}$ be a subfamily of $\mathscr{F}$ that is linearly ordered by inclusion, and write E_1 for the set $\bigcap_{\gamma \in \Gamma} E_\gamma$. Clearly E_1 is nonvoid. Assume that $\iota(E_1) < \beta$. Then there is an open set U containing E_1 such that $\iota(U) < \beta$. However, we have $\varnothing = U' \cap E_1 = U' \cap (\bigcap_{\gamma \in \Gamma} E_\gamma)$ and since the E_γ's are compact, there are $\gamma_1, \ldots, \gamma_m$ in Γ for which $\bigcap_{j=1}^{m} E_{\gamma_j} \subset U$. Since $\{E_\gamma\}_{\gamma \in \Gamma}$ is linearly ordered, we have some $E_\gamma \subset U$, a contradiction. It follows by ZORN's lemma that there is a minimal compact subset E_0 of E such that $\iota(E_0) \geqq \beta$. Assume that $\iota(E_0) > \beta$. Let x be any point of E_0. Since $\iota(\{x\}) = 0$, we can find a neighborhood U of x such that $\iota(U) < \iota(E_0) - \beta$. We find

$$\iota(E_0 \cap U') = \iota(E_0) - \iota(E_0 \cap U) > \iota(E_0) - \iota(U) > \beta,$$

and this contradicts the minimality of E_0.]

(b) The hypotheses of (a) obviously hold for every open subset U of X and also for every σ-finite ι-measurable subset of X. The conclusion of (a) must fail for all locally null nonnull sets [see (11.33)].

(11.45) Measures on subspaces. Throughout this example, let X be a locally compact Hausdorff space and $X^\bullet$ a nonvoid closed subset of X. Let $I^\bullet$ be a nonnegative linear functional on $\mathfrak{C}_{00}(X^\bullet)$ and let $\iota^\bullet$ be the set function defined on all subsets of $X^\bullet$ as in (11.20). We will show how to construct from $I^\bullet$ a functional on $\mathfrak{C}_{00}(X)$ and will compute the corresponding measure.

(a) For a function f on X, let $f^\bullet$ denote f with its domain restricted to $X^\bullet$. If f is in $\mathfrak{C}_{00}(X)$, then $f^\bullet$ is in $\mathfrak{C}_{00}(X^\bullet)$. Also, every function g in $\mathfrak{C}_{00}(X^\bullet)$ has the form $f^\bullet$ for some f in $\mathfrak{C}_{00}(X)$. [It is easy to prove that $f^\bullet$ is in $\mathfrak{C}_{00}(X^\bullet)$ whenever f is in $\mathfrak{C}_{00}(X)$. The second statement is equivalent to the assertion that every function g in $\mathfrak{C}_{00}(X^\bullet)$ admits an extension f over X such that f is in $\mathfrak{C}_{00}(X)$. If X is compact, f exists by the TIETZE extension theorem [KELLEY [2], p. 242]. If X is noncompact, let X_∞ be the 1-point compactification of X, and let ∞ be the adjoined point. Let E be a compact subset of $X^\bullet$ such that g vanishes on $X^\bullet \cap E'$. Then E is also compact as a subset of X, and there is an open subset U of X such that $E \subset U$ and U^- is compact. It is obvious that $Y = X^\bullet \cup (U' \cap X) \cup \{\infty\}$ is a closed subset of X_∞. Let f_1 be the function on Y such that $f_1(x) = g(x)$ for $x \in X^\bullet$ and $f_1(x) = 0$ for $x \in (U' \cap X) \cup \{\infty\}$. Then f_1 is continuous on Y, and by the TIETZE extension theorem there is a continuous complex-valued function f_2 on X_∞ such that f_2 is an extension of f_1. Let f be the restriction of f_2 to the space X. Note that if $g(X^\bullet) \subset [0, 1]$, then the inclusion $f(X) \subset [0, 1]$ can be realized.]

(b) For $f \in \mathfrak{C}_{00}(X)$, let $I(f) = I^\bullet(f^\bullet)$. Then I is evidently a nonnegative linear functional on $\mathfrak{C}_{00}(X)$. Let ι denote the set function on all subsets of X defined for I as in (11.20). Then for every subset A of X, we have $\iota(A) = \iota(A \cap X^\bullet) = \iota^\bullet(A \cap X^\bullet)$. [It is easy to see that $\iota(X^{\bullet\prime}) = 0$, and since $X^\bullet$ is ι-measurable, the equality $\iota(A)=\iota(A\cap X^\bullet)$ holds for all $A\subset X$. Now consider an open subset U of X. By (11.11) and part (a), we have

$$\begin{aligned}\iota(U) &= \sup\{I(f): f\in\mathfrak{C}_{00}^+(X),\ f\leqq\xi_U\}\\ &= \sup\{I^\bullet(f^\bullet): f\in\mathfrak{C}_{00}^+(X),\ f\leqq\xi_U\}\\ &\leqq \sup\{I^\bullet(g): g\in\mathfrak{C}_{00}^+(X^\bullet),\ g\leqq\xi_{U\cap X^\bullet}\}\\ &= \iota^\bullet(U\cap X^\bullet).\end{aligned}$$

To show that $\iota(U)\geqq\iota^\bullet(U\cap X^\bullet)$, consider any function g in $\mathfrak{C}_{00}^+(X^\bullet)$ such that $g\leqq\xi_{U\cap X^\bullet}$. Let h be the function on the closed set $Y=X^\bullet\cup(U'\cap X)$ such that $h(x)=g(x)$ for $x\in X^\bullet$ and $h(x)=0$ for $x\in U'\cap X$. Then h belongs to $\mathfrak{C}_{00}^+(Y)$ and by part (a), there is a function f in $\mathfrak{C}_{00}^+(X)$ such that $f(X)\subset[0,1]$ and f is an extension of h. Clearly $f^\bullet=g$ and $f\leqq\xi_U$. Thus $\iota(U)\geqq I(f)=I^\bullet(f^\bullet)=I^\bullet(g)$, and, as g is arbitrary, we have $\iota(U)\geqq\iota^\bullet(U\cap X^\bullet)$. Thus $\iota(U)=\iota^\bullet(U\cap X^\bullet)$.

Finally, let A be any subset of $X^\bullet$. By (11.22) and the preceding paragraph, we have

$$\begin{aligned}\iota^\bullet(A) &= \inf\{\iota^\bullet(U\cap X^\bullet): U \text{ is open in } X,\ U\cap X^\bullet\supset A\}\\ &= \inf\{\iota(U): U \text{ is open in } X,\ U\cap X^\bullet\supset A\}\\ &= \inf\{\iota(U): U \text{ is open in } X,\ U\supset A\}\\ &= \iota(A).\end{aligned}$$

This completes the proof.]

Notes

The theory of integration as formulated in this section is the work of many hands. An extended discussion of the historical development of the subject would be out of place here; we mention only a few highlights[1]. Corollary (11.37) for $X=[0,1]$ and the integral a Riemann-Stieltjes integral is of course due to FRÉDÉRIC RIESZ [1]. It is fascinating to see how RIESZ chose the *right* way to represent bounded linear functionals on $\mathfrak{C}([0,1])$, in contrast to earlier representations which he refers to [*loc. cit.*, p. 974].

The basic idea of extending a functional and so producing a measure is due to P. J. DANIELL [1]. The construction of a measure directly from a nonnegative functional on $\mathfrak{C}(X)$ and Corollary (11.37) for a compact Hausdorff space X appear in VON NEUMANN [5], which is the earliest statement known to us of this technique and result. The same construction, with more details, appears in KAKUTANI [3]. It is remarked with no proofs or constructions in WEIL [4], p. 32, that "il y a donc

[1] An interesting historical discussion appears in BOURBAKI [3], pp. 113—126.

correspondance biunivoque entre les mesures de Radon et les fonctionnelles $I(f)$". Unquestionably WEIL was in possession of a construction similar to that of the present section, although he leaves everything to the reader, referring only to the fascicule [2] of BOURBAKI, which appeared in print eleven years after WEIL [4]. A construction similar to but not identical with our construction of $\bar{I}$ and $\bar{\bar{I}}$ is sketched in H. CARTAN [2]. CARTAN'S process is developed with full details by EDWARDS [1].

Measures, whether countably or finitely additive, which yield integrals reproducing given linear functionals in various abstract settings, have been considered by several writers within recent years. One such treatment is in HEWITT [1], where references to earlier literature also appear. [Theorem 2.17 of HEWITT [1] is false, as was pointed out by I. GLICKSBERG: an additional separation property of sets in $\mathscr{P}$ by functions in $\mathfrak{F}$ is required.] An alternative treatment is given by LOOMIS [3]. All of these constructions lead to essentially the same result: the choice of one or another set of axioms and mode of procedure seems to be largely a matter of taste. Countable additivity is obviously essential for the application of the most powerful tools of integration theory, *viz.*, LEBESGUE'S theorem on dominated convergence and FUBINI'S theorem. The appearance of this phenomenon is discussed in GLICKSBERG [1].

The construction of $\bar{I}$, $\bar{\bar{I}}$, and ι given in the present section is very like [although not quite the same as] that found in NAĬMARK [1], §6. We have also borrowed freely from BOURBAKI [2]. Locally null sets were apparently first introduced by BOURBAKI [2], with a definition formally different from but actually equivalent to ours.

§ 12. The spaces $\mathfrak{L}_p(X)$ $(1 \leqq p \leqq \infty)$

Throughout this section, X, I, ι, $\mathscr{M}_\iota$, and $\mathscr{N}_\iota$ are as in §11. We take the elementary properties of measurable functions as known. See for example HALMOS [2], pp. 73—86 or SAKS [1], pp. 12—15.

(12.1) Definition. Let p be a positive real number. Let f be an extended real- or complex-valued measurable function on X such that $\int_X |f|^p \, d\iota < \infty$. Then we say that $f \in \mathfrak{L}_p(X, \iota)$; where no confusion can arise we will write $\mathfrak{L}_p$ for $\mathfrak{L}_p(X, \iota)$. The *norm* $\|f\|_p$ is defined by

(i) $$\|f\|_p = \left[\int_X |f|^p \, d\iota\right]^{1/p}.$$

Given two functions $f, g \in \mathfrak{L}_p$, we have $\|f-g\|_p = 0$ if and only if $f(x) = g(x)$ almost everywhere on X. **Two functions such that $\|f-g\|_p = 0$ are taken to be identical in $\mathfrak{L}_p$.** That is, $\mathfrak{L}_p$ actually consists of equivalence classes of functions. We will frequently refer to a function f in $\mathfrak{L}_p$, etc. It will be clear from the context in every case whether we mean the fixed function f or the equivalence class in $\mathfrak{L}_p$ containing f. The real functions in $\mathfrak{L}_p$ will be denoted by $\mathfrak{L}_p^r$. We shall be concerned with $\mathfrak{L}_p$ for values of $p \geqq 1$ almost exclusively.

The pathological character of ι on sets that are locally null but not null [see (11.33)] permits us to improve a little upon (12.1).

(12.2) Theorem. *Let f be a complex-valued function or an extended real-valued function on X that is locally null but not null. Then $\int_X |f|^p d\iota = \infty$ $(1 \leqq p < \infty)$. Hence if $f, g \in \mathfrak{L}_p$ and $f(x) = g(x)$ locally almost everywhere, then $\int_X |f-g|^p d\iota = 0$. Thus f and g are equal almost everywhere and define the same element of $\mathfrak{L}_p$.*

Proof. As pointed out in (11.33), if N is a locally null set but not a null set, then $\iota(N) = \infty$. For some $\alpha > 0$, we have $\iota(\{x \in X: |f(x)|^p > \alpha\}) \neq 0$, and hence $\iota(\{x \in X: |f(x)|^p > \alpha\}) = \infty$. It follows that $\int_X |f|^p d\iota = \infty$. For $f, g \in \mathfrak{L}_p$, we have $|f-g|^p \leqq (|f|+|g|)^p \leqq [2\max(|f|, |g|)]^p \leqq 2^p(|f|^p + |g|^p)$. Hence $\int_X |f-g|^p d\iota$ is finite and is therefore 0. □

It is obvious that if $f = f'$ locally almost everywhere and $g = g'$ locally almost everywhere, then $f+g = f'+g'$ locally almost everywhere. Thus addition is defined unambiguously in $\mathfrak{L}_p$. Similarly αf is defined unambiguously if $f \in \mathfrak{L}_p$ and $\alpha \in K$. Consequently, $\mathfrak{L}_p$ is a complex linear space. The norm $\|f\|_p$ has the usual properties ascribed to a norm [see (B.7)]. That is: $\|f\|_p$ is nonnegative and is zero if and only if $f = 0$ [we repeat, in the sense of $\mathfrak{L}_p$]; $\|f+g\|_p \leqq \|f\|_p + \|g\|_p$; and $\|\alpha f\|_p = |\alpha| \cdot \|f\|_p$ for $\alpha \in K$. Note that these properties are obvious for the case $p = 1$. A close examination of certain inequalities implying $\|f+g\|_p \leqq \|f\|_p + \|g\|_p$ is required for the detailed study of $\mathfrak{L}_p$, and we therefore digress to set them down.

Throughout (12.3) to (12.6) inclusive, p will be any number such that $1 < p < \infty$, and p' will be the number $\frac{p}{p-1}$, so that $\frac{1}{p} + \frac{1}{p'} = 1$. Note that $2' = 2$.

(12.3) Lemma. *Let a and b be nonnegative real numbers. Then*

(i) $ab \leqq \dfrac{a^p}{p} + \dfrac{b^{p'}}{p'}$,

and equality holds if and only if $a^p = b^{p'}$.

Proof. For $u \geqq 0$, the function u^{p-1} and its inverse $u^{\frac{1}{p-1}}$ are strictly increasing and continuous. From this it is obvious upon inspecting the areas under the curves $v = u^{p-1}$ and $u = v^{\frac{1}{p-1}}$ that

$$ab < \int_0^a u^{p-1} du + \int_0^b v^{\frac{1}{p-1}} dv = \frac{a^p}{p} + \frac{b^{p'}}{p'},$$

unless $b = a^{p-1}$, which is equivalent to $a^p = b^{p'}$. Plainly equality holds in (i) if $a^p = b^{p'}$. ▯

(12.4) Theorem [HÖLDER'S inequality]. *Let $f \in \mathfrak{L}_p$ and $g \in \mathfrak{L}_{p'}$. Then $fg \in \mathfrak{L}_1$; in fact*

(i) $\left| \int_X f g \, d\iota \right| \leqq \int_X |fg| \, d\iota$,

and

(ii) $\int_X |fg| \, d\iota \leqq \|f\|_p \|g\|_{p'}$.

Equality holds in (i) *if and only if* $\operatorname{sgn}[f(x)\, g(x)]$ *is constant almost everywhere on the set where $fg \neq 0$. Equality holds in* (ii) *if and only if there are nonnegative numbers A and B not both 0 such that $A|f(x)|^p = B|g(x)|^{p'}$ almost everywhere. Thus*

(iii) $\left| \int_X f g \, d\iota \right| \leqq \|f\|_p \|g\|_{p'}$,

and equality holds in (iii) *if and only if* $\operatorname{sgn}[f(x)\, g(x)]$ *is constant almost everywhere on the set where $fg \neq 0$ and $A|f(x)|^p = B|g(x)|^{p'}$ almost everywhere*[1].

Proof. Inequality (i) holds for any function $h \in \mathfrak{L}_1$:

$$\left| \int_X h \, d\iota \right| \leqq \int_X |h| \, d\iota. \tag{1}$$

This inequality is proved just as (11.5) was proved. Let $E = \{x \in X : h(x) \neq 0\}$. If $\operatorname{sgn} h(x) = \exp(i\alpha)$ $(\alpha \in R)$ almost everywhere on E, then we have

$$\int_X h \, d\iota = \int_E \exp(i\alpha) \, |h| \, d\iota = \exp(i\alpha) \int_X |h| \, d\iota,$$

so that equality obtains in (1). Suppose conversely that equality obtains in (1), so that $\exp(i\beta) \int_X h \, d\iota = \int_X |h| \, d\iota$, for some $\beta \in R$. Write $\exp(i\beta)\, h = \varphi = \varphi_1 + i\varphi_2$, where φ_1 and φ_2 are real-valued functions. Then we have $\int_X \varphi \, d\iota = \int_X \varphi_1 \, d\iota + i \int_X \varphi_2 \, d\iota = \int_X [\varphi_1^2 + \varphi_2^2]^{\frac{1}{2}} \, d\iota$, so that $\int_X \varphi_1 \, d\iota \leqq \int_X |\varphi_1| \, d\iota \leqq \int_X [\varphi_1^2 + \varphi_2^2]^{\frac{1}{2}} \, d\iota = \int_X \varphi_1 \, d\iota$. This implies that $\varphi_1(x) \geqq 0$ almost everywhere and $\varphi_2(x) = 0$ almost everywhere. Hence $\varphi_1(x) = |\varphi(x)| = |h(x)|$ almost everywhere, and $h(x) = \exp(-i\beta)\, |h(x)|$ almost everywhere.

We turn now to (ii). If f or g vanishes almost everywhere, then equality trivially holds in (ii). If neither f nor g vanishes almost everywhere, then (12.3) shows that

$$\frac{|f(x)\, g(x)|}{\|f\|_p \|g\|_{p'}} \leqq \frac{1}{p} \frac{|f(x)|^p}{\|f\|_p^p} + \frac{1}{p'} \frac{|g(x)|^{p'}}{\|g\|_{p'}^{p'}}, \tag{2}$$

[1] Inequality (iii) is *HÖLDER's inequality*; sometimes (ii) is also referred to as HÖLDER's inequality. For $p = p' = 2$, (iii) is called *CAUCHY's inequality* or *BUNYAKOVSKIĬ's inequality*.

with equality holding if and only if

$$\|g\|_{p'}^{p'}\,|f(x)|^p = \|f\|_p^p\,|g(x)|^{p'}. \tag{3}$$

We integrate (2) over X, obtaining (ii), and note that if (3) fails on a set of positive measure, then strict inequality obtains in (ii). □

(12.5) Corollary. *Let $f_1, \ldots, f_n$ be nonnegative functions in $\mathfrak{L}_1$ and let $\alpha_1, \ldots, \alpha_n$ be positive numbers such that $\alpha_1+\alpha_2+\cdots+\alpha_n=1$. Then $f_1^{\alpha_1}f_2^{\alpha_2}\cdots f_n^{\alpha_n}\in\mathfrak{L}_1$ and*

(i) $\int\limits_X f_1^{\alpha_1} f_2^{\alpha_2}\cdots f_n^{\alpha_n}\,d\iota \leqq \|f_1\|_1^{\alpha_1}\|f_2\|_1^{\alpha_2}\cdots\|f_n\|_1^{\alpha_n}$.

Proof. Use (12.4) and induction on n. □

(12.6) Theorem [MINKOWSKI'S inequality]. *Let f and g be functions in $\mathfrak{L}_p$ $(1<p<\infty)$. Then*

(i) $\|f+g\|_p \leqq \|f\|_p + \|g\|_p$,

and equality obtains if and only if $Af=Bg$ for nonnegative numbers A and B not both 0.

Proof. We have $|f+g|^p \leqq 2^p(|f|^p+|g|^p)$ as in (12.2), so that $f+g\in\mathfrak{L}_p$. Applying (12.4), we have

$$\left.\begin{aligned} \|f+g\|_p^p &= \int\limits_X |f+g|^p\,d\iota = \int\limits_X |f+g|\cdot|f+g|^{p-1}\,d\iota \\ &\leqq \int\limits_X |f|\cdot|f+g|^{p-1}\,d\iota + \int\limits_X |g|\cdot|f+g|^{p-1}\,d\iota \\ &\leqq \|f\|_p\Big[\int\limits_X |f+g|^{(p-1)p'}\,d\iota\Big]^{1/p'} \\ &\quad + \|g\|_p\Big[\int\limits_X |f+g|^{(p-1)p'}\,d\iota\Big]^{1/p'} = (\|f\|_p+\|g\|_p)\,\|f+g\|_p^{p-1}. \end{aligned}\right\} \tag{1}$$

This clearly implies (i).

The condition stated for equality in (i) is obviously sufficient. Suppose conversely that equality holds in (i). Then the inequalities in (1) are both equalities. From the second of these and the condition for equality in (12.4.ii), we have

$$A'\,|f(x)|^p = B'\,|f(x)+g(x)|^p \quad\text{and}\quad A''\,|g(x)|^p = B''\,|f(x)+g(x)|^p \tag{2}$$

almost everywhere, where A', B', A'', B'' are nonnegative real numbers and $(A'^2+B'^2)(A''^2+B''^2)>0$. If $f+g=0$ or $f=0$ or $g=0$ [in $\mathfrak{L}_p$] we plainly have $Af=Bg$. If not, (2) implies that $|g(x)|=\gamma|f(x)|=\delta|f(x)+g(x)|$ almost everywhere, where $\gamma, \delta>0$. Since the first inequality in (1) is also an equality, we have

$$|f(x)+g(x)| = |f(x)|+|g(x)| = (1+\gamma)\,|f(x)| \tag{3}$$

almost everywhere on the set $E=\{x\in X: f(x)+g(x)\neq 0\}$. We thus have

$$\left|1+\frac{g(x)}{f(x)}\right| = 1+\gamma = 1+\left|\frac{g(x)}{f(x)}\right| \tag{4}$$

almost everywhere on E, so that also $g(x)=\gamma f(x)$ almost everywhere on E. Thus $g(x)=\gamma f(x)$ almost everywhere. □

(12.7) Theorem. *Let f and g be functions in $\mathfrak{L}_1$. Then $\|f+g\|_1 \leqq \|f\|_1+\|g\|_1$, and equality holds if and only if there is a positive measurable function ϱ on X such that $g(x)=\varrho(x) f(x)$ almost everywhere on the set where $f(x)\, g(x) \neq 0$.*

Proof. The inequality is obvious:

$$\int_X |f+g|\, d\iota \leqq \int_X |f|\, d\iota + \int_X |g|\, d\iota. \tag{1}$$

Equality obtains in (1) if and only if

$$|f(x)+g(x)| = |f(x)| + |g(x)| \tag{2}$$

almost everywhere on X. The equality (2) is no restriction unless $f(x)\, g(x) \neq 0$. If this occurs, then (2) is equivalent to the condition that $\varrho(x) = \frac{g(x)}{f(x)}$ be real and positive. □

(12.8) Theorem. *For $1 \leqq p < \infty$, the spaces $\mathfrak{L}_p$ are complex Banach spaces, and $\mathfrak{L}_2$ is a complex Hilbert space with inner product*

(i) $\langle f, g\rangle = \int_X f\bar{g}\, d\iota$.[1]

Proof. Theorems (12.6) and (12.7) show that $\|f\|_p$ is a norm for $\mathfrak{L}_p$ [see (B.7)]. To show that $\mathfrak{L}_p$ is complete, let $\{f_n\}_{n=1}^{\infty}$ be a Cauchy sequence in $\mathfrak{L}_p$ and let $n_1 < n_2 < \cdots < n_k < \cdots$ be a sequence of integers such that $\sum_{k=1}^{\infty} \|f_{n_{k+1}} - f_{n_k}\|_p = \alpha < \infty$. Let $g_k = |f_{n_1}| + |f_{n_2} - f_{n_1}| + \cdots + |f_{n_{k+1}} - f_{n_k}|$ $(k = 1, 2, \ldots)$. From (12.7) and (12.6), we have

$$\|g_k^p\|_1 = \|g_k\|_p^p \leqq \Big[\|f_{n_1}\|_p + \sum_{j=1}^{k} \|f_{n_{j+1}} - f_{n_j}\|_p\Big]^p \leqq [\|f_{n_1}\|_p + \alpha]^p < \infty.$$

Writing $g = \lim_{k\to\infty} g_k$, we apply the theorem on monotone convergence and find

$$\int_X g^p\, d\iota \leqq [\|f_{n_1}\|_p + \alpha]^p.$$

This implies that the series $f_{n_1}(x) + \sum_{k=1}^{\infty} \big(f_{n_{k+1}}(x) - f_{n_k}(x)\big)$ converges to a complex number, say $f(x)$, for almost all $x \in X$. Thus $\lim_{k\to\infty} f_{n_k}(x) = f(x)$ almost everywhere. For $\varepsilon > 0$, choose l so large that for $m \geqq n_l$ and $k \geqq l$, we have $\|f_m - f_{n_k}\|_p^p < \varepsilon$. Then by FATOU's lemma on integrating sequences of nonnegative functions, we have

$$\int_X |f - f_m|^p\, d\iota = \int_X \varliminf_{k\to\infty} |f_{n_k} - f_m|^p\, d\iota \leqq \varliminf_{k\to\infty} \int_X |f_{n_k} - f_m|^p\, d\iota \leqq \varepsilon$$

for $m \geqq n_l$. Hence $f \in \mathfrak{L}_p$ and $\lim_{n\to\infty} \|f - f_n\|_p = 0$.

---

[1] Horizontal bars denote complex conjugates throughout this section, as they do everywhere in the text except for (19.2) *et seq.*, where certain functions are written with horizontal bars.

To prove that $\mathfrak{L}_2$ is a Hilbert space, we need only note that $\langle f, g\rangle = \int_X f\bar{g}\,d\iota$ is a legitimate inner product [see (B.39)]. □

The concepts and results of (12.1) and (12.3)—(12.8) hold for the spaces $\mathfrak{L}_p$ defined for *any* measure space: plainly (12.2) requires the special X and ι discussed in §11. Another special result follows.

(12.9) Definition. Given X, I, ι, and $\mathcal{M}_\iota$ as in §11, let $\mathfrak{S}$ denote the class of functions on X of the form $\sum_{j=1}^{m} \alpha_j \xi_{A_j}$, where the α_j's are complex numbers and the sets A_j are measurable and have finite measure. In other words, $\mathfrak{S}$ is the class of complex-valued measurable functions on X that assume only finitely many values and vanish outside of sets of finite measure.

(12.10) Theorem. *For $1 \leq p < \infty$, the linear space $\mathfrak{C}_{00}$ is a dense subspace of $\mathfrak{L}_p$, and $\mathfrak{C}_{00}^r$ is a dense subspace of $\mathfrak{L}_p^r$.*

Proof. We carry out the proof for $\mathfrak{C}_{00}$ and $\mathfrak{L}_p$; the spaces $\mathfrak{C}_{00}^r$ and $\mathfrak{L}_p^r$ are dealt with similarly. Since $0 \leq I(|f|^p) = \int_X |f|^p\,d\iota$ for $f \in \mathfrak{C}_{00}$ (11.36), it is obvious that $\mathfrak{C}_{00} \subset \mathfrak{L}_p$. To show that $\mathfrak{C}_{00}$ is dense in $\mathfrak{L}_p$, we first prove that $\mathfrak{S}$ [which is obviously a linear subspace of $\mathfrak{L}_p$] is dense in $\mathfrak{L}_p$.[1] Suppose that $f \in \mathfrak{L}_p$ and that f is real and nonnegative. For every positive integer n, and for $k = 1, 2, \ldots, n \cdot 2^n - 1$, let $A_{k,n} = \left\{x \in X : \frac{k}{2^n} \leq f(x) < \frac{k+1}{2^n}\right\}$. Define the function σ_n by $\sigma_n = \sum_{k=1}^{n2^n-1} \frac{k}{2^n} \xi_{A_{k,n}}$. Then we have $\sigma_n \in \mathfrak{S}$, $\sigma_1 \leq \sigma_2 \leq \cdots \leq \sigma_n \leq \cdots$, and $\lim_{n\to\infty} \sigma_n = f$ everywhere on X. Furthermore, we have

$$|f - \sigma_n|^p \leq (f + \sigma_n)^p \leq (2f)^p.$$

Since $(2f)^p \in \mathfrak{L}_1$, we apply LEBESGUE's theorem on dominated convergence to infer that $\lim_{n\to\infty} \int_X |f - \sigma_n|^p\,d\iota = \int_X \lim_{n\to\infty} |f - \sigma_n|^p\,d\iota = 0$. This obviously implies that $\lim_{n\to\infty} \|f - \sigma_n\|_p = 0$. Every function in $\mathfrak{L}_p$ can be written as $f_1 - f_2 + i(f_3 - f_4)$, where f_1, f_2, f_3, f_4 are real, nonnegative, and in $\mathfrak{L}_p$. Hence $\mathfrak{S}$ is dense in $\mathfrak{L}_p$.

To prove that $\mathfrak{C}_{00}$ is dense in $\mathfrak{L}_p$, it thus obviously suffices to show the following. Let A be a measurable subset of X such that $0 < \iota(A) < \infty$. Then there is an $f \in \mathfrak{C}_{00}^+$ such that $\int_X |f - \xi_A|^p\,d\iota < \varepsilon$, where ε is an arbitrary positive number. This assertion follows at once from (11.32). Indeed, choose a compact set F and an open set U such that $F \subset A \subset U$ and $\iota(U \cap F') < \varepsilon$. Then select any function $f \in \mathfrak{C}_{00}$ such that $f(F) = 1$, $f(U') = 0$, and $f(X) \subset [0, 1]$. It is obvious that $\int_X |f - \xi_A|^p\,d\iota \leq \iota(U \cap F') < \varepsilon$. □

[1] This assertion is valid for $\mathfrak{L}_p$ defined for an arbitrary measure space.

In connection with Theorem (12.10), note that $\|f\|_p$ may well be zero for a nonzero f in $\mathfrak{C}_{00}$. It is the *equivalence classes* containing elements of $\mathfrak{C}_{00}$ that are dense in $\mathfrak{L}_p$.

We now define the space of "essentially bounded" measurable functions. Our definition is motivated by Theorem (12.2) and differs somewhat from the classical one.

(12.11) Definition. Let $\mathfrak{M}_\iota$ denote the space of all bounded, complex-valued, measurable functions on X. For $f \in \mathfrak{M}_\iota$, let $\|f\|_u = \sup\{|f(x)| : x \in X\}$, as usual. Let $\mathfrak{N}_\iota$ denote the space of all locally null functions in $\mathfrak{M}_\iota$. Plainly $\mathfrak{M}_\iota$ is a Banach space under the norm $\|f\|_u$ and pointwise linear operations, and it is easy to see that $\mathfrak{N}_\iota$ is a closed linear subspace of $\mathfrak{M}_\iota$. We define $\mathfrak{L}_\infty(X, \mathscr{M}_\iota)$ [or $\mathfrak{L}_\infty$] as the quotient space $\mathfrak{M}_\iota/\mathfrak{N}_\iota$. Thus we may consider $\mathfrak{L}_\infty$ as the space of all bounded, measurable, complex-valued functions on X, two functions being regarded as equal if they differ only on a locally null set. The norm of $f \in \mathfrak{L}_\infty$ is defined as $\inf\{\|f - f'\|_u : f' \in \mathfrak{N}_\iota\}$; we write this number as $\|f\|_\infty$.

It is easy to show that $\|f\|_\infty = \inf\{\alpha \in R : \alpha \geqq 0$ and $\{x \in X : |f(x)| > \alpha\}$ is locally null$\}$. Furthermore, this infimum is attained, *i.e.*, there is a $\beta \geqq 0$ such that $\{x \in X : |f(x)| > \beta\}$ is locally null, and such that if $0 \leqq \gamma < \beta$, then $\{x \in X : |f(x)| > \gamma\}$ fails to be locally null.

(12.12) Theorem. *The space $\mathfrak{L}_\infty$ is a Banach space.*

Proof. The only property to be verified is completeness. This is a property of all quotient spaces of Banach spaces by closed linear subspaces (B.17). □

We now prove a theorem which will be used in identifying the conjugate spaces of $\mathfrak{L}_p$ $(1 \leqq p < \infty)$.

As before, if $1 < p < \infty$, then p' denotes $\frac{p}{p-1}$; we also define $1'$ as ∞ and ∞' as 1.

(12.13) Theorem. *For $f \in \mathfrak{L}_p$ $(1 \leqq p \leqq \infty)$, we have*

(i) $\|f\|_p = \sup\left\{\left|\int_X f\varphi \, d\iota\right| : \varphi \in \mathfrak{C}_{00}, \ \|\varphi\|_{p'} \leqq 1\right\}$.

A similar assertion holds for $\mathfrak{L}_p^r$ and $\mathfrak{C}_{00}^r$.

Proof. (I) Suppose that $1 < p < \infty$. Then for $f \in \mathfrak{L}_p$, Hölder's inequality (12.4) shows that the right side of (i) does not exceed the left side. To prove equality for $f \in \mathfrak{L}_p$, we may plainly restrict ourselves to the case $\|f\|_p > 0$. First let $h = \|f\|_p^{1-p} |f|^{p-1} \operatorname{sgn} \bar{f}$. Then it is obvious that $h \in \mathfrak{L}_{p'}$ $\left[\text{recall that } p' = \frac{p}{p-1}\right]$ and that $\|h\|_{p'} = 1$. Furthermore, $\int_X fh \, d\iota = \|f\|_p^{1-p} \int_X f|f|^{p-1} \operatorname{sgn} \bar{f} \, d\iota = \|f\|_p^{1-p} \int_X |f|^p \, d\iota = \|f\|_p$. Now, using (12.10), choose $\varphi_1 \in \mathfrak{C}_{00}$ such that $\|h - \varphi_1\|_{p'} < \frac{\varepsilon}{2\|f\|_p}$, where ε is an arbitrary positive number less than $2\|f\|_p$. Then we have $|\|\varphi_1\|_{p'} - 1| = |\|\varphi_1\|_{p'} - \|h\|_{p'}| \leqq \|\varphi_1 - h\|_{p'} < \frac{\varepsilon}{2\|f\|_p}$. Let $\varphi = \frac{1}{\|\varphi_1\|_{p'}} \varphi_1$; note that $\|\varphi_1\|_{p'} > 0$. Then $\|\varphi\|_{p'} = 1$; applying HÖLDER's inequality and the foregoing, we obtain $\left|\int_X f\varphi\, d\iota - \|f\|_p\right| = \left|\int_X f\varphi\, d\iota - \int_X fh\, d\iota\right| \leqq \|f\|_p \cdot \|\varphi - h\|_{p'} \leqq \|f\|_p(\|\varphi - \varphi_1\|_{p'} + \|\varphi_1 - h\|_{p'}) = \|f\|_p(|\|\varphi_1\|_{p'} - 1| + \|\varphi_1 - h\|_{p'}) < \varepsilon$.

(II) Suppose that $p = 1$. For $f \in \mathfrak{L}_1$ and $h \in \mathfrak{L}_\infty$, we obviously have

$$\left|\int_X fh\, d\iota\right| \leqq \|f\|_1 \cdot \|h\|_\infty. \tag{1}$$

Consider any function $\varphi \in \mathfrak{C}_{00}$. Let F be a compact set such that $\varphi(F') = 0$, and for $\alpha > 0$ let $E_\alpha = \{x \in X : |\varphi(x)| \geqq \alpha\}$. Let $\omega_\alpha \in \mathfrak{C}_{00}$ be such that $\omega_\alpha(E_\alpha) = 1$, $\omega_\alpha(E'_{\alpha/2}) = 0$, and $\omega_\alpha(X) \subset [0, 1]$. Let $\psi_\alpha = \operatorname{sgn} \bar{\varphi} \cdot \omega_\alpha$. Then it is clear that $\psi_\alpha \in \mathfrak{C}_{00}$ and $\|\psi_\alpha\|_\infty \leqq 1$ for all α. It is easy to see that

$$\lim_{\alpha \to 0} \int_X \varphi\psi_\alpha\, d\iota = \int_F |\varphi|\, d\iota = \|\varphi\|_1. \tag{2}$$

Now consider an arbitrary $f \in \mathfrak{L}_1$, and let ε be an arbitrary positive number. By (12.10), there is a $\varphi \in \mathfrak{C}_{00}$ such that $\|\varphi - f\|_1 < \frac{\varepsilon}{3}$. For α sufficiently small, (2) and (1) and elementary norm inequalities give us

$$\left|\|f\|_1 - \int_X f\psi_\alpha\, d\iota\right| \leqq \left|\|f\|_1 - \|\varphi\|_1\right| + \left|\|\varphi\|_1 - \int_X \varphi\psi_\alpha\, d\iota\right| + \left|\int_X \varphi\psi_\alpha\, d\iota - \int_X f\psi_\alpha\, d\iota\right| < \frac{2\varepsilon}{3} + \|\varphi - f\|_1 \cdot \|\psi_\alpha\|_\infty < \varepsilon.$$

(III) Suppose finally that $p = \infty$. In considering (i), we may ignore the trivial case $\|f\|_\infty = 0$. Let β be any number such that $0 < \beta < \|f\|_\infty$. Then (12.11) and (11.26) imply that there is a set $A \in \mathscr{M}_\iota$ such that $0 < \iota(A) < \infty$ and $|f(x)| > \beta$ for all $x \in A$. Let $g = \frac{1}{\iota(A)} \cdot \operatorname{sgn} \bar{f} \cdot \xi_A$. Then we have $g \in \mathfrak{L}_1$, $\|g\|_1 = 1$, and

$$\int_X fg\, d\iota = \frac{1}{\iota(A)} \int_A |f|\, d\iota > \beta. \tag{3}$$

Since $\|g - \varphi\|_1$ can be made arbitrarily small with $\varphi \in \mathfrak{C}_{00}$, (3) and (1) with a short computation yield (i) for $p = \infty$.

The proof for $\mathfrak{L}_p^r$ is similar and is omitted. □

We now prove a partial converse of (12.13), also used in finding the conjugate spaces of $\mathfrak{L}_p$.

(12.14) Theorem. *Let $p \geqq 1$ be a real number. If f is a measurable, nonnegative, extended real-valued function on X such that*

(i) $\sup\left\{\int\limits_F f^p\,d\iota : F \textit{ is compact, } F \subset X\right\} = \infty,$[1]

then

(ii) $\sup\left\{\int\limits_X f\varphi\,d\iota : \varphi \in \mathfrak{C}_{00}^+,\ \|\varphi\|_{p'} \leqq 1\right\} = \infty.$

For $p = \infty$, (ii) *holds for any measurable, nonnegative, extended real-valued function on* X *that is not in* $\mathfrak{L}_\infty$ [*here* $p' = 1$].

Proof. Suppose that $1 \leqq p < \infty$ and that (i) holds. Let β be an arbitrary positive number and let $F \subset X$ be a compact set such that $\int\limits_F f^p\,d\iota > \beta^p$. For $n = 1, 2, \ldots$, let $f_n = \min(f, n)$; then $\lim\limits_{n\to\infty} \int\limits_F f_n^p\,d\iota = \int\limits_F f^p\,d\iota > \beta^p$. Let n_0 be chosen so that $\int\limits_F f_{n_0}^p\,d\iota > \beta^p$. Since $\iota(F) < \infty$, we have $f_{n_0} \cdot \xi_F \in \mathfrak{L}_p$ and obviously $\|f_{n_0} \cdot \xi_F\|_p > \beta$. Hence by (12.13), there is a $\varphi \in \mathfrak{C}_{00}$, $\|\varphi\|_{p'} \leqq 1$, such that $\left|\int\limits_X f_{n_0} \cdot \xi_F\,\varphi\,d\iota\right| > \beta$. Moreover,

$$\int\limits_X f\,|\varphi|\,d\iota \geqq \left|\int\limits_F f_{n_0}\,\varphi\,d\iota\right| > \beta.$$

Since $|\varphi| \in \mathfrak{C}_{00}^+$ and $\||\varphi|\|_{p'} \leqq 1$, this proves (ii).

Suppose now that f is nonnegative and measurable and that $f \notin \mathfrak{L}_\infty$. For every $\beta > 0$, (11.33) implies that there is a compact set F such that $\iota(\{x \in F : f(x) > 2\beta\}) > 0$. Let $f_0 = \min(f, 2\beta)\,\xi_F$; clearly $f_0 \in \mathfrak{L}_\infty$ and $\|f_0\|_\infty = 2\beta$. By (12.13), there is a $\varphi \in \mathfrak{C}_{00}$ such that $\|\varphi\|_1 \leqq 1$ and $\left|\int\limits_X f_0\varphi\,d\iota\right| > \beta$. Then we have

$$\int\limits_X f\,|\varphi|\,d\iota \geqq \left|\int\limits_X f_0\,\varphi\,d\iota\right| > \beta. \quad \square$$

(12.15) Discussion. Our next aim is to show that the conjugate space (B.21) of $\mathfrak{L}_p$ $(1 \leqq p < \infty)$ can be identified with $\mathfrak{L}_{p'}$. Hölder's inequality (12.4.iii) and its trivial analogue (12.13.1) for $\mathfrak{L}_1$ and $\mathfrak{L}_\infty$ show that every element g of $\mathfrak{L}_{p'}$ defines a bounded linear functional on $\mathfrak{L}_p$:

(i) $f \to \int\limits_X fg\,d\iota = \Phi_g(f)$.

Theorem (12.13) implies as a special case that $\|\Phi_g\|$ [see (B.8)] is equal to $\|g\|_{p'}$. We wish to show that *every* bounded linear functional on $\mathfrak{L}_p$ has the form Φ_g for some $g \in \mathfrak{L}_{p'}$ $(1 \leqq p < \infty)$. To do this, we require another result, known as the Lebesgue-Radon-Nikodým theorem, which is of fundamental importance in its own right.

(12.16) Suppose that I, ι, $\mathscr{M}_\iota$, and $\mathscr{N}_\iota$ are as in §11. Consider another nonnegative linear functional J on $\mathfrak{C}_{00}(X)$, and construct for it just as for I the extended functionals $\bar{J}$ and $\bar{\bar{J}}$ and the Carathéodory outer

[1] This condition holds, for example, if $f \notin \mathfrak{L}_p$ and f vanishes outside of a set which is the union of countably many sets of finite measure. It *fails* for ξ_N if N is locally null, although ξ_N may not be in $\mathfrak{L}_p$.

measure $\overline{\overline{J}}(\xi_A)$, which we now write as $\eta(A)$. One way to form a functional J from I is to select a nonnegative ι-measurable function g such that $g\cdot\xi_F$ is in $\mathfrak{L}_1(X, \iota)$ for every compact set $F\subset X$: call such a function g a *weight function*. We can then define $J(\varphi)=\int\limits_X \varphi g\, d\iota$ for all $\varphi\in\mathfrak{C}_{00}$. We shall show that this description of J is equivalent with several relations between I and J and ι and η.

(12.17) Theorem [Lebesgue-Radon-Nikodým]. *Let I, J, ι and η be as in* (12.16) *and let $\mathscr{M}_\iota$, $\mathscr{M}_\eta$, $\mathscr{N}_\iota$, $\mathscr{N}_\eta$ be the classes of sets defined for ι and η as in* (11.28) *and* (11.26). *The following statements are equivalent.*

(i) *There is a weight function g on X such that*

$$J(\varphi)=\int\limits_X \varphi g\, d\iota \quad \textit{for all} \quad \varphi\in\mathfrak{C}_{00}^+.$$

(ii) *Let $A\subset X$ have the form $A=\bigcup\limits_{n=1}^{\infty} A_n$, where each A_n is ι-measurable and η-measurable and $\iota(A_n)<\infty$, $\eta(A_n)<\infty$. Then $\eta(A)=\int\limits_A g\, d\iota$, where g is a weight function depending only on η and ι.*

(iii) *If f is an η-measurable complex- or extended real-valued function on X such that $\int\limits_X |f|\, d\eta<\infty$, then there is an ι-measurable function f' on X such that $|f'|\leqq|f|$, $f=f'$ η-almost everywhere, and $\int\limits_X f\, d\eta=\int\limits_X f' g\, d\iota$. Again g is a weight function depending only on η and ι.*

(iv) $\mathscr{N}_\iota\subset\mathscr{N}_\eta$.

(v) *For every $\varphi\in\mathfrak{C}_{00}^+$ and for every positive number ε, there exists a positive number δ such that $\psi\in\mathfrak{C}_{00}^+$, $0\leqq\psi\leqq\varphi$, and $I(\psi)<\delta$ imply $J(\psi)<\varepsilon$.*

Proof. We begin by proving that (i) implies (ii). Suppose first that F is a compact subset of X. Using (11.22), we find open sets U_n and a compact set E such that

$$F\subset\cdots\subset U_{n+1}\subset U_n\subset\cdots\subset U_1\subset E,$$

$\lim\limits_{n\to\infty}\iota(U_n)=\iota(F)$, and $\lim\limits_{n\to\infty}\eta(U_n)=\eta(F)$. For $n=1, 2, \ldots$, let $\varphi_n\in\mathfrak{C}_{00}^+$ be such that $\varphi_n(F)=1$, $\varphi_n(U_n')=0$, and $\varphi_n(X)\subset[0, 1]$. Then $\lim\limits_{n\to\infty}\varphi_n(x)=\xi_F(x)$ η-almost everywhere and the functions φ_n are dominated by ξ_E, which is in $\mathfrak{L}_1(X, \eta)$. Also $\lim\limits_{n\to\infty}\varphi_n(x)\, g(x)=\xi_F(x)\, g(x)$ ι-almost everywhere and the functions $\varphi_n g$ are dominated by $\xi_E\cdot g\in\mathfrak{L}_1(X, \iota)$. Hence we may apply (i) and Lebesgue's theorem on dominated convergence to obtain

$$\eta(F)=\int\limits_X \xi_F\, d\eta=\lim_{n\to\infty}\int\limits_X \varphi_n\, d\eta=\lim_{n\to\infty}\int\limits_X \varphi_n g\, d\iota=\int\limits_X \xi_F g\, d\iota=\int\limits_F g\, d\iota. \quad (1)$$

Now let $A=\bigcup\limits_{n=1}^{\infty} A_n$, where each A_n is ι-measurable and η-measurable and $\iota(A_n)<\infty$, $\eta(A_n)<\infty$. By virtue of (11.32), $A=\left(\bigcup\limits_{n=1}^{\infty} F_n\right)\cup N$ where each F_n is compact and $\iota(N) = \eta(N) = 0$. We may obviously suppose that $F_n \subset F_{n+1}$ for all $n = 1, 2, \ldots$. Writing $D = \bigcup_{n=1}^{\infty} F_n$ and using (1), we see that

$$\eta(A) = \eta(D) = \lim_{n\to\infty} \eta(F_n) = \lim_{n\to\infty} \int_{F_n} g\,d\iota = \int_D g\,d\iota = \int_A g\,d\iota. \tag{2}$$

This completes the proof that (i) implies (ii).

We now show that (ii) implies (iii). We may obviously restrict ourselves to the case $f \geqq 0$. Apply (11.40) to f and the measure η to obtain f'. For every positive integer n and $k = 1, 2, \ldots, n2^n - 1$, let $A_{k,n} = \left\{x \in X : \frac{k}{2^n} \leqq f'(x) < \frac{k+1}{2^n}\right\}$. Each $A_{k,n}$ is plainly ι- and η-measurable and is the union of a countable family of sets for which ι and η are finite. By (ii), we have

$$\sum_{k=1}^{n2^n-1} \frac{k}{2^n}\,\eta(A_{k,n}) = \sum_{k=1}^{n2^n-1} \frac{k}{2^n} \int_{A_{k,n}} g\,d\iota = \int_X \left[\sum_{k=1}^{n2^n-1} \frac{k}{2^n}\,\xi_{A_{k,n}}\right] g\,d\iota. \tag{3}$$

The functions $\sum_{k=1}^{n2^n-1} \frac{k}{2^n}\,\xi_{A_{k,n}}$ increase monotonically to f', and so (3) implies that $\int_X f\,d\eta = \int_X f'\,d\eta = \int_X f'\,g\,d\iota$. Thus (iii) holds.

We next show that (iii) implies (iv). Let A be a set that is locally ι-null: $\iota(A \cap F) = 0$ for every compact set $F \subset X$. For each such F, there is a sequence $U_1 \supset U_2 \supset \cdots \supset U_n \supset \cdots \supset A \cap F$ of open sets such that U_1^- is compact and $\iota(U_n) < \frac{1}{n}$. Let $E = \bigcap_{n=1}^{\infty} U_n$; plainly $E \supset A \cap F$ and $\iota(E) = 0$. The function ξ_{U_n} is ι-measurable and $\eta(U_n) \leqq \eta(U_1^-) < \infty$. By (iii) there is a nonnegative function $f_n' \leqq \xi_{U_n}$ that is ι-measurable and for which $\eta(U_n) = \int_X f_n'\,g\,d\iota$. The functions $f_n'\,g$ are dominated by $\xi_{U_1} g$, which is in $\mathfrak{L}_1(X, \iota)$, and so, by LEBESGUE'S theorem on dominated convergence, we have $0 = \int_E \overline{\lim_{n\to\infty}}\, f_n'\,g\,d\iota \geqq \overline{\lim_{n\to\infty}} \int_X f_n'\,g\,d\iota = \overline{\lim_{n\to\infty}}\, \eta(U_n)$. Thus $\lim_{n\to\infty} \eta(U_n) = 0$, and so $\eta(A \cap F) = 0$. This shows that A is in $\mathscr{N}_\eta$, and (iv) is established.

We next prove that (iv) implies (v). Suppose in fact that (v) fails. Then there are a $\varphi_0 \in \mathfrak{C}_{00}^+$ and a positive number ε_0 such that for $n = 1, 2, \ldots$, there is a function $\psi_n \in \mathfrak{C}_{00}^+$ for which $0 \leqq \psi_n \leqq \varphi_0$, $I(\psi_n) < 2^{-n}$, and $J(\psi_n) \geqq \varepsilon_0$. Now set $h_n = \sup\{\psi_n, \psi_{n+1}, \ldots, \psi_{n+l}, \ldots\}$ $(n = 1, 2, \ldots)$, and $h = \lim_{n\to\infty} h_n$. Then we have $\int_X h_n\,d\iota \leqq \int_X \left[\sum_{k=n}^{\infty} \psi_k\right] d\iota = \sum_{k=n}^{\infty} \int_X \psi_k\,d\iota < 2 \cdot 2^{-n}$, and $\varepsilon_0 \leqq \int_X \psi_n\,d\eta \leqq \int_X h_n\,d\eta$ $(n = 1, 2, \ldots)$. Since $h_n \leqq \varphi_0$, LEBESGUE'S theorem on dominated convergence implies that

$$\int_X h\,d\iota = \lim_{n\to\infty} \int_X h_n\,d\iota = 0, \tag{4}$$

and

$$\int_X h\,d\eta = \lim_{n\to\infty} \int_X h_n\,d\eta \geqq \varepsilon_0 > 0. \tag{5}$$

From (4), (11.27), (5), and (11.32), we infer that there is a compact set $E \subset \{x \in X : h(x) > 0\}$ such that $\eta(E) > 0$ and $\iota(E) = 0$. The set E is plainly in $\mathcal{N}_\iota$ and not in $\mathcal{N}_\eta$. Thus (iv) fails.

It remains to prove that (v) implies (i). This proof requires several steps. Suppose first that $J \leqq I$ and that $\eta(X) < \infty$. Then for every $\varphi \in \mathfrak{C}_{00}^r$, CAUCHY's inequality (12.4.iii) implies that

$$\left|\int_X \varphi\,d\eta\right| \leqq \eta(X)^{\frac{1}{2}} \left[\int_X |\varphi|^2\,d\eta\right]^{\frac{1}{2}} \leqq \eta(X)^{\frac{1}{2}} \left[\int_X |\varphi|^2\,d\iota\right]^{\frac{1}{2}}.$$

These inequalities show that the mapping $\varphi \to \int_X \varphi\,d\eta$ is a linear functional on $\mathfrak{C}_{00}^r$ that is continuous in the norm $\|\varphi\|_2$ of $\mathfrak{L}_2^r(X, \iota)$. Since $\mathfrak{C}_{00}^r$ is dense in $\mathfrak{L}_2^r(X, \iota)$ (12.10), this mapping can be extended in one and only one way so as to be a bounded linear [real-valued!] functional Φ on $\mathfrak{L}_2^r(X, \iota)$ (B.11). By (B.45)[1], there is a function $g \in \mathfrak{L}_2^r(X, \iota)$ such that $\Phi(f) = \int_X f g\,d\iota$ for all $f \in \mathfrak{L}_2^r(X, \iota)$. For $\varphi \in \mathfrak{C}_{00}^r$, we thus have

$$\int_X \varphi g\,d\iota = \Phi(\varphi) = \int_X \varphi\,d\eta.$$

For $\varphi \in \mathfrak{C}_{00}^+$, the inequalities $0 \leqq J \leqq I$ imply that

$$0 \leqq \int_X \varphi g\,d\iota \leqq \int_X \varphi\,d\iota.$$

It is easy to see from these inequalities that $0 \leqq g(x) \leqq 1$ for all $x \in X$ except for an ι-null set. We may thus suppose that $0 \leqq g(x) \leqq 1$ for *all* $x \in X$. In particular, $g \cdot \xi_E \in \mathfrak{L}_1^r(X, \iota)$ for all compact subsets E of X. This proves (i) in the very special case $J \leqq I$ and $\eta(X) < \infty$.

Second, suppose that $J \leqq I$ and that $\eta(X) = \infty$. Let $\mathcal{F}$ be a family of compact sets constructed for the measure ι as in (11.39). For each $F \in \mathcal{F}$, let I_F and J_F be the linear functionals on $\mathfrak{C}_{00}$ defined by $I_F(\varphi) = \int_F \varphi\,d\iota$ and $J_F(\varphi) = \int_F \varphi\,d\eta$, $\varphi \in \mathfrak{C}_{00}$. We now show that $J_F \leqq I_F$ for $F \in \mathcal{F}$. As in the first paragraph of the proof, there is a dominated sequence $\{\varphi_n\}_{n=1}^\infty$ of functions in $\mathfrak{C}_{00}^+$ such that $\lim_{n\to\infty} \varphi_n(x) = \xi_F(x)$ ι-almost every-

[1] Since (B.45) applies to *complex* Hilbert spaces, a little explanation is called for. Extend Φ to a complex linear functional Φ_c on $\mathfrak{L}_2(X, \iota)$ (B.38). Then apply (B.45) to obtain a function $g \in \mathfrak{L}_2(X, \iota)$ such that $\Phi_c(f) = \int_X f g\,d\iota$ for $f \in \mathfrak{L}_2(X, \iota)$. It is easy to see that g must be real-valued.

where and η-almost everywhere. Hence if $\psi \in \mathfrak{C}_{00}^+$, then

$$J_F(\psi) = \int_F \psi \, d\eta = \int_X \xi_F \psi \, d\eta = \lim_{n\to\infty} \int_X \varphi_n \psi \, d\eta$$
$$\leqq \lim_{n\to\infty} \int_X \varphi_n \psi \, d\iota = \int_X \xi_F \psi \, d\iota = I_F(\psi).$$

Write ι_F and η_F for the measures corresponding to I_F and J_F, respectively. If $\psi \in \mathfrak{C}_{00}^+$ and $\psi \leqq 1$, then $J_F(\psi) = \int_F \psi \, d\eta \leqq \eta(F)$. Hence by (11.11), $\eta_F(X)$ is finite. Consequently the preceding case applies to the functionals J_F and I_F. Thus for every $F \in \mathscr{F}$, we find a function $g_F \in \mathfrak{L}_2^r(X, \iota_F)$ such that $0 \leqq g_F(x) \leqq 1$ everywhere on X and for all $\varphi \in \mathfrak{C}_{00}$,

$$\int_F \varphi \, d\eta = \int_F \varphi g_F \, d\iota.$$

We may obviously suppose that $g_F(x) = 0$ for $x \in F'$. Now define the function g as $\sum_{F \in \mathscr{F}} g_F$. That is, $g(x) = g_F(x)$ for $x \in F$ $(F \in \mathscr{F})$, and $g(x) = 0$ for $x \in (\bigcup \{F \in \mathscr{F}\})'$. It is easy to see that g is an ι-measurable function. In fact, let $A = \{x \in X : g(x) > \alpha\}$ where $\alpha \geqq 0$. Then $A \cap F = \{x \in X : g_F(x) > \alpha\}$ is ι-measurable for each $F \in \mathscr{F}$, and $A = \bigcup_{F \in \mathscr{F}} (A \cap F)$. If U is an open set of finite positive ι-measure, we have $U \cap F \neq \varnothing$ for only countably many $F \in \mathscr{F}$, say $F_1, F_2, \ldots$ (11.39.iii). Hence

$$\iota(U \cap A) + \iota(U \cap A') = \iota\Big(U \cap \Big(\bigcup_{n=1}^{\infty} (A \cap F_n)\Big)\Big) + \iota\Big(U \cap \Big(\bigcup_{n=1}^{\infty} (A \cap F_n)\Big)'\Big) = \iota(U).$$

By (11.31), A is ι-measurable. It is obvious that $g \cdot \xi_E \in \mathfrak{L}_1(X, \iota)$ for all compact sets $E \subset X$.

Now let φ be any function in $\mathfrak{C}_{00}$ and let E be a compact subset of X such that $\varphi(E') = 0$. It follows at once from (11.39.iii) that E has nonvoid intersection with only countably many $F \in \mathscr{F}$, say $F_1, F_2, \ldots, F_n, \ldots$. Write $N = (\bigcup \{F \in \mathscr{F}\})'$. As noted in (11.39), N is locally ι-null, and since $J \leqq I$, it follows that N is locally η-null. Hence we have

$$J(\varphi) = \int_X \varphi \, d\eta = \int_E \varphi \, d\eta = \int_{\bigcup_{n=1}^{\infty} F_n} \varphi \, d\eta + \int_{N \cap E} \varphi \, d\eta = \sum_{n=1}^{\infty} \int_{F_n} \varphi \, d\eta$$
$$= \sum_{n=1}^{\infty} \int_{F_n} \varphi g \, d\iota = \int_X \varphi g \, d\iota.$$

Thus we have proved (i) if $J \leqq I$.

Consider finally the general case. For every positive integer n, let $J_n = \min(nI, J)$ where $\min(nI, J)$ is the linear functional on $\mathfrak{C}_{00}$ defined in (B.34). We first show that $J(\varphi) = \lim_{n\to\infty} J_n(\varphi)$ for each $\varphi \in \mathfrak{C}_{00}^+$.[1] That is, we assert that for an arbitrary $\alpha > 0$,

$$\inf\{nI(\psi) + J(\varphi - \psi) : \psi \in \mathfrak{C}_{00}^+ \text{ and } \psi \leqq \varphi\} \geqq J(\varphi) - \alpha \tag{6}$$

[1] The reader will note that (v) is used in the proof only to establish this equality.

if n is sufficiently large. We restate (6) as:

$$nI(\psi) \geqq J(\psi) - \alpha \tag{7}$$

if $0 \leqq \psi \leqq \varphi$, $\psi \in \mathfrak{C}_{00}^+$, and n is sufficiently large. For $J(\varphi) \leqq \alpha$, (7) is no restriction. For $J(\varphi) > \alpha$, we use (v) to assert the existence of a $\delta > 0$ such that if $0 \leqq \psi \leqq \varphi$, $\psi \in \mathfrak{C}_{00}^+$, and $J(\psi) \geqq \alpha$, then $I(\psi) \geqq \delta$. For every integer $n > \dfrac{J(\varphi) - \alpha}{\delta}$, we find that if $J(\psi) - \alpha \geqq 0$, then $nI(\psi) > J(\varphi) - \alpha \geqq J(\psi) - \alpha$. Thus (7) and hence (6) hold.

Plainly we have $J_n \leqq nI$ for $n = 1, 2, \ldots$, and so by the preceding case there is an ι-measurable function g_n such that $0 \leqq g_n \leqq n$ and $J_n(\varphi) = \int_X \varphi g_n \, d\iota$ for all $\varphi \in \mathfrak{C}_{00}$. Since $J_1 \leqq J_2 \leqq \cdots$, the sequence of functions $g_1, g_2, \ldots$ is nondecreasing [ignoring as usual a locally ι-null set], and hence we have

$$J(\varphi) = \lim_{n \to \infty} J_n(\varphi) = \int_X \varphi \left[\lim_{n \to \infty} g_n\right] d\iota,$$

for $\varphi \in \mathfrak{C}_{00}^+$. Thus we set $g = \lim_{n \to \infty} g_n$ [locally ι-almost everywhere], and the proof is complete. □

Our first application of the LEBESGUE-RADON-NIKODÝM theorem follows.

(12.18) Theorem. *For $1 \leqq p < \infty$, the conjugate space of $\mathfrak{L}_p = \mathfrak{L}_p(X, \iota)$ is $\mathfrak{L}_{p'}$, in the sense that for every bounded linear functional Φ on $\mathfrak{L}_p$ there is a function $g \in \mathfrak{L}_{p'}$ such that*

(i) *$\Phi(f) = \int_X fg \, d\iota$ for all $f \in \mathfrak{L}_p$,*

and $\|\Phi\| = \|g\|_{p'}$. [Recall that we take $1' = \infty$.]

Proof. We begin with the space $\mathfrak{L}_p^r$. Let Ψ be any *nonnegative* bounded linear functional on $\mathfrak{L}_p^r$. Since $\mathfrak{C}_{00}^r$ is a [real linear] subspace of $\mathfrak{L}_p^r$, we may regard Ψ as a linear functional on $\mathfrak{C}_{00}^r$ that is nonnegative on $\mathfrak{C}_{00}^+$: hence (12.17) may be applied, if relevant, to Ψ and the original functional I in terms of which $\mathfrak{L}_p^r(X, \iota)$ was defined. Consider any function φ in $\mathfrak{C}_{00}^+$. Then, if $0 \leqq \psi \leqq \varphi$ and $\psi \in \mathfrak{C}_{00}^+$, we have

$$\Psi(\psi) \leqq \|\Psi\| \left[\int_X \psi^p \, d\iota\right]^{1/p} \leqq \|\Psi\| \cdot \|\varphi\|_u^{1/p'} \left[\int_X \psi \, d\iota\right]^{1/p}.$$

This implies that Ψ and I satisfy condition (12.17.v), and hence by Theorem (12.17), there is an ι-measurable function $g \geqq 0$ on X such that $g \cdot \xi_F$ is in $\mathfrak{L}_1(X, \iota)$ for all compact $F \subset X$ and

$$\Psi(\varphi) = \int_X \varphi g \, d\iota \quad \text{for all} \quad \varphi \in \mathfrak{C}_{00}^r. \tag{1}$$

For $\varphi \in \mathfrak{C}_{00}^r$ and $\|\varphi\|_p \leqq 1$, we thus have

$$\left|\int_X \varphi g \, d\iota\right| \leqq \int_X |\varphi| \, g \, d\iota \leqq \|\Psi\|.$$

If F is a compact subset of X, then

$$\left|\int_F \varphi g\, d\iota\right| \leqq \int_F |\varphi|\, g\, d\iota \leqq \|\Psi\|.$$

By (12.14), therefore, we have $g\xi_F \in \mathfrak{L}_{p'}$ and by (12.13), $\|g\xi_F\|_{p'} \leqq \|\Psi\|$. Let $\alpha = \sup\{\|g\xi_F\|_{p'} : F$ is compact and $F \subset X\}$. Let $F_1 \subset F_2 \subset \cdots \subset F_n \subset \cdots$ be a sequence of compact sets such that $\lim\limits_{n\to\infty} \|g\xi_{F_n}\|_{p'} = \alpha$. Let $D = \bigcup\limits_{n=1}^{\infty} F_n$. Then g is locally almost everywhere 0 on D'. If this were not the case, there would be a compact subset E of D' such that $\iota(E) > 0$ and $g(x) > 0$ for $x \in E$ [see (11.33)]. Then we have $\int_E g^{p'}\, d\iota > 0$ and $F_n \cup E$ is compact for $n = 1, 2, \ldots$. Since $\lim\limits_{n\to\infty} \int_{F_n} g^{p'}\, d\iota = \alpha^{p'}$, we then obtain

$$\int_{F_n \cup E} g^{p'}\, d\iota = \int_{F_n} g^{p'}\, d\iota + \int_E g^{p'}\, d\iota > \alpha^{p'}$$

ior n sufficiently large. Redefining g to be 0 on D', we obtain a function g fn $\mathfrak{L}_{p'}$ for which (1) holds. From (12.13), we have $\|\Psi\| = \|g\|_{p'}$. Since $\mathfrak{C}_{00}^r$ is dense in $\mathfrak{L}_p^r$ (12.10), the representation (1) holds for all $f \in \mathfrak{L}_p^r$.

For an arbitrary bounded linear functional Φ on $\mathfrak{L}_p^r$, we use (B.37) to write Φ as $\Psi_1 - \Psi_2$, where both Ψ_1 and Ψ_2 are nonnegative. If g_1 and g_2 are as in (1) for Ψ_1 and Ψ_2, respectively, then we plainly obtain (i) with $g = g_1 - g_2$, for [real] functionals on $\mathfrak{L}_p^r$. The representation (i) and Theorem (12.13) show that $\|\Phi\| = \|g\|_{p'}$.

We now consider an arbitrary bounded linear functional Φ on the complex Banach space $\mathfrak{L}_p$. Then for each $f \in \mathfrak{L}_p$, we have $\Phi(f) = \Phi_1(f) + i\Phi_2(f)$, where $\Phi_1(f)$ and $\Phi_2(f)$ are real. It is easy to see that: $\Phi_j(f + f') = \Phi_j(f) + \Phi_j(f')$ and $\Phi_j(\alpha f) = \alpha\Phi_j(f)$ for $f, f' \in \mathfrak{L}_p$, $\alpha \in R$, and $j = 1, 2$; $\Phi_2(f) = -\Phi_1(if)$; and $\sup\{|\Phi_j(f)| : \|f\|_p \leqq 1\} \leqq \|\Phi\|$ $(j = 1, 2)$. Considered on the [real] linear space $\mathfrak{L}_p^r$, therefore, Φ_1 is a bounded linear functional, and by the previous discussion admits a representation $\Phi_1(\varphi) = \int_X \varphi g\, d\iota$ for $\varphi \in \mathfrak{L}_p^r$. Similarly Φ_1 admits an integral representation on the [real] subspace of $\mathfrak{L}_p$ consisting of all functions $i\psi$ with $\psi \in \mathfrak{L}_p^r$: hence we have $\Phi_1(i\psi) = \int_X \psi h\, d\iota$, where $h \in \mathfrak{L}_{p'}^r$. Thus for $f = \varphi + i\psi \in \mathfrak{L}_p$, we have

$$\begin{aligned}\Phi(f) &= \Phi_1(f) + i\Phi_2(f) = \Phi_1(f) - i\Phi_1(if)\\ &= \Phi_1(\varphi) + \Phi_1(i\psi) - i[-\Phi_1(\psi) + \Phi_1(i\varphi)] = \int_X [\varphi + i\psi][g - ih]\, d\iota.\end{aligned}$$

Since g and h are in $\mathfrak{L}_{p'}^r$, $g - ih$ is in $\mathfrak{L}_{p'}$, so that (i) is established. A final appeal to Theorem (12.13) shows that $\|g - ih\|_{p'} = \|\Phi\|$. □

Notes

This section necessarily departs somewhat from classical treatments of $\mathfrak{L}_p$ spaces as found for example in Hardy, Littlewood, and Pólya [1] and Dunford and Schwartz [1]. This of course arises because of the possible existence of locally null, nonnull sets; (12.2), (12.11), (12.12), (12.14), (12.17), and (12.18) all differ from the corresponding classical assertions. The very useful Theorem (12.10) obviously has no analogue for $\mathfrak{L}_p$ spaces defined for abstract measure spaces.

Lemma (12.3) is due to W. H. YOUNG [1], and the proof of (12.6) to F. RIESZ [3], p. 45. Theorems (12.13) and (12.14) are adapted from BOURBAKI [2], Ch. IV, §6, N° 4, Remarque 2). The classical case of (12.13) and (12.14) is due to F. RIESZ [2]. Theorem (12.17) is of vital importance in harmonic analysis. For $X=R$ and $\iota=$Lebesgue measure, it is due to LEBESGUE [1]; for X an interval in R^n and general set functions, to RADON [1]; for abstract measure spaces, to NIKODÝM [1]. Our treatment is based on BOURBAKI [3], Ch. V, §5, N° 5, Théorème 2. For $p>1$, Theorem (12.18) is the translation to our setting of a classical result of F. RIESZ [2], p. 475. For $p=1$, the classical case of (12.18) is due to H. STEINHAUS [1].

§ 13. Integration on product spaces

(13.1) Discussion. Throughout (13.1)—(13.14), X and Y will denote arbitrary nonvoid locally compact Hausdorff spaces, and I and J will denote arbitrary nonnegative, nonzero, linear functionals on $\mathfrak{C}_{00}(X)$ and $\mathfrak{C}_{00}(Y)$ respectively. For I and J, we construct the extended functionals $\bar{I}, \bar{\bar{I}}, \bar{J}$, and $\bar{\bar{J}}$, and the corresponding outer measures ι and η. The families of sets $\mathscr{M}_\iota$, $\mathscr{N}_\iota$ and $\mathscr{M}_\eta$, $\mathscr{N}_\eta$ are then defined as in §11. Note that $\mathscr{M}_\iota, \mathscr{N}_\iota \subset \mathscr{P}(X)$ and $\mathscr{M}_\eta, \mathscr{N}_\eta \subset \mathscr{P}(Y)$.

We are concerned with the product space $X\times Y$ and with functionals, measures, and integration on $X\times Y$ defined by means of I, J and ι, η. Let f be a complex- or extended real-valued function on $X\times Y$ such that for some $x_0 \in X$, the function $\varphi\colon y \rightarrow f(x_0, y)$ is in $\mathfrak{C}_{00}(Y)$. We then write $J_y(f(x_0, y))$ for $J(\varphi)$; the expressions $I_x(f(x, y_0))$, $\bar{I}_x(f(x, y_0))$, $\bar{J}_y(f(x_0, y))$, $\bar{\bar{I}}_x(f(x, y_0))$, and $\bar{\bar{J}}_y(f(x_0, y))$ are defined analogously. For functions f on X and g on Y, let fg denote the function $(x, y) \rightarrow f(x)\,g(y)$ defined on $X\times Y$. [The ranges of f and g will always be such that the definition is meaningful.]

(13.2) Theorem. *For every* $f\in\mathfrak{C}_{00}(X\times Y)$, *we have* $J_y(f(x, y))\in\mathfrak{C}_{00}(X)$ *and* $I_x(f(x, y))\in\mathfrak{C}_{00}(Y)$. *Also the equality*

(i) $I_x[J_y(f(x, y))] = J_y[I_x(f(x, y))]$

obtains. The functional defined by (i) *is a nonnegative, nonzero, linear functional on* $\mathfrak{C}_{00}(X\times Y)$.

Proof. We may suppose that f is real-valued. For $f\in\mathfrak{C}_{00}^r(X\times Y)$, there are obviously open sets $U\subset X$ and $V\subset Y$ such that U^- and V^- are compact and $f(x, y)=0$ for $(x, y)\in(U\times V)'$. [Consider the projections of the set $\{(x, y) \in X \times Y : f(x, y) \neq 0\}$ onto X and Y.] The open subset $U \times V$ of $X \times Y$ is evidently a locally compact Hausdorff space in its relative topology. Consider the space $\mathfrak{H}$ of all functions h on $U \times V$ of the form $\sum_{j=1}^{m} \varphi_j \psi_j$ where $\varphi_1, \ldots, \varphi_m \in \mathfrak{C}_{00}^r(U)$ and $\psi_1, \ldots, \psi_m \in \mathfrak{C}_{00}^r(V)$. It is clear that $\mathfrak{H}$ is a subalgebra of $\mathfrak{C}_0^r(U \times V)$ that separates points: if (u_1, v_1) and (u_2, v_2) are in $U \times V$ and $u_1 \neq u_2$, there is a φ such that $\varphi(u_1) = 1$, $\varphi(u_2) = 0$ and a ψ such that $\psi(v_1) = 1$. The STONE-WEIERSTRASS theorem implies that $\mathfrak{H}$ is uniformly dense in $\mathfrak{C}_0^r(U \times V)$.[1] Now the function f, with its domain restricted to $U \times V$, is plainly in $\mathfrak{C}_0^r(U \times V)$. Hence f can be arbitrarily uniformly approximated on $U \times V$ by functions in $\mathfrak{H}$. We extend the functions $\sum_{j=1}^{m} \varphi_j \psi_j$ in $\mathfrak{H}$ by letting $\varphi_j(x) = 0$ for $x \in X \cap U'$ and $\psi_j(y) = 0$ for $y \in Y \cap V'$ $(j = 1, \ldots, m)$. Then for every $\varepsilon > 0$, there are functions $\varphi_1, \ldots, \varphi_m \in \mathfrak{C}_{00}^r(X)$ vanishing on U' and functions $\psi_1, \ldots, \psi_m \in \mathfrak{C}_{00}^r(Y)$ vanishing on V' such that

$$\left| f(x, y) - \sum_{j=1}^{m} \varphi_j(x)\, \psi_j(y) \right| < \varepsilon \quad \text{for all} \quad (x, y) \in X \times Y. \tag{1}$$

By Theorem (11.6), there are positive numbers α and β such that $|I(\varphi)| \leqq \alpha \|\varphi\|_u$ if $\varphi \in \mathfrak{C}_{00}(X)$ and $\varphi(U^{-\prime}) = 0$, and $|J(\psi)| \leqq \beta \|\psi\|_u$ if $\psi \in \mathfrak{C}_{00}(Y)$ and $\psi(V^{-\prime}) = 0$.

Now consider a *fixed* $x \in X$. The function

$$f(x, y) - \sum_{j=1}^{m} \varphi_j(x)\, \psi_j(y),$$

defined for $y \in Y$, is obviously in $\mathfrak{C}_{00}^r(Y)$ and vanishes on $V' \supset V^{-\prime}$. Hence (1) shows that

$$\left| J_y(f(x, y)) - \sum_{j=1}^{m} \varphi_j(x)\, J(\psi_j) \right| < \beta \varepsilon. \tag{2}$$

[1] There are several forms of the STONE-WEIERSTRASS theorem. A set $\mathfrak{F}$ of functions on a set X is said to *separate points of* X if for $x, y \in X$, $x \neq y$, there is an $f \in \mathfrak{F}$ such that $f(x) \neq f(y)$. Several versions of the STONE-WEIERSTRASS theorem follow.

(a) For a compact Hausdorff space X, a subalgebra of $\mathfrak{C}^r(X)$ that separates points of X and vanishes identically at no point of X is uniformly dense in $\mathfrak{C}^r(X)$.

(b) For a compact Hausdorff space X, a subalgebra of $\mathfrak{C}(X)$ that separates points, vanishes identically at no point of X, and is closed under complex conjugation, is uniformly dense in $\mathfrak{C}(X)$.

(c) For a locally compact Hausdorff space X, a subalgebra of $\mathfrak{C}_0^r(X)$ that separates points and vanishes identically at no point of X is uniformly dense in $\mathfrak{C}_0^r(X)$.

(d) For a locally compact Hausdorff space X, a subalgebra of $\mathfrak{C}_0(X)$ that separates points, vanishes identically at no point of X, and is closed under complex conjugation, is uniformly dense in $\mathfrak{C}_0(X)$.

For a detailed proof of (a) and sketches of proofs of (b)—(d), see for example HEWITT and STROMBERG [2], (7.30) and (7.37).

The estimate (2) is independent of $x \in X$. Thus the function $J_y(f(x,y))$ is the uniform limit on X of functions $\sum_{j=1}^{m} J(\psi_j)\varphi_j \in \mathfrak{C}_{00}^r(X)$, each vanishing on $U' \supset U^{-\prime}$ It follows that $J_y(f(x,y))$ is in $\mathfrak{C}_{00}^r(X)$ and vanishes on U'. From (2) we also infer that

$$\left| I_x[J_y(f(x,y))] - \sum_{j=1}^{m} I(\varphi_j)\, J(\psi_j) \right| < \alpha\beta\varepsilon. \tag{3}$$

Precisely the same argument shows that $I_x(f(x,y))$, *qua* function of y, is in $\mathfrak{C}_{00}^r(Y)$ and vanishes on V'; also

$$\left| J_y[I_x(f(x,y))] - \sum_{j=1}^{m} I(\varphi_j)\, J(\psi_j) \right| < \beta\alpha\varepsilon. \tag{4}$$

Since ε is arbitrary and α and β are fixed as soon as f is fixed, (3) and (4) establish (i). The linearity and nonnegativity of the functional defined by (i) are obvious. If $I(\varphi) \neq 0$ and $J(\psi) \neq 0$, then $I_x[J_y(\varphi(x)\psi(y))] = I(\varphi)\, J(\psi) \neq 0$. □

(13.3) Definition. The linear functional defined by (13.2.i) is called the *product* of the functionals I and J and is written $I \times J$. The outer measure on $\mathscr{P}(X \times Y)$ defined as in §11 from $I \times J$ is called the *product* of the measures ι and η and is denoted by $\iota \times \eta$.

(13.4) Theorem. *For* $g \in \mathfrak{M}^+(X \times Y)$, *we have* $\bar{J}_y(g(x,y)) \in \mathfrak{M}^+(X)$ *and* $\bar{I}_x(g(x,y)) \in \mathfrak{M}^+(Y)$. *Furthermore*

(i) $\overline{I \times J}(g) = \bar{I}_x[\bar{J}_y(g(x,y))] = \bar{J}_y[\bar{I}_x(g(x,y))]$.

Proof. Let $\mathfrak{U}$ be the set of functions $f \in \mathfrak{C}_{00}^+(X \times Y)$ such that $0 \leq f \leq g$. As shown in (13.2), each function $J_y(f(x,y))$ is in $\mathfrak{C}_{00}^+(X)$. By (11.10.v), we have $g = \sup\{f : f \in \mathfrak{U}\}$. Hence for each fixed $x_0 \in X$, we have $g(x_0, y) = \sup\{f(x_0, y) : f \in \mathfrak{U}\}$, for all $y \in Y$. Each function $f(x_0, y)$ is in $\mathfrak{C}_{00}^+(Y)$, and Theorem (11.13) implies that $\bar{J}_y(g(x_0, y)) = \sup\{J_y(f(x_0, y)) : f \in \mathfrak{U}\}$. This being true for all $x_0 \in X$, (11.10.iii) implies that $\bar{J}_y(g(x,y))$ *qua* function of x is in $\mathfrak{M}^+(X)$. The argument for $\bar{I}_x(g(x,y))$ is the same. Using (11.11), (13.3), and (11.13) twice, we have

$$\begin{aligned}\overline{I \times J}(g) &= \sup\{I \times J(f) : f \in \mathfrak{U}\} = \sup\{I_x[J_y(f(x,y))] : f \in \mathfrak{U}\} \\ &= \bar{I}_x[\sup\{J_y(f(x,y)) : f \in \mathfrak{U}\}] = \bar{I}_x[\bar{J}_y(g(x,y))].\end{aligned}$$

The equality $\overline{I \times J}(g) = \bar{J}_y[\bar{I}_x(g(x,y))]$ is proved similarly. □

(13.5) Corollary. *If h is any nonnegative extended real-valued function on* $X \times Y$, *then*

(i) $\overline{\overline{I \times J}}(h) \geq \max\{\bar{\bar{I}}_x[\bar{\bar{J}}_y(h(x,y))], \bar{\bar{J}}_y[\bar{\bar{I}}_x(h(x,y))]\}$.

Proof. If $g \in \mathfrak{M}^+(X \times Y)$ and $g \geq h$, then we have

$$\overline{I \times J}(g) = \bar{I}_x[\bar{J}_y(g(x,y))] = \bar{J}_y[\bar{I}_x(g(x,y))].$$

The monotonicity of $\overline{\overline{I}}$ and $\overline{\overline{J}}$ (11.17.ii) implies at once that

$$\overline{\overline{I}}_x[\overline{\overline{J}}_y(g(x,y))] \geqq \overline{\overline{I}}_x[\overline{\overline{J}}_y(h(x,y))].$$

Now take the infimum over all g. □

For functions of the form $f(x)\,g(y)$, we can improve upon (13.5).

(13.6) Theorem. *Let f and g be arbitrary nonnegative extended real-valued functions defined on X and Y respectively. Then*

(i) $\overline{\overline{I\times J}}(fg)=\overline{\overline{I}}(f)\,\overline{\overline{J}}(g)$
unless one of $\overline{\overline{I}}(f), \overline{\overline{I}}(g)$ *is* ∞ *and the other is* 0.

Proof. The inequality

$$\overline{\overline{I\times J}}(fg)\leqq\overline{\overline{I}}(f)\,\overline{\overline{J}}(g) \tag{1}$$

is all that needs to be proved, in view of (13.5). The only case to be verified is that in which $\overline{\overline{I}}(f)<\infty$, $\overline{\overline{J}}(g)<\infty$. Let β and γ be any real numbers greater than $\overline{\overline{I}}(f)$ and $\overline{\overline{J}}(g)$, respectively. Then there are functions $f'\in\mathfrak{M}^+(X)$ and $g'\in\mathfrak{M}^+(Y)$ such that $f\leqq f'$, $g\leqq g'$, $\bar{I}(f')<\beta$, and $\bar{I}(g')<\gamma$. The function $f'g'$ is evidently in $\mathfrak{M}^+(X\times Y)$ [recall that $0\cdot\infty=0$] and so we have $\beta\gamma>\bar{I}(f')\,\bar{J}(g')=\bar{I}_x[\bar{J}_y(f'(x)\;g'(y))]=\overline{I\times J}(f'\,g')\geqq\overline{\overline{I\times J}}(fg)$. This establishes (1). □

We now take up the problem of computing integrals with respect to $\iota\times\eta$ in terms of integrals with respect to ι and integrals with respect to η.

(13.7) Theorem. *Let A be a subset of $X\times Y$. If A is $\iota\times\eta$-null, then the set $A_x=\{y\in Y:(x,y)\in A\}$ is η-null for ι-almost all $x\in X$ and the set $A_y=\{x\in X:(x,y)\in A\}$ is ι-null for η-almost all $y\in Y$.*[1]

Proof. We consider only A_x. If $\iota\times\eta(A)=0$, (13.5) implies that $0=\overline{\overline{I\times J}}(\xi_A)\geqq\overline{\overline{I}}_x[\overline{\overline{J}}_y(\xi_A(x,y))]=\overline{\overline{I}}_x(\eta(A_x))\geqq 0$. Theorem (11.27) shows that $\eta(A_x)$ vanishes for ι-almost all $x\in X$. □

(13.8) Theorem [FUBINI'S theorem]. *Let f be in $\mathfrak{L}_1(X\times Y,\iota\times\eta)$. Then $f(x,y)$* qua *function of x is in $\mathfrak{L}_1(X,\iota)$ for η-almost all $y\in Y$, and $f(x,y)$* qua *function of y is in $\mathfrak{L}_1(Y,\eta)$ for ι-almost all $x\in X$. Furthermore the function of x defined by $x\rightarrow\int\limits_Y f(x,y)\,d\eta(y)$ where the integral exists and 0 elsewhere is in $\mathfrak{L}_1(X,\iota)$;*[2] *similarly for $\int\limits_X f(x,y)\,d\iota(x)$. Finally, we have*

---

[1] The sets A_x and A_y are called *sections* of A.

[2] Throughout this book, the function $x\rightarrow\int\limits_Y f(x,y)\,d\eta(y)$ will be defined everywhere on X by this convention; similarly for $y\rightarrow\int\limits_X f(x,y)\,d\iota(x)$.

(i) $\int\limits_{X\times Y} f(x,y)\, d\iota\times\eta(x,y) = \int\limits_X\int\limits_Y f(x,y)\, d\eta(y)\, d\iota(x)$
$= \int\limits_Y\int\limits_X f(x,y)\, d\iota(x)\, d\eta(y)$.[1]

Proof. Consider first a *fixed* $\iota\times\eta$-measurable function f for which $\int\limits_{X\times Y} |f|\, d\iota\times\eta < \infty$. There is no loss of generality in supposing throughout the present proof that f is nonnegative. We first show that the function $x\to f(x,y)$ is ι-measurable for η-almost all y. By (11.40), there is a function f' defined on $X\times Y$ such that $\int\limits_{X\times Y} f'(x,y)\, d\iota\times\eta(x,y) = \int\limits_{X\times Y} f(x,y)\, d\iota\times\eta(x,y)$, $f'\leqq f$, and the set $\{(x,y)\in X\times Y: f'(x,y)>\alpha\}$ is σ-compact for every $\alpha\geqq 0$. Thus $f=f'+h$, where h is a nonnegative function on $X\times Y$ that vanishes except for a set N of $\iota\times\eta$-measure 0. Let P be the set of $y\in Y$ for which $\iota\{x\in X:(x,y)\in N\}=0$; by (13.7), we have $\eta(P')=0$. Now if $y\in P$ and $\alpha\geqq 0$, we have

$$\left.\begin{aligned}\{x\in X: f(x,y)>\alpha\} = \{x\in X: f'(x,y)>\alpha\}\cup\{x\in X:(x,y)\in N \\ \text{and } h(x,y)>\alpha - f'(x,y)\}.\end{aligned}\right\} \quad (1)$$

The first set on the right side of (1) is σ-compact and the second set is ι-null, so that their union is ι-measurable. Similarly the function $y\to f(x,y)$ is η-measurable for ι-almost all x.

Now by (12.10), there is a sequence $\varphi_1, \varphi_2, \ldots$ of functions in $\mathfrak{C}_{00}^{r}(X\times Y)$ such that $\lim\limits_{n\to\infty} \int\limits_{X\times Y} |f(x,y)-\varphi_n(x,y)|\, d\iota\times\eta(x,y)=0$. From (13.5) and (11.36), we can then write

$$\lim_{n\to\infty} \bar{\bar{I}}_x\left[\int\limits_Y |f(x,y)-\varphi_n(x,y)|\, d\eta(y)\right] = 0, \quad (2)$$

where we replace $\int\limits_Y |f(x,y)-\varphi_n(x,y)|\, d\eta(y)$ by 0 if this integral is undefined, *i.e.*, if $f(x,y)-\varphi_n(x,y)$ is not η-measurable. By (11.27), there is a subsequence of the φ_n's [which we again write as $\varphi_1, \varphi_2, \ldots$] such that

$$\lim_{n\to\infty} \int\limits_Y |f(x,y)-\varphi_n(x,y)|\, d\eta(y) = 0 \ \iota\text{-almost everywhere on } X. \quad (3)$$

Since $\int\limits_Y \varphi_n(x,y)\, d\eta(y) = J_y(\varphi_n(x,y))$ [use (11.36)] is in $\mathfrak{C}_{00}^{r}(X)$ as a function of x (13.2), we infer from (3) that $\int\limits_Y f(x,y)\, d\eta(y)$ is ι-measurable: it is the limit ι-almost everywhere of functions in $\mathfrak{C}_{00}^{r}(X)$. It now follows

[1] Here and throughout this book we use $\int \ldots d$ like parentheses. Thus the right member of (i) is to be interpreted as

$$\int\limits_Y \left[\int\limits_X f(x,y)\, d\iota(x)\right] d\eta(y).$$

from (2) that

$$\lim_{n\to\infty} \int_X \left| \int_Y f(x, y)\, d\eta(y) - \int_Y \varphi_n(x, y)\, d\eta(y) \right| d\iota(x) = 0, \tag{4}$$

and so the function $x \to \int_Y f(x, y)\, d\eta(y)$ is in $\mathfrak{L}_1(X, \iota)$. Finally, taking note of (13.3), we have

$$\begin{aligned} &\left| \int_{X\times Y} f(x, y)\, d\iota\times\eta(x, y) - \int_X \int_Y f(x, y)\, d\eta(y)\, d\iota(x) \right| \\ &\quad \leqq \int_{X\times Y} |f(x, y) - \varphi_n(x, y)|\, d\iota\times\eta(x, y) \\ &\quad + \int_X \left| \int_Y f(x, y)\, d\eta(y) - \int_Y \varphi_n(x, y)\, d\eta(y) \right| d\iota(x). \end{aligned}$$

This implies the first equality in (i). The remaining equalities are obvious.

To complete the proof, it suffices to notice that if $g(x, y) = f(x, y)$ $\iota\times\eta$-almost everywhere, then for ι-almost all x, the equality $f(x, y) = g(x, y)$ holds for η-almost all y, and for η-almost all y, $f(x, y) = g(x, y)$ for ι-almost all x. This follows immediately from (13.7). □

(13.9) Theorem. *Let f be a nonnegative extended real-valued function defined on $X\times Y$ that is $\iota\times\eta$-measurable and vanishes $\iota\times\eta$-almost everywhere outside of a set $\bigcup_{n=1}^{\infty} A_n$, where each A_n is $\iota\times\eta$-measurable and $\iota\times\eta(A_n) < \infty$. Then the function $x \to f(x, y)$ is ι-measurable for η-almost all $y \in Y$, and the function $y \to \int_X f(x, y)\, d\iota(x)$ is η-measurable; similarly for $y \to f(x, y)$ and $x \to \int_Y f(x, y)\, d\eta(y)$. Furthermore, we have*

$$\text{(i)} \quad \int_{X\times Y} f(x, y)\, d\iota\times\eta(x, y) = \int_X \int_Y f(x, y)\, d\eta(y)\, d\iota(x) = \int_Y \int_X f(x, y)\, d\iota(x)\, d\eta(y).$$

Proof. We may clearly suppose that $A_1 \subset A_2 \subset \cdots \subset A_n \subset \cdots$. Let $f_n = \min(n, f\xi_{A_n})$ $(n = 1, 2, \ldots)$. Then it is clear that $f_1 \leqq f_2 \leqq \cdots \leqq f_n \leqq \cdots$, $\lim_{n\to\infty} f_n = f$ $\iota\times\eta$-almost everywhere, and $f_n \in \mathfrak{L}_1(X\times Y, \iota\times\eta)$ for $n = 1, 2, \ldots$. Let P be the set of $y \in Y$ for which $\lim_{n\to\infty} f_n(x,y) = f(x,y)$ ι-almost everywhere and for which the function $x \to f_n(x, y)$ is ι-measurable $(n = 1, 2, \ldots)$. By (13.7) and (13.8), we have $\eta(P') = 0$. Clearly the function $x \to \lim_{n\to\infty} f_n(x, y) = f(x, y)$ is ι-measurable for $y \in P$. Also each function $y \to \int_X f_n(x, y)\, d\iota(x)$ is η-measurable and is in $\mathfrak{L}_1(Y, \eta)$. By the theorem on monotone convergence, we have $\int_X f(x, y)\, d\iota(x) = \lim_{n\to\infty} \int_X f_n(x, y)\, d\iota(x)$ for $y \in P$. Note too that the functions $y \to \int_X f_n(x, y)\, d\iota(x)$ are monotone increasing with

n $(y \in P)$. Combining these observations with (13.8), we obtain

$$\int_{X\times Y} f(x,y)\,d\iota\times\eta(x,y) = \lim_{n\to\infty}\int_{X\times Y} f_n(x,y)\,d\iota\times\eta(x,y)$$
$$= \lim_{n\to\infty}\int_Y\int_X f_n(x,y)\,d\iota(x)\,d\eta(y) = \int_Y \lim_{n\to\infty}\int_X f_n(x,y)\,d\iota(x)\,d\eta(y)$$
$$= \int_Y\int_X f(x,y)\,d\iota(x)\,d\eta(y).$$

The argument for $\int_X\int_Y f(x,y)\,d\eta(y)\,d\iota(x)$ is the same. □

(13.10) Theorem. *Let f be a complex- or extended real-valued, $\iota\times\eta$-measurable function on $X\times Y$ vanishing $\iota\times\eta$-almost everywhere outside of a set $\bigcup_{n=1}^{\infty} A_n$, where each A_n is $\iota\times\eta$-measurable and $\iota\times\eta(A_n)<\infty$. The three integrals*

(i) $\int_{X\times Y} f(x,y)\,d\iota\times\eta(x,y)$, $\int_X\int_Y f(x,y)\,d\eta(y)\,d\iota(x)$,
$\int_Y\int_X f(x,y)\,d\iota(x)\,d\eta(y)$

are finite and equal to each other if and only if one of the integrals

(ii) $\int_{X\times Y} |f(x,y)|\,d\iota\times\eta(x,y)$, $\int_X\int_Y |f(x,y)|\,d\eta(y)\,d\iota(x)$,
$\int_Y\int_X |f(x,y)|\,d\iota(x)\,d\eta(y)$

is finite.

Proof. We may write f as $f_1-f_2+i(f_3-f_4)$ where $0\leqq f_j\leqq|f|$ for $j=1,2,3,4$. Then $\int_{X\times Y} |f(x,y)|\,d\iota\times\eta(x,y)$ is finite if and only if each $\int_{X\times Y} f_j(x,y)\,d\iota\times\eta(x,y)$ is finite $(j=1,2,3,4)$; similarly for the iterated integrals. Thus the assertion follows immediately from (13.9). □

(13.11) Theorem. *Let $A\subset X$ and $B\subset Y$ and suppose that $\iota(A)=0$ and B is the union of an increasing sequence $B_1, B_2, \ldots, B_n, \ldots$ of [not necessarily η-measurable] sets of finite η-measure. Then we have $\iota\times\eta(A\times B)=0$. If B is any set, and A is locally ι-null, then $A\times B$ is locally $\iota\times\eta$-null. Similar assertions hold if the rôles of A and B are interchanged.*

Proof. By (13.6), we have $\overline{\overline{I\times J}}(\xi_{A\times B_n}) = \overline{\overline{I\times J}}(\xi_A\,\xi_{B_n}) = \overline{\overline{I}}(\xi_A)\,\overline{\overline{J}}(\xi_{B_n}) = \iota(A)\,\eta(B_n)=0$. Thus by (11.18), we have $\iota\times\eta(A\times B) = \overline{\overline{I\times J}}(\xi_{A\times B}) = \lim_{n\to\infty}\overline{\overline{I\times J}}(\xi_{A\times B_n})=0$.

Now suppose that A is locally ι-null and B is arbitrary. If F is any compact subset of $X\times Y$, then F is contained in a set $D\times E$, where D is a compact subset of X and E is a compact subset of Y. Then we have $(A\times B)\cap F\subset(A\cap D)\times(B\cap E)$, and $\iota\times\eta((A\cap D)\times(B\cap E))=0$ by the first case. □

(13.12) Theorem. *Let f be in $\mathfrak{L}_1(X, \iota)$ and g in $\mathfrak{L}_1(Y, \eta)$. Then the function $(x, y) \to f(x)\,g(y)$ [written as fg] is in $\mathfrak{L}_1(X \times Y, \iota \times \eta)$ and $\int\limits_{X \times Y} f(x)\,g(y)\,d\iota \times \eta(x, y) = \int\limits_X f(x)\,d\iota(x) \int\limits_Y g(y)\,d\eta(y)$.*

Proof. In view of (13.6) and (11.36), we need only to prove that fg is $\iota \times \eta$-measurable. Using (12.10), choose a sequence $\varphi_1, \varphi_2, \ldots, \varphi_n, \ldots$ of functions in $\mathfrak{C}_{00}(X)$ and a sequence of functions $\psi_1, \psi_2, \ldots, \psi_n, \ldots$ in $\mathfrak{C}_{00}(Y)$ such that $\lim\limits_{n\to\infty} \|f - \varphi_n\|_1 = 0$ and $\lim\limits_{n\to\infty} \|g - \psi_n\|_1 = 0$. Elementary properties of $\overline{\overline{I \times J}}$, (13.6), and (11.36) imply that

$$\begin{aligned}\overline{\overline{I \times J}}(|fg - \varphi_n \psi_n|) &\leq \overline{\overline{I \times J}}(|fg - f\psi_n| + |f\psi_n - \varphi_n \psi_n|)\\ &\leq \overline{\overline{I \times J}}(|fg - f\psi_n|) + \overline{\overline{I \times J}}(|f\psi_n - \varphi_n \psi_n|)\\ &= \overline{\overline{I}}(|f|)\,\overline{\overline{J}}(|g - \psi_n|) + \overline{\overline{I}}(|f - \varphi_n|)\,\overline{\overline{J}}(|\psi_n|)\\ &= \|f\|_1 \|g - \psi_n\|_1 + \|f - \varphi_n\|_1 \|\psi_n\|_1.\end{aligned}$$

As the sequence $\{\|\psi_n\|_1\}_{n=1}^{\infty}$ is bounded, the last expression goes to 0 as $n \to \infty$. Hence by (11.27) a subsequence of $\{|fg - \varphi_n \psi_n|\}_{n=1}^{\infty}$ converges to 0 $\iota \times \eta$-almost everywhere on $X \times Y$. Since each $\varphi_n \psi_n$ is in $\mathfrak{C}_{00}(X \times Y)$, it follows that fg is $\iota \times \eta$-measurable. □

(13.13) Theorem. *Let $A \subset X$ and $B \subset Y$ be measurable and σ-finite. Then $A \times B$ is $\iota \times \eta$-measurable and $\iota \times \eta(A \times B) = \iota(A) \cdot \eta(B)$.*

Proof. Let $A_1, A_2, \ldots,$ be a monotone increasing sequence of measurable subsets of X of finite ι-measure such that $\bigcup\limits_{n=1}^{\infty} A_n = A$; similarly for $B_1, B_2, \ldots$. Then by (13.12), $\xi_{A_n \times B_n} = \xi_{A_n} \xi_{B_n} \in \mathfrak{L}_1(X \times Y, \iota \times \eta)$, and $\iota \times \eta(A_n \times B_n) = \iota(A_n)\,\eta(B_n)$. Now take limits as $n \to \infty$. □

A result similar to (13.13) holds for arbitrary sets A and B so long as the product $\iota(A) \cdot \eta(B)$ does not have the form $0 \cdot \infty$ or $\infty \cdot 0$.

(13.14) Theorem. *Let $A \subset X$ and $B \subset Y$ be arbitrary sets such that $\iota(A) \cdot \eta(B)$ is not of the form $0 \cdot \infty$ or $\infty \cdot 0$. Then $\iota \times \eta(A \times B) = \iota(A) \cdot \eta(B)$.*

Proof. By (13.6), we have

$$\iota \times \eta(A \times B) = \overline{\overline{I \times J}}(\xi_{A \times B}) = \overline{\overline{I}}(\xi_A)\,\overline{\overline{J}}(\xi_B) = \iota(A) \cdot \eta(B). \quad \square$$

(13.15) We now take up the theory of integration on arbitrary products of *compact* Hausdorff spaces. For the remainder of the present section, $\{X_\gamma\}_{\gamma \in \Gamma}$ will denote an arbitrary nonvoid family of nonvoid compact Hausdorff spaces and X will denote $\underset{\gamma \in \Gamma}{\mathsf{P}} X_\gamma$. For each $\gamma \in \Gamma$, I_γ will denote an arbitrary nonnegative linear functional on $\mathfrak{C}(X_\gamma)$ for which $I_\gamma(1) = 1$. We will shortly consider a linear functional I on $\mathfrak{C}(X)$. For all of these functionals, $\overline{I}, \overline{I}_\gamma, \overline{\overline{I}}, \overline{\overline{I}}_\gamma, \iota$, and ι_γ will have the meanings established in §11.

For $\gamma_0 \in \Gamma$ and a function g on X_{γ_0}, we can consider the function $g \circ \pi_\gamma$ on the product space X. [Recall that π_{γ_0} is the projection that carries $(x_\gamma) \in X$ onto x_{γ_0}.] It is obvious that if $g \in \mathfrak{C}(X_{\gamma_0})$, then $g \circ \pi_{\gamma_0} \in \mathfrak{C}(X)$. Throughout the remainder of this section, we will write $g \circ \pi_{\gamma_0}$ as g where no confusion can arise.

Also, consider any function $f \in \mathfrak{C}(X)$. Select a $\gamma_0 \in \Gamma$ and a point $a_\gamma \in X_\gamma$ for all $\gamma \neq \gamma_0$. Write $\boldsymbol{a}'$ for the point $(a_\gamma) \in \underset{\gamma \neq \gamma_0}{\mathsf{P}} X_\gamma$. The function $x_{\gamma_0} \to f((\boldsymbol{a}', x_{\gamma_0}))$ is obviously continuous on X_{γ_0}. Applying the functional I_{γ_0} to this function, we obtain a complex-valued function on $\underset{\gamma \neq \gamma_0}{\mathsf{P}} X_\gamma$. Its value at $\boldsymbol{a}'$ is written as $I_{\gamma_0}[f(\boldsymbol{a}')]$. It is convenient to define $I_{\gamma_0}[f]$ on all of X by the rule that $I_{\gamma_0}[f(\boldsymbol{a})]$ for $\boldsymbol{a} \in X$ is equal to $I_{\gamma_0}[f(\boldsymbol{a}')]$, where $\boldsymbol{a}' = (a_\gamma)_{\gamma \neq \gamma_0}$. That is to say, $I_{\gamma_0}[f(\boldsymbol{a})]$ does not depend on the γ_0-th coordinate of $\boldsymbol{a}$. Finally, we will abbreviate $I_{\gamma_1}[I_{\gamma_2}[\cdots[I_{\gamma_n}[f]\cdots]]]$ as $I_{\gamma_1,\ldots,\gamma_n}(f)$; similarly for $\bar{I}_{\gamma_1,\ldots,\gamma_n}(f)$ and $\bar{\bar{I}}_{\gamma_1,\ldots,\gamma_n}(f)$.

We will adopt similar conventions with regard to integration. Consider any complex- or extended real-valued function h on X. Select a $\gamma_0 \in \Gamma$ and a point $a_\gamma \in X_\gamma$ for all $\gamma \neq \gamma_0$. Write $\boldsymbol{a}'$ for the point $(a_\gamma) \in \underset{\gamma \neq \gamma_0}{\mathsf{P}} X_\gamma$. Suppose that the function $x_{\gamma_0} \to h((\boldsymbol{a}', x_{\gamma_0}))$ belongs to $\mathfrak{L}_1(X_{\gamma_0})$. Then integrating this function with respect to ι_{γ_0}, we obtain a complex-valued function defined for some or all $\boldsymbol{a}'$. Its value at $\boldsymbol{a}'$ is written as $\int_{X_{\gamma_0}} h((\boldsymbol{a}', x_{\gamma_0}))\, d\iota_{\gamma_0}(x_{\gamma_0})$ or more briefly $\int_{X_{\gamma_0}} h\, d\iota_{\gamma_0}$. As in the case of functionals, we define $\int_{X_{\gamma_0}} h\, d\iota_{\gamma_0}$ at $(\boldsymbol{a}', a_{\gamma_0}) \in \underset{\gamma \in \Gamma}{\mathsf{P}} X_\gamma$ as $\int_{X_{\gamma_0}} h((\boldsymbol{a}', x_{\gamma_0}))\, d\iota_{\gamma_0}(x_{\gamma_0})$ for all choices of $a_{\gamma_0} \in X_{\gamma_0}$.

(13.16) Theorem. *For $f \in \mathfrak{C}(X)$ and $\gamma_0 \in \Gamma$, the function $I_{\gamma_0}(f)$ is also in $\mathfrak{C}(X)$. Furthermore, if $\gamma_1, \ldots, \gamma_p$ are distinct elements of Γ, and $\{n_1, \ldots, n_p\}$ is any permutation of $\{1, \ldots, p\}$, we have*

(i) $I_{\gamma_1,\ldots,\gamma_p}(f) = I_{\gamma_{n_1},\ldots,\gamma_{n_p}}(f)$.

Proof. Let $\mathfrak{H}$ denote the space of all functions on X of the form $h = \sum_{j=1}^{m} \prod_{k=1}^{n} \varphi_{j,k}$ where each $\varphi_{j,k}$ belongs to $\mathfrak{C}(X_{\delta_k})$ and $\delta_1, \ldots, \delta_n$ are distinct elements of Γ. Let γ be any element of Γ. Then $I_\gamma(h)$ is in $\mathfrak{H}$. In fact, if γ is δ_{k_0} for some $k_0 = 1, \ldots, n$, then

$$I_\gamma(h) = I_{\delta_{k_0}}(h) = \sum_{j=1}^{m} \left(\prod_{\substack{k=1 \\ k \neq k_0}}^{n} \varphi_{j,k} \right) I_{\delta_{k_0}}(\varphi_{j,k_0})$$

and if γ is distinct from all the δ_k, then

$$I_\gamma(h) = h.$$

These expressions for $I_\gamma(h)$ show that $I_\gamma(h)$ is in $\mathfrak{H}$. It also follows that both sides of (i) are well defined and belong to $\mathfrak{H}$ if $f \in \mathfrak{H}$.

Next consider an arbitrary $f \in \mathfrak{C}(X)$ and an arbitrary positive number ε. The function space $\mathfrak{H}$ is a subalgebra of $\mathfrak{C}(X)$ that is closed under complex conjugation, separates points of X, and contains 1. The STONE-WEIERSTRASS theorem implies that there is an $h \in \mathfrak{H}$ for which

$$\|f - h\|_u < \varepsilon .$$

For $\gamma_0 \in \Gamma$, it is clear that $\|I_{\gamma_0}(f) - I_{\gamma_0}(h)\|_u < \varepsilon$. Since $I_{\gamma_0}(h) \in \mathfrak{H}$, we see that $I_{\gamma_0}(f)$ is the uniform limit on X of functions in $\mathfrak{H}$, and thus $I_{\gamma_0}(f) \in \mathfrak{C}(X)$. Therefore both sides of (i) are defined for $f \in \mathfrak{C}(X)$, and both sides are functions in $\mathfrak{C}(X)$.

With all coordinates x_γ fixed except for $x_{\gamma_1}, \ldots, x_{\gamma_p}$, we can consider f as a [continuous] function on $X_{\gamma_1} \times \cdots \times X_{\gamma_p}$. Then (13.2) shows that the two sides of (i) are equal. □

(13.17) Definition. For a finite subset $\Delta = \{\gamma_1, \ldots, \gamma_p\}$ of Γ and a function $f \in \mathfrak{C}(X)$, we define $I_\Delta(f)$ to be the function in $\mathfrak{C}(X)$ defined by (13.16.i).

(13.18) Theorem. *Let $\{\Delta\}$ be the set of all finite subsets of Γ, directed by inclusion. Then for every $f \in \mathfrak{C}(X)$, the net $I_\Delta(f)$ of functions in $\mathfrak{C}(X)$ converges uniformly to a constant function, which we denote by $I(f)$. The functional I on $\mathfrak{C}(X)$ is linear and nonnegative, and $I(1) = 1$.*

Proof. Let ε be a positive number. As pointed out in the proof of (13.16), there is a function $h = \sum_{j=1}^{m} \prod_{k=1}^{n} \varphi_{j,k} \in \mathfrak{H}$ [$\varphi_{j,k} \in \mathfrak{C}(X_{\delta_k})$, where $\delta_1, \ldots, \delta_n$ are distinct elements of Γ] such that $\|f - h\|_u < \varepsilon$. Let $\Delta_0 = \{\delta_1, \ldots, \delta_n\}$. If $\Delta \supset \Delta_0$, then

$$I_\Delta(h) = \sum_{j=1}^{m} \prod_{k=1}^{n} I_{\delta_k}(\varphi_{j,k}) = I_{\Delta_0}(h). \qquad (1)$$

Thus all of the functions $I_\Delta(h)$, $\Delta \supset \Delta_0$, are the constant function whose value is $I_{\Delta_0}(h)$. We also have

$$\|I_\Delta(f) - I_\Delta(h)\|_u < \varepsilon \quad \text{for all} \quad \Delta. \qquad (2)$$

From (1) and (2), we obtain

$$\|I_\Delta(f) - I_{\Delta_0}(h)\|_u < \varepsilon \quad \text{for all} \quad \Delta \supset \Delta_0 .$$

Since $I_{\Delta_0}(h)$ is a constant function, whose value can be regarded as depending only upon ε [once $h = h_\varepsilon$ has been chosen], an elementary argument shows that $I_\Delta(f)$ converges uniformly to a constant function. The linearity and nonnegativity of I follow from the linearity and nonnegativity of each I_Δ. □

For the remainder of this section, I will denote the linear functional on $\mathfrak{C}(X)$ given in (13.18).

(13.19) Corollary. *Let $f=f_1 f_2 \cdots f_m$ where $f_j \in \mathfrak{C}(X_{\delta_j})$ and $\delta_1, \ldots, \delta_m$ are distinct elements of Γ. Then*

$$I(f) = \prod_{j=1}^{m} I_{\delta_j}(f_j).$$

(13.20) Theorem. *Let $g=g_1 g_2 \cdots g_m$ where $g_j \in \mathfrak{M}^+(X_{\delta_j})$ and $\delta_1, \ldots, \delta_m$ are distinct elements of Γ. Then*

(i) $\bar{I}(g) = \prod_{j=1}^{m} \bar{I}_{\delta_j}(g_j)$.

Proof. We first observe that if $A_1, \ldots, A_m$ are nonvoid subsets of $[0, \infty]$, then

$$\prod_{j=1}^{m} \sup\{a : a \in A_j\} = \sup\{a_1 a_2 \cdots a_m : a_j \in A_j\}. \tag{1}$$

We omit the proof of (1), which is elementary; the validity of (1) depends upon our convention that $0 \cdot \infty = 0$.

Since $g_j = \sup\{f_j : f_j \in \mathfrak{C}^+(X_{\delta_j})$ and $f_j \leqq g_j\}$ (11.10.v), the equality (1) implies that

$$g = \sup\{f_1 f_2 \cdots f_m : f_j \in \mathfrak{C}^+(X_{\delta_j}) \quad \text{and} \quad f_j \leqq g_j,\ j = 1, \ldots, m\}.$$

Applying (11.13), (13.19), (1) again, and (11.11), we have

$$\begin{aligned} \bar{I}(g) &= \sup\{I(f_1 f_2 \cdots f_m) : f_j \in \mathfrak{C}^+(X_{\delta_j}) \text{ and } f_j \leqq g_j\} \\ &= \sup\left\{\prod_{j=1}^{m} I_{\delta_j}(f_j) : f_j \in \mathfrak{C}^+(X_{\delta_j}) \text{ and } f_j \leqq g_j\right\} \\ &= \prod_{j=1}^{m} \sup\{I_{\delta_j}(f_j) : f_j \in \mathfrak{C}^+(X_{\delta_j}) \text{ and } f_j \leqq g_j\} = \prod_{j=1}^{m} \bar{I}_{\delta_j}(g_j). \quad \square \end{aligned}$$

(13.21) Theorem. *Let $A = \underset{\gamma \in \Gamma}{\mathsf{P}} A_\gamma$ be a subset of $X = \underset{\gamma \in \Gamma}{\mathsf{P}} X_\gamma$ such that each A_γ is ι_γ-measurable and such that $A_\gamma = X_\gamma$ for all but finitely many coordinates $\gamma \in \Gamma$. Then A is ι-measurable and*

(i) $\iota(A) = \prod_{\gamma \in \Gamma} \iota_\gamma(A_\gamma)$

[*this is really a finite product*].

Proof. Let Δ be a finite subset of Γ containing $\{\gamma \in \Gamma : A_\gamma \neq X_\gamma\}$ and suppose that $\overline{\overline{\Delta}} \geqq 2$.

We first prove (i) in the case that each A_γ is open or closed. Let $\Delta_c = \{\delta \in \Delta : A_\delta$ is not open$\}$; we prove (i) by induction on $\overline{\overline{\Delta}}_c$. If $\overline{\overline{\Delta}}_c = 0$, then each $\xi_{A_\delta} \in \mathfrak{M}^+(X_\delta)$ since A_δ is open, and therefore

$$\iota(A) = \bar{I}(\xi_A) = \bar{I}\Big(\prod_{\delta \in \Delta} \xi_{A_\delta}\Big) = \prod_{\delta \in \Delta} \bar{I}_\delta(\xi_{A_\delta}) = \prod_{\delta \in \Delta} \iota_\delta(A_\delta),$$

by (13.20). Suppose now that (i) is valid and that A is ι-measurable whenever $\overline{\overline{\Delta}}_c < n$, and suppose that $\overline{\overline{\Delta}}_c = n$, $n \geqq 1$. Choose δ_0 in Δ_c.

Writing $\Delta_0=\Delta\cap\{\delta_0\}'$ and $Y=\underset{\gamma\in\Gamma\cap\Delta'}{\mathsf{P}}X_\gamma$, we see that

$$X_{\delta_0}\times\underset{\delta\in\Delta_0}{\mathsf{P}}A_\delta\times Y=(A_{\delta_0}\times\underset{\delta\in\Delta_0}{\mathsf{P}}A_\delta\times Y)\cup(A'_{\delta_0}\times\underset{\delta\in\Delta_0}{\mathsf{P}}A_\delta\times Y). \tag{1}$$

The inductive hypothesis applies to $X_{\delta_0}\times\underset{\delta\in\Delta_0}{\mathsf{P}}A_\delta\times Y$ and also to $A'_{\delta_0}\times\underset{\delta\in\Delta_0}{\mathsf{P}}A_\delta\times Y$. Therefore $A=A_{\delta_0}\times\underset{\delta\in\Delta_0}{\mathsf{P}}A_\delta\times Y$ is ι-measurable, and as ι is finite, we use (1) and our inductive hypothesis to write

$$\prod_{\delta\in\Delta_0}\iota_\delta(A_\delta)=\iota(A)+\iota_{\delta_0}(A'_{\delta_0})\prod_{\delta\in\Delta_0}\iota_\delta(A_\delta)$$

and

$$\iota(A)=[1-\iota_{\delta_0}(A'_{\delta_0})]\prod_{\delta\in\Delta_0}\iota_\delta(A_\delta)=\prod_{\delta\in\Delta}\iota_\delta(A_\delta).$$

This proves (i) in case each A_δ is open or closed.

Consider now the general case. Suppose that $\iota_{\delta_0}(A_{\delta_0})=0$ for some $\delta_0\in\Delta$. Then for $\varepsilon>0$, there is an open set $U_{\delta_0}\supset A_{\delta_0}$ for which $\iota_{\delta_0}(U_{\delta_0})<\varepsilon$. Hence $\iota(A)\leqq\iota\big(U_{\delta_0}\times\underset{\gamma\neq\delta_0}{\mathsf{P}}X_\gamma\big)=\iota_{\delta_0}(U_{\delta_0})<\varepsilon$. Suppose finally that $\iota_\delta(A_\delta)>0$ for all $\delta\in\Delta$ and that $\overline{\overline{\Delta}}=m$. Let α be an arbitrary number such that $0<\alpha<1$. Then there are open sets U_δ and compact sets F_δ such that $F_\delta\subset A_\delta\subset U_\delta$ and $\sqrt[m]{\alpha}\,\iota_\delta(U_\delta)<\iota_\delta(A_\delta)<\frac{1}{\sqrt[m]{\alpha}}\iota_\delta(F_\delta)$. Then

$$\left.\begin{aligned}\alpha\,\iota(A)&\leqq\alpha\,\iota\Big(\underset{\delta\in\Delta}{\mathsf{P}}U_\delta\times\underset{\gamma\in\Delta'}{\mathsf{P}}X_\gamma\Big)=\prod_{\delta\in\Delta}\sqrt[m]{\alpha}\,\iota_\delta(U_\delta)<\prod_{\delta\in\Delta}\iota_\delta(A_\delta)<\prod_{\delta\in\Delta}\frac{1}{\sqrt[m]{\alpha}}\iota_\delta(F_\delta)\\&=\frac{1}{\alpha}\iota\Big(\underset{\delta\in\Delta}{\mathsf{P}}F_\delta\times\underset{\gamma\in\Delta'}{\mathsf{P}}X_\gamma\Big)\leqq\frac{1}{\alpha}\iota(A).\end{aligned}\right\}\tag{2}$$

Since α is arbitrary, we have $\iota(A)=\prod_{\delta\in\Delta}\iota_\delta(A_\delta)$.

The inequalities (2) also show that A is the union of a σ-compact set and a set of ι-measure 0 if all $\iota_\delta(A_\delta)$ are positive. Hence A itself is ι-measurable. □

(13.22) Theorem. *Let A_γ be a subset of X_γ for each $\gamma\in\Gamma$. If all but a countable number of the A_γ are equal to X_γ and all A_γ are ι_γ-measurable, then $\underset{\gamma\in\Gamma}{\mathsf{P}}A_\gamma$ is ι-measurable and $\iota\Big(\underset{\gamma\in\Gamma}{\mathsf{P}}A_\gamma\Big)=\prod_{\gamma\in\Gamma}\iota_\gamma(A_\gamma)$.*[1] *If $\iota_\gamma(A_\gamma)<1$ for an uncountable number of the indices γ, then $\iota\Big(\underset{\gamma\in\Gamma}{\mathsf{P}}A_\gamma\Big)=0$.*[2]

Proof. The first statement follows at once from (13.21) and the countable additivity of ι. To prove the second, note that there is a number $\alpha<1$ such that $\iota_\gamma(A_\gamma)<\alpha$ for an infinite number of indices γ, and so there are open sets U_γ in these X_γ such that $A_\gamma\subset U_\gamma$ and $\iota_\gamma(U_\gamma)<\alpha$.

---

[1] This product is defined as $\inf\{\iota_{\gamma_1}(A_{\gamma_1})\cdots\iota_{\gamma_n}(A_{\gamma_n}):\{\gamma_1,\ldots,\gamma_n\}$ is a finite subset of $\Gamma\}$.

[2] If $\iota_\gamma(A_\gamma)=1$ and $A_\gamma\neq X_\gamma$ for an uncountable number of the indices γ, then $\underset{\gamma\in\Gamma}{\mathsf{P}}A_\gamma$ need not be measurable; see (16.13.f).

Consequently we have $\iota\left(\underset{\gamma\in\Gamma}{\mathsf{P}} A_\gamma\right) \leqq \alpha^n$ for $n=1, 2, 3, \ldots$, so that $\iota\left(\underset{\gamma\in\Gamma}{\mathsf{P}} A_\gamma\right) = 0$. □

We now take up the analogue of Theorem (13.8) for infinite products. Several preliminaries are needed.

(13.23) Let $X = \underset{\gamma\in\Gamma}{\mathsf{P}} X_\gamma$ and suppose that $\{\Gamma_\theta\}_{\theta\in\Theta}$ is a partition of Γ. If for each $\theta\in\Theta$, we define $Y_\theta = \underset{\gamma\in\Gamma_\theta}{\mathsf{P}} X_\gamma$, then X is homeomorphic with the product $Y = \underset{\theta\in\Theta}{\mathsf{P}} Y_\theta$. In fact, the mapping τ of X onto Y given by $\tau((x_\gamma)) = (y_\theta)$, where $y_\theta = (x_\gamma)_{\gamma\in\Gamma_\theta}$, is obviously a homeomorphism. For each $\theta\in\Theta$, let J_θ be the functional on $\mathfrak{C}(Y_\theta)$ formed as in (13.18) from the functionals $\{I_\gamma\}_{\gamma\in\Gamma_\theta}$, and let J be the functional on $\mathfrak{C}(Y)$ formed as in (13.18) from the functionals J_θ. Let η_θ be the measure on Y_θ associated as in §11 with J_θ. If I denotes the functional on $\mathfrak{C}(X)$ formed as in (13.18) from the functionals $\{I_\gamma\}_{\gamma\in\Gamma}$, it is natural to expect I and J to be essentially the same functional. The following theorem shows this to be the case.

(13.24) Theorem. *Let $\{\Gamma_\theta\}_{\theta\in\Theta}$, Y_θ, Y, τ, J_θ, and J be as in* (13.23). *Then*

(i) $I(f) = J(f\circ\tau^{-1})$ *for all* $f\in\mathfrak{C}(X)$.

Proof. Since τ is a homeomorphism, it is clear that f belongs to $\mathfrak{C}(X)$ if and only if $f\circ\tau^{-1}$ belongs to $\mathfrak{C}(Y)$. It is also easy to see that the mapping $f\to J(f\circ\tau^{-1})$ is a nonnegative [and hence continuous] linear functional on $\mathfrak{C}(X)$. We omit the argument.

Let $\mathfrak{H}\subset\mathfrak{C}(X)$ be as in the proof of (13.16). Since $\mathfrak{H}$ is dense in the uniform topology of $\mathfrak{C}(X)$, it suffices to establish (i) for functions in $\mathfrak{H}$. Since I and J are linear functionals, it suffices to establish (i) for functions of the form $h = \prod_{k=1}^{n} \varphi_k$ where each φ_k belongs to $\mathfrak{C}(X_{\delta_k})$ and $\delta_1, \ldots, \delta_n$ are distinct elements of Γ.

Let $\Delta_0 = \{\delta_1, \ldots, \delta_n\}$ and $\Theta_0 = \{\theta\in\Theta : \Delta_0\cap\Gamma_\theta \neq \varnothing\}$; clearly Θ_0 is finite. For $\theta\in\Theta_0$, we define $\Delta_\theta = \Delta_0\cap\Gamma_\theta$ and $\psi_\theta = \prod_{\delta_k\in\Delta_\theta}\varphi_k$; we regard ψ_θ as defined on $Y_\theta = \underset{\gamma\in\Gamma_\theta}{\mathsf{P}} X_\gamma$. Finally, we observe that the function $\psi = \prod_{\theta\in\Theta_0}\psi_\theta$ defined on Y is equal to $h\circ\tau^{-1}$. For $\theta\in\Theta_0$, we have $J_\theta(\psi_\theta) = (J_\theta)_{\Delta_\theta}\left(\prod_{\delta_k\in\Delta_\theta}\varphi_k\right) = \prod_{\delta_k\in\Delta_\theta} I_{\delta_k}(\varphi_k)$. Therefore we infer that

$$\begin{aligned} J(h\circ\tau^{-1}) &= J(\psi) = J_{\Theta_0}\Big(\prod_{\theta\in\Theta_0}\psi_\theta\Big) = \prod_{\theta\in\Theta_0} J_\theta(\psi_\theta) \\ &= \prod_{\theta\in\Theta_0}\ \prod_{\delta_k\in\Delta_\theta} I_{\delta_k}(\varphi_k) = \prod_{\delta_k\in\Delta_0} I_{\delta_k}(\varphi_k) = I_{\Delta_0}\Big(\prod_{k=1}^{n}\varphi_k\Big) = I(h). \quad \square \end{aligned}$$

In further consideration of X, Y, I, and J as in (13.23), we will identify X and Y and I and J. Theorem (13.24) shows that this is justified.

We now state and prove the $\mathfrak{L}_p$ analogue of Theorem (13.16).

(13.25) Theorem. *Let $1 \leqq p < \infty$. For $f \in \mathfrak{L}_p(X)$ and $\gamma_0 \in \Gamma$, the function $\int\limits_{X_{\gamma_0}} f \, d\iota_{\gamma_0}$ is defined and is complex-valued for ι-almost all $(x_\gamma) \in X$. Where otherwise undefined, set $\int\limits_{X_{\gamma_0}} f \, d\iota_{\gamma_0} = 0$. The function $\int\limits_{X_{\gamma_0}} f \, d\iota_{\gamma_0}$ is in $\mathfrak{L}_p(X)$ and*

(i) $\left\| \int\limits_{X_{\gamma_0}} f \, d\iota_{\gamma_0} \right\|_p \leqq \|f\|_p$.

Furthermore, let $\gamma_1, \ldots, \gamma_m$ be distinct elements of Γ, and let $\{n_1, \ldots, n_m\}$ be any permutation of $\{1, \ldots, m\}$. Then we have

(ii) $\int\limits_{X_{\gamma_m}} \int\limits_{X_{\gamma_{m-1}}} \cdots \int\limits_{X_{\gamma_1}} f \, d\iota_{\gamma_1} \cdots d\iota_{\gamma_m} = \int\limits_{X_{\gamma_{n_m}}} \int\limits_{X_{\gamma_{n_{m-1}}}} \cdots \int\limits_{X_{\gamma_{n_1}}} f \, d\iota_{\gamma_{n_1}} \cdots d\iota_{\gamma_{n_m}}$,

this equality holding ι-almost everywhere in X.

Proof. Since $\iota(X) = 1$, we have $\mathfrak{L}_p(X) \subset \mathfrak{L}_1(X)$. Now, to apply (13.24), write $\Theta = \{1, 2\}$, $\Gamma_1 = \{\gamma_0\}$, $\Gamma_2 = \Gamma \cap \{\gamma_0\}'$. Then in the notation of (13.23), we have $Y_1 = X_{\gamma_0}$, $J_1 = I_{\gamma_0}$, and $\eta_1 = \iota_{\gamma_0}$. By (13.24), we have $X = X_{\gamma_0} \times Y_2$ and $I = I_{\gamma_0} \times J_2$. By (13.3), then, we have $\iota = \iota_{\gamma_0} \times \eta_2$.

We now apply FUBINI's theorem (13.8). We have $f((x_\gamma)) = f(x_{\gamma_0}, (x_\gamma)_{\gamma \in \Gamma_2}) = f(x_{\gamma_0}, y_2)$, where $y_2 \in Y_2$. As a function of x_{γ_0}, $f(x_{\gamma_0}, y_2)$ is in $\mathfrak{L}_1(X_{\gamma_0})$ for all $y_2 \in A_2'$, where $A_2 \subset Y_2$ and $\eta_2(A_2) = 0$. Thus we can write $\int\limits_{X_{\gamma_0}} f(x_{\gamma_0}, y_2) \, d\iota_{\gamma_0}(x_{\gamma_0})$ and obtain a complex number for all $y_2 \in A_2'$. The convention in (13.15) under which $\int\limits_{X_{\gamma_0}} f \, d\iota_{\gamma_0}$ is defined shows that $\int\limits_{X_{\gamma_0}} f \, d\iota_{\gamma_0}$ is defined for $(x_{\gamma_0}, y_2) \notin X_{\gamma_0} \times A_2$. By (13.14), this set has ι-measure 0. Thus the first statement of the present theorem is proved.

The function $\int\limits_{X_{\gamma_0}} |f(x_{\gamma_0}, y_2)|^p \, d\iota_{\gamma_0}(x_{\gamma_0})$ also is defined everywhere on X by the same rule.

HÖLDER's inequality (12.4.iii) for $p > 1$ and an obvious estimate for $p = 1$ give us

$$\left| \int\limits_{X_{\gamma_0}} f \, d\iota_{\gamma_0} \right|^p \leqq \left[\left(\int\limits_{X_{\gamma_0}} |f|^p \, d\iota_{\gamma_0} \right)^{1/p} \left(\int\limits_{X_{\gamma_0}} 1 \, d\iota_{\gamma_0} \right)^{1/p'} \right]^p = \int\limits_{X_{\gamma_0}} |f|^p \, d\iota_{\gamma_0}. \tag{1}$$

The relations (1) hold ι-almost everywhere on X.

The functions $\int\limits_{X_{\gamma_0}} f \, d\iota_{\gamma_0}$ and $\int\limits_{X_{\gamma_0}} |f|^p \, d\iota_{\gamma_0}$ are in $\mathfrak{L}_1(X)$, as (13.8) and (13.12) show [recall the definition of these functions in (13.15)]. Using (1),

(13.12), and (13.8), we obtain

$$\int_X \left|\int_{X_{\gamma_0}} f\,d\iota_{\gamma_0}\right|^p d\iota \leqq \int_X \int_{X_{\gamma_0}} |f|^p\,d\iota_{\gamma_0}\,d\iota = \int_{Y_2} \int_{X_{\gamma_0}} |f|^p\,d\iota_{\gamma_0}\,d\eta_2 = \int_X |f|^p\,d\iota. \tag{2}$$

Taking p-th roots in (2), we have

$$\left\|\int_{X_{\gamma_0}} f\,d\iota_{\gamma_0}\right\|_p \leqq \|f\|_p. \tag{3}$$

This establishes (i) and of course the fact that $\int_{X_{\gamma_0}} f\,d\iota_{\gamma_0} \in \mathfrak{L}_p(X)$.

Both sides of (ii) are well defined and are complex numbers ι-almost everywhere, since at each step the integrand belongs to $\mathfrak{L}_1(X)$, as we have just shown. Now break up Γ into $\Gamma_1 = \{\gamma_1, \ldots, \gamma_m\}$ and $\Gamma_2 = \Gamma \cap \Gamma_1'$. For any fixed $y_2 \in Y_2$ for which $\int_{Y_1} f(y_1, y_2)\,d\eta_1(y_1)$ exists and is finite, we can apply (13.8) to obtain (ii). The set of such y_2's is all $y_2 \in Y_2$ except for a set of η_2-measure 0, as (13.8) shows. Then (13.11) completes the proof of (ii). □

(13.26) Definition. For a finite subset $\Delta = \{\gamma_1, \ldots, \gamma_m\}$ of Γ and a function $f \in \mathfrak{L}_1(X)$, $\int_\Delta f$ will denote the function defined by (13.25.ii).

We now state and prove the analogue of (13.8) for infinite products.

(13.27) Theorem. *Let $\{\Delta\}$ be the set of all finite subsets of Γ, directed by inclusion, and let $f \in \mathfrak{L}_p(X)$, $1 \leqq p < \infty$. Then*

$$\lim_\Delta \left\|\int_\Delta f - \int_X f\,d\iota\right\|_p = 0.$$

Proof. Let $\varepsilon > 0$ and choose $g \in \mathfrak{C}(X)$ such that $\|f - g\|_p < \frac{\varepsilon}{3}$ (12.10). By (13.18), there is a finite subset Δ_0 of Γ such that

$$\|I(g) - I_\Delta(g)\|_u < \frac{\varepsilon}{3} \quad \text{for all} \quad \Delta \supset \Delta_0.$$

Repeated application of (13.25.i) shows that $\left\|\int_\Delta f - \int_\Delta g\right\|_p < \frac{\varepsilon}{3}$ for all finite sets Δ. Using MINKOWSKI's inequality (12.6) and (12.7), we have, if $\Delta \supset \Delta_0$,

$$\begin{aligned}
\left\|\int_\Delta f - \int_X f\,d\iota\right\|_p &\leqq \left\|\int_\Delta f - \int_\Delta g\right\|_p + \left\|\int_\Delta g - \int_X g\,d\iota\right\|_p + \left\|\int_X g\,d\iota - \int_X f\,d\iota\right\|_p \\
&< \frac{\varepsilon}{3} + \|I_\Delta(g) - I(g)\|_p + \left|\int_X g\,d\iota - \int_X f\,d\iota\right| \\
&\leqq \frac{\varepsilon}{3} + \|I_\Delta(g) - I(g)\|_u + \|g - f\|_1 \\
&< \frac{2\varepsilon}{3} + \|g - f\|_p < \varepsilon.
\end{aligned}$$

[The inequality $\|g - f\|_1 \leqq \|g - f\|_p$ follows from HÖLDER's inequality.] □

Theorem (13.27) shows that the integral $\int_X f\, d\iota$ is the limit in the $\mathfrak{L}_p$ metric of the functions $\int_\Delta f$, for $f \in \mathfrak{L}_p(X)$. If Γ is *countably* infinite, this result can be sharpened considerably: the functions $\int_\Delta f$ converge ι-almost everywhere to $\int_X f\, d\iota$. To show this, we first prove a lemma.

(13.28) Lemma. *Suppose that* $\Gamma = \{1, 2, \ldots\}$; *for each positive integer* n, *let* $\Delta_n = \{1, 2, \ldots, n\}$. *For* $f \in \mathfrak{L}_1(X)$ *and* $\delta > 0$, *let* $A_\delta = \left\{x \in X : \sup_{k \geq 1} \left\{\left|\left(\int_{\Delta_k} f\right)(x)\right|\right\} > \delta\right\}$. *Then*

(i) $\delta\iota(A_\delta) \leq \int_{A_\delta} |f|\, d\iota$.

Proof. Let

$$B_n = \left\{x \in X : \sup_{k \geq n} \left\{\left|\left(\int_{\Delta_k} f\right)(x)\right|\right\} > \delta\right\}$$

and $C_n = B_n \cap B'_{n+1}$, for $n = 1, 2, \ldots$. Then $\{C_n\}_{n=1}^\infty$ partitions A_δ into measurable sets. [Theorem (13.25) shows that the functions $\int_{\Delta_k} f$ are ι-measurable.] Therefore to establish (i) it suffices by virtue of countable additivity to show that

$$\delta\iota(C_n) \leq \int_{C_n} |f|\, d\iota \tag{1}$$

for $n = 1, 2, \ldots$. In order to apply (13.24), let $\Theta = \{1, 2\}$, $\Gamma_1 = \Delta_n$, and $\Gamma_2 = \Gamma \cap \Delta'_n$. Then $Y_1 = \underset{\gamma \in \Delta_n}{\mathsf{P}} X_\gamma$ and $Y_2 = \underset{\gamma \in \Delta'_n}{\mathsf{P}} X_\gamma$. For $k \geq n$, the function $\int_{\Delta_k} f$ does not depend upon the coordinates $x_1, \ldots, x_k$, and therefore ξ_{B_n} does not depend upon the coordinates $x_1, \ldots, x_n$. It follows that ξ_{C_n} does not depend upon the coordinates $x_1, \ldots, x_n$. Equivalently, $\xi_{C_n}(y_1, y_2)$ does not depend upon the y_1-coordinate. Using this fact along with (13.8) and (13.24), we obtain

$$\begin{aligned}\int_{C_n} |f|\, d\iota &= \int_X \xi_{C_n} |f|\, d\iota = \int_{Y_2}\int_{Y_1} \xi_{C_n} |f|\, d\eta_1\, d\eta_2 \\ &= \int_{Y_2} \xi_{C_n} \int_{Y_1} |f|\, d\eta_1\, d\eta_2 = \int_{Y_2} \xi_{C_n} \left(\int_{\Delta_n} |f|\right) d\eta_2 \geq \int_{Y_2} \xi_{C_n} \left|\int_{\Delta_n} f\right| d\eta_2.\end{aligned}$$

Recalling that $\eta_1(Y_1) = 1$ and that $C_n \subset \left\{x \in X : \left|\left(\int_{\Delta_n} f\right)(x)\right| > \delta\right\}$, we see that

$$\int_{C_n} |f|\, d\iota \geq \int_{Y_1}\int_{Y_2} \xi_{C_n} \left|\int_{\Delta_n} f\right| d\eta_2\, d\eta_1 = \int_X \xi_{C_n} \left|\int_{\Delta_n} f\right| d\iota \geq \delta\iota(C_n).$$

This proves (1). □

(13.29) Theorem. *Let* Γ *and* Δ_n *be as in* (13.28). *Then for* $f \in \mathfrak{L}_1(X)$, *we have*

$$\lim_{n \to \infty} \int_{\Delta_n} f = \int_X f\, d\iota \qquad \iota\text{-almost everywhere}.$$

Proof. For $x \in X$, let $r(x)$ be defined by

$$r(x) = \lim_{p\to\infty}\left[\sup\left\{\left|\left(\int_{\Delta_n} f\right)(x) - \left(\int_{\Delta_m} f\right)(x)\right| : m, n \geqq p\right\}\right].$$

Let δ and ε be positive real numbers. By (12.10), there is a $g \in \mathfrak{C}(X)$ such that $\|f - g\|_1 < \frac{\varepsilon}{2}$ and, as shown in the proof of (13.16), there is a function $h \in \mathfrak{H}$ such that $\|g - h\|_u < \frac{\varepsilon}{2}$ and $\int_{\Delta_n} h = \int_X h\, d\iota$ for all sufficiently large n. Evidently $\|f - h\|_1 < \varepsilon$; writing ψ for $f - h$, we have

$$\begin{aligned} r(x) &= \lim_{p\to\infty}\left[\sup\left\{\left|\left(\int_{\Delta_n} \psi\right)(x) - \left(\int_{\Delta_m} \psi\right)(x)\right| : m, n \geqq p\right\}\right] \\ &\leqq 2 \sup\left\{\left|\left(\int_{\Delta_n} \psi\right)(x)\right| : n = 1, 2, \ldots\right\}. \end{aligned}$$

This inequality implies that

$$\{x \in X : r(x) > 2\delta\} \subset \left\{x \in X : \sup_n \left\{\left|\left(\int_{\Delta_n} \psi\right)(x)\right|\right\} > \delta\right\}.$$

Applying (13.28) to the function ψ, we therefore have

$$\delta \cdot \iota(\{x \in X : r(x) > 2\delta\}) \leqq \int_X |\psi|\, d\iota = \|\psi\|_1 < \varepsilon.$$

Since ε is arbitrary, we have $\iota(\{x \in X : r(x) > 2\delta\}) = 0$, and since δ is arbitrary, it follows that $r(x) = 0$ ι-almost everywhere. This implies that $\lim_{n\to\infty} \int_{\Delta_n} f$ exists ι-almost everywhere. By FATOU's lemma and (13.27), we have

$$\int_X \left|\lim_{n\to\infty} \int_{\Delta_n} f - \int_X f\, d\iota\right| d\iota \leqq \varliminf_{n\to\infty} \int_X \left|\int_{\Delta_n} f - \int_X f\, d\iota\right| d\iota = 0.$$

Consequently, $\lim_{n\to\infty} \int_{\Delta_n} f = \int_X f\, d\iota$ ι-almost everywhere (11.27). □

Notes

For the early history of FUBINI's theorem, see SAKS [1], p. 77 and BOURBAKI [3], pp. 119—120. In (13.1)—(13.14), we have followed in part BOURBAKI [3], Ch. V, §8 and also in part NAĬMARK [1], §6, N° 18. For the history of (13.15)—(13.29), see DUNFORD and SCHWARTZ [1], p. 235. A different treatment of integration on product spaces, and a number of sharper results for countable products, are given in HEWITT and STROMBERG [2], Ch. VI.

§ 14. Complex measures

Throughout this section, X will denote as usual an arbitrary nonvoid locally compact Hausdorff space. Where no confusion is to be feared, we will write $\mathfrak{C}_0$ for $\mathfrak{C}_0(X)$; similarly for $\mathfrak{C}_{00}$, $\mathfrak{C}_0^r$, etc. We are concerned with the conjugate space $\mathfrak{C}_0^*$ of $\mathfrak{C}_0$. That is, we are concerned with complex-valued linear functionals Φ on $\mathfrak{C}_0$ such that $|\Phi(f)| \leqq \beta \|f\|_u$ for all $f \in \mathfrak{C}_0$, where β is a constant depending upon Φ. As noted in (B.8), there is a least nonnegative number β for which this inequality holds: it is the same as $\sup\{|\Phi(f)| : \|f\|_u \leqq 1,\ f \in \mathfrak{C}_0\}$, is written as $\|\Phi\|$, and is called the *norm* of Φ.

Since $\mathfrak{C}_{00}$ is a dense linear subspace of $\mathfrak{C}_0$, it is obvious that each $\Phi \in \mathfrak{C}_0^*$ is completely determined by its values on $\mathfrak{C}_{00}$. Thus we can attempt to carry over the techniques and results of §§11–13 to the present case. There are two important differences. First, functionals Φ in $\mathfrak{C}_0^*$ can assume arbitrary complex values for functions in $\mathfrak{C}_0^+$. This apparent drawback is offset by the second difference, which is that the functionals in $\mathfrak{C}_0^*$ are bounded, while nonnegative functionals on $\mathfrak{C}_{00}$ may well be unbounded.

(14.1) Theorem. *Let I be a nonnegative linear functional on $\mathfrak{C}_{00}$, and let ι be constructed from I as in* §11. *The functional I is bounded if and only if $\iota(X) < \infty$, and then $\|I\| = \iota(X)$.*

Proof. Since $\iota(X) = \sup\{I(f) : 0 \leqq f \leqq 1,\ f \in \mathfrak{C}_{00}^+\}$, it is plain that I is unbounded if $\iota(X) = \infty$. Since $|I(f)| \leqq I(|f|)$ (11.5), it is also clear that I is bounded if $\iota(X) < \infty$. The same inequality shows that $\|I\| = \iota(X)$. □

(14.2) Note. It is obvious from (14.1) that every nonnegative linear functional on $\mathfrak{C}$ is bounded if X is compact. If X is not countably compact, then X contains a closed, infinite, discrete subset A. Every compact subset of X meets A in a finite set. For $f \in \mathfrak{C}_{00}$, therefore, the sum $\sum_{x \in A} f(x)$ has only a finite number of nonzero terms, and the functional $I(f) = \sum_{x \in A} f(x)$ is a nonnegative, unbounded linear functional on $\mathfrak{C}_{00}$.[1]

We next show how to represent functionals in $\mathfrak{C}_0^*$ as integrals with respect to bounded complex measures.

(14.3) Theorem. *Let Φ be a functional in $\mathfrak{C}_0^*$. Then there are nonnegative [bounded] linear functionals I_1, I_2, I_3, I_4 in $\mathfrak{C}_0^*$ such that*

(i) $\Phi = I_1 - I_2 + i(I_3 - I_4)$.

The functionals I_1, I_2 and I_3, I_4 can be chosen so that $\min(I_1, I_2) = \min(I_3, I_4) = 0$.[2] *Under this added restriction, the resolution* (i) *is unique.*

Proof. Suppose that

$$\Phi(f) = I_1(f) - I_2(f) + i[I_3(f) - I_4(f)] \tag{1}$$

[1] We have no reasonable characterization of the locally compact Hausdorff spaces X for which $\mathfrak{C}_{00}(X)$ admits an unbounded, nonnegative, linear functional.

[2] For the definition of $\min(I, J)$, see (B.34).

for all $f \in \mathfrak{C}_0^+$. Then the same holds for *all* $f \in \mathfrak{C}_0$, as (B.38) shows. Hence we may confine ourselves to functions in $\mathfrak{C}_0^+$ or $\mathfrak{C}_0^r$. Clearly we have $\Phi = \Phi_1 + i\Phi_2$, where Φ_1 and Φ_2 are real-valued, additive, and real-homogeneous. It is obvious that Φ_1 and Φ_2 are bounded. Now restrict Φ_j to the space $\mathfrak{C}_0^r$ $(j=1, 2)$. Clearly $\mathfrak{C}_0^r$ is a vector space over R such that $|f| \in \mathfrak{C}_0^r$ if $f \in \mathfrak{C}_0^r$. Also, if $f \in \mathfrak{C}_0^+$, $g \in \mathfrak{C}_0^r$, and $|g| \leq f$, we have $\|g\|_u \leq \|f\|_u$, and so $|\Phi_j(g)| \leq \|\Phi\| \cdot \|f\|_u$. Hence $\mathfrak{C}_0^r$, Φ_1, and Φ_2 satisfy the hypotheses of Theorem (B.37). Therefore, there are unique nonnegative bounded linear functionals I_1, I_2 on $\mathfrak{C}_0^r$ such that $\Phi_1 = I_1 - I_2$ and $\min(I_1, I_2) = 0$; similarly $\Phi_2 = I_3 - I_4$. ☐

(14.4) Theorem. *Let Φ be a functional in $\mathfrak{C}_0^*$. There is a complex measure μ defined on a σ-algebra $\mathscr{M}_\mu$ of subsets of X containing the Borel sets in X and such that*

(i) $\Phi(f) = \int_X f(x)\,d\mu(x)$ *for all* $f \in \mathfrak{C}_0$.

Proof. For the nonnegative bounded functionals I_j of (14.3), construct the measures ι_j as in §11 $(j=1, 2, 3, 4)$. By (11.36), we have

$$I_j(f) = \int_X f\,d\iota_j \quad \text{for} \quad f \in \mathfrak{C}_0^+ \tag{1}$$

and by definition of the integral, also for all $f \in \mathfrak{C}_0$. Since I_j is bounded, $\iota_j(X)$ is finite and $0 \leq \iota_j(A) \leq \iota_j(X)$ for all $A \subset X$. Now write $\mu = \iota_1 - \iota_2 + i(\iota_3 - \iota_4)$. Clearly $\mu(A)$ is a complex number for all $A \subset X$, and $|\mu(A)| \leq \sum_{j=1}^{4} \iota_j(A)$. On the σ-algebra $\mathscr{M}_\mu = \mathscr{M}_{\iota_1} \cap \mathscr{M}_{\iota_2} \cap \mathscr{M}_{\iota_3} \cap \mathscr{M}_{\iota_4}$, μ is obviously countably additive. The representation (i) now follows from (14.3.i) and the definition of $\int_X f\,d\mu$. ☐

We next find a nonnegative linear functional J that majorizes $|\Phi(\varphi)|$ for every $\varphi \in \mathfrak{C}_0$, in the sense that $J(|\varphi|) \geq |\Phi(\varphi)|$.

(14.5) Theorem. *Let Φ be any functional in $\mathfrak{C}_0^*$. For every $\varphi \in \mathfrak{C}_0^+$, let*

(i) $|\Phi|(\varphi) = \sup\{|\Phi(\psi)| : \psi \in \mathfrak{C}_0, |\psi| \leq \varphi\}$.

Then $|\Phi|$ can be extended to a nonnegative linear functional in $\mathfrak{C}_0^*$, and is the least nonnegative majorant of Φ.

Proof. Suppose that J is any nonnegative majorant of Φ in $\mathfrak{C}_0^*$. Then if $\varphi \in \mathfrak{C}_0^+$, $\psi \in \mathfrak{C}_0$, and $|\psi| \leq \varphi$, we have $J(\varphi) \geq J(|\psi|) \geq |\Phi(\psi)|$. Hence to show that $|\Phi|$ is the least nonnegative majorant of Φ in $\mathfrak{C}_0^*$, we need only to show that $|\Phi|$ is a bounded linear functional.

First, it is obvious that $|\Phi|(\varphi) \leq \|\Phi\| \cdot \|\varphi\|_u$ for all $\varphi \in \mathfrak{C}_0^+$: if $|\psi| \leq \varphi$, then $\|\psi\|_u \leq \|\varphi\|_u$, and $|\Phi(\psi)| \leq \|\Phi\| \cdot \|\psi\|_u \leq \|\Phi\| \cdot \|\varphi\|_u$. This shows that $|\Phi|(\varphi)$ is a nonnegative real number for all $\varphi \in \mathfrak{C}_0^+$ and also that $|\Phi|$ is a bounded functional. It is also obvious that $|\Phi|(\alpha\varphi) = \alpha|\Phi|(\varphi)$ if α is a nonnegative real number and $\varphi \in \mathfrak{C}_0^+$.

Next let φ_1, φ_2 be any functions in $\mathfrak{C}_0^+$. For $\varepsilon>0$, let ψ_j be a function in $\mathfrak{C}_0$ such that $|\psi_j|\leqq\varphi_j$ and $|\Phi(\psi_j)|+\frac{\varepsilon}{2}>|\Phi|(\varphi_j)$ $(j=1,2)$. Write $\Phi(\psi_j)=\exp(i\theta_j)|\Phi(\psi_j)|$ $(0\leqq\theta_j<2\pi)$, and write γ for $\exp(i(\theta_2-\theta_1))$. Then we have

$$|\Phi|(\varphi_1)+|\Phi|(\varphi_2)<|\Phi(\psi_1)|+|\Phi(\psi_2)|+\varepsilon$$
$$=|\gamma\,\Phi(\psi_1)+\Phi(\psi_2)|+\varepsilon=|\Phi(\gamma\,\psi_1+\psi_2)|+\varepsilon\leqq|\Phi|(\varphi_1+\varphi_2)+\varepsilon.$$

Thus we have $|\Phi|(\varphi_1)+|\Phi|(\varphi_2)\leqq|\Phi|(\varphi_1+\varphi_2)$.

To prove the reversed inequality, suppose that $\psi\in\mathfrak{C}_0$ and $|\psi|\leqq\varphi_1+\varphi_2$. Let $\varrho_1=\min(\varphi_1,|\psi|)$ and $\varrho_2=|\psi|-\varrho_1$. It is plain that $\varrho_1+\varrho_2=|\psi|$, $\varrho_j\leqq\varphi_j$, and $\varrho_j\in\mathfrak{C}_0^+$ $(j=1,2)$. Now let $\psi_j=\varrho_j\operatorname{sgn}\psi$ $(j=1,2)$. It is easy to see that $\psi_j\in\mathfrak{C}_0$, $|\psi_j|=\varrho_j\leqq\varphi_j$, and $\psi_1+\psi_2=\psi$. Hence we have

$$|\Phi(\psi)|=|\Phi(\psi_1+\psi_2)|\leqq|\Phi(\psi_1)|+|\Phi(\psi_2)|\leqq|\Phi|(\varphi_1)+|\Phi|(\varphi_2).$$

Consequently $|\Phi|$ is additive on $\mathfrak{C}_0^+$. We now extend $|\Phi|$ in the usual way (B.38) to be a complex linear functional on $\mathfrak{C}_0$. □

(14.6) Theorem. *Let Φ be an element of $\mathfrak{C}_0^*$. Let μ be the complex measure corresponding to Φ and write $|\mu|$ for the nonnegative measure corresponding to $|\Phi|$. Then $|\mu|(X)=\|\Phi\|$.*

Proof. Recall first that $|\mu|(X)=\sup\{|\Phi|(\varphi):\varphi\in\mathfrak{C}_0^+,\ \|\varphi\|_u\leqq1\}$. If $\varphi\in\mathfrak{C}_0$ and $\|\varphi\|_u\leqq1$, then we have $|\Phi(\varphi)|\leqq|\Phi|(|\varphi|)\leqq|\mu|(X)$. It follows that $\|\Phi\|\leqq|\mu|(X)$. On the other hand, for $\varphi\in\mathfrak{C}_0^+$ such that $\|\varphi\|_u\leqq1$, and for $\varepsilon>0$, there is a $\psi\in\mathfrak{C}_0$ such that $|\psi|\leqq\varphi$ and $|\Phi(\psi)|+\varepsilon>|\Phi|(\varphi)$. This implies that $\|\Phi\|\geqq|\mu|(X)$. □

(14.7) Theorem. *Let Φ be an element of $\mathfrak{C}_0^*$, and let μ and $|\mu|$ be the measures corresponding to Φ and $|\Phi|$, respectively. Then*

(i) $\left|\int\limits_X f\,d\mu\right|\leqq\int\limits_X|f|\,d|\mu|$ for $f\in\mathfrak{C}_0$.

This follows immediately from (14.5.i).

(14.8) Definition. If I and J are nonnegative linear functionals on $\mathfrak{C}_0$ and ι and η are their respective measures, then we write $\min(\iota,\eta)$ for the measure corresponding to $\min(I,J)$. If $I\leqq J$, then we write $\iota\leqq\eta$.

Suppose that I, J, ι, and η are as above. Then it is easy to see that the following assertions are equivalent:

$$I\leqq J, \tag{1}$$

$$\iota(A)\leqq\eta(A)\ \text{for all subsets } A \text{ of } X, \tag{2}$$

$$\iota(F)\leqq\eta(F)\ \text{for all compact subsets } F \text{ of } X. \tag{3}$$

(14.9) Definition. Let $\boldsymbol{M}(X)$ denote the set of all complex measures μ obtained as in (14.4) from linear functionals Φ in $\mathfrak{C}_0^*(X)$. For $\mu \in \boldsymbol{M}(X)$, we define the norm $\|\mu\|$ of μ as $|\mu|(X)$.

(14.10) Theorem. *The mapping $\Phi \to \mu$ given by (14.4) is a one-to-one, linear, norm-preserving, order-preserving*[1] *mapping of $\mathfrak{C}_0^*(X)$ onto $\boldsymbol{M}(X)$.*

Proof. The mapping $\Phi \to \mu$ is plainly order-preserving and onto; it is norm-preserving by (14.6); and it is one-to-one by formula (14.4.i). We need to show that the mapping $\Phi \to \mu$ is linear.

Let Ψ be another member of $\mathfrak{C}_0^*$ and let ν be its corresponding measure; also let α be any complex number. Let η be the measure corresponding to $\alpha\Phi + \Psi$; then we must show that $\eta = \alpha\mu + \nu$ [this implies linearity of $\Phi \to \mu$]. We have $\alpha = \alpha_1 - \alpha_2 + i(\alpha_3 - \alpha_4)$ where $\alpha_j \geqq 0$ ($j = 1, 2, 3, 4$). We also have

$$\Phi = I_1 - I_2 + i(I_3 - I_4), \quad \Psi = J_1 - J_2 + i(J_3 - J_4), \tag{1}$$

and

$$\alpha\Phi + \Psi = L_1 - L_2 + i(L_3 - L_4), \tag{2}$$

where the functionals I_j, J_j, and L_j are chosen as in (14.3). Denote the measures corresponding to I_j, J_j, and L_j by μ_j, ν_j, and η_j, respectively ($j = 1, 2, 3, 4$). Substituting the identities (1) into (2) and equating real parts, we obtain

$$\alpha_1 I_1 + \alpha_2 I_2 + \alpha_4 I_3 + \alpha_3 I_4 + J_1 + L_2 = \alpha_2 I_1 + \alpha_1 I_2 + \alpha_3 I_3 + \alpha_4 I_4 + J_2 + L_1.$$

It follows from (11.38) that the measures

$$\alpha_1 \mu_1 + \alpha_2 \mu_2 + \alpha_4 \mu_3 + \alpha_3 \mu_4 + \nu_1 + \eta_2 \tag{3}$$

and

$$\alpha_2 \mu_1 + \alpha_1 \mu_2 + \alpha_3 \mu_3 + \alpha_4 \mu_4 + \nu_2 + \eta_1 \tag{4}$$

agree on all Borel sets. For any *finite* measures ι and ι_0 constructed as in §11 from nonnegative linear functionals on $\mathfrak{C}_{00}^+$, and every set $A \subset X$, the identity

$$\left.\begin{aligned} &\inf\{\iota(U) : U \text{ is open and } U \supset A\} \\ &+ \inf\{\iota_0(U) : U \text{ is open and } U \supset A\} \\ &= \inf\{\iota(U) + \iota_0(U) : U \text{ is open and } U \supset A\}, \end{aligned}\right\} \tag{5}$$

is easily verified. This depends upon (11.21.ii) and the trivial fact that if U and V are open, $U \supset A$, and $V \supset A$, then $U \cap V$ is also open and $U \cap V \supset A$. Since the measures (3) and (4) agree on Borel sets, (5) together with (11.22) implies that the outer measures (3) and (4) agree

[1] That is, Φ is a nonnegative functional if and only if the corresponding measure μ is nonnegative.

on all subsets of X. This means that

$$(\alpha_1 - \alpha_2)(\mu_1 - \mu_2) - (\alpha_3 - \alpha_4)(\mu_3 - \mu_4) + (\nu_1 - \nu_2) = \eta_1 - \eta_2,$$

and a similar argument shows that the imaginary parts of $\alpha\mu+\nu$ and η are equal. Hence we have $\alpha\mu+\nu=\eta$. □

We now collect some elementary facts about $\boldsymbol{M}(X)$.

(14.11) Theorem. *Let Φ and Ψ be in $\mathfrak{C}_0^*$, and let μ and ν, respectively, be the corresponding measures in $\boldsymbol{M}(X)$. Then we have*

(i) $|\alpha\mu| = |\alpha| \cdot |\mu|$ *for* $\alpha \in K$,

and

(ii) $|\mu+\nu| \leqq |\mu| + |\nu|$.

Also, if $|\mu| \leqq |\nu|$, then every $|\nu|$-measurable set is $|\mu|$-measurable, i.e.,

(iii) $\mathscr{M}_{|\nu|} \subset \mathscr{M}_{|\mu|}$.

Proof. From (14.5.i) we have $|\alpha\Phi| = |\alpha| \cdot |\Phi|$ for $\alpha \in K$, and this implies that $|\alpha\mu| = |\alpha| \cdot |\mu|$.

For $f \in \mathfrak{C}_0^+$ and $\psi \in \mathfrak{C}_0$, $|\psi| \leqq f$, we have $|\Phi(\psi) + \Psi(\psi)| \leqq |\Phi(\psi)| + |\Psi(\psi)| \leqq |\Phi|(|\psi|) + |\Psi|(|\psi|) \leqq |\Phi|(f) + |\Psi|(f)$. It follows that $|\Phi+\Psi|(f) \leqq |\Phi|(f) + |\Psi|(f)$ and hence $|\Phi+\Psi| \leqq |\Phi| + |\Psi|$. This is equivalent to (ii).

Suppose finally that $|\mu| \leqq |\nu|$ and that A is $|\nu|$-measurable. Then $A = B \cup N$ where B is a σ-compact set, N is $|\nu|$-measurable, and $|\nu|(N) = 0$. Hence $|\mu|(N) = 0$ and N is $|\mu|$-measurable. Consequently, A is $|\mu|$-measurable. □

(14.12) Theorem. *Let Φ be an element of $\mathfrak{C}_0^*$, and let μ and $|\mu|$ be as in* (14.4) *and* (14.6), *respectively. There is an $|\mu|$-measurable function g such that:*

(i) $|g(x)| = 1$ *for all* $x \in X$;

(ii) $\Phi(f) = \int_X f(x)\, g(x)\, d|\mu|(x)$ *for* $f \in \mathfrak{C}_0$.

Proof. Consider the nonnegative functionals I_1, I_2, I_3, I_4 of (14.3), the nonnegative functional $|\Phi|$ of (14.5), and the measures $\iota_1, \iota_2, \iota_3, \iota_4$, and $|\mu|$ corresponding to these functionals. Let Φ_1 and Φ_2 be as in the proof of (14.3). We prove first that $I_j \leqq |\Phi|$ $(j = 1, 2, 3, 4)$. Consider I_1 and $|\Phi|$. For $\varphi \in \mathfrak{C}_{00}^+$, we have

$$\begin{aligned}
|\Phi|(\varphi) &= \sup\{|\Phi(\psi)| : \psi \in \mathfrak{C}_0, |\psi| \leqq \varphi\} \\
&= \sup\{|\Phi_1(\psi) + i\Phi_2(\psi)| : \psi \in \mathfrak{C}_0, |\psi| \leqq \varphi\} \\
&\geqq \sup\{|\Phi_1(\psi)| : \psi \in \mathfrak{C}_0, |\psi| \leqq \varphi\} \\
&\geqq \sup\{\Phi_1(\psi) : \psi \in \mathfrak{C}_0^+, 0 \leqq \psi \leqq \varphi\} = I_1(\varphi).
\end{aligned}$$

[The last equality is just the definition of $I_1=\max(\Phi_1, 0)$: see (14.3) and (B.34).] We also have $I_2=I_1-\Phi_1$. Hence for $\varphi\in\mathfrak{C}_0^+$, we have

$$\begin{aligned} I_2(\varphi) &= \sup\{\Phi_1(\psi):\psi\in\mathfrak{C}_0^+,\ 0\leqq\psi\leqq\varphi\}-\Phi_1(\varphi) \\ &= \sup\{\Phi_1(\psi-\varphi):\psi\in\mathfrak{C}_0^+,\ 0\leqq\psi\leqq\varphi\} \\ &= \sup\{\Phi_1(\omega):\omega\in\mathfrak{C}_0^r,\ -\varphi\leqq\omega\leqq 0\} \\ &\leqq \sup\{|\Phi_1(\psi)|:\psi\in\mathfrak{C}_0,\ |\psi|\leqq\varphi\}\leqq|\Phi|(\varphi). \end{aligned}$$

The same argument shows that $I_3\leqq|\Phi|$ and $I_4\leqq|\Phi|$.

In the proof of the LEBESGUE-RADON-NIKODÝM theorem (12.17), the following is proved. The inequality $I_j\leqq|\Phi|$ implies that there is an $|\mu|$-measurable weight function g_j such that $0\leqq g_j(x)\leqq 1$ for all $x\in X$ and such that

$$I_j(\varphi)=\int_X \varphi g_j\,d|\mu| \quad \text{for} \quad \varphi\in\mathfrak{C}_0. \tag{1}$$

Thus we have

$$\Phi(\varphi)=\int_X \varphi[g_1-g_2+i(g_3-g_4)]\,d|\mu| \quad \text{for} \quad \varphi\in\mathfrak{C}_0. \tag{2}$$

Write $g=g_1-g_2+i(g_3-g_4)$. Plainly (2) is equivalent with (ii).

It remains to show that $|g(x)|=1$ $|\mu|$-almost everywhere. Assume that $|g(x)|>1$ on a [measurable] set F of positive $|\mu|$-measure. We may suppose that F is compact, since $|\mu|$ is finite for all subsets of X (14.6). There is a sequence $\varphi_1\geqq\varphi_2\geqq\cdots\geqq\varphi_n\geqq\cdots$ of functions in $\mathfrak{C}_{00}^+$ such that $\varphi_n(X)\subset[0,1]$ and $\varphi_n(F)=1$ for $n=1,2,\ldots$, and $\lim\limits_{n\to\infty}\int_X \varphi_n\,d|\mu|=|\mu|(F)$. Now consider the function $\operatorname{sgn}\bar{g}$. This is plainly a function in $\mathfrak{L}_1(X,|\mu|)$, and so by (12.10), there is a sequence of functions $\psi_1,\psi_2,\ldots,\psi_n,\ldots$ in $\mathfrak{C}_{00}$ such that $\lim\limits_{n\to\infty}\int_X|\psi_n-\operatorname{sgn}\bar{g}|\,d|\mu|=0$. Now let $\omega_n=\psi_n\cdot\min\left(\frac{1}{|\psi_n|},1\right)$ [we agree that $\frac{1}{0}>1$]. It is obvious that $\omega_n\in\mathfrak{C}_{00}$ and that $|\omega_n|\leqq 1$. Since $|\operatorname{sgn}\bar{g}(x)|$ is 0 or 1, an elementary argument shows that $|\omega_n(x)-\operatorname{sgn}\bar{g}(x)|\leqq|\psi_n(x)-\operatorname{sgn}\bar{g}(x)|$ for all $x\in X$. Hence we have $\lim\limits_{n\to\infty}\int_X|\omega_n-\operatorname{sgn}\bar{g}|\,d|\mu|=0$.

The original definition of $|\Phi|$ in (14.5.i) and (2) show that

$$\int_X \varphi_n\,d|\mu|=|\Phi|(\varphi_n)\geqq|\Phi(\varphi_n\omega_n)|=\left|\int_X \varphi_n\omega_n g\,d|\mu|\right|. \tag{3}$$

We also have

$$\left.\begin{aligned} &\left|\int_X \varphi_n\omega_n g\,d|\mu|-\int_X \xi_F|g|\,d|\mu|\right| \\ &\leqq\left|\int_X \varphi_n\omega_n g\,d|\mu|-\int_X \varphi_n|g|\,d|\mu|\right|+\left|\int_X \varphi_n|g|\,d|\mu|-\int_X \xi_F|g|\,d|\mu|\right| \\ &\leqq\|\varphi_n g\|_\infty\cdot\int_X|\omega_n-\operatorname{sgn}\bar{g}|\,d|\mu|+\|g\|_\infty\left[\int_X \varphi_n\,d|\mu|-|\mu|(F)\right]. \end{aligned}\right\} \tag{4}$$

Since $|g|$ is bounded, we infer from (4) that $\lim_{n\to\infty} \int_X \varphi_n \omega_n g \, d|\mu| = \int_X \xi_F |g| \, d|\mu|$. Taking the limit as $n\to\infty$ on both sides of (3), we obtain the relations $|\mu|(F) \geqq \left|\int_X \xi_F |g| \, d|\mu|\right| = \int_X \xi_F |g| \, d|\mu|$. Since $|g(x)| > 1$ for $x \in F$, this last inequality is impossible. Thus we have

$$|g(x)| \leqq 1 \quad |\mu|\text{-almost everywhere.} \tag{5}$$

Finally, assume that $|g(x)| < 1$ on a compact set F such that $|\mu|(F) > 0$. Choose a sequence $\varphi_1 \geqq \varphi_2 \geqq \cdots \geqq \varphi_n \geqq \cdots$ of functions in $\mathfrak{C}_{00}^+$ such that $\lim_{n\to\infty} \varphi_n = \xi_E$, where E is an $|\mu|$-measurable set containing F such that $|\mu|(E \cap F') = 0$. Then we have $\lim_{n\to\infty} |\Phi|(\varphi_n) = \lim_{n\to\infty} \int_X \varphi_n \, d|\mu| = |\mu|(F)$. Using the definition of $|\Phi|$ in (14.5.i), choose functions $\psi_n \in \mathfrak{C}_{00}$ such that $|\psi_n| \leqq \varphi_n$ and $|\Phi(\psi_n)| + \frac{1}{n} > |\Phi|(\varphi_n)$ $(n = 1, 2, \ldots)$. Using (2) and (5) and obvious integral inequalities, we now have

$$|\Phi(\psi_n)| = \left|\int_X \psi_n g \, d|\mu|\right| \leqq \int_X |\psi_n| \, |g| \, d|\mu| \leqq \int_X |\psi_n| \, d|\mu| \leqq \int_X \varphi_n \, d|\mu| . \tag{6}$$

Taking limits as $n\to\infty$ in (6), we find that

$$|\mu|(F) = \lim_{n\to\infty} \int_X |\psi_n g| \, d|\mu| . \tag{7}$$

LEBESGUE's theorem on dominated convergence and the fact that $|g(x)| < 1$ on F give us

$$\left.\begin{aligned} \lim_{n\to\infty} \int_X |\psi_n g| \, d|\mu| &\leqq \int_X \left(\overline{\lim_{n\to\infty}} |\psi_n|\right) |g| \, d|\mu| \leqq \int_X \left(\lim_{n\to\infty} \varphi_n\right) |g| \, d|\mu| \\ &= \int_E |g| \, d|\mu| = \int_F |g| \, d|\mu| < |\mu|(F) . \end{aligned}\right\} \tag{8}$$

Relations (7) and (8) give an obvious contradiction. Thus $|g(x)| = 1$ for $|\mu|$-almost all x. We replace g by a function equal to it $|\mu|$-almost everywhere to obtain (i). ☐

(14.13) Theorem. *Let A be a subset of X. Then $|\mu|(A) = 0$ if and only if $\iota_1(A) = \iota_2(A) = \iota_3(A) = \iota_4(A) = 0$. A subset of X is $|\mu|$-measurable if and only if it is ι_j-measurable for $j = 1, 2, 3, 4$, and for $|\mu|$-measurable sets A, we have*

(i) $\mu(A) = \int_A g \, d|\mu|$.

Furthermore, a complex- or extended real-valued function f on X is in $\mathfrak{L}_1(X, |\mu|)$ if and only if it is in $\mathfrak{L}_1(X, \iota_j)$ for $j = 1, 2, 3, 4$, and then

(ii) $\int_X f \, d\mu = \int_X f g \, d|\mu|$.

Proof. As shown in the proof of (14.12), we have $I_j \leqq |\Phi|$ $(j=1, 2, 3, 4)$, and this immediately implies that $\iota_j \leqq |\mu|$. Hence if A is $|\mu|$-measurable, then by (14.11.iii) A is ι_j-measurable for $j=1, 2, 3, 4$. The functional $I_1+I_2+I_3+I_4$ is easily shown to be a nonnegative majorant of Φ, and (14.5) thus implies that $|\Phi| \leqq I_1+I_2+I_3+I_4$, so that $|\mu| \leqq \iota_1+\iota_2+\iota_3+\iota_4$. Now if A is ι_j-measurable for $j=1, 2, 3, 4$, then A is clearly $\iota_1+\iota_2+\iota_3+\iota_4$-measurable, and (14.11.iii) implies that A is $|\mu|$-measurable.

Consider a complex- or extended real-valued function f on X. From the previous paragraph, it is clear that $f \in \mathfrak{L}_1(X, |\mu|)$ if and only if $f \in \mathfrak{L}_1(X, \iota_j)$ for $j=1, 2, 3, 4$. The linear functionals $f \rightarrow \int\limits_X f\, d\mu$ and $f \rightarrow \int\limits_X f g\, d|\mu|$ on $\mathfrak{L}_1(X, |\mu|)$ are clearly bounded in the $\mathfrak{L}_1$ norm. By (14.12.ii), they agree on $\mathfrak{C}_0$, which is dense in $\mathfrak{L}_1(X, |\mu|)$ by (12.10). It follows that these functionals agree on $\mathfrak{L}_1(X, |\mu|)$. Hence (ii) holds; (i) is a special case of (ii). □

Finally we characterize $|\mu|$ in terms of μ.

(14.14) Theorem. *Let A be an $|\mu|$-measurable set. Then*

(i) $|\mu|(A) = \sup\left\{\sum_{j=1}^{m} |\mu(E_j)| : \{E_1, \ldots, E_m\}\right.$ *is a partition of A into $|\mu|$-measurable sets*$\left.\right\}$.

Proof. For $\{E_1, \ldots, E_m\}$ as in (i), we have

$$\sum_{j=1}^{m} |\mu(E_j)| = \sum_{j=1}^{m} \left|\int\limits_{E_j} g\, d|\mu|\right| \leqq \sum_{j=1}^{m} \int\limits_{E_j} |g|\, d|\mu| = \sum_{j=1}^{m} |\mu|(E_j) = |\mu|(A).$$

Hence $|\mu|(A)$ is greater than or equal to the right side of (i).

To prove the reversed inequality, let ε be a positive number. By (12.10), there is a function $h = \sum_{j=1}^{m} \alpha_j \xi_{E_j}$ such that the E_j are pairwise disjoint $|\mu|$-measurable subsets of X and $\int\limits_X \left|h - \frac{\xi_A}{g}\right| d|\mu| < \varepsilon$. We may suppose that the sets $\{E_j\}_{j=1}^{m}$ partition A. Let $\beta_j = \alpha_j \cdot \min\left(\frac{1}{|\alpha_j|}, 1\right)$ [again $\frac{1}{0} > 1$], and write $h_0 = \sum_{j=1}^{m} \beta_j \xi_{E_j}$. Then an elementary argument shows that $\left|h_0 - \frac{\xi_A}{g}\right| \leqq \left|h - \frac{\xi_A}{g}\right|$ and $|\beta_j| \leqq 1$. In particular, we have $\int\limits_X |h_0 g - \xi_A|\, d|\mu| < \varepsilon$. It follows that

$$\begin{aligned} |\mu|(A) - \varepsilon &= \int\limits_X \xi_A\, d|\mu| - \varepsilon < \left|\int\limits_X h_0 g\, d|\mu|\right| \\ &= \left|\sum_{j=1}^{m} \beta_j \int\limits_X \xi_{E_j} g\, d|\mu|\right| = \left|\sum_{j=1}^{m} \beta_j \mu(E_j)\right| \leqq \sum_{j=1}^{m} |\mu(E_j)|. \end{aligned}$$

As ε is arbitrary, the proof is complete. □

(14.15) Corollary. *Let E be an $|\mu|$-measurable subset of X. Then the following assertions are equivalent:*

(i) $|\mu|(E)=0$;

(ii) $\mu(A)=0$ *for all* $|\mu|$*-measurable subsets* A *of* E;

(iii) $\mu(F)=0$ *for all compact subsets* F *of* E.

This follows immediately from (14.14).

We next identify $\mathfrak{C}_0^*(X)$ for *discrete* spaces X [which are trivially locally compact].

(14.16) Theorem. *Let X be a discrete space, and let I be any nonnegative linear functional on $\mathfrak{C}_{00}$. Then there is a nonnegative real-valued function w defined on X such that*

(i) $I(f)=\sum\limits_{x\in X} f(x)\,w(x)$ *for all* $f\in\mathfrak{C}_{00}$

[the sum is actually finite]. Conversely, every nonnegative real-valued function w on X defines a nonnegative linear functional on $\mathfrak{C}_{00}$ by the formula (i). *Let Φ be a bounded complex linear functional on $\mathfrak{C}_0$. Then there is a complex-valued function w defined on X such that*

(ii) $\sum\limits_{x\in X}|w(x)|<\infty$ [1]

and

(iii) $\Phi(f)=\sum\limits_{x\in X} f(x)\,w(x)$ *for all* $f\in\mathfrak{C}_0$.

Conversely, every $w\in l_1(X)$ defines a functional in $\mathfrak{C}_0^*$ by the formula (iii).

Proof. For every $x\in X$, let δ_x be the function on X defined by $\delta_x(y)=\begin{cases}1 & \text{if } y=x\\ 0 & \text{if } y\neq x\end{cases}$. Define w by the equation $w(x)=I(\delta_x)$, $x\in X$. Then the function w satisfies (i). The converse proposition is obvious.

Let Φ be a nonzero bounded complex linear functional on $\mathfrak{C}_0$, let δ_x be as above ($x\in X$), and let $w(x)=\Phi(\delta_x)$ for $x\in X$. If (ii) fails, we can find a sequence $x_1,\ldots,x_n$ of elements in X such that $\sum\limits_{k=1}^{n}|w(x_k)|>\|\Phi\|$. Then if f_0 is equal to $\sum\limits_{k=1}^{n}\operatorname{sgn}\overline{w(x_k)}\,\delta_{x_k}$, we have $\Phi(f_0)=\sum\limits_{k=1}^{n}\operatorname{sgn}\overline{w(x_k)}\,w(x_k)=\sum\limits_{k=1}^{n}|w(x_k)|>\|\Phi\|$ and $\|f_0\|_u=1$, which is impossible. Hence $\sum\limits_{x\in X}|w(x)|\leq\|\Phi\|<\infty$ and (ii) holds.

Linearity implies that (iii) holds for all $f\in\mathfrak{C}_{00}$. Let $f\in\mathfrak{C}_0$ be arbitrary ($f\neq 0$), and let ε be a positive number. Choose a function f_0 in $\mathfrak{C}_{00}$ such that $\|f-f_0\|_u<\frac{\varepsilon}{2\|\Phi\|}$. Then $\left|\Phi(f)-\sum\limits_{x\in X} f(x)\,w(x)\right|\leq|\Phi(f)-\Phi(f_0)|+$

---

[1] That is, w is in $l_1(X)$: see § 1 for the definition.

$\left|\Phi(f_0) - \sum_{x\in X} f_0(x)\, w(x)\right| + \left|\sum_{x\in X} f_0(x)\, w(x) - \sum_{x\in X} f(x)\, w(x)\right| \leqq \|\Phi\| \cdot \|f - f_0\|_u +$
$0 + \|f - f_0\|_u \sum_{x\in X} |w(x)| < \varepsilon.$

The proof of the last statement of the theorem is simple and is omitted. □

In the next two theorems, we investigate a special class of measures in $\boldsymbol{M}(X)$.

(14.17) Theorem. *Let I be a nonnegative linear functional on $\mathfrak{C}_{00}$, and let ι be the corresponding measure as in* §11. *Suppose that w is a function in $\mathfrak{L}_1(X, \iota)$, and let*

(i) $\Phi(f) = \int_X f w\, d\iota$ *for* $f \in \mathfrak{C}_0$.

Then Φ is a functional in $\mathfrak{C}_0^*$. Write μ for the complex measure corresponding to Φ as in (14.4), *and write E for the set $\{x \in X : w(x) \neq 0\}$. A subset A of X is $|\mu|$-measurable if and only if $A \cap E$ is ι-measurable. For such sets, we have*

(ii) $\mu(A) = \int_{A\cap E} w\, d\iota$

and

(iii) $|\mu|(A) = \int_{A\cap E} |w|\, d\iota.$

In particular, $\|\Phi\| = \|w\|_1$ and if $\iota(A) = 0$, then $\mu(A) = |\mu|(A) = 0$. Also for $f \in \mathfrak{L}_1(X, |\mu|)$, we have $f w \in \mathfrak{L}_1(X, \iota)$,

(iv) $\int_X f\, d\mu = \int_E f w\, d\iota,$

and

(v) $\int_X f\, d|\mu| = \int_E f |w|\, d\iota.$

Proof. It is easy to see that Φ is a functional in $\mathfrak{C}_0^*$ and that $\|\Phi\| \leqq \|w\|_1$. To prove (ii), suppose initially that w is nonnegative. We prove (ii) first for open sets U. By (11.11), we have

$$\mu(U) = \sup\left\{\int_X f w\, d\iota : f \leqq \xi_U,\ f \in \mathfrak{C}_{00}^+\right\}, \tag{1}$$

so that $\mu(U) \leqq \int_U w\, d\iota$. The set E is clearly a countable union of ι-measurable sets having finite ι-measure. Therefore $U \cap E$ can be written as $\left(\bigcup_{n=1}^{\infty} F_n\right) \cup N$ where $\iota(N) = 0$ and $\{F_n\}_{n=1}^{\infty}$ is an increasing sequence of compact sets. Choose f_n in $\mathfrak{C}_{00}^+$ such that $f_n(F_n) = 1$, $f_n(U') = 0$, and $f_n(X) \subset [0, 1]$. Then $f_n \leqq \xi_U$ for $n = 1, 2, \ldots$, and $\lim_{n\to\infty} f_n w = \xi_U w$ ι-almost everywhere. Therefore using (1), we have

$$\mu(U) \geqq \lim_{n\to\infty} \int_X f_n w\, d\iota = \int_X \lim_{n\to\infty} f_n w\, d\iota = \int_U w\, d\iota = \int_{U\cap E} w\, d\iota.$$

Thus (ii) holds for all open sets. A simple computation shows that (ii) holds for all sets in the σ-algebra of Borel sets.

We now show that

$$\text{if } A \subset X,\ A \cap E \in \mathcal{M}_\iota, \text{ and } \iota(A \cap E) = 0, \text{ then } A \in \mathcal{M}_\mu \text{ and } \mu(A) = 0. \tag{2}$$

We note that E has the form $B_0 \cup N_0$ where B_0 is a Borel set, $N_0 \in \mathcal{M}_\iota$, and $\iota(N_0) = 0$. Also there is a Borel set B such that $A \cap E \subset B$ and $\iota(B) = 0$. Then $A \subset B \cup B_0'$, and

$$\mu(B \cup B_0') = \int_{(B \cup B_0') \cap E} w\, d\iota = \int_{B \cap E} w\, d\iota = 0.$$

It follows that $A \in \mathcal{M}_\mu$ and $\mu(A) = 0$.

Now suppose that A is any set such that $A \cap E \in \mathcal{M}_\iota$. Then $A \cap E = D \cup N$ where D is a Borel set and $\iota(N) = 0$. By (2), we have $N \in \mathcal{M}_\mu$ and $A \cap E' \in \mathcal{M}_\mu$, so that $A = D \cup N \cup (A \cap E') \in \mathcal{M}_\mu$.

Suppose that A is in $\mathcal{M}_\mu$. Let B and C be Borel sets such that $B \subset A \subset C$ and $\mu(B) = \mu(A) = \mu(C)$. Then

$$\int_{C \cap B' \cap E} w\, d\iota = \mu(C \cap B') = 0,$$

and as $w(x) > 0$ for $x \in E$, we see by (11.27) that $C \cap B' \cap E$ is ι-null. Thus $A \cap E = (B \cap E) \cup (A \cap B' \cap E)$ is ι-measurable.

Finally suppose that $\mu(A) \neq \int_{A \cap E} w\, d\iota$ for some $A \in \mathcal{M}_\mu$. Since $\mu(A) + \mu(A') = \mu(X) = \int_X w\, d\iota = \int_{A \cap E} w\, d\iota + \int_{A' \cap E} w\, d\iota$, either

$$\mu(A) < \int_{A \cap E} w\, d\iota \tag{3}$$

or

$$\mu(A') < \int_{A' \cap E} w\, d\iota \tag{4}$$

obtains; we may suppose that (3) holds. Choose an open set $U \supset A$ such that $\mu(U) < \int_{A \cap E} w\, d\iota$. Then we have $\mu(U) < \int_{A \cap E} w\, d\iota \leqq \int_{U \cap E} w\, d\iota$, which is impossible. This completes the proof of (ii) in the case that $w \geqq 0$.

For arbitrary $w \in \mathfrak{L}_1(X, \iota)$, we write $w = w_1 - w_2 + i(w_3 - w_4)$, where the w_j's are functions in $\mathfrak{L}_1^+(X, \iota)$ and $\min(w_1, w_2) = \min(w_3, w_4) = 0$. For $f \in \mathfrak{C}_0$ and $j = 1, 2, 3, 4$, we define $\Phi_j(f) = \int_X f w_j\, d\iota$. An elementary computation shows that $\min(\Phi_1, \Phi_2) = \min(\Phi_3, \Phi_4) = 0$ [1] and therefore (14.3) implies that $\Phi = \Phi_1 - \Phi_2 + i(\Phi_3 - \Phi_4)$ is the resolution of (14.3). Write $E^{(j)} = \{x \in X : w_j(x) > 0\}$; then $\mu_j(A) = \int_A w_j\, d\iota$ for all sets A such that $A \cap E^{(j)}$ is ι-measurable ($j = 1, 2, 3, 4$). If A is a subset of X, then $A \cap E$ is ι-measurable if and only if $A \cap E^{(j)}$ is ι-measurable for $j = 1, 2, 3, 4$, since $E = E^{(1)} \cup E^{(2)} \cup E^{(3)} \cup E^{(4)}$. It follows that $A \cap E$ is ι-measurable if

[1] These minima are the minima of nonnegative linear functionals, defined as in (B.34).

and only if A is $|\mu|$-measurable, and in this case $\mu(A)=\mu_1(A)-\mu_2(A)+i(\mu_3(A)-\mu_4(A))=\int\limits_{A\cap E} w\,d\iota$.

We now prove (iv). Let $\mathfrak{S}$ consist of all functions σ having the form $\sum_{j=1}^{m}\alpha_j\xi_{A_j}$ where $\alpha_j\in K$ and A_j is $|\mu|$-measurable $(j=1,\ldots,m)$. Because of (ii), (iv) obviously holds for $\sigma\in\mathfrak{S}$. Suppose now that $f\in\mathfrak{L}_1^+(X,|\mu|)$ and that $w\geqq 0$. Then, as is shown in the proof of (12.10), there is an increasing sequence $\{\sigma_n\}_{n=1}^{\infty}$ of nonnegative functions in $\mathfrak{S}$ such that $\lim\limits_{n\to\infty}\sigma_n=f$ everywhere on X. Using LEBESGUE's theorem on monotone convergence, we have

$$\int\limits_X f\,d\mu=\lim_{n\to\infty}\int\limits_X\sigma_n\,d\mu=\lim_{n\to\infty}\int\limits_E\sigma_n w\,d\iota=\int\limits_E fw\,d\iota.$$

Relation (iv) for arbitrary $f\in\mathfrak{L}_1(X,|\mu|)$ and $w\in\mathfrak{L}_1(X,\iota)$ is now proved by writing f and w as sums of functions in $\mathfrak{L}_1^+(X,|\mu|)$ and $\mathfrak{L}_1^+(X,\iota)$, respectively, and applying the case just considered.

Next we prove (iii); it suffices to prove that $|\Phi|(f)=\int\limits_X f|w|\,d\iota$ for $f\in\mathfrak{C}_0$, because then (ii) can be applied to $|\mu|$. The functional defined by $f\to\int\limits_X f|w|\,d\iota$ for $f\in\mathfrak{C}_0$ is obviously a nonnegative linear functional on $\mathfrak{C}_0$, and is a nonnegative majorant of Φ: $|\Phi(f)|=\left|\int\limits_X fw\,d\iota\right|\leqq\int\limits_X|f|\,|w|\,d\iota$. Thus by (14.5), it suffices to prove that

$$\int\limits_X f|w|\,d\iota\leqq|\Phi|(f) \tag{5}$$

for all $f\in\mathfrak{C}_0^+$. Let $\{\omega_n\}_{n=1}^{\infty}$ be a sequence of functions in $\mathfrak{C}_{00}$ such that $|\omega_n|\leqq 1$ for all n, and $\lim\limits_{n\to\infty}\int\limits_X|\omega_n-\operatorname{sgn}\overline{w}|\,d|\mu|=0$. [This sequence is constructed just as in the proof of (14.12) for the function $\operatorname{sgn}\overline{g}$.] For $f\in\mathfrak{C}_0^+$, we have $\left|\int\limits_X\omega_n f\,d\mu-\int\limits_X\operatorname{sgn}\overline{w}\,f\,d\mu\right|\leqq\|f\|_u\int\limits_X|\omega_n-\operatorname{sgn}\overline{w}|\,d|\mu|$, so that $\lim\limits_{n\to\infty}\int\limits_X\omega_n f\,d\mu=\int\limits_X\operatorname{sgn}\overline{w}\,f\,d\mu=\int\limits_X f|w|\,d\iota$; the last equality follows from (iv) since $\operatorname{sgn}\overline{w}\,f\in\mathfrak{L}_1(X,|\mu|)$. It follows from (14.5.i) that $|\Phi|(f)\geqq\int\limits_X f|w|\,d\iota$. This proves (5) and hence (iii).

Assertion (v) follows from (iv) and the fact that $|\Phi|(f)=\int\limits_X f|w|\,d\iota$ for $f\in\mathfrak{C}_0$. Finally we observe that $\|\Phi\|=|\mu|(X)=\int\limits_X|w|\,d\iota=\|w\|_1$ by (14.6) and (iii). □

(14.18) For $w\in\mathfrak{L}_1(X,\iota)$ and μ as defined from (14.17.i), we write $d\mu=w\,d\iota$, meaning that (14.17.i)—(14.17.v) hold for μ and $|\mu|$.

Let ν be a complex measure as in (14.4), and let w be in $\mathfrak{L}_1(X,|\nu|)$. By (14.12), there is an $|\nu|$-measurable function g on X such that $d\nu=g\,d|\nu|$. Let μ be the measure such that $d\mu=wg\,d|\nu|$ as in (14.17).

We then have $\int_X f\,d\mu = \int_X fw\,d\nu$ for $f \in \mathfrak{C}_0$, and we write $d\mu = w\,d\nu$. A theorem analogous to (14.17) is now valid.

(14.19) Theorem. *Let I be a nonnegative linear functional on $\mathfrak{C}_{00}$ and let ι be the corresponding measure as in § 11. Let μ be a complex measure as constructed in (14.4) with the property that $|\mu|(F) = 0$ for all compact sets $F \subset X$ that have ι-measure 0. Then there is a Borel measurable function $w \in \mathfrak{L}_1(X, \iota)$ such that $d\mu = w\,d\iota$.*

Proof. Let A be any locally ι-null set (11.26). Consider any compact subset F of X. Then $\iota(A \cap F) = 0$ and hence there is a Borel set E such that $A \cap F \subset E$ and $\iota(E) = 0$ [E can be taken to be a countable intersection of open sets]. Using (11.32), we see that $|\mu|(E) = 0$ and hence $|\mu|(A \cap F) = 0$. Thus A is locally $|\mu|$-null. [Note that $|\mu|(A) = 0$ if A is *locally* $|\mu|$-null, since $|\mu|$ is finite on all subsets of X.] Thus we have $\mathcal{N}_\iota \subset \mathcal{N}_{|\mu|}$, and we apply (12.17) to assert the existence of a nonnegative ι-measurable function w_0 such that $w_0\,\xi_F$ is in $\mathfrak{L}_1(X, \iota)$ for all compact subsets F of X and such that

$$\int_X f\,d|\mu| = \int_X f w_0\,d\iota \quad \text{for all} \quad f \in \mathfrak{C}_{00}. \tag{1}$$

Although w_0 need not be in $\mathfrak{L}_1(X, \iota)$, we will find a function $w_1 \in \mathfrak{L}_1(X, \iota)$ such that (1) holds with w_0 replaced by w_1. It is easy to see that there is a sequence $\{U_n\}_{n=1}^\infty$ of open subsets of X such that U_n^- is compact, $U_n^- \subset U_{n+1}$ for $n = 1, 2, \ldots$, and $\lim_{n\to\infty} |\mu|(U_n) = |\mu|(X)$. For each n, let $\psi_n \in \mathfrak{C}_{00}^+$ be taken so that $\psi_n(U_n^-) = 1$, $\psi_n(U'_{n+1}) = 0$, and $\psi_n(X) \subset [0, 1]$. Write $\bigcup_{n=1}^\infty U_n$ as E, and let $w_1 = \xi_E w_0$. Plainly $\{\psi_n\}_{n=1}^\infty$ is an increasing sequence of functions, $\lim_{n\to\infty} \psi_n = \xi_E$ everywhere, and $|\mu|(E') = 0$. Now for $f \in \mathfrak{C}_0^+$, (1) implies that

$$\left.\begin{aligned} \int_X f\,d|\mu| &= \int_X \xi_E f\,d|\mu| = \lim_{n\to\infty} \int_X \psi_n f\,d|\mu| \\ &= \lim_{n\to\infty} \int_X \psi_n f w_0\,d\iota = \int_X f \xi_E w_0\,d\iota = \int_X f w_1\,d\iota. \end{aligned}\right\} \tag{2}$$

Plainly (2) also holds for all $f \in \mathfrak{C}_0$. Moreover, $\int_X w_1\,d\iota = \lim_{n\to\infty} \int_X \psi_n w_0\,d\iota = \lim_{n\to\infty} \int_X \psi_n\,d|\mu| = |\mu|(E) < \infty$, so that $w_1 \in \mathfrak{L}_1(X, \iota)$. By (11.41), we may suppose that w_1 is Borel measurable.

Let g be the $|\mu|$-measurable function for μ defined in (14.12); then

$$\int_X f\,d\mu = \int_X fg\,d|\mu| \quad \text{for} \quad f \in \mathfrak{C}_0. \tag{3}$$

By (11.41), we may also suppose that g is Borel measurable. Since (2) holds for $f \in \mathfrak{C}_0$, we can apply (14.17.iv) to the function $fg \in \mathfrak{L}_1(X, |\mu|)$;

this yields

$$\int_X f\,d\mu = \int_X fg\,d|\mu| = \int_X fg\,w_1\,d\iota \quad \text{for} \quad f\in\mathfrak{C}_0.$$

Now let $w = gw_1$. □

(14.20) Definition. Let μ and ι be set functions defined on a subfamily of $\mathscr{P}(X)$ satisfying the hypotheses of Theorem (14.19). Then μ is said to be *absolutely continuous with respect to* ι. [We also define absolute continuity for measures ι and η of the kind considered in (12.17), saying that η is *absolutely continuous with respect to* ι if the conditions (12.17.i)—(12.17.v) obtain. It is clear that the second definition is consistent with the first.]

We now define and investigate a property of measures that is in a sense the antithesis of absolute continuity.

(14.21) Definition. Let I be a nonnegative linear functional on $\mathfrak{C}_{00}$ and let ι be the corresponding measure as in §11. Let μ be a complex measure as constructed in (14.4), and suppose that there is a Borel set $B\subset X$ such that $\iota(B)=0$ and $|\mu|(B')=0$. [We say that $|\mu|$ is *concentrated on* B.] Then μ is said to be *singular with respect to* ι.

(14.22) Theorem. *Let I and ι be as in* (14.21) *and let μ be any complex measure as constructed in* (14.4). *Then there are complex measures μ_1 and μ_2 as in* (14.4) *such that*

(i) $\mu = \mu_1 + \mu_2$,

(ii) *μ_1 is absolutely continuous with respect to ι,*

and

(iii) *μ_2 is singular with respect to ι.*

If ν_1 and ν_2 are any complex measures [as in (14.4)] *with properties* (i)—(iii), *then $\mu_1=\nu_1$ and $\mu_2=\nu_2$. Also*

(iv) $|\mu| = |\mu_1| + |\mu_2|$, *and* $\|\mu\| = \|\mu_1\| + \|\mu_2\|$.

Proof. Let Φ be the complex linear functional on $\mathfrak{C}_0$ in terms of which μ is defined and let $|\Phi|$ be as in (14.5). It is obvious that $|\Phi| \leqq I + |\Phi|$ and that the [nonnegative] measure corresponding to $I+|\Phi|$ is $\iota+|\mu|$. It is trivial that $|\mu|(F)=0$ if F is compact and $(\iota+|\mu|)(F)=0$. Thus applying (14.19) to μ, we find a complex-valued Borel measurable function w in $\mathfrak{L}_1(X, \iota+|\mu|)$ for which

$$\int_X f\,d\mu = \int_X fw\,d(\iota+|\mu|) = \int_X fw\,d\iota + \int_X fw\,d|\mu| \tag{1}$$

for all $f\in\mathfrak{C}_0$. Let $E=\{x\in X: w(x) \neq 0\}$. Then E is a Borel set and by (14.17) we have

$$\mu(A) = \int_{A\cap E} w\,d\iota + \int_{A\cap E} w\,d|\mu| \tag{2}$$

and

$$|\mu|(A) = |\mu|(A \cap E) = \int_{A\cap E} |w|\, d\iota + \int_{A\cap E} |w|\, d|\mu| \tag{3}$$

for all subsets A of X such that $A \cap E$ is $(\iota + |\mu|)$-measurable. Let $C = \{x \in X : |w(x)| > 1\}$ and $B = \{x \in X : |w(x)| = 1\}$. Then from (3) we have

$$|\mu|(C) = \int_C |w|\, d\iota + \int_C |w|\, d|\mu|,$$

so that $|\mu|(C) = 0$ and $\iota(C) = 0$. We may therefore suppose that $|w(x)| \leqq 1$ for all $x \in X$. Setting $A = B$ in (3), we have

$$|\mu|(B) = \iota(B) + |\mu|(B),$$

so that

$$\iota(B) = 0. \tag{4}$$

Now let Φ_1 and Φ_2 be the functionals in $\mathfrak{C}_0^*$ such that

$$\Phi_1(f) = \int_X f\xi_{B'} w\, d\iota + \int_X f\xi_{B'} w\, d|\mu|, \qquad \Phi_2(f) = \int_X f\xi_B w\, d|\mu| \tag{5}$$

for $f \in \mathfrak{C}_0$, and let μ_1 and μ_2, respectively, be the corresponding complex measures. It follows from (1), (4), and (5) that $\Phi = \Phi_1 + \Phi_2$ and $\mu = \mu_1 + \mu_2$. It follows from (14.17.iii) that $|\mu_2|(B') = 0$; this equality and (4) show that μ_2 is singular with respect to ι. Suppose that F is compact and that $\iota(F) = 0$. Setting $A = F$ in (3), we have

$$|\mu|(F) = \int_{F\cap E} |w|\, d\iota + \int_{F\cap E} |w|\, d|\mu| = \int_{F\cap E} |w|\, d|\mu| \leqq \int_F |w|\, d|\mu| \leqq |\mu|(F).$$

Thus $0 = \int_F (1 - |w|)\, d|\mu| = \int_{F\cap B'} (1 - |w|)\, d|\mu|$, and consequently $|\mu|(F \cap B') = 0$, since $|w| < 1$ on $F \cap B'$. Applying (14.17.iii), we have

$$|\mu_1|(F) = \int_{F\cap B'} |w|\, d\iota + \int_{F\cap B'} |w|\, d|\mu| = 0.$$

That is, μ_1 is absolutely continuous with respect to ι.

Finally, let ν_1 and ν_2 be as in the statement of the theorem. Then we have $\mu_1 - \nu_1 = \nu_2 - \mu_2$. For every compact set F such that $\iota(F) = 0$, we have $|\nu_2 - \mu_2|(F) = |\mu_1 - \nu_1|(F) \leqq |\mu_1|(F) + |\nu_1|(F) = 0$. Thus $\nu_2 - \mu_2$ is absolutely continuous with respect to ι. On the other hand, if $|\nu_2|(B') = |\mu_2|(C') = 0$, where $\iota(B) = \iota(C) = 0$, then $|\nu_2 - \mu_2|(B' \cap C') \leqq |\nu_2|(B') + |\mu_2|(C') = 0$, and $\iota(B \cup C) = 0$. Hence $\nu_2 - \mu_2$ is both singular and absolutely continuous with respect to ι, and is therefore 0. Property (iv) follows from (5), (1), and (14.17). □

(14.23) Remarks. Let Φ be in $\mathfrak{C}_0^*$ and let μ be the measure corresponding to Φ as in (14.4). LEBESGUE's theorem on dominated convergence is valid for integrals with respect to μ. More precisely, if f is an $|\mu|$-almost everywhere limit of a sequence $\{f_n\}_{n=1}^{\infty}$ of $|\mu|$-measurable functions on X and if there is a function $s \in \mathfrak{L}_1(X, |\mu|)$ such that $|f_n(x)| \leqq s(x)$ for

$|\mu|$-almost all $x\in X$, then $\lim_{n\to\infty}\int_X f_n\,d\mu$ exists and is equal to $\int_X f\,d\mu$. To see this, let g be as defined in (14.12). Then the sequence $\{|f_n g|\}_{n=1}^{\infty}$ of functions is dominated by s, $\lim_{n\to\infty} f_n g = fg$, and the usual dominated convergence theorem, together with (14.13.ii), gives us

$$\int_X f\,d\mu = \int_X fg\,d|\mu| = \lim_{n\to\infty}\int_X f_n g\,d|\mu| = \lim_{n\to\infty}\int_X f_n\,d\mu.$$

We now discuss product measures $\mu\times\nu$ where the measures μ and ν are complex.

Let X and Y be locally compact Hausdorff spaces. Suppose that I and J are bounded, nonnegative, linear functionals on $\mathfrak{C}_{00}(X)$ and $\mathfrak{C}_{00}(Y)$, respectively, and that ι and η are their respective measures. Then $I\times J$ is a bounded, nonnegative, linear functional on $\mathfrak{C}_{00}(X\times Y)$ and can be extended uniquely to a nonnegative [and hence bounded] linear functional on $\mathfrak{C}_0(X\times Y)$.

Let Φ and Ψ be in $\mathfrak{C}_0^*(X)$ and $\mathfrak{C}_0^*(Y)$, respectively, and let μ and ν be the corresponding measures. Let $\Phi = \sum_{j=1}^{4}\alpha_j I_j$ and $\Psi = \sum_{k=1}^{4}\beta_k J_k$ be the resolutions of Φ and Ψ, respectively, given in (14.3). We then define $\Phi\times\Psi$ to be the functional $\sum_{j=1}^{4}\sum_{k=1}^{4}\alpha_j\beta_k I_j\times J_k$; $\Phi\times\Psi$ belongs to $\mathfrak{C}_0^*(X\times Y)$. If μ_j and ν_k are the measures corresponding to I_j and J_k, respectively ($j, k = 1, 2, 3, 4$), then it is easy to see that the measure corresponding to $\Phi\times\Psi$ is $\sum_{j=1}^{4}\sum_{k=1}^{4}\alpha_j\beta_k\mu_j\times\nu_k$, and we define $\mu\times\nu$ to be this measure.

(14.24) Theorem. *Let μ, ν, and $\mu\times\nu$ be as in* (14.23). *Then we have* $|\mu\times\nu| = |\mu|\times|\nu|$.

Proof. Let g_1 and g_2 be measurable functions on X and Y, respectively, such that $|g_1| = 1$, $|g_2| = 1$,

$$\int_X f\,d\mu = \int_X fg_1\,d|\mu| \quad \text{for all} \quad f\in\mathfrak{C}_0(X),$$

and

$$\int_Y f\,d\nu = \int_Y fg_2\,d|\nu| \quad \text{for all} \quad f\in\mathfrak{C}_0(Y);$$

see (14.12). By (11.41), we may suppose that g_1 and g_2 are Borel measurable. We first show that

$$\Phi\times\Psi(f) = \int_{X\times Y} f(x, y)\, g_1(x)\, g_2(y)\, d|\mu|\times|\nu|\,(x, y), \tag{1}$$

for all $f\in\mathfrak{C}_0(X\times Y)$. It suffices to verify (1) for all functions f having the form $f(x, y) = \varphi(x)\,\psi(y)$ where $\varphi\in\mathfrak{C}_0(X)$ and $\psi\in\mathfrak{C}_0(Y)$, since linear combinations of functions of this sort are uniformly dense in $\mathfrak{C}_0(X\times Y)$

[see the proof of (13.2)]. For such f, we use (13.12) twice to obtain

$$\begin{aligned}\int_{X\times Y} f(x, y)\, g_1(x)\, g_2(y)\, d|\mu|\times|\nu|(x, y) &= \int_{X\times Y} \varphi(x)\, g_1(x)\, \psi(y)\, g_2(y)\, d|\mu|\times|\nu|(x, y)\\ &= \int_X \varphi g_1\, d|\mu| \int_Y \psi g_2\, d|\nu| = \int_X \varphi\, d\mu \int_Y \psi\, d\nu\\ &= \sum_{j=1}^{4}\sum_{k=1}^{4} \alpha_j \beta_k \int_X \varphi\, d\mu_j \int_Y \psi\, d\nu_k\\ &= \sum_{j=1}^{4}\sum_{k=1}^{4} \alpha_j \beta_k \int_{X\times Y} \varphi(x)\, \psi(y)\, d\mu_j\times\nu_k(x, y)\\ &= \sum_{j=1}^{4}\sum_{k=1}^{4} \alpha_j \beta_k\, I_j\times J_k(f) = \Phi\times\Psi(f).\end{aligned}$$

In view of (1), we can apply (14.17.v) and see that

$$|\Phi\times\Psi|(f) = \int_{X\times Y} f(x, y)\, |g_1(x)\, g_2(y)|\, d|\mu|\times|\nu|(x, y) = |\Phi|\times|\Psi|(f)$$

for all $f\in\mathfrak{C}_0(X\times Y)$. Therefore we have $|\Phi\times\Psi| = |\Phi|\times|\Psi|$ and $|\mu\times\nu| = |\mu|\times|\nu|$. □

We now prove Fubini's theorem (13.8) for complex measures.

(14.25) Theorem. *Let μ, ν, and $\mu\times\nu$ be as in (14.23), and let f be in $\mathfrak{L}_1(X\times Y, |\mu\times\nu|)$. Then $f(x, y)$* qua *function of x is in $\mathfrak{L}_1(X, |\mu|)$ for $|\nu|$-almost all $y\in Y$, so that the integral $\int_X f(x, y)\, d\mu(x)$ exists for $|\nu|$-almost all $y\in Y$. Define this integral as 0 where it does not already exist; then the function of y defined by $y\to\int_X f(x, y)\, d\mu(x)$ belongs to $\mathfrak{L}_1(Y, |\nu|)$. Similar remarks apply to $f(x, y)$* qua *function of y and the function $x\to\int_Y f(x,y)\, d\nu(y)$. Also we have*

(i) $\int_{X\times Y} f\, d\mu\times\nu = \int_X\int_Y f(x, y)\, d\nu(y)\, d\mu(x) = \int_Y\int_X f(x, y)\, d\mu(x)\, d\nu(y)$.

Proof. Let g_1 and g_2 be the Borel measurable functions used in the proof of (14.24). The function $(x, y)\to f(x, y)\, g_1(x)\, g_2(y)$ clearly belongs to $\mathfrak{L}_1(X\times Y, |\mu\times\nu|) = \mathfrak{L}_1(X\times Y, |\mu|\times|\nu|)$ (14.24). All statements of the present theorem except the last follow at once from (13.8). By (14.24.1) [for $f\in\mathfrak{C}_0(X\times Y)$], and (14.17.iv), we have

$$\int_{X\times Y} f\, d\mu\times\nu = \int_{X\times Y} f(x, y)\, g_1(x)\, g_2(y)\, d|\mu|\times|\nu|(x, y).$$

Hence by (13.8), we have

$$\begin{aligned}\int_{X\times Y} f\, d\mu\times\nu &= \int_X\int_Y f(x, y)\, g_1(x)\, g_2(y)\, d|\nu|(y)\, d|\mu|(x)\\ &= \int_X\int_Y f(x, y)\, d\nu(y)\, d\mu(x).\end{aligned}$$

The remaining equality of (i) is shown similarly. □

Notes

The contents of this section are well known to many functional analysts; for example, see W. RUDIN [2], p. 228 and P. J. COHEN [1], p. 193, for assertions related to (14.6) and (14.12), respectively. Curiously enough, no coherent treatment seems to exist in print. Some of the results appear in HEWITT [3]. Complex measures are also defined in BOURBAKI [4], Ch. VI, §2, N° 8, but are treated from an altogether different point of view.

Chapter Four

Invariant functionals

Invariant functionals, measures, and integrals are a vital tool in studying representations of locally compact groups and in establishing the detailed structure of locally compact Abelian groups. They also provide the function algebras and function spaces that are studied in harmonic analysis. The subject of invariant functionals is large, and we cannot treat it with any completeness. In §15, we construct the Haar integral, which is essential for all of our subsequent work. In §16, we give some technical but interesting facts about Haar measure, and in §§17 and 18, we follow some interesting byways.

§ 15. The Haar integral

We begin with some definitions.

(15.1) Definition. Let G be any group, E any nonvoid set, and f any function with domain G and range contained in E. For a fixed element $a \in G$, let ${}_af$ $[f_a]$ be the function on G such that ${}_af(x) = f(ax)$ $[f_a(x) = f(xa)]$ for all $x \in G$. Then ${}_af$ $[f_a]$ is called the *left translate* [*right translate*] *of f by a*. Let f^* be the function on G such that $f^*(x) = f(x^{-1})$ for all $x \in G$.

(15.2) Definition. Let G be any group and let $\mathfrak{F}$ be a set of functions on G. Suppose that $f \in \mathfrak{F}$ and $a \in G$ imply ${}_af \in \mathfrak{F}$ $[f_a \in \mathfrak{F}]$. Let I be any function on $\mathfrak{F}$ such that $I(f) = I({}_af)$ $[I(f) = I(f_a)]$ for all $f \in \mathfrak{F}$ and $a \in G$. Then I is said to be *left invariant* or *invariant under left translations* [*right invariant* or *invariant under right translations*]. If I is both left and right invariant, it is said to be *two-sided invariant*. Suppose that $f \in \mathfrak{F}$ implies $f^* \in \mathfrak{F}$. If $I(f) = I(f^*)$ for all $f \in \mathfrak{F}$, then I is said to be *inversion invariant*, or to be *invariant under inversion*.

Of course every function space $\mathfrak{F}$ of the sort described in (15.2) admits a function that is two-sided invariant and inversion invariant:

namely, any constant function. This trivial example is of no value to us. We shall be concerned throughout the present chapter with the construction and description of nonzero ***linear functionals*** on a variety of function spaces which are invariant under left or right translations or inversion.

There is an intimate relationship between certain linear functionals and measures; this relationship is worked out for $\mathfrak{C}_{00}$ and $\mathfrak{C}_0$ in Chapter Three. Frequently when an invariant functional can be represented as the integral with respect to a measure, the corresponding measure will enjoy invariance properties like those of the functional. We accordingly make a formal definition of invariance for measures.

(15.3) Definition. Let G be a group and let $\mathscr{A}$ be a family of subsets of G. Let E be any nonvoid set, and let λ be a function with domain $\mathscr{A}$ and range contained in E. Suppose that $A\in\mathscr{A}$ and $x\in G$ imply $xA\in\mathscr{A}$ [$Ax\in\mathscr{A}$]. If $\lambda(xA)=\lambda(A)$ for all $x\in G$ and $A\in\mathscr{A}$ [$\lambda(Ax)=\lambda(A)$ for all $x\in G$ and $A\in\mathscr{A}$], then λ is said to be ***left invariant*** [***right invariant***]. Suppose that $A\in\mathscr{A}$ implies $A^{-1}\in\mathscr{A}$. If $\lambda(A^{-1})=\lambda(A)$ for all $A\in\mathscr{A}$, then λ is said to be ***inversion invariant.***

The first and most important function space for which we construct an invariant functional is $\mathfrak{C}_{00}(G)$, where G is a locally compact group. As in Chapter Three, we will write $\mathfrak{C}_{00}$ for $\mathfrak{C}_{00}(G)$, etc., when no confusion can arise.

We first make a simple observation about $\mathfrak{C}_0$.

(15.4) Theorem. *Let G be any topological group, and let f be any function in $\mathfrak{C}_0(G)$. Then for every $\varepsilon>0$, there is a neighborhood U of e in G such that $|f(x)-f(y)|<\varepsilon$ for all $x,y\in G$ such that $x^{-1}y\in U$ [f is left uniformly continuous]. Similarly, there is a neighborhood V of e in G such that $|f(x)-f(y)|<\varepsilon$ for all $x, y\in G$ such that $yx^{-1}\in V$ [f is right uniformly continuous].*

Proof. This follows immediately from (4.15), if we take H to be the additive group of complex numbers. □

We now state the fundamental theorem on invariant functionals. On this theorem all of our subsequent analysis depends.

(15.5) Theorem. *Let G be a locally compact group. Then there is a functional I on $\mathfrak{C}_{00}^+$ such that:*

(i) *$I(f)$ is real and positive for $f\neq 0$;*

(ii) *$I(f+g)=I(f)+I(g)$ for all $f, g\in\mathfrak{C}_{00}^+$;*

(iii) *$I(\alpha f)=\alpha I(f)$ for $\alpha\geqq 0$ and $f\in\mathfrak{C}_{00}^+$;*

(iv) *$I({}_af)=I(f)$ for all $f\in\mathfrak{C}_{00}^+$ and $a\in G$.*

Furthermore, if J is any nonnegative functional on $\mathfrak{C}_{00}^{+}$ satisfying conditions (ii)—(iv) *and not vanishing identically, then $J = cI$ for some positive number c.*

(15.6) Remarks. Before proceeding to the proof, we make a few remarks. The functional I described in (15.5) is not a linear functional, since it is not defined on a linear space. However, properties (ii) and (iii) are as close to linearity as can be achieved on $\mathfrak{C}_{00}^{+}$. We call I a *left Haar integral on* $\mathfrak{C}_{00}^{+}$. It is clear that a right invariant Haar integral can also be constructed on $\mathfrak{C}_{00}^{+}$, and that it too will be unique up to a multiplicative constant.

Proof of (15.5). The proof, as is to be expected, requires a number of steps.

(I) We begin with any two functions f and φ in $\mathfrak{C}_{00}^{+}$ such that $\varphi \neq 0$, and seek some estimate of the size of f relative to φ. Consider all finite sequences $\{s_1, s_2, \ldots, s_m\}$ of elements of G and all sequences $\{c_1, c_2, \ldots, c_m\}$ of positive numbers such that $f \leqq \sum_{j=1}^{m} c_j\, {}_{s_j}\varphi$; that is,

$$f(x) \leqq \sum_{j=1}^{m} c_j\, \varphi(s_j x) \quad \text{for all} \quad x \in G. \tag{1}$$

Define the expression $(f:\varphi)$ as the infimum of all sums $\sum_{j=1}^{m} c_j$ for which the inequality (1) obtains. We note first that such sums $\sum_{j=1}^{m} c_j$ exist. Let F be a compact subset of G such that $f(F') = 0$, and let $\alpha = \|f\|_u$. Let a be a point of G, U a neighborhood of e in G, and μ a positive number such that $\varphi(x) \geqq \mu$ for all $x \in aU$. By compactness, there are a finite number of sets $y_1 U, y_2 U, \ldots, y_{m_0} U$ that cover F. It is then easy to see that $f(x) \leqq \sum_{j=1}^{m_0} \frac{\alpha}{\mu}\, \varphi(a y_j^{-1} x)$ for all $x \in G$. That is, $(f:\varphi)$ exists and does not exceed $\frac{m_0 \alpha}{\mu}$. Next we note that if (1) holds, then $\|f\|_u \leqq \sum_{j=1}^{m} c_j \|\varphi\|_u$, so that we have

$$\frac{\|f\|_u}{\|\varphi\|_u} \leqq (f:\varphi) \leqq \frac{m_0 \|f\|_u}{\mu}. \tag{2}$$

We next list three obvious properties of $(f:\varphi)$:

$$({}_a f:\varphi) = (f:{}_a\varphi) = (f:\varphi) \quad \text{for all} \quad a \in G; \tag{3}$$

$$(\alpha f:\varphi) = \alpha (f:\varphi) \quad \text{for all} \quad \alpha \geqq 0; \tag{4}$$

$$(f_1 + f_2:\varphi) \leqq (f_1:\varphi) + (f_2:\varphi) \quad \text{for all} \quad f_1, f_2 \in \mathfrak{C}_{00}^{+}. \tag{5}$$

If φ and ψ are nonzero functions in $\mathfrak{C}_{00}^{+}$, then we also have

$$(f:\psi) \leqq (f:\varphi)(\varphi:\psi). \tag{6}$$

To establish (6), we need only note that if $f(x) \leqq \sum_{j=1}^{m} c_j \varphi(s_j x)$ for all $x \in G$ and $\varphi(y) \leqq \sum_{k=1}^{n} d_k \psi(t_k y)$ for all $y \in G$, then $f(x) \leqq \sum_{j=1}^{m} c_j \left(\sum_{k=1}^{n} d_k \psi(t_k s_j x)\right) = \sum_{j=1}^{m} \sum_{k=1}^{n} c_j d_k \psi(t_k s_j x)$ for all $x \in G$. Thus $(f:\psi) \leqq \sum_{j=1}^{m} \sum_{k=1}^{n} c_j d_k = \left(\sum_{j=1}^{m} c_j\right)\left(\sum_{k=1}^{n} d_k\right)$.

Next choose once and for all a function $f_0 \in \mathfrak{C}_{00}^{+}$ such that $f_0 \neq 0$, and define

$$I_\varphi(f) = \frac{(f:\varphi)}{(f_0:\varphi)}, \tag{7}$$

for nonzero functions φ in $\mathfrak{C}_{00}^{+}$. Then $I_\varphi(0) = 0$ since $(0:\varphi) = 0$, and for $f \neq 0$, we see from (6) that

$$\frac{1}{(f_0:f)} \leqq I_\varphi(f) \leqq (f:f_0). \tag{8}$$

From (3), (4), and (5), respectively, we also have:

$$I_\varphi({}_a f) = I_\varphi(f) \quad \text{for all} \quad a \in G \quad \text{and} \quad f \in \mathfrak{C}_{00}^{+}; \tag{9}$$

$$I_\varphi(\alpha f) = \alpha I_\varphi(f) \quad \text{for} \quad \alpha \geqq 0 \quad \text{and} \quad f \in \mathfrak{C}_{00}^{+}; \tag{10}$$

$$I_\varphi(f_1 + f_2) \leqq I_\varphi(f_1) + I_\varphi(f_2) \quad \text{for all} \quad f_1, f_2 \in \mathfrak{C}_{00}^{+}; \tag{11}$$

$$I_\varphi(f_1) \leqq I_\varphi(f_2) \quad \text{if} \quad f_1, f_2 \in \mathfrak{C}_{00}^{+} \quad \text{and} \quad f_1 \leqq f_2. \tag{12}$$

The idea of the proof is now to consider all of the functionals I_φ, where φ vanishes outside of a neighborhood U_φ of e. Then as U_φ "decreases", it will be shown that I_φ converges to a limit functional I, which enjoys properties (i)—(iv). The uniqueness of I is also an immediate consequence. To carry out this program, we must establish two technical lemmas.

(II) Let $f_1, \ldots, f_m$ be nonzero functions in $\mathfrak{C}_{00}^{+}$ and let δ and Δ be any positive numbers. Then there is a neighborhood U of e such that

$$\sum_{j=1}^{m} \lambda_j I_\varphi(f_j) \leqq I_\varphi\left(\sum_{j=1}^{m} \lambda_j f_j\right) + \delta \tag{13}$$

if $\varphi \in \mathfrak{C}_{00}^{+}$, $\varphi \neq 0$, $\varphi(U') = 0$, and $0 \leqq \lambda_j \leqq \Delta$ $(j = 1, 2, \ldots, m)$.

To prove (II), let E be a compact subset of G such that $f_j(E') = 0$ for $j = 1, 2, \ldots, m$, and let V be a neighborhood of e such that V^- is compact. Choose a function g in $\mathfrak{C}_{00}^{+}$ having the property that $g(EV^-) = 1$ and $g(G) \subset [0, 1]$. Let M denote $\Delta m \cdot \max\{\|f_1\|_u, \ldots, \|f_m\|_u\}$. Let ε be an arbitrary positive number less than M. The functions $f_1, f_2, \ldots, f_m$, and g are left uniformly continuous, as pointed out in (15.4). Therefore there is a symmetric neighborhood U of e in G such that $U \subset V$ and

$$|f_j(s) - f_j(x)| < \frac{\varepsilon^3}{4Mm\Delta} \quad \text{if} \quad s^{-1}x \in U \quad (j = 1, 2, \ldots, m) \tag{14}$$

and

$$|g(s) - g(x)| < \frac{\varepsilon^2}{4M} \quad \text{if} \quad s^{-1}x \in U. \tag{15}$$

Now let φ be in $\mathfrak{C}_{00}^+$ where $\varphi \neq 0$ and $\varphi(U') = 0$, and let $\lambda_1, \lambda_2, \ldots, \lambda_m$ be taken so that $0 \leqq \lambda_j \leqq \Delta$ $(j = 1, 2, \ldots, m)$. Let Φ denote the function $\sum_{j=1}^{m} \lambda_j f_j + \varepsilon g$, and define the functions $h_1, h_2, \ldots, h_m$ by

$$h_j(x) = \begin{cases} \dfrac{\lambda_j f_j(x)}{\Phi(x)} & \text{for} \quad x \in E, \\ 0 & \text{for} \quad x \notin E. \end{cases}$$

It is obvious that $\Phi \in \mathfrak{C}_{00}^+$, that $h_j \Phi = \lambda_j f_j$ for $j = 1, 2, \ldots, m$, and that $\sum_{j=1}^{m} h_j \leqq 1$. If $s^{-1}x \in U$, then by (14) and (15), we have

$$\left. \begin{aligned} |\Phi(s) - \Phi(x)| &\leqq \sum_{j=1}^{m} \lambda_j |f_j(s) - f_j(x)| + \varepsilon |g(s) - g(x)| \\ &< m\Delta \frac{\varepsilon^3}{4Mm\Delta} + \varepsilon \frac{\varepsilon^2}{4M} = \frac{\varepsilon^3}{2M}. \end{aligned} \right\} \tag{16}$$

It is evident that $\|\Phi\|_u \leqq \sum_{j=1}^{m} \lambda_j \|f_j\|_u + \varepsilon \leqq M + \varepsilon < 2M$. We now show that

$$|h_j(s) - h_j(x)| < \frac{\varepsilon}{m} \quad \text{if} \quad s^{-1}x \in U \quad (j = 1, 2, \ldots, m). \tag{17}$$

In fact, if x and s also belong to EV^-, then using (14) and (16) we have

$$|h_j(s) - h_j(x)| = \left| \frac{\lambda_j f_j(s)}{\Phi(s)} - \frac{\lambda_j f_j(x)}{\Phi(x)} \right| = \left| \frac{\lambda_j f_j(s)\, \Phi(x) - \lambda_j f_j(x)\, \Phi(s)}{\Phi(s)\, \Phi(x)} \right|$$

$$\leqq \frac{\Delta}{\varepsilon^2} \{ |f_j(s)\, \Phi(x) - f_j(s)\, \Phi(s)| + |f_j(s)\, \Phi(s) - f_j(x)\, \Phi(s)| \}$$

$$< \frac{\Delta}{\varepsilon^2} \left\{ \|f_j\|_u \frac{\varepsilon^3}{2M} + \|\Phi\|_u \frac{\varepsilon^3}{4Mm\Delta} \right\} < \frac{\Delta}{\varepsilon^2} \left\{ \frac{M}{\Delta m} \frac{\varepsilon^3}{2M} + \frac{2M\varepsilon^3}{4Mm\Delta} \right\} = \frac{\varepsilon}{m}.$$

If $x \notin EV^-$ and $s^{-1}x \in U$, then $s \notin E$, since otherwise we would have $x = ss^{-1}x \in EU \subset EV^-$. If $s \notin EV^-$ and $s^{-1}x \in U$, then $x \notin E$, since otherwise $s = xx^{-1}s \in EU^{-1} \subset EV^-$. Therefore if either x or s fails to be in EV^-, the left side of (17) is zero.

Let us now estimate $(\Phi : \varphi)$. If we have

$$\Phi(x) \leqq \sum_{k=1}^{n} c_k\, \varphi(s_k x) \quad \text{for all} \quad x \in G,$$

then, since $\varphi(U') = 0$, we also have, for fixed $x \in G$,

$$\Phi(x) \leqq \sum_{k=1}^{n}{}' c_k\, \varphi(s_k x), \tag{18}$$

the summation being restricted to those indices k for which $x \in s_k^{-1} U$. From (17) and (18), we obtain

$$\Phi(x)\, h_j(x) \leqq \sum_{k=1}^{n}{}' c_k\, \varphi(s_k x) \left[h_j(s_k^{-1}) + \frac{\varepsilon}{m} \right]$$

and hence

$$\lambda_j f_j(x) = \Phi(x)\, h_j(x) \leqq \sum_{k=1}^{n} c_k \left[h_j(s_k^{-1}) + \frac{\varepsilon}{m}\right] \varphi(s_k x)$$

for all $x \in G$. This implies that

$$(\lambda_j f_j : \varphi) \leqq \sum_{k=1}^{n} c_k \left[h_j(s_k^{-1}) + \frac{\varepsilon}{m}\right];$$

summing over $j = 1, 2, \ldots, m$ and recalling that $\sum_{j=1}^{m} h_j \leqq 1$, we obtain

$$\sum_{j=1}^{m} (\lambda_j f_j : \varphi) \leqq (1+\varepsilon)\left(\sum_{k=1}^{n} c_k\right).$$

Since $\sum_{k=1}^{n} c_k$ can be taken as close as we wish to $(\Phi : \varphi)$, we have

$$\sum_{j=1}^{m} (\lambda_j f_j : \varphi) \leqq (1+\varepsilon)(\Phi : \varphi).$$

Dividing by $(f_0 : \varphi)$, we get

$$\sum_{j=1}^{m} I_\varphi(\lambda_j f_j) \leqq (1+\varepsilon)\, I_\varphi(\Phi). \tag{19}$$

Applying (10), (11), the definition of Φ, and (8) to (19), we obtain

$$\begin{aligned}\sum_{j=1}^{m} \lambda_j I_\varphi(f_j) &\leqq (1+\varepsilon)\left[I_\varphi\left(\sum_{j=1}^{m} \lambda_j f_j\right) + \varepsilon I_\varphi(g)\right] \\ &\leqq I_\varphi\left(\sum_{j=1}^{m} \lambda_j f_j\right) + \varepsilon \sum_{j=1}^{m} \lambda_j I_\varphi(f_j) + \varepsilon(1+\varepsilon)\, I_\varphi(g) \\ &\leqq I_\varphi\left(\sum_{j=1}^{m} \lambda_j f_j\right) + \varepsilon \Delta \sum_{j=1}^{m} (f_j : f_0) + \varepsilon(1+\varepsilon)(g : f_0).\end{aligned}$$

These inequalities show that (13) holds if ε is sufficiently small, and thus (II) is established[1].

Our second lemma follows.

(III) Let f be any function in $\mathfrak{C}_{00}^{+}$, $f \neq 0$, ε any positive number, and U a neighborhood of e in G such that $|f(x) - f(y)| < \varepsilon$ if $x, y \in G$ and $y^{-1}x \in U$. Let f vanish outside of the compact set $E \subset G$. Let g be a function in $\mathfrak{C}_{00}^{+}$ such that $g \neq 0$ and $g(U') = 0$. Then for every $\alpha > \varepsilon$, there are elements $t_1, t_2, \ldots, t_m \in E^{-1}$ and nonnegative numbers $c_1, c_2, \ldots, c_m$

[1] At this point, we could shorten the proof by appealing to Tihonov's theorem, and by selecting some point in a large Cartesian product of closed intervals. This selection, naturally, involves the axiom of choice, and leaves the uniqueness of the left Haar integral to be proved separately. We prefer to give the remainder of the proof in a strictly constructive form, which simultaneously demonstrates the existence and uniqueness of the left Haar integral and does not depend upon the axiom of choice. For a sketch of the shorter, nonconstructive, proof, see (15.25).

such that $\sum_{j=1}^{m} c_j > 0$ and

$$\left| f(x) - \sum_{j=1}^{m} c_j\, g(t_j x) \right| \leqq \alpha \quad \text{for all} \quad x \in G. \tag{20}$$

To prove (III), we note first that

$$[f(x) - \varepsilon]\, g(s^{-1}x) \leqq f(s)\, g(s^{-1}x) \leqq [f(x) + \varepsilon]\, g(s^{-1}x) \tag{21}$$

for all $s, x \in G$. [If $s^{-1}x \in U$, then $|f(s) - f(x)| < \varepsilon$; if $s^{-1}x \notin U$, then $g(s^{-1}x) = 0$.] Now choose $\eta > 0$ so small that $(f:g^\star)\,\eta < \alpha - \varepsilon$. Let V be a neighborhood of e in G with compact closure such that $|g(u) - g(v)| < \eta$ for all $u, v \in G$ such that $uv^{-1} \in V$. [Notice that we use the *right* uniform continuity of g.] Since f vanishes outside of the compact set E, there are elements $s_1, s_2, \ldots, s_m \in E$ such that $\bigcup_{j=1}^{m} s_j V \supset E \supset \{x \in G : f(x) \neq 0\}$. By (3.1), there are functions $h_1, h_2, \ldots, h_m \in \mathfrak{C}_{00}^+$ such that $h_j((s_j V)') = 0$ $(j = 1, 2, \ldots, m)$ and $\sum_{j=1}^{m} h_j(x) = 1$ if $f(x) \neq 0$. Now we have

$$h_j(s)\, f(s)\, [g(s^{-1}x) - \eta] \leqq h_j(s)\, f(s)\, g(s_j^{-1}x) \leqq h_j(s)\, f(s)\, [g(s^{-1}x) + \eta] \tag{22}$$

for all $s, x \in G$ and $j = 1, 2, \ldots, m$. [If $s_j^{-1}s = (s_j^{-1}x)(s^{-1}x)^{-1} \in V$, then $|g(s^{-1}x) - g(s_j^{-1}x)| < \eta$; if $s \notin s_j V$, then $h_j(s) = 0$.] Summing (22) over $j = 1, 2, \ldots, m$ and taking cognizance of (21), we obtain

$$\begin{aligned} [f(x) - \varepsilon]\, g(s^{-1}x) - \eta\, f(s) &\leqq \sum_{j=1}^{m} h_j(s)\, f(s)\, g(s_j^{-1}x) \\ &\leqq [f(x) + \varepsilon]\, g(s^{-1}x) + \eta\, f(s). \end{aligned}$$

Now consider any $\psi \in \mathfrak{C}_{00}^+$, $\psi \neq 0$. Using (9), (10), (11), and (12), we infer from the last inequalities that

$$\left.\begin{aligned} [f(x) - \varepsilon]\, I_\psi(g^\star) - \eta\, I_\psi(f) &\leqq I_\psi\left[\sum_{j=1}^{m} g(s_j^{-1}x)\, h_j f\right] \\ &\leqq [f(x) + \varepsilon]\, I_\psi(g^\star) + \eta\, I_\psi(f). \end{aligned}\right\} \tag{23}$$

Now (7) and (6) imply that $\frac{I_\psi(f)}{I_\psi(g^\star)} \leqq (f:g^\star) = \frac{\beta - \varepsilon}{\eta}$, where $\beta < \alpha$. Thus, dividing (23) by $I_\psi(g^\star)$ and using (10), we have

$$f(x) - \beta \leqq I_\psi\left[\sum_{j=1}^{m} \frac{g(s_j^{-1}x)}{I_\psi(g^\star)}\, h_j f\right] \leqq f(x) + \beta. \tag{24}$$

We now apply (II) with $\delta = \alpha - \beta$, $\Delta = (f_0 : g^\star)\, \|g\|_u$ [see (8)], $\lambda_j = \frac{g(s_j^{-1}x)}{I_\psi(g^\star)}$, and $f_j = h_j f$. This, together with (10) and (11), tells us that there is a neighborhood W of e in G such that if our ψ vanishes on W', then

$$\left| I_\psi\left[\sum_{j=1}^{m} \frac{g(s_j^{-1}x)}{I_\psi(g^\star)}\, h_j f\right] - \sum_{j=1}^{m} \frac{I_\psi(h_j f)}{I_\psi(g^\star)}\, g(s_j^{-1}x) \right| \leqq \alpha - \beta. \tag{25}$$

Combining (24) and (25), we have (20), with $t_j = s_j^{-1}$ and

$$c_j = \frac{I_\psi(h_j f)}{I_\psi(g^\star)} \qquad (j = 1, 2, \ldots, m). \tag{26}$$

(IV) We now complete the proof of the existence of a left Haar integral. For every neighborhood U of e in G, choose a function $\varphi = \varphi_U$ in $\mathfrak{C}_{00}^+$ such that $\varphi \neq 0$ and $\varphi(U') = 0$. Given $\varphi_1 = \varphi_{U_1}$ and $\varphi_2 = \varphi_{U_2}$, write $\varphi_2 \geqq \varphi_1$ if $U_1 \supset U_2$. The set $\{\varphi\}$ is then directed by $\geqq$. We will show that $\lim_\varphi I_\varphi(f) = I(f)$ exists for every $f \in \mathfrak{C}_{00}^+$. We may suppose that $f \neq 0$. By (II), this limit functional I will satisfy (ii). It will obviously satisfy (i), (iii), and (iv), in view of (8), (10), and (9), respectively. To show that $\lim_\varphi I_\varphi(f)$ exists, it obviously suffices to show that for every $\varepsilon > 0$, there is a φ_0 such that $|I_{\varphi_1}(f) - I_{\varphi_2}(f)| < \varepsilon$ for $\varphi_1 \geqq \varphi_0$ and $\varphi_2 \geqq \varphi_0$.

First let U_0 be any neighborhood of e in G such that U_0^- is compact, and let ω be a function in $\mathfrak{C}_{00}^+$ that assumes the value 1 everywhere on the compact set $F = \{x \in G : f(x) + f_0(x) \neq 0\}^- \cdot U_0^-$. Let ε be any number such that $0 < \varepsilon < 1$, and let

$$\gamma = \frac{\varepsilon}{4[1 + (\omega : f_0)][1 + (f : f_0)]}.$$

Let U be a neighborhood of e in G such that $U \subset U_0$, $|f(x) - f(y)| < \frac{\gamma}{2}$ if $y^{-1}x \in U$, and $|f_0(x) - f_0(y)| < \frac{\gamma}{2}$ if $y^{-1}x \in U$. Then choose a nonzero function g in $\mathfrak{C}_{00}^+$ such that $g(U') = 0$ and take $\alpha = \gamma$. Applying (III), we obtain $t_1, \ldots, t_m \in (\{x \in G : f(x) \neq 0\}^-)^{-1}$ and nonnegative numbers $c_1, \ldots, c_m$ such that

$$\left| f(x) - \sum_{j=1}^m c_j g(t_j x) \right| \leqq \gamma \quad \text{for all} \quad x \in G.$$

Since f and all of the functions ${}_{t_j}g$ vanish on F', we have

$$\left| f(x) - \sum_{j=1}^m c_j g(t_j x) \right| \leqq \gamma\, \omega(x) \quad \text{for all} \quad x \in G. \tag{27}$$

For every $\varphi \in \mathfrak{C}_{00}^+$, $\varphi \neq 0$, this last inequality implies that

$$\left| I_\varphi(f) - I_\varphi\left(\sum_{j=1}^m c_j\, {}_{t_j}g \right) \right| \leqq \gamma\, I_\varphi(\omega) \leqq \gamma\, (\omega : f_0). \tag{28}$$

[Simply write (27) as two inequalities and apply (12) and (11), followed by (8).] Now we apply (II). Each c_j is given by (26), and since $h_j f \leqq f$, we have $c_j \leqq \frac{I_\psi(f)}{I_\psi(g^\star)} \leqq (f : g^\star)$. From (II) we infer that there is a neighborhood V of e in G such that if $\varphi \in \mathfrak{C}_{00}^+$, $\varphi \neq 0$, and $\varphi(V') = 0$, then

$$\left| I_\varphi\left(\sum_{j=1}^m c_j\, {}_{t_j}g \right) - I_\varphi(g) \left(\sum_{j=1}^m c_j \right) \right| < \gamma. \tag{29}$$

Combining (28) and (29), and writing $c = \sum_{j=1}^{m} c_j$ $(c>0)$, we obtain

$$|I_\varphi(f) - c I_\varphi(g)| < \gamma[1 + (\omega : f_0)]. \tag{30}$$

Apply precisely the same argument to the "base" function f_0, and note that $I_\varphi(f_0) = 1$. We find a neighborhood V_0 of e in G and a $d>0$ such that

$$|1 - d I_\varphi(g)| < \gamma[1 + (\omega : f_0)], \tag{31}$$

for nonzero $\varphi \in \mathfrak{C}_{00}^+$ such that $\varphi(V_0') = 0$. Combining (30) and (31) in an obvious way, we obtain

$$\left| I_\varphi(f) - \frac{c}{d} \right| < \gamma[1 + (\omega : f_0)]\left(1 + \frac{c}{d}\right) \tag{32}$$

if $\varphi \in \mathfrak{C}_{00}^+$, $\varphi \neq 0$, and $\varphi(V' \cup V_0') = 0$. Inequality (32) implies at once that

$$\frac{c}{d}\left(1 - \gamma[1 + (\omega : f_0)]\right) < I_\varphi(f) + \gamma[1 + (\omega : f_0)],$$

so that $\frac{c}{d} + 1 < \frac{I_\varphi(f) + 1}{1 - \gamma[1 + (\omega : f_0)]} \leqq 2[1 + (f : f_0)]$. This estimate, combined with (32), yields

$$\left| I_\varphi(f) - \frac{c}{d} \right| < \gamma[1 + (\omega : f_0)]\, 2[1 + (f : f_0)] = \frac{\varepsilon}{2}.$$

Thus if φ_1 and φ_2 are any functions in $\mathfrak{C}_{00}^+$, $\varphi_j \neq 0$, $\varphi_j(V' \cup V_0') = 0$ $(j = 1, 2)$, then we have

$$|I_{\varphi_1}(f) - I_{\varphi_2}(f)| < \varepsilon.$$

As explained above, this proves that $\lim_\varphi I_\varphi(f)$ exists for all $f \in \mathfrak{C}_{00}^+$, and thus a left Haar integral exists.

(V) It remains only to show that two left Haar integrals on $\mathfrak{C}_{00}^+$ differ only by a constant factor. Let J be any nonnegative functional on $\mathfrak{C}_{00}^+$ satisfying (ii), (iii), and (iv), and such that $J(f_1) \neq 0$ for some $f_1 \in \mathfrak{C}_{00}^+$. For any $f \in \mathfrak{C}_{00}^+$, $f \neq 0$, we have $f_1 \leqq \sum_{j=1}^{m} c_j\, {}_{s_j}f$ for appropriate real c_j and $s_j \in G$. Hence $J(f_1) \leqq \left(\sum_{j=1}^{m} c_j\right) J(f)$, and thus $J(f)$ is positive for all nonzero f in $\mathfrak{C}_{00}^+$. For any $f_1 \in \mathfrak{C}_{00}^+$ and nonzero g in $\mathfrak{C}_{00}^+$, suppose that $f_1 \leqq \sum_{j=1}^{m} c_j\, {}_{s_j}g$. Then $J(f_1) \leqq \left(\sum_{j=1}^{m} c_j\right) J(g)$, so that we have

$$(f_1 : g) \geqq \frac{J(f_1)}{J(g)}. \tag{33}$$

Consider any nonzero f in $\mathfrak{C}_{00}^+$; f vanishes outside of a compact set E in G. Let V be a neighborhood of e with V^- compact and select a *fixed* $\omega \in \mathfrak{C}_{00}^+$ so that $\omega = 2$ on EV^-. Now for any $\varepsilon > 0$, apply (III) with $\alpha = 2\varepsilon$ to obtain a neighborhood $U \subset V$ of e with the following property. If g is any nonzero function in $\mathfrak{C}_{00}^+$ vanishing outside of U, then there are points

$t_1, \ldots, t_m$ in G and nonnegative numbers $c_1, \ldots, c_m$ such that

$$f(x) \leqq \varepsilon\,\omega(x) + \sum_{j=1}^{m} c_j\, g(t_j x), \tag{34}$$

and

$$f(x) + \varepsilon\,\omega(x) \geqq \sum_{j=1}^{m} c_j\, g(t_j x), \tag{35}$$

for all $x \in G$. It follows at once from (34) that

$$(f:g) \leqq \varepsilon(\omega:g) + \sum_{j=1}^{m} c_j. \tag{36}$$

Applying J to both sides of (35), we have

$$J(f) + \varepsilon J(\omega) \geqq \Big(\sum_{j=1}^{m} c_j\Big) J(g).$$

Using (36) and (6), we have

$$J(f) + \varepsilon J(\omega) \geqq \left[1 - \varepsilon \frac{(\omega:g)}{(f:g)}\right] (f:g)\, J(g) \geqq [1 - \varepsilon(\omega:f)]\,(f:g)\, J(g). \tag{37}$$

Let f_1 be any nonzero function in $\mathfrak{C}_{00}^{+}$. Divide (37) by $J(f_1)$ and use (33). This gives us

$$\frac{J(f)}{J(f_1)} + \varepsilon \frac{J(\omega)}{J(f_1)} \geqq [1 - \varepsilon(\omega:f)] \frac{(f:g)}{(f_1:g)} = [1 - \varepsilon(\omega:f)] \frac{I_g(f)}{I_g(f_1)}. \tag{38}$$

Now take the limit in (38) over g and then as $\varepsilon \to 0$. This gives

$$\frac{J(f)}{J(f_1)} \geqq \frac{I(f)}{I(f_1)}.$$

Interchanging f_1 and f, we have

$$J(f) = \frac{J(f_1)}{I(f_1)} I(f),$$

for all $f \in \mathfrak{C}_{00}^{+}$. □

(15.7) Note. There is a unique extension of I from $\mathfrak{C}_{00}^{+}$ to the complex linear space $\mathfrak{C}_{00}$ that is a complex linear functional on $\mathfrak{C}_{00}$. This is a special case of (B.38). The extension is necessarily left invariant. We will write the extended functional as I, and we will call I a *left Haar integral on* $\mathfrak{C}_{00}$, or *on* G.

(15.8) Remarks. We now use the construction of §11 to associate a set function with our left Haar integral I. We denote this set function by λ. Thus λ is a nonnegative, extended real-valued set function defined for all subsets of G, and is a measure on the σ-algebra $\mathscr{M}_\lambda$ of λ-measurable subsets of G. Every open set is in $\mathscr{M}_\lambda$. The set function λ has the following properties:

(i) $0 < \lambda(U)$ for all nonvoid open sets U;

(ii) $\lambda(U) < \infty$ for some open set U;

(iii) $\lambda(aB) = \lambda(B)$ for all $B \subset G$ and $a \in G$ [λ is left invariant in the sense of (15.3)].

To prove (i), we use (15.5.i) and the definition of $\lambda(U)$ in (11.11) and (11.20). Given $x\in U$, there is a function $\varphi\in\mathfrak{C}_{00}^{+}$ such that $\varphi(x)=1$ and $\varphi\leqq\xi_U$. Hence we have $\lambda(U)\geqq I(\varphi)>0$. Property (ii) is shared by λ with all of the set functions ι constructed in §11: $\lambda(A)$ is finite if A^- is compact. Property (15.5.iv) implies (iii), as follows. Since $I({}_af)=I(f)$ for all $f\in\mathfrak{C}_{00}^{+}$ and since $\varphi\leqq\psi$ if and only if ${}_a\varphi\leqq{}_a\psi$ for any extended real-valued functions φ and ψ on G, Definitions (11.11) and (11.16) show that $\overline{\overline{I}}({}_ah)=\overline{\overline{I}}(h)$ for all $a\in G$ and all nonnegative extended real-valued functions h on G. We thus have $\lambda(B)=\overline{\overline{I}}(\xi_B)=\overline{\overline{I}}\left({}_{a^{-1}}(\xi_B)\right)=\overline{\overline{I}}(\xi_{aB})=\lambda(aB)$. The measure λ constructed from a left Haar integral on $\mathfrak{C}_{00}$ will be called a *left Haar measure on G.*

The measure λ on $\mathscr{M}_\lambda$ gives us an integral $\int_G f\,d\lambda$ defined for all nonnegative λ-measurable functions f on G, the integral giving rise to the function spaces $\mathfrak{L}_p(G,\lambda)$ $(1\leqq p<\infty)$; λ itself gives us the function space $\mathfrak{L}_\infty(G,\lambda)$. The left invariance of λ shows at once that $\int_G {}_af\,d\lambda=\int_G f\,d\lambda$ for all $f\in\mathfrak{L}_1(G,\lambda)$, that $\langle {}_af, {}_ag\rangle=\langle f,g\rangle$ if f,g are in $\mathfrak{L}_2(G,\lambda)$, and that $\|{}_af\|_p=\|f\|_p$ for $f\in\mathfrak{L}_p(G,\lambda)$ $(1\leqq p\leqq\infty)$: see also Theorem (20.1).

Where no confusion can arise, we will often write $\int_G f\,d\lambda$ as $\int_G f(x)\,dx$, $\int_G f(y)\,dy$, and so on. Thus $dx, dy, \ldots$ will always indicate integration with respect to a left Haar measure on G. The identity $\int_G {}_af\,d\lambda=\int_G f\,d\lambda$ thus receives the suggestive form $\int_G f(x)\,dx=\int_G f(ax)\,dx$. We will call $\int_G f\,d\lambda$ a *left Haar integral of f* whenever it is defined. This terminology is justified by the identity $\overline{\overline{I}}(f)=\int_G f\,d\lambda$, valid for all nonnegative λ-measurable functions f on G [see (11.36)].

It is to be expected from (15.5) that left Haar measures, like left Haar integrals on $\mathfrak{C}_{00}$, are unique up to a multiplicative constant. This is indeed the case, under certain restrictions on the domain of definition of the measures. The exact result is the following. Let $\mathscr{B}$ denote the family of Borel sets in G. Let μ be any measure on $\mathscr{B}$ such that:

(iv) $\mu(F)<\infty$ if F is compact;

(v) $\mu(U)>0$ for some open set U;

(vi) $\mu(aB)=\mu(B)$ for all $B\in\mathscr{B}$ and $a\in G$;

(vii) μ is regular in the sense of (11.34).

Then if λ is any left Haar measure as described above, there is a positive number c such that $\mu(B)=c\,\lambda(B)$ for all $B\in\mathscr{B}$.

To prove this fact, we construct a functional J from μ in the obvious way: $J(\varphi)=\int_G \varphi\,d\mu$ for $\varphi\in\mathfrak{C}_{00}$. It is easy to verify that J enjoys all of the properties (15.5.i)—(15.5.iv). Therefore, the uniqueness of the Haar integral implies that $J=cI$, where I is an arbitrary Haar integral, and c is a positive number. From (11.38), we infer that $\mu(B)=c\lambda(B)$ for all $B\in\mathscr{B}$.

(15.9) Theorem. *Let G be a locally compact group with a left Haar measure λ. Then $\lambda(G)$ is finite if and only if G is compact. [If G is compact, we will always normalize left Haar measure by the requirement that $\lambda(G)=1$.]*

Proof. If G is compact, then $1\in\mathfrak{C}=\mathfrak{C}_{00}$. Thus $I(1)$ exists and is a positive real number. Since $I(1)=\lambda(G)$, it follows that $\lambda(G)$ is finite. If G is noncompact, let U be any neighborhood of e in G such that U^- is compact. Then no finite collection of sets $x_1U, x_2U, \ldots, x_mU$ can cover G, since if it did, G would be the union of finitely many compact sets and would be compact. Thus we can choose an infinite sequence of points $\{x_1, x_2, \ldots, x_n, \ldots\}$ in G such that $x_{n+1}\notin\bigcup_{j=1}^{n} x_jU$ for all positive integers n. Let V be any symmetric neighborhood of e such that $V^2\subset U$. Then the sets $x_1V, x_2V, \ldots, x_nV, \ldots$ are pairwise disjoint, and we have $\lambda(G)\geqq\sum_{j=1}^{n}\lambda(x_jV)=n\cdot\lambda(V)$ for all positive integers n. This implies that $\lambda(G)=+\infty$. □

(15.10) Right Haar integral and measure. The constructions and arguments used in (15.5) and (15.6)—(15.9) are plainly symmetric wtih respect to right and left translations of functions on the group G. Therefore we can state with no further ado that there exists a functional I' on $\mathfrak{C}_{00}^+$ with properties (15.5.i)—(15.5.iii) and $I'(f_a)=I'(f)$ for all $f\in\mathfrak{C}_{00}^+$ and $a\in G$. The functional I' can be extended over $\mathfrak{C}_{00}$ as in (15.7), and I' is unique up to a multiplicative constant. There is also a right Haar measure λ' on G with properties (15.8.i)—(15.8.ii) and

$$\lambda'(Ba)=\lambda'(B) \quad \text{for all} \quad a\in G \quad \text{and all sets} \quad B\subset G.$$

We shall see in (15.15) how to obtain a right Haar integral from a left Haar integral.

We next introduce an important function.

(15.11) Theorem. *Let G be a locally compact group, and let I be a left Haar integral on $\mathfrak{C}_{00}$. For $f\in\mathfrak{C}_{00}^+$, $f\neq 0$, and for $x\in G$, let $\Delta(x)=\frac{I(f_{x^{-1}})}{I(f)}$. Then Δ depends only upon x, and not upon I or f. The function Δ is continuous, positive throughout G, and satisfies the functional equation*

(i) $\Delta(xy)=\Delta(x)\,\Delta(y)$ *for all* $x, y\in G$.

Proof. Consider the functional J_x on $\mathfrak{C}_{00}^+$ defined by $J_x(f)=I(f_{x^{-1}})$. It is obvious that J_x is a left Haar integral on $\mathfrak{C}_{00}$, and so by (15.5), there is for each $x \in G$ a positive number $\Delta(x)$ such that $J_x(f) = I(f_{x^{-1}}) = \Delta(x)\, I(f)$ for all $f \in \mathfrak{C}_{00}^+$. Thus the notation $\Delta(x)$ is justified. It is clear from (15.5) that Δ does not depend upon the particular choice of I. We also have, for $x, y, u \in G$,

$$(f_{x^{-1}})_{y^{-1}}(u) = f_{x^{-1}}(u y^{-1}) = f(u y^{-1} x^{-1}) = f\big(u (x y)^{-1}\big) = f_{(xy)^{-1}}(u).$$

Hence we have

$$\Delta(x y) = \frac{I(f_{(xy)^{-1}})}{I(f)} = \frac{I((f_{x^{-1}})_{y^{-1}})}{I(f)} = \frac{I((f_{x^{-1}})_{y^{-1}})}{I(f_{x^{-1}})} \cdot \frac{I(f_{x^{-1}})}{I(f)} = \Delta(y)\, \Delta(x).$$

Thus (i) holds.

It remains to show that Δ is continuous. Since Δ is a homomorphism of G into the multiplicative group of positive real numbers, we need only show that Δ is continuous at e (5.40.a). Let U be any neighborhood of e in G such that U^- is compact, and let f be a nonzero function in $\mathfrak{C}_{00}^+$. Let ω be a function in $\mathfrak{C}_{00}^+$ such that $\omega\big(\{s \in G : f(s) > 0\}^- \cdot U^-\big) = 1$, let ε be a positive number, and let V be a neighborhood of e such that $V \subset U$ and $|f(u) - f(v)| < \frac{\varepsilon I(f)}{I(\omega)}$ whenever $u^{-1} v \in V$. Then if $x \in V$, we have $|f(s x^{-1}) - f(s)| \leqq \frac{\varepsilon I(f)}{I(\omega)} \omega(s)$ for all $s \in G$, and so $|I(f_{x^{-1}}) - I(f)| \leqq \varepsilon I(f)$. Thus

$$|\Delta(x) - 1| \leqq \varepsilon \quad \text{if} \quad x \in V. \quad \square$$

(15.12) Remark. The function Δ is called the *modular function* of the locally compact group G. If $\Delta = 1$ on G, then G is called *unimodular*. [Note that $\Delta = 1$ if and only if the class of left Haar integrals is the same as the class of right Haar integrals.] It is obvious that every locally compact Abelian group is unimodular.

(15.13) Theorem. *Every compact group is unimodular*[1].

First proof. The set $\Delta(G)$ is a compact subgroup of the multiplicative group $]0, \infty[$. Plainly $\{1\}$ is the only such subgroup of $]0, \infty[$, so that $\Delta(G) = 1$, *i.e.*, $\Delta(x) = 1$ for all $x \in G$.

Second proof. It is worth while to note that the present theorem can be proved without using the modular function. Let I be the normalized left Haar integral on $\mathfrak{C}$. For $f \in \mathfrak{C}$ and $x \in G$, let $f'(x) = I\big(({}_x f)^*\big)$. The uniform continuity of f implies that $f' \in \mathfrak{C}$. Now write $J(f) = I(f')$. It is obvious that J has properties (15.5.i)—(15.5.iii) and that $J(1) = 1$.

We prove first that J is left invariant: $J({}_a f) = J(f)$. We have $({}_a f)'(x) = I\big(({}_x({}_a f))^*\big) = I\big(({}_{ax} f)^*\big) = f'(ax) = {}_a(f')(x)$. Thus $J({}_a f) = I\big({}_a(f')\big) = I(f') = J(f)$. Since $J(1) = I(1) = 1$, we have $J = I$ by the uniqueness property of I.

[1] See also (19.28).

To prove that J is right invariant, we note first that $(f_a)^*(y) = f_a(y^{-1}) = f(y^{-1}a) = f((a^{-1}y)^{-1}) = {}_{a^{-1}}(f^*)(y)$. That is,

$$(f_a)^* = {}_{a^{-1}}(f^*). \tag{1}$$

Similarly, we have

$$({}_af)^* = (f^*)_{a^{-1}}. \tag{2}$$

Then we have $(f_a)'(x) = I[({}_x(f_a))^*] = I[(({}_xf)_a)^*] = I[{}_{a^{-1}}(({}_xf)^*)] = I[({}_xf)^*] = f'(x)$. Hence $J(f_a) = I((f_a)') = I(f') = J(f)$. Since $I = J$, our second proof is complete. □

(15.14) Theorem. *Let I be a left Haar integral. For every $f \in \mathfrak{C}_{00}$, we have $I(f) = I\left(f^* \frac{1}{\Delta}\right)$.*

Proof. Write $J(f) = I\left(f^* \frac{1}{\Delta}\right)$. We plainly have $\Delta_{a^{-1}} = \Delta(a^{-1})\Delta$. Using (15.13.2), we thus have

$$J({}_af) = I\left[({}_af)^* \frac{1}{\Delta}\right] = I\left[(f^*)_{a^{-1}} \frac{1}{\Delta}\right] = \frac{\Delta(a^{-1})}{\Delta(a^{-1})} I\left[(f^*)_{a^{-1}} \frac{1}{\Delta}\right]$$
$$= \Delta(a^{-1})\, I\left[(f^*)_{a^{-1}} \frac{1}{\Delta_{a^{-1}}}\right] = \Delta(a^{-1})\,\Delta(a)\, I\left(f^* \frac{1}{\Delta}\right) = J(f).$$

That is, J is a left invariant functional. Plainly J also satisfies (15.5.i)—(15.5.iii) [note that $f\frac{1}{\Delta} \in \mathfrak{C}_{00}^+$ if $f \in \mathfrak{C}_{00}^+$ and is zero only if f is zero]. The uniqueness of left Haar integrals shows that $J = cI$, where c is a positive number.

To show that $c = 1$, let ε be a positive number and let U be a neighborhood of e in G such that $\left|\frac{1}{\Delta(x)} - 1\right| < \varepsilon$ for all $x \in U$ [recall that Δ is continuous and that $\Delta(x^{-1}) = \frac{1}{\Delta(x)}$]. Let g be a nonzero function in $\mathfrak{C}_{00}^+$ such that $g = g^*$ and $g(U') = 0$. Then we have $\left|g(x) - g(x)\frac{1}{\Delta(x)}\right| \leqq \varepsilon g(x)$ for all $x \in G$ and therefore $\left|I(g) - I\left(g\frac{1}{\Delta}\right)\right| \leqq \varepsilon I(g)$. Thus $|1 - c| \leqq \varepsilon$, so that $c = 1$. Consequently, $I(f) = I\left(f^* \frac{1}{\Delta}\right)$ for all $f \in \mathfrak{C}_{00}$. □

We now construct a right Haar integral from a left Haar integral.

(15.15) Theorem. *Let I be a left Haar integral on $\mathfrak{C}_{00}(G)$, where G is a locally compact group. Let $J_1(f) = I(f^*)$ and $J_2(f) = I\left(f\frac{1}{\Delta}\right)$ for all $f \in \mathfrak{C}_{00}$. Then J_1 and J_2 are right Haar integrals on G, and furthermore, $J_1 = J_2$.*

Proof. By (15.14), we have

$$J_1(f) = I(f^*) = I\left((f^*)^* \frac{1}{\Delta}\right) = I\left(f\frac{1}{\Delta}\right) = J_2(f),$$

for $f \in \mathfrak{C}_{00}$; thus $J_1 = J_2$. It is obvious that J_1 satisfies conditions (15.5.i)–(15.5.iii). To check the right invariance of J_1, we use (15.13.1) to write

$$J_1(f_a) = I\left((f_a)^\star\right) = I\left({}_{a^{-1}}(f^\star)\right) = I(f^\star) = J_1(f). \quad \square$$

(15.16) Corollary. *Let G be a locally compact group and I a left Haar integral on $\mathfrak{C}_{00}$. Then G is unimodular if and only if I is inversion invariant.*

Proof. Suppose that $I(f) = I(f^\star) = J_1(f)$ for all $f \in \mathfrak{C}_{00}$. Then by (15.15), I is right invariant, *i.e.*, G is unimodular. If I is right invariant, then $\Delta = 1$, and again by (15.15), $I(f^\star) = I\left(f \frac{1}{\Delta}\right) = I(f)$. $\square$

(15.17) Examples of Haar integrals and measures. (a) Let G be a finite group, $\overline{\overline{G}} = r$. Then it is obvious that the functional $I(f) = \frac{1}{r} \sum_{x \in G} f(x)$ is the left and right normalized Haar integral for the space of all complex-valued functions on G. The corresponding measure on subsets of G is given by $\lambda(A) = \frac{1}{r} \overline{\overline{A}}$.

(b) Let G be any discrete infinite group. Then $\mathfrak{C}_{00}^+$ consists of the nonnegative functions on G that vanish outside of finite sets. Let δ denote the function such that $\delta(e) = 1$ and $\delta(x) = 0$ for $x \neq e$. Then a nonzero function f in $\mathfrak{C}_{00}^+$ can be written uniquely in the form $f = \sum_{j=1}^{m} \alpha_j \,{}_{a_j^{-1}}\delta = \sum_{j=1}^{m} \alpha_j \,\delta_{a_j^{-1}}$ $[f(a_j) = \alpha_j,\ j = 1, 2, \ldots, m$, and $f(x) = 0$ otherwise$]$. Thus a left or right Haar integral I on $\mathfrak{C}_{00}$ has the form $I(f) = \left(\sum_{j=1}^{m} \alpha_j\right) I(\delta)$. The group G is plainly unimodular. It is convenient in dealing with infinite discrete groups to take $I(\delta) = 1$, and this we will always do in the sequel. The measure λ corresponding to I is countably additive on *all* subsets of G: $\lambda(A) = \overline{\overline{A}}$ if A is a finite subset of G and $\lambda(A) = \infty$ if A is an infinite subset of G.

If G is locally compact and nondiscrete, then $\lambda(A) = 0$ for all countable subsets A of G. It suffices to prove that $\lambda(\{e\}) = 0$. Clearly $\lambda(\{x\}) = \lambda(\{e\})$ for $x \in G$. Let U be an open set in G such that $0 < \lambda(U) < \infty$. Since $\overline{\overline{U}}$ is infinite, we would have $\lambda(U) = \infty$ if $\lambda(\{e\}) \neq 0$.

(c) Consider the additive group R. A Haar integral on $\mathfrak{C}_{00}(R)$ is $I(f) = \int_{-\infty}^{\infty} f(x)\,dx$, the integral being an ordinary Riemann integral, which for each f is in reality extended only over a finite interval. The measure corresponding to I is ordinary Lebesgue measure on R. There is no single normalization for the Haar integral on $\mathfrak{C}_{00}(R)$ that is suitable for all purposes. We shall find it convenient in some subsequent chapters to use $\frac{1}{\sqrt{2\pi}} \int\limits_{-\infty}^{\infty} f(x)\,dx$.

(d) Consider the compact multiplicative group T. Here the normalized Haar integral is $\frac{1}{2\pi} \int\limits_{0}^{2\pi} f(\exp(it))\,dt$, for all $f \in \mathfrak{C}(T)$, where again the integral is the Riemann integral on the interval $[0,2\pi]$. The corresponding measure is $\frac{1}{2\pi}$ times Lebesgue measure on $[0,2\pi]$.

(e) Let G be any topological group with the following three properties. First, as a topological space, G is an open subset of some real Euclidean space R^n. For $\boldsymbol{x} = (x_1, \ldots, x_n)$ and $\boldsymbol{y} = (y_1, \ldots, y_n)$ in G, the product $\boldsymbol{x}\boldsymbol{y}$ is thus a function $F(x_1, \ldots, x_n, y_1, \ldots, y_n)$ mapping $G \times G \subset R^{2n}$ onto $G \subset R^n$. Our second hypothesis is that all of the partial derivatives

$$\frac{\partial F_j}{\partial x_k}, \frac{\partial F_j}{\partial y_k},$$

exist and are continuous throughout $G \times G$ $(j, k = 1, \ldots, n)$ [F_j is the projection of F onto the j-th coordinate]. For each $\boldsymbol{a} \in G$, let $\sigma_{\boldsymbol{a}}$ [$\delta_{\boldsymbol{a}}$] be the transformation of G onto itself defined by $\sigma_{\boldsymbol{a}}(\boldsymbol{x}) = \boldsymbol{a}\boldsymbol{x}$ [$\delta_{\boldsymbol{a}}(\boldsymbol{x}) = \boldsymbol{x}\boldsymbol{a}$] for all $\boldsymbol{x} \in G$. That is, $\sigma_{\boldsymbol{a}}$ and $\delta_{\boldsymbol{a}}$ are left and right translation by the element $\boldsymbol{a}$. The third property that we require of G is that the Jacobian of each of the transformations $\sigma_{\boldsymbol{a}}$ and $\delta_{\boldsymbol{a}}$ be a constant; that is, they depend only upon $\boldsymbol{a}$.[1]

We denote the Jacobian of a transformation τ by the symbol $J(\tau)$. For $\boldsymbol{a} \in G$, we define $S(\boldsymbol{a})$ to be $|J(\sigma_{\boldsymbol{a}})|$ and $D(\boldsymbol{a})$ to be $|J(\delta_{\boldsymbol{a}})|$. Since $\sigma_{\boldsymbol{a}} \circ \sigma_{\boldsymbol{b}} = \sigma_{\boldsymbol{a}\boldsymbol{b}}$, we use the familiar Jacobian identity $J(\sigma_{\boldsymbol{a}} \circ \sigma_{\boldsymbol{b}}) = J(\sigma_{\boldsymbol{a}}) \cdot J(\sigma_{\boldsymbol{b}})$ to point out that $S(\boldsymbol{a}\boldsymbol{b}) = S(\boldsymbol{a})\,S(\boldsymbol{b})$, for all $\boldsymbol{a}, \boldsymbol{b} \in G$. Similarly, we have $\delta_{\boldsymbol{a}\boldsymbol{b}} = \delta_{\boldsymbol{b}} \circ \delta_{\boldsymbol{a}}$, and therefore $D(\boldsymbol{a}\boldsymbol{b}) = D(\boldsymbol{a})\,D(\boldsymbol{b})$. Write $\boldsymbol{e}$ for the identity element of G. Then $\sigma_{\boldsymbol{e}}$ and $\delta_{\boldsymbol{e}}$ are the identity transformation on G, so that $S(\boldsymbol{e}) = D(\boldsymbol{e}) = 1$. It follows that S and D are continuous homomorphisms of G into the multiplicative group $]0, \infty[$.

We form the left and right Haar integrals on $\mathfrak{C}_{00}$ with the functions S and D, respectively, together with Riemann integration in R^n. Given a continuous complex-valued function φ defined on G and vanishing outside of a compact set, let $\int\limits_G \varphi(\boldsymbol{x})\,d\boldsymbol{x}$ denote the ordinary n-dimensional

[1] These conditions are fulfilled, for example, if each F_j has the form

$$F_j(x_1, \ldots, x_n, y_1, \ldots, y_n) = \sum_{k=1}^{n} \sum_{l=1}^{n} c_{kl}^{(j)} x_k y_l + d_j,$$

where the real numbers $c_{kl}^{(j)}$ and d_j do not depend upon $\boldsymbol{x}$ and $\boldsymbol{y}$.

Riemann integral of φ. Then we claim that

$$I_s(f) = \int_G f(x) \frac{1}{S(x)} dx$$

is a left Haar integral on $\mathfrak{C}_{00}$ and that

$$I_d(f) = \int_G f(x) \frac{1}{D(x)} dx$$

is a right Haar integral on $\mathfrak{C}_{00}$. To prove these assertions, we make use of the familiar formula for transformation of multiple integrals. Applied to G and σ_a, this formula is:

$$\int_{\sigma_a(G)} \varphi(x)\, dx = \int_G (\varphi \circ \sigma_a)(y)\, |J(\sigma_a)(y)|\, dy.$$

Here φ is, let us say, any function in $\mathfrak{C}_{00}$. For $\varphi = {}_{a^{-1}}f \frac{1}{S}$, we obtain

$$\begin{aligned} I_s({}_{a^{-1}}f) &= \int_G {}_{a^{-1}}f(x) \frac{1}{S(x)} dx = \int_G (({}_{a^{-1}}f) \circ \sigma_a)(y) \frac{1}{(S \circ \sigma_a)(y)} S(a)\, dy \\ &= \int_G f(y) \frac{1}{S(y)} dy = I_s(f). \end{aligned}$$

Since I_s obviously satisfies the properties (15.5.i)—(15.5.iii), we have shown that I_s is a left Haar integral. The proof that I_d is a right Haar integral is almost the same, and is omitted.

It is of interest to compute the modular function for G. We have

$$\begin{aligned} I_s(f_{a^{-1}}) &= \int_G f \circ \delta_{a^{-1}}(x) \frac{1}{S(x)} dx = \int_G f \circ \delta_{a^{-1}} \circ \delta_a(y) \frac{1}{S \circ \delta_a(y)} D(a)\, dy \\ &= \int_G f(y) \frac{1}{S(y)} \frac{D(a)}{S(a)} dy = \frac{D(a)}{S(a)} I_s(f). \end{aligned}$$

Hence the modular function Δ for G is D/S.

(f) We first apply (e) to the multiplicative group $R \cap \{0\}'$ of nonzero real numbers. The transformation $x \to ax$ has Jacobian a, and so a Haar integral has the form

$$\int_{-\infty}^{\infty} \frac{f(x)}{|x|} dx,$$

where for each $f \in \mathfrak{C}_{00}(R \cap \{0\}')$ the integral is actually extended over $[-\beta, -\alpha] \cup [\alpha, \beta]$ for real numbers α and β, $0 < \alpha < \beta$. Similarly a Haar integral for functions on the multiplicative group of all nonzero complex numbers is

$$\int_{-\infty}^{\infty} \int_{-\infty}^{\infty} \frac{f(x+iy)}{x^2+y^2} dx\, dy.$$

(g) We next apply (e) to the group G of all matrices

$$\begin{pmatrix} x & y \\ 0 & 1 \end{pmatrix},$$

with $x, y \in R$ and $x \neq 0$. For convenience we write the elements of G as (x, y), with $(x, y)(u, v) = (xu, xv + y)$. Topologize G as a subset of R^2. Then G is plainly a topological group satisfying the hypotheses of (e). For $(a, b) \in G$, the transformation $\sigma_{(a,b)}$ has the form $\sigma_{(a,b)}(x, y) = (ax, ay + b)$. The Jacobian of $\sigma_{(a,b)}$ is a^2. Thus a left Haar integral on $\mathfrak{C}_{00}(G)$ is provided by

$$\int_{-\infty}^{\infty}\int_{-\infty}^{\infty} \frac{f(x,y)}{x^2}\, dx\, dy,$$

where the integral is, for each fixed f, actually extended over a bounded closed subset of R^2 that does not intersect the line $\{(x, y) \in R^2 : x = 0\}$.

The transformation $\delta_{(a,b)}$ on G has the form $\delta_{(a,b)}(x, y) = (ax, bx + y)$; the Jacobian of $\delta_{(a,b)}$ is a. Thus

$$\int_{-\infty}^{\infty}\int_{-\infty}^{\infty} \frac{f(x, y)}{|x|}\, dx\, dy$$

is a right Haar integral on $\mathfrak{C}_{00}(G)$.

This group is one of the simplest examples of a locally compact group in which the right and left Haar integrals are essentially different. The modular function is $\Delta(x, y) = \dfrac{D(x, y)}{S(x, y)} = \dfrac{1}{|x|}$.

(h) Consider next the group $\mathfrak{G}\mathfrak{L}(2, R)$. For $A = \begin{pmatrix} a_{11} & a_{12} \\ a_{21} & a_{22} \end{pmatrix}$ in $\mathfrak{G}\mathfrak{L}(2, R)$, a simple computation shows that $J(\sigma_A) = J(\delta_A) = (a_{11}a_{22} - a_{12}a_{21})^2 = (\det A)^2$. Thus the right and left Haar integrals are identical, and are given by

$$\int_{-\infty}^{\infty}\int_{-\infty}^{\infty}\int_{-\infty}^{\infty}\int_{-\infty}^{\infty} \frac{f(x_{11}, x_{12}, x_{21}, x_{22})}{(x_{11}x_{22} - x_{12}x_{21})^2}\, dx_{11}\, dx_{12}\, dx_{21}\, dx_{22}.$$

(i) Let $G_1, G_2, \ldots, G_m$ be locally compact groups, and let I_j be a left Haar integral on G_j ($j = 1, 2, \ldots, m$). The product functional $I_1 \times I_2 \times \cdots \times I_m$ is a left Haar integral on $G_1 \times G_2 \times \cdots \times G_m$. This follows immediately from (13.3) and (13.2).

(j) Let $\{G_\gamma\}_{\gamma \in \Gamma}$ be an arbitrary nonvoid family of *compact* groups; let I_γ be the normalized Haar integral on G_γ for each $\gamma \in \Gamma$ [*i.e.*, $I_\gamma(1) = 1$]; and let I be the functional on $\mathfrak{C}(\underset{\gamma\in\Gamma}{\mathsf{P}} G_\gamma)$ defined in (13.18). Then I is the normalized Haar integral on $\underset{\gamma\in\Gamma}{\mathsf{P}} G_\gamma$.

(k) We now describe normalized Haar measure λ on the groups $\Delta_{\boldsymbol{a}}$ (10.2). Notation is as in (10.2). Topologically, $\Delta_{\boldsymbol{a}}$ is the Cartesian product $\mathop{\mathsf{P}}\limits_{k=0}^{\infty}\{0, 1, \ldots, a_k - 1\}$, where each finite factor space is discrete. The finite space $\{0, 1, \ldots, a_k - 1\}$ admits the measure λ_k such that $\lambda_k(A) = \frac{\overline{\overline{A}}}{a_k}$ for all $A \subset \{0, 1, \ldots, a_k - 1\}$. Haar measure on $\Delta_{\boldsymbol{a}}$ is the product measure of these measures $\lambda_0, \lambda_1, \ldots, \lambda_k, \ldots$, as constructed in (13.15)–(13.22)[1]. [Given subsets A_k of $\{0, 1, \ldots, a_k - 1\}$ for $k = 0, 1, \ldots$, let $C(A_0, A_1, \ldots, A_n)$ be the set of all $\boldsymbol{x} \in \Delta_{\boldsymbol{a}}$ such that each x_k belongs to A_k for $k = 0, 1, \ldots, n$. For $n = 0, 1, \ldots$, let $\boldsymbol{u}^{(n)}$ be the element of $\Delta_{\boldsymbol{a}}$ such that $u_k^{(n)} = \delta_{nk}$ $(k = 0, 1, \ldots)$. We will compute $\lambda\left(\mathop{\mathsf{P}}\limits_{k=0}^{\infty} A_k\right)$. Consider first $C(\{0\}_0) = \{\boldsymbol{x} \in \Delta_{\boldsymbol{a}} : x_0 = 0\}$. Since $\Delta_{\boldsymbol{a}} = \bigcup\limits_{k=0}^{a_0-1} \left(k\boldsymbol{u}^{(0)} + C(\{0\}_0)\right)$ and the translates just written are pairwise disjoint, the invariance of λ implies that $\lambda\left(C(\{0\}_0)\right) = \frac{1}{a_0}$. A set $C(A_0)$ is the union of $\overline{\overline{A}}_0$ pairwise disjoint translates of $C\left(\{0\}_0\right)$. Hence $\lambda\left(C(A_0)\right) = \frac{\overline{\overline{A}}_0}{a_0}$. Similarly, there are a_1 pairwise disjoint translates of $C\left(A_0, \{0\}_1\right)$ whose union is $C(A_0)$, from which we infer that $\lambda\left(C(A_0, A_1)\right) = \prod\limits_{k=0}^{1} \frac{\overline{\overline{A}}_k}{a_k}$. Finite induction shows that $\lambda\left(C(A_0, A_1, \ldots, A_n)\right) = \prod\limits_{k=0}^{n} \frac{\overline{\overline{A}}_k}{a_k}$ for $n = 2, 3, \ldots$. The countable additivity of λ implies that $\lambda\left(\mathop{\mathsf{P}}\limits_{k=0}^{\infty} A_k\right) = \lim\limits_{n\to\infty} \lambda\left(C(A_0, A_1, \ldots, A_n)\right) = \prod\limits_{k=0}^{\infty} \frac{\overline{\overline{A}}_k}{a_k}$. Characteristic functions of sets $C(A_0, A_1, \ldots, A_n)$ span a linear subspace of $\mathfrak{C}(\Delta_{\boldsymbol{a}})$ that is dense in $\mathfrak{C}(\Delta_{\boldsymbol{a}})$. [This follows from the STONE-WEIERSTRASS theorem (13.2), since the product of two of these characteristic functions is another of the same sort, and these characteristic functions separate points of $\Delta_{\boldsymbol{a}}$.] The foregoing and (13.22) show that $\int\limits_{\Delta_{\boldsymbol{a}}} f\, d\lambda$ agrees with $\int\limits_{\Delta_{\boldsymbol{a}}} f\, d\iota$, where ι is the product measure for $\lambda_0, \lambda_1, \ldots, \lambda_n, \ldots$, for f in a dense subspace of $\mathfrak{C}(\Delta_{\boldsymbol{a}})$. Hence λ is the product measure.]

(l) Using (k), it is easy to compute Haar measure λ on $\Omega_{\boldsymbol{a}}$. We normalize λ by requiring that $\lambda(\Delta_{\boldsymbol{a}}) = 1$. For each $k \in Z$, let λ_k be the measure on $\{0, 1, \ldots, a_k - 1\}$ defined in (k). For each $n \in Z$, let μ_n denote the product measure of these λ_k's on the space $\mathop{\mathsf{P}}\limits_{k=n}^{\infty}\{0, 1, \ldots, a_k - 1\}$;

[1] It is curious that Haar measure on $\Delta_{\boldsymbol{a}}$ is identical with Haar measure on $\mathop{\mathsf{P}}\limits_{k=0}^{\infty}\{0, 1, \ldots, a_k - 1\}$, with the topology and group operation as the direct product of finite groups of order a_k. See (j) *supra*.

this space is homeomorphic with the set Λ_n defined in (10.4). If A is a λ-measurable subset of Λ_n where $n<0$, then $\lambda(A)=a_n\, a_{n+1} \cdots a_{-1}\cdot\mu_n(A)$. For an arbitrary λ-measurable subset A of $\Omega_{\mathbf{a}}$, we of course have $\lambda(A)= \lim\limits_{n\to-\infty} a_n\, a_{n+1} \cdots a_{-1}\cdot\mu_n(A\cap\Lambda_n)$.

(15.18) We will now investigate relatively invariant functionals on quotient spaces. Throughout (15.18)–(15.24), we will suppose that G is a locally compact group and that H is a closed subgroup of G. Given any function ψ on G/H and any $a\in G$, let ${}_a\psi$ be the function on G/H defined by ${}_a\psi(xH)=\psi(axH)$. The space G/H is locally compact (5.22) and Hausdorff (5.21), so that the function space $\mathfrak{C}_{00}(G/H)$ is nontrivial if $H\subsetneqq G$. We are concerned here with finding nonzero functionals J on $\mathfrak{C}_{00}(G/H)$ such that:

(i) $0\leqq J(\psi)<\infty$ for $\psi\in\mathfrak{C}_{00}^{+}(G/H)$;

(ii) $J(\psi_1+\psi_2)=J(\psi_1)+J(\psi_2)$ for $\psi_1,\psi_2\in\mathfrak{C}_{00}(G/H)$;

(iii) $J(\alpha\psi)=\alpha J(\psi)$ for $\alpha\in K$ and $\psi\in\mathfrak{C}_{00}(G/H)$;

(iv) $J({}_a\psi)=\chi(a)\, J(\psi)$ for all $a\in G$ and $\psi\in\mathfrak{C}_{00}(G/H)$.

For $a\in G$, the number $\chi(a)$ is defined by (iv) and is to be independent of the choice of ψ. Such a functional J is called *relatively invariant*.

Let J be as above. As in part (V) of the proof of (15.5), it is easy to see that $J(\psi)>0$ for all nonzero $\psi\in\mathfrak{C}_{00}^{+}(G/H)$. Property (iv) is responsible for this. Since ${}_{ab}\psi={}_b({}_a\psi)$ for $a, b\in G$, the function χ satisfies the functional equation

(v) $\chi(ab)=\chi(a)\,\chi(b)$ for all $a, b\in G$,

and also has the property that $\chi(G)\subset\,]0,\infty[$. A less trivial fact is that χ is continuous. To prove this we need a lemma.

(15.19) Lemma. *Let ψ be a continuous complex-valued function on G/H such that for every $\delta>0$ there is a compact subset $\{xH: x\in F\}$ of G/H outside of which $|\psi|$ is less than δ. Then for every $\varepsilon>0$, there is a symmetric neighborhood V of e in G such that if $x, y\in G$ and $yx^{-1}\in V$, then $|\psi(xH)-\psi(yH)|<\varepsilon$.*

Proof. This lemma and its proof are similar to but not identical with (4.15). Consider $\varepsilon>0$ and choose $\{zH: z\in F\}$ to be a compact set such that $|\psi(wH)|<\frac{\varepsilon}{2}$ for $wH\notin\{zH: z\in F\}$. Now for every $x\in F$, there is a neighborhood U_x of e in G such that if $yH\in\{zH: z\in U_x x\}$, then $|\psi(yH)-\psi(xH)|<\frac{\varepsilon}{2}$. Let V_x be a symmetric neighborhood of e in G such that $V_x^2\subset U_x$. A finite number of the open sets $\{zH: z\in V_x x\}$ cover the compact set $\{zH:z\in F\}$; that is,

$$\{zH:z\in F\}\subset\bigcup_{k=1}^{m}\{zH:z\in V_{x_k}x_k\},$$

for some $x_1,\ldots,x_m\in F$. Let $V=\bigcap_{k=1}^{m}V_{x_k}$.

Now consider any $x, y\in G$ such that $yx^{-1}\in V$. If $xH\in\{zH:z\in F\}$, then we have $x\in V_{x_k}x_kH$ for some $k=1,\ldots,m$. Therefore $y=yx^{-1}x\in VV_{x_k}x_kH\subset U_{x_k}x_kH$, and $yH\in\{zH:z\in U_{x_k}x_k\}$. Obviously xH also belongs to $\{zH:z\in U_{x_k}x_k\}$ and hence

$$|\psi(yH)-\psi(xH)|\leqq|\psi(yH)-\psi(x_kH)|+|\psi(x_kH)-\psi(xH)|<\varepsilon.$$

Similarly, if $yH\in\{zH:z\in F\}$, then $|\psi(yH)-\psi(xH)|<\varepsilon$. If neither yH nor xH lies in $\{zH:z\in F\}$, then we have $|\psi(yH)-\psi(xH)|\leqq|\psi(yH)|+|\psi(xH)|<\varepsilon$. □

(15.20) Theorem. *Let J be a relatively invariant functional on $\mathfrak{C}_{00}(G/H)$ and let χ be as in (15.18.iv). Then χ is continuous.*

Proof. Let ψ be a nonzero function in $\mathfrak{C}_{00}^{+}(G/H)$, and let $\{xH:x\in F\}$ be a compact subset of G/H outside of which ψ vanishes. By (5.24.b), we may suppose that F is compact. Let U be a neighborhood of e in G such that U^{-} is compact. Then the set $\{uxH:u\in U^{-}, x\in F\}$ is compact in G/H since $U^{-}F$ is compact in G. Let ω be a function in $\mathfrak{C}_{00}^{+}(G/H)$ such that $\omega(\{uxH:u\in U^{-}, x\in F\})=1$. For $\varepsilon>0$, use (15.19) to choose a symmetric neighborhood V of e in G such that $V\subset U$ and

$$|\psi(vxH)-\psi(xH)|<\varepsilon\frac{J(\psi)}{J(\omega)}$$

for all $v\in V$ and $x\in G$. Then we have

$$|\psi(vxH)-\psi(xH)|\leqq\varepsilon\frac{J(\psi)}{J(\omega)}\omega(xH)\quad\text{for all}\quad xH\in G/H\quad\text{and}\quad v\in V;$$

applying J to the last inequality, we have

$$|\chi(v)\,J(\psi)-J(\psi)|\leqq\varepsilon J(\psi)\quad\text{for all}\quad v\in V.$$

This implies that χ is continuous at e, and since χ is a homomorphism of G onto a topological group, χ is continuous throughout G. □

We will ultimately establish a relationship between the function χ and the modular functions for the groups G and H. We first establish the following theorem.

(15.21) Theorem. *Let L denote a left Haar integral on $\mathfrak{C}_{00}(H)$. Then the correspondence $f\to f'$ given by*

(i) $f'(xH)=L({}_xf)$

defines a linear mapping of $\mathfrak{C}_{00}(G)$ onto $\mathfrak{C}_{00}(G/H)$. [In the expression $L({}_xf)$, ${}_xf$ is regarded as the function ${}_xf$ restricted to H.]

Proof. Let $f \in \mathfrak{C}_{00}(G)$; then f vanishes outside of a compact set $F \subset G$. If we restrict the domain of ${}_xf$ to H, we see at once that ${}_xf$ is continuous on H and that ${}_xf$ vanishes outside of the compact set $(x^{-1}F) \cap H$. Thus $L({}_xf)$ is defined for $x \in G$ and $f \in \mathfrak{C}_{00}(G)$. If $h \in H$, then we have $L({}_{xh}f) = L\big({}_h({}_xf)\big) = L({}_xf)$. Thus the function $L({}_xf)$, defined for all $x \in G$, is constant on the cosets xH; we can therefore define f' on G/H by the equality (i). The uniform continuity of f guarantees that f' is continuous on G/H. We omit the details of this argument, which should by now be familiar to the reader. Also f' vanishes outside of the compact set $\{xH : x \in F\} \subset G/H$. Thus (i) defines a mapping of $\mathfrak{C}_{00}(G)$ into $\mathfrak{C}_{00}(G/H)$. It is obvious that $(f_1+f_2)' = f_1' + f_2'$, $(\alpha f)' = \alpha f'$ $(\alpha \in K)$, that $f' \geqq 0$ if $f \geqq 0$, and that $f' = 0$ if and only if $f = 0$ [for $f \geqq 0$]. In particular, $f \to f'$ is a linear mapping.

Now let ψ be any function in $\mathfrak{C}_{00}(G/H)$. Suppose that ψ vanishes outside of the compact set $\{xH : x \in F\}$. By (5.24.b), we may suppose that F is compact in G. Let g be a function in $\mathfrak{C}_{00}^+(G)$ such that $g(F) = 1$. Let $U = \{x \in G : g(x) > 0\} \cdot H$. Then it is clear that $g'(uH) > 0$ if and only if $u \in U$ [recall that L is a strictly positive functional]. Let φ be the natural mapping of G onto G/H. Let Ψ be the function on G defined by

$$\Psi(x) = \begin{cases} \dfrac{g(x)}{g' \circ \varphi(x)} \psi \circ \varphi(x) & \text{if} \quad g(x) \neq 0, \\ 0 & \text{otherwise}. \end{cases}$$

It is clear that $\Psi \in \mathfrak{C}_{00}(G)$ [Ψ is continuous since $\Psi = \frac{g}{g' \circ \varphi} \psi \circ \varphi$ on the open set U, $\Psi = 0$ on the open set $(FH)'$, and $G = U \cup (FH)'$]. For $x \in U$, we have

$${}_x\Psi(y) = \frac{{}_xg(y)\, \psi(xH)}{L({}_xg)} \quad \text{for all} \quad y \in H,$$

and therefore $L({}_x\Psi) = \psi(xH)$. If $x \notin U$, then clearly $L({}_x\Psi) = 0 = \psi(xH)$. That is, $\psi(xH) = \Psi'(xH)$ for all $xH \in G/H$. Hence the mapping $f \to f'$ carries $\mathfrak{C}_{00}(G)$ *onto* $\mathfrak{C}_{00}(G/H)$. □

(15.22) Theorem. *Let J be a relatively invariant functional on* $\mathfrak{C}_{00}(G/H)$ *as in* (15.18) *and let χ be as in* (15.18.iv). *Let Δ and δ denote the modular functions of G and H, respectively. Then we have*

(i) $\chi(h) = \dfrac{\Delta(h)}{\delta(h)}$ *for all* $h \in H$.[1]

Proof. Let L be a left Haar integral on $\mathfrak{C}_{00}(H)$ and let the mapping $f \to f'$ be given by (15.21.i). For $f \in \mathfrak{C}_{00}(G)$, let $K(f) = J(f')$. For all $x, s \in G$, it is evident that $({}_sf)'(xH) = L\big({}_x({}_sf)\big) = L({}_{sx}f) = f'(sxH)$. Thus $K({}_sf) = J\big({}_s(f')\big) = \chi(s)\, J(f') = \chi(s)\, K(f)$. This tells us that K is a "relatively

[1] Thus, if there is a relatively invariant functional J, the homomorphism Δ/δ of H into the multiplicative group $]0, \infty[$ admits an extension over G which is a continuous homomorphism of G into $]0, \infty[$.

invariant" [in an obvious sense], strictly positive, linear functional on $\mathfrak{D}_{00}(G)$. Consider the functional $I(f)=K(\chi f)$ on $\mathfrak{C}_{00}(G)$. We have

$$\chi(s)\, I(f) = \chi(s)\, K(\chi f) = K\left({}_s(\chi f)\right) = K\left(\chi(s)\,\chi\cdot({}_sf)\right)$$
$$= \chi(s)\, K\left(\chi\cdot({}_sf)\right) = \chi(s)\, I({}_sf),$$

for $f\in\mathfrak{C}_{00}(G)$ and $s\in G$. Since I is plainly linear, it follows that I is a left Haar integral on $\mathfrak{C}_{00}(G)$. Thus we have

$$J(f') = K(f) = I(\chi^{-1}f).^{1} \tag{1}$$

Now take any $h\in H$, and consider the function $f_{h^{-1}}$. For $x\in G$, we have

$$(f_{h^{-1}})'(xH) = L\left({}_x(f_{h^{-1}})\right) = L\left(({}_xf)_{h^{-1}}\right) = \delta(h)\, L({}_xf) = \delta(h)\, f'(xH).$$

Thus

$$J\left((f_{h^{-1}})'\right) = \delta(h)\, J(f'). \tag{2}$$

We also have

$$\chi^{-1}(x)\, f(xh^{-1}) = \chi(h^{-1})\, \chi^{-1}(xh^{-1})\, f(xh^{-1}),$$

so that

$$I\left(\chi^{-1}(f_{h^{-1}})\right) = \chi(h^{-1})\, I\left((\chi^{-1}f)_{h^{-1}}\right) = \chi(h^{-1})\, \Delta(h)\, I(\chi^{-1}f). \tag{3}$$

Applying (2), (1), (3), and then (1) again, we find

$$\delta(h)\, J(f') = \chi(h^{-1})\, \Delta(h)\, J(f').$$

Hence the equality (i) holds. □

(15.23) Remark. Suppose that H is a closed normal subgroup of G, so that G/H is a locally compact group. For J, let us take a left Haar integral on $\mathfrak{C}_{00}(G/H)$. Since $(aH)(xH)=axH$ for all $a, x\in G$, we have ${}_{(aH)}\psi={}_a\psi$ for all functions ψ on G/H. Thus J satisfies (15.18.i)—(15.18.iv) with $\chi(a)=1$ for all $a\in G$. From (15.22), we infer that $\delta(h)=\Delta(h)$ for $h\in H$, where Δ and δ are the modular functions for G and H, respectively.

The converse of (15.22) also holds.

(15.24) Theorem. *Let Δ and δ be the modular functions for G and H respectively, and suppose that there is a continuous homomorphism χ of G into $]0, \infty[$ such that*

(i) $\chi(h) = \dfrac{\Delta(h)}{\delta(h)}$ *for all* $h\in H$.

Then there is a functional J on $\mathfrak{C}_{00}(G/H)$ satisfying (15.18.i)—(15.18.iii) *and* (15.18.iv) *with the given homomorphism* χ.

Proof. Let I and L denote left Haar integrals on G and H, respectively, and for $f\in\mathfrak{C}_{00}(G)$, let f' be defined by (15.21.i). Define the functional K on $\mathfrak{C}_{00}(G)$ by

$$K(f)=I(\chi^{-1}f). \tag{1}$$

---

[1] Here χ^{-1} designates the function $1/\chi$; it is not to be confused with the inverse mapping.

We will show that K can be regarded as a functional J on $\mathfrak{C}_{00}(G/H)$. Since the mapping $f \to f'$ is linear and onto (15.21), this will be accomplished if we show that $f'=0$ implies $K(f)=0$. [We will then define $J(f')=K(f)$.] Suppose then that $L({}_xf)=0$ for all $x \in G$. By (15.14), we then have $L\left(({}_xf)\star\frac{1}{\delta}\right)=0$ for all $x \in G$. Now let g be any function in $\mathfrak{C}_{00}^+(G)$ such that $g'(xH)=1$ for all $x \in \{y \in G : f(y) \neq 0\}^-$. That is,

$$L({}_xg)=1 \quad \text{whenever} \quad f(x) \neq 0. \tag{2}$$

Let us use the notation I_x $[L_y]$ to indicate that the functional I $[L]$ is being applied to a function of $x \in G$ $[y \in H]$. We then have

$$\begin{aligned} 0 = I(\chi^{-1}g \cdot 0) &= I_x\left[\chi^{-1}(x)\, g(x)\, L\left(({}_xf)\star\frac{1}{\delta}\right)\right] \\ &= I_x\left[\chi^{-1}(x)\, g(x)\, L_y\left(f(xy^{-1})\,\frac{1}{\delta(y)}\right)\right]. \end{aligned}$$

Applying FUBINI's theorem[1] to this equality, we obtain

$$L_y\left[\frac{1}{\delta(y)}\, I_x\left(g(x)\,\chi^{-1}(x)\, f(xy^{-1})\right)\right]=0. \tag{3}$$

By (15.11), we have

$$I(g\chi^{-1}f_{y^{-1}}) = \Delta(y)\, I\left(g_y\,\frac{1}{\chi_y}\,f\right) = \frac{\Delta(y)}{\chi(y)}\, I\left(g_y\,\frac{1}{\chi}\,f\right).$$

Substituting this into (3) and using (i), we find

$$L_y\left[I_x\left(g(xy)\,\chi^{-1}(x)\, f(x)\right)\right]=0.$$

Applying FUBINI's theorem (13.2) again, we have

$$I_x\left[L({}_xg)\,\chi^{-1}(x)\, f(x)\right]=0. \tag{4}$$

In view of (2), (4) implies that $I(\chi^{-1}f)=0$. That is, $K(f)=0$. Thus $J(f')=K(f)$ defines a nonzero functional on $\mathfrak{C}_{00}(G/H)$ satisfying (15.18.i) to (15.18.iii). Furthermore,

$$\begin{aligned} J\left({}_a(f')\right) = J\left(({}_af)'\right) = K({}_af) &= I\left(\frac{1}{\chi}\,{}_af\right) \\ &= I\left(\frac{\chi(a)}{{}_a\chi}\,{}_af\right) = \chi(a)\, I\left(\frac{1}{\chi}\,f\right) = \chi(a)\, J(f'), \end{aligned}$$

and thus J is relatively invariant on $\mathfrak{C}_{00}(G/H)$ with the given homomorphism χ appearing in (15.18.iv). □

[1] Since $f \in \mathfrak{C}_{00}(G)$ and $g \in \mathfrak{C}_{00}(G)$, the function $(x, y) \to \chi^{-1}(x)\, g(x)\, f(xy^{-1})\,\frac{1}{\delta(y)}$ is in $\mathfrak{C}_{00}(G \times H)$. Thus we need only the simplest version of FUBINI's theorem (13.2) to obtain (3).

Miscellaneous theorems and examples

(15.25) We sketch here a proof of the existence of the Haar integral using steps (I) and (II) of (15.5) and TIHONOV's theorem. [For each nonzero $f \in \mathfrak{C}_{00}^+$, let $X_f = \left[\frac{1}{(f_0:f)}, (f:f_0)\right]$, and let $X = \mathsf{P}_f X_f$, where f runs through all nonzero f in $\mathfrak{C}_{00}^+$. For nonzero $\varphi \in \mathfrak{C}_{00}^+$, the functional I_φ can be regarded as the point (x_f) of X where $x_f = I_\varphi(f)$. Let $\mathscr{U}$ consist of all neighborhoods of e in G. For each $U \in \mathscr{U}$, choose exactly one $\varphi_U \in \mathfrak{C}_{00}^+$ such that $\varphi_U(U') = 0$. The set $\mathscr{U}$ is directed by inclusion $\subset$, *i.e.*, $U \geqq V$ if $U \subset V$. Thus I_{φ_U}, $U \in \mathscr{U}$, is a net in the compact space X. Consequently, there is a subnet I_β, $\beta \in E$, of I_{φ_U} and an $I \in X$ such that $\lim_\beta I_\beta = I$. Then I is a Haar integral; the linearity follows from step (II) of (15.5).]

(15.26) Haar integrals and automorphisms of G (BRACONNIER [1]). Let G be a locally compact group and let τ be a topological automorphism of G *onto* G; *i.e.* let τ be an automorphism and a homeomorphism. Let I be a left Haar integral on $\mathfrak{C}_{00}$. The mapping $f \to f \circ \tau$ is plainly a one-to-one linear transformation of $\mathfrak{C}_{00}$ onto itself that preserves nonnegativity and also preserves the norm $\|f\|_u$. The linear functional $I_\tau(f) = I(f \circ \tau)$ thus obviously satisfies conditions (15.5.i)—(15.5.iii). For $a \in G$, we have $({}_a f) \circ \tau = {}_{\tau^{-1}(a)}(f \circ \tau)$, as is easy to verify. Hence we have $I_\tau({}_a f) = I\left({}_{\tau^{-1}(a)}(f \circ \tau)\right)$, and as I is a left Haar integral, we obtain $I_\tau({}_a f) = I_\tau(f)$. That is, I_τ is also a left Haar integral, and by the uniqueness of left Haar integrals (15.5), there is a positive number $\Delta(\tau)$ such that $I_\tau(f) = \Delta(\tau)\, I(f)$ for all $f \in \mathfrak{C}_{00}$. Given τ_1 and τ_2, it is easy to see that $\Delta(\tau_1 \circ \tau_2) = \Delta(\tau_1)\, \Delta(\tau_2)$. That is, the mapping $\tau \to \Delta(\tau)$ is a homomorphism of the group of all τ's into the multiplicative group $]0, \infty[$. We obtain the function $x \to \Delta(x)$ of (15.11) by considering the inner automorphisms ϱ_x: $y \to x y x^{-1}$. Then plainly $\Delta(\varrho_x)$ in our present notation is $\Delta(x)$ in the notation of (15.11)[1].

Similarly, if J is a right Haar integral on $\mathfrak{C}_{00}$, we define $J_\tau(f) = J(f \circ \tau)$ and find $J_\tau(f_a) = J\left((f_a) \circ \tau\right) = J\left((f \circ \tau)_{\tau^{-1}(a)}\right) = J(f \circ \tau) = J_\tau(f)$ and hence $J_\tau(f) = \delta(\tau)\, J(f)$, where $\delta(\tau) > 0$ and $\delta(\tau_1 \circ \tau_2) = \delta(\tau_1)\, \delta(\tau_2)$.

(15.27) Comments on right Haar integrals. (a) Let G be a locally compact group, J a right Haar integral on $\mathfrak{C}_{00}$, and Δ the modular function on G. Then the mappings $f \to J(f\Delta)$ and $f \to J(f^\star)$ are identical and are left Haar integrals on $\mathfrak{C}_{00}$. [This follows from (15.15). Given a left Haar integral I, $f \to I\left(f \frac{1}{\Delta}\right)$ is a right Haar integral. Thus we can write $J(f) = I\left(f \frac{1}{\Delta}\right)$ and $J(f\Delta) = I\left(f\Delta \frac{1}{\Delta}\right) = I(f)$. Moreover, we have by (15.15) that

$$J(f^\star) = I\left(f^\star \frac{1}{\Delta}\right) = I\left((f\Delta)^\star\right) = I\left((f\Delta)\frac{1}{\Delta}\right) = I(f) = J(f\Delta).]$$

---

[1] For more facts about $\Delta(\tau)$, see (26.21).

(b) Let G, Δ, and J be as in (a). Then $J({}_a f) = \Delta(a)\, J(f)$ for all $f \in \mathfrak{C}_{00}$ and $a \in G$. [We have $J(f) = I\left(f\frac{1}{\Delta}\right)$, where I is a left Haar integral. Thus

$$J({}_a f) = I\left({}_a f\frac{1}{\Delta}\right) = I\left(\Delta(a)\,{}_a\!\left(f\frac{1}{\Delta}\right)\right) = \Delta(a)\, I\left(f\frac{1}{\Delta}\right) = \Delta(a)\, J(f).]$$

(15.28) More examples of Haar integrals. (a) The construction of (15.17.h) can be generalized to $\mathfrak{G}\mathfrak{L}(n, R)$. For $A \in \mathfrak{G}\mathfrak{L}(n, R)$, we have $J(\delta_A) = J(\sigma_A) = (\det A)^n$: this is shown by an elementary computation. Thus $\mathfrak{G}\mathfrak{L}(n, R)$ is unimodular, and the Haar integral on $\mathfrak{C}_{00}(\mathfrak{G}\mathfrak{L}(n, R))$ has the form

$$\int \frac{f(X)}{|\det(X)|^n}\, dX,$$

the integral in each case being extended over a compact subset of R^{n^2} that is disjoint from the set $\{(x_{11}, \ldots, x_{nn}) : \det(X) = 0\}$.

(b) Consider the group G of all triangular $n \times n$ matrices

$$\begin{pmatrix} x_{11} & x_{12} & x_{13} & \cdots & x_{1n} \\ 0 & x_{22} & x_{23} & \cdots & x_{2n} \\ 0 & 0 & x_{33} & \cdots & x_{3n} \\ \vdots & \vdots & \vdots & & \vdots \\ 0 & 0 & 0 & \cdots & x_{nn} \end{pmatrix}$$

such that $x_{jk} \in R$ $(1 \leqq j \leqq k \leqq n)$ and $x_{11}\, x_{22} \cdots x_{nn} \neq 0$. Plainly G can be regarded as an open subset of $R^{\frac{n(n+1)}{2}}$, and satisfies the hypotheses of (15.17.e). A left Haar integral on $\mathfrak{C}_{00}(G)$ is the integral

$$\int_{-\infty}^{\infty} \cdots \int_{-\infty}^{\infty} f(x_{11}, x_{12}, \ldots, x_{nn}) \frac{1}{|x_{11}^n\, x_{22}^{n-1} \cdots x_{nn}|}\, dx_{11}\, dx_{12} \cdots dx_{nn}.$$

A right Haar integral on $\mathfrak{C}_{00}(G)$ is the integral

$$\int_{-\infty}^{\infty} \cdots \int_{-\infty}^{\infty} f(x_{11}, x_{12}, \ldots, x_{nn}) \frac{1}{|x_{11}\, x_{22}^2 \cdots x_{nn}^n|}\, dx_{11}\, dx_{12} \cdots dx_{nn}.$$

(c) Let G be the group of matrices $\begin{pmatrix} x & y \\ 0 & x^{-1} \end{pmatrix}$ $[x, y \in R,\ x \neq 0]$, with its topology as an open subset of R^2. For $A = \begin{pmatrix} a & b \\ 0 & a^{-1} \end{pmatrix} \in G$, we have $J(\sigma_A) = a^2$ and $J(\delta_A) = 1$. [We use x and y as parameters to describe G.] Thus

$$\int_{-\infty}^{\infty} \int_{-\infty}^{\infty} \frac{f(x, y)}{x^2}\, dx\, dy$$

is a left Haar integral on $\mathfrak{C}_{00}(G)$ and

$$\int_{-\infty}^{\infty}\int_{-\infty}^{\infty} f(x, y)\, dx\, dy$$

is a right Haar integral on $\mathfrak{C}_{00}(G)$. The modular function is $\Delta(x, y) = \frac{1}{x^2}$.

(d) Consider the group of nonzero quaternions. We describe the quaternions informally as the set of all linear combinations $x+\boldsymbol{i}y+\boldsymbol{j}z+\boldsymbol{k}w$, where x, y, z, and w are arbitrary real numbers and the quaternion units $1, -1, \boldsymbol{i}, -\boldsymbol{i}, \boldsymbol{j}, -\boldsymbol{j}, \boldsymbol{k}, -\boldsymbol{k}$ form a group: 1 is the identity, $(-1)^2=1$, $\boldsymbol{i}^2=\boldsymbol{j}^2=\boldsymbol{k}^2=-1$, $\boldsymbol{ij}=-\boldsymbol{ji}=\boldsymbol{k}$, $\boldsymbol{jk}=-\boldsymbol{kj}=\boldsymbol{i}$, $\boldsymbol{ki}=-\boldsymbol{ik}=\boldsymbol{j}$. The product $(x+\boldsymbol{i}y+\boldsymbol{j}z+\boldsymbol{k}w)(x'+\boldsymbol{i}y'+\boldsymbol{j}z'+\boldsymbol{k}w')$ is formed by multiplying out, assuming the validity of the usual axioms for an algebra over a field, and applying the above multiplication table for the quaternion units. Thus

$$\begin{aligned}(x+\boldsymbol{i}y+\boldsymbol{j}z+\boldsymbol{k}w)&(x'+\boldsymbol{i}y'+\boldsymbol{j}z'+\boldsymbol{k}w')\\ &=(xx'-yy'-zz'-ww')+\boldsymbol{i}(xy'+yx'+zw'-wz')\\ &\quad+\boldsymbol{j}(xz'-yw'+zx'+wy')+\boldsymbol{k}(xw'+yz'-zy'+wx').\end{aligned}$$

The quaternion $1+\boldsymbol{i}0+\boldsymbol{j}0+\boldsymbol{k}0$ is an identity for multiplication, and $0+\boldsymbol{i}0+\boldsymbol{j}0+\boldsymbol{k}0$ is a zero for multiplication. The nonzero quaternions form a [non-Abelian] group under multiplication. [The inverse of $x+\boldsymbol{i}y+\boldsymbol{j}z+\boldsymbol{k}w$ is $\frac{1}{\alpha}(x-\boldsymbol{i}y-\boldsymbol{j}z-\boldsymbol{k}w)$, where $\alpha=x^2+y^2+z^2+w^2$.] Let us denote this group by G. We can evidently regard G topologically as R^4 without the origin $(0, 0, 0, 0)$, and with this topology G is a locally compact group. Left and right Haar integrals on $\mathfrak{C}_{00}(G)$ are the same, and are given by

$$\int_{-\infty}^{\infty}\int_{-\infty}^{\infty}\int_{-\infty}^{\infty}\int_{-\infty}^{\infty} \frac{f(x, y, z, w)}{[x^2+y^2+z^2+w^2]^2}\, dx\, dy\, dz\, dw.$$

(15.29) Haar integrals for semidirect products. Let G and H be locally compact groups, and let $h\to\tau_h$ be a homomorphism of H into the group of automorphisms of G. Suppose also that the mapping $(x, h)\to\tau_h(x)$ is a continuous mapping of $G\times H$ onto G. Then the semidirect product $G\,\circledS\,H$, with the Cartesian product topology, is a locally compact group [see (2.6) and (6.20)].

(a) Let J and K be *right* Haar integrals for G and H respectively. Then the product functional $J\times K$ is a right Haar integral for $G\,\circledS\,H$. [Let f be any function in $\mathfrak{C}_{00}(G\,\circledS\,H)$. Then, using FUBINI's theorem (13.2), we have

$$\begin{aligned}J\times K[f_{(x_0, h_0)}] &= K_h[J_x(f(x\tau_h(x_0), hh_0))] = K_h[J_x(f(x, hh_0))]\\ &= J_x[K_h(f(x, hh_0))] = J_x[K_h(f(x, h))] = J\times K(f).]\end{aligned}$$

(b) We next compute the left Haar integrals for $G \circledS H$. While there is no necessity for doing so, we find it convenient to use the right Haar integrals J and K of (a). Let Δ_G and Δ_H be the modular functions for the groups G and H, respectively. For fixed $h \in H$, the functional $J(f \circ \tau_h)$ on $\mathfrak{C}_{00}(G)$ has the form $J(f \circ \tau_h) = \delta(h)\, J(f)$; moreover, $\delta(h_1 h_2) = \delta(h_1)\, \delta(h_2)$, $\delta(h) > 0$, and δ is continuous on H. [All of these properties of δ except for continuity follow at once from (15.26). It suffices to check continuity at the identity e' of H. Let U be any neighborhood of the identity e in G such that U^- is compact, and let f be any nonzero function in $\mathfrak{C}_{00}^+(G)$. Let A denote the set $\{y \in G : f(y) > 0\}^-$, and let $\omega \in \mathfrak{C}_{00}^+(G)$ be such that $\omega(A \cdot U^-) = 1$. For $\varepsilon > 0$, let U_0 be a neighborhood of e such that $U_0 \subset U$ and

$$|f(u) - f(v)| < \varepsilon \frac{J(f)}{J(\omega)} \text{ whenever } u^{-1} v \in U_0.$$

Finally, let U_1 be a symmetric neighborhood of e such that $U_1^2 \subset U_0$. For each $x \in A$, joint continuity at $\tau_{e'}(x) = x$ implies that there exist symmetric neighborhoods V_x and W_x of e' and e, respectively, such that $W_x \subset U_1$ and

$$\tau_h(y) \in x U_1 \text{ whenever } h \in V_x \text{ and } y \in x W_x.$$

Let $x_1 W_{x_1}, \ldots, x_m W_{x_m}$ be a finite covering of A and define V as $\bigcap_{j=1}^{m} V_{xj}$. If $h \in V$ and $y \in A$, then $y^{-1} \tau_h(y) \in U_0$ and hence $|f(y) - f \circ \tau_h(y)| < \varepsilon \frac{J(f)}{J(\omega)}$. It follows that

$$|f(y) - f \circ \tau_h(y)| \leqq \varepsilon \frac{J(f)}{J(\omega)} \omega(y)$$

for $h \in V$ and all $y \in G$. Applying J to this inequality, we obtain

$$|J(f) - \delta(h)\, J(f)| \leqq \varepsilon\, J(f) \quad \text{for} \quad h \in V.]$$

We now have:

$$\begin{aligned} J \times K\left[{}_{(x_0, h_0)}f\right] &= K_h\left[J_x\left(f(x_0 \tau_{h_0}(x), h_0 h)\right)\right] \\ &= K_h\left[J_x\left(f(x, h_0 h)\right) \delta(h_0)\, \Delta_G(x_0)\right]^{1} \\ &= \delta(h_0)\, \Delta_G(x_0)\, J_x\left[K_h\left(f(x, h_0 h)\right)\right] \\ &= \delta(h_0)\, \Delta_G(x_0)\, \Delta_H(h_0)\, J \times K(f). \end{aligned}$$

That is, the modular function for the semidirect product $G \circledS H$ is $\Delta(x, h) = \delta(h)\, \Delta_H(h)\, \Delta_G(x)$. Comment (15.27.a) now gives us the left Haar integrals on $\mathfrak{C}_{00}(G \circledS H)$: they all have the form $f \rightarrow J \times K(f \delta \Delta_H \Delta_G)$.

(c) The group of matrices $\begin{pmatrix} x & y \\ 0 & 1 \end{pmatrix}$ discussed in (15.17.g) is a semidirect product of the multiplicative group $]-\infty, 0[\cup]0, \infty[$ [which we take

[1] Here we use the identities $(({}_{x_0}\varphi) \circ \tau_{h_0})(x) = \varphi(x_0 \tau_{h_0}(x))$ and $J({}_a\varphi) = \Delta(a)\, J(\varphi)$ [see (15.27.b)].

as H] and the additive group R [which we take as G]. For real $x \neq 0$ and real y, let $\tau_x(y) = xy$. Then $(y, x)(y', x') = (y + xy', xx')$; this corresponds precisely to the product of the matrices $\begin{pmatrix} x & y \\ 0 & 1 \end{pmatrix}$ and $\begin{pmatrix} x' & y' \\ 0 & 1 \end{pmatrix}$. Applying (15.17.f) and (b) above, we see once again that $\int_{-\infty}^{\infty} \int_{-\infty}^{\infty} \frac{f(x, y)}{|x|} dx\, dy$ is a right Haar integral for this group. It is also easy to see that $\delta(x) = \frac{1}{|x|}$, which gives us from (b) another verification that $\int_{-\infty}^{\infty} \int_{-\infty}^{\infty} \frac{f(x, y)}{x^2} dx\, dy$ is a left Haar integral for our group.

(d) Let H be any locally compact group with modular function Δ_H, and let G be the additive real numbers R. For $h \in H$ and $x \in R$, let $\tau_h(x) = \Delta_H(h)\, x$. Plainly we define in this manner a semidirect product of R and H. To compute δ, we write $J(f \circ \tau_h) = \int_{-\infty}^{\infty} f(\Delta_H(h)\; x)\, dx = \frac{1}{\Delta_H(h)} \int_{-\infty}^{\infty} f(x)\, dx = \frac{1}{\Delta_H(h)} J(f) = \delta(h)\, J(f)$. Thus (b) gives the modular function for $R \circledS H$ as $\frac{1}{\Delta_H(h)} \Delta_H(h) \cdot 1 = 1$. Therefore every locally compact group admits a semidirect product with R that is unimodular.

(15.30) Relatively invariant integrals. (a) A simple example of a relatively invariant but not invariant integral is the following. Consider the additive group R^2 and its closed subgroup $H = \{(x, 0) : x \in R\}$. Let α be any nonzero real number. Then $\int_{-\infty}^{\infty} \int_{-\infty}^{\infty} \exp(\alpha y)\, f(x, y)\, dx\, dy$ defines as in (15.24) a relatively invariant integral on $\mathfrak{C}_{00}(R^2/H)$, for which the proportionality factor χ in (15.18.iv) is $\chi(x, y) = \exp(-\alpha y)$. [One needs only to note that R^2 is Abelian so that $\Delta/\delta = 1$ on H, and that χ is a continuous homomorphism of R^2 onto $]0, \infty[$ for which $\chi(H) = 1$.]

(b) Consider the special linear group $\mathfrak{S}\mathfrak{L}(2, R)$, and its subgroup $\mathfrak{H}$ consisting of all matrices $\begin{pmatrix} x & y \\ 0 & x^{-1} \end{pmatrix}$, with $x, y \in R$ and $x \neq 0$. The group $\mathfrak{S}\mathfrak{L}(2, R)$ is unimodular, as (15.17.h) and (15.23) show. There is no relatively invariant functional on $\mathfrak{C}_{00}(\mathfrak{S}\mathfrak{L}(2, R)/\mathfrak{H})$ in the sense of (15.18). [The modular function δ of $\mathfrak{H}$ has the form $\delta\begin{pmatrix} x & y \\ 0 & x^{-1} \end{pmatrix} = \frac{1}{x^2}$, as (15.28.c) shows. If there were a relatively invariant functional on $\mathfrak{C}_{00}(\mathfrak{S}\mathfrak{L}(2, R)/\mathfrak{H})$, there would be a continuous positive function χ on $\mathfrak{S}\mathfrak{L}(2, R)$ such that $\chi(XY) = \chi(X)\, \chi(Y)$ for $X, Y \in \mathfrak{S}\mathfrak{L}(2, R)$ and $\chi(X) = \frac{\Delta(X)}{\delta(X)} = \frac{1}{\delta(X)}$ for $X \in \mathfrak{H}$ (15.22). It clearly suffices to show that the only positive function χ on $\mathfrak{S}\mathfrak{L}(2, R)$ such that $\chi(XY) = \chi(X)\, \chi(Y)$

is the function 1. On the subgroup of all matrices $\begin{pmatrix}1 & x\\ 0 & 1\end{pmatrix}$, write $\chi\begin{pmatrix}1 & x\\ 0 & 1\end{pmatrix} = f(x)$. Then $f(x+y) = f(x)\,f(y)$. Similarly if we write $g(y) = \chi\begin{pmatrix}1 & 0\\ y & 1\end{pmatrix}$, we get $g(x+y) = g(x)\,g(y)$. On the subgroup of all matrices $\begin{pmatrix}z & 0\\ 0 & z^{-1}\end{pmatrix}$, write $\chi\begin{pmatrix}z & 0\\ 0 & z^{-1}\end{pmatrix} = h(z)$. Since

$$\begin{pmatrix}z & 0\\ 0 & z^{-1}\end{pmatrix}\begin{pmatrix}1 & x\\ 0 & 1\end{pmatrix} = \begin{pmatrix}1 & z^2x\\ 0 & 1\end{pmatrix}\begin{pmatrix}z & 0\\ 0 & z^{-1}\end{pmatrix} = \begin{pmatrix}z & zx\\ 0 & z^{-1}\end{pmatrix},$$

we infer that $h(z)\,f(x) = f(z^2x)\,h(z)$, so that $f(x) = f(z^2x)$ for all real x and all real $z \neq 0$. Hence $f(x)$ is constant for $x > 0$, and since $f(2) = f(1)^2$, we get $f(x) = 1$ for $x > 0$. Since $f(-x) = \frac{1}{f(x)}$ and $f(0) = 1$, it follows that $f = 1$. Similarly we see that $g = 1$ $\left[\text{note that } \begin{pmatrix}z & 0\\ 0 & z^{-1}\end{pmatrix}\begin{pmatrix}1 & 0\\ z^2x & 1\end{pmatrix} = \begin{pmatrix}1 & 0\\ x & 1\end{pmatrix}\begin{pmatrix}z & 0\\ 0 & z^{-1}\end{pmatrix}\right]$. If $ad - bc = 1$ and $b \neq 0$, then

$$\begin{pmatrix}a & b\\ c & d\end{pmatrix} = \begin{pmatrix}1 & 0\\ (d-1)\,b^{-1} & 1\end{pmatrix}\begin{pmatrix}1 & b\\ 0 & 1\end{pmatrix}\begin{pmatrix}1 & 0\\ (a-1)\,b^{-1} & 1\end{pmatrix},$$

and therefore $\chi\begin{pmatrix}a & b\\ c & d\end{pmatrix} = 1$. If $ad - bc = 1$ and $b = 0$, then $a \neq 0$, $d = a^{-1}$, and

$$\begin{pmatrix}a & 0\\ c & a^{-1}\end{pmatrix} = \begin{pmatrix}0 & a\\ -a^{-1} & c\end{pmatrix}\begin{pmatrix}0 & -1\\ 1 & 0\end{pmatrix},$$

so that $\chi\begin{pmatrix}a & 0\\ c & a^{-1}\end{pmatrix} = 1$. Hence we have $\chi = 1$. That is, the group $\mathfrak{S}\mathfrak{L}(2, R)$ admits *no* nonconstant homomorphism into the multiplicative group $]0, \infty[$.]

Notes

Invariant integration on one or another special class of groups has long been known and used. A detailed computation of the invariant integral on $\mathfrak{S}\mathfrak{O}(n)$ was given in 1897 by HURWITZ [1]. SCHUR and FROBENIUS in the years 1900—1920 made frequent use of averages over finite groups; for references, see the notes in WEYL [3]. SCHUR in [1] computed and applied intensively the invariant integrals for $\mathfrak{S}\mathfrak{O}(n)$ and $\mathfrak{O}(n)$. WEYL in [1] computed the invariant integrals for $\mathfrak{U}(n)$, $\mathfrak{O}(n)$, the unitary subgroup of the symplectic group, and [more or less explicitly] for certain other compact Lie groups. WEYL and PETER in [1] showed the existence of an invariant integral for any compact Lie group.

The decisive step in founding modern harmonic analysis was taken by A. HAAR [3] in 1933. He proved directly the existence [but not the uniqueness] of left Haar measure on a locally compact group with a countable open basis. His construction was reformulated in terms of linear functionals and extended to arbitrary locally compact groups by A. WEIL [1], [2], and [4], pp. 33—38. KAKUTANI [2] pointed out also that HAAR'S construction can be extended to all locally compact groups. Theorem (15.5) as stated is thus due to WEIL. The proof we present is due to H. CARTAN [1].

For an arbitrary compact group G, VON NEUMANN [5] proved the existence and uniqueness of the Haar integral, as well as its two-sided and inversion invariance. In [6], VON NEUMANN proved the uniqueness of left Haar measure for locally compact G with a countable open basis; a special case was also established by SZ.-NAGY [1]. WEIL [4], pp. 37—38, proved the uniqueness of the left Haar integral for all locally compact groups. A generalization was proved by RAĬKOV [2]; see also AUBERT [1]. All of these uniqueness proofs function only *ex post facto*, when the axiom of choice has already been invoked to prove existence. The virtue of CARTAN'S existence proof is that it demonstrates uniqueness and existence at one and the same time.

Theorem (15.9) is due to WEIL [4], p. 38; VON NEUMANN [3] pointed out that $\lambda(G)$ is finite if G is compact. It is interesting to note that WEYL [1], Kap. I, §5, considered an invariant measure of total measure 1 on the group $\mathfrak{S}\mathfrak{L}(n,R)$, and concluded that the existence of such a measure was doubtful, because $\mathfrak{S}\mathfrak{L}(n, R)$ is noncompact.

Theorem (15.11) appears in VON NEUMANN [6] and is attributed by VON NEUMANN to A. HAAR. Theorem (15.13) is due to VON NEUMANN [5], and Theorem (15.14) to WEIL [4], pp. 39—40. Example (15.17.g) is immensely popular; so far as we know, it was first computed by VON NEUMANN [6]. The other examples of (15.17) are widely known. It is difficult, as well as of little interest, to trace their history.

Explicit computation of the Haar integral for groups like $\mathfrak{S}\mathfrak{L}(n, R)$, $\mathfrak{S}\mathfrak{L}(n, K)$, $\mathfrak{U}(n)$, and $\mathfrak{Q}(n)$, which are not given as open subsets of some R^l [in which case the simple process of (15.17.e) does not apply] requires some sort of parametrization of the group. The computation, still in abstract form, is given for all Lie groups by CHEVALLEY [1], pp. 167—170. The Haar integral for $\mathfrak{U}(n)$ and other compact classical groups is described in WEYL [3], pp. 194—198, 218, 225. A description of the Haar integral for $\mathfrak{U}(n)$, the symplectic unitary group, and $\mathfrak{O}(n)$ is also given by TÔYAMA [1]. An illuminating discussion of the rôle of invariant integration in general appears in WEYL [3], pp. 185—194.

All of (15.18)—(15.24) is taken from WEIL [4], pp. 42—45; see also WEIL [1].

We now mention some further results which the reader interested in invariant measures may wish to look into. A. D. ALEKSANDROV [1] has sketched the construction and properties of a *finitely* additive analogue of Haar measure for groups that are not necessarily locally compact but only locally bounded. [A topological group G is *locally bounded* if there is a neighborhood U of e such that for every neighborhood V of e, a finite set $\{x_1, \ldots, x_n\} \subset G$ exists such that $\bigcup_{k=1}^{n} x_k V \supset U$.] OXTOBY [1] has made a careful examination of the existence and strange properties of countably additive invariant Borel measures on complete metric groups. H. F. DAVIS [1] has characterized Haar measure as a *measure-theoretic* product of n-dimensional Lebesgue measure, Haar measure on a compact group, and "counting" measure on a discrete set. LOOMIS [1] has studied invariant measures in uniform structures. ROSEN [1] and SCHWARZ [2] have examined invariant means on $\mathfrak{C}(S)$ for a compact *semi*group S.

§ 16. More about Haar measure

In this section, we present a remarkable theorem of KAKUTANI and OXTOBY on extensions of Haar measure and also some facts about nonmeasurable sets.

(16.1) Let G be any infinite compact metric group. Let λ be normalized Haar measure on G, and let $\mathscr{M}$ be the σ-algebra of λ-measurable subsets of G. By (4.26), we have $\overline{\overline{G}} \geqq \mathfrak{c}$. Since G contains a countable dense subset D and every element of G is the limit of a sequence of elements of D, we see that $\overline{\overline{G}} = \mathfrak{c}$. Throughout (16.1)–(16.11), G, λ, and $\mathscr{M}$ will be as defined here. We will have occasion to use the notation E^{ε}, where ε is either 1 or $'$: this will denote E or E', respectively.

It is known that there is no countably additive measure defined for *all* subsets of G, agreeing with λ on $\mathscr{M}$, and invariant under, say, left translations. [For a discussion, see (16.13).] Nevertheless, λ can be extended to a very much larger measure space.

(16.2) Definition. Let X be any set, $\mathscr{S}$ any σ-algebra of subsets of X, and μ any measure defined on $\mathscr{S}$. Then the *character* of the measure space $(X, \mathscr{S}, \mu)$ is the smallest cardinal number $\mathfrak{m}$ for which there is a subfamily $\mathscr{A}$ of $\mathscr{S}$ such that $\overline{\overline{\mathscr{A}}} = \mathfrak{m}$ and such that for every $S \in \mathscr{S}$ and every $\varepsilon > 0$, there exists a set $A \in \mathscr{A}$ satisfying $\mu(S \triangle A) < \varepsilon$. Any such subfamily $\mathscr{A}$ of $\mathscr{S}$ will be called a *basis* for $(X, \mathscr{S}, \mu)$.

We first note that the measure space $(G, \mathscr{M}, \lambda)$ has character $\aleph_0$. There is a countable open basis for the topology of G; let $\mathscr{A}$ consist of finite unions of sets from this countable open basis. If $M \in \mathscr{M}$ and $\varepsilon > 0$, then there is a compact set F and an open set U such that $F \subset M \subset U$

and $\lambda(U\cap F')<\varepsilon$. Since F is compact, there is a set $A\in\mathscr{A}$ such that $F\subset A\subset U$. Then clearly $\lambda(M\triangle A)<\varepsilon$. Since $\overline{\overline{\mathscr{A}}}=\aleph_0$, $(G,\mathscr{M},\lambda)$ has character $\aleph_0$. [It is easy to see that $(G,\mathscr{M},\lambda)$ cannot have finite character.]

Given a set X and two measure spaces $(X,\mathscr{M}_1,\mu_1)$ and $(X,\mathscr{M}_2,\mu_2)$, we say that the second is an *extension* of the first if $\mathscr{M}_2\supset\mathscr{M}_1$ and $\mu_2(A)=\mu_1(A)$ for all $A\in\mathscr{M}_1$.

We will prove the following assertion.

(16.3) Theorem. *There is an extension $(G,\mathscr{M}^*,\lambda^*)$ of $(G,\mathscr{M},\lambda)$ such that the character of $(G,\mathscr{M}^*,\lambda^*)$ is $2^{\mathfrak{c}}$ and λ^* is invariant under left translations, right translations, and inversion.*

The proof of this theorem is long, and is broken up into a sequence of lemmas. We shall repeatedly use well ordering and transfinite induction, but we do *not* use the continuum hypothesis. In order to avoid using the continuum hypothesis, we first extend $(G,\mathscr{M},\lambda)$ to a measure space $(G,\mathscr{M}^\dagger,\lambda^\dagger)$ such that every set $A\subset G$ for which $\overline{\overline{A}}<\mathfrak{c}$ belongs to $\mathscr{M}^\dagger$ and has $\lambda^\dagger$-measure zero. We will then extend $(G,\mathscr{M}^\dagger,\lambda^\dagger)$ to the desired measure space $(G,\mathscr{M}^*,\lambda^*)$.

Let $\mathfrak{T}$ consist of the transformations $x\to ax$, $x\to xa$, and $x\to x^{-1}$ of G onto itself $(a\in G)$. To check that a measure is [left and right] translation and inversion invariant, it suffices to check that if M is measurable, then each $\tau(M)$ is measurable $(\tau\in\mathfrak{T})$ and that the measures of M and $\tau(M)$ are equal.

(16.4) Lemma. *The measure space $(G,\mathscr{M},\lambda)$ can be extended to a measure space $(G,\mathscr{M}^\dagger,\lambda^\dagger)$ such that every set $A\subset G$ for which $\overline{\overline{A}}<\mathfrak{c}$ belongs to $\mathscr{M}^\dagger$ and has $\lambda^\dagger$-measure zero. Furthermore, $\lambda^\dagger$ is translation and inversion invariant and $(G,\mathscr{M}^\dagger,\lambda^\dagger)$ is complete; that is, if $N\in\mathscr{M}^\dagger$, $\lambda^\dagger(N)=0$, and $N_1\subset N$, then $N_1\in\mathscr{M}^\dagger$.*

Proof. Let $\mathscr{P}$ be the family of subsets P of G such that $\overline{\overline{P}}<\mathfrak{c}$. It is not known whether $\mathscr{P}\subset\mathscr{M}$; however, the continuum hypothesis obviously implies this inclusion, since $\mathscr{M}$ contains all countable sets. At any rate, if $P\in\mathscr{P}\cap\mathscr{M}$, then $\lambda(P)=0$.[1] We also have

$$P_1,P_2,\ldots,P_n,\ldots\in\mathscr{P} \text{ implies } \bigcup_{n=1}^{\infty}P_n\in\mathscr{P}. \tag{1}$$

[1] If $P\in\mathscr{M}$ and $\lambda(P)>0$, then P contains an uncountable compact set. Every compact subset F of G is the union of a countable set C and a perfect set F_0: let C be the set of all $x\in F$ such that $U\cap F$ is countable, for some neighborhood U of x. All nonvoid perfect subsets of compact metric spaces have cardinal number $\mathfrak{c}$. This is shown by the fact that every nonvoid perfect set F_0 contains a homeomorphic image of $\{0,1\}^{\aleph_0}$; such a subset of F_0 can be constructed by trivially modifying the proof of (4.26). The set F_0 plays the rôle of U, and each $U_{j_1\cdots j_m}$ is chosen to have diameter less than $1/m$. Then for any sequence $\{j_m\}_{m=1}^{\infty}$ of 0's and 1's, the set $\bigcap_{m=1}^{\infty}U_{j_1\cdots j_m}^{-}$ consists of exactly one point $x_{j_1\cdots j_m\cdots}$ and this point belongs to F_0. The set consisting of all of the points $x_{j_1\cdots j_m\cdots}$ is homeomorphic with $\{0,1\}^{\aleph_0}$.

For, KÖNIG's theorem [see HAUSDORFF [1], p. 34] states that if $\{\mathfrak{m}_\gamma\}_{\gamma\in\Gamma}$ and $\{\mathfrak{n}_\gamma\}_{\gamma\in\Gamma}$ are sets of cardinal numbers such that $\mathfrak{m}_\gamma < \mathfrak{n}_\gamma$ for all $\gamma\in\Gamma$, then $\sum_{\gamma\in\Gamma}\mathfrak{m}_\gamma < \prod_{\gamma\in\Gamma}\mathfrak{n}_\gamma$. Therefore, since $\overline{\overline{P}}_n < \mathfrak{c}$ for all n, we have $\overline{\overline{\bigcup_{n=1}^{\infty} P_n}} < \prod_{n=1}^{\infty}\mathfrak{c} = \mathfrak{c}$. That is, $\bigcup_{n=1}^{\infty} P_n \in \mathscr{P}$.

Let $\mathscr{M}^\dagger$ consist of all subsets $M^\dagger$ of G such that the symmetric difference $M^\dagger \triangle M$ is in $\mathscr{P}$ for some $M\in\mathscr{M}$. In other words, a set lies in $\mathscr{M}^\dagger$ if and only if it has the form $(M\cap A')\cup B$ where $M\in\mathscr{M}$ and $A, B\in\mathscr{P}$. We first verify that $\mathscr{M}^\dagger$ is a σ-algebra. If $M^\dagger\in\mathscr{M}^\dagger$ and $M^\dagger\triangle M\in\mathscr{P}$ where $M\in\mathscr{M}$, then $(M^\dagger)'\triangle M' = M^\dagger\triangle M\in\mathscr{P}$ so that $(M^\dagger)'\in\mathscr{M}^\dagger$. Also, if $\{M_n^\dagger\}_{n=1}^{\infty}\subset\mathscr{M}^\dagger$ and $M_n^\dagger\triangle M_n\in\mathscr{P}$, then $\left(\bigcup_{n=1}^{\infty} M_n^\dagger\right)\triangle\left(\bigcup_{n=1}^{\infty} M_n\right)\subset\bigcup_{n=1}^{\infty}(M_n^\dagger\triangle M_n)\in\mathscr{P}$ by (1). Thus $\bigcup_{n=1}^{\infty} M_n^\dagger\in\mathscr{M}^\dagger$.

For $M^\dagger\in\mathscr{M}^\dagger$ and $M^\dagger\triangle M\in\mathscr{P}$, $M\in\mathscr{M}$, we define

$$\lambda^\dagger(M^\dagger) = \lambda(M).$$

If $M^\dagger\triangle M_1\in\mathscr{P}$ and $M^\dagger\triangle M_2\in\mathscr{P}$, where M_1 and M_2 are in $\mathscr{M}$, then we have $(M_1\triangle M_2)\subset(M_1\triangle M^\dagger)\cup(M_2\triangle M^\dagger)\in\mathscr{P}$, so that $\lambda(M_1\triangle M_2)=0$ and $\lambda(M_1)=\lambda(M_2)$. Therefore $\lambda^\dagger$ is well defined on the σ-algebra $\mathscr{M}^\dagger$. To prove that $\lambda^\dagger$ is countably additive, suppose that $\{M_n^\dagger\}_{n=1}^{\infty}$ is a pairwise disjoint sequence in $\mathscr{M}^\dagger$ and that $M_n^\dagger\triangle M_n\in\mathscr{P}$ where $M_n\in\mathscr{M}$. Let $Q_1 = M_1$ and $Q_n = M_n\cap\left(\bigcap_{k=1}^{n-1} M_k'\right)$ for $n=2,3,\ldots$. Then $\{Q_n\}_{n=1}^{\infty}$ is a pairwise disjoint sequence in $\mathscr{M}$, $M_n^\dagger\triangle Q_n\subset\bigcup_{k=1}^{n}(M_k^\dagger\triangle M_k)\in\mathscr{P}$,[1] and $\left(\bigcup_{n=1}^{\infty} M_n^\dagger\right)\triangle\left(\bigcup_{n=1}^{\infty} Q_n\right)\in\mathscr{P}$. Thus $\lambda^\dagger\left(\bigcup_{n=1}^{\infty} M_n^\dagger\right)=\lambda\left(\bigcup_{n=1}^{\infty} Q_n\right)=\sum_{n=1}^{\infty}\lambda(Q_n)=\sum_{n=1}^{\infty}\lambda^\dagger(M_n^\dagger)$, and we have proved that $\lambda^\dagger$ is a measure on $\mathscr{M}^\dagger$.

We next show that $(G,\mathscr{M}^\dagger,\lambda^\dagger)$ is complete. Suppose that $N\in\mathscr{M}^\dagger$, $\lambda^\dagger(N)=0$, and $N_1\subset N$. Then $N\triangle M\in\mathscr{P}$ for some $M\in\mathscr{M}$ such that $\lambda(M)=0$. Hence $N_1\cap M\in\mathscr{M}$, $\lambda(N_1\cap M)=0$, and $N_1\triangle(N_1\cap M)\subset N\triangle M\in\mathscr{P}$, so that $N_1\in\mathscr{M}^\dagger$.

Finally, we check that $\lambda^\dagger$ is translation and inversion invariant on $\mathscr{M}^\dagger$. Clearly $P\in\mathscr{P}$ implies that $\tau(P)\in\mathscr{P}$ for all $\tau\in\mathfrak{T}$. Thus if $M^\dagger\in\mathscr{M}^\dagger$, $M^\dagger\triangle M\in\mathscr{P}$ and $M\in\mathscr{M}$, then $\tau(M^\dagger)\triangle\tau(M)=\tau(M^\dagger\triangle M)\in\mathscr{P}$. Consequently, $\tau(M^\dagger)\in\mathscr{M}^\dagger$ and $\lambda(\tau(M^\dagger))=\lambda(\tau(M))=\lambda(M)=\lambda(M^\dagger)$ for all $\tau\in\mathfrak{T}$. □

---

[1] To verify this inclusion, suppose that $x\in Q_n$ and $x\notin M_n^\dagger$; then $x\in M_n$, so that $x\in M_n^\dagger\triangle M_n$. Suppose that $x\in M_n^\dagger$ and $x\notin Q_n$. If $x\notin M_n$, then clearly $x\in M_n^\dagger\triangle M_n$. Otherwise we have $x\in M_k$ for some $k<n$. Since $x\in M_n^\dagger$ and $M_k^\dagger\cap M_n^\dagger=\varnothing$, we have $x\in(M_k^\dagger)'$, so that $x\in M_k^\dagger\triangle M_k$.

(16.5) Lemma. *The set G contains exactly $\mathfrak{c}$ distinct compact subsets having cardinal number $\mathfrak{c}$. Let $\omega_{\mathfrak{c}}$ denote the smallest ordinal having cardinal number $\mathfrak{c}$, and let $\{F_\alpha : 1 \leqq \alpha < \omega_{\mathfrak{c}}\}$ be any well ordering of all compact subsets of G having cardinal number $\mathfrak{c}$. Then for every $M^\dagger \in \mathscr{M}^\dagger$ such that $\lambda^\dagger(M^\dagger) > 0$ and for every ordinal number $\beta < \omega_{\mathfrak{c}}$, there is an ordinal α such that $\alpha > \beta$ and $F_\alpha \subset M^\dagger$.*

Proof. Every open subset of G is a union of sets in a countable open basis, and thus the number of compact subsets of G having cardinal number $\mathfrak{c}$ does not exceed $\mathfrak{c}$. In the next paragraph, we will see that G contains exactly $\mathfrak{c}$ compact sets F such that $\overline{\overline{F}} = \mathfrak{c}$.

To prove the lemma, it suffices to show that every $M^\dagger \in \mathscr{M}^\dagger$ with $\lambda^\dagger(M^\dagger) > 0$ contains $\mathfrak{c}$ compact sets having cardinal number $\mathfrak{c}$. Let $M \in \mathscr{M}$ have the property that $M^\dagger \triangle M \in \mathscr{P}$. Then we have $\lambda(M) > 0$ and $\overline{\overline{M^\dagger \triangle M}} < \mathfrak{c}$. It thus suffices to prove that M contains $\mathfrak{c}$ pairwise disjoint compact subsets each of cardinal number $\mathfrak{c}$, because then $\mathfrak{c}$ of these compact sets are necessarily disjoint from $M^\dagger \triangle M$ and are hence contained in $M^\dagger$. The set M contains an uncountable compact set, and therefore M contains a nonvoid perfect set. As shown in the proof of (16.4), this set contains a subset homeomorphic with $\{0, 1\}^{\aleph_0}$. This space is plainly homeomorphic with $\{0, 1\}^{\aleph_0} \times \{0, 1\}^{\aleph_0} = \bigcup \{\{0, 1\}^{\aleph_0} \times \{x\} : x \in \{0, 1\}^{\aleph_0}\}$; this is a compact space which is the union of $\mathfrak{c}$ pairwise disjoint compact subsets each having cardinal number $\mathfrak{c}$. □

(16.6) Definitions. For every subset A of G, we define the *outer $\lambda^\dagger$-measure* $\mu(A)$ of A to be

$$\inf\{\lambda^\dagger(M^\dagger) : M^\dagger \in \mathscr{M}^\dagger \text{ and } A \subset M^\dagger\}.$$

A subset A of G is said to be *absolutely invariant* [*with respect to $\lambda^\dagger$ and $\mathfrak{T}$*] if for every $\tau \in \mathfrak{T}$, $\tau(A) \triangle A \in \mathscr{M}^\dagger$ and $\lambda^\dagger(\tau(A) \triangle A) = 0$.

It is easy to verify that the family of absolutely invariant subsets of G is a σ-algebra of subsets of G [recall that $(G, \mathscr{M}^\dagger, \lambda^\dagger)$ is complete].

(16.7) Lemma. *There exists a family $\{X_\nu\}_{\nu \in \mathsf{N}}$ of subsets X_ν of G with the following properties:*

(i) $\overline{\overline{\mathsf{N}}} = \mathfrak{c}$;

(ii) *the sets X_ν, $\nu \in \mathsf{N}$, are pairwise disjoint;*

(iii) $\mu(X_\nu) = 1$ *for* $\nu \in \mathsf{N}$;

(iv) *for every subset N_0 of N, the union $\bigcup\limits_{\nu \in \mathsf{N}_0} X_\nu$ is absolutely invariant.*

Proof. Let $\{F_\alpha : 1 \leqq \alpha < \omega_{\mathfrak{c}}\}$ be as in (16.5), and let $\{\tau_\alpha : 1 \leqq \alpha < \omega_{\mathfrak{c}}\}$ be any well ordering of $\mathfrak{T}$ with order type $\omega_{\mathfrak{c}}$. For each $x \in G$ and each ordinal $\alpha < \omega_{\mathfrak{c}}$, let $C_\alpha(x)$ be the set of points of G that can be written as $\tau_{\beta_1}^{\varepsilon_1} \circ \cdots \circ \tau_{\beta_n}^{\varepsilon_n}(x)$ where $1 \leqq \beta_k \leqq \alpha$ and ε_k is 1 or -1, $k = 1, \ldots, n$, and

$n=1, 2, 3, \ldots$. Clearly we have $x \in C_\alpha(x)$ and for $x \in G$ and $1 \leqq \beta \leqq \alpha < \omega_{\mathfrak{c}}$, we have $\tau_\beta(C_\alpha(x)) = C_\alpha(x)$. We also have $\overline{\overline{C_\alpha(x)}} \leqq \max(\bar{\bar{\alpha}}, \aleph_0) < \mathfrak{c}$[1] and therefore $C_\alpha(x) \in \mathscr{M}^\dagger$.

We shall show that there exists a transfinite double sequence $\{x_\beta^\alpha : 1 \leqq \beta \leqq \alpha < \omega_{\mathfrak{c}}\}$ of elements of G such that:

$$x_\beta^\alpha \in F_\alpha \quad \text{for} \quad 1 \leqq \beta \leqq \alpha < \omega_{\mathfrak{c}}; \tag{1}$$

$$\text{the sets} \quad \{C_\alpha(x_\beta^\alpha) : 1 \leqq \beta \leqq \alpha < \omega_{\mathfrak{c}}\} \quad \text{are pairwise disjoint.} \tag{2}$$

If we agree that $(\gamma, \delta) < (\alpha, \beta)$ whenever $\gamma < \alpha$ or whenever $\gamma = \alpha$ and $\delta < \beta$ [lexicographic ordering], then $\{(\alpha, \beta) : 1 \leqq \beta \leqq \alpha < \omega_{\mathfrak{c}}\}$ is a well-ordered set. We shall define the sequence $\{x_\beta^\alpha : 1 \leqq \beta \leqq \alpha < \omega_{\mathfrak{c}}\}$ by transfinite induction. Let x_1^1 be an arbitrary point of F_1. Suppose that $1 \leqq \beta \leqq \alpha < \omega_{\mathfrak{c}}$ and that x_δ^γ has already been defined for all pairs $(\gamma, \delta) < (\alpha, \beta)$, $1 \leqq \delta \leqq \gamma$. Consider the union $D(\alpha, \beta)$ of the sets $C_\alpha(x_\delta^\gamma)$ as (γ, δ) runs over all pairs $(\gamma, \delta) < (\alpha, \beta)$. Then $\overline{\overline{D(\alpha, \beta)}} \leqq (\bar{\bar{\alpha}})^2 \cdot \max(\bar{\bar{\alpha}}, \aleph_0) = \max(\bar{\bar{\alpha}}, \aleph_0) < \mathfrak{c}$. Since $\overline{\overline{F_\alpha}} = \mathfrak{c}$, the set $F_\alpha \cap (D(\alpha, \beta))'$ is nonvoid. Let x_β^α be an arbitrary point of $F_\alpha \cap (D(\alpha, \beta))'$. Then $C_\alpha(x_\beta^\alpha)$ is disjoint from every $C_\gamma(x_\delta^\gamma)$, $(\gamma, \delta) < (\alpha, \beta)$. Otherwise we would have

$$\tau_{\beta_1}^{\varepsilon_1} \circ \cdots \circ \tau_{\beta_n}^{\varepsilon_n}(x_\beta^\alpha) = \tau_{\delta_1}^{\eta_1} \circ \cdots \circ \tau_{\delta_m}^{\eta_m}(x_\delta^\gamma)$$

where $\beta_k \leqq \alpha$, $\delta_j \leqq \gamma \leqq \alpha$, ε_k is 1 or -1, and η_j is 1 or -1, $k = 1, \ldots, n$, $j = 1, \ldots, m$. Hence

$$x_\beta^\alpha = \tau_{\beta_n}^{-\varepsilon_n} \circ \cdots \circ \tau_{\beta_1}^{-\varepsilon_1} \circ \tau_{\delta_1}^{\eta_1} \circ \cdots \circ \tau_{\delta_m}^{\eta_m}(x_\delta^\gamma) \in C_\alpha(x_\delta^\gamma) \subset D(\alpha, \beta);$$

and this contradicts the choice of x_β^α.

Now let $\mathsf{N} = \{\nu : \nu \text{ is an ordinal and } 1 \leqq \nu < \omega_{\mathfrak{c}}\}$ and let

$$X_\nu = \bigcup \{C_\alpha(x_\nu^\alpha) : \nu \leqq \alpha < \omega_{\mathfrak{c}}\}, \qquad \nu \in \mathsf{N}. \tag{3}$$

Property (i) is obvious and (ii) follows immediately from (2). To prove (iii), assume that $\mu(X_\nu) < 1$ for some $\nu \in \mathsf{N}$. Then there is an $M^\dagger \in \mathscr{M}^\dagger$ such that $X_\nu \subset M^\dagger$ and $\lambda^\dagger(M^\dagger) < 1$. Therefore Lemma (16.5) applies to the set $(M^\dagger)'$ and there exists an ordinal α, $\nu < \alpha < \omega_{\mathfrak{c}}$, such that $F_\alpha \subset (M^\dagger)'$. This is impossible since $x_\nu^\alpha \in F_\alpha$ by (1) and $x_\nu^\alpha \in C_\alpha(x_\nu^\alpha) \subset X_\nu \subset M^\dagger$ by (3). This proves (iii).

To prove (iv), it suffices to show that $\tau_\gamma(\bigcup_{\nu \in \mathsf{N}_0} X_\nu) \triangle (\bigcup_{\nu \in \mathsf{N}_0} X_\nu)$ has cardinal number less than $\mathfrak{c}$ for all $\gamma < \omega_{\mathfrak{c}}$ and $\mathsf{N}_0 \subset \mathsf{N}$. Since

$$\bigcup_{\nu \in \mathsf{N}_0} X_\nu = \bigcup \{C_\alpha(x_\nu^\alpha) : \nu \in \mathsf{N}_0, \ \nu \leqq \alpha < \omega_{\mathfrak{c}}\}$$

---

[1] If α is an ordinal, we write $\bar{\bar{\alpha}}$ for the cardinal number of the set of ordinals $\{\beta : 1 \leqq \beta \leqq \alpha\}$.

is a disjoint union by (2), and since $\tau_\gamma(C_\alpha(x_\nu^\alpha)) = C_\alpha(x_\nu^\alpha)$ whenever $\alpha \geqq \gamma$, we have

$$\tau_\gamma\Big(\bigcup_{\nu\in\mathsf{N}_0} X_\nu\Big) \triangle \Big(\bigcup_{\nu\in\mathsf{N}_0} X_\nu\Big) \subset \bigcup\{C_\alpha(x_\nu^\alpha) \cup \tau_\gamma(C_\alpha(x_\nu^\alpha)) : \nu\in\mathsf{N}_0,\ \nu\leqq\alpha<\gamma\}. \quad (4)$$

Since $\overline{\overline{C_\alpha(x_\nu^\alpha) \cup \tau_\gamma(C_\alpha(x_\nu^\alpha))}} \leqq \max(\overline{\overline{\alpha}}, \aleph_0)$ for $\nu \leqq \alpha$ and since $\overline{\overline{\{(\alpha,\nu):\nu\in\mathsf{N}_0,\ \nu\leqq\alpha<\gamma\}}} \leqq (\overline{\overline{\gamma}})^2$, the right side of (4) has cardinal number less than or equal to $\max(\overline{\overline{\gamma}}, \aleph_0)\cdot(\overline{\overline{\gamma}})^2 = \max(\overline{\overline{\gamma}}, \aleph_0) < \mathfrak{c}$. □

The following purely set-theoretical lemma is a special case of a lemma of A. Tarski [1].

(16.8) Lemma. *Let* $\mathsf{N} = \{\nu\}$ *be an arbitrary set having cardinal number* $\mathfrak{c}$. *Then there exists a family* $\{\mathsf{N}_\theta\}_{\theta\in\Theta}$ *of distinct subsets of* N *such that:*

(i) $\overline{\overline{\Theta}} = 2^{\mathfrak{c}}$;

(ii) *the set* $\bigcap_{n=1}^{\infty} \mathsf{N}_{\theta_n}^{\varepsilon_n}$ *is nonvoid for every sequence* $\{\theta_n\}_{n=1}^\infty$ *of distinct elements from* Θ *and every sequence* $\{\varepsilon_n\}_{n=1}^\infty$ *with* $\varepsilon_n = 1$ *or* $'$, $n = 1, 2, \ldots$.

Proof. It suffices to prove the lemma for some particular set N such that $\overline{\overline{\mathsf{N}}} = \mathfrak{c}$. Let Θ be the family of all subsets of $[0, 1[$; clearly $\overline{\overline{\Theta}} = 2^{\mathfrak{c}}$. For each $\theta\in\Theta$, let $B(\theta) = \theta \cup ([1, 2[\cap (\theta+1)')$; here $\theta+1 = \{x+1 : x\in\theta\}$. We have

$$\theta_1, \theta_2\in\Theta \quad\text{and}\quad \theta_1 \neq \theta_2 \quad\text{imply}\quad B(\theta_1)\cap B(\theta_2)' \neq \varnothing. \quad (1)$$

In fact, if $x\in\theta_1\cap\theta_2'$, then $x\in B(\theta_1)\cap B(\theta_2)'$, and if $x\in\theta_2\cap\theta_1'$, then $x+1\in B(\theta_1)\cap B(\theta_2)'$.

Let $\mathscr{C}$ be the family of all countable subsets of $[0, 2[$, and let N be the family of all countable subsets of $\mathscr{C}$. Then $\overline{\overline{\mathscr{C}}} = \overline{\overline{\mathsf{N}}} = \mathfrak{c}$. For each $\theta\in\Theta$, define $\mathscr{C}_\theta$ as the family of all countable subsets of $B(\theta)\subset[0, 2[$, and N_θ as the family of all countable subsets of $\mathscr{C}\cap(\mathscr{C}_\theta)'$. Then we obviously have $\mathscr{C}_\theta\subset\mathscr{C}$ and $\mathsf{N}_\theta\subset\mathsf{N}$ for $\theta\in\Theta$.

Let $\{\theta_n\}_{n=1}^\infty$ and $\{\varepsilon_n\}_{n=1}^\infty$ be given sequences as described in assertion (ii). Choose any element $\theta_0\in\Theta$ such that $\theta_0\neq\theta_n$ for $n=1, 2, \ldots$, and put ε_0 equal to 1 or $'$ so that $\varepsilon_0\neq\varepsilon_1$. Let I be the set of nonnegative integers i such that $\varepsilon_i = 1$ and let J be the set of nonnegative integers j such that $\varepsilon_j = '$. Then I and J are disjoint nonvoid sets whose union is $\{0, 1, 2, \ldots\}$. [The reason for adjoining θ_0 is to ensure that both I and J are nonvoid.] For every $i\in I$ and $j\in J$, we have $\theta_i\neq\theta_j$ and therefore $B(\theta_j)\cap B(\theta_i)'\neq\varnothing$ by (1). For every such pair $(i, j)\in I\times J$, choose an element x_{ij} in the set $B(\theta_j)\cap B(\theta_i)'$. For $j\in J$, let $C_j = \{x_{ij} : i\in I\}$ and let $\nu = \{C_j : j\in J\}$. Then obviously $C_j\in\mathscr{C}$ and $\nu\in\mathsf{N}$. For $i\in I$ and $j\in J$, we have $x_{ij}\in B(\theta_j)$ and so $C_j\subset B(\theta_j)$. Hence each C_j belongs to $\mathscr{C}_{\theta_j}$ and thus $\nu\notin\mathsf{N}_{\theta_j}$. That is,

$$\nu\in\bigcap_{j\in J}\mathsf{N}_{\theta_j}^{\varepsilon_j}. \quad (2)$$

Likewise, for $i\in I$ and $j\in J$, we have $x_{ij}\notin B(\theta_i)$ and so $C_j\not\subset B(\theta_i)$. Therefore $C_j\notin\mathscr{C}_{\theta_i}$ for all $j\in J$ and thus $\nu\subset\mathscr{C}\cap(\mathscr{C}_{\theta_i})'$. This implies that $\nu\in\mathsf{N}_{\theta_i}$ and, as i is arbitrary, it follows that $\nu\in\bigcap\limits_{i\in I}\mathsf{N}_{\theta_i}^{\varepsilon_i}$. From this and (2) we infer that $\bigcap\limits_{n=1}^{\infty}\mathsf{N}_{\theta_n}^{\varepsilon_n}\neq\varnothing$.

From (ii) it follows that the members of N are distinct. □

We are now in a position to state our crucial lemma, which gives us a family $\{E_\theta\}_{\theta\in\Theta}$ of $2^{\mathfrak{c}}$ subsets of G satisfying certain properties. Later in the construction we will find that the sets E_θ are all λ^*-measurable and that $\lambda^*(E_{\theta_1}\triangle E_{\theta_2})=\frac{1}{2}$ whenever $\theta_1\neq\theta_2$. This property implies immediately that the character of $(G,\mathscr{M}^*,\lambda^*)$ is $2^{\mathfrak{c}}$.

(16.9) Lemma. *There exists a family $\{E_\theta\}_{\theta\in\Theta}$ of distinct subsets of G such that*

(i) $\overline{\overline{\Theta}}=2^{\mathfrak{c}}$;

(ii) $\bigcap\limits_{n=1}^{\infty}E_{\theta_n}^{\varepsilon_n}$ *is absolutely invariant for every sequence $\{\theta_n\}_{n=1}^{\infty}$ of [not necessarily distinct] elements from Θ and every sequence $\{\varepsilon_n\}_{n=1}^{\infty}$ where ε_n is 1 or $'$, $n=1,2,\ldots$;*

(iii) $\mu\left(\bigcap\limits_{n=1}^{\infty}E_{\theta_n}^{\varepsilon_n}\right)=1$ *for every sequence $\{\theta_n\}_{n=1}^{\infty}$ of distinct elements from Θ and every sequence $\{\varepsilon_n\}_{n=1}^{\infty}$ where ε_n is 1 or $'$, $n=1,2,\ldots$.*

Proof. Let $\{X_\nu\}_{\nu\in\mathsf{N}}$ be the family of subsets of G obtained in (16.7), and let $\{\mathsf{N}_\theta\}_{\theta\in\Theta}$ be the family of subsets of N obtained in (16.8). For $\theta\in\Theta$, we define E_θ to be $\bigcup\{X_\nu:\nu\in\mathsf{N}_\theta\}$. The sets E_θ are all distinct because the sets X_ν are pairwise disjoint. Property (i) is exactly (16.8.i).

Each set E_θ is absolutely invariant by (16.7.iv). Property (ii) follows from this and the fact that the family of absolutely invariant subsets of G is a σ-algebra.

To prove (iii), we first choose any ν_0 from $\bigcap\limits_{n=1}^{\infty}\mathsf{N}_{\theta_n}^{\varepsilon_n}$ (16.8.ii). By definition we have $E_\theta=\bigcup\{X_\nu:\nu\in\mathsf{N}_\theta\}$, and since the sets $\{X_\nu\}_{\nu\in\mathsf{N}}$ are pairwise disjoint (16.7.ii), we also have $E_\theta'\supset\bigcup\{X_\nu:\nu\in\mathsf{N}_\theta'\}$. That is, $E_\theta^\varepsilon\supset\bigcup\{X_\nu:\nu\in\mathsf{N}_\theta^\varepsilon\}$ when ε is 1 or $'$. Hence we have the relations

$$\bigcap_{n=1}^{\infty}E_{\theta_n}^{\varepsilon_n}\supset\bigcap_{n=1}^{\infty}\left(\bigcup\{X_\nu:\nu\in\mathsf{N}_{\theta_n}^{\varepsilon_n}\}\right)\supset\bigcup\left\{X_\nu:\nu\in\bigcap_{n=1}^{\infty}\mathsf{N}_{\theta_n}^{\varepsilon_n}\right\}\supset X_{\nu_0}.$$

Since $\mu(X_{\nu_0})=1$ (16.7.iii), it follows that $\mu\left(\bigcap\limits_{n=1}^{\infty}E_{\theta_n}^{\varepsilon_n}\right)=1$. □

(16.10) Let $\{E_\theta\}_{\theta\in\Theta}$ be the family of subsets of G obtained in (16.9). Consider the family $\mathscr{E}$ of all subsets E of G having the form

(i) $E=\bigcup_{\{\varepsilon_1,\ldots,\varepsilon_n\}}\left(\left(\bigcap\limits_{k=1}^{n}E_{\theta_k}^{\varepsilon_k}\right)\cap M_{\varepsilon_1\cdots\varepsilon_n}^{\dagger}\right)$

where n is a positive integer, $\{\theta_1, \ldots, \theta_n\}$ is a finite subset of Θ, and for each sequence $\{\varepsilon_1, \ldots, \varepsilon_n\}$ with $\varepsilon_k = 1$ or $'$, $k = 1, \ldots, n$, $M^\dagger_{\varepsilon_1 \cdots \varepsilon_n}$ is an element of $\mathscr{M}^\dagger$. Here $\bigcup_{\{\varepsilon_1, \ldots, \varepsilon_n\}}$ means the union over the 2^n distinct sequences $\{\varepsilon_1, \ldots, \varepsilon_n\}$ where $\varepsilon_k = 1$ or $'$. Similarly $\Sigma_{\{\varepsilon_1, \ldots, \varepsilon_n\}}$ will denote the sum over all of the distinct sequences $\{\varepsilon_1, \ldots, \varepsilon_n\}$ [n is fixed].

(16.11) Lemma. *The family $\mathscr{E}$ is an algebra of subsets of G and $\mathscr{M}^\dagger \subset \mathscr{E}$. Let*

(i) $\lambda^*(E) = \frac{1}{2^n} \Sigma_{\{\varepsilon_1, \ldots, \varepsilon_n\}} \lambda^\dagger(M^\dagger_{\varepsilon_1 \cdots \varepsilon_n})$

whenever E has the form (16.10.i). *Then λ^* is well defined and finitely additive on $\mathscr{E}$ and $\lambda^*(M^\dagger) = \lambda^\dagger(M^\dagger)$ for all $M^\dagger \in \mathscr{M}^\dagger$. Finally, if $E \in \mathscr{E}$ and $\tau \in \mathfrak{T}$, then $\tau(E) \in \mathscr{E}$ and $\lambda^*(\tau(E)) = \lambda^*(E)$.*

Proof. We first observe that if E has the form

$$E = \bigcup_{\{\varepsilon_1, \ldots, \varepsilon_n\}} \left(\left(\bigcap_{k=1}^{n} E_{\theta_k}^{\varepsilon_k} \right) \cap M^\dagger_{\varepsilon_1 \cdots \varepsilon_n} \right), \tag{1}$$

and $\{\theta_{n+1}, \ldots, \theta_m\}$ is a finite subset of Θ disjoint from $\{\theta_1, \ldots, \theta_n\}$, then

$$E = \bigcup_{\{\varepsilon_1, \ldots, \varepsilon_m\}} \left(\left(\bigcap_{j=1}^{m} E_{\theta_j}^{\varepsilon_j} \right) \cap M^\dagger_{\varepsilon_1 \cdots \varepsilon_m} \right)$$

where $M^\dagger_{\varepsilon_1 \cdots \varepsilon_m}$ is defined to be $M^\dagger_{\varepsilon_1 \cdots \varepsilon_n}$ for all sequences $\{\varepsilon_1, \ldots, \varepsilon_m\}$ where $\varepsilon_j = 1$ or $'$. Moreover, we have

$$\frac{1}{2^n} \Sigma_{\{\varepsilon_1, \ldots, \varepsilon_n\}} \lambda^\dagger(M^\dagger_{\varepsilon_1 \cdots \varepsilon_n}) = \frac{1}{2^m} \Sigma_{\{\varepsilon_1, \ldots, \varepsilon_m\}} \lambda^\dagger(M^\dagger_{\varepsilon_1 \cdots \varepsilon_m}).$$

These observations imply that whenever in the present proof we have two sets E and F in $\mathscr{E}$, we may suppose that E has the form (1) and F has the form

$$F = \bigcup_{\{\varepsilon_1, \ldots, \varepsilon_n\}} \left(\left(\bigcap_{k=1}^{n} E_{\theta_k}^{\varepsilon_k} \right) \cap M^{\dagger\dagger}_{\varepsilon_1 \cdots \varepsilon_n} \right). \tag{2}$$

That is, the sets $\{\theta_1, \ldots, \theta_n\}$ in (1) and (2) are the same.

If $E \in \mathscr{E}$ has the form (1), then

$$E' = \bigcup_{\{\varepsilon_1, \ldots, \varepsilon_n\}} \left(\left(\bigcap_{k=1}^{n} E_{\theta_k}^{\varepsilon_k} \right) \cap (M^\dagger_{\varepsilon_1 \cdots \varepsilon_n})' \right) \in \mathscr{E}.$$

Also if E and F have the forms (1) and (2), respectively, then

$$E \cup F = \bigcup_{\{\varepsilon_1, \ldots, \varepsilon_n\}} \left(\left(\bigcap_{k=1}^{n} E_{\theta_k}^{\varepsilon_k} \right) \cap (M^\dagger_{\varepsilon_1 \cdots \varepsilon_n} \cup M^{\dagger\dagger}_{\varepsilon_1 \cdots \varepsilon_n}) \right) \in \mathscr{E}.$$

Hence $\mathscr{E}$ is an algebra of subsets of G. It is obvious that $\mathscr{M}^\dagger \subset \mathscr{E}$.

Suppose that E has the forms (1) and (2); *i.e.*, $E = F$. To prove that λ^* is well defined on $\mathscr{E}$, we need only to show that

$$\lambda^\dagger(M^\dagger_{\varepsilon_1 \cdots \varepsilon_n}) = \lambda^\dagger(M^{\dagger\dagger}_{\varepsilon_1 \cdots \varepsilon_n}) \tag{3}$$

for all sequences $\{\varepsilon_1, \ldots, \varepsilon_n\}$. From (1) and (2), we see that $\left(\bigcap_{k=1}^{n} E_{\theta_k}^{\varepsilon_k}\right) \cap M^{\dagger}_{\varepsilon_1 \cdots \varepsilon_n} = \left(\bigcap_{k=1}^{n} E_{\theta_k}^{\varepsilon_k}\right) \cap M^{\dagger\dagger}_{\varepsilon_1 \cdots \varepsilon_n}$. Consequently, we have

$$(M^{\dagger}_{\varepsilon_1 \cdots \varepsilon_n} \triangle M^{\dagger\dagger}_{\varepsilon_1 \cdots \varepsilon_n}) \subset \left(\bigcap_{k=1}^{n} E_{\theta_k}^{\varepsilon_k}\right)' . \tag{4}$$

Since $\mu\left(\bigcap_{k=1}^{n} E_{\theta_k}^{\varepsilon_k}\right) = 1$ (16.9.iii), every $\lambda^{\dagger}$-measurable subset of $\left(\bigcap_{k=1}^{n} E_{\theta_k}^{\varepsilon_k}\right)'$ has $\lambda^{\dagger}$-measure 0. Thus (4) implies (3).

We now show that λ^* is finitely additive on $\mathscr{E}$. Suppose that E and F are disjoint sets having the forms (1) and (2). Then $\left(\bigcap_{k=1}^{n} E_{\theta_k}^{\varepsilon_k}\right) \cap M^{\dagger}_{\varepsilon_1 \cdots \varepsilon_n} \cap M^{\dagger\dagger}_{\varepsilon_1 \cdots \varepsilon_n} = \varnothing$ so that $M^{\dagger}_{\varepsilon_1 \cdots \varepsilon_n} \cap M^{\dagger\dagger}_{\varepsilon_1 \cdots \varepsilon_n} \subset \left(\bigcap_{k=1}^{n} E_{\theta_k}^{\varepsilon_k}\right)'$. As in the previous paragraph, it follows that $\lambda^{\dagger}(M^{\dagger}_{\varepsilon_1 \cdots \varepsilon_n} \cap M^{\dagger\dagger}_{\varepsilon_1 \cdots \varepsilon_n}) = 0$. Then we have $\lambda^{\dagger}(M^{\dagger}_{\varepsilon_1 \cdots \varepsilon_n} \cup M^{\dagger\dagger}_{\varepsilon_1 \cdots \varepsilon_n}) = \lambda^{\dagger}(M^{\dagger}_{\varepsilon_1 \cdots \varepsilon_n}) + \lambda^{\dagger}(M^{\dagger\dagger}_{\varepsilon_1 \cdots \varepsilon_n})$, and it follows at once from (i) that $\lambda^*(E \cup F) = \lambda^*(E) + \lambda^*(F)$. The fact that $\lambda^*(M^{\dagger}) = \lambda^{\dagger}(M^{\dagger})$ for all $M^{\dagger} \in \mathscr{M}^{\dagger}$ is obvious.

Finally, let E have the form (1) and let τ be in $\mathfrak{T}$. Then

$$\tau(E) = \bigcup_{\{\varepsilon_1, \ldots, \varepsilon_n\}} \left(\tau\left(\bigcap_{k=1}^{n} E_{\theta_k}^{\varepsilon_k}\right) \cap \tau(M^{\dagger}_{\varepsilon_1 \cdots \varepsilon_n})\right).$$

If we let

$$E_0 = \bigcup_{\{\varepsilon_1, \ldots, \varepsilon_n\}} \left(\left(\bigcap_{k=1}^{n} E_{\theta_k}^{\varepsilon_k}\right) \cap \tau(M^{\dagger}_{\varepsilon_1 \cdots \varepsilon_n})\right),$$

we find that

$$\tau(E) \triangle E_0 \subset \bigcup_{\{\varepsilon_1, \ldots, \varepsilon_n\}} \left(\tau\left(\bigcap_{k=1}^{n} E_{\theta_k}^{\varepsilon_k}\right) \triangle \left(\bigcap_{k=1}^{n} E_{\theta_k}^{\varepsilon_k}\right)\right).$$

Each term of this union belongs to $\mathscr{M}^{\dagger}$ and has $\lambda^{\dagger}$-measure 0 by (16.9.ii). Thus $\tau(E) \triangle E_0 \in \mathscr{M}^{\dagger} \subset \mathscr{E}$ and obviously we have $E_0 \in \mathscr{E}$. Hence $\tau(E) \in \mathscr{E}$ [recall that $(G, \mathscr{M}^{\dagger}, \lambda^{\dagger})$ is complete]; then using (16.4) we obtain

$$\begin{aligned} \lambda^*(\tau(E)) &= \lambda^*(E_0) = \frac{1}{2^n} \Sigma_{\{\varepsilon_1, \ldots, \varepsilon_n\}} \lambda^{\dagger}(\tau(M^{\dagger}_{\varepsilon_1 \cdots \varepsilon_n})) \\ &= \frac{1}{2^n} \Sigma_{\{\varepsilon_1, \ldots, \varepsilon_n\}} \lambda^{\dagger}(M^{\dagger}_{\varepsilon_1 \cdots \varepsilon_n}) = \lambda^*(E). \quad \square \end{aligned}$$

Proof of Theorem (16.3). Let $\mathscr{M}^*$ be the σ-algebra of subsets of G generated by $\mathscr{E}$. To show that λ^* can be extended to a [countably additive] measure on $\mathscr{M}^*$, it suffices to prove that if $\{E_p\}_{p=1}^{\infty}$ is a decreasing sequence of sets in $\mathscr{E}$ and $\lambda^*(E_p) \geqq \alpha > 0$ for some positive number α and $p = 1, 2, \ldots$, then $\bigcap_{p=1}^{\infty} E_p \neq \varnothing$. [See Theorem 13.A, HALMOS [2]; the above property implies immediately that λ^* is a measure on the algebra $\mathscr{E}$ in HALMOS's sense.] The observations of the first paragraph in the proof of (16.11) show that we lose no generality if we suppose that we have a sequence $\{\theta_n\}_{n=1}^{\infty}$ of distinct elements of Θ, a strictly increasing sequence $\{n_p\}_{p=1}^{\infty}$ of positive integers, and elements $M^{\dagger}_{\varepsilon_1\cdots\varepsilon_{n_p}}$ in $\mathscr{M}^{\dagger}$ for each sequence $\{\varepsilon_1, \ldots, \varepsilon_{n_p}\}$ such that

$$E_p = \bigcup_{\{\varepsilon_1,\ldots,\varepsilon_{n_p}\}}\left(\left(\bigcap_{k=1}^{n_p} E_{\theta_k}^{\varepsilon_k}\right)\cap M^{\dagger}_{\varepsilon_1\cdots\varepsilon_{n_p}}\right). \tag{1}$$

If in (1), each set $M^{\dagger}_{\varepsilon_1\cdots\varepsilon_{n_p}}$ were replaced by the set $\bigcap_{q=1}^{p} M^{\dagger}_{\varepsilon_1\cdots\varepsilon_{n_q}}$, the equality would not be destroyed [recall that $\{E_p\}_{p=1}^{\infty}$ is a decreasing sequence of sets]. Hence we may also suppose that

$$M^{\dagger}_{\varepsilon_1\cdots\varepsilon_{n_{p+1}}} \subset M^{\dagger}_{\varepsilon_1\cdots\varepsilon_{n_p}} \tag{2}$$

for $p=1, 2, \ldots$ and every sequence $\{\varepsilon_1, \ldots, \varepsilon_{n_{p+1}}\}$. By hypothesis, we have $\lambda^*(E_p) = \frac{1}{2^{n_p}} \Sigma_{\{\varepsilon_1,\ldots,\varepsilon_{n_p}\}} \lambda^{\dagger}(M^{\dagger}_{\varepsilon_1\cdots\varepsilon_{n_p}}) \geqq \alpha$ for each p and therefore

$$\lambda^{\dagger}\left(M^{\dagger}_{\varepsilon_1^{(p)}\cdots\varepsilon_{n_p}^{(p)}}\right) \geqq \alpha, \tag{3}$$

for some sequence $\{\varepsilon_1^{(p)}, \ldots, \varepsilon_{n_p}^{(p)}\}$. There exists some sequence $\{\delta_1, \ldots, \delta_{n_1}\}$ where $\delta_k = 1$ or $'$ for which the set

$$Z_1 = \{q : \varepsilon_k^{(q)} = \delta_k \text{ for all } k = 1, \ldots, n_1\}$$

of positive integers is infinite. Having defined $\{\delta_1, \ldots, \delta_{n_{p-1}}\}$ and an infinite set Z_{p-1} of positive integers, we choose $\{\delta_{n_{p-1}+1}, \ldots, \delta_{n_p}\}$ in such a manner that

$$Z_p = \{q \in Z_{p-1} : q \geqq p \text{ and } \varepsilon_k^{(q)} = \delta_k \text{ for all } k = n_{p-1}+1, \ldots, n_p\}$$

is infinite. For $p = 1, 2, \ldots$, there is a $q \geqq p$ such that

$$\lambda^{\dagger}\left(M^{\dagger}_{\delta_1\cdots\delta_{n_p}}\right) = \lambda^{\dagger}\left(M^{\dagger}_{\varepsilon_1^{(q)}\cdots\varepsilon_{n_p}^{(q)}}\right) \geqq \lambda^{\dagger}\left(M^{\dagger}_{\varepsilon_1^{(q)}\cdots\varepsilon_{n_q}^{(q)}}\right) \geqq \alpha > 0, \tag{4}$$

the first inequality following from (2) and the second from (3). Since $\{M^{\dagger}_{\delta_1\cdots\delta_{n_p}}\}_{p=1}^{\infty}$ is a decreasing sequence of sets, we have

$$\lambda^{\dagger}\left(\bigcap_{p=1}^{\infty} M^{\dagger}_{\delta_1\cdots\delta_{n_p}}\right) \geqq \alpha > 0.$$

Since $\mu\left(\bigcap_{k=1}^{\infty} E_{\theta_k}^{\delta_k}\right) = 1$, it follows that $\bigcap_{p=1}^{\infty} M^{\dagger}_{\delta_1\cdots\delta_{n_p}}$ is not disjoint from $\bigcap_{k=1}^{\infty} E_{\theta_k}^{\delta_k}$. Thus we have

$$\bigcap_{p=1}^{\infty} E_p \supset \bigcap_{p=1}^{\infty}\left(\left(\bigcap_{k=1}^{n_p} E_{\theta_k}^{\delta_k}\right)\cap M^{\dagger}_{\delta_1\cdots\delta_{n_p}}\right) = \left(\bigcap_{k=1}^{\infty} E_{\theta_k}^{\delta_k}\right)\cap\left(\bigcap_{p=1}^{\infty} M^{\dagger}_{\delta_1\cdots\delta_{n_p}}\right) \neq \varnothing,$$

and we have shown that λ^* can be extended to a measure, also denoted by λ^*, on $\mathscr{M}^*$.

We now show that $(G, \mathscr{M}^*, \lambda^*)$ has character $2^{\mathfrak{c}}$. Let $\{E_\theta\}_{\theta\in\Theta}$ be as in (16.9). From (16.11.i) we have $\lambda^*(E_{\theta_1}\cap E'_{\theta_2}) = \frac{1}{4}$ and therefore $\lambda^*(E_{\theta_1}\triangle E_{\theta_2}) = \frac{1}{2}$ for $\theta_1, \theta_2 \in \Theta$ and $\theta_1 \neq \theta_2$. Suppose that $\mathscr{A} \subset \mathscr{M}^*$ is a basis for $(G, \mathscr{M}^*, \lambda^*)$. Then for every $\theta \in \Theta$, there is an $A_\theta \in \mathscr{A}$ such that

$\lambda^*(E_\theta \triangle A_\theta) < \frac{1}{4}$. If $\theta_1 \neq \theta_2$ and $A_{\theta_1} = A_{\theta_2}$, then $\frac{1}{2} = \lambda^*(E_{\theta_1} \triangle E_{\theta_2}) \leqq \lambda^*(E_{\theta_1} \triangle A_{\theta_1}) + \lambda^*(A_{\theta_2} \triangle E_{\theta_2}) < \frac{1}{2}$, which is impossible. Thus $\{A_\theta : \theta \in \Theta\}$ consists of distinct elements and $\mathscr{A}$ has cardinal number $2^{\mathfrak{c}}$.

It remains to prove that $M^* \in \mathscr{M}^*$ implies that $\tau(M^*) \in \mathscr{M}^*$ and $\lambda^*(\tau(M^*)) = \lambda^*(M^*)$ for all $\tau \in \mathfrak{T}$. Let $\mathscr{B}$ consist of those sets $B \in \mathscr{M}^*$ such that $\tau(B) \in \mathscr{M}^*$ and $\lambda^*(\tau(B)) = \lambda^*(B)$ for all $\tau \in \mathfrak{T}$. Since $\mathscr{E} \subset \mathscr{B}$ by (16.11), Theorem 6.B of HALMOS [2] shows that $\mathscr{B} = \mathscr{M}^*$ if $\mathscr{B}$ is a monotone class; that is, if $\mathscr{B}$ is closed under the formation of unions of increasing sequences of sets and of intersections of decreasing sequences of sets. If $\{B_n\}_{n=1}^\infty$ is an increasing sequence of sets in $\mathscr{B}$ and $\tau \in \mathfrak{T}$, then $\tau\left(\bigcup_{n=1}^\infty B_n\right) = \bigcup_{n=1}^\infty \tau(B_n) \in \mathscr{M}^*$ and

$$\lambda^*\left(\tau\left(\bigcup_{n=1}^\infty B_n\right)\right) = \lambda^*\left(\bigcup_{n=1}^\infty \tau(B_n)\right) = \lim_{n\to\infty} \lambda^*(\tau(B_n))$$
$$= \lim_{n\to\infty} \lambda^*(B_n) = \lambda^*\left(\bigcup_{n=1}^\infty B_n\right),$$

so that $\bigcup_{n=1}^\infty B_n \in \mathscr{B}$. Similarly $\bigcap_{n=1}^\infty B_n \in \mathscr{B}$ if $\{B_n\}_{n=1}^\infty$ is a decreasing sequence of sets in $\mathscr{B}$. □

Miscellaneous theorems and examples

(16.12) Character of a measure space and dimension of $\mathfrak{L}_2$. Let $(X, \mathscr{M}, \mu)$ be a measure space such that $\mu(X) = 1$. Let $\mathfrak{d}$ be the dimension of $\mathfrak{L}_2(X, \mathscr{M}, \mu)$ [1] [we will write $\mathfrak{L}_2$ for $\mathfrak{L}_2(X, \mathscr{M}, \mu)$] and let $\mathfrak{m}$ be the character of $(X, \mathscr{M}, \mu)$ in the sense of (16.2). If $\mathfrak{d}$ is finite, then $\mathfrak{m} = 2^{\mathfrak{d}}$. If $\mathfrak{d}$ is infinite, then $\mathfrak{m} = \mathfrak{d}$. [Suppose first that μ assumes only finitely many values; let α_1 be the least positive value of μ. Let E_1 be a set in $\mathscr{M}$ such that $\mu(E_1) = \alpha_1$. Then for every set $A \in \mathscr{M}$ such that $A \subset E_1$, we have $\mu(A) = \alpha_1$ or $\mu(A) = 0$. Let α_2 be the smallest positive value assumed by μ on subsets of E_1', if there are any such values. Note that $\alpha_2 \geqq \alpha_1$. Choosing $E_2 \subset E_1'$ such that $\mu(E_2) = \alpha_2$, we see that E_2 is like E_1; induction gives us a partition of X into sets $E_1, \ldots, E_n$, where $\mu(E_k) = \alpha_k$, and every subset A of E_k, $A \in \mathscr{M}$, has measure α_k or 0 $(k = 1, 2, \ldots, n)$. It is then easy to see that $\{\xi_{E_k}\}_{k=1}^n$ is an orthogonal basis in $\mathfrak{L}_2$, so that $\mathfrak{L}_2$ has dimension n. It is also easy to see that the character of the measure space $(X, \mathscr{M}, \mu)$ is 2^n. In fact, the sets $\varnothing$ and all unions $E_{k_1} \cup \cdots \cup E_{k_l}$ for nonvoid subsets $\{k_1, \ldots, k_l\}$ of $\{1, 2, \ldots, n\}$ form a basis; and no basis can have fewer elements. That is, $\mathfrak{m} = 2^{\mathfrak{d}} = 2^n$ in this case.

[1] The space $\mathfrak{L}_2(X, \mathscr{M}, \mu)$ is defined as usual as the normed linear space of all $\mathscr{M}$-measurable complex-valued functions f on X such that $\|f\|_2^2 = \int_X |f|^2 \, d\mu < \infty$; see §12 for a special case. With the inner product $\langle f, g\rangle = \int_X f \bar{g} \, d\mu$, $\mathfrak{L}_2(X, \mathscr{M}, \mu)$ is a Hilbert space.

Suppose next that μ assumes infinitely many values. Then there is an infinite family $\{A_n\}_{n=1}^{\infty}\subset\mathscr{M}$ of pairwise disjoint sets of positive measure: this is shown by an elementary argument, which we omit. The functions ξ_{A_n} $(n=1, 2, \ldots)$ are in $\mathfrak{L}_2$ and are plainly orthogonal, so that $\mathfrak{d}$ is infinite[1].

It remains to show that $\mathfrak{m}=\mathfrak{d}$ if $\mathfrak{d}$ is infinite. Let $\mathfrak{p}$ be the least cardinal number of a dense subset of $\mathfrak{L}_2$; it is easy to see that $\mathfrak{d}=\mathfrak{p}$. Let $\mathscr{A}$ be a basis for $(X,\mathscr{M},\mu)$ of cardinal number $\mathfrak{m}$. Functions of the form $\sum_{j=1}^{m}(\alpha_j+i\beta_j)\,\xi_{M_j}$ $(\alpha_j,\beta_j\in Q,\ M_j\in\mathscr{M})$ are dense in $\mathfrak{L}_2$, and every function ξ_B can be approximated arbitrarily in the $\mathfrak{L}_2$ metric by a function ξ_A with $A\in\mathscr{A}$, since $\|\xi_A-\xi_B\|_2=\mu(A\triangle B)^{\frac{1}{2}}$. This implies that $\mathfrak{m}\geqq\mathfrak{p}=\mathfrak{d}$. To show that $\mathfrak{m}\leqq\mathfrak{d}$, select a dense subset $\mathfrak{F}$ of $\mathfrak{L}_2$ such that $\overline{\overline{\mathfrak{F}}}=\mathfrak{d}$. For every $M\in\mathscr{M}$ and positive integer n, there is an $f\in\mathfrak{F}$ such that $\|f-\xi_M\|_2<\frac{1}{n}$. For every f and n for which an $M\in\mathscr{M}$ exists satisfying the last inequality, choose one such M; let $\mathscr{A}$ be the family of all the sets M so selected. It is obvious that $\overline{\overline{\mathscr{A}}}\leqq\aleph_0\cdot\mathfrak{d}=\mathfrak{d}$. Now, given $M\in\mathscr{M}$ and a positive integer n, there is an $f\in\mathfrak{F}$ such that $\|f-\xi_M\|_2<\frac{1}{n}$ and an $A\in\mathscr{A}$ such that $\|f-\xi_A\|_2<\frac{1}{n}$. Therefore, using MINKOWSKI's inequality (12.6), we have $\mu(A\triangle M)^{\frac{1}{2}}=\|\xi_A-\xi_M\|_2\leqq\|\xi_A-f\|_2+\|f-\xi_M\|_2<\frac{2}{n}$. That is, $\mathscr{A}$ is a basis for $(X,\mathscr{M},\mu)$.]

(16.13) Nonmeasurable sets. (a) Every nondiscrete locally compact group G contains a subset E that is nonmeasurable under every left translation invariant measure μ that is an extension of left Haar measure λ. [Let U be a neighborhood of e such that $\lambda(U)<\infty$ and let V be a symmetric neighborhood of e such that $V^3\subset U$. Let D be any countably infinite subset of V and let H be the subgroup of G generated by D. Clearly H is countable. Denote the set of distinct *right* cosets of H by $\{Hx_\alpha\}_{\alpha\in A}$; let $A_0=\{\alpha\in A:(Hx_\alpha)\cap V\neq\varnothing\}$. For $\alpha\in A_0$, choose exactly one element y_α in $(Hx_\alpha)\cap V$ and define E to be the set $\{y_\alpha:\alpha\in A_0\}$. Assume that E is μ-measurable. Since $H\cap V^2$ is countably infinite, the set $\bigcup\{xE:x\in H\cap V^2\}=(H\cap V^2)E$ is a disjoint union of a countably infinite number of left translates of E. It follows that $(H\cap V^2)E$ must have either μ-measure 0 or μ-measure ∞. Since $\lambda(V)>0$ and $\lambda(U)<\infty$, this assertion will be contradicted when we have established the following inclusions:

$$V\subset(H\cap V^2)E\subset U.$$

[1] For every cardinal number $\mathfrak{n}>0$ there is a compact group G, with Haar measure λ, such that $\dim\big(\mathfrak{L}_2(G,\lambda)\big)=\mathfrak{n}$. See (24.16) *infra*.

If $v \in V$, then $v \in Hx_\alpha = Hy_\alpha$ for some $\alpha \in A_0$; thus $v = hy_\alpha$ for some $h \in H$. Since $h = vy_\alpha^{-1} \in VV^{-1} \subset V^2$, we have $v = hy_\alpha \in (H \cap V^2) E$. The inclusions $(H \cap V^2) E \subset V^2 E \subset V^2 V \subset U$ are obvious.]

(b) Let G be a group, and let H be a subgroup of G [not necessarily normal] such that the [left] coset space G/H is countably infinite: $G = \bigcup_{n=1}^{\infty} a_n H$, where the sets $a_n H$ are pairwise disjoint. Assume that there is a left invariant measure μ on an algebra $\mathscr{A}$ of subsets of G such that $H \in \mathscr{A}$ and $\mu(G) = 1$. If $\mu(H)$ is 0, then we have $\mu(G) = \sum_{n=1}^{\infty} \mu(a_n H) = \sum_{n=1}^{\infty} \mu(H) = 0$. If $\mu(H)$ is positive, then we have $\mu(G) = \sum_{n=1}^{\infty} \mu(H) = \infty$. It is therefore impossible for H to belong to $\mathscr{A}$. In particular, if G is a compact group with normalized Haar measure λ, then H is not λ-measurable, and there is no left invariant extension of Haar measure under which H is measurable.

(c) Let G be an infinite Abelian group. Then G contains a subgroup H such that G/H is countably infinite. [Case I: G is a torsion group. Since G is the weak direct product of a countable number of primary groups (A.3), it suffices to prove our assertion for the case in which G is p-primary for some prime p. By (A.24), G contains a subgroup B such that G/B is divisible and B is isomorphic with a weak direct product of cyclic groups. If $B = G$, then our assertion clearly holds. If $B \neq G$, then G/B is a weak direct product of copies of $Z(p^\infty)$ (A.14). It is clear that G/B admits a subgroup J such that $(G/B)/J$ is countably infinite, and hence G does also.

Case II: G is not a torsion group. Let $a \in G$ have infinite order, let L be the group $\{a^n\}_{n=-\infty}^{\infty}$, and let M be a subgroup of G such that $M \cap L = \{e\}$ and M is maximal with respect to the property $M \cap L = \{e\}$. Then LM is a subgroup of G, and $G/(LM)$ is a torsion group. If $G/(LM)$ is countable, then G/M is countably infinite. If $G/(LM)$ is uncountable, we can apply Case I.]

(d) Every infinite compact Abelian group contains a subgroup H that is nonmeasurable under every translation invariant measure that is an extension of Haar measure. [Simply apply (b) and (c).]

(e) Every infinite compact metric group G contains $\mathfrak{c}$ pairwise disjoint subsets that are not measurable with respect to Haar measure. [Consider the $\mathfrak{c}$ sets X_ν constructed in (16.7), let μ be the set function of (16.6), and let λ denote the set function defined from the Haar integral as in (11.20). It is obvious from (11.22) and (16.6) that $\lambda(X_\nu) \geqq \mu(X_\nu)$ for all ν. By (16.7), we have $\lambda(X_\nu) = 1$ for all ν. If any X_{ν_0} were λ-measurable, we would have $\lambda(X_\nu) \leqq \lambda(X'_{\nu_0}) = 0$ for all $\nu \neq \nu_0$.]

(f) Relatively simple nonmeasurable sets can be obtained by resorting to large product spaces. Let Γ be an uncountable index set, and for each $\gamma\in\Gamma$, let G_γ be an infinite compact group. Let λ_γ denote normalized Haar measure on G_γ and let λ denote normalized Haar measure on $G=\underset{\gamma\in\Gamma}{\mathsf{P}}G_\gamma$. Then λ is the product measure (13.18) of the measures λ_γ (15.17.j). For each $\gamma\in\Gamma$, let A_γ be a proper λ_γ-measurable subset of G_γ such that $\lambda_\gamma(A_\gamma)=1$. Then $A=\underset{\gamma\in\Gamma}{\mathsf{P}}A_\gamma$ is nonmeasurable. [Assume that A is measurable. Let F be a compact subset of A, and for $\gamma\in\Gamma$, let E_γ be the projection of F into G_γ. Then E_γ is a compact subset of A_γ and $\lambda_\gamma(E_\gamma)<1$. Thus $\lambda\big(\underset{\gamma\in\Gamma}{\mathsf{P}}E_\gamma\big)=0$ (13.22), and since $F\subset\underset{\gamma\in\Gamma}{\mathsf{P}}E_\gamma$, we have $\lambda(F)=0$. Thus $\lambda(A)=0$ by (11.32).

The Baire sets of G are defined in (11.1). According to (19.30.b) *infra* [which is of course proved without recourse to the present discussion], there is a Baire set B such that $B\supset A$ and $\lambda(B)=\lambda(A)=0$. Since G is normal, the Baire sets in G are the smallest σ-algebra of subsets of G containing every open set that is the union of a countable family of closed sets. Now consider the subsets D of G for which there is a countable subset Γ_D of Γ such that

$$(x_\gamma)\in D,\quad (y_\gamma)\in G,\quad \text{and}\quad x_\gamma=y_\gamma\quad \text{for}\quad \gamma\in\Gamma_D\quad \text{imply}\quad (y_\gamma)\in D.$$

This family of sets is a σ-algebra containing every open subset of G that is a union of a countable family of closed sets, since such a set is a countable union of basic open sets. It follows that there is a countable subset Γ_0 of Γ such that

$$(x_\gamma)\in B,\quad (y_\gamma)\in G,\quad \text{and}\quad x_\gamma=y_\gamma\quad \text{for}\quad \gamma\in\Gamma_0\quad \text{imply}\quad (y_\gamma)\in B.$$

We infer from this and the inclusion $B\supset A$ that $B\supset\underset{\gamma\in\Gamma_0}{\mathsf{P}}A_\gamma\times\underset{\gamma\in\Gamma_0'}{\mathsf{P}}G_\gamma$. Theorem (13.22) now implies that $\lambda(B)=1$, which is a contradiction.]

(16.14) Locally null sets. Every nondiscrete locally compact topological group G that is not σ-compact contains a locally λ-null subset that is not λ-null; compare (11.33). [By (5.7), G contains an open and closed σ-compact subgroup H. Plainly the quotient space G/H is uncountable. Choose exactly one element from each left coset of H, and let A denote the set of elements so chosen. The set A is locally λ-null because the only compact subsets of A are finite and hence have λ-measure zero. We now show that $\lambda(A)=\infty$. If U is an open set containing A, then $\lambda\big((xH)\cap U\big)>0$ for all left cosets xH of H. Since G/H is uncountable, this implies that $\lambda(U)=\infty$ (11.25). Hence by (11.22), we have $\lambda(A)=\infty$.]

Notes

The Theorem (16.3) of KAKUTANI and OXTOBY [1] is remarkable in showing that Haar measure does not, so to say, tell the whole story regarding invariant integration on compact metric groups. This theorem opens the door to possible extensions of harmonic analysis; we will return to this matter at an appropriate place in the sequel.

There are many other important and interesting facts about invariant measures. Some of these are taken up in §§17 and 18. Both §17 and §18 deal with what may be called "finitely additive" situations. In §§15 and 16, we have treated only countably additive invariant measures, which are by far the most important invariant measures for harmonic analysis in its present form. In these sections, we have by no means exhausted the useful known facts about invariant measures[1]. Limitations of both space and time prevent us from treating them all.

Several topics are important enough for certain specialized applications to warrant at least mention. The first of these is A. WEIL'S converse to the construction of Haar measure, proved in WEIL [4], pp. 140—146. A slightly stronger version appears in HALMOS [2], pp. 266—276. The basic idea is simple. As is shown in (20.17) *infra*, every Borel set A of finite positive measure in a locally compact group G has the property that $A(A^{-1})$ contains a neighborhood of e. Now let G be a group with no topology but with a left invariant measure μ on a left invariant σ-ring $\mathscr{A}$ of subsets such that the mapping $(x,y)\rightarrow(x,xy)$ of $G\times G$ onto itself preserves measurability for the product measure $\mu\times\mu$,[2] and such that every A in $\mathscr{A}$ is σ-finite with respect to μ. Suppose also that for $x\in G$ and $x\neq e$, there is an $A\in\mathscr{A}$ such that $0<\mu(A)<\infty$ and $\mu\big((xA)\bigtriangleup A\big)>0$. Then taking sets $\{x\in G:\mu\big((xA)\bigtriangleup A\big)<\varepsilon\}$ as an open basis at e, one obtains a topology on G under which it is a topological group. There is a locally compact group G_0 in which G is a dense subgroup, and in which Haar measure is closely related to μ. For the details, we refer the reader to HALMOS, *loc. cit.* For groups in which the σ-ring $\mathscr{A}$ is a σ-algebra of a special sort, this result has been greatly sharpened by MACKEY [3]. The details are rather technical and will not be reproduced here.

There is an extensive literature on the subject of $\lambda(AB)$ for subsets A and B of a locally compact group. See for example KEMPERMAN [1], KNESER [1], MACBEATH [1]. Isolated but interesting theorems on Haar measure appear in ŚWIERCZKOWSKI [1] and [2], and URBANIK [1].

---

[1] Our selection of topics for §§16—18 was governed partly by sheer personal preference.

[2] Since G need not be locally compact and μ need not be as constructed in §11, we cannot define $\mu\times\mu$ by the procedure of §13. We must use the abstract construction found, for example, in HALMOS [2], ch. VII.

§ 17. Invariant means defined for all bounded functions

As is pointed out in Definition (15.2), the notion of invariance for a linear functional defined on a space of real or complex functions on a group need not be limited to the case of a locally compact group G and the Haar integrals defined on $\mathfrak{C}_{00}(G)$. In this section, we take up the interesting [and so far only partially solved] problem of finding the groups for which there is an invariant mean defined for *all* bounded functions on the group.

Many of our previous definitions given for groups also have meaning for semigroups. For example, if S is a semigroup, if a is in S, and f is a function on S with values in any nonvoid set, then ${}_af$ and f_a have exactly the same meanings as in Definition (15.1). In the cases where definitions are carried over *verbatim* from groups to semigroups, we will usually not spell out the semigroup definitions in detail. A number of the theorems of the present section can be stated and proved for semigroups just as easily as for groups. We shall make this extension where it is possible.

We proceed to exact definitions.

(17.1) Definition. Let X be any nonvoid set. Let $\mathfrak{B}(X)$ denote the set of all bounded complex-valued functions on X; $\mathfrak{B}^r(X)$ and $\mathfrak{B}^+(X)$ are defined as in (11.4). For $f\in\mathfrak{B}(X)$, let $\|f\|_u=\sup\{|f(x)|:x\in X\}$.

Clearly $\mathfrak{B}(X)$ and $\mathfrak{B}^r(X)$ are complex and real linear spaces, respectively, under pointwise addition and scalar multiplication. With the norm $\|\ \|_u$, $\mathfrak{B}(X)$ and $\mathfrak{B}^r(X)$ are obviously complex and real Banach spaces, respectively.

(17.2) Definition. Let X be any nonvoid set and let $\mathfrak{F}$ be a real linear subspace of $\mathfrak{B}^r(X)$. A *mean* M on $\mathfrak{F}$ is a real linear functional on $\mathfrak{F}$ having the property that

(i) $\inf\{f(x):x\in X\}\leqq M(f)\leqq\sup\{f(x):x\in X\}$ for all $f\in\mathfrak{F}$.

Note that if M is a real linear functional on $\mathfrak{F}$ and if $\mathfrak{F}$ contains the constant functions, then property (i) holds if and only if

$$M(f)\geqq 0 \quad \text{whenever} \quad f\in\mathfrak{F} \quad \text{and} \quad f\geqq 0 \tag{1}$$

and

$$M(1)=1. \tag{2}$$

(17.3) Definition. Let S be a semigroup and let $\mathfrak{F}$ be a linear subspace of $\mathfrak{B}^r(S)$ such that $x\in S$ and $f\in\mathfrak{F}$ imply ${}_xf\in\mathfrak{F}$ [$f_x\in\mathfrak{F}$]. A *left* [*right*] *invariant mean* M on $\mathfrak{F}$ is a mean such that $M({}_xf)=M(f)$ [$M(f_x)=M(f)$] for all $f\in\mathfrak{F}$ and $x\in S$. A mean M that satisfies $M({}_xf)=M(f_x)=M(f)$ for $f\in\mathfrak{F}$ and $x\in S$ is said to be a *two-sided invariant mean* or simply an *invariant mean*[1].

[1] Note the close connection with Definition (15.2).

(17.4) Theorem. *Let S be a semigroup and let $\mathfrak{F}$ be a linear subspace of $\mathfrak{B}^r(S)$ such that $x \in S$ and $f \in \mathfrak{F}$ imply ${}_x f \in \mathfrak{F}$. Let $\mathfrak{H}_l$ consist of all functions $h \in \mathfrak{F}$ having the form*

$$h = \sum_{k=1}^{n} [f_k - {}_{a_k} f_k]$$

for some $f_1, \ldots, f_n \in \mathfrak{F}$ and $a_1, \ldots, a_n \in S$. Then there exists a left invariant mean for $\mathfrak{F}$ if and only if

(i) $\sup\{h(x) : x \in S\} \geqq 0$ *for all $h \in \mathfrak{H}_l$.*

Suppose further that $x \in S$ and $f \in \mathfrak{F}$ imply $f_x \in \mathfrak{F}$. Let $\mathfrak{H}$ consist of all functions $h \in \mathfrak{F}$ having the form

$$h = \sum_{k=1}^{n} [f_k - {}_{a_k} f_k] + \sum_{j=1}^{m} [g_j - (g_j)_{b_j}],$$

for some $f_1, \ldots, f_n, g_1, \ldots, g_m \in \mathfrak{F}$ and $a_1, \ldots, a_n, b_1, \ldots, b_m \in S$. Then there exists a two-sided invariant mean for $\mathfrak{F}$ if and only if

(ii) $\sup\{h(x) : x \in S\} \geqq 0$ *for all $h \in \mathfrak{H}$.*

Proof. We will prove the second assertion; the proof of the first one is the same. If M is a two-sided invariant mean for $\mathfrak{F}$, then $M(h) = 0$ for all $h \in \mathfrak{H}$ and thus

$$\sup\{h(x) : x \in S\} \geqq M(h) = 0.$$

Suppose now that (ii) holds. Clearly $\mathfrak{H}$ is a linear subspace of $\mathfrak{F}$. For $h \in \mathfrak{H}$, we define

$$M_0(h) = 0.$$

Then clearly we have

$$M_0(h) = 0 \leqq \sup\{h(x) : x \in S\}$$

for $h \in \mathfrak{H}$. Letting $p(f) = \sup\{f(x) : x \in S\}$ for $f \in \mathfrak{F}$ and applying (B.13), we see that M_0 can be extended to a linear functional M on $\mathfrak{F}$ satisfying

$$M(f) \leqq \sup\{f(x) : x \in S\}$$

for $f \in \mathfrak{F}$. Also we have

$$-M(f) = M(-f) \leqq \sup\{-f(x) : x \in S\} = -\inf\{f(x) : x \in S\}$$

so that

$$\inf\{f(x) : x \in S\} \leqq M(f).$$

Finally if $a \in S$ and $f \in \mathfrak{F}$, then $f - {}_a f$ and $f - f_a$ belong to $\mathfrak{H}$ so that $M(f) = M({}_a f) = M(f_a)$. □

(17.5) Theorem. *Let S be a commutative semigroup. Then there is an invariant mean M on $\mathfrak{B}^r(S)$. [In particular, every linear subspace $\mathfrak{F}$ of $\mathfrak{B}^r(S)$ such that ${}_x f \in \mathfrak{F}$ whenever $x \in S$ and $f \in \mathfrak{F}$ has an invariant mean.]*

Proof. Let $f_1, \ldots, f_n$ be elements of $\mathfrak{B}^r(S)$, let $a_1, \ldots, a_n$ be elements of S, and let

$$h = \sum_{k=1}^{n} [f_k - {}_{a_k}f_k].$$

By virtue of (17.4), it suffices to prove that $\sup\{h(x) : x \in S\} \geqq 0$. Assume then that for some $\varepsilon > 0$, we have

$$\sup\{h(x) : x \in S\} = -\varepsilon. \tag{1}$$

Let p be any positive integer. Let Λ consist of all functions λ with domain $\{1, \ldots, n\}$ and range contained in $\{1, \ldots, p\}$. Evidently Λ contains exactly p^n elements. Let τ be the mapping of Λ into S defined by $\tau(\lambda) = a_1^{\lambda(1)} a_2^{\lambda(2)} \cdots a_n^{\lambda(n)}$. For a fixed k, $k = 1, \ldots, n$, we wish to estimate the sum

$$\sum_{\lambda \in \Lambda} [f_k(\tau(\lambda)) - f_k(a_k \tau(\lambda))]. \tag{2}$$

It is easy to see that all of the terms in (2) cancel each other except possibly those $f_k(\tau(\lambda))$ such that $\lambda(k) = 1$ and those $f_k(a_k \tau(\lambda))$ such that $\lambda(k) = p$. The number of these terms is $2p^{n-1}$. Therefore

$$\sum_{\lambda \in \Lambda} [f_k(\tau(\lambda)) - f_k(a_k \tau(\lambda))] \geqq -2p^{n-1} \|f_k\|_u.$$

Applying (1), we have

$$\begin{aligned} -\varepsilon p^n \geqq \sum_{\lambda \in \Lambda} h(\tau(\lambda)) &= \sum_{\lambda \in \Lambda} \sum_{k=1}^{n} [f_k(\tau(\lambda)) - f_k(a_k \tau(\lambda))] \\ &= \sum_{k=1}^{n} \sum_{\lambda \in \Lambda} [f_k(\tau(\lambda)) - f_k(a_k \tau(\lambda))] \\ &\geqq -\sum_{k=1}^{n} 2p^{n-1} A = -2np^{n-1} A, \end{aligned}$$

where $A = \max\{\|f_k\|_u : k = 1, \ldots, n\}$. Consequently we have $\varepsilon p \leqq 2nA$. Since p can be chosen arbitrarily, this leads to a contradiction. Hence (1) cannot hold and there is an invariant mean on $\mathfrak{B}^r(S)$. □

We now state an elementary but useful theorem.

(17.6) Theorem. *Let S be a semigroup and let $\mathfrak{F}$ be a linear subspace of $\mathfrak{B}^r(S)$. Suppose that $\mathscr{T}$ is a family of subsemigroups of S such that*

(i) *if $T_1, T_2 \in \mathscr{T}$, then there is a $T_3 \in \mathscr{T}$ such that $T_3 \supset T_1 \cup T_2$;*

(ii) $\cup\{T : T \in \mathscr{T}\} = S$.

For each $T \in \mathscr{T}$, let $\mathfrak{F}_T$ be the linear space of all functions in $\mathfrak{F}$ with their domains of definition restricted to T. If ${}_xf \in \mathfrak{F}$ whenever $x \in S$ and $f \in \mathfrak{F}$ and if there exists a left invariant mean on each $\mathfrak{F}_T$, then there exists a left invariant mean on $\mathfrak{F}$. If also $f_x \in \mathfrak{F}$ whenever $x \in S$ and $f \in \mathfrak{F}$ and if there exists a two-sided invariant mean on each $\mathfrak{F}_T$, then there exists a two-sided invariant mean on $\mathfrak{F}$.

Proof. We prove the first assertion; the proof of the second is similar. Let $f_1, \ldots, f_n \in \mathfrak{F}$, $a_1, \ldots, a_n \in S$, and consider $h = \sum_{k=1}^{n} [f_k - {}_{a_k}f_k]$. For some $T_0 \in \mathscr{T}$, we have $a_1, \ldots, a_n \in T_0$. By (17.4) applied to $\mathfrak{F}_{T_0}$, we see that $\sup\{h(x) : x \in T_0\} \geqq 0$. Then obviously $\sup\{h(x) : x \in S\} \geqq 0$. By (17.4) again, we see that there is a left invariant mean on $\mathfrak{F}$. ☐

Suppose that S is a *topological* semigroup, *i.e.*, a semigroup that is a topological space in which the mapping $(x, y) \rightarrow xy$ is continuous on $S \times S$ into S. Then $\mathfrak{C}^r(S)$ is plainly a translation invariant linear subspace of $\mathfrak{B}^r(S)$, so that (17.4)—(17.6) hold for $\mathfrak{F} = \mathfrak{C}^r(S)$.

We now list three corollaries of (17.6).

(17.7) Corollary. *Let S be a semigroup. There exists a [left, right, two-sided] invariant mean on $\mathfrak{B}^r(S)$ if there exists a [left, right, two-sided] invariant mean on $\mathfrak{B}^r(T)$ for all finitely generated subsemigroups T of S.*

(17.8) Corollary. *Let G be a group. There exists a two-sided invariant mean on $\mathfrak{B}^r(G)$ if there exists a two-sided invariant mean on $\mathfrak{B}^r(H)$ for all finitely generated subgroups H of G.*[1]

(17.9) Corollary. *Let G be a group such that every finite subset of G generates a finite group. Then $\mathfrak{B}^r(G)$ admits a two-sided invariant mean.*

Corollary (17.9) follows from the fact that if H is a finite group, then $\mathfrak{B}^r(H)$ has a two-sided invariant mean, namely the Haar integral.

(17.10) Theorem. *Let S be a semigroup and suppose that there exist left and right invariant means for $\mathfrak{B}^r(S)$. Then there exists a two-sided invariant mean for $\mathfrak{B}^r(S)$.*

Proof. Let M and M' be left and right invariant means, respectively, for $\mathfrak{B}^r(S)$. For $f \in \mathfrak{B}^r(S)$ and $x \in S$, let $f'(x) = M'({}_xf)$; clearly $f' \in \mathfrak{B}^r(S)$. Now let

$$M_0(f) = M(f').$$

Since the mapping $f \rightarrow f'$ is linear on $\mathfrak{B}^r(S)$, we see that M_0 is linear. Moreover $M_0(1) = 1$, and $M_0(f) \geqq 0$ whenever $f \geqq 0$. For $a, x \in S$, we have $({}_af)'(x) = M'({}_x({}_af)) = M'({}_{ax}f) = f'(ax) = {}_a(f')(x)$. Therefore

$$M_0({}_af) = M(({}_af)') = M({}_a(f')) = M(f') = M_0(f).$$

For $a, x \in S$, we have $(f_a)'(x) = M'({}_x(f_a)) = M'(({}_xf)_a) = M'({}_xf) = f'(x)$. It follows that

$$M_0(f_a) = M((f_a)') = M(f') = M_0(f).$$

Consequently M_0 is a two-sided invariant mean for $\mathfrak{B}^r(S)$. ☐

[1] We omit here the result for left and right invariant means, because of (17.11) [*q.v.*].

(17.11) Theorem. *Let G be a group and suppose that there exists a left invariant mean M for $\mathfrak{B}^r(G)$. Then there exists a two-sided invariant mean for $\mathfrak{B}^r(G)$.*

Proof. As in Definition (15.1), $f^\star$ denotes the function on G such that $f^\star(x)=f(x^{-1})$. For $f\in\mathfrak{B}^r(G)$, let $M'(f)=M(f^\star)$. Then M' is a right invariant mean for $\mathfrak{B}^r(G)$; now apply (17.10). □

(17.12) Theorem. *Let G be a group and let H be a subgroup of G. If $\mathfrak{B}^r(G)$ admits a two-sided invariant mean M, then so does $\mathfrak{B}^r(H)$.*

Proof. For each *right* coset Hx of H, let $\tau(Hx)$ be an arbitrary but fixed element of Hx. Every element x of G then has a unique representation $x=h_x\tau(Hx)$, where $h_x\in H$. For $f\in\mathfrak{B}^r(H)$ and $x\in G$, we define

$$f'(x)=f(h_x).$$

Clearly $f\to f'$ is a linear mapping of $\mathfrak{B}^r(H)$ into $\mathfrak{B}^r(G)$. For $f\in\mathfrak{B}^r(H)$, let $M_0(f)=M(f')$; obviously M_0 is linear, $M_0(1)=1$, and $M_0(f)\geqq 0$ whenever $f\geqq 0$. We will prove that M_0 is left invariant on $\mathfrak{B}^r(H)$. For $x\in G$ and $a\in H$, we have $ax=ah_x\tau(Hx)$ and also $ax=h_{ax}\tau(Hax)=h_{ax}\tau(Hx)$. Thus $ah_x=h_{ax}$, and for any $f\in\mathfrak{B}^r(H)$, the equalities

$$({}_af)'(x)={}_af(h_x)=f(ah_x)=f(h_{ax})=f'(ax)={}_a(f')(x)$$

hold. Thus we have

$$M_0({}_af)=M\big(({}_af)'\big)=M\big({}_a(f')\big)=M(f')=M_0(f).$$

Since M_0 is a left invariant mean for $\mathfrak{B}^r(H)$, Theorem (17.11) shows that there is a two-sided invariant mean for $\mathfrak{B}^r(H)$. □

(17.13) Corollary. *Let G be a group. Then $\mathfrak{B}^r(G)$ admits a two-sided invariant mean if and only if $\mathfrak{B}^r(H)$ admits a two-sided invariant mean for all finitely generated subgroups H of G.*

Proof. This follows from (17.8) and (17.12). □

(17.14) Theorem. *Let G be a group and let H be a normal subgroup of G. Then $\mathfrak{B}^r(G)$ admits an invariant mean if and only if both $\mathfrak{B}^r(H)$ and $\mathfrak{B}^r(G/H)$ admit invariant means*[1].

Proof. Suppose that M is an invariant mean for $\mathfrak{B}^r(G)$. By (17.12), $\mathfrak{B}^r(H)$ admits an invariant mean. Let φ denote the natural mapping of G onto G/H. For $f\in\mathfrak{B}^r(G/H)$, let $M_0(f)=M(f\circ\varphi)$. It is easy to show that M_0 is linear on $\mathfrak{B}^r(G/H)$, that $M_0(1)=1$, and that $M_0(f)\geqq 0$ whenever $f\geqq 0$. For $f\in\mathfrak{B}^r(G/H)$ and $x, y\in G$, we see that $\big(({}_{xH}f)\circ\varphi\big)(y)=f(xyH)=$

[1] Note the analogy between this theorem and Theorems (5.25) and (7.14).

${}_x(f\circ\varphi)(y)$ and consequently

$$M_0({}_{xH}f)=M\left(({}_{xH}f)\circ\varphi\right)=M\left({}_x(f\circ\varphi)\right)=M(f\circ\varphi)=M_0(f).$$

Similarly M_0 is right invariant on $\mathfrak{B}^r(G/H)$.

Next suppose that M and M' are invariant means for $\mathfrak{B}^r(H)$ and $\mathfrak{B}^r(G/H)$, respectively. For $f\in\mathfrak{B}^r(G)$ and $x\in G$, let $\bar{f}(x)=M({}_xf)$; here the domain of ${}_xf$ is restricted to H. We first prove that if x and y are in the same coset of H, then $\bar{f}(x)=\bar{f}(y)$. Thus suppose that $x=yh_0$ where $h_0\in H$. For $h\in H$, we have ${}_xf(h)=f(xh)=f(yh_0h)={}_{h_0}({}_yf)(h)$. Therefore $\bar{f}(x)=M({}_xf)=M\left({}_{h_0}({}_yf)\right)=M({}_yf)=\bar{f}(y)$. Since $\bar{f}$ is constant on the cosets of H, we may define the function f' on G/H by the rule $f'(xH)=\bar{f}(x)$. Finally, we set $M_0(f)=M'(f')$ for $f\in\mathfrak{B}^r(G)$. The linearity of M_0, the equality $M_0(1)=1$, and the nonnegativity of M_0 are obvious. For $a,\ x\in G$ and $f\in\mathfrak{B}^r(G)$, we obtain

$$({}_af)'(xH)=M\left({}_x({}_af)\right)=M({}_{ax}f)=f'(axH)={}_{aH}(f')(xH),$$

which implies that

$$M_0({}_af)=M'\left(({}_af)'\right)=M'(f')=M_0(f).$$

Thus M_0 is left invariant and, by (17.11), $\mathfrak{B}^r(G)$ has a two-sided invariant mean. □

(17.15) Theorem. *Let S be a semigroup and let $\mathfrak{H}_l$ consist of all $h\in\mathfrak{B}^r(S)$ having the form*

$$h=\sum_{k=1}^{n}[f_k-{}_{a_k}f_k]$$

for $f_1,\ldots,f_n\in\mathfrak{B}^r(S)$ and $a_1,\ldots,a_n\in S$. Then the following statements are equivalent:

(i) *there is a left invariant mean M on $\mathfrak{B}^r(S)$;*

(ii) $\inf\{\|1-h\|_u : h\in\mathfrak{H}_l\}=1$.

If S satisfies the left cancellation law $[xy=xz$ implies $y=z]$, then (i) *and* (ii) *are equivalent to*

(iii) *the uniform closure of $\mathfrak{H}_l$ is not equal to $\mathfrak{B}^r(S)$.*

Proof. Suppose that (i) holds. Then by (17.4), we have $\sup\{h(x):x\in S\}\geqq 0$ for all $h\in\mathfrak{H}_l$. This implies that $\|1-h\|_u\geqq 1$ for all $h\in\mathfrak{H}_l$, and hence that $\inf\{\|1-h\|_u : h\in\mathfrak{H}_l\}\geqq 1$. Since $\|1\|_u=1$, we infer that (ii) holds.

Suppose now that $\inf\{\|1-h\|_u : h\in\mathfrak{H}_l\}=1$. Then by (B.15), there is a linear functional M on $\mathfrak{B}^r(S)$ such that $M(1)=1$, $\|M\|=1$, and $M(h)=0$ for all $h\in\mathfrak{H}_l$. Recall that $\|M\|=\sup\{|M(f)| : \|f\|_u=1$ and $f\in\mathfrak{B}^r(S)\}$. Since $M(\mathfrak{H}_l)=0$, M is left invariant. To show that M is a mean, we need only show that $f\geqq 0$ implies $M(f)\geqq 0$. Assume that $f\geqq 0$ and $M(f)<0$. If

$g = \|f\|_u - f$, then $\|g\|_u \leqq \|f\|_u$ and $|M(g)| = \|f\|_u - M(f) > \|f\|_u \geqq \|g\|_u$. This implies that $\|M\| > 1$, contrary to the choice of M. Therefore M is a left invariant mean on $\mathfrak{B}^r(S)$.

Obviously (ii) implies (iii). Suppose finally that (iii) holds and that S satisfies the left cancellation law. Choose a function f_0 in $\mathfrak{B}^r(S) \cap (\mathfrak{H}_l^-)'$. By (B.15), there is a bounded linear functional L on $\mathfrak{B}^r(S)$ such that $L(f_0) = 1$ and $L(\mathfrak{H}_l^-) = 0$. By (B.37), L has the form $L_+ - L_-$, where $L_+ = \max(L, 0)$, and L_+ and L_- are nonnegative linear functionals. Note that by definition,

$$L_+(f) = \sup\{L(g) : 0 \leqq g \leqq f\} \tag{1}$$

for nonnegative f. We now prove that L_+ is left invariant; we already have $L({}_xf) = L(f)$ for all f since $L(\mathfrak{H}_l) = 0$. Suppose that $f \geqq 0$ and $x \in S$. Then $g \leqq f$ implies ${}_xg \leqq {}_xf$. Moreover, if $h \leqq {}_xf$, then $h = {}_xg$ for some $g \in \mathfrak{B}^r(S)$ such that $0 \leqq g \leqq f$. [To see this, note that every $y \in xS$ has the form xs, where s is unique by the left cancellation law. Then let $g(y) = h(s)$ for $y = xs$ and let $g(y) = 0$ for $y \in S \cap (xS)'$.] It follows from (1) that $L_+({}_xf) = L_+(f)$. Since every function in $\mathfrak{B}^r(S)$ is the difference of two nonnegative functions, we have $L_+({}_xf) = L_+(f)$ for all $f \in \mathfrak{B}^r(S)$. Since L and L_+ are left invariant, so is L_-. We have either $L_+(f_0) \neq 0$ or $L_-(f_0) \neq 0$, say $L_+(f_0) \neq 0$. Since $|L_+(f_0)| \leqq L_+(\|f_0\|_u) = \|f_0\|_u L_+(1)$, we have $L_+(1) > 0$. Defining $M(f) = \dfrac{L_+(f)}{L_+(1)}$, we obtain a left invariant mean for $\mathfrak{B}^r(S)$. □

We now prove that for some groups G, $\mathfrak{B}^r(G)$ admits no left invariant mean.

(17.16) Theorem. *Let G be a group containing a free subgroup having two generators. Then there is no left invariant mean on $\mathfrak{B}^r(G)$.*

Proof. In view of (17.11) and (17.12), it suffices to show that $\mathfrak{B}^r(F)$ admits no left invariant mean, where F is the free group having two generators a and b. Each element of F has a unique representation as a reduced word (2.8). Let A be the set of elements of F that begin with a or a^{-1} when written as reduced words, and let $f(x) = 1$ for $x \in A$ and $f(x) = 0$ for $x \notin A$. By (17.4), it suffices to show that the function

$$h = \left({}_{ba^{-1}}f - {}_{ab^{-1}a}({}_{ba^{-1}}f)\right) + \left((-f) - {}_{b^{-1}a^{-1}}(-f)\right)$$

has the property that $\sup\{h(x) : x \in F\} < 0$.

We have

$$h(x) = f(ba^{-1}x) + f(b^{-1}a^{-1}x) - f(ax) - f(x),$$

for $x \in F$. For all $x \in F$, either x or ax belongs to A. Therefore $h(x) \leqq -1$ except possibly in the cases that $ba^{-1}x$ or $b^{-1}a^{-1}x$ belongs to A. For one of these situations to hold, x must have the form $ab^{-1}a^{\varepsilon}y$ or $aba^{\varepsilon}y$ where

$\varepsilon = \pm 1$ and $a^{\varepsilon} y$ is reduced. In each of these cases, both x and ax belong to A whereas only one of the elements $ba^{-1}x$ and $b^{-1}a^{-1}x$ belongs to A. Thus in these cases we have $h(x) = -1$. Hence $\sup\{h(x): x \in F\} \leqq -1$. □

Miscellaneous theorems and examples

(17.17) (ROBISON [1].) Theorem (17.4) can be generalized as follows. Let X be any nonvoid set, $\mathfrak{F}$ any linear subspace of $\mathfrak{B}^r(X)$ containing some function $f \neq 0$, and $\mathscr{T}$ any family of linear transformations of $\mathfrak{F}$ into $\mathfrak{F}$. A necessary and sufficient condition that there exist a nonnegative linear functional M on $\mathfrak{F}$ satisfying

(i) $M(Tf) = M(f)$ for each $T \in \mathscr{T}$ and $f \in \mathfrak{F}$,

and

(ii) $\inf\{f(x): x \in X\} \leqq M(f) \leqq \sup\{f(x): x \in X\}$,

is that for every finite set of pairs (f_k, T_k), $f_k \in \mathfrak{F}$, $T_k \in \mathscr{T}$ $(k = 1, \ldots, n)$, we have

(iii) $\sup\left\{\sum_{k=1}^{n} (f_k - T_k f_k)(x): x \in X\right\} \geqq 0.$

(17.18) Invariant means for $\mathfrak{B}^r(S)$. (a) Let S and T be semigroups and let φ be a homomorphism of S onto T. If there is a left [two-sided] invariant mean for $\mathfrak{B}^r(S)$, there is also a left [two-sided] invariant mean for $\mathfrak{B}^r(T)$. [If M is the mean for $\mathfrak{B}^r(S)$, and $f \in \mathfrak{B}^r(T)$, define $M_0(f) = M(f \circ \varphi)$ and show that M_0 is the desired mean for $\mathfrak{B}^r(T)$.]

(b) Let S and T be semigroups and suppose that M and M' are left [two-sided] invariant means for $\mathfrak{B}^r(S)$ and $\mathfrak{B}^r(T)$, respectively. Then $\mathfrak{B}^r(S \times T)$ also admits a left [two-sided] invariant mean[1]. [For $f \in \mathfrak{B}^r(S \times T)$, let

$$M_0(f) = M_s(M'_t(f(s, t)))$$

where M_s [M'_t] indicates that the variable is $s \in S$ [$t \in T$]. It is evident that M_0 is well defined, linear, and nonnegative. Also we have $M_0(1) = 1$. Suppose that M and M' are left invariant. Then if $(a, b) \in S \times T$, we have

$$M_0({}_{(a,b)}f) = M_s(M'_t(f(as, bt))) = M_s(M'_t(f(s, t))) = M_0(f)$$

by applying the left invariance of M' and then that of M.]

(c) Suppose that $\{S_\iota\}_{\iota \in I}$ is a family of semigroups each having an identity e_ι and let $S = \mathsf{P}^*_{\iota \in I} S_\iota$. If each $\mathfrak{B}^r(S_\iota)$ admits a left [two-sided] invariant mean, then so does $\mathfrak{B}^r(S)$. [Each S_ι may be regarded as a

[1] The definition of the direct product of a family of semigroups is exactly like that of groups (2.3). If we choose a fixed element in each of a family of semigroups [the identity if there is one], then the weak direct product of semigroups can also be defined by analogy with (2.3).

subsemigroup of S in a natural way. In fact, $x \in S_{\iota_0}$ corresponds to the element $s \in S$, where

$$s_\iota = \begin{cases} x & \text{if} \quad \iota = \iota_0, \\ e_\iota & \text{if} \quad \iota \neq \iota_0. \end{cases}$$

By (b), all finite products $\mathop{\mathsf{P}}_{\iota \in I_0} S_\iota$ have the property that $\mathfrak{B}^r(\mathop{\mathsf{P}}_{\iota \in I_0} S_\iota)$ admits an invariant mean. If we let $\mathcal{T}$ consist of all such $\mathop{\mathsf{P}}_{\iota \in I_0} S_\iota$, then (17.6) shows that $\mathfrak{B}^r(S)$ admits an invariant mean.]

(d) The result of (c) does not necessarily hold if we take S to be $\mathop{\mathsf{P}}_{\iota \in I} S_\iota$ rather than $\mathop{\mathsf{P}^*}_{\iota \in I} S_\iota$. The analogous statement for groups does not even hold. [Let F be the free group generated by a and b with identity $\boldsymbol{e}$, and let $F_0 = F \cap \{\boldsymbol{e}\}'$. Let $\boldsymbol{x} = x_1^{\varepsilon_1} x_2^{\varepsilon_2} \cdots x_n^{\varepsilon_n}$ be any reduced word in F_0; each x_k is a or b and each ε_k is 1 or -1. By (2.9), there exist an integer l and elements P_a and P_b in the permutation group $\mathfrak{S}_l$ such that $P_{x_1}^{\varepsilon_1} \circ P_{x_2}^{\varepsilon_2} \circ \cdots \circ P_{x_n}^{\varepsilon_n}$ is different from the identity permutation. We define $G_{\boldsymbol{x}}$ as $\mathfrak{S}_l$, $a_{\boldsymbol{x}}$ as P_a, and $b_{\boldsymbol{x}}$ as P_b. Having made these definitions for each $\boldsymbol{x} \in F_0$, we define G to be $\mathop{\mathsf{P}}_{\boldsymbol{x} \in F_0} G_{\boldsymbol{x}}$. Since each $G_{\boldsymbol{x}}$ is finite, $\mathfrak{B}^r(G_{\boldsymbol{x}})$ admits a two-sided invariant mean for all $\boldsymbol{x} \in F_0$. The elements $(a_{\boldsymbol{x}})$ and $(b_{\boldsymbol{x}})$ of G generate a free group. That is, G contains a free group on two generators. Thus by (17.16), $\mathfrak{B}^r(G)$ does not admit a left invariant mean. Note the analogy between this result and Theorem (8.8).]

(e) Let S be any semigroup whatever; $\mathfrak{B}^r(S)$ may or may not admit invariant means. Let $S^* = S \cup \{0\}$ where $s0 = 0s = 0$ for all $s \in S^*$. Then $\mathfrak{B}^r(S^*)$ admits a two-sided invariant mean. [For $f \in \mathfrak{B}^r(S^*)$, let $M(f) = f(0)$.]

(f) (Day [5].) In view of (e), the following result is interesting. Let S be a semigroup and let M be a left invariant mean for $\mathfrak{B}^r(S)$. Suppose that T is a subsemigroup of S such that $M(\xi_T) > 0$, where ξ_T is as usual the characteristic function of T. Then $\mathfrak{B}^r(T)$ admits a left invariant mean. [For $f \in \mathfrak{B}^r(T)$, let $f' \in \mathfrak{B}^r(S)$ be defined by $f'(x) = f(x)$ for $x \in T$ and $f'(x) = 0$ for $x \in T'$, and let $M_0(f) = \dfrac{M(f')}{M(\xi_T)}$. Then M_0 is a mean on $\mathfrak{B}^r(T)$. To show that M_0 is left invariant, let $x \in T$, $f \in \mathfrak{B}^r(T)$, and consider $g = ({}_x f)' - {}_x(f')$. It is easy to see that $g = g\xi_E$ where $E = \{y \in S : y \notin T$ and $xy \in T\}$. For any $s \in S$, one can show easily that at most one of the elements $\{x^k s\}_{k=1}^{\infty}$ belongs to E. Hence for every positive integer n, we have

$$\sum_{k=1}^{n} {}_{x^k}(\xi_E)(s) \leqq 1 \quad \text{for all} \quad s \in S.$$

Thus $n M(\xi_E) = M\left(\sum_{k=1}^{n} {}_{x^k}(\xi_E)\right) \leqq M(1) < \infty$, so that $M(\xi_E) = 0$. Now we have $|M(g)| = |M(g\xi_E)| \leqq M(\|g\|_u \xi_E) = 0$, and so $M(g) = 0$. We have thus shown that $M(({}_x f)') = M({}_x(f'))$ for all $x \in T$ and $f \in \mathfrak{B}^r(T)$. This implies immediately that M_0 is left invariant.]

(17.19) Invariant means for $\mathfrak{B}^r(G)$. (a) The two-sided invariant mean for $\mathfrak{B}^r(G)$ in (17.11) can be taken to be inversion invariant. [Let M be the two-sided invariant mean and for $f\in\mathfrak{B}^r(G)$, define $M_0(f)=\frac{1}{2}[M(f)+M(f^*)]$.]

(b) Let G be the group generated by a and b and satisfying the relations $a^2=b^2=e$ and no others. Then $a^{-1}=a$, $b^{-1}=b$, and $G=\{e, a, ab, aba, \ldots, b, ba, bab, \ldots\}$. Let H consist of all of the elements of even length; that is, $H=\{e, ab, abab, \ldots, ba, baba, \ldots\}$. Then H is an infinite cyclic normal subgroup of G and G/H consists of two elements. Thus $\mathfrak{B}^r(H)$ and $\mathfrak{B}^r(G/H)$ admit invariant means. By (17.14), $\mathfrak{B}^r(G)$ also admits invariant means.

There is a left invariant mean on $\mathfrak{B}^r(G)$ that is not right invariant. [Let A consist of all elements in G ending in a; that is, $A=\{a, ba, aba, baba, \ldots\}$. For $p<q$, let $A^{(p,q)}$ be the set of elements $x\in A$ such that $p\leqq$ length $x\leqq q$. Consider an arbitrary element y of G whose length is m and an $x\in A$ whose length is $j>m$. Then yx is also in A and it is either the element of length $j-m$ or the element of length $j+m$. Thus if p and q satisfy $m<p$ and $p+2m<q$, we see that $y^{-1}A^{(p+m,q-m)}\subset A^{(p,q)}$, so that $yA^{(p,q)}$ contains the set $A^{(p+m,q-m)}$. Thus if $f\in\mathfrak{B}^r(G)$ and we consider the sum $\sum_{x\in A^{(p,q)}}[f(x)-f(yx)]$, we see that all except at most $4m$ terms cancel. It follows that

$$\sum_{x\in A^{(p,q)}}[f(x)-f(yx)]\leqq 4m\,\|f\|_u. \tag{1}$$

Consider $h=\sum_{k=1}^{n}[f_k-{}_{y_k}f_k]$ where $f_1,\ldots,f_n\in\mathfrak{B}^r(G)$ and $y_1,\ldots,y_n\in G$. We assert that

$$\inf\{h(x):x\in A\}\leqq 0. \tag{2}$$

Assume that $\inf\{h(x):x\in A\}=\alpha>0$. Let m_0 be the maximum of the lengths of $y_1,\ldots,y_n$ and let $B=\max\{\|f_1\|_u,\ldots,\|f_n\|_u\}$. Choose p and q so that $m_0<p$ and $p+2m_0<q$. Now apply (1): this yields the relations

$$\begin{aligned}\alpha(q-p+1)&\leqq\sum_{x\in A^{(p,q)}}\sum_{k=1}^{n}[f_k(x)-f_k(y_kx)]\\&=\sum_{k=1}^{n}\sum_{x\in A^{(p,q)}}[f_k(x)-f_k(y_kx)]\leqq\sum_{k=1}^{n}4m_0B=4nm_0B.\end{aligned}$$

Since $q-p$ can be taken arbitrarily large, we have a contradiction. Thus (2) holds.

The function $\xi_A-(\xi_A)_a$ is clearly equal to 1 on the set A. Hence by (2), it is not in the uniform closure of all functions of the form $h=\sum_{k=1}^{n}[f_k-{}_{y_k}f_k]$. An argument like that used in proving that (17.15.iii) implies (17.15.i) now shows that there is a left invariant mean M on

$\mathfrak{B}_r(G)$ such that $M((\xi_A)-(\xi_A)_a)\neq 0$. This mean is obviously not right invariant.]

(c) Let G be a noncompact locally compact topological group and suppose that M is a mean for $\mathfrak{B}^r(G)$. Then $M(f)=0$ for all functions f that are arbitrarily small outside of compact sets. [As in (15.9), there is an open set V in G and a sequence $x_1, \ldots, x_n, \ldots$ in G such that V^- is compact and $\{x_k V\}_{k=1}^{\infty}$ is a pairwise disjoint sequence of sets. Then $\sum_{k=1}^{n} \xi_{x_k V} = \sum_{k=1}^{n} {}_{x_k^{-1}}(\xi_V) \leqq 1$ for all n so that $nM(\xi_V) \leqq M(1)$. Thus $M(\xi_V)=0$. Now suppose that $|f(x)|<\varepsilon$ for $x\in F'$, where F is compact. Then for some $y_1, \ldots, y_m \in G$ we have $\bigcup_{j=1}^{m} y_j V \supset F$. Thus we have $|f| \leqq \|f\|_u \sum_{j=1}^{m} \xi_{y_j V} + \varepsilon$ and hence $|M(f)| \leqq M(\varepsilon)=\varepsilon$.]

(17.20) Invariant means for $\mathfrak{B}^r(R)$. (a) Let M be an invariant mean for $\mathfrak{B}^r(R)$. Suppose that the limits $\lim_{t\to\infty} f(t)$ and $\lim_{t\to-\infty} f(t)$ exist. Then the number $M(f)$ lies between these two limits. In particular, if the above limits are equal, then $M(f)$ is equal to the common limit. [Let $g(x)=\lim_{t\to\infty} f(t)$ for $x\geqq 0$ and $g(x)=\lim_{t\to-\infty} f(t)$ for $x<0$. Then $\lim_{t\to\infty}(g(t)-f(t)) = \lim_{t\to-\infty}(g(t)-f(t))=0$, and so (17.19.c) implies that $M(g-f)=0$. Since $M(f)=M(g)$, it is clear that $M(f)$ lies between $\lim_{t\to\infty} f(t)$ and $\lim_{t\to-\infty} f(t)$.]

(b) Theorems (17.5) and (17.4) show us that $\sup\{h(x): x\in R\}\geqq 0$ for all $h=\sum_{k=1}^{n} [f_k - {}_{a_k}f_k]$ where $f_k\in\mathfrak{B}^r(R)$ and $a_k\in R$ $(k=1, \ldots, n)$. By modifying the proof of (17.5), one can prove that $\sup\{h(x): x\geqq N\}\geqq 0$ for any positive number N. [We may suppose that each a_k is positive; otherwise replace $f_k - {}_{a_k}f_k$ by $(-{}_{a_k}f_k) - {}_{-a_k}(-{}_{a_k}f_k)$. Let p be any positive integer and let Λ consist of all functions λ with domain $\{1, \ldots, n\}$ and range contained in $\{1, \ldots, p\}$. Let τ be the mapping of Λ into $[N, \infty[$ defined by $\tau(\lambda)=N+\sum_{k=1}^{n}\lambda(k)\, a_k$. Assume that $\sup\{h(x): x\geqq N)=\alpha<0$. Using this and taking into account the terms that cancel, we see that

$$\alpha p^n \geqq \sum_{\lambda\in\Lambda}\sum_{k=1}^{n}[f_k(\tau(\lambda)) - f_k(a_k+\tau(\lambda))]$$
$$\geqq \sum_{k=1}^{n} -2p^{n-1}\max\{\|f_1\|_u, \ldots, \|f_n\|_u\} = -2np^{n-1}\max\{\|f_1\|_u, \ldots, \|f_n\|_u\}.$$

Since p is arbitrary, the inequality $\alpha<0$ is impossible.]

(c) There is an invariant mean M for $\mathfrak{B}^r(R)$ such that $M(f)=\lim_{t\to\infty} f(t)$ whenever this limit exists. [Let $\mathfrak{H}_0$ consist of those functions h in $\mathfrak{B}^r(R)$ of the form

$$h=g+\sum_{k=1}^{n}[f_k - {}_{a_k}f_k], \tag{1}$$

where $g, f_1, \ldots, f_n \in \mathfrak{B}^r(R)$, $a_1, \ldots, a_n \in R$, and $\lim\limits_{t\to\infty} g(t)=0$. Clearly $\mathfrak{H}_0$ is a linear subspace of $\mathfrak{B}^r(R)$. Let h be as in (1) and choose N so large that

$$|g(t)| < \tfrac{1}{2} \quad \text{for} \quad t \geqq N. \tag{2}$$

By virtue of (b), we have

$$\sup\left\{\sum_{k=1}^{n}[f_k(t) - f_k(a_k + t)] : t \geqq N\right\} \geqq 0. \tag{3}$$

Combining (1), (2), and (3), we have $\sup\{h(t) : t \geqq N\} \geqq -\frac{1}{2}$. Thus $\|h - (-1)\|_u \geqq \frac{1}{2}$ for all $h \in \mathfrak{H}_0$. As in (17.15), there is a mean M on $\mathfrak{B}^r(R)$ such that $M(\mathfrak{H}_0) = 0$. Then clearly M is invariant. Also if $\lim\limits_{t\to\infty} f(t) = \alpha$, then $f - \alpha \in \mathfrak{H}_0$, so that $M(f) = M(f - \alpha) + M(\alpha) = \alpha$.]

(d) Part (c) remains valid if R is replaced by any additive subsemigroup S of R having the property that $x \in S$ implies $|x| \in S$. In particular, it is valid if S is Z, Q, $\{x \in R : x \geqq N\}$, $\{x \in Z : x \geqq N\}$, or $\{x \in Q : x \geqq N\}$ where $N \geqq 0$. [Repeat the proofs of (b) and (c) *verbatim*, using S instead of R.]

(17.21) Uniqueness of means (Day [5], §7). If G is a finite group, it is trivial that $\mathfrak{B}^r(G)$ has exactly one invariant mean. We will show in (c) that for infinite Abelian groups, $\mathfrak{B}^r(G)$ always admits more than one invariant mean.

(a) Let G be a group such that $\mathfrak{B}^r(G)$ admits an invariant mean, and suppose that H is a normal subgroup of G such that $\mathfrak{B}^r(H)$ admits more than one left invariant mean. Then $\mathfrak{B}^r(G)$ admits more than one left invariant mean. [By (17.14) and (17.11), $\mathfrak{B}^r(G/H)$ has an invariant mean M'. Let M_1 and M_2 be distinct left invariant means for $\mathfrak{B}^r(H)$; we construct $M_0^{(1)}$ and $M_0^{(2)}$ from M_1 and M' and M_2 and M', respectively, exactly as in the last paragraph of the proof of (17.14), using the notation of this paragraph with obvious modifications. We will show that $M_0^{(1)} \neq M_0^{(2)}$. Choose $g \in \mathfrak{B}^r(H)$ such that $M_1(g) \neq M_2(g)$. Write the distinct elements of G/H as $\{x_\alpha H\}_{\alpha\in A}$, where the x_α are fixed elements of G. Thus each $y \in G$ can be uniquely written as $x_\alpha h$ for $\alpha \in A$ and $h \in H$; define $f \in \mathfrak{B}^r(G)$ by $f(y) = g(h)$. For all $\alpha \in A$, we have $\bar{f}_1(x_\alpha) = M_1({}_{x_\alpha}f) = M_1(g)$ and $\bar{f}_2(x_\alpha) = M_2(g)$. Hence $f_1'(x_\alpha H) = M_1(g)$ and $f_2'(x_\alpha H) = M_2(g)$ for all $\alpha \in A$. It follows that $M_0^{(1)}(f) = M'(M_1(g)) = M_1(g)$ and $M_0^{(2)}(f) = M_2(g)$.]

(b) Suppose that G is a group and that $G = \bigcup\limits_{n=1}^{\infty} G_n$, where $\{G_n\}_{n=1}^{\infty}$ is a strictly increasing sequence of finite subgroups of G. Then $\mathfrak{B}^r(G)$ admits more than one left invariant mean. [Let k_n denote the number of elements in G_n, $n = 1, 2, \ldots$. By taking a subsequence if necessary, we may suppose that $k_n \geqq 3k_{n-1}$ for $n = 2, 3, \ldots$. For $f \in \mathfrak{B}^r(G)$, let $p(f) = \overline{\lim\limits_{n\to\infty}} \frac{1}{k_n} \sum\limits_{x\in G_n} f(x)$. Let $\mathfrak{A}$ be the closed linear subspace of $\mathfrak{B}^r(G)$ consisting of all functions f such that

$$\lim_{n\to\infty}\left[\frac{1}{k_n}\sum_{x\in G_n} f(x)\right]=0.$$

Consider f in $\mathfrak{B}^r(G)$ and a in G. Then $a\in G_m$ for some m. Clearly we have $aG_n=G_n$ for $n\geqq m$, and therefore

$$\frac{1}{k_n}\sum_{x\in G_n}[{}_af(x)-f(x)]=0$$

for $n\geqq m$. It follows that $\mathfrak{A}$ contains all linear combinations of functions of the form ${}_af-f$.

Let f_0 be defined on G by $f_0(x)=1$ if $x\in G_1\cup(G_3\cap G_2')\cup\cdots\cup(G_{2n+1}\cap G_{2n}')\cup\cdots$ and $f_0(x)=0$ otherwise. Since

$$p(f_0)=\overline{\lim_{n\to\infty}}\frac{1}{k_n}\sum_{x\in G_n} f_0(x)\geqq\overline{\lim_{n\to\infty}}\frac{1}{k_{2n+1}}\sum_{x\in G_{2n+1}} f_0(x)\geqq\frac{2}{3},$$

f_0 does not belong to $\mathfrak{A}$. Note also that

$$-p(-f_0)=\underline{\lim_{n\to\infty}}\frac{1}{k_n}\sum_{x\in G_n} f_0(x)\leqq\underline{\lim_{n\to\infty}}\frac{1}{k_{2n}}\sum_{x\in G_{2n}} f_0(x)\leqq\frac{1}{3}.$$

Let $M(f)=0$ for $f\in\mathfrak{A}$. It is easy to show that

$$\sup\{-p(-f-f_0)-M(f):f\in\mathfrak{A}\}=-p(-f_0)\leqq\tfrac{1}{3}$$

and

$$\inf\{p(f+f_0)-M(f):f\in\mathfrak{A}\}=p(f_0)\geqq\tfrac{2}{3}.$$

Theorems (B.12) and (B.13) show that for any real number α such that $\frac{1}{3}\leqq\alpha\leqq\frac{2}{3}$, there is a linear functional M_α on $\mathfrak{B}^r(G)$ with the following properties: $M_\alpha(\mathfrak{A})=0$; $M_\alpha(f_0)=\alpha$; and $M_\alpha(f)\leqq p(f)$ for all $f\in\mathfrak{B}^r(G)$. Since $-p(-f)\leqq M_\alpha(f)\leqq p(f)$ for all $f\in\mathfrak{B}^r(G)$, it follows that $M_\alpha(1)=1$ and that M_α is nonnegative. Since $M_\alpha(f)=0$ for all $f\in\mathfrak{A}$, M_α is also left invariant. Therefore $\mathfrak{B}^r(G)$ admits at least $\mathfrak{c}$ distinct left invariant means.]

(c) Let G be an infinite Abelian group. Then $\mathfrak{B}^r(G)$ admits more than one invariant mean. [If G contains an element of infinite order, then G contains a subgroup H isomorphic with Z. By (17.20.c) and (17.20.d), there are distinct invariant means for $\mathfrak{B}^r(Z)$; such means M_1 and M_2 may be chosen so that $M_1(f)=\lim\limits_{n\to\infty}f(n)$ and $M_2(f)=\lim\limits_{n\to\infty}f(-n)$ whenever these limits exist. Now apply (a). If all elements of G have finite order, there is a strictly increasing sequence $\{H_n\}_{n=1}^{\infty}$ of finite subgroups of G. Let $H=\bigcup\limits_{n=1}^{\infty}H_n$ and apply (b) and (a).]

(17.22) Invariant finitely additive measures. (a) Let G be any group such that $\mathfrak{B}^r(G)$ admits an invariant mean M [which in view of (17.11) may be taken to be two-sided invariant]. For any subset A of G, define $\nu(A) = M(\xi_A)$. Then ν is a finitely additive two-sided invariant measure.

(b) For any subset A of R, let $\nu(A) = M(\xi_A)$ where M is an invariant mean for $\mathfrak{B}^r(R)$ as in (17.20.c). Then ν is a finitely additive set function satisfying

(i) $0 \leqq \nu(A) \leqq 1$ for all $A \subset R$;

(ii) $\nu(A) = 1$ if $[\alpha, \infty[\subset A$ for some $\alpha \in R$;

(iii) $\nu(A) = 0$ if A is bounded above;

(iv) $\nu(A + a) = \nu(A)$ for all $A \subset R$ and $a \in R$.

(c) (Adapted from VON NEUMANN [2].) Let G be a locally compact group such that $\mathfrak{B}^r(G)$ admits a [two-sided] invariant mean M. Let λ denote left Haar measure on G. Then there is a finitely additive set function μ, defined for all subsets of G, having the following properties:

(i) $0 \leqq \mu(A) \leqq +\infty$ for all $A \subset G$;

(ii) $\mu(xA) = \mu(A)$ for all $x \in G$ and $A \subset G$;

(iii) $\mu(A) = \lambda(A)$ for all Haar measurable subsets A of G.

That is, we obtain a *finitely* additive measure defined for all subsets of G that is invariant under left translations, and which agrees with λ on measurable subsets of G.

[The proof is carried out in two steps.

(I) There is a finitely additive set function ν, defined for all subsets of G, satisfying (i) and (iii)[1].

This assertion depends upon the axiom of choice. Let $\mathscr{M}$ be the family of all Haar measurable subsets of G. Since the theorem is trivial when $\mathscr{M}$ consists of all subsets of G, we suppose that $\mathscr{M}$ is different from $\mathscr{P}(G)$, the family of all subsets of G. Well order all of the subsets of G, putting all of the sets in $\mathscr{M}$ first. Thus the family $\mathscr{P}(G)$ is written as

$$A_1, A_2, \ldots, A_\alpha, \ldots, \qquad \alpha < \varDelta,$$

where $\varDelta$ is [say] the smallest ordinal with corresponding cardinal $2^{\overline{\overline{G}}}$, and $A_\alpha \in \mathscr{M}$ for $\alpha < \Gamma$; here $\Gamma < \varDelta$. Our purpose is to define ν in such a way that if $a_1, \ldots, a_m, b_1, \ldots, b_n$ are positive real numbers and $\alpha_1, \ldots, \alpha_m, \beta_1, \ldots, \beta_n$ are ordinals ($\alpha_j < \varDelta$, $\beta_k < \varDelta$), then

$$\sum_{j=1}^{m} a_j \xi_{A_{\alpha_j}} \leqq \sum_{k=1}^{n} b_k \xi_{A_{\beta_k}} \quad \text{implies} \quad \sum_{j=1}^{m} a_j \nu(A_{\alpha_j}) \leqq \sum_{k=1}^{n} b_k \nu(A_{\beta_k}). \tag{1}$$

Condition (1) implies that ν is finitely additive; in fact, if $A \cap B = \varnothing$, then $\xi_A + \xi_B = \xi_{A \cup B}$. For $\alpha < \Gamma$, let $\nu(A_\alpha) = \lambda(A_\alpha)$. Let us verify (1) in the case that $\alpha_j < \Gamma$ and $\beta_k < \Gamma$, $j = 1, \ldots, m$, $k = 1, \ldots, n$. Integrating

[1] For a different proof of this fact, see BIRKHOFF [2], pp. 185—186.

the left inequality of (1) with respect to λ gives

$$\sum_{j=1}^{m} a_j \nu(A_{\alpha_j}) = \sum_{j=1}^{m} a_j \lambda(A_{\alpha_j}) = \int_G \Big[\sum_{j=1}^{m} a_j \xi_{A_{\alpha_j}}(x)\Big] d\lambda(x)$$
$$\leqq \int_G \Big[\sum_{k=1}^{n} b_k \xi_{A_{\beta_k}}(x)\Big] d\lambda(x) = \sum_{k=1}^{n} b_k \nu(A_{\beta_k}).$$

We now extend the definition of ν by transfinite induction. Suppose that $\nu(A_\alpha)$ has been defined for all $\alpha < \eta$ in such a way that (1) holds, where $\Gamma \leqq \eta < \Delta$. If

$$\inf\left\{\sum_{k=1}^{n} b_k \nu(A_{\beta_k}) : \sum_{k=1}^{n} b_k \xi_{A_{\beta_k}} \geqq \xi_{A_\eta},\ \text{all } \beta_k < \eta\right\} = +\infty,$$

we define $\nu(A_\eta) = +\infty$. Then it is simple to verify (1) for all ordinals α_j and β_k less than or equal to η. Suppose now that $\inf\left\{\sum_{k=1}^{n} b_k \nu(A_{\beta_k}) : \sum_{k=1}^{n} b_k \xi_{A_{\beta_k}} \geqq \xi_{A_\eta}\right\} < +\infty$. Then there exist positive real numbers a_j and b_k and ordinals α_j and β_k less than η such that

$$\sum_{j=1}^{m} a_j \xi_{A_{\alpha_j}} + \xi_{A_\eta} \leqq \sum_{k=1}^{n} b_k \xi_{A_{\beta_k}}; \tag{2}$$

$$\sum_{k=1}^{n} b_k \nu(A_{\beta_k}) < \infty. \tag{3}$$

We define $\nu(A_\eta) = \inf\left\{\sum_{k=1}^{n} b_k \nu(A_{\beta_k}) - \sum_{j=1}^{m} a_j \nu(A_{\alpha_j}) : \text{(2) and (3) hold}\right\}$. To show that (1) holds for all ordinals α, $\alpha \leqq \eta$, it suffices to consider the following two cases:

$$\sum_{j=1}^{m} a_j \xi_{A_{\alpha_j}} + \xi_{A_\eta} \leqq \sum_{k=1}^{n} b_k \xi_{A_{\beta_k}}; \tag{4}$$

$$\sum_{j=1}^{m'} a'_j \xi_{A_{\alpha'_j}} \leqq \sum_{k=1}^{n'} b'_k \xi_{A_{\beta'_k}} + \xi_{A_\eta}. \tag{5}$$

If (4) holds, then clearly

$$\sum_{j=1}^{m} a_j \nu(A_{\alpha_j}) + \nu(A_\eta) \leqq \sum_{k=1}^{n} b_k \nu(A_{\beta_k}).$$

Suppose finally that (5) holds. Consider *any* a_j, b_k, α_j, and β_k for which (2) and (3) hold. Adding (2) and (5) and applying (1) and the induction hypothesis, we obtain

$$\sum_{j=1}^{m} a_j \nu(A_{\alpha_j}) + \sum_{j=1}^{m'} a'_j \nu(A_{\alpha'_j}) \leqq \sum_{k=1}^{n} b_k \nu(A_{\beta_k}) + \sum_{k=1}^{n'} b'_k \nu(A_{\beta'_k}).$$

Taking the infimum over all a_j, b_k, α_j, and β_k satisfying (2) and (3), we get

$$\sum_{j=1}^{m'} a'_j \nu(A_{\alpha'_j}) \leqq \nu(A_\eta) + \sum_{k=1}^{n'} b'_k \nu(A_{\beta'_k}).$$

This completes the proof of the existence of ν.

(II) We now use the measure ν of (I) and the mean M to construct a finitely additive measure μ satisfying (i)—(iii). For each subset A of G, we define a function $\widetilde{A}$ on G by $\widetilde{A}(x) = \nu(xA)$. The function $\widetilde{A}$ is nonnegative, but may be unbounded and may also assume the value $+\infty$. Now let $\mu(A) = \lim_{n\to\infty} M(\min(\widetilde{A}, n))$. Conditions (i) and (iii) are obvious. To show that μ is additive, let A and B be disjoint subsets of G. Then clearly $\widetilde{A \cup B} = \widetilde{A} + \widetilde{B}$. The inequalities

$$\min(\widetilde{A}+\widetilde{B}, n) \leqq \min(\widetilde{A}, n) + \min(\widetilde{B}, n) \leqq \min(\widetilde{A}+\widetilde{B}, 2n)$$

then imply that $\mu(A \cup B) = \lim_{n\to\infty} M(\min(\widetilde{A}+\widetilde{B}, n)) = \lim_{n\to\infty} M(\min(\widetilde{A}, n)) + \lim_{n\to\infty} M(\min(\widetilde{B}, n)) = \mu(A) + \mu(B)$. To verify (ii), let $x \in G$ and $A \subset G$. Then we have $\widetilde{xA} = (\widetilde{A})_x$, and therefore

$$\begin{aligned} \mu(xA) &= \lim_{n\to\infty} M(\min(\widetilde{xA}, n)) = \lim_{n\to\infty} M(\min((\widetilde{A})_x, n)) \\ &= \lim_{n\to\infty} M((\min(\widetilde{A}, n))_x) = \lim_{n\to\infty} M(\min(\widetilde{A}, n)) = \mu(A).] \end{aligned}$$

Notes

The theory of invariant means on $\mathfrak{B}(G)$ was founded by VON NEUMANN [2], who established [for groups] the main results of the present section. Invariant means have exerted a strange fascination for mathematicians in recent years, and a startling number of writers on the subject have published previously published results. We make no effort to trace all of these repetitions.

Theorem (17.4) is due to DIXMIER [1]. Theorem (17.5) for groups appears in VON NEUMANN [2], and for semigroups in DAY [2]. Our proof of (17.5) is taken from DIXMIER [1]. DAY proved (17.6), (17.7), and (17.9) [4], as well as (17.11) in slightly disguised form [4], Lemma 7. Theorem (17.12) was announced without proof by DAY [3]; the first published proof was given by FØLNER [3]. For left invariant means, (17.14) was proved by VON NEUMANN [2], and for invariant means by DAY [4], Theorem 6. Theorems similar to (17.15) appear in DAY [4], FØLNER [4], and RAIMI [1]. Theorem (17.16) is due to VON NEUMANN [2].

The reader wishing to pursue the subject further should consult: DAY [5] for a useful survey; FØLNER [3] and KESTEN [1] for other conditions under which invariant means exist; LORENTZ [1] and LUTHAR [1] for uniqueness questions; SILVERMAN [1], [2], [3].

§ 18. Invariant means on almost periodic functions

In § 17, we obtained invariant means on $\mathfrak{B}(G)$ and $\mathfrak{C}(G)$ in various cases where the HAHN-BANACH theorem can be applied. In the present section, we prove the existence of invariant means in other function classes by the use of *compactness* arguments. This seems altogether fitting, since compactness and well ordering are two of the most powerful concepts used for existence proofs in analysis. As we shall show in Vol. II, some of the results presented here can be subsumed under the construction of Haar measure on compact groups. Others cannot, and in any case it seems worth while to give a completely elementary construction, as we will do.

We begin with a description of the class of almost periodic functions on a group G. For a complex function f on G and $a \in G$, let $D_a f$ be the function on $G \times G = G^2$ such that $D_a f(x, y) = f(xay)$.

(18.1) Theorem. *Let G be a group and let f be a function in $\mathfrak{B}(G)$. Let $^-$ denote closure in the uniform topology for $\mathfrak{B}(G)$. The following properties of f are equivalent:*

(i) $\{f_a : a \in G\}^-$ *is compact in* $\mathfrak{B}(G)$;

(ii) $\{{}_af : a \in G\}^-$ *is compact in* $\mathfrak{B}(G)$;

(iii) $\{{}_bf_a : a, b \in G\}^-$ *is compact in* $\mathfrak{B}(G)$;

(iv) $\{D_a f : a \in G\}^-$ *is compact in* $\mathfrak{B}(G^2)$.

Proof. The space $\mathfrak{B}(G)$ under the metric $\|f - g\|_u$ is a complete metric space; a subset $\mathfrak{A}$ of $\mathfrak{B}(G)$ is thus compact if and only if it is closed and totally bounded, and $\mathfrak{A}$ has compact closure if and only if it is totally bounded [see (3.7)]. Hence we have only to prove the equivalence of total boundedness for the sets $\{f_a : a \in G\}$, $\{{}_af : a \in G\}$, $\{{}_bf_a : a, b \in G\}$, $\{D_a f : a \in G\}$.

It is obvious that (iii) implies (i), (iii) implies (ii), (iv) implies (i), and (iv) implies (ii). We will prove that (i) implies (iv), (ii) implies (iv), (i) implies (iii), and (ii) implies (iii). This will complete the proof.

Suppose, then, that $\{f_a : a \in G\}$ is totally bounded. Given $\varepsilon > 0$, there is a finite subset $\{a_1, a_2, \ldots, a_m\}$ of G such that $\{f_{a_j}\}_{j=1}^m$ is an $\varepsilon/4$-mesh in $\{f_a : a \in G\}$.[1] For $j = 1, 2, \ldots, m$, let $A_j = \{a \in G : \|f_a - f_{a_j}\|_u < \varepsilon/4\}$. Consider the family of all sets $(A_{l_1} a_1^{-1}) \cap (A_{l_2} a_2^{-1}) \cap \cdots \cap (A_{l_m} a_m^{-1})$, where $l_j \in \{1, 2, \ldots, m\}$ for $j = 1, 2, \ldots, m$. Write the nonvoid sets in this family as $B_1, B_2, \ldots, B_n$, and choose $b_k \in B_k$ for $k = 1, 2, \ldots, n$. It is obvious that $\bigcup_{k=1}^{n} B_k = G$. Now consider any $c \in G$; let B_{k_0} be a set B_k such that $c \in B_{k_0}$. Let (x, y) be an arbitrary point in G^2; select j_0 ($j_0 = 1, 2, \ldots, m$) such that $y \in A_{j_0}$. Then we have

$$\left.\begin{aligned} &|f(xcy) - f(xb_{k_0}y)| \leqq |f(xcy) - f(xca_{j_0})| + |f(xca_{j_0}) - f(xb_{k_0}a_{j_0})| \\ &+ |f(xb_{k_0}a_{j_0}) - f(xb_{k_0}y)| \leqq \|f_y - f_{a_{j_0}}\|_u + \|f_{ca_{j_0}} - f_{b_{k_0}a_{j_0}}\|_u + \|f_{a_{j_0}} - f_y\|_u . \end{aligned}\right\} \quad (1)$$

The first and third summands on the right side of (1) are less than $\varepsilon/4$ because $y \in A_{j_0}$. The second summand is less than $\varepsilon/2$ because c and b_{k_0}

[1] For the definition of α-mesh, see (3.7).

are both in $A_{l_{j_0}} a_{j_0}^{-1}$ for some l_{j_0}. Since (x, y) is arbitrary, the functions $D_{b_1}f, D_{b_2}f, \ldots, D_{b_n}f$ form an ε-mesh in $\{D_c f : c \in G\}$. Thus (iv) holds.

The proof that (ii) implies (iv) is very like the proof just given. Without writing it out, therefore, we may assert that (i), (ii), and (iv) are equivalent.

Finally, suppose that (i) holds. Then (ii) holds also. Let $\{f_{a_j}\}_{j=1}^m$ and $\{{}_{b_k}f\}_{k=1}^n$ be $\varepsilon/2$-meshes in $\{f_a : a \in G\}$ and $\{{}_b f : b \in G\}$, respectively. Then for all x, a, b in G, we have

$$|f(bxa) - f(b_k x a_j)| \leqq |f(bxa) - f(b_k x a)| + |f(b_k x a) - f(b_k x a_j)| < \varepsilon$$

for some k and j. That is, $\{{}_{b_k}f_{a_j}\}_{j=1}^m{}_{,}{}_{k=1}^n$ is an ε-mesh in $\{{}_b f_a : a, b \in G\}$. □

(18.2) Definition. Let G be a group. A function f in $\mathfrak{B}(G)$ satisfying one [and hence all] of the conditions (18.1.i)—(18.1.iv) is said to be *almost periodic*. The set of all almost periodic functions on G is denoted by $\mathfrak{A}(G)$. For a topological group G, the set of all continuous functions in $\mathfrak{A}(G)$ is denoted by $\mathfrak{A}_c(G)$.

(18.3) Theorem. *Let G be a group.*

(i) *Every constant function is in $\mathfrak{A}(G)$.*

(ii) *If f is in $\mathfrak{A}(G)$, then also* Re f, Im f, *and $\bar{f}$ are in $\mathfrak{A}(G)$.*

(iii) *If f, g are in $\mathfrak{A}(G)$, then $f+g$ and fg are in $\mathfrak{A}(G)$.*

(iv) *If $f^{(1)}, f^{(2)}, f^{(3)}, \ldots$ are in $\mathfrak{A}(G)$ and $\lim\limits_{n\to\infty} \|f^{(n)} - f\|_u = 0$, then $f \in \mathfrak{A}(G)$.*

(v) *If f is in $\mathfrak{A}(G)$ and a, b are in G, then ${}_a f$, f_a, and ${}_b f_a$ are in $\mathfrak{A}(G)$.*

Similar assertions hold for $\mathfrak{A}_c(G)$.

Proof. Assertions (i), (ii), and (v) are obvious. To prove (iii), consider first $f+g$. Let ε be any positive number; let $\{f_{a_j}\}_{j=1}^m$ be an $\varepsilon/4$-mesh for $\{f_a : a \in G\}$ and $\{g_{b_k}\}_{k=1}^n$ be an $\varepsilon/4$-mesh for $\{g_b : b \in G\}$. Let $A_j = \{a \in G : \|f_a - f_{a_j}\|_u < \varepsilon/4\}$ $(j = 1, 2, \ldots, m)$ and $B_k = \{b \in G : \|g_b - g_{b_k}\|_u < \varepsilon/4\}$ $(k = 1, 2, \ldots, n)$. Let the nonvoid sets $A_j \cap B_k$ be written as $C_1, C_2, \ldots, C_p$, and choose $c_l \in C_l$ $(l = 1, 2, \ldots, p)$. Then it is easy to see that $\{(f+g)_{c_l}\}_{l=1}^p$ is an ε-mesh for $\{(f+g)_c : c \in G\}$. To show that $fg \in \mathfrak{A}(G)$, suppose that $\|f\|_u > 0$ and $\|g\|_u > 0$, and form an $\frac{\varepsilon}{4\|g\|_u}$-mesh for $\{f_a : a \in G\}$ and an $\frac{\varepsilon}{4\|f\|_u}$-mesh for $\{g_b : b \in G\}$. Then, as above, the sets C_l and points $c_l \in C_l$ yield an ε-mesh for $\{(fg)_c : c \in G\}$.

To prove (iv), note that if $\{f_{a_1}^{(n)}, \ldots, f_{a_p}^{(n)}\}$ is an $\varepsilon/3$-mesh for $\{f_a^{(n)} : a \in G\}$, and $\|f - f^{(n)}\|_u < \varepsilon/3$, then $\{f_{a_1}, \ldots, f_{a_p}\}$ is an ε-mesh for $\{f_a : a \in G\}$. □

We will show that $\mathfrak{A}(G)$ admits a unique, strictly positive, two-sided invariant mean. We first prove a combinatorial lemma [in somewhat greater generality than we actually need].

(18.4) Lemma. *Let P and Q be nonvoid sets. Let ϱ be a function that maps P into the family of nonvoid subsets of Q. Suppose that*

(i) $\overline{\overline{\bigcup\{\varrho(p): p \in P_1\}}} \geqq \overline{\overline{P_1}}$

for all finite subsets P_1 of P. Suppose also that P is finite or that $\varrho(p)$ is finite for all $p \in P$. Then there is a one-to-one function σ mapping P into Q with the property that $\sigma(p) \in \varrho(p)$ for all $p \in P$.[1]

Proof. Suppose that P is finite, $\overline{\overline{P}} = n$. If $n = 1$, the result is trivial. For $n > 1$, there are two cases to consider. Suppose first that the left side of (i) is greater than $\overline{\overline{P_1}}$ if $P_1 \subset P$ and $0 < \overline{\overline{P_1}} < n$. Choose any $p_0 \in P$ and let $\sigma(p_0)$ be any element of $\varrho(p_0)$. Then $P \cap \{p_0\}'$ and the function $p \to \varrho(p) \cap \{\sigma(p_0)\}'$ satisfy the inductive hypothesis, and $\overline{\overline{P \cap \{p_0\}'}}$ is $n-1$. Hence σ can be constructed in this case.

Second, suppose that there is a $P_0 \subset P$ such that $0 < \overline{\overline{P_0}} < n$ and $\overline{\overline{\bigcup_{p \in P_0} \varrho(p)}} = \overline{\overline{P_0}}$. Then P_0 satisfies the inductive hypothesis and so σ can be defined on P_0 so as to be one-to-one, and such that $\sigma(p) \in \varrho(p)$ for $p \in P_0$. Note that $\sigma(P_0) = \bigcup_{p \in P_0} \varrho(p)$. Now look at the set $P \cap P_0'$. If there were a subset P_2 of $P \cap P_0'$ such that $\overline{\overline{\bigcup\{\varrho(p) \cap \sigma(P_0)': p \in P_2\}}} < \overline{\overline{P_2}}$, then we would have $\overline{\overline{\bigcup_{p \in P_0 \cup P_2} \varrho(p)}} = \overline{\overline{\left(\sigma(P_0) \cup \bigcup_{p \in P_2} \varrho(p)\right)}} = \overline{\overline{P_0}} + \overline{\overline{\bigcup_{p \in P_2} \varrho(p) \cap \sigma(P_0)'}} < \overline{\overline{P_0 \cup P_2}}$. That is, (i) would fail for $P_0 \cup P_2$. Hence we can apply the inductive hypothesis to $P \cap P_0'$ and the mapping $\varrho^*(p) = \varrho(p) \cap \sigma(P_0)'$, and define σ on $P \cap P_0'$.

The case in which P is infinite and all $\varrho(p)$ are finite is handled by a compactness argument. With the discrete topology, each $\varrho(p)$ is compact, and so the Cartesian product $X = \underset{p \in P}{\mathsf{P}} \varrho(p)$ is compact. For a finite nonvoid subset F of P, let H_F be the set of all $(q_p) \in X$ such that all q_p are distinct for $p \in F$. By the first case, H_F is nonvoid. It is also closed. Since $H_{F_1} \cap \cdots \cap H_{F_n} \supset H_{F_1 \cup \cdots \cup F_n}$, the intersection $\bigcap\{H_F : F$ is a finite nonvoid subset of $P\}$ is nonvoid. Choose any point (q_p) in this intersection, and let $\sigma(p) = q_p$ for all $p \in P$. Plainly this σ has all of the required properties. □

Lemma (18.4) can be used to establish a useful fact about metric spaces.

(18.5) Lemma. *Let X be a metric space with metric d, and suppose that $\{x_1, x_2, \ldots, x_n\}$ is an ε-mesh in X for some $\varepsilon > 0$, such that the number n of elements of the mesh is as small as possible for the given ε. Let Y be any subset of X such that for each $x \in X$ there is a $y \in Y$ such that $d(x, y) < \varepsilon$.*[2] *Then there are a one-to-one mapping σ of $\{x_1, x_2, \ldots, x_n\}$ into Y and a sequence $\{z_1, z_2, \ldots, z_n\}$ of elements of X such that $d(x_k, z_k) < \varepsilon$ and $d(z_k, \sigma(x_k)) < \varepsilon$, $k = 1, \ldots, n$.*

---

[1] This assertion is called "the marriage lemma" for an obvious reason.

[2] Thus Y is an ε-mesh if it is finite.

Proof. For every $k=1, 2, \ldots, n$, let $\varrho(x_k)=\{y\in Y: d(x_k, z)<\varepsilon$ and $d(y, z)<\varepsilon$ for some $z\in X\}$. Note that all $x_1, \ldots, x_n$ are distinct. Consider an arbitrary nonvoid subset $\{x_{j_1}, \ldots, x_{j_r}\}$ of $\{x_1, \ldots, x_n\}$ and write the remaining elements [if any] as $x_{j_{r+1}}, \ldots, x_{j_n}$. We will prove that

$$\overline{\overline{\varrho(x_{j_1})\cup\cdots\cup\varrho(x_{j_r})}}\geqq r. \tag{1}$$

If the set $\varrho(x_{j_1})\cup\cdots\cup\varrho(x_{j_r})$ is infinite, there is nothing to prove. Suppose then that it is finite: we write $\varrho(x_{j_1})\cup\cdots\cup\varrho(x_{j_r})=\{y_1, y_2, \ldots, y_s\}$, where the y_l's are distinct. Form the set $A=\{y_1, y_2, \ldots, y_s, x_{j_{r+1}}, \ldots, x_{j_n}\}$. Some y_l's may be equal to x_j's in A, but at any rate we have $\overline{\overline{A}}\leqq s+n-r$. Also A is an ε-mesh. For, if $z\in X$ and $d(z, x_{j_p})\geqq\varepsilon$ for $p=r+1, \ldots, n$, then $d(z, x_{j_p})<\varepsilon$ for some $p=1, 2, \ldots, r$. There is also a $y\in Y$ such that $d(z, y)<\varepsilon$, and by the definition of ϱ, we have $y\in\varrho(x_{j_p})$; *i.e.*, $y=y_l$ for some $l=1, 2, \ldots, s$. By the choice of n, we have $s+n-r\geqq\overline{\overline{A}}\geqq n$, so that $s\geqq r$. This establishes (1).

Now apply (18.4) with $P=\{x_1, \ldots, x_n\}$ and $Q=Y$.[1] □

With the aid of (18.5) it is easy to construct an invariant mean on $\mathfrak{A}(G)$.

(18.6) Theorem. *Let G be a group, f a function in $\mathfrak{A}(G)$, and ε a positive number. Let $\{D_{a_1}f, \ldots, D_{a_n}f\}$ and $\{D_{b_1}f, \ldots, D_{b_n}f\}$ be ε-meshes in $\{D_a f: a\in G\}$ both having the least cardinal number n among all ε-meshes. Then we have*

(i) $\left\|\frac{1}{n}\sum_{k=1}^{n} D_{a_k}f-\frac{1}{n}\sum_{k=1}^{n} D_{b_k}f\right\|_u<2\varepsilon.$

Proof. We apply Lemma (18.5) to the ε-meshes $\{D_{a_k}f\}$ and $\{D_{b_k}f\}$ and the metric space $\{D_a f: a\in G\}$. Thus we can find $c_1, c_2, \ldots, c_n\in G$ and a one-to-one mapping σ of $\{1, 2, \ldots, n\}$ onto itself such that

$$\|D_{a_k}f-D_{c_k}f\|_u<\varepsilon, \qquad \|D_{c_k}f-D_{b_{\sigma(k)}}f\|_u<\varepsilon \qquad (k=1, 2, \ldots, n).$$

Thus we have

$$\|D_{a_k}f-D_{b_{\sigma(k)}}f\|_u<2\varepsilon \qquad (k=1, 2, \ldots, n). \tag{1}$$

Adding the inequalities (1) from 1 to n, dividing by n, using elementary properties of the norm, and noting that σ carries $\{1, 2, \ldots, n\}$ onto itself, we obtain (i). □

(18.7) Theorem. *Let G, f, ε, and $\{D_{a_1}f, \ldots, D_{a_n}f\}$ be as in (18.6). Then*

(i) $\left\|\frac{1}{n}\sum_{k=1}^{n} f(a_k)-\frac{1}{n}\sum_{k=1}^{n} D_{a_k}f\right\|_u\leqq 2\varepsilon.$

[1] Observe that only the elementary part of (18.4) is used in this proof: TIHONOV's theorem is not required.

Proof. Let u, v be arbitrary elements of G. Then $\{D_{ua_1v}f, \ldots, D_{ua_nv}f\}$ is an ε-mesh in $\{D_a f : a \in G\}$. For, given $a \in G$, we have $\|D_{u^{-1}av^{-1}}f - D_{a_k}f\|_u < \varepsilon$ for some $k = 1, 2, \ldots, n$, and so $\|D_a f - D_{ua_kv}f\|_u < \varepsilon$. For $\{D_{a_k}f\}$ and $\{D_{b_k}f\}$ with $b_k = ua_kv$, compute the function in (18.6.i) at the point $(e, e) \in G^2$. This gives (i). □

(18.8) Theorem. *Let G be a group. There is a complex linear functional M on $\mathfrak{A}(G)$ such that:*

(i) $M({}_bf_a) = M(f)$ *for* $f \in \mathfrak{A}(G)$ *and* $a, b \in G$;

(ii) $M(f) > 0$ *if* $f \in \mathfrak{A}^+(G)$ *and* $f \neq 0$;

(iii) $M(1) = 1$.[1]

Proof. Let $f \in \mathfrak{A}(G)$ and let ε be a positive number. Let E_ε be the set of all complex numbers z such that for some sequence $\{c_1, \ldots, c_p\}$ of elements of G, the inequality

$$\left\| z - \frac{1}{p} \sum_{j=1}^{p} D_{c_j} f \right\|_u < \varepsilon \tag{1}$$

holds. Theorem (18.7) shows that E_ε is nonvoid. Suppose that z_1 and z_2 are in E_ε. That is,

$$\left| z_1 - \frac{1}{p} \sum_{j=1}^{p} f(x c_j y) \right| < \varepsilon \quad \text{for all} \quad (x, y) \in G^2 \tag{2}$$

and

$$\left| z_2 - \frac{1}{q} \sum_{k=1}^{q} f(x d_k y) \right| < \varepsilon \quad \text{for all} \quad (x, y) \in G^2. \tag{3}$$

Setting $x = e$ and $y = d_k$ in (2), adding over $k = 1, 2, \ldots, q$, and dividing by q, we obtain

$$\left| z_1 - \frac{1}{pq} \sum_{k=1}^{q} \sum_{j=1}^{p} f(c_j d_k) \right| < \varepsilon.$$

Setting $y = e$ and $x = c_j$ in (3), we obtain similarly

$$\left| z_2 - \frac{1}{pq} \sum_{j=1}^{p} \sum_{k=1}^{q} f(c_j d_k) \right| < \varepsilon.$$

Hence we have

$$|z_1 - z_2| < 2\varepsilon. \tag{4}$$

It is easy to see that if $z \in E_\varepsilon$, then

$$|z| < \|f\|_u + \varepsilon.$$

---

[1] On $\mathfrak{A}^r(G)$, M is thus a two-sided invariant mean in the sense of (17.3). We could of course define M only on $\mathfrak{A}^r(G)$ and then extend it by (B.38), but there is no advantage in doing this.

Let F_ε be the closure [in the ordinary topology of the complex plane] of E_ε. Since $F_{\varepsilon_1} \cap \cdots \cap F_{\varepsilon_m} \supset F_{\min(\varepsilon_1, \ldots, \varepsilon_m)}$, compactness shows that the set $\bigcap_{\varepsilon>0} F_\varepsilon$ is nonvoid. The relation (4) shows that $\bigcap_{\varepsilon>0} F_\varepsilon$ contains exactly one point. This number we take as $M(f)$. Since we have $F_\varepsilon \subset E_{\varepsilon'}$ for all $\varepsilon, \varepsilon'$ such that $0<\varepsilon<\varepsilon'$, it follows that $M(f)$ is the unique point lying in all E_ε. Thus we have:

$M(f)$ is the unique complex number such that for every $\varepsilon>0$, there is a sequence $\{a_1, \ldots, a_n\}$ of group elements for which

$$\left\| M(f) - \frac{1}{n} \sum_{k=1}^{n} D_{a_k} f \right\|_u < \varepsilon. \tag{5}$$

It is obvious that $M(\alpha f) = \alpha M(f)$ for all $f \in \mathfrak{A}(G)$ and complex numbers α, that $M({}_bf_a) = M(f)$ for all $a, b \in G$, and that $M(1) = 1$.

We now prove that $M(f+g) = M(f) + M(g)$ for $f, g \in \mathfrak{A}(G)$. For $\varepsilon > 0$, apply (5) to choose sequences $\{a_1, \ldots, a_m\}$ and $\{b_1, \ldots, b_n\}$ of elements of G such that

$$\left\| M(f) - \frac{1}{m} \sum_{j=1}^{m} D_{a_j} f \right\|_u < \frac{\varepsilon}{2} \tag{6}$$

and

$$\left\| M(g) - \frac{1}{n} \sum_{k=1}^{n} D_{b_k} g \right\|_u < \frac{\varepsilon}{2}. \tag{7}$$

From (6) we see that

$$\left| M(f) - \frac{1}{m} \sum_{j=1}^{m} f(x a_j b_k y) \right| < \frac{\varepsilon}{2}$$

for all x, y in G. Summing over k and dividing by n, we get

$$\left| M(f) - \frac{1}{mn} \sum_{k=1}^{n} \sum_{j=1}^{m} f(x a_j b_k y) \right| < \frac{\varepsilon}{2}$$

for all x, y in G. That is,

$$\left\| M(f) - \frac{1}{mn} \sum_{k=1}^{n} \sum_{j=1}^{m} D_{a_j b_k} f \right\|_u < \frac{\varepsilon}{2}. \tag{8}$$

Similarly we find

$$\left\| M(g) - \frac{1}{mn} \sum_{j=1}^{m} \sum_{k=1}^{n} D_{a_j b_k} g \right\|_u < \frac{\varepsilon}{2}. \tag{9}$$

Adding (8) and (9), we have

$$\left\| M(f) + M(g) - \frac{1}{mn} \sum_{k=1}^{n} \sum_{j=1}^{m} D_{a_j b_k}(f+g) \right\|_u < \varepsilon.$$

Since ε is arbitrary, (5) shows that $M(f) + M(g) = M(f+g)$.

We next establish (ii). It is obvious that $M(f) \geqq 0$ for $f \in \mathfrak{A}^+(G)$. Suppose that $f \in \mathfrak{A}^+(G)$ and that $f(b) > 0$ for some $b \in G$. Let $\{D_{a_1}f, \ldots, D_{a_n}f\}$ be an $f(b)/2$-mesh in $\{D_a f : a \in G\}$. Then for all $x, y, a \in G$, we have

$$f(x a_1 y) + \cdots + f(x a_n y) \geqq \max[f(x a_1 y), \ldots, f(x a_n y)] > f(x a y) - \frac{f(b)}{2}.$$

Setting $y = e$ and $a = x^{-1} b$, we obtain

$$f(x a_1) + \cdots + f(x a_n) > \frac{f(b)}{2} \text{ for all } x \in G.$$

Hence we have $M(f_{a_1} + \cdots + f_{a_n}) = n M(f) \geqq f(b)/2$, since M is linear and is nonnegative on $\mathfrak{A}^+(G)$. □

(18.9) Theorem. *The linear functional M constructed in* (18.8) *is uniquely determined by properties* (18.8.i)—(18.8.iii). *In fact, suppose that G is a topological group and that M' is any complex linear functional on $\mathfrak{A}_c(G)$ such that*

(i$_l$) *$M'({}_a f) = M'(f)$ for all $a \in G$ and $f \in \mathfrak{A}_c(G)$*

or

(i$_r$) *$M'(f_a) = M'(f)$ for all $a \in G$ and $f \in \mathfrak{A}_c(G)$,*

(ii) *$M'(f) \geqq 0$ for $f \in \mathfrak{A}_c^+(G)$,*

(iii) *$M'(1) = 1$.*

Then $M'(f) = M(f)$ for all $f \in \mathfrak{A}_c(G)$.

Proof. Since M and M' are linear on $\mathfrak{A}_c(G)$, we need only prove $M'(f) = M(f)$ for $f \in \mathfrak{A}_c^r(G)$. By (18.8.5), we have

$$-\varepsilon < M(f) - \frac{1}{n} \sum_{k=1}^{n} f(x a_k y) < \varepsilon \tag{1}$$

for all $x, y \in G$, where ε is an arbitrary positive number and $a_1, \ldots, a_n$ are appropriate elements of G. If M' satisfies (i$_l$), we set $x = e$ in (1), and obtain

$$-\varepsilon < M(f) - \frac{1}{n} \sum_{k=1}^{n} {}_{a_k}f < \varepsilon. \tag{2}$$

Applying M' to (2), we get

$$-\varepsilon \leqq M(f) - \frac{1}{n} \sum_{k=1}^{n} M'(f) \leqq \varepsilon.$$

Hence $M'(f) = M(f)$ for $f \in \mathfrak{A}_c(G)$. The case in which (i$_r$) holds is dealt with similarly. □

In some cases, the mean value M on $\mathfrak{A}_c(G)$ has a particularly simple form. We begin with a general theorem.

(18.10) Theorem. *Let G be a locally compact group with a left Haar measure λ. Suppose that there is a sequence $\{H_n\}_{n=1}^{\infty}$ of subsets of G such that $0 < \lambda(H_n) < \infty$ for all n and such that for each $x \in G$,*

(i) $\lim_{n\to\infty} \frac{\lambda((xH_n)\cap H_n')}{\lambda(H_n)} = 0$.

Then for every $f\in\mathfrak{A}_c(G)$, *we have*

(ii) $M(f) = \lim_{n\to\infty} \frac{1}{\lambda(H_n)} \int\limits_{H_n} f(x)\,dx$.

Proof. We first prove that for $f\in\mathfrak{C}(G)$ and $b\in G$, we have

$$\lim_{n\to\infty} \frac{1}{\lambda(H_n)} \int\limits_{H_n} ({}_bf(x) - f(x))\,dx = 0. \tag{1}$$

Clearly we have

$$\left.\begin{aligned} &\left|\int\limits_{H_n} {}_bf(x)\,dx - \int\limits_{H_n} f(x)\,dx\right| = \left|\int\limits_{bH_n} f(x)\,dx - \int\limits_{H_n} f(x)\,dx\right| \\ &\quad \leqq \int\limits_{(bH_n)\triangle H_n} |f(x)|\,dx \leqq \|f\|_u [\lambda((bH_n)\cap H_n') + \lambda(H_n\cap(bH_n)')]. \end{aligned}\right\} \tag{2}$$

We also have

$$\lambda(H_n\cap(bH_n)') = \lambda(b^{-1}(H_n\cap(bH_n)')) = \lambda((b^{-1}H_n)\cap H_n'). \tag{3}$$

Substituting (3) in (2), dividing the result by $\lambda(H_n)$, and applying (i), we obtain (1).

We shall use (1) to establish (ii). It obviously suffices to prove (ii) for $f\in\mathfrak{A}_c^r(G)$. For such f, let

$$p(f) = \overline{\lim_{n\to\infty}} \frac{1}{\lambda(H_n)} \int\limits_{H_n} f(x)\,dx.$$

It is evident that $p(f+g)\leqq p(f)+p(g)$ and $p(\alpha f)=\alpha p(f)$ for real nonnegative numbers α. By the HAHN-BANACH theorem (B.13), there is a linear functional M_0 on $\mathfrak{A}_c^r(G)$ such that

$$-p(-f)\leqq M_0(f)\leqq p(f) \quad \text{for all} \quad f\in\mathfrak{A}_c^r(G). \tag{4}$$

By (1) we have $p({}_bf-f) = -p(-{}_bf+f) = 0$ for all $f\in\mathfrak{A}_c^r(G)$ and all $b\in G$. This and (4) show that $M_0(f) = M_0({}_bf)$. It is also clear that M_0 is nonnegative and $M_0(1)=1$. From (18.9), we infer that $M_0(f) = M(f)$ for $f\in\mathfrak{A}_c^r(G)$. Now if there were a function $f\in\mathfrak{A}_c^r(G)$ such that

$$-p(-f) = \underline{\lim_{n\to\infty}} \frac{1}{\lambda(H_n)} \int\limits_{H_n} f(x)\,dx < \overline{\lim_{n\to\infty}} \frac{1}{\lambda(H_n)} \int\limits_{H_n} f(x)\,dx = p(f),$$

Theorem (B.12) would imply the existence of two *distinct* functionals M_0 and M_1 both satisfying (4). They would both be equal to M on $\mathfrak{A}_c(G)$, and a contradiction would result. Consequently the limit in (ii) exists and (ii) itself is established. □

Sets $H_1, H_2, \ldots, H_n, \ldots$ for which (18.10.ii) holds need not exist. However, if G is locally compact, σ-compact, and Abelian, they do. We give the construction in a sequence of lemmas.

(18.11) Lemma. *Let G be a locally compact group with left Haar measure λ, let A be a λ-measurable subset of G such that $0<\lambda(A)<\infty$, and let x be any element of G. Write $A_0=A$, and $A_n=A\cup xA\cup\cdots\cup x^nA$ $(n=1, 2, 3, \ldots)$. Then*

(i) $\dfrac{\lambda(xA_n\cap A'_n)}{\lambda(A_n)}\leqq\dfrac{1}{n+1}\qquad (n=0, 1, 2, \ldots)$.

Proof. Let $B_n=A_n\cap A'_{n-1}$ $(n=1, 2, \ldots)$. For $n=2, 3, 4, \ldots$, we have

$$B_n=x^nA\cap(A\cup xA\cup\cdots\cup x^{n-1}A)'\subset x^nA\cap(xA\cup\cdots\cup x^{n-1}A)'$$
$$=x\left(x^{n-1}A\cap(A\cup xA\cup\cdots\cup x^{n-2}A)'\right)=xB_{n-1}.$$

It follows that

$$\lambda(A)\geqq\lambda(B_1)\geqq\lambda(B_2)\geqq\cdots\geqq\lambda(B_n)\geqq\cdots. \tag{1}$$

It is also clear that

$$\frac{\lambda(xA_n\cap A'_n)}{\lambda(A_n)}=\frac{\lambda(B_{n+1})}{\lambda(A)+\lambda(B_1)+\cdots+\lambda(B_n)} \tag{2}$$

$(n=1, 2, 3, \ldots)$. If $\lambda(B_{n+1})$ is zero, (2) shows that (i) holds. If $\lambda(B_{n+1})$ is positive, then (2) and (1) yield

$$\frac{\lambda(xA_n\cap A'_n)}{\lambda(A_n)}\leqq\frac{\lambda(B_{n+1})}{(n+1)\,\lambda(B_n)}\leqq\frac{\lambda(B_n)}{(n+1)\,\lambda(B_n)}=\frac{1}{n+1}$$

[note that $\lambda(B_n)<\infty$]. □

(18.12) Lemma. *Let G be a locally compact Abelian group with Haar measure λ. Let U and V be neighborhoods of e in G such that U^- and V^- are compact, and let ε be a positive number. Then there is an open subset H of G such that H^- is compact, $V\subset H$, and*

(i) $\dfrac{\lambda((HU)\cap H')}{\lambda(H)}<\varepsilon$.

Proof. Since V^-U^- is compact (4.4), there are elements $t_1, t_2, \ldots, t_r\in G$ such that $VU\subset V^-U^-\subset\bigcup_{j=1}^{r}t_jV$. For $n=1, 2, \ldots$, let $W_n=\bigcup\{t_1^{\alpha_1}t_2^{\alpha_2}\cdots t_r^{\alpha_r}V\}$, the union being extended over all ordered r-tuples $(\alpha_1, \alpha_2, \ldots, \alpha_r)$ of nonnegative integers such that $\alpha_j\leqq n$ $(j=1, 2, \ldots, r)$. It is obvious that

$$W_nU\subset\bigcup_{j=1}^{r}t_jW_n\qquad(n=1, 2, \ldots). \tag{1}$$

Now consider any $j=1, 2, \ldots, r$. We have $W_n=\bigcup_{\alpha_j=0}^{n}t_j^{\alpha_j}A$, where $A=\bigcup\{t_1^{\alpha_1}\cdots t_{j-1}^{\alpha_{j-1}}t_{j+1}^{\alpha_{j+1}}\cdots t_r^{\alpha_r}V\colon 0\leqq\alpha_k\leqq n,\ k=1, \ldots, r,\ k\neq j\}$. Plainly we have $0<\lambda(A)<\infty$; (18.11) shows that

$$\frac{\lambda(t_jW_n\cap W'_n)}{\lambda(W_n)}\leqq\frac{1}{n+1}\qquad(n=1, 2, \ldots). \tag{2}$$

Using (1) and (2), we make the following estimates:

$$\frac{\lambda((W_n U) \cap W_n')}{\lambda(W_n)} \leqq \frac{\lambda\left(\left(\bigcup_{j=1}^{r} t_j W_n\right) \cap W_n'\right)}{\lambda(W_n)}$$
$$\leqq \sum_{j=1}^{r} \frac{\lambda((t_j W_n) \cap W_n')}{\lambda(W_n)}$$
$$\leqq \sum_{j=1}^{r} \frac{1}{n+1} = \frac{r}{n+1}.$$

Since r is fixed, we obtain (i) by taking $H = W_n$ with any $n > \frac{r}{\varepsilon} - 1$. □

(18.13) Lemma. *Let G be a locally compact, σ-compact, Abelian group. There is a sequence $\{H_n\}_{n=1}^{\infty}$ of subsets of G with the following properties:*

(i) *each H_n is open and H_n^- is compact;*

(ii) $H_1 \subset H_2 \subset \cdots \subset H_n \subset \cdots$;

(iii) $\bigcup_{n=1}^{\infty} H_n = G$;

(iv) *for each $x \in G$,* $\lim_{n\to\infty} \frac{\lambda((xH_n) \cap H_n')}{\lambda(H_n)} = 0.$

Proof. Let $\{F_n\}_{n=1}^{\infty}$ be an increasing sequence of compact subsets of G such that $e \in F_1$ and $\bigcup_{n=1}^{\infty} F_n = G$. Let W be any neighborhood of e with compact closure, and let $U_n = F_n W$ $(n = 1, 2, 3, \ldots)$. We define the sets H_n by induction. Use (18.12) to find an open set H_1 with compact closure such that $H_1 \supset U_1$ and

$$\frac{\lambda((H_1 U_1) \cap H_1')}{\lambda(H_1)} < 1.$$

Suppose that $H_1, H_2, \ldots, H_{n-1}$ have been defined. Use (18.12) to find an open set H_n with compact closure such that $H_n \supset H_{n-1} \cup U_n$ and

$$\frac{\lambda((H_n U_n) \cap H_n')}{\lambda(H_n)} < \frac{1}{n}.$$

Thus $\{H_n\}_{n=1}^{\infty}$ is defined. Properties (i), (ii), and (iii) are evident. Since $\bigcup_{n=1}^{\infty} U_n = G$ and $U_1 \subset U_2 \subset \cdots \subset U_n \subset \cdots$, we can choose for each $x \in G$ a positive integer n_0 such that $x \in U_n$ for $n \geqq n_0$. For all such n, we have

$$\frac{\lambda((xH_n) \cap H_n')}{\lambda(H_n)} \leqq \frac{\lambda((H_n U_n) \cap H_n')}{\lambda(H_n)} < \frac{1}{n}.$$

This implies property (iv). □

(18.14) Theorem. *Let G be a locally compact, σ-compact, Abelian group. There is an increasing sequence $\{H_n\}_{n=1}^{\infty}$ of open sets having compact closure such that*

(i) $M(f) = \lim_{n\to\infty} \frac{1}{\lambda(H_n)} \int_{H_n} f(x)\,dx$

for all $f \in \mathfrak{A}_c(G)$.

Proof. Let $\{H_n\}_{n=1}^{\infty}$ be as in (18.13) and apply (18.10). □

Miscellaneous theorems and examples

(18.15) Applications of Theorems (18.10) and (18.14). (a) Consider the additive group R. For $f \in \mathfrak{A}_c(R)$, we have $M(f) = \lim_{T\to\infty} \frac{1}{2T} \int_{-T}^{T} f(x)\,dx = \lim_{T\to\infty} \frac{1}{(T-a)} \int_{a}^{T} f(x)\,dx = \lim_{T\to\infty} \frac{1}{(T+b)} \int_{-T}^{b} f(x)\,dx$, where a and b are arbitrary real numbers. [Consider the subadditive functional $p(f) = \overline{\lim_{T\to\infty}} \frac{1}{2T} \int_{-T}^{T} f(x)\,dx$, and argue as in (18.10); similarly for the second and third equalities.]

(b) Consider the additive group Z. For $f \in \mathfrak{A}(Z)$, we have

$$M(f) = \lim_{m\to\infty} \frac{1}{2m+1} \sum_{j=-m}^{m} f(j) = \lim_{m\to\infty} \frac{1}{m-a+1} \sum_{j=a}^{m} f(j)$$
$$= \lim_{m\to\infty} \frac{1}{m+b+1} \sum_{j=-m}^{b} f(j).$$

[See (18.10).]

(c) Let G be a compact group. For $f \in \mathfrak{A}_c(G)$, we have $M(f) = I(f)$, where I is the Haar integral on $\mathfrak{C}(G)$.[1] [This is obvious from (15.5) and (18.9).]

(d) Let G and $\widetilde{G}$ be locally compact groups with left Haar measures λ and λ_0, respectively. Suppose that $\{H_n\}_{n=1}^{\infty}$ and $\{\widetilde{H}_n\}_{n=1}^{\infty}$ are sequences of subsets of G and $\widetilde{G}$, respectively, that satisfy (18.10.i). Then for every $(x, y) \in G \times \widetilde{G}$, we have

$$\lim_{n\to\infty} \frac{\lambda\times\lambda_0\left((x,y)\cdot(H_n\times\widetilde{H}_n)\cap(H_n\times\widetilde{H}_n)'\right)}{\lambda(H_n)\,\lambda_0(\widetilde{H}_n)} = 0.$$

(e) Consider a topological group of the form $G_0 \times R^a \times Z^b$, where G_0 is a compact group and a and b are nonnegative integers. For $n = 1, 2, \ldots$, let H_n be the set of all $(x, y_1, \ldots, y_a, z_1, \ldots, z_b) \in G_0 \times R^a \times Z^b$ such that $y_j \in R$ and $|y_j| < n$, $z_k \in Z$ and $|z_k| \leqq n$ ($j = 1, \ldots, a$; $k = 1, \ldots, b$), and x is arbitrary in G_0. Then for $f \in \mathfrak{A}_c(G_0 \times R^a \times Z^b)$, we have

$$M(f) = \lim_{n\to\infty} (2n)^{-a}(2n+1)^{-b} \int_{H_n} f(w)\,d\lambda(w),$$

[1] It is easy to show that $\mathfrak{C}(G) = \mathfrak{A}_c(G)$ if G is compact; see Vol. II, (33.26.b). This fact, while interesting, is unimportant for our present purpose.

where λ is an appropriately normalized Haar measure on $G_0 \times R^a \times Z^b$. [Apply (d) and (18.10).]

(18.16) Miscellaneous means (adapted from MAAK [2]). (a) Let S be an arbitrary semigroup. For a function f in $\mathfrak{B}(S)$ and $a \in S$, let $D_a f$ be the function on $S \times S = S^2$ such that $D_a f(x, y) = f(x a y)$. For a positive number ε, let E_ε be the set of all complex numbers z such that $\left\| z - \frac{1}{m} \sum_{j=1}^{m} D_{a_j} f \right\|_u < \varepsilon$ for some sequence $\{a_1, a_2, \ldots, a_m\}$ of elements of S. If $z_1, z_2 \in E_\varepsilon$, then $|z_1 - z_2| < 2\varepsilon$. If all sets E_ε are nonvoid, then $\bigcap_{\varepsilon > 0} E_\varepsilon$ consists of a single complex number, which we write as $M(f)$. [Suppose that

$$\left| z_1 - \frac{1}{m} \sum_{j=1}^{m} f(x a_j y) \right| < \varepsilon \quad \text{and} \quad \left| z_2 - \frac{1}{n} \sum_{k=1}^{n} f(x b_k y) \right| < \varepsilon$$

for all $(x, y) \in S^2$. Set $x = a_{j'} b_k$ and $y = b_{k'}$ in the first inequality, sum over j', k, k', and divide by $m n^2$. This yields

$$\left| z_1 - \frac{1}{m^2 n^2} \sum_{j, j', k, k'} f(a_{j'} b_k a_j b_{k'}) \right| < \varepsilon .$$

This relation with z_1 replaced by z_2 is obtained by setting $x = a_{j'}$ and $y = a_j b_{k'}$ in the second inequality above. The argument now follows the proof of (18.8).]

(b) Suppose that $f \in \mathfrak{B}(S)$ and that $M(f)$ exists. Then $M(\alpha f) = \alpha M(f)$ for all $\alpha \in K$, and $M(f) = M({}_a f) = M(f_b) = M({}_a f_b)$ for all $a, b \in S$. If $f \in \mathfrak{B}^+(S)$ and $M(f)$ exists, then $M(f) \geqq 0$. If $\|f^{(n)} - f\|_u \to 0$ as $n \to \infty$ and $M(f^{(n)})$ exists for all n, then $M(f)$ exists and is equal to $\lim_{n \to \infty} M(f^{(n)})$. If $M(f)$ and $M(g)$ exist, then $M(f+g) = M(f) + M(g)$. [All assertions but the last are immediate. We sketch a proof of the last. If $\left| z - \frac{1}{m} \sum_{j=1}^{m} f(x a_j y) \right| < \frac{\varepsilon}{8}$ and $\left| w - \frac{1}{n} \sum_{k=1}^{n} g(x b_k y) \right| < \frac{\varepsilon}{8}$ for all $(x, y) \in S^2$, then $\left| z - \frac{1}{mn} \sum_{j,k} f(x a_j b_k y) \right| < \frac{\varepsilon}{8}$ and $\left| w - \frac{1}{mn} \sum_{j,k} g(x a_j b_k y) \right| < \frac{\varepsilon}{8}$. This implies that $\left| (z + w) - \frac{1}{mn} \sum_{j,k} (f+g)(x a_j b_k y) \right| < \frac{\varepsilon}{4}$, and so E_ε is nonvoid for the function $f + g$. We also have $|M(f) - z| < \frac{\varepsilon}{4}$, $|M(g) - w| < \frac{\varepsilon}{4}$, and $|M(f+g) - (z+w)| < \frac{\varepsilon}{2}$, so that $|M(f) + M(g) - M(f+g)| < \varepsilon$.]

(c) A *zero element* or *zero* of a semigroup S is an element $0 \in S$ such that $0x = x0 = 0$ for all $x \in S$. If S has a zero, then $M(f)$ exists and is equal to $f(0)$ for all $f \in \mathfrak{B}(S)$. The functional $f \to f(0)$ is the only invariant mean on $\mathfrak{B}(S)$. See also (17.18.e).

(d) Let S be an infinite *cancellation* semigroup: $ax=ay$ implies $x=y$ and $xa=ya$ implies $x=y$, for all $x, y, a\in S$. If $f\in\mathfrak{B}(S)$ and $\{x\in S: |f(x)|>\varepsilon\}$ is finite for all $\varepsilon>0$, then $M(f)$ exists and is zero. [In view of (b), it suffices to show that $M(\xi_{\{b\}})=0$ for all $b\in S$. If $a_1, a_2, \ldots, a_m$ are distinct elements of S and x, y are elements of S, then the equality $xa_jy=b$ can hold for at most one j, $j=1, 2, \ldots, m$. Thus we have $\left|\frac{1}{m}\sum_{j=1}^{m}\xi_{\{b\}}(xa_jy)\right|\leq\frac{1}{m}$ for all $(x, y)\in S^2$.]

(e) Let G be a locally compact, noncompact topological group for which the uniform structures $\mathscr{S}_l(G)$ and $\mathscr{S}_r(G)$ are equivalent. Let f be a function in $\mathfrak{B}(G)$ such that the set $\{x\in G: |f(x)|>\varepsilon\}^-$ is compact for all $\varepsilon>0$. Then $M(f)$ exists and is equal to zero. [We will show that there is a neighborhood W of e in G such that $M(\xi_W)$ exists and is equal to zero. Note also that $0\leq f\leq g$ and $M(g)=0$ imply $M(f)=0$. These facts together with (b) suffice to prove (e). Let V and W be neighborhoods of e in G such that V^- is compact, W is symmetric (4.6), and $xW^2x^{-1}\subset V$ for all $x\in G$ [(4.14.g) and (4.5.i)]. Define a sequence $\{a_j\}_{j=1}^{\infty}$ of elements of G by induction. Let a_1 be arbitrary. When $a_1, a_2, \ldots, a_{n-1}$ have been defined, let a_n be any element not in $(a_1V)\cup(a_2V)\cup\cdots\cup(a_{n-1}V)$. Now consider the function $h=\sum_{j=1}^{n}D_{a_j}(\xi_W)$. Its value at every $(x, y)\in G^2$ is either 1 or 0. In fact, if $xa_jy\in W$ and $xa_ky\in W$ where $j\geq k$, then we have $(xa_ky)^{-1}(xa_jy)=y^{-1}a_k^{-1}a_jy\in W^2$, $a_k^{-1}a_j\in yW^2y^{-1}\subset V$, so that $a_j\in a_kV$. This implies that $j\leq k$; and so $j=k$. It follows that $\left\|\frac{1}{n}h\right\|_u=\frac{1}{n}$, and that $M(\xi_W)=0$.]

Let G be a locally compact, noncompact topological group and suppose that there is a continuous homomorphism τ of G *onto* some noncompact Abelian group H.[1] If f is a function in $\mathfrak{B}(G)$ and the sets $\{x\in G: |f(x)|>\varepsilon\}^-$ are compact for $\varepsilon>0$, then $M(f)$ exists and is equal to zero. [It suffices to prove that $M(\xi_F)=0$ for all nonvoid compact subsets F of G. Let $a_1=e$. Since FF^{-1} is compact, $\tau(FF^{-1})$ is a compact subset of H and since H is Abelian, we have $\tau(\bigcup_{x\in G}xFF^{-1}x^{-1})=\tau(FF^{-1})$. It follows that $\bigcup_{x\in G}xFF^{-1}x^{-1}\neq G$; let $a_2\notin\bigcup_{x\in G}xFF^{-1}x^{-1}$. Having defined $a_1, \ldots, a_{n-1}$, we find that $\bigcup_{k=1}^{n-1}(\bigcup_{x\in G}xFF^{-1}x^{-1})a_k\neq G$ and choose a_n in $G\cap\left(\bigcup_{k=1}^{n-1}(\bigcup_{x\in G}xFF^{-1}x^{-1})a_k\right)'$. We then have $\left\|\frac{1}{n}\sum_{k=1}^{n}D_{a_k}(\xi_F)\right\|_u=\frac{1}{n}$ for $n=1, 2, \ldots$.][2]

---

[1] This property holds whenever G is not unimodular [use the modular function Δ (15.11)] and also holds for some unimodular groups such as $\mathfrak{GL}(n, K)$ [use the function $A\rightarrow\det A$].

[2] We do not know if the result of (18.16.e) holds for all locally compact groups G and all $f\in\mathfrak{B}(G)$ such that $\{x\in G: |f(x)|>\varepsilon\}^-$ is compact for all $\varepsilon>0$. The question is open even for $f\in\mathfrak{C}_0(G)$.

(f) Let G be a group and H a normal subgroup of G such that G/H is infinite. Then $M(\xi_H)=0$. [If $a_1, a_2, \ldots, a_n$ lie in distinct cosets of H, then $\left\|\frac{1}{n}\sum_{k=1}^{n} D_{a_k}(\xi_H)\right\|_u = \frac{1}{n}$.]

(g) Let G be a group and H an infinite normal subgroup of G. Let B be a subset of G such that $\overline{\overline{B\cap(xH)}}$ is less than or equal to p (a positive integer) for all $x\in G$. Then $M(\xi_B)=0$. [If $a_1, a_2, \ldots, a_n$ are distinct elements of H, then $\left\|\frac{1}{n}\sum_{k=1}^{n} D_{a_k}(\xi_B)\right\|_u \leqq \frac{p}{n}$.]

(h) Let G be an infinite Abelian group. There is a function f in $\mathfrak{B}^+(G)$ such that $f(x)>0$ for all $x\in G$ and $M(f)=0$. [By (16.13.c), G has a subgroup H such that G/H is countably infinite. Let $A_1, A_2, \ldots, A_n, \ldots$ be the distinct cosets of H. Then $M(\xi_{A_n})=M(\xi_H)=0$ as (b) and (f) show. Now set $f=\sum_{n=1}^{\infty} 2^{-n}\xi_{A_n}$ and apply (b) once more[1].]

(18.17) ω-Functions (KEINER [1] and MAAK [2]). (a) We now describe a class of functions f in $\mathfrak{B}(S)$ [S is an arbitrary semigroup] for which the mean value defined in (18.16.a) exists. Suppose that for every $\varepsilon>0$, there are subsets $A_1, A_2, \ldots, A_n$ and E of S with the following properties:

$$E=S\cap(A_1\cup\cdots\cup A_n)'; \tag{1}$$

$$M(\xi_E)=0; \tag{2}$$

if $a, b\in S$ and the set $B_{a,b}=\{(a,b)\}\cup\{(xay, xby): x, y\in S\}$ has nonvoid intersection with some $A_k\times A_k$, then $|f(z)-f(w)|<\varepsilon$ for all $(z,w)\in B_{a,b}\cap(E'\times E')$. (3)

Then f is called an *ω-function*. We will prove that $M(f)$ [in the sense of (18.16)] exists for ω-functions f. We use a sequence of lemmas.

(b) Let f be an ω-function; let $A_1, A_2, \ldots, A_n$ and E be as in (a) for a given $\varepsilon>0$; and suppose that n has the smallest possible value for this ε. Let x, y be any fixed elements of S. Then there is a permutation τ of $\{1, 2, \ldots, n\}$ such that $A_k\cap(xA_{\tau(k)}y)\neq\varnothing$ for $k=1, 2, \ldots, n$. [Like (18.7), this result depends upon (18.4). For $k=1, 2, \ldots, n$, let $B_k=xA_ky$ and let $\varrho(k)=\{j: A_j\cap B_k\neq\varnothing,\ 1\leqq j\leqq n\}$. Given a set $\{k_1, k_2, \ldots, k_r\}$ where $1\leqq k_1<k_2<\cdots<k_r\leqq n$, let $\{j_1, j_2, \ldots, j_s\}$ denote the set $\varrho(k_1)\cup\cdots\cup\varrho(k_r)$, where $1\leqq j_1<j_2<\cdots<j_s\leqq n$. It is obvious that

$$B_{k_1}\cup\cdots\cup B_{k_r}\subset A_{j_1}\cup\cdots\cup A_{j_s}\cup E.$$

[1] This construction shows spectacularly that there is no analogue of LEBESGUE'S theorem on dominated convergence for the functional M. In fact, writing $h_n=\xi_{A_1}+\cdots+\xi_{A_n}$, we have $\lim_{n\to\infty} h_n=1$, $h_1\leqq h_2\leqq\cdots\leqq h_n\leqq\cdots$, $M(1)=1$, and $M(h_n)=0$ for $n=1, 2, 3, \ldots$.

Now let $C_l=\{z\in S: xzy\in A_{j_l}\}$ for $l=1, 2, \ldots, s$, and let $F=\{z\in S: xzy\in E\}$. Clearly we have $M(\xi_F)=M({}_x(\xi_E)_y)=0$ and hence $M(\xi_{E\cup F})=0$. Write the sets A_k different from $A_{k_1}, \ldots, A_{k_r}$ as $A_{k_{r+1}}, \ldots, A_{k_n}$, and consider the family of sets

$$C_1\cap(E\cup F)', \ldots, C_s\cap(E\cup F)', A_{k_{r+1}}\cap(E\cup F)', \ldots, A_{k_n}\cap(E\cup F)', E\cup F.$$

It is easy to see that this family of sets satisfies (1) and (3), and so $s+n-r\geqq n$, $s\geqq r$. Thus (18.4) can be applied, and (b) follows.]

(c) Let f be an ω-function; let $A_1, A_2, \ldots, A_n$ and E be as in (a) for a given $\varepsilon>0$; and suppose that n is as small as possible for this ε. For every $\delta>0$, there is a positive integer N and in each set A_k there are elements $v_{k,1}, v_{k,2}, \ldots, v_{k,N}$ such that

$$\left\|\frac{1}{N}\sum_{j=1}^{N} D_{v_{k,j}}(\xi_E)\right\|_u<\delta \qquad (k=1, 2, \ldots, n). \tag{4}$$

[Let $\alpha_k=\inf\left\{\left\|\frac{1}{p}\sum_{l=1}^{p} D_{a_l}(\xi_{A_k})\right\|_u : \{a_1, \ldots, a_p\}\text{ is a finite sequence of elements of } S\right\}$, for $k=1, 2, \ldots, n$. Since n is as small as possible, all α_k's are positive. Let $\alpha=\frac{1}{2}\min(\alpha_1, \alpha_2, \ldots, \alpha_n)$. Choose elements $u_1, u_2, \ldots, u_L$ in S so that $\left\|\frac{1}{L}\sum_{l=1}^{L} D_{u_l}(\xi_E)\right\|_u<\alpha\delta$. For $k=1, 2, \ldots, n$, there are elements x_k and y_k in S such that

$$\frac{1}{L}\sum_{l=1}^{L}\xi_{A_k}(x_k u_l y_k)\geqq\frac{\alpha_k}{2}\geqq\alpha, \tag{5}$$

while at the same time

$$\frac{1}{L}\sum_{l=1}^{L}\xi_E(x x_k u_l y_k y)<\alpha\delta \quad\text{for all}\quad (x, y)\in S^2. \tag{6}$$

Let N be the smallest integer such that $N\geqq\alpha L$. Because of (5), at least N of the elements $x_k u_1 y_k, \ldots, x_k u_L y_k$ lie in A_k: take N of these elements as $v_{k,1}, v_{k,2}, \ldots, v_{k,N}$. On account of (6), we have (4).]

(d) Let f be an ω-function on S. Then $M(f)$ exists. More precisely, for every $\varepsilon>0$ there are elements $w_1, \ldots, w_m$ in S such that

$$\left\|\frac{1}{m}\sum_{j=1}^{m} f(w_j)-\frac{1}{m}\sum_{j=1}^{m} D_{w_j}f\right\|_u\leqq\varepsilon. \tag{7}$$

[Let ε be a positive number, and let $A_1, A_2, \ldots, A_n, E$ be as in (a), with ε replaced by $\varepsilon/3$, and with the smallest possible value of n. Let $\delta=\frac{\varepsilon}{6\|f\|_u}$, and choose N and the elements $v_{k,l}$ ($k=1, 2, \ldots, n$; $l=1, 2, \ldots, N$) as in (c), for this value of δ. Let x, y be arbitrary elements of S, let τ be as in (b), and let c_k be an element of $A_k\cap(xA_{\tau(k)}y)$ ($k=$

1, 2, ..., n). We make the following estimates:

$$\left.\begin{aligned}&\left|\frac{1}{nN}\sum_{k,l} f(v_{k,l}) - \frac{1}{nN}\sum_{k,l} f(x v_{k,l} y)\right| \\ &\quad = \frac{1}{nN}\left|\sum_{k,l}\left(f(v_{k,l}) - f(x v_{\tau(k),l} y)\right)\right| \leqq \frac{1}{nN}\sum_{k,l}|f(v_{k,l}) - f(c_k)| \\ &\quad + \frac{1}{nN}\sum_{k,l}|f(c_k) - f(x v_{\tau(k),l} y)| = \Sigma_1 + \Sigma_2.\end{aligned}\right\} \tag{8}$$

Since $v_{k,l} \in A_k$, $c_k \in A_k$, (3) shows that

$$\Sigma_1 < \frac{\varepsilon}{3}. \tag{9}$$

To estimate Σ_2, consider the values of k and l such that $x v_{\tau(k),l} y \notin E$. We have $c_k = x s y$, where $s \in A_{\tau(k)}$, and again (3) shows that

$$\frac{1}{nN}\Sigma' |f(c_k) - f(x v_{\tau(k),l} y)| < \frac{\varepsilon}{3}, \tag{10}$$

the sum being taken over all k and l such that $x v_{\tau(k),l} y \notin E$. Let Σ'' denote the sum over all k and l such that $x v_{\tau(k),l} y \in E$. From (4), we see that

$$\frac{1}{nN}\Sigma'' |f(c_k) - f(x v_{\tau(k),l} y)| \leqq \frac{1}{nN} n N \delta \cdot 2\|f\|_u = \frac{\varepsilon}{3}. \tag{11}$$

Combining (11), (10), (9), and (8), we get (7) with the w_j's equal to the $v_{k,l}$'s.]

Notes

The literature on almost periodic functions is enormous: for a survey see MAAK [1], pp. 222—235. Lemma (18.4) is due to HALMOS and VAUGHAN [1]; our language is regrettably less colorful than theirs. The construction of M and the proof of its uniqueness (18.6)—(18.9) are taken from MAAK [1], pp. 31—42. Theorems (18.10)—(18.14) are based on LYUBARSKIĬ [1] and an unpublished manuscript of A. BEURLING and E. HEWITT. See also KAWADA [1] and STRUBLE [1] for related results. One-parameter families of sets resembling the sets $\{H_n\}_{n=1}^{\infty}$ of (18.13) were used by CALDERÓN [1] to prove ergodic theorems.

Chapter Five

Convolutions and group representations

In the present chapter, we initiate our study of harmonic analysis proper. The basic operation in harmonic analysis is *convolution*; in §19, we give a reasonably general definition of convolutions and develop with some care the fundamental properties of convolutions of measures. In §20, we examine explicit formulas for convolutions of measures and functions. In §21, we present some facts about representations of groups and algebras. In §22, we prove the existence of irreducible representations of locally compact groups.

Our main goal in this chapter is to prove the GEL'FAND-RAĬKOV theorem (22.12): this theorem is one of the most important in the entire book and incidentally is needed at once in the theory of locally compact Abelian groups, which we treat in Chapter Six. It would be perfectly reasonable to postpone a detailed discussion of convolutions until we have treated representations of topological groups and the theory of locally compact Abelian groups. However, we wish to treat representations of locally compact groups by means of the *algebra* $\mathfrak{L}_1(G, \lambda)$. To define multiplication in $\mathfrak{L}_1(G, \lambda)$ [which is convolution], we need to define convolutions in general. Our discussion of convolutions in the present chapter is only introductory: several subsequent chapters are devoted to a detailed study of various convolution algebras.

§ 19. Introduction to convolutions

As usual, we begin with a definition.

(19.1) Definition. Let S be a semigroup [not necessarily topological] and let $\mathfrak{F}$ be a linear space of real- or complex-valued functions on S. We suppose throughout that $\mathfrak{F}$ is left invariant: ${}_xf\in\mathfrak{F}$ if $f\in\mathfrak{F}$ and $x\in S$. For a linear functional M defined on $\mathfrak{F}$ and $f\in\mathfrak{F}$, let $\overline{M}f$ be the function on S such that $\overline{M}f(x)=M({}_xf)$ for all $x\in S$. Suppose further that the linear functional M is such that $\overline{M}f\in\mathfrak{F}$ for all $f\in\mathfrak{F}$. Then if L is any linear functional on $\mathfrak{F}$, the functional whose value at $f\in\mathfrak{F}$ is $L(\overline{M}f)=(L\circ\overline{M})f$ is well defined. It is called the *convolution* of L and M and is written $L*M$. For a *right* invariant space of functions $\mathfrak{F}$, a linear functional L on $\mathfrak{F}$, and $f\in\mathfrak{F}$, we define $\underline{L}f$ as the function $\underline{L}f(x)=L(f_x)$.

(19.2) Theorem. (i) *The convolution $L*M$ is a linear functional on $\mathfrak{F}$.*

(ii) *For a [real or complex] number α, the identities $\alpha(L*M)=(\alpha L)*M=L*(\alpha M)$ hold.*

(iii) *If $L*M$ and $L*N$ exist, then $L*(M+N)$ exists and $L*(M+N)=L*M+L*N$; similarly $(M+N)*L=M*L+N*L$.* [1]

(iv) *If L, M, N are linear functionals on $\mathfrak{F}$ and $\overline{M}$ and $\overline{N}$ carry $\mathfrak{F}$ into itself, then $L*(M*N)=(L*M)*N$.*

---

[1] Properly we should write $(L*M)+(L*N)$; but we regard convolution as a sort of multiplication and use the universal convention that $ax+by=(ax)+(by)$ in any ring.

Proof. Assertions (i), (ii), and (iii) are immediate consequences of (19.1): we omit their proofs. To prove (iv), note first that $(L*M)*N$ obviously exists. For $x, y \in S$, we have $\overline{N}({}_x f)(y) = N({}_y({}_x f)) = N({}_{xy} f) = \overline{N} f(xy) = {}_x(\overline{N} f)(y)$.[1] Thus $\overline{N}({}_x f) = {}_x(\overline{N} f)$. From this we infer that $(\overline{M*N}) f(x) = (M*N)({}_x f) = M(\overline{N}({}_x f)) = M({}_x(\overline{N} f)) = \overline{M}(\overline{N} f)(x)$. That is, $\overline{M*N} = \overline{M} \circ \overline{N}$. This shows that $L*(M*N)$ exists, and we can write $L*(M*N) = L \circ (\overline{M*N}) = L \circ (\overline{M} \circ \overline{N}) = (L \circ \overline{M}) \circ \overline{N} = (L*M)*N$. □

(19.3) Definition. Let $\mathfrak{F}$ be as in (19.1) and let $\boldsymbol{L}$ be a linear space [real or complex according as $\mathfrak{F}$ is a real or complex linear space] of linear functionals on $\mathfrak{F}$ such that $L*M$ exists and is an element of $\boldsymbol{L}$ for all $L, M \in \boldsymbol{L}$. Then $\boldsymbol{L}$ is called a *convolution algebra* [with $*$ as multiplication and addition defined as usual]. It follows at once from (19.2) that $\boldsymbol{L}$ is an associative algebra over the real or complex field.

Convolution algebras abound in both analysis and algebra. Many have been exhaustively studied; others, of equal interest and possibly of equal importance, have hardly been noticed. We first give an example.

(19.4) The l_1-algebra of a semigroup. Let S be any semigroup and let $\mathfrak{F}$ be the linear space $\mathfrak{B}(S)$ of all bounded complex-valued functions on S. For each $a \in S$, let E_a be the linear functional $E_a(f) = f(a)$. Then $\overline{E}_a f = f_a$, as is easy to see, and $E_a * E_b(f) = E_a(f_b) = f_b(a) = f(ab) = E_{ab}(f)$. Thus the set of linear functionals E_a forms a semigroup under convolution that is isomorphic with S. By forming the set of all complex linear combinations $\sum_{k=1}^{n} \alpha_k E_{a_k}$, we obtain an algebra containing an isomorph of S as a multiplicative subsemigroup. One can go a step further and consider all of the linear functionals $\sum_{k=1}^{\infty} \alpha_k E_{a_k}$ such that $\alpha_k \neq 0$, $\sum_{k=1}^{\infty} |\alpha_k| < \infty$ and $a_1, a_2, \ldots, a_n, \ldots$ are distinct elements of S. [The zero functional must be treated separately, but is trivial.] For an obvious reason, we call this *the l_1-algebra of* S, and denote it [although this is not really accurate] by $l_1(S)$. It is clear that every $\sum_{k=1}^{\infty} \alpha_k E_{a_k}$ is a bounded linear functional on $\mathfrak{B}(S)$: $\left(\sum_{k=1}^{\infty} \alpha_k E_{a_k}\right)(f) = \sum_{k=1}^{\infty} \alpha_k f(a_k)$. It is obvious [take $f(a_k) = \operatorname{sgn} \bar{\alpha}_k$] that $\left\| \sum_{k=1}^{\infty} \alpha_k E_{a_k} \right\| = \sum_{k=1}^{\infty} |\alpha_k|$. To verify that $l_1(S)$ is a convolution algebra, and so to justify the expression "l_1-algebra", we compute as follows. For $f \in \mathfrak{B}(S)$, $A = \sum_{k=1}^{\infty} \alpha_k E_{a_k} \in l_1(S)$, and $B = \sum_{l=1}^{\infty} \beta_l E_{b_l} \in l_1(S)$, we have $\overline{B} f = \sum_{l=1}^{\infty} \beta_l f_{b_l}$, which is a function in $\mathfrak{B}(S)$, and hence

$$(A * B)(f) = \sum_{k=1}^{\infty} \alpha_k \left[\sum_{l=1}^{\infty} \beta_l f(a_k b_l) \right]. \tag{1}$$

---

[1] The associativity of S is needed only to prove the identity ${}_y({}_x f) = {}_{xy} f$.

Since

$$\sum_{k=1}^{\infty}|\alpha_k|\left[\sum_{l=1}^{\infty}|\beta_l|\,|f(a_k b_l)|\right]\leqq\|A\|\cdot\|B\|\cdot\|f\|_u, \tag{2}$$

we may rearrange the left side of (2) and the right side of (1) in any way we like without changing their values. Let $\{x_1, x_2, \ldots, x_n, \ldots\}$ be the [finite or countably infinite] set of all *distinct* products $a_k b_l$. Then we have

$$(A*B)(f)=\sum_{n=1}^{\infty}\left[\sum_{a_k b_l=x_n}\alpha_k\beta_l\right]f(x_n).$$

To show that $A*B\in l_1(S)$, we must show that

$$\sum_{n=1}^{\infty}\left|\sum_{a_k b_l=x_n}\alpha_k\beta_l\right|<\infty.$$

Since $\sum_{k=1}^{\infty}\sum_{l=1}^{\infty}|\alpha_k|\cdot|\beta_l|=\|A\|\cdot\|B\|<\infty$, this sum can be rearranged arbitrarily. For each pair of positive integers (k, l), we have $a_k b_l=x_n$ for precisely one n. Therefore

$$\sum_{n=1}^{\infty}\left|\sum_{a_k b_l=x_n}\alpha_k\beta_l\right|\leqq\sum_{n=1}^{\infty}\sum_{a_k b_l=x_n}|\alpha_k\beta_l|=\sum_{k=1}^{\infty}\sum_{l=1}^{\infty}|\alpha_k|\cdot|\beta_l|=\|A\|\cdot\|B\|.$$

That is, $A*B$ is in $l_1(S)$ and $\|A*B\|\leqq\|A\|\cdot\|B\|$.

Various other examples of convolution algebras, of only ancillary interest, are given in (19.23)—(19.24).

We now take up the most important of the convolution algebras to be studied in this book, namely, the algebra of all bounded linear functionals on $\mathfrak{C}_0(G)$ for a locally compact group G. Throughout (19.5)—(19.28) inclusive, G will denote a locally compact group. From this point on, the symbol $\mathfrak{L}_p(G)$ $(1\leqq p\leqq\infty)$ will denote the Banach space $\mathfrak{L}_p(G, \lambda)$, where λ is a fixed but arbitrary *left* Haar measure on G.

(19.5) Lemma. *Let f be a function in $\mathfrak{C}_0(G)$, and let L be any element of $\mathfrak{C}_0^*(G)$. Then $\overline{L}f$ and $\underline{L}f$ are in $\mathfrak{C}_0(G)$.*

Proof. We carry out the proof only for $\overline{L}f$. If $L=0$ or $f=0$, the lemma is trivial: $\overline{O}(f)=0$ and $\overline{L}(0)=0$. First of all, it is obvious that ${}_xf$ and f_x are in $\mathfrak{C}_0(G)$ for every $f\in\mathfrak{C}_0(G)$ and $x\in G$. Let ε be any positive number, and choose a neighborhood V of the identity in G such that $|f(u)-f(v)|<\frac{\varepsilon}{\|L\|}$ for all $u, v\in G$ such that $uv^{-1}\in V$. This can be done by (15.4) [f is right uniformly continuous]. Then if x and y in G are such that $xy^{-1}\in V$, we have $\|{}_xf-{}_yf\|_u<\frac{\varepsilon}{\|L\|}$ and so $|\overline{L}f(x)-\overline{L}f(y)|<\varepsilon$. This shows that $\overline{L}f$ is continuous [indeed, right uniformly continuous].

To show that $|\overline{L}f|$ is arbitrarily small outside of a suitably chosen compact set, we first write $f=f_1-f_2+i(f_3-f_4)$, where the f_j's are in $\mathfrak{C}_0^+(G)$. Thus we may suppose that f itself is in $\mathfrak{C}_0^+(G)$ and not zero, since $\overline{L}$ is plainly a linear operator. Let μ be the complex measure that represents L as in (14.4) and let $|\mu|$ be the nonnegative measure defined in (14.6). Then as noted in (14.6), $|\mu|(G)$ is finite. Let ε be an arbitrary positive number. By (11.23), there is a compact set $E\subset G$ such that $|\mu|(E')<\frac{\varepsilon}{2\|f\|_u}$. There is also a compact set $F\subset G$ such that $|f(z)|<\frac{\varepsilon}{2|\mu|(G)}$ for $z\in F'$. We now have

$$\left.\begin{aligned} |\overline{L}f(x)| = |L({}_xf)| &\leqq |L|({}_xf) = \int_G f(xy)\,d|\mu|(y) \\ &= \int_E f(xy)\,d|\mu|(y) + \int_{E'} f(xy)\,d|\mu|(y). \end{aligned}\right\} \tag{1}$$

It is clear that

$$\int_{E'} f(xy)\,d|\mu|(y) \leqq \|f\|_u \cdot |\mu|(E') < \frac{\varepsilon}{2}. \tag{2}$$

Also, if $f(xy) \geqq \frac{\varepsilon}{2|\mu|(G)}$, we have $xy\in F$; and if in addition $y\in E$, we have $x\in F(y^{-1})\subset F(E^{-1})$. Thus if $x\in \big(F(E^{-1})\big)'$, we have

$$\int_E f(xy)\,d|\mu|(y) \leqq |\mu|(E)\,\frac{\varepsilon}{2|\mu|(G)} \leqq \frac{\varepsilon}{2}. \tag{3}$$

Combining (3) and (2) with (1) and noting that $F(E^{-1})$ is compact [(4.2) and (4.4)], we see that $\overline{L}f\in\mathfrak{C}_0(G)$. □

(19.6) Theorem. *The conjugate space $\mathfrak{C}_0^*(G)$ is a convolution algebra under Definition* (19.3). *With the norm $\|L\|$, it is a Banach algebra in the sense of* (C.1). *The functionals E_a defined by $E_a(f)=f(a)$ for $f\in\mathfrak{C}_0(G)$ satisfy the relation*

(i) $E_a * E_b = E_{ab}$ *for* $a, b\in G$.

The functional E_e is the unit of $\mathfrak{C}_0^*(G)$:

(ii) $E_e * L = L * E_e = L$ *for all* $L\in\mathfrak{C}_0^*(G)$.

The algebra $\mathfrak{C}_0^*(G)$ is commutative if and only if G is Abelian.

Proof. To show that $\mathfrak{C}_0^*(G)$ is a convolution algebra, we need only to prove that $L*M$ is a *bounded* linear functional on $\mathfrak{C}_0(G)$ if L and M are [see (19.5)]. Let μ and ν be the complex measures that represent L and M, respectively, as in (14.4). Applying (14.7), we have $|M(f)| = \left|\int_G f\,d\nu\right| \leqq \int_G |f|\,d|\nu|$. Hence, for every $x\in G$, we have

$$|\overline{M}f(x)| = |M({}_xf)| \leqq \int_G |{}_xf|\,d|\nu| \leqq \|f\|_u\,|\nu|(G) = \|f\|_u\cdot\|M\|.$$

This implies that $\|\overline{M}f\|_u \leqq \|f\|_u\cdot\|M\|$, and this inequality in turn allows us to write

$$|L*M(f)| = |L(\overline{M}f)| \leqq \|L\|\cdot\|\overline{M}f\|_u \leqq \|L\|\cdot\|M\|\cdot\|f\|_u. \tag{1}$$

The relations (1) obviously show that $L*M$ is a bounded linear functional on $\mathfrak{C}_0(G)$ and also that

$$\|L*M\| \leq \|L\| \cdot \|M\|. \tag{2}$$

By (B.22), the space $\mathfrak{C}_0^*(G)$ is complete in the metric $\|L-M\|$. This fact and (2) show that $\mathfrak{C}_0^*(G)$ is a Banach algebra.

It is obvious that the functionals E_a are in $\mathfrak{C}_0^*(G)$ and that $\|E_a\|=1$, for all $a \in G$. To verify (i), we write

$$(E_a*E_b)(f) = E_a(\overline{E_b}f) = E_a(f_b) = f_b(a) = f(ab) = E_{ab}(f).$$

To verify (ii), we note that $\overline{E_e}f(x) = E_e({}_xf) = f(xe) = f(x)$. That is, $\overline{E_e}f = f$, and so $(L*E_e)(f) = L(f)$. Also we have

$$(E_e*L)(f) = E_e(\overline{L}f) = \overline{L}f(e) = L({}_ef) = L(f).$$

This proves (ii).

If G is non-Abelian, then (i) shows that $\mathfrak{C}_0^*(G)$ is noncommutative. To prove the converse, suppose that G is Abelian, and that L and M are elements of $\mathfrak{C}_0^*(G)$. Let μ and ν be the measures[1] representing L and M, respectively. Let f be in $\mathfrak{C}_0(G)$. The function $(x,y) \to f(xy)$ is continuous [and bounded] on $G \times G$, and so belongs to $\mathfrak{L}_1(G \times G, |\mu \times \nu|)$. Hence by (14.25), we have

$$\left.\begin{aligned} L*M(f) &= \int_G \int_G f(xy)\, d\nu(y)\, d\mu(x) = \int_G \int_G f(xy)\, d\mu(x)\, d\nu(y) \\ &= \int_G \int_G f(yx)\, d\mu(x)\, d\nu(y) = M*L(f). \quad \square \end{aligned}\right\} \tag{3}$$

(19.7) Note. FUBINI's theorem plays the essential rôle in proving (19.6.3), and hence in showing that $\mathfrak{C}_0^*(G)$ is commutative if G is. In situations where FUBINI's theorem fails, convolution algebras can be noncommutative even though the underlying group or semigroup is commutative (19.24).

(19.8) Definition. For L, M in $\mathfrak{C}_0^*(G)$, let μ and ν be the complex measures that represent L and M, respectively, as in (14.4). Then $\mu*\nu$ will denote the complex measure that represents $L*M$ as in (14.4). We call $\mu*\nu$ the *convolution* of the two measures μ and ν.

We now make some computations involving μ, ν, and $\mu*\nu$.

(19.9) Theorem. *Let L, M, μ, and ν be as in* (19.8). *Then*

$$|\mu*\nu| \leq |\mu| * |\nu|.$$

---

[1] Where there is no danger of confusion, we use "measure" to mean "complex measure as constructed in (14.4)".

Proof. Let $\varphi \in \mathfrak{C}_0^+$ and suppose that $f \in \mathfrak{C}_0$ and $|f| \leqq \varphi$. Then using (14.7) twice, we have

$$\begin{aligned} |L*M(f)| &= \left|\int_G \int_G f(xy)\, d\nu(y)\, d\mu(x)\right| \leqq \int_G \left|\int_G f(xy)\, d\nu(y)\right| d|\mu|(x) \\ &\leqq \int_G \int_G |f(xy)|\, d|\nu|(y)\, d|\mu|(x) \\ &\leqq \int_G \int_G \varphi(xy)\, d|\nu|(y)\, d|\mu|(x) = |L| * |M|(\varphi). \end{aligned}$$

Thus by (14.5.i), we have $|L*M|(\varphi) \leqq |L|*|M|(\varphi)$ so that $|\mu*\nu| \leqq |\mu|*|\nu|$. □

(19.10) Theorem. *Suppose that L and M are functionals in $\mathfrak{C}_0^*(G)$ and that μ and ν, respectively, are the complex measures representing them. Let τ denote the mapping $(x, y) \to xy$ of $G \times G$ onto G. If f is a function on G that vanishes almost everywhere for $|\mu| * |\nu|$, then $f \circ \tau$ vanishes almost everywhere on $G \times G$ for $|\mu \times \nu|$. Thus if $f \in \mathfrak{L}_1(G, |\mu| * |\nu|)$, the function $f \circ \tau$ is defined $|\mu \times \nu|$-almost everywhere on $G \times G$. Furthermore, $f \circ \tau$ is in $\mathfrak{L}_1(G \times G, |\mu \times \nu|)$, and*

$$\text{(i)} \quad \int_G f\, d\mu*\nu = \int_{G\times G} (f\circ\tau)\, d\mu\times\nu = \int_G \int_G f(xy)\, d\nu(y)\, d\mu(x) = \int_G \int_G f(xy)\, d\mu(x)\, d\nu(y).$$

Proof. Let F be a compact subset of G. Then $\tau^{-1}(F)$ is closed [τ is continuous!] and hence $|\mu| \times |\nu|$-measurable. Let ε be a positive number. By (11.22), there is an open set U such that $F \subset U$ and $|\mu| * |\nu|(U) < |\mu| * |\nu|(F) + \varepsilon$. Let $f \in \mathfrak{C}_{00}^+(G)$ be taken so that $f(F) = 1$, $f(U') = 0$, and $f(G) \subset [0, 1]$. Then

$$\begin{aligned} \int_{G\times G} \xi_F \circ \tau\, d|\mu| \times |\nu| &= \int_G \int_G \xi_F(xy)\, d|\nu|(y)\, d|\mu|(x) \\ &\leqq \int_G \int_G f(xy)\, d|\nu|(y)\, d|\mu|(x) = |L| * |M|(f) = \int_G f\, d|\mu| * |\nu| \\ &\leqq \int_G \xi_U\, d|\mu| * |\nu| = |\mu| * |\nu|(U) < |\mu| * |\nu|(F) + \varepsilon. \end{aligned}$$

Thus we have

$$\int_{G\times G} \xi_F \circ \tau\, d|\mu| \times |\nu| \leqq |\mu| * |\nu|(F) \tag{1}$$

for compact sets $F \subset G$. It follows immediately that (1) holds for any σ-compact subset of G. Next consider an $|\mu| * |\nu|$-null set A in G. Let D be a countable intersection of open sets in G such that $A \subset D$ and $|\mu|*|\nu|(D) = 0$. We will show that $\xi_A \circ \tau$ is an $|\mu| \times |\nu|$-null function, by showing that $\xi_D \circ \tau = \xi_{\tau^{-1}(D)}$ is an $|\mu| \times |\nu|$-null function. Since $\tau^{-1}(D)$ is a Borel set and hence $|\mu| \times |\nu|$-measurable, it suffices to show that $|\mu| \times |\nu|(E) = 0$ for all compact subsets E of $\{(x, y) \in G \times G : xy \in D\}$. Clearly

$\tau(E)$ is a compact subset of D and $\xi_E \leqq \xi_{\tau(E)} \circ \tau$. Hence by (1), we have

$$|\mu| \times |\nu| (E) = \int\limits_{G \times G} \xi_E \, d|\mu| \times |\nu| \leqq \int\limits_{G \times G} \xi_{\tau(E)} \circ \tau \, d|\mu| \times |\nu|$$
$$\leqq |\mu| * |\nu| (\tau(E)) \leqq |\mu| * |\nu| (D) = 0.$$

Consequently, $\xi_D \circ \tau$ and $\xi_A \circ \tau$ are $|\mu| \times |\nu|$-null functions. Thus if A is an $|\mu| * |\nu|$-null set in G, then $\tau^{-1}(A)$ is an $|\mu| \times |\nu|$-null set in $G \times G$. This implies the first statement of the theorem.

Next consider any $|\mu| * |\nu|$-measurable subset B of G. Then $B = F \cup A$ where $F \cap A = \varnothing$, F is σ-compact, and A is an $|\mu| * |\nu|$-null set. Then $\xi_B \circ \tau = \xi_A \circ \tau + \xi_F \circ \tau$ is $|\mu| \times |\nu|$-measurable and

$$\int\limits_{G \times G} \xi_B \circ \tau \, d|\mu| \times |\nu| \leqq |\mu| * |\nu| (B).$$

It now follows that if f in $\mathfrak{L}_1(G, |\mu| * |\nu|)$ has the form $\sum_{k=1}^{n} \alpha_k \xi_{B_k}$ for $|\mu| * |\nu|$-measurable sets B_k and positive numbers α_k, then $f \circ \tau \in \mathfrak{L}_1(G \times G, |\mu| \times |\nu|)$ and

$$\int\limits_{G \times G} f \circ \tau \, d|\mu| \times |\nu| \leqq \int\limits_{G} f \, d|\mu| * |\nu|. \tag{2}$$

Now consider an arbitrary nonnegative function f in $\mathfrak{L}_1(G, |\mu| * |\nu|)$; we may suppose that f is real-valued. For all positive integers n and k, let $A_{k,n} = \left\{x \in G : \frac{k}{2^n} \leqq f(x) < \frac{k+1}{2^n}\right\}$. Let $f_n = \sum_{k=1}^{n 2^n - 1} \frac{k}{2^n} \xi_{A_{k,n}}$. Then we have $f_1 \leqq f_2 \leqq \cdots \leqq f_n \leqq \cdots \leqq f$, and $\lim_{n \to \infty} f_n(x) = f(x)$ for all $x \in G$. Similarly, we have $f_1 \circ \tau \leqq f_2 \circ \tau \leqq \cdots \leqq f_n \circ \tau \leqq \cdots$ and $\lim_{n \to \infty} f_n \circ \tau(x, y) = f \circ \tau(x, y)$ for all $(x, y) \in G \times G$. Since each $f_n \circ \tau$ is $|\mu| \times |\nu|$-measurable, $f \circ \tau$ is $|\mu| \times |\nu|$-measurable. The monotone convergence theorem and (2) now imply that

$$\left.\begin{aligned} \int\limits_{G \times G} f \circ \tau \, d|\mu| \times |\nu| &= \lim_{n \to \infty} \int\limits_{G \times G} f_n \circ \tau \, d|\mu| \times |\nu| \\ &\leqq \lim_{n \to \infty} \int\limits_{G} f_n \, d|\mu| * |\nu| = \int\limits_{G} f \, d|\mu| * |\nu|. \end{aligned}\right\} \tag{3}$$

From the foregoing it is clear that $f \circ \tau \in \mathfrak{L}_1(G \times G, |\mu| \times |\nu|)$ whenever $f \in \mathfrak{L}_1(G, |\mu| * |\nu|)$. For $f \in \mathfrak{L}_1(G, |\mu| * |\nu|)$, we write $\|f\|_1$ for $\int\limits_G |f| \, d|\mu| * |\nu|$; it is obvious from (19.9) that

$$\left| \int\limits_G f \, d\mu * \nu \right| \leqq \|f\|_1. \tag{4}$$

Moreover, using (3), we see that

$$\left.\begin{aligned} \left| \int\limits_{G \times G} (f \circ \tau) \, d\mu \times \nu \right| &\leqq \int\limits_{G \times G} |f \circ \tau| \, d|\mu| \times |\nu| \\ &= \int\limits_{G \times G} |f| \circ \tau \, d|\mu| \times |\nu| \leqq \int\limits_G |f| \, d|\mu| * |\nu| = \|f\|_1. \end{aligned}\right\} \tag{5}$$

Now consider the linear functionals $f \to \int_G f\,d\mu * \nu$ and $f \to \int_{G\times G} (f \circ \tau)\,d\mu \times \nu$ on $\mathfrak{L}_1(G, |\mu| * |\nu|)$. These functionals agree on $\mathfrak{C}_0(G)$ since $\int_G f\,d\mu * \nu = L * M(f) = L \circ \overline{M}(f) = \int_G \int_G f(xy)\,d\nu(y)\,d\mu(x) = \int_{G\times G} f \circ \tau\,d\mu \times \nu$, and they are bounded, by (4) and (5). By (12.10) and (B.11), these functionals agree on all of $\mathfrak{L}_1(G, |\mu| * |\nu|)$. This proves the first equality in (i). The remaining equalities are immediate consequences of (14.25). ☐

(19.11) Theorem. *Let L, M, μ, ν, and τ be as in* (19.10). *Then for every $|\mu| * |\nu|$-measurable set A, $\tau^{-1}(A)$ is $|\mu \times \nu|$-measurable, and*

(i) $\mu * \nu(A) = \mu \times \nu(\tau^{-1}(A)) = \int_G \nu(x^{-1}A)\,d\mu(x) = \int_G \mu(Ay^{-1})\,d\nu(y)$. [1]

Proof. This follows immediately from (19.10). ☐

(19.12) Let $\boldsymbol{M}(G)$ denote the set of all complex measures μ obtained as in (14.4) for linear functionals L in $\mathfrak{C}_0^*(G)$. By (14.10), the mapping $L \to \mu$ of $\mathfrak{C}_0^*(G)$ onto $\boldsymbol{M}(G)$ is one-to-one, linear, norm-preserving, and order-preserving. We have defined $\mu * \nu$ [where $L \to \mu$ and $M \to \nu$] in (19.8) so that the mapping preserves convolution. In our further study of $\mathfrak{C}_0^*(G)$, we will replace it by the equivalent entity $\boldsymbol{M}(G)$, and will usually state our results in terms of $\boldsymbol{M}(G)$ alone. Where convenient, of course, we will make use of the fact that $\boldsymbol{M}(G)$ is a concrete representation of $\mathfrak{C}_0^*(G)$, so that we can use the HAHN-BANACH theorem and other devices of functional analysis.

The word "closed" in reference to $\boldsymbol{M}(G)$ means "closed in the topology for $\boldsymbol{M}(G)$ defined by the metric $\|\mu - \nu\|$".

We now describe, and characterize so far as possible, some important subsets of $\boldsymbol{M}(G)$.

(19.13) Definition. A measure μ in $\boldsymbol{M}(G)$ is said to be *purely discontinuous* if there is a countable subset E of G such that $|\mu|(E') = 0$. The measure μ is said to be *continuous* if $\mu(\{x\}) = 0$ for all $x \in G$. A continuous measure μ is said to be *singular* if μ is singular with respect to left Haar measure λ in the sense of (14.21). A measure μ is said to be *absolutely continuous* if μ is absolutely continuous with respect to left Haar measure λ in the sense of (14.20). The sets of purely discontinuous, continuous, singular, and absolutely continuous measures in $\boldsymbol{M}(G)$ are denoted by $\boldsymbol{M}_d(G)$, $\boldsymbol{M}_c(G)$, $\boldsymbol{M}_s(G)$, and $\boldsymbol{M}_a(G)$, respectively. The symbols $\boldsymbol{M}^+(G)$, $\boldsymbol{M}_d^+(G)$, etc., will designate the sets of nonnegative measures in $\boldsymbol{M}(G)$, $\boldsymbol{M}_d(G)$, etc., respectively.

---

[1] In the last expression, we mean that $\mu(Ay^{-1})$ exists for $|\nu|$-almost all y and that $\mu(Ay^{-1})$ is ν-measurable as a function of y [let $\mu(Ay^{-1})$ be 0, say, where it is not already defined]. Similar considerations apply to $\nu(x^{-1}A)$.

(19.14) Lemma. *Let A be a subset of G that is measurable for all $|\mu|$, $\mu \in \boldsymbol{M}(G)$. The set $\boldsymbol{V}_A = \{\mu \in \boldsymbol{M}(G) : |\mu|(A) = 0\}$ is a closed linear subspace of $\boldsymbol{M}(G)$.*

Proof. For $\mu, \nu \in \boldsymbol{V}_A$, we have $|\mu(A)+\nu(A)| \leqq |\mu|(A) + |\nu|(A) = 0$ (14.11.ii), so that $\mu+\nu \in \boldsymbol{V}_A$; also $|\alpha\mu|(A) = |\alpha|\,|\mu|(A) = 0$ for $\alpha \in K$ and $\mu \in \boldsymbol{V}_A$ (14.11.i). Thus $\boldsymbol{V}_A$ is a linear subspace of $\boldsymbol{M}(G)$. If $\mu_n \in \boldsymbol{V}_A$ ($n = 1, 2, 3, \ldots$), if $\mu \in \boldsymbol{M}(G)$, and $\lim_{n\to\infty} \|\mu_n - \mu\| = 0$, then for every Borel subset B of A, we have

$$|\mu(B)| = |\mu_n(B) - \mu(B)| = |(\mu_n - \mu)(B)| \leqq |\mu_n - \mu|(B) \leqq \|\mu_n - \mu\|.$$

Hence $\mu(B) = 0$, and so $|\mu|(A) = 0$ (14.15). □

(19.15) Theorem. *For $a \in G$, let ε_a be the measure such that $\varepsilon_a(P) = \xi_P(a)$ for all $P \subset G$. All subsets of G are ε_a-measurable, and ε_a is the measure in $\boldsymbol{M}(G)$ corresponding to the functional $E_a \in \mathfrak{C}_0^*(G)$ [see (19.6)]. For a sequence of complex numbers $\{\alpha_n\}_{n=1}^{\infty}$ such that $\sum_{n=1}^{\infty} |\alpha_n| < \infty$ and a sequence $\{a_n\}_{n=1}^{\infty}$ of distinct points of G, the measure*

(i) $\sum_{n=1}^{\infty} \alpha_n \varepsilon_{a_n}$ [1]

corresponds to the functional $\sum_{n=1}^{\infty} \alpha_n E_{a_n} \in \mathfrak{C}_0^*(G)$. We also have

(ii) $\left|\sum_{n=1}^{\infty} \alpha_n \varepsilon_{a_n}\right| = \sum_{n=1}^{\infty} |\alpha_n|\, \varepsilon_{a_n}$

and

(iii) $\left\|\sum_{n=1}^{\infty} \alpha_n \varepsilon_{a_n}\right\| = \sum_{n=1}^{\infty} |\alpha_n|$.

All measures of the form (i) *are in $\boldsymbol{M}_d(G)$, and every nonzero measure in $\boldsymbol{M}_d(G)$ has a unique representation* (i) *in which all α_n's are nonzero [the sum may be finite]. The set $\boldsymbol{M}_d(G)$ is a closed subalgebra of $\boldsymbol{M}(G)$. The equality $\boldsymbol{M}_d(G) = \boldsymbol{M}(G)$ obtains if and only if G is discrete.*

Proof. The first two assertions are obvious upon a little reflection; we omit their proofs. To prove (ii), write $\mu = \sum_{n=1}^{\infty} \alpha_n \varepsilon_{a_n}$. Then for $E \subset G$, it is clear that $\mu(E) = \Sigma \alpha_n$, the sum being taken over all n such that $a_n \in E$. By (14.14), we have $|\mu|(E) = \sup\left\{\sum_{j=1}^{m} |\mu(E_j)| : \{E_1, \ldots, E_m\} \text{ is a partition of } E\right\} \geqq \Sigma |\alpha_n|$, the sum again over all n such that $a_n \in E$. Hence we have $|\mu| \geqq \sum_{n=1}^{\infty} |\alpha_n|\, \varepsilon_{a_n}$. Theorem (14.14) shows that $|\mu|$ is the least nonnegative majorant of μ, and hence $|\mu| \leqq \sum_{n=1}^{\infty} |\alpha_n|\, \varepsilon_{a_n}$. The equality (iii) follows at once from (ii) and the definition of $\|\mu\|$.

[1] The definition of this measure is obvious; see also (19.4).

It is also obvious that every measure of the form (i) is in $\boldsymbol{M}_d(G)$. For $\mu \in \boldsymbol{M}_d(G)$, let $E = \{a_1, a_2, a_3, \ldots\}$ be a set [all a_n's are distinct] such that $|\mu|(E') = 0$. For every $|\mu|$-measurable set $X \subset G$, we have

$$\mu(X) = \mu(X \cap E) + \mu(X \cap E') = \mu(X \cap E) = \sum_{a_n \in X} \mu(\{a_n\}).$$

Hence $\mu = \sum_{a_n \in E} \mu(\{a_n\})\, \varepsilon_{a_n}$, as we wished to show. For $\{a_1, a_2, \ldots, a_N\} \subset E$, we have $\sum_{n=1}^{N} |\mu(\{a_n\})| \leqq |\mu|(G) = \|\mu\| < \infty$, so that $\sum_{n=1}^{\infty} |\mu(\{a_n\})| < \infty$. It is also obvious that the representation (i) is unique if the a_n's are distinct and the α_n's are different from zero.

To show that $\boldsymbol{M}_d(G)$ is norm-closed, let $\{\mu_n\}_{n=1}^{\infty}$ be a sequence of elements of $\boldsymbol{M}_d(G)$ and μ an element of $\boldsymbol{M}(G)$ such that $\lim_{n\to\infty} \|\mu - \mu_n\| = 0$. Let E_n be a countable set such that $|\mu_n|(E_n') = 0$ $(n = 1, 2, 3, \ldots)$ and let $E = \bigcup_{n=1}^{\infty} E_n$. For a compact subset X of E', we have

$$|\mu(X)| = |\mu(X) - \mu_n(X)| = |(\mu - \mu_n)(X)| \leqq |\mu - \mu_n|(X) \leqq \|\mu - \mu_n\|,$$

for each n. It follows that $\mu(X) = 0$ and so $|\mu|(E') = 0$ by (14.15).

We next show that $\boldsymbol{M}_d(G)$ is a subalgebra of $\boldsymbol{M}(G)$. It is obvious that $\boldsymbol{M}_d(G)$ is a linear subspace of $\boldsymbol{M}(G)$. If $\mu_n \to \mu$ and $\nu_n \to \nu$ in $\boldsymbol{M}(G)$, then it is clear that $\|\mu_n * \nu_n - \mu * \nu\| \leqq \|\mu_n * \nu_n - \mu_n * \nu\| + \|\mu_n * \nu - \mu * \nu\| \leqq \|\mu_n\| \cdot \|\nu_n - \nu\| + \|\nu\| \cdot \|\mu_n - \mu\|$ (19.6.2), so that $\mu_n * \nu_n \to \mu * \nu$. Thus if $\sum_{n=1}^{\infty} \alpha_n \varepsilon_{a_n}$ and $\sum_{m=1}^{\infty} \beta_m \varepsilon_{b_m}$ are in $\boldsymbol{M}_d(G)$, a simple computation based on the identity $\varepsilon_a * \varepsilon_b = \varepsilon_{ab}$ shows that

$$\left(\sum_{n=1}^{\infty} \alpha_n \varepsilon_{a_n}\right) * \left(\sum_{m=1}^{\infty} \beta_m \varepsilon_{b_m}\right) = \sum_{n=1}^{\infty} \sum_{m=1}^{\infty} \alpha_n \beta_m \varepsilon_{a_n b_m}.$$

The set of all *distinct* $a_n b_m$ is countable, and so $\boldsymbol{M}_d(G)$ is closed under convolution.

Suppose that G is discrete. Theorem (14.16) shows at once that every functional in $\mathfrak{C}_0^*(G)$ has the form $\sum_{n=1}^{\infty} \alpha_n E_{a_n}$ with $\sum_{n=1}^{\infty} |\alpha_n| < \infty$: thus $\boldsymbol{M}(G) = \boldsymbol{M}_d(G)$ if G is discrete. If G is nondiscrete, let w be any nonnegative, nonzero function in $\mathfrak{L}_1(G)$, and let μ be the measure in $\boldsymbol{M}(G)$ such that $d\mu = w\, d\lambda$, as in (14.17). By (15.17.b), we have $\lambda(A) = 0$ for all countable sets $A \subset G$, and so $\mu(A) = \int_A w\, d\lambda = 0$ if A is countable. Since μ is not the zero measure, we have $\boldsymbol{M}_d(G) \subsetneqq \boldsymbol{M}(G)$. □

(19.16) Theorem. *The set $\boldsymbol{M}_c(G)$ is a closed two-sided ideal in the algebra $\boldsymbol{M}(G)$.*

Proof. By (19.13), $\boldsymbol{M}_c(G)$ is the set $\bigcap_{x\in G} \boldsymbol{V}_{\{x\}}$, which by (19.14) is the intersection of closed linear subspaces. This implies that $\boldsymbol{M}_c(G)$ is a closed linear subspace of $\boldsymbol{M}(G)$. For $\mu\in\boldsymbol{M}_c(G)$, $\nu\in\boldsymbol{M}(G)$, and $a\in G$, (19.11.i) gives us

$$\mu*\nu(\{a\}) = \int_G \mu(\{ay^{-1}\})\, d\nu(y) = 0$$

and

$$\nu*\mu(\{a\}) = \int_G \mu(\{x^{-1}a\})\, d\nu(x) = 0.$$

Thus $\boldsymbol{M}_c(G)$ is a two-sided ideal in $\boldsymbol{M}(G)$. □

(19.17) Lemma. *For all subsets A of G and all $x\in G$, we have*

(i) $\lambda(Ax) = \Delta(x)\,\lambda(A)$.

In particular, if A is λ-null [locally λ-null], then Ax is λ-null [locally λ-null].

Proof. Consider first the case of an open set U. If $f\in\mathfrak{C}_{00}^+(G)$ and $f\leqq\xi_U$, then plainly $f_{x^{-1}}\leqq(\xi_U)_{x^{-1}}=\xi_{Ux}$. It follows immediately from (15.11) and (11.11) that $\Delta(x)\,\lambda(U)\leqq\lambda(Ux)$. Replacing x by x^{-1} and U by Ux, we also have $\Delta(x^{-1})\,\lambda(Ux)\leqq\lambda(U)$, and as $\Delta(x)\,\Delta(x^{-1})=1$, (i) follows for open sets. Given any set $A\subset G$, we have $A\subset U$ if and only if $Ax\subset Ux$, so that (i) follows for all subsets of G from (11.22).

Suppose that A is locally λ-null. If F is a compact subset of Ax, then Fx^{-1} is a compact subset of A, and so $\lambda(Fx^{-1})=0$. Thus $\lambda(F)= \Delta(x)\,\lambda(Fx^{-1})=0$; since Ax is λ-measurable, it is locally λ-null. □

(19.18) Theorem. *The set $\boldsymbol{M}_a(G)$ is a closed two-sided ideal in the algebra $\boldsymbol{M}(G)$. The mapping $\mu\to w$ carrying $\boldsymbol{M}_a(G)$ into $\mathfrak{L}_1(G)=\mathfrak{L}_1(G,\lambda)$ as in* (14.17) *and* (14.19) *is a norm-preserving linear isomorphism of $\boldsymbol{M}_a(G)$ onto $\mathfrak{L}_1(G)$.*

Proof. An element μ of $\boldsymbol{M}(G)$ is in $\boldsymbol{M}_a(G)$ if and only if $|\mu|(F)=0$ for every compact set F such that $\lambda(F)=0$. Lemma (19.14) shows that $\boldsymbol{M}_a(G)$ is a closed linear subspace of $\boldsymbol{M}(G)$. Let μ be a measure in $\boldsymbol{M}_a(G)$ and ν a measure in $\boldsymbol{M}(G)$. If F is a compact set such that $\lambda(F)=0$, then we have $\lambda(Fy^{-1})=0$ for all $y\in G$ (19.17), and hence $|\mu|(Fy^{-1})=0$ for all $y\in G$. Using (19.9) and (19.11), we obtain

$$|\mu*\nu|(F)\leqq|\mu|*|\nu|(F)=\int_G|\mu|(Fy^{-1})\,d|\nu|(y)=0,$$

so that $\mu*\nu\in\boldsymbol{M}_a(G)$. We evidently have $|\mu|(x^{-1}F)=0$ for all $x\in G$, and thus

$$|\nu*\mu|(F)\leqq|\nu|*|\mu|(F)=\int_G|\mu|(x^{-1}F)\,d|\nu|(x)=0.$$

This shows that $\nu*\mu$ is in $\boldsymbol{M}_a(G)$: that is, $\boldsymbol{M}_a(G)$ is a two-sided ideal in $\boldsymbol{M}(G)$.

For $\mu \in \boldsymbol{M}_a(G)$, Theorem (14.19) shows that there is a function $w \in \mathfrak{L}_1(G)$ such that $d\mu = w\, d\lambda$. The mapping of $\boldsymbol{M}_a(G)$ into $\mathfrak{L}_1(G)$ engendered in this way is obviously linear. Theorem (14.17) shows that this mapping is onto and norm-preserving; since it is norm-preserving, it is also one-to-one. ▯

(19.19) Theorem. *The set $\boldsymbol{M}_s(G)$ is a closed linear subspace of $\boldsymbol{M}(G)$.*

Proof. Let μ, ν be elements of $\boldsymbol{M}_s(G)$. Then $\mu + \nu$ is continuous, clearly enough, and if B and C are Borel sets such that $\lambda(B) = \lambda(C) = 0$ and $|\mu|(B') = |\nu|(C') = 0$, then $|\mu + \nu|(B' \cap C') \leqq |\mu|(B') + |\nu|(C') = 0$, and $\lambda(B \cup C) = 0$. Thus $\mu + \nu \in \boldsymbol{M}_s(G)$ and $\boldsymbol{M}_s(G)$ is a linear subspace. Suppose that $\mu_1, \mu_2, \ldots, \mu_n, \ldots$ are in $\boldsymbol{M}_s(G)$ and that $\lim\limits_{n \to \infty} \|\mu - \mu_n\| = 0$. Let B_n be a Borel set such that $\lambda(B_n) = 0$ and $|\mu_n|(B_n') = 0$. Write $B = \bigcup\limits_{n=1}^{\infty} B_n$; then $\lambda(B) = 0$. If F is a compact subset of B', then

$$|\mu(F)| = |\mu(F) - \mu_n(F)| \leqq |\mu - \mu_n|(F) \leqq \|\mu - \mu_n\|$$

for all n, so that $\mu(F) = 0$. Hence $|\mu|(B') = 0$; since μ is clearly continuous, μ is in $\boldsymbol{M}_s(G)$. ▯

(19.20) Theorem. *If G is nondiscrete, the algebra $\boldsymbol{M}(G)$ is a direct sum as a linear space:*

(i) $\boldsymbol{M}(G) = \boldsymbol{M}_d(G) \oplus \boldsymbol{M}_s(G) \oplus \boldsymbol{M}_a(G)$.

Writing $\mu \in \boldsymbol{M}(G)$ in the form $\mu = \mu_d + \mu_s + \mu_a$, we have

(ii) $|\mu| = |\mu_d| + |\mu_s| + |\mu_a|$,

and in particular

(iii) $\|\mu\| = \|\mu_d\| + \|\mu_s\| + \|\mu_a\|$.

If G is discrete, then $\boldsymbol{M}(G) = \boldsymbol{M}_d(G) = \boldsymbol{M}_a(G)$ and $\boldsymbol{M}_s(G) = \{0\}$.

Proof. Suppose that G is nondiscrete. The equality (i) asserts that every μ in $\boldsymbol{M}(G)$ can be written as a sum of measures in $\boldsymbol{M}_d(G)$, $\boldsymbol{M}_s(G)$, and $\boldsymbol{M}_a(G)$, and that this decomposition is unique. Given $\mu \in \boldsymbol{M}(G)$, the set $\{x \in G : \mu(\{x\}) \neq 0\}$ is clearly finite or countably infinite. If it is void, set $\mu_d = 0$. If it is nonvoid, enumerate it in any fashion: $\{a_1, a_2, a_3, \ldots\}$, where the a_n's are distinct. Then write $\mu_d = \sum\limits_{n=1}^{\infty} \mu(\{a_n\})\, \varepsilon_{a_n}$. It is clear that $\mu_d \in \boldsymbol{M}_d$, and that $\mu_c = \mu - \mu_d$ is a continuous measure. Now apply Theorem (14.22) to μ_c and left Haar measure λ. We obtain $\mu_c = \mu_s + \mu_a$, where $\mu_s \in \boldsymbol{M}_s(G)$ and $\mu_a \in \boldsymbol{M}_a(G)$; this decomposition is unique, and $|\mu_c| = |\mu_s| + |\mu_a|$. The equality $|\mu| = |\mu_d| + |\mu_c|$ is also easy to verify: we omit the details. This proves (ii) and (iii).

If $\mu = \mu_d' + \mu_c'$, where $\mu_d' \in \boldsymbol{M}_d(G)$ and $\mu_c' \in \boldsymbol{M}_c(G)$, then $\mu_d - \mu_d' = \mu_c' - \mu_c$, so that $0 = \mu_d - \mu_d' = \mu_c' - \mu_c$. Hence the decomposition $\mu = \mu_d + \mu_s + \mu_a$ is unique, and (i) is proved.

The last statement of the theorem follows from (15.17.b), (14.16), and (19.13). □

(19.21) Note. The subspace $\boldsymbol{M}_d(G)$ is a subalgebra of $\boldsymbol{M}(G)$; $\boldsymbol{M}_c(G)$ and $\boldsymbol{M}_a(G)$ are two-sided ideals in $\boldsymbol{M}(G)$. However, $\boldsymbol{M}_s(G)$ is not in general a subalgebra of $\boldsymbol{M}(G)$ [see (19.26)]. If G is nondiscrete, then $\boldsymbol{M}_c(G)=\boldsymbol{M}_s(G)\oplus\boldsymbol{M}_a(G)$.

Miscellaneous theorems and examples

(19.22) (a) We made an arbitrary choice in Definition (19.1). Take $\mathfrak{F}$ as a *right* invariant linear space of functions on S, and define $\underline{M}f$ as the function whose value at $x\in S$ is $M(f_x)$. If $\underline{M}$ carries $\mathfrak{F}$ into itself and L is a linear functional on $\mathfrak{F}$, define $L\nabla M$ as the functional $L\circ\underline{M}$. Then we have $\underline{N}(f_x)(y)=N((f_x)_y)=N(f_{yx})=\underline{N}f(yx)=(\underline{N}f)_x(y)$. Thus $\underline{N}(f_x)=(\underline{N}f)_x$ for all $x\in S$. Also $(\underline{M\nabla N})f(x)=(M\nabla N)(f_x)=M(\underline{N}(f_x))=M((\underline{N}f)_x)=(\underline{M}\circ\underline{N})f(x)$. Thus $L\nabla(M\nabla N)=L\circ(\underline{M\nabla N})=L\circ(\underline{M}\circ\underline{N})=(L\circ\underline{M})\circ\underline{N}=(L\nabla M)\nabla N$. That is, the operation ∇, like the operation $*$, is associative. It is clear that for every theorem about $*$, there is a corresponding theorem about ∇.

(b) Let $\mathfrak{F}=\mathfrak{B}(S)$ and E_a $(a\in S)$ be as in (19.4). Then $\underline{E_a}f={}_af$ and $(E_b\nabla E_a)f=f(ab)=E_{ab}f$. Thus the mapping $a\to E_a$ is an anti-isomorphism of S onto a semigroup under ∇ of linear functionals on $\mathfrak{B}(S)$. This fact makes our choice of $*$ for the definition of convolution somewhat more convenient than the choice of ∇, which as pointed out in (a) would have been perfectly possible.

(c) We can apply the operation ∇ to the conjugate space $\mathfrak{C}_0^*(G)$ [identified as usual with $\boldsymbol{M}(G)$], where G is a locally compact group. For L and M in $\mathfrak{C}_0^*(G)$ and $f\in\mathfrak{C}_0(G)$, (19.5) shows that $\underline{M}f\in\mathfrak{C}_0(G)$. Thus $L\nabla M$ is in $\mathfrak{C}_0^*(G)$.

For $\mu,\nu\in\boldsymbol{M}(G)$ and $f\in\mathfrak{L}_1(G,|\mu|\nabla|\nu|)$, we have

$$\int_G f\,d\mu\nabla\nu=\int_G\int_G f(yx)\,d\nu(y)\,d\mu(x). \tag{1}$$

For $|\mu|\nabla|\nu|$-measurable sets A, we have

$$\mu\nabla\nu(A)=\int_G\nu(Ax^{-1})\,d\mu(x)=\int_G\mu(y^{-1}A)\,d\nu(y). \tag{2}$$

(19.23) Other examples of convolution algebras. The choice of $\mathfrak{C}_0(G)$ and $\mathfrak{C}_0^*(G)$ in (19.5) *et seq.* is not accidental. Considerations of accessibility, availability of analytic machinery, and even personal taste influence one's choice of function spaces and spaces of linear functionals in which to study convolutions. Some natural candidates, however, suffer from serious defects, as the following examples show.

(a) Let G be a noncompact, locally compact, nondiscrete group with a countable open basis of neighborhoods at the identity. As in (15.9), select a countably infinite set $\{x_1, x_2, \ldots, x_n, \ldots\} \subset G$ and a neighborhood V of e such that V^- is compact and $(Vx_n) \cap (Vx_m) = \emptyset$ if $m \neq n$. Let $\{V_n\}_{n=1}^{\infty}$ be a decreasing sequence of neighborhoods of e such that $V_1 \subset V$ and $\bigcap_{n=1}^{\infty} V_n = \{e\}$. Let f be a function in $\mathfrak{C}(G)$ such that $f(G) \subset [0, 1]$, $f(x_n) = 1$ for $n = 1, 2, \ldots$, and $f\left(\left(\bigcup_{n=1}^{\infty} V_n x_n\right)'\right) = 0$ $\left[\bigcup_{n=1}^{\infty} \{x_n\}\right.$ is closed and G is a normal topological space$\left.\right]$. Let L be a functional in $\mathfrak{C}^*(G)$ such that $L(\varphi) = \lim_{n\to\infty} \varphi(x_n)$ whenever this limit exists [apply (B.14)]. If $y \in V$ and $y \notin V_m$, then ${}_yf(x_n) = 0$ for all $n > m$. It follows that $L(f) = 1$ and $L({}_yf) = 0$ for all $y \in V \cap \{e\}'$. That is, the function $\overline{L}f$ is discontinuous, and $\mathfrak{C}^*(G)$ is not a convolution algebra in our sense.

(b) (Suggested by BUCK [1].) Let G be any locally compact group and let $\mathfrak{C}_{ru}(G)$ $[\mathfrak{C}_{lu}(G)]$ be the linear space of all right [left] uniformly continuous, bounded, complex-valued functions on G. Let $\mathfrak{C}_u(G) = \mathfrak{C}_{lu}(G) \cap \mathfrak{C}_{ru}(G)$. Under the uniform norm $\|f\|_u$, these spaces are Banach spaces [closed subspaces of $\mathfrak{C}(G)$]. For $\varphi \in \mathfrak{C}_{ru}(G)$ and $\varepsilon > 0$, there is a neighborhood U of e such that $|\varphi(ux) - \varphi(vx)| < \varepsilon$ for all $x \in G$ provided that $uv^{-1} = (ux)(vx)^{-1} \in U$. For $M \in \mathfrak{C}_{ru}^*(G)$, therefore, we have $\overline{M}\varphi \in \mathfrak{C}_{ru}(G)$ and so $L * M \in \mathfrak{C}_{ru}^*(G)$ for $L, M \in \mathfrak{C}_{ru}^*(G)$. That is, $\mathfrak{C}_{ru}^*(G)$ is a convolution algebra in the sense of (19.3). Similarly $L \nabla M$ [see (19.22.a)] is in $\mathfrak{C}_{lu}^*(G)$ if L, M are in $\mathfrak{C}_{lu}^*(G)$. Both of these algebras are highly legitimate objects of study. They present grave difficulties, however, because of the great complexity of their structure. There is no known analogue for $\mathfrak{C}_{ru}(G)$ of Theorem (14.4), which represents linear functionals in $\mathfrak{C}_0^*(G)$ as integrals with respect to *countably* additive complex measures. It seems unlikely, indeed, that all functionals in $\mathfrak{C}_{ru}^*(G)$ can be represented as integrals with respect to *finitely* additive measures.

(c) Apparently less tractable even than $\mathfrak{C}_{ru}^*(G)$ is the convolution algebra $\mathfrak{B}^*(S)$, where S is a semigroup [no topology], $\mathfrak{B}(S)$ is as defined in (17.1), and $\mathfrak{B}(S)$ is made into a Banach space with the uniform norm, as in (17.1). It is trivial that $\underline{M}\varphi$ and $\overline{M}\varphi$ are in $\mathfrak{B}(S)$ if $\varphi \in \mathfrak{B}(S)$ and $M \in \mathfrak{B}^*(S)$. Thus $\mathfrak{B}^*(S)$ is a convolution algebra under both $*$ and ∇. It is known that a functional $L \in \mathfrak{B}^*(S)$ can be represented as an integral: $L(\varphi) = \int_S \varphi(x)\, d\mu(x)$, where μ is a finitely additive, complex measure defined for all subsets of S. However, in no nontrivial case has the analysis of $\mathfrak{B}^*(S)$ been carried out. Even for the group Z, the complexity and sheer size of $\mathfrak{B}^*(Z)$ are such that a detailed study seems very difficult indeed. [The representation theorem for $\mathfrak{B}^*(S)$ is simple: let $\mu(A) = L(\xi_A)$ for all $A \subset S$.]

(19.24) (a) The algebras $\mathfrak{C}_{ru}^*(G)$ and $\mathfrak{B}^*(S)$ of (19.23) need not be commutative even if G and S are commutative. We consider $\mathfrak{C}_{ru}^*(G)$, the argument for $\mathfrak{B}^*(S)$ being similar. Suppose that $\mathfrak{C}_{ru}(G)$ contains a closed linear subspace $\mathfrak{A}$ that is left invariant [${}_af \in \mathfrak{A}$ if $f \in \mathfrak{A}$ and $a \in G$], that $1 \in \mathfrak{A}$, and that $\mathfrak{A}$ admits two distinct bounded linear functionals N_1 and N_2 with the properties $N_j({}_af) = N_j(f)$ and $N_j(1) = 1$ $(j = 1, 2)$. By (B.14), there are extensions M_j of N_j that are bounded linear functionals on $\mathfrak{C}_{ru}(G)$. Now if $f \in \mathfrak{A}$ and $N_1(f) \neq N_2(f)$, we have $M_1 * M_2(f) = M_1 \circ \overline{M}_2(f) = M_1(1)\, M_2(f) = M_2(f)$, and $M_2 * M_1(f) = M_1(f) \neq M_2(f)$. Thus $\mathfrak{C}_{ru}^*(G)$ is noncommutative.

(b) Let $G = R$. In $\mathfrak{C}_u(R)$, consider the subspace $\mathfrak{A}$ consisting of all φ for which $N_1(\varphi) = \lim\limits_{x\to\infty} \varphi(x)$ and $N_2(\varphi) = \lim\limits_{x\to-\infty} \varphi(x)$ exist. Obviously $N_1 \neq N_2$, N_j is invariant, and $N_j(1) = 1$. Hence $\mathfrak{C}_u^*(R)$ and $\mathfrak{B}^*(R)$ are noncommutative.

(c) Let G be an infinite Abelian group. Then by (17.21.c) and (a), $\mathfrak{B}^*(G)$ is noncommutative.

(19.25) Measurability with respect to $|\mu|$, $|\nu|$, $|\mu * \nu|$, and $|\mu| * |\nu|$. (a) Let G be a locally compact group. It is possible for a subset A of G to be measurable with respect to measures $|\mu|$ and $|\nu|$, where μ and ν are elements of $\boldsymbol{M}(G)$, while at the same time A is *not* measurable with respect to $|\mu * \nu|$ or $|\mu| * |\nu|$. [As a simple example, take $G = R$, let λ be ordinary Lebesgue measure on R, and let μ be the measure such that $d\mu = \xi_{[-1,1]}\, d\lambda$. An elementary computation using (19.11) shows that $d\mu * \mu = f\, d\lambda$, where $f(x) = \max(2 - |x|, 0)$ for all $x \in R$; note that it suffices to show that $\int\limits_{-1}^{1} \int\limits_{-1}^{1} \xi_{]-\infty, a]}(x + y)\, dx\, dy = \int\limits_{-\infty}^{a} f(x)\, dx$ for all $a \in R$. If A is a subset of $[1, 2]$ that is not Lebesgue measurable, then A is $|\mu|$-measurable since $|\mu|(A) = 0$, but A is not $|\mu * \mu|$-measurable or $|\mu| * |\mu|$-measurable, as (14.17) shows. Note that $|\mu| = \mu$ and $|\mu * \mu| = |\mu| * |\mu| = \mu * \mu$.]

(b) A subset of G can be $|\mu * \nu|$-measurable and yet not $|\mu| * |\nu|$-measurable. [Consider the group T with the normalized Haar measure λ described in (15.17.d). Let μ and ν be defined by $d\mu(x) = x\, d\lambda(x)$ and $d\nu(x) = x^{-1}\, d\lambda(x)$. Then for every integer n, we have

$$\int\limits_T x^n\, d\mu * \nu(x) = \int\limits_T \int\limits_T (xy)^n\, d\mu(x)\, d\nu(y) = \int\limits_T x^{n+1}\, d\lambda(x) \int\limits_T y^{n-1}\, d\lambda(y)$$

$$= \frac{1}{4\pi^2} \int\limits_0^{2\pi} \exp\big(i(n+1)t\big)\, dt \int\limits_0^{2\pi} \exp\big(i(n-1)t\big)\, dt = 0.$$

Since the set $\mathfrak{T}$ of all linear combinations of functions x^n $(n = 0, \pm 1, \pm 2, \ldots)$ is a dense subspace of $\mathfrak{C}(T)$,[1] we infer that $\mu * \nu = 0$. From

[1] The group T is compact and $\mathfrak{T}$ is a subalgebra of $\mathfrak{C}(T)$ closed under complex conjugation and separating points of T. See the footnote to (13.2). This is a special case of a phenomenon occurring in *all* compact groups. For the Abelian case, see (23.20). The non-Abelian case will be taken up in Vol. II, (27.39).

(14.17) we see that $|\mu| = |\nu| = \lambda$, and from (19.11.i) that $\lambda * \lambda = \lambda$. Hence every subset of T is $|\mu * \nu|$-measurable, while (16.13.d) shows that there are subgroups of T that are not λ-measurable.]

(c) The hypothesis in (19.11) that A be $|\mu| * |\nu|$-measurable is needed in order to ensure that $\tau^{-1}(A)$ is $|\mu \times \nu|$-measurable. [Take $G = T$ and μ and ν the measures described in (b). Let H be a subgroup of T such that T/H is countably infinite (16.13.c). Then H is $|\mu * \nu|$-measurable, since $\mu * \nu = 0$. Theorem (2.2) shows that $(T \times T)/\tau^{-1}(H)$ is isomorphic with T/H. Thus $\tau^{-1}(H)$ comes under (16.13.b) and is not measurable with respect to Haar measure on $T \times T$. If λ is normalized Haar measure on T, then $\lambda \times \lambda$ is normalized Haar measure on $T \times T$ (15.17.j). Also by (14.24) and (b) above we have $|\mu \times \nu| = |\mu| \times |\nu| = \lambda \times \lambda$. That is, $\tau^{-1}(H)$ is not $|\mu \times \nu|$-measurable.]

(19.26) Properties of $\boldsymbol{M}_s(G)$.

(a) Let G be a nondiscrete locally compact group. Then $\boldsymbol{M}_s(G)$ is different from $\{0\}$. In fact, every nonvoid open set in G contains a compact set which is the support of a nonnegative measure μ in $\boldsymbol{M}_s(G)$ for which $\mu(G) = 1$. [Suppose first that G is metrizable, *i.e.*, that $\{e\}$ is the intersection of a countable family of open sets (8.5). Let U be any neighborhood of e. Then the construction used in (4.26) can be applied to G and U to produce a subset F of U such that $\lambda(F) = 0$ and F is homeomorphic with $\{0, 1\}^{\aleph_0}$. Thus F is homeomorphic with the product of $\aleph_0$ 2-element groups. Normalized Haar measure on this product group can be transferred to a measure on F and thence by (11.45) to a measure μ on G which is certainly continuous and singular. If G is not metrizable, (5.14) and (8.7) show that every neighborhood of e in G contains a compact infinite subgroup N such that $\lambda(N) = 0$. Normalized Haar measure on N can be used with (11.45) to produce once again a continuous singular measure μ on G as required. The invariance of Haar measure on G shows that an arbitrary nonvoid open subset of G can be used instead of a neighborhood of e.]

(b) The subspace $\boldsymbol{M}_s(G)$ need not be a subalgebra of $\boldsymbol{M}(G)$. [Let G_1 and G_2 be infinite compact groups, let λ_1 and λ_2 be normalized Haar measures on G_1 and G_2, respectively, and let $G = G_1 \times G_2$. Denote by e_j the identity element in G_j. If $f \in \mathfrak{C}(G)$, then the function $x \to f(x, e_2)$ belongs to $\mathfrak{C}(G_1)$ and $f \to \int_{G_1} f(x, e_2)\, d\lambda_1(x)$ defines an element of $\mathfrak{C}^*(G)$. Let $\lambda_1^\dagger$ be the measure in $\boldsymbol{M}(G)$ such that $\int_G f\, d\lambda_1^\dagger = \int_{G_1} f(x, e_2)\, d\lambda_1(x)$ for $f \in \mathfrak{C}(G)$; similarly let $\lambda_2^\dagger \in \boldsymbol{M}(G)$ be taken so that $\int_G f\, d\lambda_2^\dagger = \int_{G_2} f(e_1, y)\, d\lambda_2(y)$. Then it is clear that $\lambda_1^\dagger$ and $\lambda_2^\dagger$ are in $\boldsymbol{M}_s(G)$ and also that $\lambda_1^\dagger * \lambda_2^\dagger = \lambda_2^\dagger * \lambda_1^\dagger = \lambda$, where λ is normalized Haar measure on G.]

(19.27) Let μ be a measure in $\boldsymbol{M}(G)$. Suppose that for every compact subset F of G for which $\lambda(F)=0$, either the function $x\to\mu(xF)$ or the function $x\to\mu(Fx)$ is continuous at e. Then μ is absolutely continuous. [If the function $x\to\mu(xF)$ is continuous at e, so is the function $x\to\mu(x^{-1}F)$. Let U be a neighborhood of e such that $\lambda(U)<\infty$ and let $\nu\in\boldsymbol{M}_a(G)$ be defined by $d\nu=\xi_U\,d\lambda$. By (19.18), $|\mu|*\nu$ and $\nu*|\mu|$ are in $\boldsymbol{M}_a(G)$. By (19.11), we have $0=\nu*|\mu|(F)=\int_G|\mu|(x^{-1}F)\,d\nu(x)=\int_U|\mu|(x^{-1}F)\,dx\geqq\int_U|\mu(x^{-1}F)|\,dx$. Since $|\mu(x^{-1}F)|$ is nonnegative, we have $\mu(x^{-1}F)=0$ for λ-almost all $x\in U$; and hence $\mu(e^{-1}F)=\mu(F)=0$. The argument if $x\to\mu(Fx)$ is continuous at e is similar.]

(19.28) (a) (BRACONNIER [1].) Using (19.17), we can generalize (15.12) and (15.13). Every locally compact group G having equivalent left and right uniform structures is unimodular. [Assume that G is not unimodular and that the left and right uniform structures are equivalent. Let U be a neighborhood of e such that $\lambda(U)<\infty$. By (4.14.g), there is a neighborhood V of e such that $xVx^{-1}\subset U$ for all $x\in G$. Choose x_0 in G such that $\Delta(x_0^{-1})\,\lambda(V)>\lambda(U)$. Then $x_0Vx_0^{-1}\subset U$ and $\lambda(x_0Vx_0^{-1})=\Delta(x_0^{-1})\,\lambda(x_0V)=\Delta(x_0^{-1})\,\lambda(V)>\lambda(U)$.]

(b) The converse of (a) is false, as the groups $\mathfrak{GL}(n,R)$ show. By (15.28.a), these groups are unimodular; by (4.24), their left and right uniform structures are inequivalent.

(19.29) Convolutions involving unbounded functionals. (a) Let G be a locally compact group; let L and M be nonnegative linear functionals on $\mathfrak{C}_{00}(G)$ such that the measure corresponding to M has compact support E (11.25); let μ and ν be the measures corresponding to L and M respectively as in §11. Thus ν is in $\boldsymbol{M}^+(G)$ and μ is arbitrary. For every $f\in\mathfrak{C}_{00}(G)$, Lemma (19.5) shows that the function $\overline{M}f$ defined by

$$\overline{M}f(x)=M({}_xf)=\int_G f(xy)\,d\nu(y)$$

is continuous: observe that it vanishes outside of the compact set $F=\{u\in G: f(u)\neq 0\}^-\cdot E^{-1}$, and is nonnegative if f is nonnegative. Hence the mapping

$$f\to L(\overline{M}f)=\int_G\int_E f(xy)\,d\nu(y)\,d\mu(x)$$

is a nonnegative [and obviously linear] functional on the linear space $\mathfrak{C}_{00}(G)$. We write $L*M$ for this functional, and $\mu*\nu$ for the measure corresponding to $L*M$ as in §11. Thus $L*M$ exists if L and M are nonnegative linear functionals on $\mathfrak{C}_{00}(G)$ **and the measure corresponding to M has compact support.**

(b) Consider a left Haar integral I on $\mathfrak{C}_{00}(G)$ and any functional M as described in (a). Let E, ν, and F [defined for $f \in \mathfrak{C}_{00}(G)$] be as in (a). Let λ as usual denote the Haar measure corresponding to I, and let dx stand for $d\lambda(x)$. Then we have $I * M = cI$, where $c = \int\limits_E \Delta(y^{-1})\, d\nu(y)$. [It is easy to see that $\overline{M}({}_a f) = {}_a(\overline{M} f)$; hence $I * M({}_a f) = I\left({}_a(\overline{M} f)\right) = I(\overline{M} f) = I * M(f)$. By (15.5), therefore, there is a nonnegative number c such that $I * M = cI$. To evaluate c, we note that

$$\begin{aligned} I * M(f) &= \int\limits_G \int\limits_E f(xy)\, d\nu(y)\, dx \\ &= \int\limits_F \int\limits_E f(xy)\, d\nu(y)\, dx. \end{aligned}$$

Hence (13.8) can be applied, and we have

$$\begin{aligned} I * M(f) &= \int\limits_E \int\limits_F f(xy)\, dx\, d\nu(y) \\ &= \int\limits_E \int\limits_G f(xy)\, dx\, d\nu(y) \\ &= \int\limits_E I(f)\, \Delta(y^{-1})\, d\nu(y).] \end{aligned}$$

(c) Let μ, ν, and $\mu * \nu$ be as in (a), and suppose further that the support of μ is σ-compact. Let f be a function in $\mathfrak{L}_1(G, \mu * \nu)$ and let τ be as defined in (19.10). Then we have $\int\limits_G f\, d\mu * \nu = \int\limits_{G \times G} (f \circ \tau)\, d\mu \times \nu = \int\limits_G \int\limits_E f(xy)\, d\nu(y)\, d\mu(x) = \int\limits_E \int\limits_G f(xy)\, d\mu(x)\, d\nu(y)$. [The proof is carried out by repeating *mutatis mutandis* the proof of (19.10). An important difference occurs only in showing that $\mu \times \nu\left(\tau^{-1}(D)\right) = 0$ [notation is as in the proof of (19.10)]. Here it is needful to recall that the support of $\mu \times \nu$ is σ-compact so that $\tau^{-1}(D)$ is $\mu \times \nu$-null if every compact subset of $\tau^{-1}(D)$ is $\mu \times \nu$-null.]

(d) Let μ, ν, and $\mu * \nu$ be as in (a), and suppose again that the support of μ is σ-compact. Let A be a $\mu * \nu$-measurable subset of G. Then we have $\mu * \nu(A) = \int\limits_{G \times G} (\xi_A \circ \tau)\, d\mu \times \nu = \int\limits_G \nu\left((x^{-1}A) \cap E\right) d\mu(x) = \int\limits_E \mu(Ay^{-1})\, d\nu(y)$. [If $\mu * \nu(A) < \infty$, this is a special case of (c). An elementary argument using the monotone convergence theorem proves the result for arbitrary A.]

(e) Let G be a locally compact, σ-compact group, let H be a compact subgroup of G, let ν be any nonzero measure in $\boldsymbol{M}^+(G)$ whose support is contained in H, and let λ be a left Haar measure on G. Then for every λ-measurable set $A \subset G$, we have $\lambda(A) = \frac{1}{\nu(G)} \int\limits_G \nu\left((x^{-1}A) \cap H\right) dx$. [This follows from (d), (b), and the obvious fact that $\Delta(y^{-1}) = 1$ for all $y \in H$.]

(19.30) Baire sets and Haar measure (KAKUTANI and KODAIRA [1]; proof based on HALMOS [2], pp. 282—289). For the definition of Baire sets, see (11.1). Throughout this example, λ will denote a fixed left Haar measure on whatever group G is under consideration.

(a) Let G be a compactly generated, locally compact group, and let U be an open subset of G such that $0<\lambda(U)<\infty$. Then U contains a set D such that:

(i) D is a Baire set;

(ii) $\lambda(U\cap D')=0$;

(iii) there is a compact normal subgroup N of G such that $UN=\{x\in G: f(x)>0\}$ for some $f\in\mathfrak{C}^r(G)$, and $\lambda(UN)=\lambda(U)$.

[Define by finite induction a sequence of compact sets F_n, each of the form $\{x\in G: f(x)=0\}$ for some $f\in\mathfrak{C}^r(G)$, and a sequence of neighborhoods V_n of e such that: $\lambda(F_n)+\frac{1}{n}>\lambda(U)$; $F_n\cdot V_n^-$ is a compact subset of U; and $F_{n+1}\supset F_n\cdot V_n^-$ for $n=1, 2, \ldots$. Theorems (4.10) and (11.23) and obvious properties of continuous real-valued functions show that this construction is possible.

By (8.7), there is a compact normal subgroup N of G such that $N\subset\bigcap_{n=1}^{\infty}V_n$ and G/N has a countable basis for open sets; plainly G/N is metrizable. Let φ denote the natural mapping of G onto G/N, and let $D=\bigcup_{n=1}^{\infty}F_n$. It is easy to see that $D=DN$, and it is obvious that (i) and (ii) hold for D.

We now note that $\varphi(U)$ is open in G/N (5.17). Since G/N is a metric space, there is a function $g\in\mathfrak{C}^r(G/N)$ such that $g(xN)>0$ for $xN\in\varphi(U)$ and $g(xN)=0$ for $xN\notin\varphi(U)$. The function $g\circ\varphi$ is continuous on G and positive exactly on UN.

Next let ν denote normalized Haar measure on the compact group N. From (11.45) and (19.29.e) we infer that

$$\left.\begin{aligned}0=\lambda(U\cap D')&=\int_G \nu\big((x^{-1}(U\cap D'))\cap N\big)\,dx\\ &\geqq\int_{(UN)\cap D'}\nu\big((x^{-1}(U\cap D'))\cap N\big)\,dx.\end{aligned}\right\}\quad(1)$$

If $x\in(UN)\cap D'$, then we have

$$\big(x^{-1}(U\cap D')\big)\cap N=(x^{-1}U)\cap N\neq\varnothing,$$

as is easy to see, and thus the integrand $\nu\big((x^{-1}(U\cap D'))\cap N\big)$ in (1) is positive throughout $(UN)\cap D'$. It follows from (11.27) that $\lambda\big((UN)\cap D'\big)=0$. Since $D\subset U\subset UN$, we have proved that $\lambda(UN)=\lambda(U)$. Thus (iii) is established.]

(b) Let G be as in (a) and let A be any λ-measurable subset of G. Then there are Baire sets C and D such that $D \subset A \subset C$ and $\lambda(C \cap D') = 0$. [Suppose first that $\lambda(A) < \infty$. Let $\{U_n\}_{n=1}^{\infty}$ be a decreasing sequence of open sets such that $U_n \supset A$ and $\lim_{n\to\infty} \lambda(U_n) = \lambda(A)$. Using (a), choose for each n a set V_n of the form $\{x \in G : f(x) > 0\}$ for some $f \in \mathfrak{C}^r(G)$ such that $V_n \supset U_n$ and $\lambda(V_n) = \lambda(U_n)$. Let $C = \bigcap_{n=1}^{\infty} V_n$. It is obvious that C is a Baire set, that $C \supset A$, and that $\lambda(C \cap A') = 0$. Let $\{W_n\}_{n=1}^{\infty}$ be a decreasing sequence of open sets such that $W_n \supset C \cap A'$ and $\lim_{n\to\infty} \lambda(W_n) = 0$; let Y_n be an open Baire set containing W_n such that $\lambda(Y_n) = \lambda(W_n)$; let $Y = \bigcap_{n=1}^{\infty} Y_n$; and finally let $D = C \cap Y'$. Then D is a Baire set, and the relations $D \subset A \subset C$ and $\lambda(C \cap D') = 0$ are evident.

Suppose now that A is any λ-measurable subset of G. Being compactly generated, G is σ-compact, and so $A = \bigcup_{n=1}^{\infty} A_n$, where $\{A_n\}_{n=1}^{\infty}$ is an increasing sequence of λ-measurable sets of finite λ-measure. For each n, choose Baire sets D_n and C_n such that $D_n \subset A_n \subset C_n$ and $\lambda(C_n \cap D_n') = 0$. Then set $D = \bigcup_{n=1}^{\infty} D_n$ and $C = \bigcup_{n=1}^{\infty} C_n$.]

(c) Let G be an arbitrary locally compact group and let A be any λ-measurable subset of G. Then there are Baire sets C and D such that $D \subset A \subset C$ and $C \cap D'$ is a locally λ-null set. [Let H be a compactly generated open subgroup of G (5.14) and let $\{xH : x \in X\}$ be the family of all distinct left cosets of H in G. We obtain a left Haar measure on H by restricting λ to subsets of H; and each coset xH is a topological and λ-measure-theoretic replica of H. Thus for each x one can use (b) to choose Baire sets D_x and C_x as in (b) such that $D_x \subset A \cap (xH) \subset C_x \subset xH$ and $\lambda(C_x \cap D_x') = 0$. Now set $D = \bigcup_{x \in X} D_x$ and $C = \bigcup_{x \in X} C_x$. It is clear that $C \cap D'$ is locally λ-null; and it is easy to see from the construction in (b) that C and D are Baire sets.]

Notes

The notion of convolution [Fr. *Produit de composition*, Ger. *Faltung*, Russ. Свертка] is a venerable one in both analysis and algebra. The convolution

$$\frac{1}{\pi} \int_{-\pi}^{\pi} \varphi(t) \, \frac{\sin\left(\left(n + \frac{1}{2}\right)(t - x)\right)}{2 \sin \frac{1}{2}(t - x)} \, dt$$

is found in Dirichlet's original memoir [1]. Weierstrass's original proof [1] of his famous approximation theorem uses the convolution

$$\frac{1}{\delta \sqrt{\pi}} \int_{-\infty}^{\infty} f(t) \exp\left[-\left(\frac{t - x}{\delta}\right)^2\right] dt.$$

For $f \in \mathfrak{C}(R)$, the limit of this expression as $\delta \downarrow 0$ is $f(x)$; the convergence is uniform on compact sets; and this expression is also evidently the uniform limit of polynomials, on each compact subset of R. This proof of WEIERSTRASS's approximation theorem retains a certain charm, despite the many other proofs that have been given.

Fractional integration and differentiation are defined by means of convolutions: see for example ZYGMUND [1], Vol. II, pp. 133–142. The central rôle played by convolutions in classical harmonic analysis is in evidence throughout ZYGMUND's treatise [1]. Convolutions of functions of finite variation on R [equivalently, convolutions of complex Borel measures on R] were discussed by BOCHNER [1], §§18–21. An *algebraic* viewpoint concerning convolutions of measures on R was manifested in the papers of WIENER and PITT [1] and BEURLING [1]. GEL'FAND in [1] pointed out that $\mathfrak{L}_1(R)$ and $\boldsymbol{M}(R)$ are Banach algebras and initiated the study of their structure that has continued up to the present day.

It is worth pointing out also that the WIENER-HOPF equation

$$\int_0^\infty K(x-y)\,\varphi(y)\,dy = \varphi(x)$$

concerns a convolution on the additive semigroup $[0, \infty[$. For a very detailed survey of results on this equation, see M. G. KREĬN [1]. Convolutions on $[0, \infty[$ were also considered by VOLTERRA and PÉRÈS [1].

Convolutions have also been used for many years in algebra and number theory. For a finite group G, the classical group algebra [over K] consists of all "formal complex linear combinations" $\sum\limits_{x \in G} \alpha_x x$, with componentwise addition and scalar multiplication and with the product

$$\Big(\sum_{x\in G} \alpha_x x\Big) * \Big(\sum_{y\in G} \beta_y y\Big) = \sum_{x\in G}\sum_{y\in G} \alpha_x \beta_y xy = \sum_{z\in G}\Big(\sum_{x\in G} \alpha_x \beta_{x^{-1}z}\Big) z = \sum_{z\in G}\Big(\sum_{y\in G} \alpha_{zy^{-1}} \beta_y\Big) z.$$

It is plain that this algebra is isomorphic with $\boldsymbol{M}(G)$ [and obviously also with $\mathfrak{L}_1(G)$]. The group algebra of a finite group G is an important tool in studying the representations of G, as is brought out for locally compact G in §22. For the history of this technique, see the notes to Chapter III of WEYL [3]. Convolutions of functions on semigroups that are not groups also have a long history. DAUBLEBSKY VON STERNECK [1] wrote down and used for number-theoretic purposes convolutions of functions on $\{1, 2, 3, \ldots\}$ under multiplication. Many years later, D. H. LEHMER [1] used the same formulas and also convolution formulas based on other semigroup operations on $\{1, 2, 3, \ldots\}$. The l_1-algebras of various semigroups have been studied by W. D. MUNN [1] and HEWITT and ZUCKERMAN [2], [3], [4].

The decisive step in uniting the algebraic and analytic lines of development in the theory of convolutions was taken by H. WEYL.

In [2], he defined the convolution of two continuous almost periodic functions on R and used this operation together with the theory of integral equations to establish the principal facts about continuous almost periodic functions on R. WEYL and PETER in [1] defined convolution of two continuous functions on a compact Lie group G and used exactly the same technique as in WEYL [2] to study the representations of G. In [4], VON NEUMANN exploited WEYL's convolution of almost periodic functions to establish the theory of almost periodic functions on an arbitrary group.

Convolutions of measures in $\mathfrak{C}_0^*(G)$ as formulated in (19.6)—(19.12) were defined [in a different form] by WEIL [4], pp. 46—48; a definition almost identical with ours for $\mathfrak{C}_0^*(G)$ appeared in H. CARTAN [2]. The general definition of convolutions in (19.1)—(19.3) is due to HEWITT and ZUCKERMAN [2] and [3]. An independent and somewhat different formulation has been presented by R. C. BUCK [1] and [2].

The construction in (19.4) is due to HEWITT and ZUCKERMAN [4]. The notions introduced in (19.13) go back to VITALI and LEBESGUE for the real line: see for example LEBESGUE [1]. Their adaptation to the case under study is widely known. Theorems (19.16)—(19.20) are announced in HEWITT [5], p. 141.

§20. Convolutions of functions and measures

In this section we present a number of definitions and formulas dealing with convolutions. These are of some interest *per se* and are also essential in our study of representations of locally compact groups. Throughout this section, G will always denote a locally compact group. We begin with some technicalities [(20.1)—(20.4)].

(20.1) Theorem. *Let f be a nonnegative λ-measurable function on G and let a be any element of G. Then ${}_af$ and f_a are λ-measurable,*

(i) $\int_G {}_af \, d\lambda = \int_G f \, d\lambda$,

and

(ii) $\int_G f_a \, d\lambda = \Delta(a^{-1}) \int_G f \, d\lambda$.

If f is a function in $\mathfrak{L}_1(G)$, then ${}_af$ and f_a are also in $\mathfrak{L}_1(G)$ and (i) *and* (ii) *hold for f.*

Proof. Let f be nonnegative and λ-measurable, and let $\{\sigma^{(n)}\}_{n=1}^{\infty}$ be an increasing sequence of λ-measurable functions, each assuming only finitely many values, such that $\lim_{n\to\infty} \sigma^{(n)}(x) = f(x)$ everywhere on G. We plainly have $\lim_{n\to\infty} {}_a\sigma^{(n)}(x) = {}_af(x)$ for all $x \in G$, and so $\lim_{n\to\infty} \int_G {}_a\sigma^{(n)} \, d\lambda = \int_G {}_af \, d\lambda$, $\lim_{n\to\infty} \int_G \sigma^{(n)} \, d\lambda = \int_G f \, d\lambda$. Since $\int_G {}_a\sigma^{(n)} \, d\lambda = \int_G \sigma^{(n)} \, d\lambda$ for all n, (i) holds. In view of (19.17), we have $\int_G \sigma_a^{(n)}\,d\lambda = \Delta(a^{-1})\int_G \sigma^{(n)}\,d\lambda$ for all n. Formula (ii) is proved from this equality just as (i) was proved.

To prove the last assertion, note first that if $f(x)=g(x)$ λ-almost everywhere, then ${}_af(x)={}_ag(x)$ and $f_a(x)=g_a(x)$ λ-almost everywhere. Now write f as $f_1-f_2+i(f_3-f_4)$ where each $f_j\in\mathfrak{L}_1^+(G)$, and apply the previous case to these functions. □

(20.2) Theorem. *Let f be a nonnegative λ-measurable function on G. Then $f^\star$ is also λ-measurable,*

(i) $\displaystyle\int_G f^\star\,d\lambda=\int_G f\frac{1}{\Delta}\,d\lambda,$

and

(ii) $\displaystyle\int_G f\,d\lambda=\int_G f^\star\frac{1}{\Delta}\,d\lambda.$

If $f\in\mathfrak{L}_1(G)$, then $f^\star\frac{1}{\Delta}\in\mathfrak{L}_1(G)$, (ii) *holds, and $\|f\|_1=\left\|f^\star\frac{1}{\Delta}\right\|_1$.*

Proof. Consider first a λ-null set A. We will prove that A^{-1} is λ-null. For $n=1,2,\ldots$, let $D_n=\left\{x\in G:\frac{1}{n}<\Delta(x)<n\right\}$. To prove that A^{-1} is λ-null, it suffices to prove that $A^{-1}\cap D_n$ is λ-null for each n. Choose $\varepsilon>0$ and choose an open set U containing A such that $\lambda(U)<\frac{\varepsilon}{n}$. It is obvious that $U^{-1}\cap D_n$ is open and that $A^{-1}\cap D_n\subset U^{-1}\cap D_n$. Now consider any function $f\in\mathfrak{C}_{00}^+(G)$ such that $f\leqq\xi_{U^{-1}\cap D_n}$. Then $f^\star\leqq\xi_{U\cap D_n}$ and (15.14) implies that

$$I(f)=I\left(f^\star\frac{1}{\Delta}\right)=\int_G f^\star\frac{1}{\Delta}\,d\lambda\leqq\int_{U\cap D_n}\frac{1}{\Delta}\,d\lambda\leqq n\lambda(U\cap D_n)\leqq n\lambda(U)<\varepsilon.$$

Since f is arbitrary, we infer that $\lambda(U^{-1}\cap D_n)\leqq\varepsilon$ and hence also $\lambda(A^{-1}\cap D_n)\leqq\varepsilon$. As ε is arbitrary, $\lambda(A^{-1}\cap D_n)=0$ and so $A^{-1}\cap D_n$ is a λ-null set.

Consider next a λ-measurable set E such that $\lambda(E)<\infty$. Then E is the union of a Borel set B and a λ-null set A. Since B^{-1} is also a Borel set and A^{-1} is λ-null, $E^{-1}=B^{-1}\cup A^{-1}$ is λ-measurable. Finally, consider an arbitrary λ-measurable set C. If F is compact, then $F^{-1}\cap C$ is λ-measurable. By the preceding case, $F\cap C^{-1}$ is also λ-measurable. As F is arbitrary, (11.31) implies that C^{-1} is λ-measurable. Since C^{-1} is λ-measurable whenever C is, it follows immediately that $f^\star$ is λ-measurable whenever f is.

We now prove (ii). For functions f in $\mathfrak{C}_{00}^+(G)$, Theorem (15.14) and Corollary (11.37) imply that $\int_G f\,d\lambda=\int_G f^\star\frac{1}{\Delta}\,d\lambda$. Suppose that $g\in\mathfrak{M}^+(G)$; *i.e.*,

g is a nonnegative lower semicontinuous function on G. Then by (11.11),

$$\left.\begin{aligned}\int_G g\,d\lambda &= \sup\left\{\int_G f\,d\lambda : f\in\mathfrak{C}_{00}^+(G) \text{ and } f\leqq g\right\}\\ &= \sup\left\{\int_G f^\star\frac{1}{\Delta}\,d\lambda : f\in\mathfrak{C}_{00}^+(G) \text{ and } f\leqq g\right\}\leqq\int_G g^\star\frac{1}{\Delta}\,d\lambda.\end{aligned}\right\}\quad(1)$$

Replacing g by $g^\star\frac{1}{\Delta}$ in (1), we obtain

$$\int_G g^\star\frac{1}{\Delta}\,d\lambda\leqq\int_G\left(g^\star\frac{1}{\Delta}\right)^\star\frac{1}{\Delta}\,d\lambda=\int_G g\,d\lambda.$$

Thus we have $\int_G g\,d\lambda=\int_G g^\star\frac{1}{\Delta}\,d\lambda$.

Now let f be an arbitrary nonnegative λ-measurable function on G. Then by (11.36) and (11.16), we have

$$\begin{aligned}\int_G f\,d\lambda &= \inf\left\{\int_G g\,d\lambda : g\in\mathfrak{M}^+(G) \text{ and } g\geqq f\right\}\\ &= \inf\left\{\int_G g^\star\frac{1}{\Delta}\,d\lambda : g\in\mathfrak{M}^+(G) \text{ and } g\geqq f\right\}\geqq\int_G f^\star\frac{1}{\Delta}\,d\lambda,\end{aligned}$$

and replacing f by $f^\star\frac{1}{\Delta}$ in this inequality, we also have

$$\int_G f^\star\frac{1}{\Delta}\,d\lambda\geqq\int_G f\,d\lambda.$$

The remaining statements of the theorem follow at once from (ii). □

(20.3) Note. Theorem (20.2) does *not* imply that $f^\star\in\mathfrak{L}_1(G)$ if $f\in\mathfrak{L}_1(G)$: see (20.29.a).

(20.4) Theorem. *Let p be a number such that $1\leqq p<\infty$ and let f be a function in $\mathfrak{L}_p(G)$. For every $\varepsilon>0$, there is a neighborhood U of e in G such that*

(i) $\|{}_sf-{}_tf\|_p<\varepsilon$ *if* $s,t\in G$ *and* $st^{-1}\in U$.

That is, the mapping $x\to{}_xf$ of G into $\mathfrak{L}_p(G)$ is right uniformly continuous. For every fixed $t\in G$ and every $\varepsilon>0$, there is a neighborhood V of e such that

(ii) $\|f_s-f_t\|_p<\varepsilon$ *if* $s\in tV$.

That is, the mapping $x\to f_x$ of G into $\mathfrak{L}_p(G)$ is continuous.

Proof. Let φ be a function in $\mathfrak{C}_{00}(G)$ such that $\|f-\varphi\|_p<\frac{\varepsilon}{3}$ (12.10). Let F be a compact set such that $\varphi(F')=0$ and let W be a symmetric neighborhood of e such that W^- is compact. Let U be a neighborhood of e such that $U\subset W$ and such that $ab^{-1}\in U$ implies that $|\varphi(a)-\varphi(b)|<$

$\frac{\varepsilon}{3}\lambda(WF)^{-\frac{1}{p}}$ (15.4). If $st^{-1}\in U$, we then have by (20.1)

$$\int_G |\varphi(sx) - \varphi(tx)|^p\, dx = \int_G |\varphi(st^{-1}x) - \varphi(x)|^p\, dx$$
$$= \int_{WF} |\varphi(st^{-1}x) - \varphi(x)|^p\, dx < \left(\frac{\varepsilon}{3}\right)^p;$$

that is, $\|_s\varphi - {}_t\varphi\|_p < \frac{\varepsilon}{3}$. Using (20.1), (12.6), and (12.7), we now can write

$$\|_sf - {}_tf\|_p \leqq \|_sf - {}_s\varphi\|_p + \|_s\varphi - {}_t\varphi\|_p + \|_t\varphi - {}_tf\|_p$$
$$= 2\|f - \varphi\|_p + \|_s\varphi - {}_t\varphi\|_p < 2\frac{\varepsilon}{3} + \frac{\varepsilon}{3} = \varepsilon.$$

This proves (i).

To prove (ii), choose $\psi\in\mathfrak{C}_{00}(G)$ such that $\|\psi - f\|_p < \frac{\varepsilon}{4}\Delta(t)^{\frac{1}{p}}$. Let W be a symmetric neighborhood of e such that W^- is compact and $W\subset\{x\in G:\Delta(x)<2^p\}$. Let E be a compact set such that $\psi(E')=0$ and let V be a symmetric neighborhood of e in G such that $V\subset W$ and such that $b^{-1}a\in V$ implies $|\psi(b)-\psi(a)|<(\varepsilon/4)\,\lambda(EW)^{-\frac{1}{p}}\,\Delta(t)^{\frac{1}{p}}$. Then if $s=tv$, where $v\in V$, we have from (20.1) that

$$\|\psi_s - \psi_t\|_p = \|\psi_{tv} - \psi_t\|_p = \Delta(t^{-1})^{\frac{1}{p}}\|\psi_v - \psi\|_p$$
$$= \Delta(t^{-1})^{\frac{1}{p}}\left[\int_{EW} |\psi(xv) - \psi(x)|^p\, dx\right]^{\frac{1}{p}} < \frac{\varepsilon}{4}.$$

As above, we obtain

$$\|f_s - f_t\|_p = \|f_{tv} - f_t\|_p \leqq \|f_{tv} - \psi_{tv}\|_p + \|\psi_{tv} - \psi_t\|_p + \|\psi_t - f_t\|_p$$
$$= \Delta(v^{-1})^{\frac{1}{p}}\Delta(t^{-1})^{\frac{1}{p}}\|f - \psi\|_p + \|\psi_{tv} - \psi_t\|_p + \Delta(t^{-1})^{\frac{1}{p}}\|\psi - f\|_p$$
$$< 2\cdot\frac{\varepsilon}{4} + \frac{\varepsilon}{4} + \frac{\varepsilon}{4} = \varepsilon. \quad \square$$

Our basic definition follows.

(20.5) Definition. Let μ be a measure in $\boldsymbol{M}(G)$ and ν a measure in $\boldsymbol{M}_a(G)$. Let f be the function in $\mathfrak{L}_1(G)$ as in (19.18) such that $d\nu = f\,d\lambda$. By (19.18), the measures $\mu*\nu$ and $\nu*\mu$ are in $\boldsymbol{M}_a(G)$. Let $\mu*f$ and $f*\mu$, respectively, denote the functions in $\mathfrak{L}_1(G)$ such that $d(\mu*\nu) = (\mu*f)\,d\lambda$ and $d(\nu*\mu) = (f*\mu)\,d\lambda$. If μ is also in $\boldsymbol{M}_a(G)$, so that $d\mu = g\,d\lambda$ for $g\in\mathfrak{L}_1(G)$, then we write $f*g$ for the function in $\mathfrak{L}_1(G)$ such that $d(\nu*\mu) = (f*g)\,d\lambda$. We call $\mu*f$, $f*\mu$, and $f*g$ *convolutions*.

We wish to obtain explicit representations for $\mu*f$, $f*\mu$, and $f*g$ as defined in (20.5), and also to extend the definition of these entities to classes of functions other than $\mathfrak{L}_1(G)$. We begin with two measure-theoretic lemmas, which are of little intrinsic interest but are essential. "Function" means "extended real- or complex-valued function" throughout this section.

(20.6) Lemma. *Let X and Y be locally compact Hausdorff spaces. Let ι and η be measures on X and Y, respectively, as constructed in § 11. Suppose that τ is a continuous mapping of $X\times Y$ into X such that*

$$\text{(i)}\quad\begin{cases}\textit{for every } \iota\textit{-null } F_\sigma \textit{ set } B\subset X \textit{ and every } y\in Y,\\ \textit{the set } N_{y,B}=\{x\in X:\tau(x,y)\in B\} \textit{ is } \iota\textit{-null.}\end{cases}$$

Then $f\circ\tau$ is an $\iota\times\eta$-measurable function on $X\times Y$ for every ι-measurable function f on X.

Proof. (I) We will first show that if E is an ι-null subset of X, then $\tau^{-1}(E)$ is $\iota\times\eta$-measurable. There is a Borel subset B of X [in fact a G_δ set] such that $E\subset B$ and $\iota(B)=0$. Write B_0 for $\tau^{-1}(B)$. Observe that B_0 is a Borel set in $X\times Y$ [in fact a G_δ set] and hence is $\iota\times\eta$-measurable, since τ is continuous. Let F be any compact subset of $X\times Y$. By FUBINI's theorem (13.9), we have

$$\iota\times\eta(F\cap B_0)=\int\limits_{X\times Y}\xi_{F\cap B_0}\,d\iota\times\eta=\int\limits_Y\int\limits_X\xi_{F\cap B_0}(x,y)\,d\iota(x)\,d\eta(y). \qquad (1)$$

For every $y\in Y$, the set $N_{y,B}$ is ι-null by (i). Since $\xi_{F\cap B_0}(x,y)\leqq\xi_{N_{y,B}}(x)$ for all (x,y) in $X\times Y$, (1) implies that

$$\iota\times\eta(F\cap B_0)\leqq\int\limits_Y\int\limits_X\xi_{N_{y,B}}(x)\,d\iota(x)\,d\eta(y)=\int\limits_Y\iota(N_{y,B})\,d\eta(y)=0.$$

Since $\iota\times\eta$ is a complete measure and $\tau^{-1}(E)\subset B_0$, we have proved that $F\cap\tau^{-1}(E)$ is $\iota\times\eta$-null for all compact sets $F\subset X\times Y$. Therefore $\tau^{-1}(E)$ is $\iota\times\eta$-locally null and hence $\iota\times\eta$-measurable by (11.30).

(II) Next we show that if E is an ι-measurable subset of X and if $\iota(E)<\infty$, then $\tau^{-1}(E)$ is $\iota\times\eta$-measurable. There is a Borel subset B of X [in fact an F_σ set] such that $B\subset E$ and $\iota(B)=\iota(E)$. Thus we can write $E=B\cup A$ where $\iota(A)=0$. Then $\tau^{-1}(E)=\tau^{-1}(B)\cup\tau^{-1}(A)$ is $\iota\times\eta$-measurable by (I) and the fact that $\tau^{-1}(B)$ is an F_σ set and hence is $\iota\times\eta$-measurable.

(III) Now we show that if E is any ι-measurable subset of X, then $\tau^{-1}(E)$ is $\iota\times\eta$-measurable. Let F be a compact subset of $X\times Y$. Since τ is continuous, the set $C=\tau(F)$ is compact in X. By (II), the set $\tau^{-1}(C\cap E)$ is $\iota\times\eta$-measurable, since $\iota(C\cap E)\leqq\iota(C)<\infty$. The obvious equality $\tau^{-1}(E)\cap F=\tau^{-1}(C\cap E)\cap F$ shows that $\tau^{-1}(E)\cap F$ is $\iota\times\eta$-measurable. This holds for all compact sets $F\subset X\times Y$ and so $\tau^{-1}(E)$ is $\iota\times\eta$-measurable by (11.31).

(IV) Finally, consider any ι-measurable extended real-valued function f on X. If B is a Borel set of real numbers, $f^{-1}(B)$ is an ι-measurable subset of X, and (III) shows that $\tau^{-1}(f^{-1}(B))$ is an $\iota\times\eta$-measurable subset of $X\times Y$. Since $\tau^{-1}(f^{-1}(B))=(f\circ\tau)^{-1}(B)$ and B is an arbitrary Borel set, we conclude that $f\circ\tau$ is $\iota\times\eta$-measurable. Complex functions f are dealt with by considering their real and imaginary parts. $\square$

(20.7) Lemma. *Let G be a locally compact group, let λ be any left Haar measure on G, and let η be a measure on G constructed as in § 11. Let f be a λ-measurable function on G. The following functions are $\lambda\times\eta$-measurable on $G\times G$:*

$$\text{(i)}\ \begin{cases} (x,y)\to f(xy), & (x,y)\to f(xy^{-1}), & (x,y)\to f(x^{-1}y),\\ (x,y)\to f(yx), & (x,y)\to f(y^{-1}x), & (x,y)\to f(yx^{-1}),\\ (x,y)\to f(yxy^{-1}), & (x,y)\to f(x), & (x,y)\to f(x^{-1}).\end{cases}$$

The following functions are $\eta\times\lambda$-measurable:

$$\text{(ii)}\ (x,y)\to f(y),\qquad (x,y)\to f(y^{-1}).$$

Proof. We will check two cases of (i), leaving the rest of the lemma to the reader. The function $(x,y)\to f(xy^{-1})$ has the form $f\circ\tau$ where $\tau(x,y)=xy^{-1}$ for $(x,y)\in G\times G$. For a subset B of G and $y\in G$, we have $\{x\in G:\tau(x,y)\in B\}=By$ and so the condition (20.6.i) follows from (19.17). Hence $(x,y)\to f(xy^{-1})$ is $\lambda\times\eta$-measurable by (20.6). If τ is the mapping $\tau(x,y)=x^{-1}$ for $(x,y)\in G\times G$, then $\{x\in G:\tau(x,y)\in B\}=B^{-1}$ for all $y\in G$, and the condition (20.6.i) follows in this case from (20.2). Hence the function $(x,y)\to f(x^{-1})$ is $\lambda\times\eta$-measurable, again by (20.6). □

We can now prove the general theorem that shows the integrability of convolutions. It too is rather technical.

(20.8) Theorem. *Suppose that $1\leqq p\leqq\infty$. Let f be a λ-measurable function on G, and let τ be either the function $(x,y)\to x^{-1}y$ or $(x,y)\to yx^{-1}$ mapping $G\times G$ onto G. Suppose that Φ is an $|\eta|$-measurable function on G, where either* (a) *η is a measure in $\boldsymbol{M}(G)$, or* (b) *η is a nonnegative measure as in § 11 and Φ vanishes outside of a σ-compact set. Finally suppose that there is a positive number γ such that*

$$\text{(i)}\ \int_G\int_G |\psi(y)\,\Phi(x)\,f\circ\tau(x,y)|\,dy\,d|\eta|(x)\leqq\|\psi\|_{p'}\,\gamma$$

for all $\psi\in\mathfrak{C}_{00}(G)$.[1] *If $1\leqq p<\infty$ and f vanishes outside of a σ-compact set S, then*

$$\text{(ii)}\ \int_G \Phi(x)\,f\circ\tau(x,y)\,d\eta(x)=h(y)$$

exists and is finite for λ-almost all $y\in G$; h belongs to $\mathfrak{L}_p(G)$; and $\|h\|_p\leqq\gamma$. If $p=\infty$ and $f\in\mathfrak{L}_\infty(G)$, then the integral in (ii) *exists and is finite for locally λ-almost all $y\in G$; h is in $\mathfrak{L}_\infty(G)$; and $\|h\|_\infty\leqq\gamma$.*

Proof. (I) First we show that the function h of (ii) exists and is λ-measurable. In case (a) we select an $|\eta|$-measurable function g such

[1] Recall our convention that $1'=\infty$ and $\infty'=1$; see page 141.

that $d\eta = g\,d|\eta|$ and $|g| = 1$ (14.12). Since there exists a σ-compact subset D of G such that $|\eta|(D') = 0$, we may suppose that g vanishes outside of a σ-compact set. In case (b) we define $g(x) = 1$ for all $x \in G$. In this case, Φ vanishes outside of a σ-compact set. Thus in either case we have $d\eta = g\,d|\eta|$ and $g\Phi$ is an $|\eta|$-measurable function that vanishes outside of a σ-compact set E.

Let ψ be in $\mathfrak{C}_{00}(G)$; then ψ vanishes outside of a compact set A_ψ. The foregoing and (20.7) show that $H(x,y) = \psi(y) f \circ \tau(x,y) \Phi(x) g(x)$ is an $|\eta| \times \lambda$-measurable function on $G \times G$ that vanishes outside of the σ-compact set $E \times A_\psi$. Thus we may apply (13.10), citing (i), to obtain

$$\int_G \int_G H(x,y)\,dy\,d|\eta|(x) = \int_G \int_G H(x,y)\,d|\eta|(x)\,dy. \tag{1}$$

By (13.10), (13.8), and [again] (i), the integral

$$\int_G H(x,y)\,d|\eta|(x) = \psi(y) \int_G f \circ \tau(x,y)\,\Phi(x) g(x)\,d|\eta|(x) \tag{2}$$

exists and is finite for λ-almost all $y \in G$, and defines a λ-measurable function in $\mathfrak{L}_1(G)$. For $y \in G$, we define

$$h(y) = \int_G f \circ \tau(x,y)\,\Phi(x) g(x)\,d|\eta|(x) \tag{3}$$

provided that the integral exists; we define $h(y) = 0$ otherwise, as suggested by footnote 2 to page 153. We see that ψh is λ-measurable for all $\psi \in \mathfrak{C}_{00}(G)$. By (11.42), h itself is λ-measurable.

(II) We next show that h belongs to $\mathfrak{L}_p(G)$ if $1 \leqq p < \infty$ and f vanishes outside of a σ-compact set S. Let $B = ES$ if $\tau(x,y) = x^{-1}y$ and let $B = SE$ if $\tau(x,y) = yx^{-1}$; plainly B is σ-compact. Hence there is a sequence $\{\psi_n\}_{n=1}^{\infty}$ in $\mathfrak{C}_{00}^+(G)$ such that $\xi_B = \lim\limits_{n\to\infty} \psi_n$ λ-almost everywhere and $\psi_n(G) \subset [0,1]$ for all n. From (2) we infer that

$$h(y) = \lim_{n\to\infty} \psi_n(y) \int_G f \circ \tau(x,y)\,\Phi(x) g(x)\,d|\eta|(x)$$

exists and is finite for λ-almost all $y \in B$. It is easy to show that $h(y) = 0$ for all $y \in B'$. Since B is σ-compact, we thus have

$$\int_G |h(y)|^p\,dy = \sup\left\{\int_F |h(y)|^p\,dy : F \text{ is compact, } F \subset G\right\}.$$

For ψ in $\mathfrak{C}_{00}(G)$, we have by (1) and (i) that

$$\left|\int_G h(y)\psi(y)\,dy\right| \leqq \|\psi\|_{p'}\,\gamma$$

and thus (12.14) and (12.13) show that $h \in \mathfrak{L}_p(G)$ and that $\|h\|_p \leqq \gamma$.

(III) Finally, suppose that f is in $\mathfrak{L}_\infty(G)$. Let N denote the set of $y \in G$ for which the integral in (3) does not exist as a finite number. Assume that there is a compact set F such that $\lambda(N\cap F)>0$. Choose $\psi\in\mathfrak{C}_{00}^{+}(G)$ such that $\psi(F)=1$. Then the integrals in (2) do not exist as finite numbers for this ψ and for $y\in N\cap F$, a contradiction. Thus N is locally λ-null, and so (3) holds locally λ-almost everywhere.

If $\{y\in G: |h(y)|>\gamma\}$ is not locally λ-null, then $\{y\in G: |h(y)|>\beta\}$ is not locally λ-null for some $\beta>\gamma$. Thus $\{y\in G: |h(y)|>\beta\}$ contains a compact set F for which $\lambda(F)>0$. Choose an open set U such that $F\subset U$ and $\lambda(U\cap F')<\left(\frac{\beta-\gamma}{\gamma}\right)\lambda(F)$. Choose ψ in $\mathfrak{C}_{00}^{+}(G)$ such that $\psi(F)=1$, $\psi(U')=0$, and $\psi(G)\subset[0,1]$. Thus by (i) and (13.10), we have

$$\begin{aligned}\|\psi\|_1\gamma&=\gamma\int_U|\psi(y)|\,dy\leqq\gamma\,\lambda(U)=\gamma\,\lambda(F)+\gamma\,\lambda(U\cap F')\\&<\gamma\,\lambda(F)+(\beta-\gamma)\,\lambda(F)=\beta\,\lambda(F)\leqq\int_G|\psi(y)\,h(y)|\,dy\\&\leqq\int_G\int_G|f\circ\tau(x,y)\,\Phi(x)\,\psi(y)|\,d|\eta|(x)\,dy\\&=\int_G\int_G|\psi(y)\,\Phi(x)\,f\circ\tau(x,y)|\,dy\,d|\eta|(x).\end{aligned}$$

This is contrary to (i). $\square$

With (20.8), we can now explicitly represent the functions $\mu*f$, $f*\mu$, and $f*g$ as defined in (20.5).

(20.9) Theorem. *For $\mu\in\boldsymbol{M}(G)$ and $\nu\in\boldsymbol{M}_a(G)$ so that $d\nu=f\,d\lambda$ with $f\in\mathfrak{L}_1(G)$, we have*

(i) $\mu*f(x)=\int_G f(y^{-1}x)\,d\mu(y)$ *for λ-almost all $x\in G$,*

and

(ii) $f*\mu(x)=\int_G\Delta(y^{-1})\,f(xy^{-1})\,d\mu(y)$ *for λ-almost all $x\in G$.*

Proof. For $\psi\in\mathfrak{C}_{00}(G)$, (19.10) shows that

$$\int_G\psi\,d\mu*\nu=\int_G\int_G\psi(xy)\,f(y)\,dy\,d\mu(x).$$

Using (20.1), we can write

$$\int_G\psi(xy)\,f(y)\,dy=\int_G{}_x[\psi(y)\,{}_{x^{-1}}f(y)]\,dy=\int_G\psi(y)\,f(x^{-1}y)\,dy,$$

for all $x\in G$. Thus

$$\int_G\psi\,d\mu*\nu=\int_G\int_G\psi(y)\,f(x^{-1}y)\,dy\,d\mu(x).\tag{1}$$

We will now apply (20.8) with $p=1$. Note that f must vanish outside of a σ-compact set, in fact outside of an open σ-compact subgroup. Let

$\tau(x, y) = x^{-1}y$, $\Phi = 1$, $\eta = \mu$, and $\gamma = \|\mu\| \cdot \|\nu\|$. For $\psi \in \mathfrak{C}_{00}(G)$, we have

$$\left.\begin{aligned} \int_G \int_G |\psi(y) f(x^{-1}y)| \, dy \, d|\mu|(x) &\leqq \|\psi\|_\infty \int_G \|{}_{x^{-1}}f\|_1 \, d|\mu|(x) \\ &= \|\psi\|_\infty \|f\|_1 \|\mu\| = \|\psi\|_\infty \|\nu\| \|\mu\|. \end{aligned}\right\} \quad (2)$$

Therefore by (20.8), the function $h(y) = \int_G f(x^{-1}y) \, d\mu(x)$ belongs to $\mathfrak{L}_1(G)$. Moreover, using (1), (2), and (13.8), we see that

$$\begin{aligned} \int_G \psi h \, d\lambda &= \int_G \int_G \psi(y) f(x^{-1}y) \, d\mu(x) \, dy \\ &= \int_G \int_G \psi(y) f(x^{-1}y) \, dy \, d\mu(x) = \int_G \psi \, d\mu * \nu = \int_G \psi \mu * f \, d\lambda \end{aligned}$$

for all $\psi \in \mathfrak{C}_{00}(G)$. Theorem (12.13) shows that $\|\mu * f - h\|_1 = 0$, and this implies (i).

For (ii) we let $\tau(x, y) = yx^{-1}$, $\Phi = \Delta^{-1}$, $\eta = \mu$, and $\gamma = \|\nu\| \cdot \|\mu\|$. Then for $\psi \in \mathfrak{C}_{00}(G)$, we have

$$\begin{aligned} &\int_G \int_G |\psi(y) \Delta^{-1}(x) f(yx^{-1})| \, dy \, d|\mu|(x) \\ &\quad = \int_G \Delta(x^{-1}) \int_G |\psi(y) f_{x^{-1}}(y)| \, dy \, d|\mu|(x) \\ &\quad \leqq \int_G \Delta(x^{-1}) \|\psi\|_\infty \int_G |f_{x^{-1}}(y)| \, dy \, d|\mu|(x) \\ &\quad = \int_G \|\psi\|_\infty \int_G |f(y)| \, dy \, d|\mu|(x) = \|\psi\|_\infty \|f\|_1 \|\mu\| = \|\psi\|_\infty \|\nu\| \cdot \|\mu\|. \end{aligned}$$

Now argue as above. ▯

(20.10) Theorem. *For $f, g \in \mathfrak{L}_1(G)$, we have*

(i) $f * g(x) = \int_G f(y) g(y^{-1}x) \, dy$,

(ii) $f * g(x) = \int_G f(xy) g(y^{-1}) \, dy$,

(iii) $f * g(x) = \int_G \Delta(y^{-1}) f(xy^{-1}) g(y) \, dy$,

(iv) $f * g(x) = \int_G \Delta(y^{-1}) f(y^{-1}) g(yx) \, dy$,

each equality holding for λ-almost all $x \in G$.

Proof. Equalities (i) and (iii) follow at once from (20.9) and the definition of $f * g$. To prove (ii), we write

$$\begin{aligned} \int_G f(y) g(y^{-1}x) \, dy &= \int_G f(y) g^\star(x^{-1}y) \, dy \\ &= \int_G {}_{x^{-1}}[{}_x f(y) \, g^\star(y)] \, dy = \int_G f(xy) g(y^{-1}) \, dy. \end{aligned}$$

To prove (iv), we write

$$\int_G \frac{1}{\Delta(y)} f(xy^{-1})\, g(y)\, dy = \int_G \frac{1}{\Delta(y)} f^\star(yx^{-1})\, g(y)\, dy$$
$$= \Delta(x^{-1}) \int_G \left[\frac{1}{\Delta(y)} f^\star(y)\, g_x(y)\right]_{x^{-1}} dy = \int_G \Delta(y^{-1})\, f(y^{-1})\, g(yx)\, dy.$$

One can also obtain (ii)—(iv) from (i) alone by applying (20.1) and (20.2). □

(20.11) Remark. Convolutions of functions and translations are connected by the following identities, in which f and g are λ-measurable functions and a and x are elements of G:

(i) $({}_af) * g(x) = f * g(ax) = {}_a(f*g)(x)$,
(ii) $f * (g_a)(x) = f * g(xa) = (f*g)_a(x)$,
(iii) $(f_a) * g(x) = \Delta(a^{-1})\, f * ({}_ag)(x)$.

Each of these identities means that whenever one of the integrals exists and is finite, the others also exist and all are equal. These identities follow at once from (20.1).

The convolution formulas (20.9.i) and (20.9.ii) for $\mu \in \boldsymbol{M}(G)$ and $f \in \mathfrak{L}_1(G)$ can be extended to $f \in \mathfrak{L}_p(G)$ $(1 < p \leqq \infty)$. The following two theorems show how this happens.

(20.12) Theorem. *Let f be a function on G that is in $\mathfrak{L}_p(G)$ $(1 \leqq p \leqq \infty)$, and let μ be a measure in $\boldsymbol{M}(G)$. Then the integral*

(i) $\int_G f(y^{-1}x)\, d\mu(y) = \mu * f(x)$

exists and is finite for all $x \in G \cap N'$, where N is λ-null if $1 \leqq p < \infty$ and N is locally λ-null if $p = \infty$. Defining $\mu * f(x)$ as 0 where it is not defined by (i), *we obtain a function in $\mathfrak{L}_p(G)$, for which*

(ii) $\|\mu * f\|_p \leqq \|\mu\| \cdot \|f\|_p$.

Proof. We apply Theorem (20.8). If p is finite, the function f vanishes outside of a σ-compact set. Let $\tau(x, y) = x^{-1}y$, $\Phi = 1$, $\eta = \mu$, and $\gamma = \|\mu\| \cdot \|f\|_p$. If $\psi \in \mathfrak{C}_{00}(G)$, then HÖLDER's inequality yields

$$\int_G |f(x^{-1}y)\, \psi(y)|\, dy \leqq \|{}_{x^{-1}}f\|_p \cdot \|\psi\|_{p'} = \|f\|_p \cdot \|\psi\|_{p'}.$$

We therefore have

$$\int_G \int_G |\psi(y)\, f(x^{-1}y)|\, dy\, d|\mu|(x) \leqq \int_G \|f\|_p \cdot \|\psi\|_{p'}\, d|\mu|(x) = \|\psi\|_{p'} \cdot \|f\|_p\, \|\mu\|.$$

The present theorem follows from (20.8). □

For the convolution $f * \mu$ with $f \in \mathfrak{L}_p(G)$, our results are less elegant than for $\mu * f$. This difference arises because of the distinction between left and right Haar measure.

(20.13) Theorem. *Let μ be in $\boldsymbol{M}(G)$ and suppose that $\int_G \Delta(y)^{-\frac{1}{p'}} d|\mu|(y)$ is finite, where $1 \leqq p \leqq \infty$ and $p' = \frac{p}{p-1}$ $[1' = \infty,\ \infty' = 1$, and $\frac{1}{\infty} = 0]$. Let f be a function in $\mathfrak{L}_p(G)$. Then the integral*

(i) $\int_G \Delta(y^{-1}) f(xy^{-1})\, d\mu(y) = f * \mu(x)$

exists and is finite for all $x \in G \cap N'$, where N is λ-null if $1 \leqq p < \infty$ and N is locally λ-null if $p = \infty$. Relation (i) *defines a function in $\mathfrak{L}_p(G)$ for which*

(ii) $\|f * \mu\|_p \leqq \|f\|_p \int_G \Delta(y)^{-\frac{1}{p'}} d|\mu|(y)$.

Proof. In order to apply Theorem (20.8), let $\tau(x, y) = yx^{-1}$, $\Phi = \Delta^{-1}$, $\eta = \mu$, and $\gamma = \|f\|_p \int_G \Delta(y)^{-\frac{1}{p'}} d|\mu|(y)$. Let ψ be in $\mathfrak{C}_{00}(G)$. By Hölder's inequality (12.4.ii) and Theorem (20.1), we have

$$\int_G |f(yx^{-1})\, \psi(y)|\, dy \leqq \|f_{x^{-1}}\|_p \|\psi\|_{p'} = \Delta(x)^{\frac{1}{p}} \|f\|_p \|\psi\|_{p'}.$$

Therefore we have

$$\begin{aligned}\int_G \int_G |\psi(y)\, \Delta^{-1}(x)\, f(yx^{-1})|\, dy\, d|\mu|(x) &\leqq \int_G \Delta(x)^{\frac{1}{p}-1} \|f\|_p \|\psi\|_{p'}\, d|\mu|(x)\\ &= \|\psi\|_{p'} \|f\|_p \int_G \Delta(x)^{-\frac{1}{p'}} d|\mu|(x). \quad \square\end{aligned}$$

(20.14) Corollary. *Let $f \in \mathfrak{L}_1(G)$ and $g \in \mathfrak{L}_p(G)$ $(1 \leqq p \leqq \infty)$. The integral*

(i) $f * g(x) = \int_G f(xy)\, g(y^{-1})\, dy = \int_G f(y)\, g(y^{-1}x)\, dy$

exists and is finite for all $x \in G \cap N'$, where N is λ-null if $p < \infty$ and is locally λ-null if $p = \infty$. The function $f * g$ is in $\mathfrak{L}_p(G)$, and

(ii) $\|f * g\|_p \leqq \|f\|_1 \|g\|_p$.

If $f \in \mathfrak{L}_p(G)$ and $\Delta^{-\frac{1}{p'}} g \in \mathfrak{L}_1(G)$, the integral

(iii) $$\begin{aligned} f * g(x) &= \int_G \Delta(y^{-1}) f(xy^{-1})\, g(y)\, dy = \int_G f(xy)\, g(y^{-1})\, dy\\ &= \int_G f(y)\, g(y^{-1}x)\, dy\end{aligned}$$

exists and is finite for all $x \in G \cap N'$, where N is as above. The function $f * g$ is in $\mathfrak{L}_p(G)$, and

(iv) $\|f * g\|_p \leqq \|f\|_p \|\Delta^{-\frac{1}{p'}} g\|_1$.

Proof. The statement for $f \in \mathfrak{L}_1(G)$ and $g \in \mathfrak{L}_p(G)$ follows at once from (20.12).

For $f \in \mathfrak{L}_p(G)$ and $\Delta^{-\frac{1}{p'}} g \in \mathfrak{L}_1(G)$, we apply Theorem (20.8). [Note that if g happens to be in $\mathfrak{L}_1(G)$, then we can simply apply (20.13) with $d\mu = g\, d\lambda$.] Let $\tau(x, y) = yx^{-1}$, $\Phi = \Delta^{-1} g$, $\eta = \lambda$, and $\gamma = \|f\|_p \|\Delta^{-\frac{1}{p'}} g\|_1$. Note that Φ is a λ-measurable function that vanishes outside of a σ-compact set. A computation similar to that in (20.13) shows that

$$\int_G \int_G |\psi(y)\, \Delta^{-1}(x)\, g(x)\, f(yx^{-1})|\, dy\, dx \leqq \|\psi\|_{p'} \|f\|_p \|\Delta^{-\frac{1}{p'}} g\|_1,$$

for $\psi \in \mathfrak{C}_{00}(G)$. □

Theorems (20.12) and (20.13) lead naturally to the following result, which expresses a continuity property of the mappings $f \to \mu * f$ and $f \to f * \mu$ for $f \in \mathfrak{L}_p(G)$.

(20.15) Theorem. *Let f be a function in $\mathfrak{L}_p(G)$ $(1 \leqq p < \infty)$ and ε a positive number. There is a neighborhood U of e in G such that*

(i) $\|\mu * f - f\|_p < \varepsilon$

for all $\mu \in \boldsymbol{M}^+(G)$ such that $\mu(G) = 1$ and $\mu(U') = 0$. There is also a neighborhood V of e such that

(ii) $\|f * \mu - f\|_p < \varepsilon$

for all $\mu \in \boldsymbol{M}^+(G)$ such that $\mu(G) = 1$ and $\mu(V') = 0$.

Proof. By (20.12), we have $\mu * f \in \mathfrak{L}_p(G)$. Let U be a neighborhood of e in G such that $\|_{y^{-1}}f - f\|_p < \frac{\varepsilon}{2}$ if $y \in U$ (20.4). Then if $\psi \in \mathfrak{C}_{00}(G)$ and $\mu(U') = 0$, the function $(x, y) \to |f(y^{-1}x) - f(x)| \cdot |\psi(x)|$ satisfies the conditions of Theorem (13.10). Thus we have

$$\begin{aligned}\int_G |\mu * f(x) - f(x)|\, |\psi(x)|\, dx &= \int_G \Big|\int_U f(y^{-1}x)\, d\mu(y) - \int_U f(x)\, d\mu(y)\Big|\, |\psi(x)|\, dx \\ &\leqq \int_G \int_U |f(y^{-1}x) - f(x)|\, d\mu(y)\, |\psi(x)|\, dx \\ &= \int_U \int_G |f(y^{-1}x) - f(x)|\, |\psi(x)|\, dx\, d\mu(y) \\ &\leqq \int_U \|_{y^{-1}}f - f\|_p \|\psi\|_{p'}\, d\mu(y) \leqq \frac{\varepsilon}{2} \|\psi\|_{p'}.\end{aligned}$$

Theorem (12.13) implies that $\|\mu * f - f\|_p < \varepsilon$.

To prove (ii), let V be a neighborhood of e in G such that $\|f\|_p\, |\Delta(y^{-1}) - 1| < \frac{\varepsilon}{2}$ for $y \in V$ and $\Delta(y^{-1})\, \|f_{y^{-1}} - f\|_p < \frac{\varepsilon}{2}$ for $y \in V$. Then if $\mu \in \boldsymbol{M}^+(G)$, $\mu(G) = 1$, and $\mu(V') = 0$, we have

$$\|f\|_p \cdot \Big|1 - \int_V \Delta(y^{-1})\, d\mu(y)\Big| \leqq \int_V \|f\|_p \cdot |1 - \Delta(y^{-1})|\, d\mu(y) < \frac{\varepsilon}{2}. \tag{1}$$

For $\psi \in \mathfrak{C}_{00}(G)$, we have

$$\begin{aligned}&\int_G \Big|f * \mu(x) - \Big(\int_V \Delta(y^{-1})\, d\mu(y)\Big) f(x)\Big|\, |\psi(x)|\, dx \\ &\quad = \int_G \Big|\int_V (f(xy^{-1}) - f(x))\, \Delta(y^{-1})\, d\mu(y)\Big|\, |\psi(x)|\, dx\end{aligned}$$

$$\leqq \int_G \int_V |f(xy^{-1}) - f(x)| \, \Delta(y^{-1}) \, d\mu(y) |\psi(x)| \, dx$$
$$= \int_V \int_G |f(xy^{-1}) - f(x)| \, |\psi(x)| \, dx \, \Delta(y^{-1}) \, d\mu(y)$$
$$\leqq \|\psi\|_{p'} \cdot \int_V \|f_{y^{-1}} - f\|_p \, \Delta(y^{-1}) \, d\mu(y) \leqq \frac{\varepsilon}{2} \|\psi\|_{p'}.$$

Theorem (12.13) implies that

$$\left\| f * \mu - \left(\int_V \Delta(y^{-1}) \, d\mu(y) \right) f \right\|_p \leqq \frac{\varepsilon}{2}. \tag{2}$$

Combining (1) and (2), we obtain (ii). □

(20.16) Theorem. *Let p be a real number such that $1 < p < \infty$, and let $p' = \frac{p}{p-1}$. If $f \in \mathfrak{L}_p(G)$ and $g^\star \in \mathfrak{L}_{p'}(G)$, then the integral*

(i) $\int_G f(xy)\, g(y^{-1})\, dy = f * g(x)$

exists for all $x \in G$ and defines a function in $\mathfrak{C}_0(G)$. If $f \in \mathfrak{L}_1(G)$ and $g \in \mathfrak{L}_\infty(G)$, then the integral (i) *exists for all $x \in G$ and defines a function in $\mathfrak{C}_{ru}(G)$.*[1] *If $f \in \mathfrak{L}_\infty(G)$ and $g^\star \in \mathfrak{L}_1(G)$, then the integral* (i) *exists for all $x \in G$ and defines a function in $\mathfrak{C}_{lu}(G)$.*[1]

Proof. Suppose first that $f \in \mathfrak{L}_p(G)$ and $g^\star \in \mathfrak{L}_{p'}(G)$, $1 < p < \infty$. By HÖLDER's inequality (12.4.ii) we have

$$\int_G |f(xy)\, g(y^{-1})| \, dy \leqq \|{}_x f\|_p \, \|g^\star\|_{p'} = \|f\|_p \, \|g^\star\|_{p'}$$

for all $x \in G$. Thus $f * g(x)$ exists and is finite for all $x \in G$, and also $\|f * g\|_u \leqq \|f\|_p \, \|g^\star\|_{p'}$. Applying HÖLDER's inequality again, we have

$$|f * g(s) - f * g(t)| \leqq \|{}_s f - {}_t f\|_p \, \|g^\star\|_{p'}$$

for all $s, t \in G$. It follows from (20.4.i) that $f * g$ is right uniformly continuous. If $f(E') = 0$ and $g(F') = 0$, where E and F are subsets of G, then $f * g((EF)') = 0$. Thus if f and g vanish outside of compact subsets of G, $f * g$ is in $\mathfrak{C}_{00}(G)$. Otherwise, let $\{f_n\}_{n=1}^\infty$ and $\{g_n\}_{n=1}^\infty$ be sequences of functions in $\mathfrak{C}_{00}(G)$ such that $\lim_{n\to\infty} \|f_n - f\|_p = 0$ and $\lim_{n\to\infty} \|g_n^\star - g^\star\|_{p'} = 0$ (12.10). Then we have

$$\|f_n * g_n - f * g\|_u \leqq \|f_n * g_n - f_n * g\|_u + \|f_n * g - f * g\|_u$$
$$\leqq \|f_n\|_p \, \|g_n^\star - g^\star\|_{p'} + \|f_n - f\|_p \, \|g^\star\|_{p'}.$$

Therefore

$$\lim_{n\to\infty} \|f_n * g_n - f * g\|_u = 0,$$

and $f * g$ accordingly is a function in $\mathfrak{C}_0(G)$.

[1] For the definitions of $\mathfrak{C}_{lu}(G)$ and $\mathfrak{C}_{ru}(G)$, see (19.23.b).

If $f \in \mathfrak{L}_1(G)$ and $g \in \mathfrak{L}_\infty(G)$, then $\int\limits_G f(xy)\, g(y^{-1})\, dy$ plainly exists [1] and is finite for all $x \in G$. For $s, t \in G$, we have

$$|f*g(s) - f*g(t)| \leqq \int\limits_G |f(sy) - f(ty)|\, |g(y^{-1})|\, dy \leqq \|_s f - {}_t f\|_1 \|g\|_\infty .$$

By (20.4.i) we see that $f*g$ is right uniformly continuous.

If $f \in \mathfrak{L}_\infty(G)$ and $g^\star \in \mathfrak{L}_1(G)$, then we have from (20.1) that

$$f*g(x) = \int\limits_G f(xy)\, g(y^{-1})\, dy = \int\limits_G f(y)\, g(y^{-1}x)\, dy = \int\limits_G f \cdot ({}_{x^{-1}}(g^\star))\, d\lambda .$$

As in the preceding paragraph we see that $|f*g(s) - f*g(t)|$ is arbitrarily small if $s^{-1}t$ lies in a suitable neighborhood of the identity. Thus $f*g$ is left uniformly continuous. ▯

(20.17) Corollary. *Let A and B be λ-measurable subsets of G such that $0 < \lambda(A) < \infty$ and $0 < \lambda(B^{-1}) < \infty$. The function $x \to \lambda(A \cap (xB^{-1}))$ is in $\mathfrak{C}_0^+(G)$ and is positive somewhere. The set AB contains a nonvoid open set, and the set $A(A^{-1})$ contains a neighborhood of the identity.*

Proof. Both of the functions ξ_A and $\xi_B^\star = \xi_{B^{-1}}$ are in $\mathfrak{L}_p(G)$ for $1 \leqq p \leqq \infty$. Applying (20.16) with $p = p' = 2$, say, we find that $\xi_A * \xi_B$ is in $\mathfrak{C}_0(G)$. The value $\xi_A * \xi_B(x)$ is just $\lambda(A \cap (xB^{-1}))$. It is also clear that $\xi_A * \xi_B(x) = 0$ if $x \in (AB)'$. Finally, using (13.9), we have

$$\begin{aligned}\int\limits_G \xi_A * \xi_B(x)\, dx &= \int\limits_G \int\limits_G \xi_A(xy)\, \xi_{B^{-1}}(y)\, dy\, dx \\ &= \int\limits_G \int\limits_G \xi_A(xy)\, dx\, \xi_{B^{-1}}(y)\, dy = \lambda(A) \int\limits_G \Delta(y^{-1})\, \xi_{B^{-1}}(y)\, dy = \lambda(A)\, \lambda(B) > 0 .\end{aligned}$$

Hence $\xi_A * \xi_B$ is positive on a set of positive λ-measure, and is accordingly positive on an open subset of AB. Since $\lambda(A^{-1}) = \int\limits_A \frac{1}{\Delta}\, d\lambda > 0$, it is easy to see that A contains a compact set E such that $\lambda(E)$ and $\lambda(E^{-1})$ are positive. We have $\xi_E * \xi_{E^{-1}}(e) = \lambda(E) > 0$, so that $E(E^{-1})$ and also $A(A^{-1})$ contain a neighborhood of e.[2] ▯

Still another result along the same lines as (20.14) and (20.16) can be obtained.

(20.18) Theorem. *Let p and q be real numbers such that $1 < p < \infty$, $1 < q < \infty$, $\frac{1}{p} + \frac{1}{q} > 1$; and let $r = \frac{pq}{p + q - pq}$, so that $\frac{1}{p} + \frac{1}{q} - \frac{1}{r} = 1$.*

---

[1] Note that if $g \in \mathfrak{L}_\infty(G)$, then $g^\star \in \mathfrak{L}_\infty(G)$ and $\|g^\star\|_\infty = \|g\|_\infty$. This is because $\lambda(F) = 0$ if and only if $\lambda(F^{-1}) = 0$, for compact subsets F of G [use (20.2)].

[2] The result in (20.17) may be false if $\lambda(A) = \infty$. To see this, consider $G = R \times R_d$ and $A = \{0\} \times R_d$. Then A is a locally null non-null λ-measurable subset of G, so that $\lambda(A) = \infty$. However, $AA^{-1} = A$ does not contain a neighborhood of $(0, 0)$.

Let f and g be λ-measurable functions on G such that $f \in \mathfrak{L}_p(G)$, $g \in \mathfrak{L}_q(G)$, $g^\star \in \mathfrak{L}_q(G)$, and $\|g\|_q = \|g^\star\|_q$. Then for λ-almost all $x \in G$, the integral

(i) $f * g(x) = \int_G f(xy)\, g(y^{-1})\, dy$

exists and is finite. The function $f * g$ is in $\mathfrak{L}_r(G)$,[1] and the inequality

(ii) $\|f * g\|_r \leq \|f\|_p \cdot \|g\|_q$

obtains.

Proof. Again the functions f and g must vanish outside of σ-compact sets. We write

$$\int_G |f(xy)\, g(y^{-1})|\, dy = \int_G \left(|f(xy)|^p\, |g(y^{-1})|^q\right)^{\frac{1}{r}} |f(xy)|^{1-\frac{p}{r}}\, |g(y^{-1})|^{1-\frac{q}{r}}\, dy. \quad (1)$$

By (20.14), the function $y \to |f(xy)|^p\, |g(y^{-1})|^q$ is in $\mathfrak{L}_1(G)$ for λ-almost all $x \in G$. For all such x, we apply (12.5) to the right side of (1), with $n = 3$, $\alpha_1 = \frac{1}{r}$, $\alpha_2 = 1 - \frac{1}{q}$, and $\alpha_3 = 1 - \frac{1}{p}$. This gives us

$$\left.\begin{aligned} &\int_G |f(xy)\, g(y^{-1})|\, dy \\ &\leq \left(\int_G |f(xy)|^p\, |g(y^{-1})|^q\, dy\right)^{\frac{1}{r}} \left(\int_G |f(xy)|^p\, dy\right)^{\frac{q-1}{q}} \left(\int_G |g(y^{-1})|^q\, dy\right)^{\frac{p-1}{p}}, \end{aligned}\right\} \quad (2)$$

since $\left(1 - \frac{p}{r}\right)\frac{q}{q-1} = p$ and $\left(1 - \frac{q}{r}\right)\frac{p}{p-1} = q$. Hence $f * g(x)$ exists and is finite for λ-almost all $x \in G$. We now show that $f * g$ is actually λ-measurable. We may, as usual, suppose that f and g are nonnegative. Then there is an increasing sequence $\{f_n\}_{n=1}^{\infty}$ of nonnegative functions such that $\lim_{n\to\infty} f_n(x) = f(x)$ for all $x \in G$ and such that each f_n is a finite linear combination of characteristic functions of λ-measurable sets having finite measure. Likewise g is the limit of an increasing sequence $\{g_n\}_{n=1}^{\infty}$ of such functions. Each function $f_n * g_n$ belongs to $\mathfrak{C}_0^+(G)$ by (20.16). The monotone convergence theorem implies that $f * g(x) = \lim_{n\to\infty} f_n * g_n(x)$ for all $x \in G$, and hence $f * g$ is λ-measurable.

We can rewrite (2) as

$$\left(\int_G |f(xy)\, g(y^{-1})|\, dy\right)^r \leq \|f\|_p^{\frac{rp(q-1)}{q}} \cdot \|g\|_q^{\frac{rq(p-1)}{p}} \int_G |f(xy)|^p\, |g(y^{-1})|^q\, dy. \quad (3)$$

It is simple to verify that the hypotheses of (13.9) are satisfied by the function $(x, y) \to |f(xy)|^p\, |g(y^{-1})|^q$; see (20.7). Thus we have

$$\left.\begin{aligned} \int_G \int_G |f(xy)|^p\, |g(y^{-1})|^q\, dy\, dx &= \int_G \int_G |f(xy)|^p\, dx\, |g(y^{-1})|^q\, dy \\ &= \int_G \Delta(y^{-1})\, \|f\|_p^p\, |g(y^{-1})|^q\, dy \\ &= \|f\|_p^p\, \|g\|_q^q. \end{aligned}\right\} \quad (4)$$

[1] Note that r is always greater than 1.

The equalities (4) show that we can integrate both sides of the inequality (3) with respect to x, over the group G, to obtain the inequalities

$$\int_G |f*g(x)|^r\,dx \leqq \int_G \Big(\int_G |f(xy)\,g(y^{-1})|\,dy\Big)^r dx \leqq \|f\|_p^{p+\frac{rp(q-1)}{q}}\,\|g\|_q^{q+\frac{rq(p-1)}{p}}.$$

Taking r^{th} roots of both sides, we find

$$\|f*g\|_r \leqq \|f\|_p\cdot\|g\|_q. \quad \square$$

(20.19) Summary of convolution formulas. For convenience in writing relations involving convolutions, we introduce some abbreviated notation. If $\mathfrak{A}$ is a set of functions on G, then $\mathfrak{A}^\star$ denotes the set $\{f^\star : f\in\mathfrak{A}\}$. If $\mathfrak{A}$ and $\mathfrak{B}$ are sets of functions on G, then $\mathfrak{A}*\mathfrak{B}$ denotes the set $\{f*g : f\in\mathfrak{A},\, g\in\mathfrak{B}\}$. If $\boldsymbol{D}$ and $\boldsymbol{E}$ are subsets of $\boldsymbol{M}(G)$, then $\boldsymbol{D}*\boldsymbol{E}$, $\boldsymbol{D}*\mathfrak{A}$, and $\mathfrak{A}*\boldsymbol{D}$ are defined analogously.

We summarize (20.12), (20.14), and (20.16) as follows:

(i) $\boldsymbol{M}(G)*\mathfrak{L}_p(G)\subset\mathfrak{L}_p(G)$ for $1\leqq p\leqq\infty$;

(ii) $\mathfrak{L}_1(G)*\mathfrak{L}_p(G)\subset\mathfrak{L}_p(G)$ for $1\leqq p\leqq\infty$;

(iii) $\mathfrak{L}_p(G)*\big(\mathfrak{L}_{p'}(G)\big)^\star\subset\mathfrak{C}_0(G)$ for $1<p<\infty$;

(iv) $\mathfrak{L}_1(G)*\mathfrak{L}_\infty(G)\subset\mathfrak{C}_{ru}(G)$;

(v) $\mathfrak{L}_\infty(G)*\big(\mathfrak{L}_1(G)\big)^\star\subset\mathfrak{C}_{lu}(G)$.

The proof of (20.18) shows that the hypothesis $\|g\|_q=\|g^\star\|_q$ is unnecessary in order to conclude that $f*g\in\mathfrak{L}_r(G)$. In view of this, we have

(vi) $\mathfrak{L}_p(G)*\big[\mathfrak{L}_q(G)\cap(\mathfrak{L}_q(G))^\star\big]\subset\mathfrak{L}_r(G)$ whenever $\frac{1}{p}+\frac{1}{q}>1$ and $\frac{1}{p}+\frac{1}{q}-\frac{1}{r}=1$.

As seen in (20.13) and (20.14), we cannot state that $\mathfrak{L}_p(G)*\boldsymbol{M}(G)\subset\mathfrak{L}_p(G)$ and $\mathfrak{L}_p(G)*\mathfrak{L}_1(G)\subset\mathfrak{L}_p(G)$. If the group G is *unimodular*, however, these relations hold. Moreover, in this case, (iii) and (vi) may be written as $\mathfrak{L}_p(G)*\mathfrak{L}_{p'}(G)\subset\mathfrak{C}_0(G)$ and $\mathfrak{L}_p(G)*\mathfrak{L}_q(G)\subset\mathfrak{L}_r(G)\left[\frac{1}{p}+\frac{1}{q}-\frac{1}{r}=1\right]$.

The inclusions (iv), (v), and (ii) for $1\leqq p<\infty$ are actually equalities. See Vol. II, (32.45) and (32.30).

(20.20) Discussion. The algebra $\boldsymbol{M}(G)$ admits an *adjoint* operation [see (21.6) and (C.27)] which is very useful in establishing properties of $\boldsymbol{M}(G)$ and $\mathfrak{L}_1(G)$. To see what this operation should be like, let us consider the mapping $f\to\mu*f$ of $\mathfrak{L}_2(G)$ into itself, for a fixed $\mu\in\boldsymbol{M}(G)$. We write $T_\mu f=\mu*f$. Then as (20.12) shows, T_μ is a bounded linear operator, for which $\|T_\mu\|\leqq\|\mu\|$, carrying $\mathfrak{L}_2(G)$ into $\mathfrak{L}_2(G)$. Like all bounded linear operators on Banach spaces, T_μ admits an adjoint operator, $(T_\mu)^\sim=T_\mu^\sim$ [see (B.53)]. Since $\mathfrak{L}_2(G)$ is a Hilbert space (12.8), we have for all $f, g\in\mathfrak{L}_2(G)$:

$$\langle T_\mu f, g\rangle=\langle f, T_\mu^\sim g\rangle,$$

i.e.,

$$\int_G \mu*f(x)\,\overline{g(x)}\,dx=\int_G f(x)\,\overline{T_\mu^\sim g(x)}\,dx,$$

or

$$\int_G f(x)\,\overline{T_\mu^\sim g(x)}\,dx = \int_G\int_G f(y^{-1}x)\,d\mu(y)\,\overline{g(x)}\,dx.$$

Using (13.10) and (20.1), which is legitimate [although at the moment we are concerned only with a heuristic argument to give us a reasonable guess as to the form of $T_\mu^\sim$], we find

$$\begin{aligned}\int_G\int_G f(y^{-1}x)\,d\mu(y)\,\overline{g(x)}\,dx &= \int_G\int_G f(y^{-1}x)\,\overline{g(x)}\,dx\,d\mu(y)\\ &= \int_G\int_G f(x)\,\overline{g(yx)}\,dx\,d\mu(y)\\ &= \int_G f(x)\int_G \overline{g(yx)}\,d\mu(y)\,dx\\ &= \int_G f(x)\,\overline{\overline{\int_G \overline{g(yx)}\,d\mu(y)}}\,dx.\end{aligned}$$

Combining these computations, we see that $T_\mu^\sim$ *ought* to be the transformation of $\mathfrak{L}_2(G)$ into itself such that

(i) $T_\mu^\sim g(x) = \overline{\int_G \overline{g(yx)}\,d\mu(y)}$.

We will show that $T_\mu^\sim$ is in fact a convolution operator; that is, $T_\mu^\sim g = \mu^\sim * g$ for a measure $\mu^\sim \in \boldsymbol{M}(G)$. We proceed to a formal definition, which it is convenient to state in terms of linear functionals.

(20.21) Definition. For $L \in \mathfrak{C}_0^*(G)$, let $L^\sim$ be the functional on $\mathfrak{C}_0(G)$ defined by $L^\sim(\varphi) = \overline{L(\overline{\varphi^\star})}$.

(20.22) Theorem. *The functional $L^\sim$ has the following properties [for all $L, M \in \mathfrak{C}_0^*(G)$]:*

(i) *$L^\sim$ is bounded and linear on $\mathfrak{C}_0(G)$;*

(ii) *$(\alpha L + \beta M)^\sim = \bar{\alpha} L^\sim + \bar{\beta} M^\sim$ for $\alpha, \beta \in K$;*

(iii) *$L^{\sim\sim} = L$;*

(iv) *$(L * M)^\sim = M^\sim * L^\sim$;*

(v) *$|L^\sim| = |L|^\sim$;*

(vi) *$\|L\| = \|L^\sim\|$.*

Proof. Assertions (i), (ii), (iii), and (vi) are very simple, and we omit their proofs. To prove (iv), let μ and ν be the measures in $\boldsymbol{M}(G)$ representing L and M respectively. Then for $\varphi \in \mathfrak{C}_0(G)$, we have

$$\left.\begin{aligned}(L*M)^\sim(\varphi) &= \overline{\int_G\int_G \overline{\varphi^\star(xy)}\,d\nu(y)\,d\mu(x)} = \overline{\int_G\int_G \overline{\varphi(y^{-1}x^{-1})}\,d\nu(y)\,d\mu(x)}\\ &= \overline{\int_G\int_G \overline{\varphi(y^{-1}x^{-1})}\,d\mu(x)\,d\nu(y)}\end{aligned}\right\} \quad (1)$$

[the last equality follows from (14.25)]. Now it is plain from (20.21) that

$$L^\sim(\psi) = \overline{\int_G \overline{\psi^\star(x)}\,d\mu(x)} = \overline{\int_G \overline{\psi(x^{-1})}\,d\mu(x)} \quad \text{for} \quad \psi \in \mathfrak{C}_0(G).$$

Thus we have

$$\left.\begin{aligned} M^{\sim} * L^{\sim}(\varphi) &= \overline{\int_G (L^{\sim}({}_y\varphi))^{\star}\, d\nu(y)} = \overline{\int_G L^{\sim}({}_{y^{-1}}\varphi)\, d\nu(y)} \\ &= \overline{\int_G \int_G \overline{\varphi(y^{-1}x^{-1})}\, d\mu(x)\, d\nu(y)}. \end{aligned}\right\} \qquad (2)$$

The identity (iv) follows at once from (1) and (2).

We now prove (v). By (14.5) and (20.21), if $\varphi \in \mathfrak{C}_0^+(G)$ we have

$$|L^{\sim}|(\varphi) = \sup\{|L(\overline{\psi^{\star}})| : \psi \in \mathfrak{C}_0(G),\ |\psi| \leqq \varphi\}.$$

For $\psi \in \mathfrak{C}_0(G)$ such that $|\psi| \leqq \varphi$, we have $|\overline{\psi^{\star}}| = |\psi^{\star}| \leqq \varphi^{\star}$, and hence $|L(\overline{\psi^{\star}})| \leqq |L|(\varphi^{\star}) = |L|^{\sim}(\varphi)$ [note that φ is nonnegative and $|L|(\varphi^{\star})$ is a nonnegative number]. It follows that $|L^{\sim}| \leqq |L|^{\sim}$ on $\mathfrak{C}_0^+(G)$. Thus

$$|L|^{\sim} = |L^{\sim\sim}|^{\sim} \leqq |L^{\sim}|^{\sim\sim} = |L^{\sim}|,$$

and so $|L|^{\sim} = |L^{\sim}|$ on $\mathfrak{C}_0^+(G)$; this implies that $|L|^{\sim} = |L^{\sim}|$ on $\mathfrak{C}_0(G)$. □

If $L \in \mathfrak{C}_0^*(G)$ and μ is the measure in $\boldsymbol{M}(G)$ representing L, we will write $\mu^{\sim}$ for the measure representing $L^{\sim}$.

(20.23) Theorem. *Let $L \in \mathfrak{C}_0^*(G)$ and let μ be the measure in $\boldsymbol{M}(G)$ corresponding to L. Then:*

(i) *a subset A of G is $|\mu^{\sim}|$-measurable if and only if A^{-1} is $|\mu|$-measurable, and in this case we have $\mu^{\sim}(A) = \overline{\mu(A^{-1})}$ and $|\mu^{\sim}|(A) = |\mu|(A^{-1})$;*

(ii) *a function f is in $\mathfrak{L}_1(G, |\mu^{\sim}|)$ if and only if $f^{\star}$ is in $\mathfrak{L}_1(G, |\mu|)$, and in this case we have $\int_G f\, d\mu^{\sim} = \overline{\int_G \overline{f^{\star}}\, d\mu}$;*

(iii) *for every $f \in \mathfrak{L}_p(G)$ $(1 \leqq p \leqq \infty)$ we have*

$$\mu^{\sim} * f(x) = \overline{\int_G \overline{f(yx)}\, d\mu(y)},$$

the integral being finite for $x \in G \cap N'$, where N is λ-null if $1 \leqq p < \infty$ and N is locally λ-null if $p = \infty$, and representing a function in $\mathfrak{L}_p(G)$;

(iv) *if $\mu \in \boldsymbol{M}_a(G)$, so that $d\mu = g\, d\lambda$ with $g \in \mathfrak{L}_1(G)$, then $\mu^{\sim}$ is also in $\boldsymbol{M}_a(G)$ and $d\mu^{\sim} = \overline{g^{\star}} \frac{1}{\Delta}\, d\lambda$.*[1]

Proof. To prove (i), consider first an open subset U of G. It is easy to see that there is an increasing sequence $\{\varphi_n\}_{n=1}^{\infty}$ of functions in $\mathfrak{C}_{00}^+(G)$ such that $\lim_{n\to\infty} \varphi_n = \xi_A$, where $A \subset U$ and $|\mu^{\sim}|(U \cap A') = 0$. Similarly there is an increasing sequence $\{\psi_n\}_{n=1}^{\infty}$ of functions in $\mathfrak{C}_{00}^+(G)$ such that $\lim_{n\to\infty} \psi_n = \xi_B$, where $B \subset U^{-1}$ and $|\mu|(U^{-1} \cap B') = 0$. For $n = 1, 2, \ldots$, let $\omega_n = \max(\varphi_n, \psi_n^{\star})$; then $\lim_{n\to\infty} \omega_n = \xi_U$ $|\mu^{\sim}|$-almost everywhere and

[1] We write $g^{\sim}$ for the function $\overline{g^{\star}} \frac{1}{\Delta}$.

$\lim_{n\to\infty} \omega_n^\star = \xi_{U^{-1}}$ $|\mu|$-almost everywhere. Thus by dominated convergence (14.23), we have

$$\mu^\sim(U) = \lim_{n\to\infty} \int_G \omega_n \, d\mu^\sim = \lim_{n\to\infty} \overline{\int_G \omega_n^\star \, d\mu} = \lim_{n\to\infty} \overline{\int_G \omega_n^\star \, d\mu} = \overline{\mu(U^{-1})}.$$

The additivity of μ and $\mu^\sim$ now shows that $\mu^\sim(E) = \overline{\mu(E^{-1})}$ for all closed subsets E of G. From (14.15) we infer that if $|\mu^\sim|(B) = 0$, then $|\mu|(B^{-1}) = 0$. An $|\mu^\sim|$-measurable set A can be written in the form $A = (\bigcup_{n=1}^{\infty} F_n) \cup B$, where $\{F_n\}_{n=1}^{\infty}$ is an increasing sequence of compact sets and B is a set of $|\mu^\sim|$-measure 0 disjoint from $\bigcup_{n=1}^{\infty} F_n$. Then we have

$$\mu^\sim(A) = \lim_{n\to\infty} \mu^\sim(F_n) = \lim_{n\to\infty} \overline{\mu(F_n^{-1})} + \overline{\mu(B^{-1})}$$
$$= \overline{\mu\left(\left(\bigcup_{n=1}^{\infty} F_n^{-1}\right) \cup B^{-1}\right)} = \overline{\mu(A^{-1})}.$$

Hence A^{-1} is $|\mu|$-measurable, and (i) holds. If A^{-1} is $|\mu|$-measurable, we use the identities $(A^{-1})^{-1} = A$ and $\mu^{\sim\sim} = \mu$ to infer that A is $|\mu^\sim|$-measurable.

To prove (ii) [1], consider first a function

$$\sigma = \sum_{j=1}^{m} \alpha_j \xi_{A_j}, \tag{1}$$

where $\alpha_j \in K$ and $A_j \in \mathscr{M}_{|\mu^\sim|}$. Using (i), we have

$$\int_G \sigma \, d\mu^\sim = \sum_{j=1}^{m} \alpha_j \mu^\sim(A_j) = \overline{\sum_{j=1}^{m} \bar{\alpha}_j \mu(A_j^{-1})} = \overline{\int_G \sigma^\star \, d\mu}.$$

If $f \in \mathfrak{L}_1(G, |\mu^\sim|)$, there is a sequence of functions $\{\sigma_n\}_{n=1}^{\infty}$ each having the form (1) such that $\lim_{n\to\infty} \sigma_n(y) = f(y)$ and $|\sigma_1(y)| \leqq |\sigma_2(y)| \leqq \cdots \leqq |f(y)|$ for all $y \in G$. [2] Dominated convergence gives us

$$\lim_{n\to\infty} \int_G \sigma_n \, d\mu^\sim = \int_G f \, d\mu^\sim,$$

and the previous case shows that

$$\int_G \sigma_n \, d\mu^\sim = \overline{\int_G \sigma_n^\star \, d\mu}.$$

Furthermore we have

$$\int_G |\sigma_n| \, d|\mu^\sim| = \int_G |\sigma_n| \, d|\mu|^\sim = \int_G |\sigma_n^\star| \, d|\mu|.$$

---

[1] Relation (ii) is obvious for $f \in \mathfrak{C}_0(G)$: it reduces to the definition of $\mu^\sim$. Like Theorem (19.10), however, (ii) is not completely obvious for the larger class of functions for which it is asserted.

[2] Write $f = f_1 - f_2 + i(f_3 - f_4)$ where the f_j are real and $\min(f_1, f_2) = \min(f_3, f_4) = 0$. Then apply the construction of (11.36) to each f_j.

The theorem on monotone convergence implies that $\lim_{n\to\infty} |\sigma_n^\star| = |f^\star|$ is in $\mathfrak{L}_1(G, |\mu|)$, and thus dominated convergence implies that

$$\lim_{n\to\infty} \overline{\int_G \sigma_n^\star \, d\mu} = \overline{\int_G f^\star \, d\mu}\,.$$

Thus (ii) holds for $f \in \mathfrak{L}_1(G, |\mu^\sim|)$. If $f^\star \in \mathfrak{L}_1(G, |\mu|)$, then $f = f^{\star\star}$ is in $\mathfrak{L}_1(G, |\mu^\sim|)$, since $\mu^{\sim\sim} = \mu$. Thus (ii) is proved.

Assertion (iii) follows at once from (ii) and (20.12).

To prove (iv), we use (14.17). As usual, g is taken to be Borel measurable. For $f \in \mathfrak{C}_0(G)$, we have

$$\int_G f\, d\mu^\sim = \overline{\int_G f^\star d\mu} = \overline{\int_G f^\star g\, d\lambda} = \int_G f^\star \bar{g}\, d\lambda = \int_G f g^\star \frac{1}{\Delta}\, d\lambda\,.$$

By (14.17), this proves (iv). □

It is convenient to put down here some further algebraic properties of the algebras $\boldsymbol{M}(G)$ and $\boldsymbol{M}_a(G)$.

(20.24) Theorem. *The following statements are equivalent:*

(i) *G is Abelian;*

(ii) *$\boldsymbol{M}(G)$ is commutative;*

(iii) *there exists a commutative $*$-subsemigroup $\boldsymbol{A}$ of $\boldsymbol{M}(G)$ such that for every nonvoid open subset U of G, there is a $\mu \in \boldsymbol{A}^+$ such that $\mu(U) = 1$ and $\mu(U') = 0$.*[1]

Proof. The equivalence of (i) and (ii) follows from (19.6). Obviously (ii) implies (iii).

Next suppose that G is non-Abelian, *i.e.*, $ac \neq ca$ for some $a, c \in G$. An elementary continuity argument shows that there is a symmetric neighborhood V of e such that

$$(VcaV) \cap (VaVc) = \varnothing\,. \tag{1}$$

Let $\boldsymbol{A}$ be any $*$-subsemigroup of $\boldsymbol{M}(G)$ such that for every nonvoid open subset U of G there is a $\mu \in \boldsymbol{A}^+$ such that $\mu(U) = 1$ and $\mu(U') = 0$. Choose a neighborhood V_1 of e such that $V_1^- \subset V$. Next choose $\nu \in \boldsymbol{A}^+$ such that $\nu(c^{-1}V_1) = 1$ and $\nu((c^{-1}V_1)') = 0$. Let $S(\nu)$ denote the support of ν (11.25). Obviously we have $S(\nu) \subset c^{-1}V$; let b be any element of $S(\nu)^{-1} \subset Vc$. Now let W be a symmetric neighborhood of e such that $W \subset V$ and $a^{-1}Wa \subset V$. Finally, let $\mu \in \boldsymbol{A}^+$ be taken so that $\mu(Wab) = 1$ and $\mu((Wab)') = 0$.

We will show that $\mu * \nu(aV) \neq \nu * \mu(aV)$; this will violate (iii) and show that (iii) implies (i). By (19.11), we have

$$\mu * \nu(aV) = \int_G \nu(x^{-1}aV)\, d\mu(x) = \int_{Wab} \nu(x^{-1}aV)\, d\mu(x).$$

[1] For example, $\boldsymbol{A}$ might be the algebra $\boldsymbol{M}_a(G)$.

If $x \in Wab$, then $x \in Wab = aa^{-1}Wab \subset aVb$, so that $b^{-1} \in x^{-1}aV$. In other words, the open set $x^{-1}aV$ intersects $S(\nu)$, so that $\nu(x^{-1}aV) > 0$. It follows that $\mu * \nu(aV) > 0$. We complete the proof by showing that $\nu * \mu(aV) = 0$. We have $\nu * \mu(aV) = \int\limits_{c^{-1}V} \mu(x^{-1}aV)\, d\nu(x)$. If $x \in c^{-1}V$, then $x^{-1} \in Vc$ and

$$(x^{-1}aV) \cap (Wab) \subset (VcaV) \cap (VaVc) = \varnothing$$

by (1). Consequently, $\mu(x^{-1}aV) = 0$ for all $x \in c^{-1}V$ and $\nu * \mu(aV) = 0$. □

If G is discrete, the algebra $\boldsymbol{M}_a(G)$ coincides with $\boldsymbol{M}(G)$ and hence has a unit, *viz.*, the measure ε_e. If G is nondiscrete, then $\boldsymbol{M}_a(G)$ and $\boldsymbol{M}_c(G)$ both lack even one-sided units. A little more is true, as follows.

(20.25) Theorem. *Suppose that G is nondiscrete. Then there is no measure μ in $\boldsymbol{M}_c(G)$ such that $\mu * \nu = \nu$ for all $\nu \in \boldsymbol{M}_a(G)$, and there is no measure μ' in $\boldsymbol{M}_c(G)$ such that $\nu * \mu' = \nu$ for all $\nu \in \boldsymbol{M}_a(G)$.*

Proof. Assume that $\mu \in \boldsymbol{M}_c(G)$ and that $\mu * \nu = \nu$ for all $\nu \in \boldsymbol{M}_a(G)$. Since $|\mu|(\{e\}) = 0$, there is a neighborhood U of e such that $|\mu|(U) < 1$. Let V be a symmetric neighborhood of e such that $V^2 \subset U$ and $\lambda(V) < \infty$; then $\xi_V \in \mathfrak{L}_1(G)$. The relation (20.9.i) implies that for λ-almost all $x \in V$, we have

$$1 = \xi_V(x) = \mu * \xi_V(x) = \int\limits_G \xi_V(y^{-1}x)\, d\mu(y) = \int\limits_U \xi_V(y^{-1}x)\, d\mu(y)$$
$$\leq \int\limits_U 1 \cdot d|\mu|(y) < 1.$$

This contradiction proves the first statement. The second is proved in like manner, starting from (20.9.ii). □

The algebra $\boldsymbol{M}_a(G)$ always contains an *approximate* unit, however, as we will now prove.

(20.26) Definition. Let S be a topological semigroup. A net $\{x_\alpha\}_{\alpha \in A}$ of elements in S is called an *approximate unit* for S if the relations $y = \lim\limits_\alpha x_\alpha y = \lim\limits_\alpha y x_\alpha$ hold for every $y \in S$.

(20.27) Theorem. *The algebra $\boldsymbol{M}_a(G)$ contains an approximate unit.*

Proof. Let $\mathfrak{U}$ denote the family of neighborhoods of e and regard $\mathfrak{U}$ as a directed set in the usual way: $U \geqq V$ if $U \subset V$. For each $U \in \mathfrak{U}$, choose a measure $\mu_U \in \boldsymbol{M}_a^+(G)$ such that $\mu_U(U) = 1$ and $\mu_U(U') = 0$. Then the net $\{\mu_U\}_{U \in \mathfrak{U}}$ is an approximate unit for $\boldsymbol{M}_a(G)$ by (20.15) applied to $\mathfrak{L}_1(G)$, which is isomorphic with $\boldsymbol{M}_a(G)$. □

(20.28) Theorem. *A closed linear subspace $\boldsymbol{A}$ of $\boldsymbol{M}_a(G)$ is a left [right] ideal of $\boldsymbol{M}_a(G)$ if and only if $\mu \in \boldsymbol{A}$ and $x \in G$ imply that $\varepsilon_x * \mu \in \boldsymbol{A}$ [$\mu * \varepsilon_x \in \boldsymbol{A}$]. In other words, if $\mathfrak{A}$ is the subspace of $\mathfrak{L}_1(G)$ corresponding to $\boldsymbol{A}$, then $f \in \mathfrak{A}$ and $x \in G$ imply ${}_xf \in \mathfrak{A}$ [$f_x \in \mathfrak{A}$].* [1]

[1] Note that if $d\mu = f\, d\lambda$ and $x \in G$, then $d\varepsilon_x * \mu = {}_{x^{-1}}f\, d\lambda$ and $d\mu * \varepsilon_x = \Delta(x^{-1})\, f_{x^{-1}}\, d\lambda$.

Proof. Let $\boldsymbol{A}$ be a closed left ideal in $\boldsymbol{M}_a(G)$ and let $\{\nu_\alpha\}$ be an approximate unit for $\boldsymbol{M}_a(G)$. Then for $\mu \in \boldsymbol{A}$ and $x \in G$ we have

$$\varepsilon_x * \mu = \lim_\alpha \varepsilon_x * (\nu_\alpha * \mu) = \lim_\alpha (\varepsilon_x * \nu_\alpha) * \mu \in \boldsymbol{A}$$

since each $\varepsilon_x * \nu_\alpha$ belongs to $\boldsymbol{M}_a(G)$ (19.18) and $\boldsymbol{A}$ is a closed left ideal.

Suppose conversely that $\mu \in \boldsymbol{A}$ and $x \in G$ imply $\varepsilon_x * \mu \in \boldsymbol{A}$. Let ν be in $\boldsymbol{M}_a(G)$ and μ_0 in $\boldsymbol{A}$. If every bounded linear functional Φ on $\boldsymbol{M}_a(G)$ for which $\Phi(\boldsymbol{A}) = 0$ also satisfies $\Phi(\nu * \mu_0) = 0$, then by the HAHN-BANACH theorem (B.15), $\nu * \mu_0$ also belongs to $\boldsymbol{A}$. Let Φ be as above. By (12.18), there is a function $g \in \mathfrak{L}_\infty(G)$ such that

$$\Phi(\mu) = \int_G g\, d\mu \quad \text{for} \quad \mu \in \boldsymbol{M}_a(G).$$

We then have

$$\begin{aligned}\Phi(\nu * \mu_0) &= \int_G g\, d\nu * \mu_0 = \int_G \int_G g(xy)\, d\mu_0(y)\, d\nu(x) \\ &= \int_G \left[\int_G \int_G g(zy)\, d\varepsilon_x(z)\, d\mu_0(y)\right] d\nu(x) \\ &= \int_G \Phi(\varepsilon_x * \mu_0)\, d\nu(x) = 0,\end{aligned}$$

since $\varepsilon_x * \mu_0 \in \boldsymbol{A}$ for $x \in G$ and $\Phi(\boldsymbol{A}) = 0$. □

Miscellaneous theorems and examples

(20.29) Measures of inverse-sets. (a) Let G be a locally compact group that is not unimodular. Then there is an open subset W of G such that $\lambda(W) < \infty$ and $\lambda(W^{-1}) = \infty$. [Let $a \in G$ be such that $\Delta(a^{-1}) \geqq 2$. Let U be a symmetric neighborhood of e in G such that $U^2 \subset \{x \in G: \frac{1}{2} < \Delta(x) < 2\}$. It is easy to see that the sets $aU, a^2U, \ldots, a^nU, \ldots$ are pairwise disjoint. Since G is nondiscrete, there are symmetric neighborhoods W_k of e such that $\lambda(W_k) \leqq 2^{-k}$ and $W_k \subset U$ $(k = 1, 2, \ldots)$. Let $\{n_k\}_{k=1}^\infty$ be an increasing sequence of positive integers such that $2^{-n_k+1} \leqq \lambda(W_k)$ for all k. If $x \in a^{n_k} W_k$, then $x = a^{n_k} w$ for some $w \in W_k$ and $\Delta(x^{-1}) = \Delta(w^{-1})\, \Delta(a^{-1})^{n_k} > 2^{n_k - 1} \geqq \lambda(W_k)^{-1}$. Now let $W = \bigcup_{k=1}^\infty a^{n_k} W_k$. We have $\lambda(W) = \sum_{k=1}^\infty \lambda(W_k) \leqq 1$ and $\lambda(W^{-1}) = \int_W \frac{1}{\Delta}\, d\lambda = \sum_{k=1}^\infty \int_{a^{n_k} W_k} \Delta(x^{-1})\, dx = \infty$.]

(b) A simple and instructive special case of (a) is provided by the group G described in (15.17.g). Let $W = \{(x, y) \in G : x > 1 \text{ and } |y| < 1\}$. Then as remarked in (15.17.g), we have

$$\lambda(W) = 2 \int_1^\infty \frac{dx}{x^2} = 2.$$

The modular function $\Delta(x, y)$ is $|x|^{-1}$, as noted in (15.17.g). Hence (20.2) shows that

$$\lambda(W^{-1}) = \int_G \xi_W \frac{1}{\Delta}\, d\lambda = \int_1^\infty \int_{-1}^1 \Delta\left((x, y)^{-1}\right) \frac{1}{x^2}\, dy\, dx = \int_1^\infty \int_{-1}^1 \frac{1}{x}\, dy\, dx = \infty.$$

Also we have $W^{-1} = \{(x, y) \in G : 0 < x < 1,\ |y| < x\}$, and

$$\lambda(W^{-1}) = \int_0^1 \int_{-x}^x \frac{1}{x^2}\, dy\, dx = \int_0^1 \frac{2}{x}\, dx = \infty.$$

(20.30) Notes on Theorem (20.4). (a) The mapping $x \to {}_x f$ of G into $\mathfrak{L}_p(G)$ is left uniformly continuous for all $f \in \mathfrak{L}_p(G)$ if and only if G has equivalent uniform structures $(1 \leqq p < \infty)$. [Suppose that G has inequivalent uniform structures. By (4.14.g), there is a neighborhood U of e such that given a neighborhood V of e there is an $x \in G$ for which $xVx^{-1} \not\subset U$. Choose W to be a symmetric neighborhood of e such that $W^2 \subset U$ and $\lambda(W) < \infty$.

Now consider any neighborhood V of e. Let x in G have the property that $xVx^{-1} \not\subset U$, and choose $z \in (xVx^{-1}) \cap U'$. It is easy to see that the sets $x^{-1}zW$ and $x^{-1}W$ are disjoint. Therefore we have

$$\|{}_{z^{-1}x}(\xi_W) - {}_x(\xi_W)\|_p = \left(2\lambda(W)\right)^{\frac{1}{p}},$$

and, moreover, $(z^{-1}x)^{-1}x = x^{-1}zx \in V$. Since V is arbitrary, the mapping $x \to {}_x(\xi_W)$ is not left uniformly continuous.]

(b) The mapping $x \to f_x$ of G into $\mathfrak{L}_p(G)$ is left uniformly continuous for all $f \in \mathfrak{L}_p(G)$ if and only if G is unimodular. [Suppose that G is unimodular. By (20.2), $f^\star$ also belongs to $\mathfrak{L}_p(G)$. Given $\varepsilon > 0$, use (20.4) to choose a neighborhood U of e such that $\|{}_s(f^\star) - {}_t(f^\star)\|_p < \varepsilon$ whenever $st^{-1} \in U$. Then $x^{-1}y \in U$ implies that $\|f_x - f_y\|_p = \|(f_x - f_y)^\star\|_p = \|{}_{x^{-1}}(f^\star) - {}_{y^{-1}}(f^\star)\|_p < \varepsilon$.

Suppose next that G is not unimodular. Let U be a symmetric neighborhood of e such that U^- is compact. Let $x_1 = e$; having chosen x_{n-1}, choose x_n in $\left(\bigcup_{k=1}^{n-1} x_k U\right)'$. Let W_1 be a symmetric neighborhood of e such that $W_1^3 \subset U$ and $\lambda(W_1) < 2^{-1}$. When W_{n-1} has been chosen, let W_n be a symmetric neighborhood of e such that $W_n^2 \subset W_{n-1}$ and $\lambda(W_n) < 2^{-n}$. Finally, let $W = \bigcup_{n=1}^{\infty} x_n W_n$. Clearly we have $\xi_W \in \mathfrak{L}_p(G)$. To show that $x \to (\xi_W)_x$ is not left uniformly continuous, let V be a symmetric neighborhood of e such that $V \subset W_1$. Choose a positive integer m so that $2^{1-m} < \lambda(V)$. Then $V \cap W'_{m-1}$ must be nonvoid; pick any element v of $V \cap W'_{m-1}$. We next show that $x_m W_m v$ is disjoint from W. Assume that $x_m W_m v \cap x_n W_n \neq \varnothing$ for some n. Then $x_n^{-1} x_m \in W_n V^{-1} W_m^{-1} \subset W_1^3 \subset U$ and hence $n = m$. Thus $v \in W_m^{-1} W_m = W_m^2 \subset W_{m-1}$, which is a contradiction. Since $x_m W_m v \subset Wv$ and $x_m W_m v$ is disjoint from W, we infer that $\|(\xi_W) - (\xi_W)_{v^{-1}}\|_p > 0$. For every $y \in G$, we have

$$\|(\xi_W)_y - (\xi_W)_{yv^{-1}}\|_p = \Delta(y^{-1})^{\frac{1}{p}} \|(\xi_W) - (\xi_W)_{v^{-1}}\|_p. \tag{1}$$

Since G is not unimodular, the right side of (1) can be made arbitrarily large. Since $(yv^{-1})^{-1}y=v\in V$, this shows that $x\to(\xi_W)_x$ is not left uniformly continuous.]

(c) Suppose that G is unimodular. The mapping $x\to f_x$ of G into $\mathfrak{L}_p(G)$ is right uniformly continuous for all $f\in\mathfrak{L}_p(G)$ if and only if G has equivalent uniform structures ($1\leqq p<\infty$). [If G has equivalent uniform structures, then (b) shows that $x\to f_x$ is left [and hence right] uniformly continuous. If G has inequivalent uniform structures, choose U and W as in part (a). For a neighborhood V of e, choose x and z so that $z\in(xVx^{-1})\cap U'$. Then $\|(\xi_W)_{x^{-1}z^{-1}}-(\xi_W)_{x^{-1}}\|_p=(2\lambda(W))^{\frac{1}{p}}$ and $x^{-1}(x^{-1}z^{-1})^{-1}$ $=x^{-1}zx\in V$.]

(20.31) Converse of (19.27). (a) Let μ be a measure in $\boldsymbol{M}_a(G)$, and let A be a Borel set in G. Then the function $x\to\mu(xA)$ is right uniformly continuous on G and the function $x\to\mu(Ax)$ is left uniformly continuous on G. [By (19.18) we have $d\mu=f\,d\lambda$ with $f\in\mathfrak{L}_1(G)$, and by (14.17)

$$\mu(xA)=\int_G\xi_{xA}(y)\,f(y)\,dy=\int_G{}_{x^{-1}}(\xi_A)(y)\,f(y)\,dy=\int_A{}_xf(y)\,dy.$$

Thus $|\mu(xA)-\mu(yA)|\leqq\int_A|{}_xf-{}_yf|\,d\lambda\leqq\|{}_xf-{}_yf\|_1$. Now apply (20.4.i). To deal with the function $x\to\mu(Ax)$, note that $\mu(Ax)=\overline{\mu^{\sim}(x^{-1}A^{-1})}$ and use (20.23.iv).]

(b) Let μ be in $\boldsymbol{M}(G)$ and suppose that for all compact sets such that $\lambda(F)=0$, the function $x\to\mu(xF)$ or the function $x\to\mu(Fx)$ is continuous at e. Then for every Borel set A, the function $x\to\mu(xA)$ [$x\to\mu(Ax)$] is right [left] uniformly continuous throughout G. [Apply (19.27) and (a).]

(20.32) Convolutions using right Haar measure. (a) Unlike some of the choices made in developing the theory of convolutions, our choice of left Haar measure instead of right Haar measure was purely arbitrary. Left Haar measure can be replaced by right Haar measure everywhere with no loss whatever. Some formulas look different, but this is the only change. Given a left Haar measure λ on G, we write $J(\psi)=\int_G\psi^\star\,d\lambda=\int_G\psi\frac{1}{\Delta}\,d\lambda$ for all functions ψ on G for which the integrals have meaning [see (15.15) and (20.2)]. The identity $(\psi_a)^\star={}_{a^{-1}}(\psi^\star)$ shows that J is right invariant. Now restrict J to $\mathfrak{C}_{00}(G)$ and construct the measure ϱ for J by the process of §11. For an open subset U of G, we have

$$\begin{aligned}\varrho(U^{-1})&=\sup\{J(f):f\in\mathfrak{C}_{00}^+(G),\ f\leqq\xi_{U^{-1}}\}\\&=\sup\left\{\int_G f^\star\,d\lambda:f\in\mathfrak{C}_{00}^+(G),\ f\leqq\xi_{U^{-1}}\right\}\\&=\sup\left\{\int_G g\,d\lambda:g\in\mathfrak{C}_{00}^+(G),\ g\leqq\xi_U\right\}=\lambda(U).\end{aligned}$$

A similar argument shows that $\varrho(A^{-1})=\lambda(A)$ for all $A\subset G$. We infer from (20.2) that $\mathscr{M}_\varrho=\mathscr{M}_\lambda$ and that $\varrho(A)=\int_A \frac{1}{\Delta}\,d\lambda$ for all $A\in\mathscr{M}_\varrho$. This implies that $\int_G f\,d\varrho=\int_G f^\star\,d\lambda=\int_G f\frac{1}{\Delta}\,d\lambda$ for all $f\in\mathfrak{L}_1(G,\varrho)$. Thus $J(f)=\int_G f\,d\varrho$ whenever either side is finite. Throughout (20.32), λ is a fixed left Haar measure and ϱ is as just defined. In (20.32), but not elsewhere, $\|f\|_p$ means the norm of f in $\mathfrak{L}_p(G,\varrho)$.

(b) For $1\leqq p\leqq\infty$, the mapping $f\to f^\star$ carries $\mathfrak{L}_p(G,\lambda)$ onto $\mathfrak{L}_p(G,\varrho)$ [as well as $\mathfrak{L}_p(G,\varrho)$ onto $\mathfrak{L}_p(G,\lambda)$]. It is linear, norm-preserving, reality-preserving, and order-preserving. The mapping $f\to f\Delta$ has all of these properties for $p=1$, but loses them for $1<p\leqq\infty$ if G is not unimodular.

(c) It is easy to see that $\varrho(A)=0$ if and only if $\lambda(A)=0$, for all $A\subset G$. Hence a measure ν in $\boldsymbol{M}_a(G)$ has the property that $d\nu=f\,d\varrho$ for some $f\in\mathfrak{L}_1(G,\varrho)$ [see (14.19) and (19.13)]. Since $\boldsymbol{M}_a(G)$ is a two-sided ideal in $\boldsymbol{M}(G)$, we can define $\mu\underset{r}{*}f$, $f\underset{r}{*}\mu$, and $f\underset{r}{*}g$ for $\mu\in\boldsymbol{M}(G)$ and $f,g\in\mathfrak{L}_1(G,\varrho)$ by analogy with (20.5). The functions $\mu\underset{r}{*}f$, $f\underset{r}{*}\mu$, and $f\underset{r}{*}g$ can be written as integrals involving ϱ instead of λ, and are of course functions in $\mathfrak{L}_1(G,\varrho)$. For ϱ-almost all $x\in G$, we have:

(i) $\mu\underset{r}{*}f(x)=\int_G f(y^{-1}x)\,\Delta(y)\,d\mu(y)$;

(ii) $f\underset{r}{*}\mu(x)=\int_G f(xy^{-1})\,d\mu(y)$;

(iii)
$$\begin{aligned} f\underset{r}{*}g(x)&=\int_G f(xy^{-1})\,g(y)\,d\varrho(y)\\ &=\int_G f(y^{-1})\,g(yx)\,d\varrho(y)\\ &=\int_G f(y)\,g(y^{-1}x)\,\Delta(y)\,d\varrho(y)\\ &=\int_G f(xy)\,g(y^{-1})\,\Delta(y)\,d\varrho(y). \end{aligned}$$

Note that $\mu\underset{r}{*}f$ is the function such that $(\mu\underset{r}{*}f)\,d\varrho=(\mu\underset{r}{*}f)\frac{1}{\Delta}\,d\lambda=\mu*\left(f\frac{1}{\Delta}\right)d\lambda$; that is, $\mu\underset{r}{*}f=\Delta\left(\mu*\left(f\frac{1}{\Delta}\right)\right)$; similarly for $f\underset{r}{*}\mu$ and $f\underset{r}{*}g$. [This can be checked by comparing (i)—(iii) with (20.9) and (20.10).]

(d) The analogues of (20.12) and (20.13) are as follows. Suppose that $f\in\mathfrak{L}_p(G,\varrho)$ $(1\leqq p\leqq\infty)$ and that $\mu\in\boldsymbol{M}(G)$. Then the integral in (ii) defines a function in $\mathfrak{L}_p(G,\varrho)$ and $\|f\underset{r}{*}\mu\|_p\leqq\|f\|_p\,\|\mu\|$. If $\int_G \Delta(y)^{\frac{1}{p'}}\,d|\mu|(y)=\alpha$ is finite, then the integral in (i) defines a function in $\mathfrak{L}_p(G,\varrho)$ and $\|\mu\underset{r}{*}f\|_p\leqq\alpha\,\|f\|_p$.

(e) Theorem (20.16) has the following analogue. Suppose that

$1<p<\infty$. If $f^\star \in \mathfrak{L}_p(G,\varrho)$ and $g\in\mathfrak{L}_{p'}(G,\varrho)$, then the integrals in (iii) define a function in $\mathfrak{C}_0(G)$ and $\|f \underset{r}{*} g\|_u \leqq \|f^\star\|_p \|g\|_{p'}$. If $f^\star\in\mathfrak{L}_1(G,\varrho)$ and $g\in\mathfrak{L}_\infty(G,\varrho)=\mathfrak{L}_\infty(G,\lambda)$, then (iii) defines a function in $\mathfrak{C}_{ru}(G)$, and $\|f \underset{r}{*} g\|_u \leqq \|f^\star\|_1 \|g\|_\infty$. If $f\in\mathfrak{L}_\infty(G,\varrho)$ and $g\in\mathfrak{L}_1(G,\varrho)$, then (iii) defines a function in $\mathfrak{C}_{lu}(G)$, and $\|f \underset{r}{*} g\|_u \leqq \|f\|_\infty \|g\|_1$.

(f) Finally we state the analogue of (20.18). Let p, q, r be as in (20.18). Let $f\in\mathfrak{L}_p(G,\varrho)\cap\mathfrak{L}_p^\star(G,\varrho)$ and let $g\in\mathfrak{L}_q(G,\varrho)$. Suppose also that $\|f\|_p=\|f^\star\|_p$. Then the integrals (iii) define a function in $\mathfrak{L}_r(G,\varrho)$, and $\|f \underset{r}{*} g\|_r \leqq \|f\|_p \|g\|_q$.

(20.33) If G is noncompact, then neither $\mathfrak{L}_1(G)*\mathfrak{L}_\infty(G)$ nor $\mathfrak{L}_\infty(G)*\mathfrak{L}_1^\star(G)$ is contained in $\mathfrak{C}_0(G)$.

(20.34) The hypothesis in (20.13) that $\int_G \Delta(y)^{-\frac{1}{p'}}\,d|\mu|(y)$ be finite is essential. For groups G that are not unimodular and for $1<p<\infty$, one can find a measure μ in $\boldsymbol{M}_a(G)$ and a function f in $\mathfrak{L}_p(G)$ such that $f*\mu$ does not belong to $\mathfrak{L}_{p'}(G)$.

[Choose a in G such that $\Delta(a^{-1})\geqq 2$ and let $\beta=\Delta(a^{-1})$. Let U, V be symmetric neighborhoods of e such that $U^2\subset\left\{x\in G:\frac{1}{\beta}<\Delta(x)<\beta\right\}$, $\lambda(U)<\infty$, and $V^2\subset U$. Note that the sets $\{a^kU\}_{k=-\infty}^{\infty}$ are pairwise disjoint. For $k=1,2,\ldots$, choose symmetric neighborhoods W_k of e such that $\lambda(W_k)\leqq 2^{-k}$ and $W_k\subset V$. Let $\{n_k\}_{k=1}^\infty$ be an increasing sequence of positive integers such that

$$(\beta^{n_k-1})^{1/p'}\geqq\frac{(\beta^{k+1})^{1/p}}{\lambda(W_k)}.$$

Define W to be $\bigcup_{k=1}^{\infty} a^{n_k}W_k$; then $\lambda(W)=\sum_{k=1}^\infty\lambda(W_k)\leqq\sum_{k=1}^\infty 2^{-k}=1$. Define μ in $\boldsymbol{M}_a(G)$ by the relation $d\mu=\xi_W\,d\lambda$. Finally, define A to be $\bigcup_{k=1}^\infty a^{n_k}Ua^k$ and f to be $\Delta^{-\frac{1}{p}}\xi_{A^{-1}}$. To show that $f\in\mathfrak{L}_p(G)$, we use (20.2.i) and (19.17) to obtain

$$\int_G f^p\,d\lambda=\int_G\frac{1}{\Delta}\xi_{A^{-1}}\,d\lambda=\int_G\xi_A\,d\lambda=\lambda(A)\leqq\sum_{k=1}^\infty\lambda(a^{n_k}Ua^k)$$
$$=\sum_{k=1}^\infty\Delta(a)^k\lambda(U)\leqq\lambda(U)\sum_{k=1}^\infty\left(\frac{1}{2}\right)^k=\lambda(U)<\infty.$$

It remains to show that $f*\mu$ is not in $\mathfrak{L}_{p'}(G)$. Since $\lambda\left(\bigcup_{k=1}^\infty a^{-k}V\right)=\infty$, it suffices to prove that $x\in a^{-k}V$ implies that $f*\mu(x)\geqq 1$. We have

$$f*\mu(x)=\int_G\Delta(y^{-1})f(xy^{-1})\,d\mu(y)=\int_G\Delta(y^{-1})\Delta(xy^{-1})^{-\frac{1}{p}}\xi_{A^{-1}}(xy^{-1})\xi_W(y)\,dy$$
$$=\int_G\Delta(y^{-1})^{1/p'}\Delta(x^{-1})^{1/p}\xi_A(yx^{-1})\xi_W(y)\,dy.$$

If $y\in a^{nk}W_k$, then $y\in W$ and $yx^{-1}\in a^{nk}W_k Va^k\subset a^{nk}Ua^k\subset A$. Since $y^{-1}\in W_k a^{-nk}$, we have $\Delta(y^{-1})>\frac{1}{\beta}\cdot\Delta(a^{-1})^{nk}=\beta^{nk-1}$ and since $x^{-1}\in Va^k$, we have $\Delta(x^{-1})>\frac{1}{\beta}\cdot\Delta(a)^k=\beta^{-k-1}$. It follows that for $x\in a^{-k}V$,

$$f*\mu(x)\geqq\int\limits_{a^{nk}W_k}\Delta(y^{-1})^{1/p'}\Delta(x^{-1})^{1/p}\,dy\geqq(\beta^{nk-1})^{1/p'}(\beta^{-k-1})^{1/p}\lambda(W_k)\geqq 1.]$$

(20.35) An alternative approach to $\mathfrak{L}_p(G)*\boldsymbol{M}(G)$. Instead of restricting the $\mu\in\boldsymbol{M}(G)$ for which we define $f*\mu$ for $f\in\mathfrak{L}_p(G)$ $(1<p\leqq\infty)$, as was done in (20.13), we can change the definition of convolution. For $f\in\mathfrak{L}_p(G)$ and $\mu\in\boldsymbol{M}(G)$, let

$$f\,\square\,\mu(x)=\int\limits_G\Delta(y)^{-\frac{1}{p}}f(xy^{-1})\,d\mu(y)$$

whenever the integral exists[1]. Then $f\,\square\,\mu(x)$ exists for λ-almost all $x\in G$ and defines a function in $\mathfrak{L}_p(G)$ for which $\|f\,\square\,\mu\|_p\leqq\|f\|_p\,\|\mu\|$. For $p=1$, we have $f\,\square\,\mu=f*\mu$. [To apply Theorem (20.8), let $\tau(x,y)=yx^{-1}$, $\Phi=\Delta^{-\frac{1}{p}}$, $\eta=\mu$, and $\gamma=\|f\|_p\,\|\mu\|$. For $\psi\in\mathfrak{C}_{00}(G)$, we have

$$\int\limits_G\int\limits_G|\psi(y)\,\Delta(x)^{-\frac{1}{p}}f(yx^{-1})|\,dy\,d|\mu|(x)\leqq\int\limits_G\|\psi\|_{p'}\|f_{x^{-1}}\|_p\,\Delta(x)^{-\frac{1}{p}}\,d|\mu|(x)$$
$$=\int\limits_G\|\psi\|_{p'}\,\Delta(x)^{\frac{1}{p}}\,\|f\|_p\,\Delta(x)^{-\frac{1}{p}}\,d|\mu|(x)=\|\psi\|_{p'}\,\|f\|_p\,\|\mu\|.]$$

(20.36) Invariant measures in $\boldsymbol{M}(G)$. (a) Let G be a compact group, and let λ be normalized Haar measure on G. Then for arbitrary $\mu\in\boldsymbol{M}(G)$ and all Borel sets $A\subset G$, we have $\lambda*\mu(A)=\mu*\lambda(A)=\mu(G)\,\lambda(A)$. Thus $\lambda*\mu=\mu*\lambda=\mu(G)\,\lambda$, and λ generates a 1-dimensional two-sided ideal in $\boldsymbol{M}(G)$.

(b) To obtain all 1-dimensional one- and two-sided ideals in $\boldsymbol{M}(G)$, we need the following lemma. Suppose that g is a λ-measurable function on any locally compact group G and that for all a in some dense subset D of G, we have $g(x)=g(ax)$ for locally λ-almost all $x\in G$. Then there is a constant β such that $g(x)=\beta$ for locally λ-almost all $x\in G$. [We may suppose that g is real-valued. If g is not a constant locally λ-almost everywhere, there exists a real number α such that the sets $\{x\in G: g(x)>\alpha\}$ and $\{x\in G: g(x)<\alpha\}$ are not locally λ-null. Hence there are compact sets E and F such that $E\subset\{x\in G: g(x)>\alpha\}$, $F\subset\{x\in G: g(x)<\alpha\}$, $\lambda(E)>0$, and $\lambda(F)>0$. Then $\lambda(F^{-1})>0$ and by (20.17), there is an $a\in D$ such that $\lambda(E\cap(a^{-1}F))>0$. If $x\in E\cap(a^{-1}F)$, then $g(x)>\alpha$ and $g(ax)<\alpha$; this is a contradiction.]

(c) Let G be a locally compact group. A measure $\mu\in\boldsymbol{M}(G)$ generates a 1-dimensional left ideal if $\nu*\mu=\alpha\mu$ for all $\nu\in\boldsymbol{M}(G)$, where α is a complex number depending of course on ν. Similarly μ generates a 1-dimensional right ideal if $\mu*\nu=\alpha\mu$ for all $\nu\in\boldsymbol{M}(G)$: see (C.1).

[1] This formula with p' instead of p appears in GODEMENT [1], p. 11; GODEMENT'S formula is equivalent to ours only for $p=2$.

Suppose that there is a nonzero μ in $\boldsymbol{M}(G)$ that generates either a 1-dimensional left ideal or a 1-dimensional right ideal. Then G is compact, μ generates a 1-dimensional two-sided ideal, and $d\mu = \beta\chi\, d\lambda$, where β is a nonzero complex number and χ is a bounded continuous nonzero complex function on G such that $\chi(xy) = \chi(x)\,\chi(y)$ for all $x, y \in G$.

[Suppose first that μ generates a 1-dimensional left ideal. Then we have $\varepsilon_a * \mu = \chi(a^{-1})\mu$ for all $a \in G$. It follows that $\int\limits_G {}_a f\, d\mu = \chi(a^{-1}) \int\limits_G f\, d\mu$ for $f \in \mathfrak{C}_0(G)$. Since $\mu \neq 0$, we also have $\chi \neq 0$. Since $\|\varepsilon_a * \mu\| = |\chi(a^{-1})| \cdot \|\mu\| \leqq \|\varepsilon_a\| \cdot \|\mu\| = \|\mu\|$, the function χ is bounded. Since $\varepsilon_a * \varepsilon_b * \mu = \varepsilon_{ab} * \mu$, we have $\chi(a)\,\chi(b) = \chi(ab)$. The continuity of χ follows from the right uniform continuity of f for $f \in \mathfrak{C}_0(G)$. Now let μ_1 be the measure in $\boldsymbol{M}(G)$ such that $d\mu_1 = \chi^{-1}\, d\mu$. It is easy to see that $\int\limits_G {}_a f\, d\mu_1 = \int\limits_G f\, d\mu_1$ for all $f \in \mathfrak{C}_0(G)$ and also that $\mu_1 \neq 0$ [see (14.17)]. We infer from (14.5) that $\int\limits_G f\, d|\mu_1| = \int\limits_G {}_a f\, d|\mu_1|$ for all $f \in \mathfrak{C}_0(G)$, and of course $|\mu_1|$ is not zero. Hence $|\mu_1|$ is a nonnegative invariant regular measure such that $|\mu_1|(G) < \infty$. Accordingly G is compact (15.9), and $d\,|\mu_1| = r\, d\lambda$, where r is a positive number. There is a Borel measurable function g, $|g| = 1$, such that $d\mu_1 = g\, d|\mu_1|$. Thus we have $d\mu = rg\chi\, d\lambda$; it follows that $g(x) = g(a^{-1}x)$ for λ-almost all x, a being any fixed element of G. Hence by (b), g can be taken as a constant function, $g(x) = \exp(i\theta)$, and we infer that $d\mu = \beta\chi\, d\lambda$ with $\beta = r\exp(i\theta)$.

It follows easily from (20.22.iv) that $\mu^{\sim}$ generates a 1-dimensional right ideal. Since $d\mu^{\sim} = \bar{\beta}\chi\, d\lambda$ by (20.23.iv), it follows that μ and $\mu^{\sim}$ generate the same 1-dimensional ideal, which is necessarily two-sided.

Suppose now that μ generates a 1-dimensional right ideal. Then $\mu^{\sim}$ generates a 1-dimensional left ideal and the previous case applies.]

(20.37) The adjoint operation $L \to L^{\sim}$ of (20.21) can be defined in many convolution algebras other than $\boldsymbol{M}(G)$ and $\mathfrak{L}_1(G)$ for a locally compact group G. Making no pretense at completeness, we cite $\mathfrak{C}_u^*(G)$ for any locally compact group G and $\mathfrak{B}^*(G)$ for any group G. If $\varphi \in \mathfrak{C}_u(G)$, then $\overline{\varphi^{\star}}$ is also in $\mathfrak{C}_u(G)$; if $\varphi \in \mathfrak{B}(G)$, then $\overline{\varphi^{\star}}$ is also in $\mathfrak{B}(G)$. Thus the mapping $\varphi \to \overline{L(\overline{\varphi^{\star}})}$ is defined if L is a bounded linear functional, and plainly is again a bounded linear functional. It is worth noting that the important relation (20.22.iv) may fail in $\mathfrak{C}_u^*(G)$ and $\mathfrak{B}^*(G)$. Let $\mathfrak{F}$ denote either $\mathfrak{C}_u(G)$ or $\mathfrak{B}(G)$. Let $\mathfrak{F}$ contain a two-sided invariant subspace $\mathfrak{A}$ such that $1 \in \mathfrak{A}$, $f^{\star} \in \mathfrak{A}$ if $f \in \mathfrak{A}$, $\bar{f} \in \mathfrak{A}$ if $f \in \mathfrak{A}$, and $\mathfrak{A}$ admits two distinct two-sided invariant bounded linear functionals N_1 and N_2 for which $N_1(1) = N_2(1) = 1$. Extend N_1 and N_2 to functionals M_1 and M_2, respectively, in $\mathfrak{F}^*$. Then for $f \in \mathfrak{A}$ we have $M_1 * M_2(f) = M_2(f)$ [as in (19.24.a)] so that $(M_1 * M_2)^{\sim}(f) = M_2^{\sim}(f)$. On the other hand, for $f \in \mathfrak{A}$ and $x \in G$, we have $M_1^{\sim}({}_x f) = \overline{M_1(\overline{({}_x f)^{\star}})} = \overline{M_1((\bar{f}^{\star})_{x^{-1}})} = M_1^{\sim}(f)$. Hence we have $M_2^{\sim} * M_1^{\sim}(f) = M_2^{\sim}(1)\, M_1^{\sim}(f) = M_1^{\sim}(f)$. Since $M_1^{\sim}$ is distinct from $M_2^{\sim}$ on $\mathfrak{A}$, we infer that $(M_1 * M_2)^{\sim} \neq M_2^{\sim} * M_1^{\sim}$. Specific examples illustrating this phenomenon can be constructed as in (19.24.b) and (19.24.c).

Notes

Most of (20.1)—(20.19) are copies of or marginal improvements on corresponding results in WEIL [4]. The second assertion of (20.4) is stated by GEL'FAND and RAĬKOV [2]. Convolutions of measures and functions in $\mathfrak{L}_1(G)$, treated in (20.5)—(20.9), appear in WEIL [4] only implicitly, on p. 48. Formula (20.9.ii) appears in GODEMENT [1], p. 13, for $f \in \mathfrak{C}_{00}^{+}(G)$. We are unsure of the first explicit mention of (20.9.i); WENDEL [1] uses it. The *form* of (20.9.i) and (20.9.ii) is obvious; although as we have seen a rigorous proof requires some care. Theorem (20.10) is found in WEIL [4], p. 49. Formula (20.10.ii) and its analogue, formula (iii) in (20.32.c), have been long used as an *ad hoc* definition of multiplication in $\mathfrak{L}_1(G)$ or $\mathfrak{L}_1(G, \varrho)$: see for example WEYL and PETER [1], GEL'FAND and RAĬKOV [1], SEGAL [1], and GEL'FAND and RAĬKOV [2]. Corollary (20.14) appears in WEIL [4], p. 49, and also in SEGAL [1]. Theorem (20.16), and also (20.17) and (20.18), are due to WEIL [4], pp. 50, 54—55.

The adjoint operation in $\boldsymbol{M}(G)$, examined in (20.20)—(20.23), was introduced at an early stage. For continuous functions on compact Lie groups, it appears in WEYL and PETER [1]. The formula $g^{\sim} = \frac{1}{\Delta}\overline{g^{\star}}$ for $g \in \mathfrak{L}_1(G)$ appears in GEL'FAND and RAĬKOV [2] and in SEGAL [2]. The adjoint operation $\mu \rightarrow \mu^{\sim}$ for measures in $\boldsymbol{M}(G)$ appears in YU. A. ŠREĬDER [1], and is utilized heavily in ŠREĬDER [2].

Theorem (20.25) for the case of a two-sided unit and $\boldsymbol{M}_a(G)$ is due to SEGAL. Theorem (20.27) appears in LOOMIS [2], p. 124, Th. 31 E. A more general assertion was proved by SEGAL [2], Lemma 1.1. Approximate units in $\boldsymbol{M}_a(G)$ have long been used: see for example HELSON [1], p. 84.

§21. Introduction to representation theory

A central technique in contemporary analysis is the study of topologico-algebraic objects of a given class by means of their continuous homomorphisms into the most elementary objects of the same class. Thus the bounded linear functionals on a Banach space E, which are just the continuous linear space homomorphisms of E into a 1-dimensional Banach space, are a very useful entity. Thus too, continuous algebra homomorphisms of commutative Banach algebras into the field K are a vital tool in analyzing the structure of such algebras. This phenomenon persists in the study of locally compact groups G and of $\boldsymbol{M}(G)$ and its subalgebras. By a proper choice of the groups and algebras that serve as the images under these homomorphisms, we obtain much information about the structure of G and $\boldsymbol{M}(G)$. We begin with some quite elementary definitions and facts.

(21.1) Definition. Let S be a semigroup, with or without a topology. Let E be a linear space over an arbitrary field F. A *representation* V *of* S is a homomorphism $x \to V_x$ of S into the semigroup of all operators on E (B.2). That is, for each $x \in S$, V_x is an operator on E, and $V_{xy} = V_x V_y$ for all $x, y \in S$. The linear space E is called the *representation space of* V. A subspace E_1 of E is said to be *invariant under the representation* V if $V_x(E_1) \subset E_1$ for all $x \in S$.

A first obvious specialization is appropriate for representations of groups.

(21.2) Theorem. *Let V be a representation of a group G as in* (21.1). *Then E is the direct sum of invariant subspaces E_0 and E_1 such that:*

(i) $V_x(E_0) = \{0\}$ *for all* $x \in G$;

(ii) $V_e(\xi) = \xi$ *for all* $\xi \in E_1$ [*e is the group identity*];

(iii) $V_x V_{x^{-1}}(\xi) = V_{x^{-1}} V_x(\xi) = \xi$ *for all* $\xi \in E_1$ *and* $x \in G$.

If E is a topological linear space and V_e is a continuous operator, then E_0 and E_1 are closed in E.

Proof. Let $E_0 = \{\xi \in E : V_e \xi = 0\}$ and let $E_1 = \{\xi \in E : V_e \xi = \xi\}$. Since $V_x V_e = V_e V_x = V_x$ for all $x \in G$, it is plain that $V_x(E_0) = \{0\}$ for all $x \in G$. Properties (ii) and (iii) follow at once from the definition of E_1. It is also obvious that $E_0 \cap E_1 = \{0\}$. For $\xi \in E$, we have

$$\xi = (\xi - V_e \xi) + V_e \xi,$$
$$V_e(\xi - V_e \xi) = V_e \xi - V_e \xi = 0,$$
$$V_e(V_e \xi) = V_e \xi.$$

Thus E is the direct sum $E_0 \oplus E_1$. It is immediate that both E_0 and E_1 are invariant and are closed if V_e is continuous. □

(21.3) Note. Theorem (21.2) shows that in studying representations of *groups*, we lose nothing in supposing that V_e is the identity operator I on E. From this point on, therefore, we shall mean by "representation V of a group" a representation such that $V_e = I$. The situation is quite different for semigroups that are not groups: representations V such that $V_x = 0$ for some elements x in the semigroup play a vital rôle in many interesting and important problems. See for example HEWITT and ZUCKERMAN [3], [4], [5].

(21.4) Definition. Let A be an algebra over a field F and let E be a linear space over the same field. A *representation* T *of* A is a homomorphism $x \to T_x$ into the algebra of operators on E (B.2). That is, for each $x \in A$, T_x is an operator on E, and for $x, y \in A$ and $\alpha \in F$, we have $T_{x+y} = T_x + T_y$, $T_{xy} = T_x T_y$, and $T_{\alpha x} = \alpha T_x$. The linear space E is called the *representation space of* T. A subspace E_1 of E is said to be *invariant under the representation* T if $T_x(E_1) \subset E_1$ for all $x \in A$.

(21.5) Note. (a) Let A be an algebra with a multiplicative unit u, and let T be a representation of A as in (21.4). Then E is the direct sum of subspaces E_0 and E_1 such that $T_x(E_0) = \{0\}$ for all $x \in A$ and $T_u \xi = \xi$ for all $\xi \in E_1$. If E is a topological linear space and T_u is a continuous operator, then E_0 and E_1 are closed subspaces. This fact is proved by the same argument as used in (21.2). From this point on, we will always suppose that $T_u = I$ if T is a representation of an algebra with unit.

(b) Suppose that T is a representation of an algebra A [or of a semigroup S] by continuous operators on a topological linear space E. If E_1 is a subspace of E that is invariant under T, then E_1^- is also invariant under T.

The algebras $\boldsymbol{M}(G)$, $\boldsymbol{M}_d(G)$, $\boldsymbol{M}_c(G)$, and $\boldsymbol{M}_a(G)$, defined in (19.13) for a locally compact group G, all admit the adjoint operation $\mu \rightarrow \mu^\sim$ described in (20.21)–(20.23). This adjoint operation is intimately connected with representations of G by unitary operators on Hilbert spaces, and is an essential tool in studying these algebras. Many facts about this adjoint operation $\sim$ can be obtained with no extra effort for arbitrary algebras over K that admit an adjoint operation. We shall therefore define and study such algebras. [The term "algebra" means "algebra over K", unless the contrary is specified.]

(21.6) Definition. Let A be an algebra, and let there be a mapping $x \rightarrow x^\sim$ of A onto A such that for all $x, y \in A$ and $\alpha \in K$ the following identities hold:

(i) $(x+y)^\sim = x^\sim + y^\sim$;

(ii) $(\alpha x)^\sim = \bar{\alpha} x^\sim$;

(iii) $(xy)^\sim = y^\sim x^\sim$;

(iv) $x^{\sim\sim} = x$.

Then A is called an *algebra with adjoint operation* or a *$\sim$-algebra*. If A is also a normed algebra (C.1) and

(v) $\|x^\sim\| = \|x\|$ for $x \in A$,

then A is called a *normed $\sim$-algebra*. A Banach algebra that is a normed $\sim$-algebra is called *Banach $\sim$-algebra*. An element $x \in A$ such that $x^\sim = x$ is said to be *Hermitian*[1]. Throughout Chapter Five, if a normed algebra has a unit u, then we suppose that $\|u\| = 1$.

(21.7) Remarks. Let A be a $\sim$-algebra. If A has a unit u, then $u^\sim = u$: for $u^\sim = u u^\sim$ and so $u = u^{\sim\sim} = (u u^\sim)^\sim = u u^\sim$. Every $x \in A$ can be written in exactly one way as the sum $x_1 + i x_2$, where x_1 and x_2 are

[1] In the literature, the symbol "*" is most commonly used for what we write as "$\sim$". We choose the latter notation to avoid confusion with "$_*$", already used to denote convolution.

Hermitian. Namely, we have $x = \frac{x + x^{\sim}}{2} + i\frac{x - x^{\sim}}{2i}$. Every element of the form $x^{\sim}x$ is Hermitian. If A has a unit and x has an inverse, then $(x^{\sim})^{-1}$ exists and is equal to $(x^{-1})^{\sim}$. If x has a quasi-inverse y (C.5), then $y^{\sim}$ is the quasi-inverse of $x^{\sim}$.

Our present concern with $\sim$-algebras stems from the fact that every subalgebra of $\boldsymbol{M}(G)$ [where G is a locally compact group] containing $\mu^{\sim}$ along with μ is a normed $\sim$-algebra.

Certain special representations of $\sim$-algebras are of great importance to us.

(21.8) Definition. Let A be a $\sim$-algebra and T a representation of A by *bounded* operators on a Hilbert space H. Suppose that $T_{x^{\sim}} = T_x^{\sim}$ for all $x \in A$ [$T_x^{\sim}$ is the adjoint operator of T_x: see (B.53)]. Then the representation T is called a *$\sim$-representation*.

Let T and T' be representations of an algebra A by bounded operators on Hilbert spaces H and H', respectively. Suppose that there is a linear isometry W (B.42) carrying H onto H' such that $WT_xW^{-1} = T'_x$ for all $x \in A$. Then T and T' are said to be *equivalent*. Equivalence of two representations V and V' of a semigroup S by bounded operators on H and H' is defined similarly.

(21.9) Theorem. *Let T be a $\sim$-representation of a $\sim$-algebra A, as in* (21.8). *If H_1 is a linear subspace of H invariant under T, then the orthogonal subspace $H_1^{\perp}$ is also invariant under T.*

Proof. For $\xi \in H_1^{\perp}$, $\zeta \in H_1$, and $x \in A$, we have

$$\langle T_x\xi, \zeta\rangle = \langle \xi, T_x^{\sim}\zeta\rangle = \langle \xi, T_{x^{\sim}}\zeta\rangle = 0.$$

Thus $T_x\xi$ is orthogonal to all $\zeta \in H_1$ and is thus in $H_1^{\perp}$. □

(21.10) Corollary. *Let A be a $\sim$-algebra and T a $\sim$-representation of A by* [*bounded*] *operators on a Hilbert space H. Then H is the direct sum of closed orthogonal invariant subspaces N and H_1 such that $T_x(N) = \{0\}$ for all $x \in A$, and for all nonzero $\xi \in H_1$ there is an $x \in A$ such that $T_x\xi \neq 0$.*

Proof. Let $N = \{\xi \in H : T_x\xi = 0$ for all $x \in A\}$. Plainly N is a closed invariant subspace of H. By (21.9), the orthogonal subspace $N^{\perp}$ is also invariant. Setting $N^{\perp} = H_1$, we have the present assertion. □

(21.11) Theorem. *Let G be a group and V a representation of G by unitary operators on a Hilbert space H. If H_1 is a linear subspace of H invariant under V, then $H_1^{\perp}$ is also invariant under V.*

Proof. We have $V_x^{\sim} = V_x^{-1}$ since V_x is unitary and $V_x^{-1} = V_{x^{-1}}$ since V is a representation. Now argue as in (21.9). □

Our goal in studying $\sim$-representations of $\sim$-algebras is to decompose them into "elementary" components so far as this can be done. Our first step in this direction is the following.

(21.12) Definition. A representation T of an algebra A by operators on a topological linear space E is said to be *cyclic* if there is a vector $\xi \in E$ such that the linear subspace $\{T_x \xi : x \in A\}$ is dense in E. A representation V of a semigroup S by operators on E is said to be *cyclic* if there is a vector $\xi \in E$ such that the smallest linear subspace containing $\{V_x \xi : x \in S\}$ is dense in E. In both cases ξ is called a *cyclic vector.*

(21.13) Theorem. *Let T be a $\sim$-representation of a $\sim$-algebra A by operators on a Hilbert space H. Then H is the direct sum of subspaces N and $\{H_\gamma\}_{\gamma \in \Gamma}$, which are closed, pairwise orthogonal, and invariant. We have $T_x(N) = \{0\}$ for all $x \in A$, and T, restricted to H_γ, is cyclic for all $\gamma \in \Gamma$. For every nonzero $\zeta \in H_\gamma$, there is an $x \in A$ such that $T_x \zeta \neq 0$. For $\xi \in H$, let ξ_γ be the projection of ξ in H_γ. Then $T_x \xi = \sum_{\gamma \in \Gamma} T_x \xi_\gamma$.*

Proof. First, let N be as in (21.10). If $N^\perp = \{0\}$, then T is the trivial representation $T_x = 0$ for all $x \in A$. If $N^\perp \neq \{0\}$, well order the nonzero elements of $N^\perp$ in any fashion:

$$\xi_1, \xi_2, \ldots, \xi_n, \ldots, \xi_\delta, \ldots, \tag{1}$$

where δ runs through all ordinals less than some ordinal Δ. Define H_1 as the closure in H of the linear subspace $\{\alpha \xi_1 + T_x \xi_1 : \alpha \in K,\ x \in A\}$. It is easy to see that H_1 is an invariant subspace of H under the representation T. We claim that the subspace $\{T_x \xi_1 : x \in A\}$ is dense in H_1. Let ζ be a vector in H_1 orthogonal to all $T_x \xi_1$ for $x \in A$. For all $y, z \in A$ and $\alpha \in K$, we have

$$0 = \langle \zeta, T_{\alpha y^\sim + y^\sim z} \xi_1 \rangle = \langle \zeta, T_{y^\sim}(\alpha \xi_1 + T_z \xi_1) \rangle = \langle T_y \zeta, \alpha \xi_1 + T_z \xi_1 \rangle.$$

Since ζ is in H_1, $T_y \zeta$ is also in H_1, and since $\{\alpha \xi_1 + T_z \xi_1 : \alpha \in K,\ z \in A\}$ is dense in H_1, we infer that $T_y \zeta = 0$ for all $y \in A$. Since ζ is in $N^\perp$, this implies that $\zeta = 0$. It follows that $\{T_x \xi_1 : x \in A\}$ is dense in H_1, so that the representation T restricted to the subspace H_1 is cyclic with cyclic vector ξ_1.

If $H_1 = N^\perp$, our construction stops. If $H_1 \subsetneqq N^\perp$, let ξ_{δ_2} be the first vector in (1) such that $\xi_{\delta_2} \in H_1^\perp$. Form the subspace $\{\alpha \xi_{\delta_2} + T_x \xi_{\delta_2} : \alpha \in K,\ x \in A\}^- = H_2$. By (21.9), $H_1^\perp \cap N^\perp$ is an invariant subspace for T, so that H_2 is an invariant subspace for T contained in $H_1^\perp \cap N^\perp$. As in the preceding paragraph, T is cyclic on H_2 with cyclic vector ξ_{δ_2}. The definition now proceeds by transfinite induction. Suppose that subspaces $H_1, H_2, \ldots, H_\gamma, \ldots$ of H and elements ξ_{δ_γ} in $N^\perp$ have been defined for all ordinals γ less than γ_0, such that the H_γ are pairwise orthogonal, are contained in $N^\perp$, and have the form $\{\alpha \xi_{\delta_\gamma} + T_x \xi_{\delta_\gamma} : \alpha \in K,\ x \in A\}^-$ $[1 = \delta_1 < \delta_2 < \cdots < \delta_\gamma < \cdots]$. If there is no nonzero vector in $N^\perp$ orthogonal to all H_γ $(\gamma < \gamma_0)$, the construction stops. Otherwise, let $\xi_{\delta_{\gamma_0}}$ be the first vector in (1) that is orthogonal to all $H_\gamma (\gamma < \gamma_0)$. Construct

$H_{\gamma_0}=\{\alpha\xi_{\delta\gamma_0}+T_x\xi_{\delta\gamma_0}:\alpha\in K,\ x\in A\}^-$. This defines the family $\{H_\gamma\}_{\gamma\in\Gamma}$, where Γ can be taken as a subset of the set of ordinal numbers less than Δ. Since no nonzero vector in $N^\perp$ is orthogonal to every H_γ, we infer that H is the direct sum of N and the subspaces H_γ (B.63).

The last statement is easily verified from the boundedness of each operator T_x: we omit the details. □

(21.14) Theorem. *Let G be any group and let V be a representation of G by unitary operators on a Hilbert space H. Then H is the direct sum of subspaces $\{H_\gamma\}_{\gamma\in\Gamma}$, which are closed, pairwise orthogonal, and invariant under V. The representation V restricted to H_γ is cyclic, for each $\gamma\in\Gamma$.*

The proof is similar to but even simpler than that of (21.13), and is omitted.

(21.15) Discussion. Suppose that A is a $\sim$-algebra, T is a $\sim$-representation of A by operators on H, and that ξ is a vector in H. The function $x\to\langle T_x\xi,\xi\rangle=p(x)$ is evidently a linear functional on A, with the property that $p(x^\sim x)=\langle T_{x^\sim x}\xi,\xi\rangle=\langle T_x\xi,T_x\xi\rangle\geqq 0$ for all $x\in A$. Functionals of this sort can be studied *in abstracto*. They turn out to have a wealth of useful properties and to be intimately connected with cyclic representations of $\sim$-algebras.

(21.16) Definition. Let A be a $\sim$-algebra. A linear functional p on A [regarded as a complex linear space] is said to be a *positive functional* if $p(x^\sim x)\geqq 0$ for all $x\in A$.

(21.17) Theorem. *Let p be a positive functional on a $\sim$-algebra A. Then for all $x, y\in A$, we have*

(i) $p(y^\sim x)=\overline{p(x^\sim y)}$

and

(ii) $|p(y^\sim x)|\leqq p(x^\sim x)^{\frac{1}{2}}p(y^\sim y)^{\frac{1}{2}}$.

Proof. For $\alpha\in K$, (21.16) and (21.6) show that

$$\left.\begin{aligned}0&\leqq p((x+\alpha y)^\sim(x+\alpha y))=p(x^\sim x+\bar\alpha y^\sim x+\alpha x^\sim y+|\alpha|^2y^\sim y)\\&=p(x^\sim x)+\bar\alpha p(y^\sim x)+\alpha p(x^\sim y)+|\alpha|^2p(y^\sim y).\end{aligned}\right\}\quad(1)$$

Since $p(x^\sim x)$ and $p(y^\sim y)$ are nonnegative, $\bar\alpha p(y^\sim x)+\alpha p(x^\sim y)$ is real. Setting $\alpha=1$ in (1), we infer from this that $\operatorname{Im}p(y^\sim x)+\operatorname{Im}p(x^\sim y)=0$. Setting $\alpha=i$ in (1), we also infer that $\operatorname{Re}p(y^\sim x)=\operatorname{Re}p(x^\sim y)$. This proves (i).

We can thus rewrite (1) as

$$0\leqq p(x^\sim x)+2\operatorname{Re}[\alpha p(x^\sim y)]+|\alpha|^2p(y^\sim y).\qquad(2)$$

If $p(x^\sim y)=0$, then (i) shows that (ii) holds. If $p(x^\sim y)\neq 0$, set

$$\alpha=-\frac{p(x^\sim x)}{p(x^\sim y)}$$

in (2). This yields

$$p(x^\sim x)\,|p(x^\sim y)|^2 \leqq p(x^\sim x)^2\, p(y^\sim y). \tag{3}$$

If $p(x^\sim x)$ is positive, then (3) implies (ii). If $p(y^\sim y)$ is positive, we repeat the argument interchanging x and y and again obtain (ii). If $p(x^\sim x)=p(y^\sim y)=0$, (2) shows that $\mathrm{Re}[\alpha p(x^\sim y)]\geqq 0$ for all $\alpha\in K$. This is impossible if $p(x^\sim y)\neq 0$, and so (ii) is established. □

(21.18) Theorem. *Let A be any $\sim$-algebra and let A_u be the algebra obtained from A by adjoining a unit u, as in* (C.3); *elements (x, α) of A_u will be written as $\alpha u + x$. For $\alpha u + x \in A_u$, let $(\alpha u + x)^\sim = \bar{\alpha} u + x^\sim$. Then A_u is a $\sim$-algebra. Let p be a positive functional on A. Then there is a positive functional $p^\dagger$ on A_u that agrees with p on A if and only if*

(i) $p(x^\sim)=\overline{p(x)}$ *for* $x\in A$

and

(ii) $|p(x)|^2\leqq a\,p(x^\sim x)$ *for* $x\in A$,

where a is a positive number independent of x. If (ii) *holds, $p^\dagger(u)$ can be taken as any number greater than or equal to a.*

Proof. If $p^\dagger$ exists, then (i) follows from setting $y=u$ in (21.17.i) and noting that $u^\sim=u$. Setting $y=u$ in (21.17.ii), we get

$$|p^\dagger(u^\sim x)| = |p(x)| \leqq p(x^\sim x)^{\frac{1}{2}}\, p^\dagger(u)^{\frac{1}{2}},$$

so that (ii) holds with $a=p^\dagger(u)$.

If (i) and (ii) hold, let b be any nonnegative number such that $|p(x)|^2\leqq b\,p(x^\sim x)$ for all $x\in A$. Then define

$$p^\dagger(\alpha u + x)=\alpha b + p(x).$$

We have

$$\begin{aligned} p^\dagger\left((\alpha u + x)^\sim(\alpha u + x)\right) &= p^\dagger\left(|\alpha|^2 u + \bar{\alpha}x + \alpha x^\sim + x^\sim x\right)\\ &= |\alpha|^2 b + \bar{\alpha}p(x) + \alpha p(x^\sim) + p(x^\sim x)\\ &= |\alpha|^2 b + 2\,\mathrm{Re}[\alpha p(x^\sim)] + p(x^\sim x)\\ &\geqq |\alpha|^2 b - 2|\alpha|\,|p(x)| + p(x^\sim x)\\ &\geqq |\alpha|^2 b - 2|\alpha|\, b^{\frac{1}{2}}\, p(x^\sim x)^{\frac{1}{2}} + p(x^\sim x)\\ &= \left(|\alpha|\, b^{\frac{1}{2}} - p(x^\sim x)^{\frac{1}{2}}\right)^2 \geqq 0. \end{aligned}$$

Since $p^\dagger$ is plainly a linear functional on A_u, the proof is complete. □

It is a remarkable fact that *all* positive functionals on a large class of $\sim$-algebras are continuous.

(21.19) Theorem. *Let A be a Banach $\sim$-algebra having a unit u. Then every positive functional p on A satisfies the relation*

(i) $|p(x)|\leqq p(u)\,\|x\|$ *for* $x\in A$.

Proof. The binomial series

$$\sum_{k=0}^{\infty}(-1)^k\binom{\frac{1}{2}}{k}\alpha^k \tag{1}$$

converges absolutely for all $\alpha\in K$ such that $|\alpha|\leqq 1$, and its sum is $\exp\left(\frac{1}{2}\operatorname{Log}(1-\alpha)\right)$ [where $-\pi<\operatorname{Im}\operatorname{Log}(1-\alpha)\leqq\pi$] if $\alpha\neq 1$; its sum is 0 for $\alpha=1$. For $x\in A$ such that $\|x\|\leqq 1$ and $n=1, 2, 3, \ldots$, let

$$y_n=u+\sum_{k=1}^{n}(-1)^k\binom{\frac{1}{2}}{k}x^k.$$

It is clear that $\lim\limits_{m,n\to\infty}\|y_n-y_m\|=0$, and since A is complete, there is a $y\in A$ such that $\lim\limits_{n\to\infty}y_n=y$. Now define

$$c_n=(-1)^n\sum_{k=0}^{n}\binom{\frac{1}{2}}{k}\binom{\frac{1}{2}}{n-k}\qquad(n=1,2,3,\ldots).$$

By MERTENS' theorem [APOSTOL [1], p. 376], we have

$$1+\sum_{n=1}^{\infty}c_n\alpha^n=(1-\alpha)$$

if $\alpha\in K$ and $|\alpha|\leqq 1$. Considering only real α, we see that $c_1=-1$ and $c_n=0$ for $n=2, 3, \ldots$. Examining the *proof* of MERTENS' theorem, we see that

$$\lim_{n\to\infty}y_n^2=\lim_{n\to\infty}\left[u+\sum_{k=1}^{n}c_k x^k\right]=u-x.$$

Since $\lim\limits_{n\to\infty}y_n^2=y^2$, we infer that $y^2=u-x$. If also x is Hermitian, then it is clear that y is Hermitian. Therefore $y^2=y^\sim y=u-x$ and

$$0\leqq p(y^\sim y)=p(u)-p(x).$$

Consequently $p(x)$ is real and

$$p(x)\leqq p(u).$$

Similarly

$$-p(x)=p(-x)\leqq p(u),$$

so that (i) holds for Hermitian x [recall that p is homogeneous]. For arbitrary $x\in A$, $x^\sim x$ is Hermitian, so that

$$p(x^\sim x)\leqq p(u)\|x^\sim x\|\leqq p(u)\|x\|^2.$$

By (21.17.ii) with $y=u$, we have

$$|p(x)|^2\leqq p(u)\,p(x^\sim x).$$

Combining the last two inequalities, we obtain (i). □

(21.20) Corollary. *Let A be a Banach $\sim$-algebra with unit u and let p be a positive functional on A. Then p is a bounded linear functional, and $\|p\|=p(u)$. If A is any Banach $\sim$-algebra and p is a positive nonzero* *functional on A for which* (21.18.i) *and* (21.18.ii) *hold, then p is bounded. Let* $M(p)=\sup\left\{\frac{|p(x)|^2}{p(x^\sim x)} : x\in A,\ p(x^\sim x)\neq 0\right\}$. *Then we have* $\|p\|\leqq M(p)$, *and if A has a unit,* $M(p)=\|p\|$. *Furthermore, if p and q are positive functionals for which* (21.18.i) *and* (21.18.ii) *hold, then* $M(p+q)\leqq M(p)+M(q)$.

Proof. The first statement is obvious from (21.19) and the equality $\|u\|=1$ (21.6). If A is any Banach $\sim$-algebra and (21.18.i) and (21.18.ii) hold for p, then (21.18) shows that p can be extended to a positive linear functional $p^\dagger$ on A_u. The number $M(p)$ is the smallest value that can be assigned as $p^\dagger(u)$: this follows from (21.18). Since $\|p^\dagger\|=p^\dagger(u)$, p must be a bounded functional on $A\subset A_u$, and as the norm does not increase on a subspace, we have $\|p\|\leqq p^\dagger(u)=M(p)$. If A has a unit, the inequality $M(p)\leqq p(u)=\|p\|$ follows from (21.17.ii). As shown in (21.18), $M(p+q)$ is the least value of $(p+q)^\dagger(u)$ for a positive extension of $p+q$ over A_u. Since $M(p)+M(q)=(p^\dagger+q^\dagger)(u)$ is such a value for $p+q$, we infer that $M(p+q)\leqq M(p)+M(q)$. □

(21.21) Corollary. *Let A be a Banach $\sim$-algebra and p a positive functional on A satisfying* (21.18.i) *and* (21.18.ii). *Let $M(p)$ be defined as in* (21.20). *Then for all $x\in A$, we have*

(i) $p(x^\sim x)\leqq M(p)\lim\limits_{n\to\infty}\|(x^\sim x)^n\|^{\frac{1}{n}}$.

Proof. In view of (21.18), there is no loss of generality in supposing that A has a unit u and that $M(p)=p(u)$. Repeated application of (21.17.ii) gives

$$\begin{aligned}|p(x)| &\leqq p(u)^{\frac{1}{2}}\,p(x^\sim x)^{\frac{1}{2}}\\ &\leqq [p(u)]^{\frac{1}{2}+\frac{1}{4}}\,[p((x^\sim x)^2)]^{\frac{1}{4}}\leqq\cdots\\ &\leqq [p(u)]^{\frac{1}{2}+\frac{1}{4}+\cdots+\frac{1}{2^n}}\,[p((x^\sim x)^{2^{n-1}})]^{\frac{1}{2^n}}.\end{aligned}$$

From (21.19), we have

$$[p((x^\sim x)^{2^{n-1}})]^{\frac{1}{2^n}}\leqq p(u)^{\frac{1}{2^n}}\,\|(x^\sim x)^{2^{n-1}}\|^{\frac{1}{2^n}},$$

so that

$$|p(x)|\leqq p(u)\,\|(x^\sim x)^{2^{n-1}}\|^{\frac{1}{2^n}}.$$

It follows that

$$p(x^\sim x)\leqq p(u)\,\|(x^\sim x)^{2^n}\|^{\frac{1}{2^n}};$$

applying (C.4), we now have

$$p(x^\sim x)\leqq p(u)\,\overline{\lim_{n\to\infty}}\,\|(x^\sim x)^n\|^{\frac{1}{n}}=p(u)\lim_{n\to\infty}\|(x^\sim x)^n\|^{\frac{1}{n}}.\quad\square$$

Corollary (21.20) has an important consequence.

(21.22) Theorem. *Every ~-representation T of a Banach ~-algebra A by [bounded] operators on a Hilbert space H is a continuous mapping of A into the space of bounded operators on H supplied with the norm topology. Indeed, we have*

(i) $\|T_x\| \leqq \|x\|$ *for all* $x \in A$.

Proof. For an arbitrary $\xi \in H$, let p be the function on A such that $p(x) = \langle T_x \xi, \xi \rangle$. Plainly p is linear, and we have

$$p(x^{\sim}x) = \langle T_{x^{\sim}x}\xi, \xi \rangle = \langle T_{x^{\sim}} T_x \xi, \xi \rangle = \langle T_x \xi, T_x \xi \rangle \geqq 0.$$

Thus p is a positive functional. Furthermore,

$$p(x^{\sim}) = \langle T_{x^{\sim}}\xi, \xi \rangle = \langle \xi, T_x \xi \rangle = \overline{\langle T_x \xi, \xi \rangle} = \overline{p(x)},$$

and

$$|p(x)|^2 = |\langle T_x \xi, \xi \rangle|^2 \leqq \|T_x \xi\|^2 \, \|\xi\|^2 = p(x^{\sim}x) \, \|\xi\|^2.$$

Thus p satisfies conditions (21.18.i) and (21.18.ii), and $M(p)$ does not exceed $\|\xi\|^2$. By (21.20), p is a bounded linear functional of norm not exceeding $\|\xi\|^2$, and so we have

$$\|T_x \xi\|^2 = \langle T_{x^{\sim}x}\xi, \xi \rangle = p(x^{\sim}x) \leqq \|p\| \cdot \|x^{\sim}x\| \leqq \|\xi\|^2 \, \|x^{\sim}x\| \leqq \|\xi\|^2 \, \|x\|^2.$$

Taking square roots, we obtain (i). □

We now take up the connection between positive functionals on a ~-algebra A and ~-representations of A, remarked on already in (21.15).

(21.23) Theorem. *Let T and T' be cyclic ~-representations of a ~-algebra A by operators on Hilbert spaces H and H', with cyclic vectors ζ and ζ', respectively. If the positive functionals $p(x) = \langle T_x \zeta, \zeta \rangle$ and $p'(x) = \langle T'_x \zeta', \zeta' \rangle$ are equal, then the representations T and T' are equivalent.*

Proof. For all $x, y \in A$, we have

$$p(y^{\sim}x) = \langle T_x \zeta, T_y \zeta \rangle = p'(y^{\sim}x) = \langle T'_x \zeta', T'_y \zeta' \rangle.$$

The mapping W defined by $W(T_x \zeta) = T'_x \zeta'$ is thus a linear, inner product preserving mapping of $\{T_x \zeta : x \in A\}$ onto $\{T'_x \zeta' : x \in A\}$. As these subspaces are dense in H and H', respectively, W can be extended to a linear isometry that carries H onto H'. For all x and y in A, we have

$$W^{-1} T'_y \zeta' = T_y \zeta,$$
$$T_x W^{-1} T'_y \zeta' = T_{xy} \zeta,$$
$$W T_x W^{-1} T'_y \zeta' = T'_{xy} \zeta' = T'_x T'_y \zeta'.$$

Thus $W T_x W^{-1} = T'_x$ on the dense subspace $\{T'_y \zeta' : y \in A\}$ of H', and so this identity persists throughout H'. That is, T and T' are equivalent (21.8). □

(21.24) Theorem. *Let A be a Banach $\sim$-algebra and let p be a positive functional on A such that*

(i) $p(x^{\sim}) = \overline{p(x)}$

and

(ii) $|p(x)|^2 \leqq a\,p(x^{\sim}x)$ $(a > 0)$

for all $x \in A$. Then there is a cyclic $\sim$-representation T of A by operators on a Hilbert space H, with cyclic vector ζ, such that

(iii) $p(x) = \langle T_x \zeta, \zeta \rangle$ *for all $x \in A$.*

Proof. Let $A_1 = A$ if A has a unit; let $A_1 = A_u$ if A has no unit. By (21.18), there is a positive functional p_1 on A_1 agreeing with p on A. We shall construct a representation T of A_1 such that (iii) holds for all $x \in A_1$. Consider first the set $B = \{x \in A_1 : p_1(x^{\sim}x) = 0\}$. For $y \in A_1$ and $x \in B$, we have $p_1(yx) = 0$, since by (21.17.ii),

$$|p_1(yx)| \leqq p_1(x^{\sim}x)^{\frac{1}{2}}\, p_1(yy^{\sim})^{\frac{1}{2}} = 0.$$

If $x_1, x_2 \in B$, then we have

$$p_1\big((x_1 + x_2)^{\sim}(x_1 + x_2)\big)$$
$$= p_1(x_1^{\sim}x_1) + p_1(x_2^{\sim}x_1) + p_1(x_1^{\sim}x_2) + p_1(x_2^{\sim}x_2) = 0,$$

and for $\alpha \in K$ and $x \in B$, $p_1\big((\alpha x)^{\sim}\alpha x\big) = |\alpha|^2 p_1(x^{\sim}x) = 0$. If $y \in A_1$ and $x \in B$, then $p_1\big((yx)^{\sim}yx\big) = p_1\big((x^{\sim}y^{\sim}y)\,x\big) = 0$. Furthermore, B is closed in A_1 since p_1 is continuous (21.19). Thus B is a closed left ideal in A_1.

Form the difference space A_1/B [see (B.2)], denoting its elements $x + B, y + B, \ldots$ by $\xi, \eta, \ldots$. If $x_1 - x_2 \in B$ and $y_1 - y_2 \in B$, then

$$p_1(y_1^{\sim}x_1) - p_1(y_2^{\sim}x_2) = p_1\big(y_1^{\sim}(x_1 - x_2)\big) + p_1\big((y_1 - y_2)^{\sim}x_2\big)$$
$$= 0 + \overline{p_1\big(x_2^{\sim}(y_1 - y_2)\big)} = 0.$$

Hence the function $(\xi, \eta) \to p_1(y^{\sim}x)$ is well defined on A_1/B: we write $\langle \xi, \eta \rangle = p_1(y^{\sim}x)$ for $x \in \xi$ and $y \in \eta$. It is obvious that

$$\langle \xi, \eta \rangle = \overline{\langle \eta, \xi \rangle}, \tag{1}$$

$$\langle \xi_1 + \xi_2, \eta \rangle = \langle \xi_1, \eta \rangle + \langle \xi_2, \eta \rangle, \tag{2}$$

$$\langle \alpha\xi, \eta \rangle = \alpha \langle \xi, \eta \rangle, \tag{3}$$

$$\langle \xi, \xi \rangle > 0 \quad \text{if} \quad \xi \neq 0, \tag{4}$$

for all $\xi, \eta \in A_1/B$ and $\alpha \in K$. Thus A_1/B is an inner product space, which in general is incomplete.

For $x \in A_1$, let T'_x be the operator on A_1/B such that $T'_x(y + B) = xy + B$. Since B is a left ideal, each T'_x is well defined and is an operator on A_1/B. It is obvious that the mapping $x \to T'_x$ is a representation of A_1. Let us show that each T'_x is a bounded operator. For this purpose, fix $y \in A_1$

and let q_1 be the function on A_1 defined by $q_1(x)=p_1(y^\sim xy)$ for $x\in A_1$. Plainly q_1 is a positive functional on A_1, and $q_1(u)=p_1(y^\sim y)$ [u is the unit of A_1]. By (21.19), we have

$$|q_1(x)|\leqq p_1(y^\sim y)\|x\| \quad \textit{for} \quad x\in A_1.$$

We can now compute

$$\begin{aligned}\langle T'_x\eta, T'_x\eta\rangle &= \langle xy+B, xy+B\rangle = p_1((xy)^\sim(xy))\\ &= q_1(x^\sim x)\leqq p_1(y^\sim y)\|x^\sim x\|\leqq p_1(y^\sim y)\|x\|^2=\langle\eta,\eta\rangle\|x\|^2.\end{aligned}$$

Thus T'_x is a bounded operator on the inner product space A_1/B, and

$$\|T'_x\|\leqq\|x\|. \tag{5}$$

Furthermore, we have

$$\left.\begin{aligned}\langle T'_x\eta,\zeta\rangle &= \langle T'_x(y+B), z+B\rangle = p_1(z^\sim(xy))\\ &= p_1((x^\sim z)^\sim y) = \langle y+B, x^\sim z+B\rangle = \langle\eta, T'_{x^\sim}\zeta\rangle.\end{aligned}\right\} \tag{6}$$

Finally, let H be the completion of the inner product space A_1/B. Every operator T'_x admits a unique extension over H, denoted by T_x, which is a bounded operator on H. It is easy to see that the mapping $x\rightarrow T_x$ is a representation of A_1 by bounded operators on H, and the identity (6), which is preserved upon passage to H and T_x, shows that this mapping is a $\sim$-representation. The inequality $\|T_x\|\leqq\|x\|$ for $x\in A_1$ follows from (5) [see also Theorem (21.22)].

It remains to show that the representation $x\rightarrow T_x$ is cyclic *when restricted to* A. Let $\zeta=u+B$. Then $\{T'_x(u+B):x\in A_1\}=\{x+B:x\in A_1\}=A_1/B$. By the definition of H, therefore, ζ is a cyclic vector for the representation $x\rightarrow T_x$ of A_1. Furthermore, we have $\langle T_x\zeta,\zeta\rangle=p_1(u^\sim x)=p(x)$ for $x\in A$. Thus the proof is complete if $A_1=A$. If A lacks a unit, so that $A_1=A_u$, we alter the definition of H. Let $H_1=\{\xi\in H: T_x\xi=0$ for all $x\in A\}$, let $H_2=H_1^\perp$, and write $\zeta=\zeta_1+\zeta_2$ where $\zeta_j\in H_j$ $(j=1,2)$. Then for all $x\in A$, we have

$$\begin{aligned}p(x) &= \langle T_x\zeta,\zeta\rangle = \langle T_x\zeta_1+T_x\zeta_2,\zeta_1+\zeta_2\rangle\\ &= \langle T_x\zeta_2,\zeta_1\rangle+\langle T_x\zeta_2,\zeta_2\rangle = \langle T_x\zeta_2,\zeta_2\rangle,\end{aligned}$$

since $T_x\zeta_2\in H_2$ by (21.9). It is easy to see that the linear subspace $\{\alpha\zeta_2+T_x\zeta_2:\alpha\in K,\ x\in A\}$ is dense in H_2, since $\|\alpha\zeta_2+T_x\zeta_2-\xi_2\|\leqq\|\alpha(\zeta_1+\zeta_2)+T_x(\zeta_1+\zeta_2)-(\xi_1+\xi_2)\|$ for all $\xi=\xi_1+\xi_2\in H$, and $A_1/B=\{\alpha\zeta+T_x\zeta:\alpha\in K,\ x\in A\}$ is dense in H. Suppose that $\eta_2\in H_2$ and that η_2 is orthogonal to all $T_x\zeta_2$ for $x\in A$. Then for $y, z\in A$ and $\alpha\in K$, we have

$$0=\langle\eta_2, T_{y^\sim(\alpha u+z)}\zeta_2\rangle=\langle T_y\eta_2,\alpha\zeta_2+T_z\zeta_2\rangle.$$

Hence $T_y\eta_2=0$ for all $y\in A$, and this implies that $\eta_2=0$. Therefore $\{T_x\zeta_2: x\in A\}$ is dense in H_2. Putting H_2 and ζ_2 for H and ζ in the statement of the theorem, we see that the present proof is complete. □

(21.25) Theorem. *Let A be a $\sim$-algebra and T a cyclic $\sim$-representation of A by operators on a nonzero Hilbert space H. Let ζ be a cyclic vector in H. Let p be the positive functional on A defined by $p(x) = \langle T_x\zeta, \zeta\rangle$. Then $M(p) = \langle\zeta, \zeta\rangle$.*

Proof. Since $H \neq \{0\}$, there is an $x \in A$ such that $T_x\zeta \neq 0$. For all such x, we have

$$\frac{|p(x)|^2}{p(x^\sim x)} = \frac{|\langle T_x\zeta, \zeta\rangle|^2}{\langle T_x\zeta, T_x\zeta\rangle} \leqq \frac{\|T_x\zeta\|^2\|\zeta\|^2}{\|T_x\zeta\|^2} = \langle\zeta, \zeta\rangle.$$

The definition of $M(p)$ (21.20) implies that $M(p) \leqq \langle\zeta, \zeta\rangle$. The reversed inequality is proved from the cyclicity of T: if $x \in A$ is chosen so that $\|T_x\zeta - \zeta\|$ is sufficiently small, then $|\langle T_x\zeta, \zeta\rangle - \langle\zeta, \zeta\rangle| \leqq \|T_x\zeta - \zeta\| \cdot \|\zeta\|$ is small, and elementary continuity considerations show that $\frac{|\langle T_x\zeta, \zeta\rangle|^2}{\|T_x\zeta\|^2}$ is arbitrarily close to $\frac{\|\zeta\|^4}{\|\zeta\|^2} = \langle\zeta, \zeta\rangle$. □

The reader who wishes to read our treatment of the spectral theorem may find it advisable at this point to read (C.36)—(C.42), since (C.42) is needed in (21.30) and some preceding theorems of §21 are needed in (C.36)—(C.42).

For reasons that will become apparent in the sequel, we are particularly concerned with representations of groups, semigroups, and algebras admitting no proper closed invariant subspaces.

(21.26) Definition. Let $\{M\}$ be a nonvoid set of continuous operators on a topological linear space E. If $\{0\}$ and E are the only *closed* subspaces of E that are invariant under all M, then $\{M\}$ is said to be *irreducible.* In the contrary case, $\{M\}$ is said to be *reducible,* and a proper closed subspace F of E invariant under all M is said to *reduce* $\{M\}$. A representation T of an algebra A is said to be *irreducible* if the set of operators $\{T_x\}_{x\in A}$ is irreducible; similarly for a representation V of a semigroup S.

The following fact is useful in studying reducibility.

(21.27) Theorem. *Let T be a $\sim$-representation of a $\sim$-algebra A by operators on a Hilbert space H. Let L be a closed linear subspace of H, and let P be the projection operator mapping H onto L. Then L reduces T if and only if*

(i) $PT_x = T_xP$ *for all* $x \in A$.

Proof. If L is invariant under T, x is in A, and ξ is a vector in H, we have $P\xi \in L$, $T_xP\xi \in L$, and so $PT_xP\xi = T_xP\xi$. That is,

$$T_xP = PT_xP. \tag{1}$$

Taking adjoints, we have

$$PT_x^\sim = PT_x^\sim P$$

and
$$PT_{x^\sim} = PT_{x^\sim} P \tag{2}$$
for all $x \in A$. Putting $x^\sim$ for x in (2), we find
$$PT_x = PT_x P,$$
which equality, combined with (1), gives (i).

Conversely, if (i) holds, $\xi \in L$, and $x \in A$, then $P(T_x \xi) = T_x(P\xi) = T_x \xi$, so that $T_x \xi \in L$. Thus L is invariant under T. ☐

(21.28) Theorem. *Let V be a representation of a group G by unitary operators on a Hilbert space H. Let L be a closed linear subspace of H, and let P be the projection operator mapping H onto L. Then L reduces V if and only if*

(i) $PV_x = V_x P$ *for all* $x \in G$.

Proof. Since $V_x^\sim = V_{x^{-1}}$, the argument of (21.27) can be repeated. ☐

(21.29) Remarks. It is trivial that every representation of a semigroup or an algebra by operators on a 1-dimensional complex linear space is irreducible. A 1-dimensional representation of an algebra A can be regarded as a multiplicative linear functional on A, in the sense of (C.12). A 1-dimensional representation of a semigroup S can be regarded as a complex function χ on S such that

(i) $\chi(xy) = \chi(x)\chi(y)$ for $x, y \in S$.

We shall explore both of these concepts in the sequel.

(21.30) Theorem. *Let T be a $\sim$-representation of a $\sim$-algebra A by operators on a Hilbert space H, different from the zero representation. The following properties are equivalent:*

(i) *T is irreducible;*

(ii) *every nonzero vector in H is a cyclic vector for T;*

(iii) *the only bounded operators on H commuting with all T_x $(x \in A)$ are of the form αI $(\alpha \in K)$.*

Properties (i)—(iii) *are also equivalent for every unitary representation V of a group G.*[1]

Proof. We carry out the proof only for A and T; it is similar for G and V. Suppose that (i) holds. Let $N = \{\xi \in H : T_x \xi = 0 \text{ for all } x \in A\}$. Then N is closed and invariant under T, and so N is $\{0\}$ or H. The equality $N = H$ means that T is the zero representation, and so we have $N = \{0\}$. Now if $\zeta \in H$ and $\zeta \neq 0$, then $H_1 = \{T_x \zeta : x \in A\}$ is a subspace of H that is invariant under T and $\{0\} \subsetneqq H_1$. Since H_1^- is also invariant under T, we have $H_1^- = H$. Hence (i) implies (ii). If (i) fails, and H_1 is a proper closed invariant subspace of H, then no vector in H_1 can be cyclic for T in H. Thus (i) and (ii) are equivalent.

---

[1] This may be viewed as a form of SCHUR's lemma; compare with (27.9), Vol. II.

We now consider (i) and (iii). If (i) fails, if H_1 is a proper closed invariant subspace of H, and if P is the projection operator onto H_1, then P is not of the form αI, and $PT_x = T_x P$ for all $x \in A$ (21.27). Hence (iii) fails if (i) fails, so that (iii) implies (i). Suppose finally that (i) holds, and let B be a nonzero Hermitian operator on H commuting with all T_x. Let $\{P_t : t \in R\}$ be the family of projections on H for the operator B given by the spectral theorem (C.42). By (C.42.ii), each P_t commutes with all T_x and so $P_t = 0$ or I (21.27). By virtue of (C.42.i), there is a real number t_0 such that $P_t = 0$ for $t < t_0$ and $P_t = I$ for $t > t_0$. It follows from (C.42.v) that $B = t_0 I$. If now B is any bounded operator commuting with all T_x, we also have $T_x B^\sim = (BT_{x^\sim})^\sim = (T_{x^\sim} B)^\sim = B^\sim T_x$ for all $x \in A$. Writing $B = \dfrac{B + B^\sim}{2} + i\dfrac{B - B^\sim}{2i}$, we see that B is a linear combination of Hermitian operators commuting with all T_x, and so $B = \alpha I$ for a complex number α. □

We are going to show that every nonzero $\sim$-representation of a Banach $\sim$-algebra A gives rise to irreducible representations of A. For this purpose we first introduce a partial ordering in the class of positive functionals on a $\sim$-algebra.

(21.31) Definition. Let A be a $\sim$-algebra and p and q positive functionals on A. If there is a complex number α such that $\alpha p - q$ is a positive functional, we write $p > q$ [or $q < p$].

Note that the α in (21.31) can always be taken to be positive. If $q(x^\sim x) = 0$ for all $x \in A$, let $\alpha = 1$; if $q(x_0^\sim x_0) > 0$ for some $x_0 \in A$, then the inequality $q(x_0^\sim x_0) \leqq \alpha p(x_0^\sim x_0)$ forces α to be positive. It is easy to show that $p < p$, that $p < q$ and $q < r$ imply $p < r$, and that $0 < p$.

(21.32) Theorem. *Let A be a Banach $\sim$-algebra and let p be a positive functional on A satisfying conditions* (21.24.i) *and* (21.24.ii). *Let $x \to T_x$, H, and ζ be as given in* (21.24). *Let q be any positive functional on A satisfying* (21.24.i) *and* (21.24.ii) *and such that $q < p$. Then there is a positive-definite Hermitian operator B on H commuting with all T_x $(x \in A)$ such that*

(i) $q(x) = \langle BT_x \zeta, \zeta \rangle$ *for all* $x \in A$.

Conversely, if B is a positive-definite Hermitian operator commuting with all T_x, then (i) defines a positive functional q on A satisfying (21.24.i) *and* (21.24.ii) *and such that $q < p$.*

Proof. It is convenient to prove the converse assertion first. By (C.35), there is a unique positive-definite Hermitian operator C on H such that $C^2 = B$; the operator C commutes with all operators in $\mathfrak{B}(H)$ that commute with B, and hence C commutes with all of the operators T_x.

Writing $q(x) = \langle BT_x\zeta, \zeta\rangle$, we have

$$\begin{aligned}|q(x)|^2 &= |\langle BT_x\zeta,\zeta\rangle|^2 = |\langle CCT_x\zeta,\zeta\rangle|^2 = |\langle T_x C\zeta, C\zeta\rangle|^2 \\ &\leqq \|T_x C\zeta\|^2 \cdot \|C\zeta\|^2 = \langle T_x C\zeta, T_x C\zeta\rangle \cdot \|C\zeta\|^2 = \langle BT_{x^\sim x}\zeta,\zeta\rangle \cdot \|C\zeta\|^2 \\ &= q(x^\sim x)\cdot\|C\zeta\|^2.\end{aligned}$$

Consequently the functional q satisfies (21.24.ii) with $a = \|C\zeta\|^2$. Property (21.24.i), linearity, and positivity of q are all obvious. It is also easy to see that $\alpha p - q$ is a positive functional if $\alpha \geqq \|B\|$ [this is the norm of B defined in (B.8)].

The direct assertion of the present theorem is more interesting and also more complicated. Suppose that q is a positive functional on A satisfying (21.24.i) and (21.24.ii) such that $q < p$. Then there is a positive number α such that

$$0 \leqq q(x^\sim x) \leqq \alpha p(x^\sim x) \quad \text{for all} \quad x \in A. \tag{1}$$

Let $H' = \{T_x\zeta : x \in A\}$. Then H' is a dense linear subspace of H (21.24 For all $x, y \in A$, let

$$\Phi(T_x\zeta, T_y\zeta) = q(y^\sim x). \tag{2}$$

The function Φ is actually well defined on $H' \times H'$. If $T_x\zeta = T_{x_1}\zeta$ and $T_y\zeta = T_{y_1}\zeta$, then $T_{x-x_1}\zeta = T_{y-y_1}\zeta = 0$, so that

$$p\big((x-x_1)^\sim (x-x_1)\big) = p\big((y-y_1)^\sim (y-y_1)\big) = 0.$$

Thus by (1) and (21.17.ii), we have

$$\begin{aligned}|q(y^\sim x) - q(y_1^\sim x_1)| &\leqq |q((y^\sim - y_1^\sim)x)| + |q(y_1^\sim(x-x_1))| \\ &\leqq q(x^\sim x)^{\frac12} q\big((y-y_1)^\sim (y-y_1)\big)^{\frac12} + q\big((x-x_1)^\sim (x-x_1)\big)^{\frac12} q(y_1^\sim y_1)^{\frac12} = 0.\end{aligned}$$

The function Φ on $H' \times H'$ is obviously a bilinear functional (B.4). Furthermore, we have

$$\begin{aligned}|\Phi(T_x\zeta, T_y\zeta)| &= |q(y^\sim x)| \leqq q(x^\sim x)^{\frac12} q(y^\sim y)^{\frac12} \\ &\leqq \alpha p(x^\sim x)^{\frac12} p(y^\sim y)^{\frac12} = \alpha\|T_x\zeta\|\cdot\|T_y\zeta\|.\end{aligned}$$

Thus Φ is continuous and can be extended in just one way to a continuous bilinear functional on $H \times H$. As noted in (B.60), there is a bounded operator B on H such that

$$\Phi(\xi, \eta) = \langle B\xi, \eta\rangle \quad \text{for all} \quad \xi, \eta \in H. \tag{3}$$

For all $x, y \in A$, we have

$$\langle BT_x\zeta, T_y\zeta\rangle = q(y^\sim x) = \overline{q(x^\sim y)} = \overline{\langle BT_y\zeta, T_x\zeta\rangle} = \langle T_x\zeta, BT_y\zeta\rangle.$$

Hence B is Hermitian. Similarly we have $\langle BT_x\zeta, T_x\zeta\rangle = q(x^\sim x) \geqq 0$. Thus B is positive-definite. For all $x, y, z \in A$, we have

$$\langle BT_xT_y\zeta, T_z\zeta\rangle = q(z^\sim xy),$$

and

$$\langle T_x B T_y \zeta, T_z \zeta \rangle = \langle B T_y \zeta, T_{x^\sim z} \zeta \rangle = q\big((x^\sim z)^\sim y\big) = q(z^\sim x y).$$

Since H' is dense in H, we infer that $\langle BT_x \xi, \eta \rangle = \langle T_x B \xi, \eta \rangle$ for all $\xi, \eta \in H$. This proves that B commutes with every T_x.

We infer from (2) and (3) that

$$q(y^\sim x) = \langle BT_x \zeta, T_y \zeta \rangle = \langle BT_{y^\sim x} \zeta, \zeta \rangle \quad \text{for all} \quad x, y \in A. \tag{4}$$

It remains to show that (i) holds [a fact requiring proof only if A lacks a unit]. To do this, let $q^\dagger$ be the functional on A defined by

$$q^\dagger(x) = \langle BT_x \zeta, \zeta \rangle \quad \text{for} \quad x \in A. \tag{5}$$

As shown above, $q^\dagger$ is a positive functional on A satisfying (21.24.i) and (21.24.ii). Thus there are a Hilbert space $H^\dagger$ and a cyclic $\sim$-representation $x \to S_x^\dagger$ of A by operators on $H^\dagger$, with cyclic vector $\zeta^\dagger$, such that

$$q^\dagger(x) = \langle S_x^\dagger \zeta^\dagger, \zeta^\dagger \rangle \quad \text{for} \quad x \in A. \tag{6}$$

Similarly q can be written as

$$q(x) = \langle S_x^{\dagger\dagger} \zeta^{\dagger\dagger}, \zeta^{\dagger\dagger} \rangle$$

where $x \to S_x^{\dagger\dagger}$ is a cyclic $\sim$-representation of A by operators on a Hilbert space $H^{\dagger\dagger}$ with cyclic vector $\zeta^{\dagger\dagger}$. We will define a linear isometry U carrying $H^\dagger$ onto $H^{\dagger\dagger}$ as follows. Let $US_x^\dagger \zeta^\dagger = S_x^{\dagger\dagger} \zeta^{\dagger\dagger}$ for all $x \in A$. Then U is a well-defined, linear, inner product preserving mapping of a dense subspace of $H^\dagger$ onto a dense subspace of $H^{\dagger\dagger}$, as (4), (5), and (6) show. It is easy to see that U can be extended to a linear isometry (B.42) of $H^\dagger$ onto $H^{\dagger\dagger}$. The definition of U implies that $U(S_x^\dagger(S_y^\dagger \zeta^\dagger)) = S_x^{\dagger\dagger}(US_y^\dagger \zeta^\dagger)$ for all $x, y \in A$; since $\{S_y^\dagger \zeta^\dagger : y \in A\}$ is dense in $H^\dagger$, we have $US_x^\dagger \xi^\dagger = S_x^{\dagger\dagger} U \xi^\dagger$ for all $x \in A$ and $\xi^\dagger \in H^\dagger$.

For all $x \in A$, we now have

$$\begin{aligned}\langle U\zeta^\dagger, S_x^{\dagger\dagger} \zeta^{\dagger\dagger} \rangle &= \langle S_{x^\sim}^{\dagger\dagger} U \zeta^\dagger, \zeta^{\dagger\dagger} \rangle = \langle US_{x^\sim}^\dagger \zeta^\dagger, \zeta^{\dagger\dagger} \rangle \\ &= \langle S_{x^\sim}^{\dagger\dagger} \zeta^{\dagger\dagger}, \zeta^{\dagger\dagger} \rangle = \langle \zeta^{\dagger\dagger}, S_x^{\dagger\dagger} \zeta^{\dagger\dagger} \rangle.\end{aligned}$$

This implies that $U\zeta^\dagger = \zeta^{\dagger\dagger}$. In consequence we have for all $x \in A$

$$\begin{aligned}q(x) &= \langle S_x^{\dagger\dagger} \zeta^{\dagger\dagger}, \zeta^{\dagger\dagger} \rangle = \langle S_x^{\dagger\dagger} U \zeta^\dagger, U\zeta^\dagger \rangle = \langle US_x^\dagger \zeta^\dagger, U\zeta^\dagger \rangle \\ &= \langle S_x^\dagger \zeta^\dagger, \zeta^\dagger \rangle = q^\dagger(x) = \langle BT_x \zeta, \zeta \rangle.\end{aligned}$$

This is just (i). □

(21.33) Theorem. *Let A be a Banach $\sim$-algebra, and let A_h be the set of all Hermitian elements of A. Then A_h is a real Banach space. Let P*

denote the set of all positive functionals p on A such that

(i) $p(x^{\sim}) = \overline{p(x)}$

and

(ii) $|p(x)| \leqq p(x^{\sim}x)^{\frac{1}{2}}$

for all $x \in A$. Each element of P, when restricted to A_h, is a real, bounded linear functional on A_h, of norm less than or equal to 1. Thus P can be regarded as a subset of A_h^*. So regarded, P is a convex set compact in the weak-$*$ topology of A_h^* (B.24).

Proof. Condition (ii) is just the condition $M(p) \leqq 1$ $(p \neq 0)$. Thus by (21.20) each $p \in P$ is a bounded linear functional on A of norm not exceeding 1. It is trivial that A_h is a real linear space, and it is simple to prove that A_h is closed in A. Thus A_h is a real Banach space. Let p be any element of P. By (i), p is real-valued on A_h, and is obviously linear. The norm of p is not increased when it is restricted to A_h, so that $\|p\| \leqq 1$ on A_h.

If we are given a real linear functional f on A_h such that $|f(x)| \leqq f(x^{\sim}x)^{\frac{1}{2}} = f(x^2)^{\frac{1}{2}}$, we may ask for conditions under which f is the restriction to A_h of a functional in P. Every element of A can be written in just one way in the form $x + iy$ with $x, y \in A_h$ (21.7). Thus the only possible linear extension of f has the form $p(x+iy) = f(x) + if(y)$. Obviously p is a complex linear functional on A. If $p \in P$, then for all $x, y \in A_h$, we have

$$|p(x+iy)|^2 \leqq p\big((x-iy)(x+iy)\big) = p\big(x^2 + y^2 + i(xy - yx)\big)$$

or equivalently

$$f^2(x) + f^2(y) \leqq f(x^2) + f(y^2) + f\big(i(xy - yx)\big).$$

Similarly, we must have

$$f^2(x) + f^2(y) \leqq f(x^2) + f(y^2) + f\big(i(yx - xy)\big),$$

and thus

$$f^2(x) + f^2(y) + \big|f\big(i(xy - yx)\big)\big| \leqq f(x^2) + f(y^2) \tag{1}$$

is a necessary condition that the function f in A_h^* be the restriction to A_h of an element of P. Conversely, if (1) holds for f, then p is a positive functional on A satisfying (i) and (ii). Thus $\|f\| \leqq 1$ for all $f \in A_h^*$ satisfying (1). It follows that P, regarded as a subset of A_h^*, is the part of the unit ball in A_h^* defined by condition (1). As such, it is obviously closed in the weak-$*$ topology and is hence compact (B.25). Finally, (i) and (ii) are equivalent to the existence of a positive extension $p^{\dagger}$ of p over A_u for which $p^{\dagger}(u) \leqq 1$. This shows that P is convex. □

(21.34) Theorem. *Let A, A_h, and P be as in* (21.33). *A nonzero element p of P is an extreme point of P if and only if $M(p) = 1$ and the representation $x \to T_x$ of A constructed as in* (21.24) *is irreducible.*

Proof. It is obvious that 0 is an extreme point of P. If $p \in P$ and $0 < M(p) < 1$, then $p = M(p)\left(\frac{p}{M(p)}\right) + (1 - M(p)) \cdot 0$, so that p is not an extreme point of P. Suppose that $p \in P$, $p \neq 0$, and that the representation $x \to T_x$ of A as constructed in (21.24) is reducible. Thus there is a closed linear subspace H_1 of H invariant under all T_x such that $\{0\} \subsetneqq H_1 \subsetneqq H$. Write H_2 for $H_1^{\perp}$. By (21.9), H_2 is also invariant under all T_x. Write the ζ of (21.24.iii) as $\zeta_1 + \zeta_2$, where $\zeta_j \in H_j$ $(j = 1, 2)$. It is easy to see that $\zeta_1 \neq 0$, $\zeta_2 \neq 0$. Also it is clear that

$$\begin{aligned} p(x) &= \langle T_x \zeta, \zeta \rangle = \langle T_x \zeta_1 + T_x \zeta_2, \zeta_1 + \zeta_2 \rangle = \langle T_x \zeta_1, \zeta_1 \rangle + \langle T_x \zeta_2, \zeta_2 \rangle \\ &= p_1(x) + p_2(x); \end{aligned}$$

furthermore, p_1 and p_2 are functionals belonging to P. We claim that $p_1 \neq 0$, $p_2 \neq 0$, and that there is no real constant $\varkappa$ such that $p_2 = \varkappa p_1$. Since T is a cyclic representation, there is an $x \in A$ such that $\|T_x \zeta - \zeta_1\| < \|\zeta_1\|$. Then we have

$$\begin{aligned} |p_1(x) - \langle \zeta_1, \zeta_1 \rangle| &= |\langle T_x \zeta_1, \zeta_1 \rangle - \langle \zeta_1, \zeta_1 \rangle| \\ &= |\langle T_x \zeta - \zeta_1, \zeta_1 \rangle| \leqq \|T_x \zeta - \zeta_1\| \cdot \|\zeta_1\| < \langle \zeta_1, \zeta_1 \rangle. \end{aligned}$$

Hence $p_1 \neq 0$; similarly we have $p_2 \neq 0$. Now assume that $p_2 = \varkappa p_1$, where $\varkappa$ is a [necessarily] positive number. Let δ be a positive number such that

$$\delta < \frac{\varkappa \|\zeta_1\|^2}{\|\zeta_2\| + \varkappa \|\zeta_1\|}. \tag{1}$$

Choose an element $a \in A$ such that $\|T_a \zeta - \zeta_1\| < \delta$. Then we have

$$\begin{aligned} |\langle T_a \zeta_1, \zeta_1 \rangle - \langle \zeta_1, \zeta_1 \rangle| &= |\langle T_a \zeta, \zeta_1 \rangle - \langle \zeta_1, \zeta_1 \rangle| \\ &\leqq \|T_a \zeta - \zeta_1\| \cdot \|\zeta_1\| < \delta \|\zeta_1\|; \end{aligned}$$

thus

$$|\langle T_a \zeta_1, \zeta_1 \rangle| > \|\zeta_1\|^2 - \delta \|\zeta_1\|. \tag{2}$$

A like computation gives

$$|\langle T_a \zeta_2, \zeta_2 \rangle| = |\langle T_a \zeta, \zeta_2 \rangle - \langle \zeta_1, \zeta_2 \rangle| \leqq \|T_a \zeta - \zeta_1\| \cdot \|\zeta_2\| < \delta \|\zeta_2\|. \tag{3}$$

From (1) we infer that

$$\delta \|\zeta_2\| < \varkappa (\|\zeta_1\|^2 - \delta \|\zeta_1\|). \tag{4}$$

Combining (2), (3), and (4), we have

$$|\langle T_a \zeta_2, \zeta_2 \rangle| < \varkappa |\langle T_a \zeta_1, \zeta_1 \rangle|,$$

which is a contradiction.

Finally, write $q_j = \frac{M(p)}{M(p_j)} p_j$ $(j = 1, 2)$ and $t = \frac{M(p_1)}{M(p)}$. By (21.25), we have

$$M(p) = \langle \zeta, \zeta \rangle = \langle \zeta_1, \zeta_1 \rangle + \langle \zeta_2, \zeta_2 \rangle = M(p_1) + M(p_2).$$

Hence we also have

$$p = p_1 + p_2 = tq_1 + (1-t)\,q_2.$$

Therefore p is not an extreme point of P.

Conversely, suppose that $M(p)=1$ and that the representation T is irreducible. Assume that p is not an extreme point of P. Then we have $p=tp_1+(1-t)p_2$, where $0<t<1$, $M(p_1)\leqq 1$, $M(p_2)\leqq 1$, $p_1 \neq p$. The inequality $M(p)\leqq tM(p_1)+(1-t)M(p_2)$ shows that $M(p_1)=M(p_2)=1$. Since $\frac{1}{t}p-p_1=\left(\frac{1-t}{t}\right)p_2$, we see that $p>p_1$, and so by (21.32) there is a positive-definite Hermitian operator B on H such that

$$p_1(x)=\langle BT_x\zeta,\zeta\rangle \quad \text{for all} \quad x\in A,$$

and such that B commutes with all T_x. Then (21.30) implies that $B=\alpha I$ for some nonnegative number α . Thus $p_1=\alpha p$; since $M(p_1)=M(p)=1$, we have $p_1=p$, and so arrive at a contradiction. □

(21.35) Theorem. *Let A be a Banach $\sim$-algebra, let p be any functional in P, let $\{x_1,\ldots,x_m\}$ be a finite subset of A, and let ε be a positive number. Then there are nonzero extreme points $p_1,\ldots,p_n$ of P and positive numbers $t_1,\ldots,t_n$ such that $\sum_{k=1}^{n} t_k\leqq 1$ for which*

(i) $|p(x_j)-(t_1p_1(x_j)+t_2p_2(x_j)+\cdots+t_np_n(x_j))|<\varepsilon \quad (j=1,\ldots,m).$

Proof. This result follows at once from the KREĬN-MIL'MAN theorem (B.30). Convex combinations of extreme points of P are dense in the weak-$*$ topology of P as a subset of A_h^*. This establishes (i) for *Hermitian* elements $x_1,\ldots,x_m$ of A. Since every x in A has the form $y+iz$ with $y,z\in A_h$, (i) is immediate for arbitrary $x_1,\ldots,x_m\in A$. □

(21.36) Corollary. *Let A be a Banach $\sim$-algebra and x an element of A for which there is a functional $p\in P$ such that $p(x)\neq 0$. Then there is an extreme point $q\in P$ such that $q(x)\neq 0$.*

This follows at once from (21.35).

We now present our main result on representations of $\sim$-algebras.

(21.37) Theorem. *Let A be a Banach $\sim$-algebra and let x be an element of A. The following conditions on x are equivalent:*

(i) *there is a functional $p_1\in P$ such that $p_1(x)\neq 0$;*

(ii) *there is a functional $p_2\in P$ such that $p_2(x^\sim x)>0$;*

(iii) *there is a $\sim$-representation T of A such that $T_x\neq 0$;*

(iv) *there is an irreducible $\sim$-representation S of A such that $S_x\neq 0$.*

Proof. The inequality $|p_1(x)|^2\leqq p_1(x^\sim x)$ shows that (i) implies (ii). If (ii) holds, then by (21.36) there is an extreme point q of P such that $q(x^\sim x)>0$. By (21.34), the $\sim$-representation S of A for which $q(x)=$

$\langle S_x\zeta, \zeta\rangle$ is irreducible. Since $q(x^{\sim}x) = \|S_x\zeta\|^2 > 0$, it follows that $S_x \neq 0$. Thus (iv) holds. It is trivial that (iv) implies (iii).

Finally, to show that (iii) implies (i), let T be a $\sim$-representation of A by operators on H for which $T_x \neq 0$. If the relation $\langle T_x\xi, \xi\rangle = 0$ held for all $\xi \in H$, then by (B.58) we would have $T_x = 0$, a contradiction. Choosing p_1 as the function $p_1(y) = \langle T_y\xi, \xi\rangle$ for some ξ such that $p_1(x) \neq 0$ and $\|\xi\| = 1$, we obtain (i). □

Miscellaneous theorems and examples

(21.38) The following examples show that the hypotheses of Theorems (21.18) and (21.19) are really necessary.

(a) Let X be any nonvoid compact Hausdorff space, and let A be the set $\mathfrak{C}(X)$. Give A the usual linear operations and the trivial multiplication $f \cdot g = 0$ for all $f, g \in A$. For $f \in A$, let $f^{\sim} = \bar{f}$ and $\|f\| = \max\{|f(x)| : x \in X\}$. Then A is a Banach $\sim$-algebra without a unit, and every linear functional on A is positive. Fix a in X, and define $p(f) = f(a)$ for $f \in A$. Then p is a continuous positive linear functional, $\|p\| = 1$, and $p(f^{\sim}) = \overline{p(f)}$ for $f \in A$. However, $|p(f)|^2 \leq \varkappa p(f^{\sim} \cdot f)$ holds for no real number $\varkappa$ and so p is not extensible to A_u.

If we define q to be ip, then q is also a continuous positive functional having norm 1 but $q(f^{\sim})$ need not equal $\overline{q(f)}$. If X is infinite, A admits discontinuous linear functionals [let $\mathfrak{B}$ be a Hamel basis in $\mathfrak{C}(X)$ over K such that $\|f\| = 1$ for $f \in \mathfrak{B}$, define q_0 on $\mathfrak{B}$ so as to be unbounded, and extend q_0 to $\mathfrak{C}(X)$]. Hence A is a Banach $\sim$-algebra with a discontinuous positive functional.

(b) Let X be a locally compact Hausdorff space such that $\mathfrak{C}_{00}(X)$ admits an unbounded nonnegative linear functional I. Such a functional will exist, for example, if X is not countably compact; see (14.2). Let A be $\mathfrak{C}_{00}(X)$; define linear operations and multiplication to be pointwise. For $f \in A$, let $f^{\sim} = \bar{f}$ and $\|f\| = \max\{|f(x)| : x \in X\}$. Then A is an incomplete normed $\sim$-algebra without unit and I is a discontinuous positive functional on A. Also the inequality $I(|f|) \leq \varkappa I(f^{\sim} f)^{\frac{1}{2}}$ holds for no real number $\varkappa$.

(c) Let A consist of all polynomials $f(z) = \sum_{k=0}^{n} \alpha_k z^k$ $(z \in K)$ having complex coefficients. Let the linear operations and multiplication in A be pointwise, and for $f(z) = \sum_{k=0}^{n} \alpha_k z^k$, let $f^{\sim}(z) = \sum_{k=0}^{n} \bar{\alpha}_k z^k$ and $\|f\| = \max\{|f(t)| : 0 \leq t \leq 1\}$. Notice that $f^{\sim} f(t) = |f(t)|^2$ for $f \in A$ and $t \in R$. Clearly A is an incomplete normed $\sim$-algebra with unit. For $f \in A$, we define $p(f) = f(1)$ and $q(f) = f(2)$. Then plainly p is a continuous positive functional on A and q is a discontinuous positive functional on A. The existence of such a functional q shows that (21.19) may fail if A is incomplete.

(d) (GODEMENT [2].) Let G be a nondiscrete locally compact group. Regard $\mathfrak{C}_{00}(G)$ as a subalgebra of $\mathfrak{L}_1(G)$, which in turn is identified with $\boldsymbol{M}_a(G)$. Thus for $f, g \in \mathfrak{C}_{00}(G)$ we define

$$f*g(x)=\int_G f(y)\,g(y^{-1}x)\,dy, \qquad f^\sim(x)=\frac{1}{\Delta(x)}\overline{f(x^{-1})}, \quad \text{and} \quad \|f\|_1=\int_G |f|\,d\lambda.$$

Then $\mathfrak{C}_{00}(G)$ is a dense subalgebra of $\mathfrak{L}_1(G)$ [the fact that $f*g$ is in $\mathfrak{C}_{00}(G)$ for $f, g\in\mathfrak{C}_{00}(G)$ is shown in the proof of (20.16)]. The mapping

$$f\to f(e)=p(f)$$

is a positive functional on $\mathfrak{C}_{00}(G)$ that is unbounded. It satisfies the condition $p(f^\sim)=\overline{p(f)}$ but not the condition $|p(f)|^2\leq \varkappa p(f^\sim * f)$. Note that $p(f^\sim * f)=\|f\|_2^2$.

(21.39) If A is a Banach $\sim$-algebra and p is a positive functional on A satisfying (21.18.i) and (21.18.ii), then $\|p\|\leq M(p)<\infty$ by (21.20). We now explore this inequality.

(a) Suppose that A is a Banach $\sim$-algebra and that for every $x\in A$ and $\varepsilon>0$, there exists a $v\in A$ such that $\|v\|\leq 1$ and $\|x-vx\|<\varepsilon$. If p is a positive nonzero functional on A satisfying (21.18.i) and (21.18.ii), then $\|p\|=M(p)$. [Let x be a fixed nonzero element of A, and let $\varepsilon>0$. Choose $v\in A$ such that $\|v\|\leq 1$ and $\|x-vx\|<\dfrac{\varepsilon}{2\|p\|^2\|x\|}$. Then

$$\begin{aligned}\big||p(x)|^2-|p(vx)|^2\big| &\leq |p(x)^2-p(vx)^2| = |p(x)+p(vx)|\cdot|p(x)-p(vx)|\\ &\leq \|p\|\{\|x\|+\|vx\|\}\|p\|\,\|x-vx\|\leq 2\|p\|^2\|x\|\,\|x-vx\|<\varepsilon,\end{aligned}$$

and therefore by (21.17.ii), we have

$$\begin{aligned}|p(x)|^2<|p(vx)|^2+\varepsilon &\leq p(x^\sim x)\,p(vv^\sim)+\varepsilon\leq p(x^\sim x)\,\|p\|\,\|vv^\sim\|+\varepsilon\\ &\leq p(x^\sim x)\|p\|+\varepsilon.\end{aligned}$$

Since ε is arbitrary, we have $|p(x)|^2\leq p(x^\sim x)\|p\|$. It follows that $M(p)\leq\|p\|$.]

(b) Let A denote all bounded complex-valued functions f defined and analytic on the open unit disk $D=\{z\in K:|z|<1\}$ and for which $f(0)=0$. We make A into an algebra by defining the linear operations and multiplication pointwise. For $f\in A$ and $z\in D$, we define $f^\sim(z)=\overline{f(\bar z)}$ and $\|f\|=\sup\{|f(z)|:z\in D\}$. Then A is a Banach $\sim$-algebra without unit. Given η, where $0<\eta<1$, we will construct a positive functional p satisfying (21.18.i), (21.18.ii), $\|p\|\leq\eta$, and $M(p)\geq 1$. [We begin by choosing a nonnegative continuous function w on $[0, 1]$ such that $\int_0^1 w(t)\,dt=1$ and $w([\eta, 1])=0$. Then we have $\int_0^1 t w(t)\,dt\leq\eta$. For $f\in A$, we define

$$p(f)=\int_0^1 f(t)\,w(t)\,dt.$$

Clearly p is a linear functional and p is positive since

$$p(f^{\sim}f) = \int_0^1 f^{\sim}(t)\, f(t)\, w(t)\, dt = \int_0^1 |f(t)|^2\, w(t)\, dt \geqq 0.$$

Condition (21.18.i) is obviously satisfied, and (21.18.ii) follows from the inequalities

$$|p(f)|^2 = \left|\int_0^1 f(t)\, w(t)\, dt\right|^2 \leqq \int_0^1 |f(t)\, w(t)|^2 dt \leqq \int_0^1 f^{\sim}(t)\, f(t)\, w(t)\, \varkappa\, dt = \varkappa\, p(f^{\sim}f),$$

where $\varkappa$ denotes the number $\max\{w(t) : 0 \leqq t \leqq 1\}$. If $f \in A$ and $\|f\| \leqq 1$, then SCHWARZ's lemma for analytic functions shows that $|f(t)| \leqq t$ for $0 \leqq t \leqq 1$ and therefore

$$|p(f)| \leqq \int_0^1 |f(t)|\, w(t)\, dt \leqq \int_0^1 t\, w(t)\, dt \leqq \eta.$$

That is, $\|p\| \leqq \eta$.

Let d denote the function on $[0, 1]$ defined by $d(t) = \begin{cases} 1 \text{ if } t > 0 \\ 0 \text{ if } t = 0 \end{cases}$. For $n = 1, 2, \ldots$, define $f_n(z) = 1 - (1 - z)^n$ for $z \in D$. For $t \in [0, 1]$, we have $\lim_{n\to\infty} f_n(t) = d(t)$. Thus by LEBESGUE's theorem on dominated convergence, we have

$$1 = \int_0^1 w(t)\, dt = \lim_{n\to\infty} \int_0^1 f_n(t)\, w(t)\, dt = \lim_{n\to\infty} p(f_n),$$

and

$$1 = \int_0^1 w(t)\, dt = \lim_{n\to\infty} \int_0^1 f_n^{\sim}(t)\, f_n(t)\, w(t)\, dt = \lim_{n\to\infty} p(f_n^{\sim} f_n).$$

Therefore $\lim_{n\to\infty} \dfrac{|p(f_n)|^2}{p(f_n^{\sim} f_n)} = 1$ and $M(p) = \sup\left\{\dfrac{|p(f)|^2}{p(f^{\sim}f)} : p(f^{\sim}f) \neq 0\right\} \geqq 1$.]

(21.40) (a) Every finite-dimensional $\sim$-representation T of a $\sim$-algebra A [unitary representation V of a group G] is the direct sum of a finite number of irreducible $\sim$-representations [unitary representations]. If the original representation is a continuous mapping of A [or G] into the algebra of all operators on the representation space, so are the direct summands. [We sketch the proof for a representation T of A; the proof for V and G is similar. Let H be the representation space, of dimension n. The proposition is true and trivial for $n = 1$. Suppose that the proposition is true for all T and H such that $\dim(H) < n$. If T is reducible, let H_1, $0 < \dim(H_1) < n$, be an invariant subspace. Then $H_1^{\perp}$ is also invariant (21.9) and $0 < \dim(H_1^{\perp}) < n$. By our inductive hypothesis, T is the direct sum of irreducible components on H_1 and $H_1^{\perp}$, and therefore also on H. Continuity of the irreducible components of T is obvious if T is continuous.]

(b) A finite-dimensional representation, even one that is continuous in the sense of (a), need not be the direct sum of irreducible representations. A simple example is the representation V,

$$z \to \begin{pmatrix} 1 & z \\ 0 & 1 \end{pmatrix},$$

of the additive group K by operators on K^2. [The subspace $H_1 = \{(\xi, 0) : \xi \in K\}$ of K^2 is invariant, so that V is reducible. The orthogonal subspace $H_1^\perp = \{(0, \zeta) : \zeta \in K\}$ is not invariant, and H_1 is easily shown to be the only 1-dimensional subspace of K^2 invariant under V. Thus V is not a direct sum of irreducible representations.]

(c) Consider the algebra $\mathfrak{A}$ of all matrices $\begin{pmatrix} \alpha & \beta \\ 0 & \alpha \end{pmatrix}$, where $\alpha, \beta \in K$, and let $\begin{pmatrix} \alpha & \beta \\ 0 & \alpha \end{pmatrix}^\sim = \begin{pmatrix} \bar{\alpha} & \bar{\beta} \\ 0 & \bar{\alpha} \end{pmatrix}$. Then $\mathfrak{A}$ is a $\sim$-algebra, and the identity mapping of $\mathfrak{A}$ onto itself is a representation that is reducible but is not the direct sum of irreducible representations. [Argue as in (b).]

(d) Uniqueness of the irreducible components of the representation V in (a) will be discussed in Vol. II, (27.30); see also (27.44) and (27.56).

Notes

The theory of representations of groups and algebras has a long history; see for example the notes to WEYL [3]. The definition of $\sim$-algebras is due to GEL'FAND and NAĬMARK [1], and much of the present section is taken from GEL'FAND and NAĬMARK [2]. Positive functionals on $\boldsymbol{M}_a(G)$ make their appearance in GEL'FAND and RAĬKOV [2], on $\sim$-algebras in GEL'FAND and NAĬMARK [1]. Cyclic $\sim$-representations of $\sim$-algebras with unit were defined by GEL'FAND and NAĬMARK [1], and of groups by GODEMENT [1]. Theorem (21.11) is due to GODEMENT [1]. Theorem (21.18) is due to GODEMENT [2]; (21.19), (21.20) and (21.22) [for the case in which A has a unit] to GEL'FAND and NAĬMARK [2]. So far as we know, the function $M(p)$ defined in (21.20) is new, as are its applications in (21.34) and (21.39). We find throughout the present section that some care is needed in going from the case in which A has a unit to the general case which we treat. Theorem (21.23) is due to GEL'FAND and NAĬMARK [2]; its analogue for cyclic unitary representations of groups to GODEMENT [1]. I. E. SEGAL [2] has obtained the main results of the present section, for norm-closed $\sim$-algebras of operators on Hilbert spaces [C^*-algebras]. See also RAĬKOV [4]. Part of the proof of (21.32) is taken from RICKART [1].

The main theorem of the present section, (21.37), is due to GEL'FAND and NAĬMARK [2] for Banach $\sim$-algebras with unit, and to I. E. SEGAL [2] for C^*-algebras. Both proofs, and of course ours as well, depend upon the KREĬN-MIL'MAN theorem.

§22. Unitary representations of locally compact groups

In the present section we study the connection between representations of a locally compact group G and representations of various subalgebras of $\boldsymbol{M}(G)$. This culminates in the famous GEL'FAND-RAĬKOV theorem (22.12), which establishes the existence of a large number of continuous irreducible unitary representations of G.

(22.1) Notation. Throughout (22.1)—(22.21), G will denote an arbitrary locally compact group. The symbol E will denote an arbitrary reflexive complex Banach space, and $E^{\sim}$ the linear space of all bounded *conjugate-linear* functionals on E. For $\omega \in E^{\sim}$ and $\xi \in E$, we write the value of ω at ξ as $\langle\omega, \xi\rangle$. As pointed out in (B.52), every functional in $E^{\sim\sim}$ has the form $\omega \rightarrow \overline{\langle\omega, \xi\rangle}$ for some fixed $\xi \in E$. We therefore write $\langle\xi, \omega\rangle = \overline{\langle\omega, \xi\rangle}$ for $\xi \in E$ and $\omega \in E^{\sim}$, and identify $E^{\sim\sim}$ with E. Except in (22.18), the symbol H will denote a Hilbert space, which may be arbitrary or may be specified in the context. We identify $H^{\sim}$ with H in the usual way, writing as always $\langle\zeta, \xi\rangle$ for the inner product of ζ and ξ in H.

For notational reasons it is convenient to consider representations of G by operators on $E^{\sim}$ instead of on E. Since we are concerned only with reflexive Banach spaces, this is a matter of convenience, nothing more. If E is a Hilbert space H, the distinction vanishes.

(22.2) Definition. Let V be a representation of G by bounded operators on $E^{\sim}$. If the function $x \rightarrow \langle V_x \omega, \xi\rangle$ is Borel measurable [λ-measurable, ...] for all $\omega \in E^{\sim}$ and $\xi \in E$, then V is said to be *weakly Borel measurable* [*weakly λ-measurable*, ...]. If the function $x \rightarrow \langle V_x \omega, \xi\rangle$ is continuous for all $\omega \in E^{\sim}$ and $\xi \in E$, then V is said to be *weakly continuous*. If the mapping $x \rightarrow V_x \omega$ is a continuous mapping of G into the Banach space $E^{\sim}$ for each $\omega \in E^{\sim}$, then V is said to be *strongly continuous*.

We first show how to obtain representations of $\boldsymbol{M}(G)$ and its subalgebras from certain representations of G.

(22.3) Theorem. *Let $\boldsymbol{A}$ be a subalgebra of $\boldsymbol{M}(G)$, and let V be a representation of G by bounded operators on $E^{\sim}$ such that:*

(i) *V is weakly $|\mu|$-measurable and weakly $|\mu| * |\nu|$-measurable for all $\mu, \nu \in \boldsymbol{A}$;*

(ii) *$\alpha = \sup\{\|V_x\| : x \in G\}$ is finite.*

Then for every $\mu \in \boldsymbol{A}$, there is a unique operator T_μ on $E^{\sim}$ such that

(iii) $\langle T_\mu \omega, \xi\rangle = \int_G \langle V_x \omega, \xi\rangle \, d\mu(x)$ *for $\omega \in E^{\sim}$ and $\xi \in E$.*

The mapping $\mu \rightarrow T_\mu$ is a representation of $\boldsymbol{A}$ by bounded operators on $E^{\sim}$, for which

(iv) $\|T_\mu\| \leqq \alpha \|\mu\|$ *for $\mu \in \boldsymbol{A}$.*

Proof. The integral in (iii) plainly exists and is a complex number for every $\omega \in E^{\sim}$ and $\xi \in E$. For a fixed $\omega \in E^{\sim}$, consider the mapping

$$\xi \rightarrow \int_G \langle V_x \omega, \xi \rangle \, d\mu(x) \tag{1}$$

of E into K. This mapping is obviously conjugate-linear, and as

$$\left| \int_G \langle V_x \omega, \xi \rangle \, d\mu(x) \right| \leqq \sup\{\|V_x\| : x \in G\} \, \|\omega\| \cdot \|\xi\| \cdot \|\mu\|, \tag{2}$$

it is bounded. Hence there is an element of $E^{\sim}$, which we write as $T_\mu \omega$, such that

$$\langle T_\mu \omega, \xi \rangle = \int_G \langle V_x \omega, \xi \rangle \, d\mu(x)$$

for all $\xi \in E$. The linearity of $\langle \omega, \xi \rangle$ in ω and the linearity of $\int_G \cdots d\mu$ show at once that the mapping $\omega \rightarrow T_\mu \omega$ is a linear mapping of $E^{\sim}$ into itself for each fixed μ. The inequality (iv) follows readily from (2). It is also clear that $T_{\mu+\nu} = T_\mu + T_\nu$ and $T_{\alpha\mu} = \alpha T_\mu$ for $\mu, \nu \in \boldsymbol{A}$ and $\alpha \in K$.

We now show that $T_{\mu * \nu} = T_\mu T_\nu$. Using (19.10), we have

$$\begin{aligned}
\langle T_{\mu*\nu} \omega, \xi \rangle &= \int_G \int_G \langle V_{xy} \omega, \xi \rangle \, d\nu(y) \, d\mu(x) \\
&= \int_G \int_G \langle V_x V_y \omega, \xi \rangle \, d\nu(y) \, d\mu(x) \\
&= \int_G \int_G \langle V_y \omega, V_x^{\sim} \xi \rangle \, d\nu(y) \, d\mu(x) = \int_G \langle T_\nu \omega, V_x^{\sim} \xi \rangle \, d\mu(x) \\
&= \int_G \langle V_x T_\nu \omega, \xi \rangle \, d\mu(x) = \langle T_\mu T_\nu \omega, \xi \rangle.
\end{aligned}$$
[1] □

(22.4) Note. If $\varepsilon_a \in \boldsymbol{A}$ for some $a \in G$, then it is clear that $T_{\varepsilon_a} = V_a$. In particular, $T_{\varepsilon_e} = V_e = I$ [see (21.3)], so that T obeys the convention of (21.5) for representations of algebras having a unit.

We can also construct the adjoint of T_μ, which naturally is an operator on E.

(22.5) Theorem. *Let $\boldsymbol{A}$ and V be as in (22.3). For all $x \in G$, let U_x be the operator $V_{x^{-1}}^{\sim}$ acting on E. The mapping $x \rightarrow U_x$ is a representation of G satisfying (22.3.ii) and (22.3.i) for the algebra $\boldsymbol{A}^{\sim} = \{\mu^{\sim} : \mu \in \boldsymbol{A}\}$. Let S be the representation of the algebra $\boldsymbol{A}^{\sim}$ constructed from U by the process of (22.3). That is, we have*

(i) $\langle S_{\mu^\sim} \xi, \omega \rangle = \int_G \langle U_x \xi, \omega \rangle \, d\mu^{\sim}(x)$ *for* $\mu \in \boldsymbol{A}$.

The relation

(ii) $(T_\mu)^{\sim} = S_{\mu^\sim}$

[1] Our hypothesis that E be reflexive is needed in this computation. Otherwise, the adjoint $V_x^{\sim}$ could not be regarded as an operator on E; see (B.52).

holds for all $\mu \in \boldsymbol{A}$. *If* E *is a Hilbert space, so that* $E = E^{\sim} = H$, *if* $\mu^{\sim} \in \boldsymbol{A}$ *whenever* $\mu \in \boldsymbol{A}$, *and if all of the operators* V_x *are unitary*[1], *then* $S_\mu = T_\mu$ *and accordingly* T *is a* $\sim$*-representation of the* $\sim$*-algebra* $\boldsymbol{A}$.

Proof. It is obvious that U is a representation of G. Since $\langle U_x \xi, \omega\rangle = \overline{\langle V_{x^{-1}}\omega, \xi\rangle}$, (20.23.i) shows that (22.3.i) holds for all $\mu^{\sim}, \nu^{\sim} \in \boldsymbol{A}^{\sim}$. Since $\|U_x\| = \|V_{x^{-1}}\|$, the relation (22.3.ii) holds with the same value of α. Noting again that $\langle \xi, \omega\rangle = \overline{\langle \omega, \xi\rangle}$, we manipulate (i) as follows for arbitrary $\mu \in \boldsymbol{A}$:

$$\begin{aligned}\langle \omega, S_{\mu^{\sim}}\xi\rangle &= \overline{\langle S_{\mu^{\sim}}\xi, \omega\rangle} = \overline{\int_G \langle U_x \xi, \omega\rangle \, d\mu^{\sim}(x)} \\ &= \overline{\int_G \overline{\langle \omega, V_{x^{-1}}^{\sim}\xi\rangle}\, d\mu^{\sim}(x)} = \overline{\int_G \overline{\langle V_{x^{-1}}\omega, \xi\rangle}\, d\mu^{\sim}(x)} = \langle T_\mu \omega, \xi\rangle.\end{aligned}$$

This proves (ii).

If $E = E^{\sim} = H$ and V_x is unitary, then $V_{x^{-1}}^{\sim} = V_x$, and thus $T_\mu = S_\mu$. Equality (ii) then implies that $T_\mu^{\sim} = T_{\mu^{\sim}}$, *i.e.*, T is a $\sim$-representation. □

(22.6) Theorem. *Let* $\boldsymbol{A}$ *be a* $\sim$*-subalgebra of* $\boldsymbol{M}(G)$, *let* V *be a representation of* G *by unitary operators on a Hilbert space* H *satisfying* (22.3.i), *and let* T *be the* $\sim$*-representation of* $\boldsymbol{A}$ *constructed as in* (22.3). *Then every closed subspace of* H *invariant under* V *is invariant under* T. *A partial converse of this holds,* viz.: *let* V *be a weakly continuous unitary representation of* G *and let* $\boldsymbol{A}$ *have the property that for every nonvoid open subset* U *of* G, *there is a nonnegative* $\mu \in \boldsymbol{A}$ *such that* $\mu(U) = 1$ *and* $\mu(U') = 0$. *Then every closed subspace of* H *invariant under* T *is also invariant under* V.

Proof. Let H_1 be a closed subspace of H invariant for all V_x, and let P be the projection operator onto H_1: thus $PV_x = V_x P$ for all $x \in G$, by (21.28). Then for all $\mu \in \boldsymbol{A}$ and $\xi, \eta \in H$, we have

$$\begin{aligned}\langle T_\mu P\xi, \eta\rangle &= \int_G \langle V_x P\xi, \eta\rangle \, d\mu(x) = \int_G \langle PV_x\xi, \eta\rangle \, d\mu(x) \\ &= \int_G \langle V_x\xi, P\eta\rangle \, d\mu(x) = \langle T_\mu\xi, P\eta\rangle = \langle PT_\mu\xi, \eta\rangle.\end{aligned}$$

Thus $T_\mu P = PT_\mu$, and by (21.27) H_1 is invariant for all T_μ.

To prove the second statement of the theorem, we consider a subspace H_1 invariant for T and the projection operator P onto H_1. Then for all $\xi, \eta \in H$ and $\mu \in \boldsymbol{A}$, we have

$$\left.\begin{aligned}\int_G \langle PV_x\xi, \eta\rangle \, d\mu(x) &= \langle T_\mu\xi, P\eta\rangle = \langle PT_\mu\xi, \eta\rangle \\ &= \langle T_\mu P\xi, \eta\rangle = \int_G \langle V_x P\xi, \eta\rangle \, d\mu(x).\end{aligned}\right\} \quad (1)$$

Consider the bounded continuous function $f(x) = \langle PV_x\xi, \eta\rangle - \langle V_x P\xi, \eta\rangle$ If $f(a) \neq 0$ and [say] $\operatorname{Re} f(a) > 0$, then $\operatorname{Re} f(x) \geqq \beta > 0$ for all x in a

[1] In this case, (22.3.ii) is automatic: $\|V_x\| = 1$ for all x.

neighborhood U of a. Choosing $\mu \in \boldsymbol{A}$ such that $\mu \geqq 0$, $\mu(U)=1$, and $\mu(U')=0$, we have

$$\int_G f(x)\, d\mu(x) = \int_U \operatorname{Re} f(x)\, d\mu(x) + i \int_U \operatorname{Im} f(x)\, d\mu(x) \neq 0,$$

which contradicts (1). Therefore $PV_x = V_x P$ for all $x \in G$; *i.e.*, H_1 is invariant under V. ▯

Theorem (22.3) does not tell the whole story on representations of $\boldsymbol{M}(G)$ and its subalgebras. In fact, if G is nondiscrete, $\boldsymbol{M}(G)$ has a $\sim$-representation *not* of the form (22.3.iii) [see (23.28.a)]. However, for certain very special subalgebras $\boldsymbol{A}$ of $\boldsymbol{M}(G)$, all $\sim$-representations of $\boldsymbol{A}$ are obtained from representations of G as in (22.3).

(22.7) Theorem. *Let $\boldsymbol{A}$ be a Banach $\sim$-subalgebra of $\boldsymbol{M}(G)$ with the following properties:*

(i) *$\mu \in \boldsymbol{A}$ and $a \in G$ imply $\varepsilon_a * \mu \in \boldsymbol{A}$;*

(ii) *every element of the conjugate space $\boldsymbol{A}^*$ of $\boldsymbol{A}$ has the form $\mu \to \int_G h\, d\mu$, where h is a bounded function on G such that $h\xi_B$ is Borel measurable for every σ-compact set $B \subset G$.*

Let $\mu \to T_\mu$ be a $\sim$-representation of $\boldsymbol{A}$ by bounded operators on a Hilbert space H such that for every $\xi \neq 0$ in H, there is a $\mu \in \boldsymbol{A}$ for which $T_\mu \xi \neq 0$. Then there is a representation V of G by unitary operators on H such that (22.3.i) *holds and such that*

(iii) *$\langle T_\mu \xi, \eta\rangle = \int_G \langle V_x \xi, \eta\rangle\, d\mu(x)$ for $\mu \in \boldsymbol{A}$ and $\xi, \eta \in H$.*

Proof. (I) Suppose first that T is a cyclic representation with cyclic vector $\zeta \in H$: thus the linear subspace $H' = \{T_\mu \zeta : \mu \in \boldsymbol{A}\}$ of H is dense in H. Consider the mapping

$$T_\mu \zeta \to T_{\varepsilon_x * \mu} \zeta \tag{1}$$

carrying H' into H', for a fixed $x \in G$. Let us show that (1) is well defined; that is, if $T_\mu \zeta = T_\nu \zeta$, then $T_{\varepsilon_x * \mu} \zeta = T_{\varepsilon_x * \nu} \zeta$. Using the fact that $(\varepsilon_x)^\sim = \varepsilon_{x^{-1}}$ (20.23), and evident properties of the representation T, we can write

$$\left.\begin{aligned} \langle T_{\varepsilon_x*(\mu-\nu)} \zeta, T_{\varepsilon_x*(\mu-\nu)} \zeta\rangle &= \langle T_{(\mu-\nu)^\sim * \varepsilon_{x^{-1}} * \varepsilon_x * (\mu-\nu)} \zeta, \zeta\rangle \\ &= \langle T_{\mu-\nu} \zeta, T_{\mu-\nu} \zeta\rangle = 0. \end{aligned}\right\} \tag{2}$$

Therefore (1) defines a [single-valued] mapping of H' into itself. We define the operator V_x' on H' by

$$V_x'(T_\mu \zeta) = T_{\varepsilon_x * \mu} \zeta. \tag{3}$$

As in (2), we see that

$$\langle V_x' T_\mu \zeta, V_x' T_\nu \zeta\rangle = \langle T_\mu \zeta, T_\nu \zeta\rangle.$$

Given $T_\mu \zeta$ in H', $T_{\varepsilon_{x^{-1}} * \mu} \zeta$ also belongs to H' and $V_x'(T_{\varepsilon_{x^{-1}} * \mu} \zeta) = T_\mu \zeta$. Since V_x' is plainly linear, V_x' is a linear isometry of H' onto itself [see

(B.42)]. The equalities $V_e' = I$ and $V_x' V_y' = V_{xy}'$ [for all x, y in G] are evident. As pointed out in (B.59), each V_x' admits a unique extension V_x over H that is a unitary operator on H. It is easy to verify that $V_e = I$, and that $V_x V_y = V_{xy}$: *i.e.*, V is a unitary representation of G.

Like all $\sim$-representations of Banach $\sim$-algebras, T is bounded (21.22), and $\|T_\mu\| \leqq \|\mu\|$. Hence for every η in H, the mapping

$$\mu \to \langle T_\mu \zeta, \eta \rangle$$

is a bounded linear functional on $\boldsymbol{A}$. Let h be a function as in (ii) such that

$$\langle T_\mu \zeta, \eta \rangle = \int_G h(y)\, d\mu(y) \quad \text{for all} \quad \mu \in \boldsymbol{A}. \tag{4}$$

By (19.10), we have

$$\langle V_x T_\mu \zeta, \eta \rangle = \langle T_{\varepsilon_x * \mu} \zeta, \eta \rangle = \int_G h(y)\, d(\varepsilon_x * \mu)(y) = \int_G h(xy)\, d\mu(y). \tag{5}$$

Let μ and ν be any measures in $\boldsymbol{M}(G)$. Let B and C be σ-compact subsets of G such that $|\mu|(B') = |\nu|(C') = 0$. Then it is clear that $|\mu| \times |\nu|((B \times C)') = 0$. Since $B \times C \subset \{(x, y) \in G \times G : xy \in BC\}$, we see that

$$h(xy) = h(xy)\, \xi_{BC}(xy)$$

for $|\mu| \times |\nu|$-almost all $(x, y) \in G \times G$. Since $h\xi_{BC}$ is Borel measurable by hypothesis (ii), we infer that the function $(x, y) \to h(xy)$ is in $\mathfrak{L}_1(|\mu| \times |\nu|)$. By (13.8), we know that the function

$$x \to \int_G h(xy)\, d\nu(y) = \langle V_x T_\nu \zeta, \eta \rangle$$

is $|\mu|$-measurable for all $\mu \in \boldsymbol{M}(G)$. Since vectors $T_\nu \zeta$ ($\nu \in \boldsymbol{A}$) are dense in H, it follows that the function $x \to \langle V_x \xi, \eta \rangle$ is $|\mu|$-measurable for all $\mu \in \boldsymbol{M}(G)$: thus (22.3.i) holds.

Let us now establish (iii). Fix $\nu \in \boldsymbol{A}$. Then for every $\mu \in \boldsymbol{A}$, h is plainly in $\mathfrak{L}_1(|\mu| * |\nu|)$. Thus we can apply (19.10) and (5) to write

$$\begin{aligned}\int_G \langle V_x T_\nu \zeta, \eta \rangle\, d\mu(x) &= \int_G \int_G h(xy)\, d\nu(y)\, d\mu(x) \\ &= \int_G h(z)\, d\mu * \nu(z) = \langle T_{\mu * \nu} \zeta, \eta \rangle = \langle T_\mu T_\nu \zeta, \eta \rangle.\end{aligned}$$

This proves (iii) for the case $\xi = T_\nu \zeta$ for some $\nu \in \boldsymbol{A}$. Such vectors being dense in H, we infer that (iii) holds for all ξ, η in H.

(II) Suppose now that T is not a cyclic representation. We apply Theorem (21.13) to T. The subspace N of (21.13) is $\{0\}$, and so H is the direct sum of closed invariant cyclic subspaces H_γ. Let $T_\mu^{(\gamma)}$ denote the operator T_μ with its domain restricted to H_γ. By (I), there is a unitary representation $V^{(\gamma)}$ of G by operators on H_γ such that

$$\langle T_\mu^{(\gamma)} \xi_\gamma, \eta_\gamma \rangle = \int_G \langle V_x^{(\gamma)} \xi_\gamma, \eta_\gamma \rangle\, d\mu(x) \tag{6}$$

for $\mu \in \boldsymbol{A}$ and all $\xi_\gamma, \eta_\gamma \in H_\gamma$. Every vector $\xi \in H$ has a unique representation $\xi = \sum_{\gamma \in \Gamma} \xi_\gamma$, where $\xi_\gamma \in H_\gamma$, only a countable number of ξ_γ are nonzero, the series is convergent in the norm of H, and $\|\xi\|^2 = \sum_{\gamma \in \Gamma} \|\xi_\gamma\|^2$. Let V_x be the operator on H such that $V_x \xi = \sum_{\gamma \in \Gamma} V_x^{(\gamma)} \xi_\gamma$. It is easy to see that $x \to V_x$ is a unitary representation of G that is weakly $|\mu|$-measurable for all $\mu \in \boldsymbol{M}(G)$ and for which (iii) holds. □

We next discuss continuity of representations V of G.

(22.8) Theorem. *Let V be a weakly λ-measurable representation of G by bounded operators on $E^\sim$ for which* (22.3.ii) *holds, and suppose that the function $x \to \langle V_x \omega, \xi \rangle$ is continuous at e for all $\omega \in E^\sim$ and $\xi \in E$. Then the representation V is strongly continuous. In fact, for every $\varrho \in E^\sim$ and $\varepsilon > 0$, there is a neighborhood U of e such that*

(i) $\|V_x \varrho - V_y \varrho\| < \varepsilon$ *for $x, y \in G$ such that $y^{-1} x \in U$.*

That is, the mapping $x \to V_x \varrho$ is a left uniformly continuous mapping of G into the metric space $E^\sim$. In particular, every weakly continuous V satisfying (22.3.ii) *is strongly continuous.*

Proof. Let T be the representation of $\boldsymbol{M}_a(G)$ constructed from the representation V as in (22.3). We first show that if $\varrho \in E^\sim$, then $\varrho \in \{T_\mu \varrho : \mu \in \boldsymbol{M}_a(G)\}^-$. Assume that $\varrho \in E^\sim$ and that $\varrho \notin \{T_\mu \varrho : \mu \in \boldsymbol{M}_a(G)\}^-$. As in (B.15), there is an element ξ of E such that

$$\langle T_\mu \varrho, \xi \rangle = 0 \quad \text{for all} \quad \mu \in \boldsymbol{M}_a(G), \tag{1}$$

and

$$\langle \varrho, \xi \rangle = 1 \tag{2}$$

[recall that E is reflexive]. From (2) and the continuity of $\langle V_x \varrho, \xi \rangle$ at e, we infer that there is a neighborhood W of e such that $\lambda(W) < \infty$ and $\operatorname{Re} \langle V_x \varrho, \xi \rangle > \frac{1}{2}$ for all $x \in W$. Let ν be the measure in $\boldsymbol{M}_a(G)$ such that $d\nu = \xi_W \, d\lambda$ [ξ_W is the characteristic function of W]. Then applying (22.3.iii), we have

$$\operatorname{Re} \langle T_\nu \varrho, \xi \rangle = \operatorname{Re} \int_W \langle V_x \varrho, \xi \rangle \, dx = \int_W \operatorname{Re} \langle V_x \varrho, \xi \rangle \, dx > \tfrac{1}{2} \lambda(W).$$

This contradicts (1).

Now consider any $x \in G$, $\varrho \in E^\sim$, $\xi \in E$ and $\mu \in \boldsymbol{M}_a(G)$. By (22.3.iii) and (19.10), we have

$$\begin{aligned} \langle T_{\varepsilon_x * \mu} \varrho, \xi \rangle &= \int_G \langle V_u \varrho, \xi \rangle \, d\varepsilon_x * \mu(u) = \int_G \langle V_{xy} \varrho, \xi \rangle \, d\mu(y) \\ &= \int_G \langle V_y \varrho, V_x^\sim \xi \rangle \, d\mu(y) = \langle T_\mu \varrho, V_x^\sim \xi \rangle = \langle V_x T_\mu \varrho, \xi \rangle . \end{aligned}$$

Thus we have

$$|\langle V_x T_\mu \varrho - T_\mu \varrho, \xi \rangle| = |\langle (T_{\varepsilon_x * \mu} - T_\mu) \varrho, \xi \rangle| \leqq \alpha \|\varrho\| \cdot \|\xi\| \, \|\varepsilon_x * \mu - \mu\|, \tag{3}$$

where the number α is as in (22.3.ii). Now if $d\mu = f\,d\lambda$, with $f \in \mathfrak{L}_1(G)$, then $d(\varepsilon_x * \mu) = {}_{x^{-1}}f\,d\lambda$, as (20.9.i) shows. From (20.4.i) we infer that there is a neighborhood U of e such that

$$\alpha \|\varrho\| \cdot \|\varepsilon_x * \mu - \mu\| < \frac{\varepsilon}{2\alpha} \tag{4}$$

for all $x \in U$. Combining (3) and (4), we infer that

$$\|V_x T_\mu \varrho - T_\mu \varrho\| \leqq \frac{\varepsilon}{2\alpha} \tag{5}$$

for all $x \in U$.

As shown already, there is a $\mu \in \boldsymbol{M}_a(G)$ such that

$$(1+\alpha)\|\varrho - T_\mu \varrho\| < \frac{\varepsilon}{2\alpha}. \tag{6}$$

If $x \in U$, where U is chosen for the μ just selected, (5) and (6) give

$$\begin{aligned}\|V_x\varrho - \varrho\| &\leqq \|V_x\varrho - V_x T_\mu \varrho\| + \|V_x T_\mu \varrho - T_\mu \varrho\| + \|T_\mu \varrho - \varrho\| \\ &\leqq (\alpha+1)\|\varrho - T_\mu \varrho\| + \frac{\varepsilon}{2\alpha} < \frac{\varepsilon}{\alpha}.\end{aligned}$$

Finally, if $x, y \in G$ and $y^{-1}x \in U$, we have

$$\|V_x\varrho - V_y\varrho\| \leqq \|V_y\| \cdot \|V_{y^{-1}x}\varrho - \varrho\| < \varepsilon. \quad \square$$

(22.9) Note. In view of (22.8) and an obvious calculation, a representation V of G satisfying (22.3.ii) is strongly continuous if and only if it is weakly continuous. We will henceforward write "continuous" instead of "strongly continuous" and "weakly continuous" in referring to such V's.

(22.10) Theorem. *Let T be a $\sim$-representation of the algebra $\boldsymbol{M}_a(G)$ by operators on a Hilbert space H such that for every $\xi \neq 0$ in H, there is a $\mu \in \boldsymbol{M}_a(G)$ for which $T_\mu \xi \neq 0$. There is a continuous unitary representation V of G by operators on H such that* (22.7.iii) *holds. The representation V is determined by* (22.7.iii) *among all continuous unitary representations.*

Proof. (I) For typographical reasons, this step of the proof is given on page 508. (II) By (20.23.iv), $\boldsymbol{M}_a(G)$ is a $\sim$-subalgebra of $\boldsymbol{M}(G)$. By (19.18), it is a closed two-sided ideal in $\boldsymbol{M}(G)$. By (12.18), every bounded linear functional on $\boldsymbol{M}_a(G)$ has the form

$$\mu \to \int_G h(y)\,d\mu(y) = \int_G h(y)\,f(y)\,dy,$$

where $h \in \mathfrak{L}_\infty(G)$ and f is the function in $\mathfrak{L}_1(G)$ such that $d\mu = f\,d\lambda$. By (I), we may suppose that $h\xi_B$ is Borel measurable for every σ-compact set $B \subset G$. Thus $\boldsymbol{M}_a(G)$ satisfies the hypotheses of (22.7). Let V be the unitary representation of G constructed in (22.7) for which (22.7.iii) holds.

We will show that V is continuous. Consider first the case in which T is cyclic with cyclic vector ζ. For $\mu \in \boldsymbol{M}_a(G)$, with $f \in \mathfrak{L}_1(G)$ as above, and $x \in G$, we have from (21.22) and (20.9.i) that

$$\|V_x T_\mu \zeta - T_\mu \zeta\| = \|T_{\varepsilon_x * \mu} \zeta - T_\mu \zeta\| \leqq \|\varepsilon_x * \mu - \mu\| \cdot \|\zeta\| = \|{}_{x^{-1}}f - f\|_1 \cdot \|\zeta\|.$$

By (20.4.i), the last expression is arbitrarily small for x in a suitable neighborhood of e. Since vectors $T_\mu \zeta$ are dense in H, we find that

$$\|V_x \xi - \xi\| < \varepsilon \tag{1}$$

for all $x \in U$, where U is a neighborhood of e that depends upon ξ and ε.

For arbitrary T, we use the construction and notation of part (II) of (22.7). For $\xi \in H$, we have $\|V_x \xi - \xi\|^2 = \sum_{\gamma \in \Gamma} \|V_x^{(\gamma)} \xi_\gamma - \xi_\gamma\|^2$. This obviously implies (1) for arbitrary T. For arbitrary $y \in G$, we have

$$\|V_{yx} \xi - V_y \xi\| = \|V_x \xi - \xi\|,$$

so that the mapping $z \to V_z \xi$ is a left uniformly continuous mapping of G into the metric space H.

Suppose that $V^{(1)}$ and $V^{(2)}$ are continuous unitary representations of G such that

$$\int_G \langle V_x^{(1)} \xi, \eta \rangle f(x)\, dx = \int_G \langle V_x^{(2)} \xi, \eta \rangle f(x)\, dx$$

for all $f \in \mathfrak{L}_1(G)$. The argument used in (22.6) shows that $\langle V_x^{(1)} \xi, \eta \rangle = \langle V_x^{(2)} \xi, \eta \rangle$ for all $\xi, \eta \in H$ and $x \in G$. Thus $V_x^{(1)} = V_x^{(2)}$ for all $x \in G$. □

We need one more preliminary to our main theorem.

(22.11) Theorem. *For $\mu \in \boldsymbol{M}(G)$ and $\varphi \in \mathfrak{L}_2(G)$, let $T_\mu \varphi = \mu * \varphi$. Each T_μ is a bounded operator on the Hilbert space $\mathfrak{L}_2(G)$, and the mapping $\mu \to T_\mu$ is a faithful*[1] *$\sim$-representation of $\boldsymbol{M}(G)$.* [2]

Proof. The linearity of T_μ on $\mathfrak{L}_2(G)$ is obvious from (20.12.i), and the boundedness of T_μ, with $\|T_\mu\| \leqq \|\mu\|$, follows from (20.12.ii). For $\varphi \in \mathfrak{L}_1(G) \cap \mathfrak{L}_2(G)$, we have

$$(\mu * \nu) * \varphi = \mu * (\nu * \varphi)$$

from (19.2.iv). Thus $T_{\mu * \nu}(\varphi) = T_\mu(T_\nu \varphi)$ for all $\varphi \in \mathfrak{L}_1(G) \cap \mathfrak{L}_2(G)$. Since $\mathfrak{C}_{00}(G) \subset \mathfrak{L}_1(G) \cap \mathfrak{L}_2(G)$, the subspace $\mathfrak{L}_1(G) \cap \mathfrak{L}_2(G)$ is dense in $\mathfrak{L}_2(G)$ [see (12.10)]. It follows that $T_{\mu*\nu} = T_\mu T_\nu$. To show that $T_\mu \neq 0$ if $\mu \neq 0$, consider an $f \in \mathfrak{C}_{00}(G)$ such that $\int_G f^\star \, d\mu \neq 0$. The inequality

$$|\mu * f(y) - \mu * f(x)| \leqq \|{}_{y^{-1}}(f^\star) - {}_{x^{-1}}(f^\star)\|_u \, \|\mu\|$$

---

[1] That is, $T_\mu \neq 0$ if $\mu \neq 0$.

[2] We call this the *regular representation of* $\boldsymbol{M}(G)$. The mapping $a \to T_{\varepsilon_a}$ of G into operators on $\mathfrak{L}_2(G)$ is called the *regular representation of* G. For $f \in \mathfrak{L}_2(G)$, we have $T_{\varepsilon_a} f = {}_{a^{-1}}f$: see (20.12).

and the right uniform continuity of $f^\star$ (4.15) show that $\mu * f$ is continuous; also we have $\mu * f(e) = \int_G f^\star \, d\mu$. Thus $\|\mu * f\|_2^2 = \int_G |\mu * f|^2 \, d\lambda$ is positive; *i.e.*, $T_\mu f$ is not the zero element of $\mathfrak{L}_2(G)$.

It remains to show that T is a $\sim$-representation. For $\varphi, \psi \in \mathfrak{C}_{00}(G)$, we have by (20.23.iii) that

$$\left.\begin{aligned} \langle \varphi, T_{\mu^\sim} \psi \rangle &= \int_G \varphi(x) \int_G \overline{\psi(yx)} \, d\mu(y) \, dx \\ &= \int_G \int_G \varphi(x) \overline{\psi(yx)} \, dx \, d\mu(y) \\ &= \int_G \int_G \varphi(y^{-1}x) \overline{\psi(x)} \, dx \, d\mu(y) \\ &= \int_G \int_G \varphi(y^{-1}x) \, d\mu(y) \overline{\psi(x)} \, dx = \langle T_\mu \varphi, \psi \rangle. \end{aligned}\right\} \tag{1}$$

The two applications of FUBINI's theorem are justified easily because φ and ψ are in $\mathfrak{C}_{00}(G)$. Since $\mathfrak{C}_{00}(G)$ is dense in $\mathfrak{L}_2(G)$, (1) holds for all φ, ψ in $\mathfrak{L}_2(G)$, and this shows that $T_\mu^\sim = T_{\mu^\sim}$. □

(22.12) Theorem [GEL'FAND-RAĬKOV]. *Let G be a locally compact group. For every a in G different from e, there is a continuous, irreducible, unitary representation V of G such that $V_a \neq I$.*[1]

Proof. Let U be a symmetric neighborhood of e such that $\lambda(U) < \infty$ and $a \notin U^2$. Let μ be the measure in $\boldsymbol{M}_a(G)$ such that $d\mu = \xi_U \, d\lambda$ [recall that ξ_U is the characteristic function of U]. Formula (20.9.i) shows at once that $d(\varepsilon_a * \mu) = \xi_{aU} \, d\lambda$. Since $a \notin U^2$, we see that $\|\varepsilon_a * \mu - \mu\| = \|\xi_{aU} - \xi_U\|_1 = 2\lambda(U)$, so that $\varepsilon_a * \mu \neq \mu$. By (22.11), we thus have $T_{\varepsilon_a * \mu} \neq T_\mu$, where T is the regular representation of $\boldsymbol{M}(G)$, restricted to the $\sim$-algebra $\boldsymbol{M}_a(G)$. By Theorem (21.37) applied to $\boldsymbol{M}_a(G)$ and T, there is an irreducible $\sim$-representation, say S, of $\boldsymbol{M}_a(G)$ such that

$$S_{\varepsilon_a * \mu} \neq S_\mu . \tag{1}$$

Theorem (22.10) shows that there is a continuous unitary representation V of G such that

$$\langle S_\nu \xi, \eta \rangle = \int_G \langle V_x \xi, \eta \rangle \, d\nu(x) \tag{2}$$

for all $\nu \in \boldsymbol{M}_a(G)$ and all ξ, η in the representation space H of S and V. From (1) and (2) we have

$$\int_G \langle V_x \xi, \eta \rangle \, d\mu(x) \neq \int_G \langle V_x \xi, \eta \rangle \, d\varepsilon_a * \mu(x) \tag{3}$$

for some vectors ξ, η in H. Theorem (19.10) shows that

$$\int_G \langle V_x \xi, \eta \rangle \, d\varepsilon_a * \mu(x) = \int_G \langle V_{ax} \xi, \eta \rangle \, d\mu(x) = \int_G \langle V_a V_x \xi, \eta \rangle \, d\mu(x).$$

[1] We abbreviate this assertion by the statement that G has *sufficiently many* representations of the sort described: similarly with other homomorphisms of G.

If V_a were the operator I, (3) would fail. By (22.6), the representation V is irreducible: any closed subspace of H invariant under V is also invariant under the irreducible representation S, and hence is $\{0\}$ or H. □

For compact groups, Theorem (22.12) becomes particularly simple, as the following theorem shows.

(22.13) Theorem. *Every irreducible continuous representation V of a compact group G by unitary operators on a Hilbert space H is finite-dimensional.*

Proof. Let ξ, η, and ζ be in H and consider the integral $\int_G \langle V_x\zeta, \eta\rangle \overline{\langle V_x\zeta, \xi\rangle}\, dx$. If we regard ζ as fixed, this is a function that is linear in ξ and conjugate-linear in η. Moreover, we have

$$\left|\int_G \langle V_x\zeta, \eta\rangle \overline{\langle V_x\zeta, \xi\rangle}\, dx\right| \leqq \|\zeta\|^2 \|\eta\| \|\xi\|.$$

Therefore by (B.60), there is a bounded operator B_ζ on H such that

$$\langle B_\zeta \xi, \eta\rangle = \int_G \langle V_x\zeta, \eta\rangle \overline{\langle V_x\zeta, \xi\rangle}\, dx.$$

We next show that B_ζ commutes with every operator V_y, $y \in G$. The left invariance of the Haar integral shows that

$$\begin{aligned}\langle B_\zeta V_y\xi, \eta\rangle &= \int_G \langle V_x\zeta, \eta\rangle \overline{\langle V_x\zeta, V_y\xi\rangle}\, dx = \int_G \langle V_{y^{-1}x}\zeta, V_{y^{-1}}\eta\rangle \overline{\langle V_{y^{-1}x}\zeta, \xi\rangle}\, dx \\ &= \int_G \langle V_x\zeta, V_{y^{-1}}\eta\rangle \overline{\langle V_x\zeta, \xi\rangle}\, dx = \langle B_\zeta\xi, V_{y^{-1}}\eta\rangle = \langle V_y B_\zeta\xi, \eta\rangle,\end{aligned}$$

and hence $B_\zeta V_y = V_y B_\zeta$. By (21.30), B_ζ is a multiple of the identity operator I: $B_\zeta = \alpha(\zeta) I$. Thus we have

$$\int_G \langle V_x\zeta, \eta\rangle \overline{\langle V_x\zeta, \xi\rangle}\, dx = \alpha(\zeta)\langle \xi, \eta\rangle$$

and in particular

$$\int_G |\langle V_x\zeta, \xi\rangle|^2\, dx = \alpha(\zeta)\, \|\xi\|^2 \tag{1}$$

for all $\zeta, \xi \in H$. Interchanging ζ and ξ in (1) and using (20.2.i) [being compact, G is unimodular], we obtain

$$\begin{aligned}\alpha(\xi)\, \|\zeta\|^2 &= \int_G |\langle V_x\xi, \zeta\rangle|^2\, dx = \int_G |\langle \zeta, V_x\xi\rangle|^2\, dx \\ &= \int_G |\langle V_{x^{-1}}\zeta, \xi\rangle|^2\, dx = \int_G |\langle V_x\zeta, \xi\rangle|^2\, dx = \alpha(\zeta)\, \|\xi\|^2.\end{aligned}$$

Hence there is a constant c such that $\alpha(\xi) = c\|\xi\|^2$ for all $\xi \in H$. Setting $\xi = \zeta$ and $\|\xi\| = 1$ in (1), we find that

$$\int_G |\langle V_x\xi, \xi\rangle|^2\, dx = \alpha(\xi)\, \|\xi\|^2 = c\, \|\xi\|^4 = c.$$

Hence c is positive, since the continuous function $x \to |\langle V_x\xi, \xi\rangle|^2$ has the value 1 at e.

Now let $\vartheta_1, \ldots, \vartheta_n$ be orthonormal vectors in H, and apply (1) with $\zeta = \vartheta_k$ and $\xi = \vartheta_1$ ($k = 1, \ldots, n$). We obtain

$$\int_G |\langle V_x \vartheta_k, \vartheta_1 \rangle|^2 \, dx = \alpha(\vartheta_k) \, \|\vartheta_1\|^2 = c. \tag{2}$$

Summing (2) from 1 to n, using the fact that $\{V_x \vartheta_k\}_{k=1}^n$ is orthonormal, and applying BESSEL's inequality, we find that

$$n c = \sum_{k=1}^{n} \int_G |\langle V_x \vartheta_k, \vartheta_1 \rangle|^2 \, dx = \int_G \sum_{k=1}^{n} |\langle V_x \vartheta_k, \vartheta_1 \rangle|^2 \, dx \leqq \int_G \|\vartheta_1\|^2 \, dx = 1.$$

Consequently, the dimension of H does not exceed $1/c$. ▯

(22.14) Corollary. *Let G be a compact group. Then G has sufficiently many irreducible, continuous representations by unitary matrices.*

Proof. Combine (22.12) and (22.13), and identify the group of unitary transformations of K^n with its concrete representative $\mathfrak{U}(n)$ [choose any orthonormal basis in K^n]. ▯

(22.15) Definition. Let S be a semigroup. Let χ be a complex-valued function on S such that

$$\chi(xy) = \chi(x)\,\chi(y) \quad \text{for all} \quad x, y \in S.$$

Then χ is called a *multiplicative function on S*. If χ is also bounded and not identically zero, χ is called a *semicharacter of S*. A semicharacter of a group is called a *character*.

(22.16) Remarks. (a) A multiplicative function is a homomorphism of a semigroup S into the multiplicative semigroup K. A semicharacter is a homomorphism of S into the multiplicative semigroup $D = \{z \in K : |z| \leqq 1\}$ that is not identically 0. A character of a group G is a homomorphism of G into the group T.

(b) Let V be a 1-dimensional unitary representation of a group G. Then the representation space H of V has the form $\{\alpha\xi : \alpha \in K\}$ where ξ has norm 1. Defining $\chi(x)$ by the relation $V_x \xi = \chi(x)\xi$, we plainly obtain a character of G. Conversely, every character χ of G defines a 1-dimensional unitary representation of G, so that there is a one-to-one correspondence between all 1-dimensional unitary representations and all characters. It is clear that V is continuous if and only if the corresponding character is continuous. Henceforth we will identify 1-dimensional unitary representations of a group with characters of the group.

(22.17) Theorem. *Let G be a locally compact Abelian group. Then G admits sufficiently many continuous characters.*

Proof. In view of (22.12) and (22.16.b), we need only to prove that every irreducible unitary representation V of G is 1-dimensional. Let a be any element of G. Then the unitary operator V_a commutes with all of the operators V_x. Theorem (21.30) shows that V_a is a multiple of I: $V_a = \chi(a) I$ for all $a \in G$. Thus every subspace of the representation space H is invariant under V, so that H has to be 1-dimensional. □

Continuity of characters follows from λ-measurability. In fact, much more can be proved.

(22.18) Theorem. *Let G be a locally compact group with left Haar measure λ. Let H be a topological group that is either σ-compact or contains a countable dense subset [H need not be locally compact]. Let τ be a homomorphism of G into H such that for some λ-measurable set $A \subset G$ for which $0 < \lambda(A) < \infty$, $\tau^{-1}(U \cap \tau(G)) \cap A$ is λ-measurable for all open subsets U of H. Then τ is continuous on G.*

Proof. Let W_0 and W be symmetric neighborhoods of the identity in H such that $W^2 \subset W_0$. The hypotheses on H ensure that there is a countable subset $\{y_1, \ldots, y_n, \ldots\}$ of H such that $\bigcup_{n=1}^{\infty} W y_n = H$. We have $\bigcup_{n=1}^{\infty} (\tau^{-1}(W y_n) \cap A) = A$, and so there is at least one value of n for which $0 < \lambda(\tau^{-1}(W y_n) \cap A) < \infty$. By (20.17), there is a neighborhood V of the identity in G such that $V \subset (\tau^{-1}(W y_n) \cap A) \cdot (\tau^{-1}(W y_n) \cap A)^{-1}$. For $x \in V$, we have $x = a b^{-1}$, where $\tau(a) = w_1 y_n$ and $\tau(b) = w_2 y_n$, and where $w_1, w_2 \in W$. Hence $\tau(x) = \tau(a)\, \tau(b)^{-1} = w_1 w_2^{-1} \in W^2 \subset W_0$. Thus τ is continuous at the identity and therefore is continuous throughout G (5.40.a). □

(22.19) Corollary. *Let G be a locally compact group and τ a homomorphism of G into $\mathfrak{GL}(n, K)$; we write $\tau(x) = (a_{jk}(x))_{j,k=1}^{n}$. Suppose that there is a λ-measurable set $A \subset G$ such that $0 < \lambda(A) < \infty$ and each function $x \to a_{jk}(x)$ is λ-measurable on A. Then τ is continuous on G. In particular, every multiplicative complex function on G that is λ-measurable on such a set A is continuous.*

Proof. It is easy to show that $\{x \in A : (a_{jk}(x))_{j,k=1}^{n} \in \mathfrak{U}\}$ is λ-measurable for all open subsets $\mathfrak{U}$ of $\mathfrak{GL}(n, K)$. Then apply (22.18). □

Miscellaneous theorems and examples

(22.20) Continuous unitary representations. (a) Let G be a locally compact group and V a representation of G by unitary operators on a Hilbert space H. The following properties of V are equivalent:

(i) V is strongly continuous;

(ii) V is weakly continuous;

(iii) the mapping $x \to V_x \xi$ of G into H is left uniformly continuous for all $\xi \in H$;

(iv) the mapping $x \to V_x \xi$ of G into H is continuous at e for all $\xi \in H$;

(v) the function $x \to \langle V_x \eta, \xi\rangle$ is left uniformly continuous for all $\xi, \eta \in H$;

(vi) the function $x \to \langle V_x \xi, \xi\rangle$ is continuous at e for all $\xi \in H$.

[The equivalence of (i) and (ii) was noted in (22.9). Taking into account trivial implications, we see that it suffices to prove that (iv) implies (iii) and (vi) implies (v).

The equality
$$\|V_x \xi - V_y \xi\| = \|V_{y^{-1}x}\xi - \xi\|$$
shows that (iv) implies (iii). Suppose that (vi) holds. Then for $\xi, \eta \in H$ and $x, y \in G$, we have
$$\begin{aligned}\|V_x\eta - V_y\eta\|^2 &= \langle V_x\eta, V_x\eta\rangle - \langle V_y\eta, V_x\eta\rangle - \langle V_x\eta, V_y\eta\rangle + \langle V_y\eta, V_y\eta\rangle \\ &= 2\langle \eta,\eta\rangle - 2\operatorname{Re}\langle V_y\eta, V_x\eta\rangle \leqq 2|\langle\eta,\eta\rangle - \langle V_y\eta, V_x\eta\rangle| \\ &= 2|\langle V_e\eta,\eta\rangle - \langle V_{x^{-1}y}\eta,\eta\rangle|,\end{aligned}$$
and therefore
$$\begin{aligned}|\langle V_x\eta,\xi\rangle - \langle V_y\eta,\xi\rangle|^2 &\leqq \|V_x\eta - V_y\eta\|^2\,\|\xi\|^2 \\ &\leqq 2\|\xi\|^2\,|\langle V_e\eta,\eta\rangle - \langle V_{x^{-1}y}\eta,\eta\rangle|.\end{aligned}$$
This last inequality and (vi) imply (v).]

(b) Let G be a locally compact group and V a representation of G by unitary operators on a separable Hilbert space H. Then the following are equivalent:

(i) V is continuous;

(ii) the function $x \to \langle V_x\eta, \xi\rangle$ is λ-measurable for all $\eta, \xi \in H$;

(iii) the mapping $x \to V_x\xi$ of G into H is λ-measurable for each fixed $\xi \in H$; *i.e.*, the inverse image of any open set in H is λ-measurable.

[Obviously (i) implies (ii). For $\xi, \zeta \in H$, we have
$$\{x \in G : \|V_x\xi - \zeta\|^2 < \delta\} = \{x \in G : 2\operatorname{Re}\langle V_x\xi, \zeta\rangle > \langle\xi,\xi\rangle + \langle\zeta,\zeta\rangle - \delta\};$$
this shows that (ii) implies (iii). Suppose that (iii) holds. To prove (i), it suffices by (a) to show that $x \to V_x\xi$ is continuous at e, where $\xi \in H$ and $\xi \neq 0$. For $\varepsilon > 0$, let $A = \left\{x \in G : \|V_x\xi - \xi\| < \frac{\varepsilon}{2}\right\}$. Clearly A is λ-measurable and symmetric. If x and y are in A, then $\|V_{xy}\xi - \xi\| \leqq \|V_{xy}\xi - V_x\xi\| + \|V_x\xi - \xi\| = \|V_y\xi - \xi\| + \|V_x\xi - \xi\| < \varepsilon$, so that $A^2 \subset \{x \in G : \|V_x\xi - \xi\| < \varepsilon\}$. It remains to prove that A^2 contains a neighborhood of e. Since H is separable and metric, the subset $\{V_x\xi : x \in G\}$ of H is also separable. Thus there exists a sequence $\{x_n\}_{n=1}^{\infty}$ of elements in G such that $\{V_{x_n}\xi\}_{n=1}^{\infty}$ is dense in $\{V_x\xi : x \in G\}$. If $y \in G$, then for some positive integer k, we have $\|V_{x_k}\xi - V_y\xi\| = \|\xi - V_{x_k^{-1}y}\xi\| < \frac{\varepsilon}{2}$, and therefore $y \in x_k A$. In other words, $G = \bigcup_{n=1}^{\infty} x_n A$. It follows that A contains a compact subset F having positive measure. By (20.17), $F(F^{-1})$ contains a neighborhood of e.]

(c) (E. THOMA, oral communication.) Let G be a nondiscrete locally compact group. Then G admits a cyclic, weakly λ-measurable, unitary representation that is not continuous. [Let $l_2(G)$ be the Hilbert space of all functions ξ on G such that $\sum_{x\in G}|\xi(x)|^2<\infty$, with pointwise linear operations and $\langle\xi,\eta\rangle=\sum_{x\in G}\xi(x)\overline{\eta(x)}$. This is of course $\mathfrak{L}_2(G_d)$, where G_d denotes G with the discrete topology. Let δ be the function in $l_2(G)$ such that $\delta(e)=1$ and $\delta(x)=0$ for $x\neq e$. For $x\in G$, let V_x be the operator on $l_2(G)$ such that $V_x\xi(y)=\xi(x^{-1}y)$. It is obvious that the mapping $x\to V_x$ is a cyclic unitary representation of G, with cyclic vector δ: $V_x\delta$ is the function on G whose value at x is 1 and elsewhere is 0. Linear combinations of such functions are dense in $l_2(G)$. For $\xi,\eta\in l_2(G)$, let A and B be countable sets outside of which ξ and η, respectively, vanish. If $\langle V_x\xi,\eta\rangle\neq 0$, then x must be in the countable set $B(A^{-1})$. Hence the representation V is weakly λ-measurable. Since G is nondiscrete, (4.26) shows that the set $\{x\in G:\langle V_x\xi,\eta\rangle\neq 0\}$ is open if and only if it is void. Thus the function $x\to\langle V_x\xi,\eta\rangle$ is continuous if and only if it vanishes identically. Therefore V is not a weakly continuous representation.]

(d) The representation $\mu\to T_\mu$ constructed in (22.3) may well be identically 0: T_μ may be the zero operator for all $\mu\in\boldsymbol{A}$. [For example, consider the algebra $\boldsymbol{M}_c(G)$ for a nondiscrete G and apply (22.3) to $\boldsymbol{M}_c(G)$ and the representation V constructed in part (c).]

(22.21) The definition of the regular representation T of $\boldsymbol{M}(G)$ given in (22.11) is by no means arbitrary. Alternative candidates exhibit serious drawbacks. A first naïve candidate might be the following. For $\mu\in\boldsymbol{M}(G)$ and $\varphi\in\mathfrak{L}_2(G)$, define $S_\mu\varphi=\varphi*\mu$. With such a definition, S will not even satisfy the relation $S_{\mu*\nu}=S_\mu S_\nu$. To correct this, we redefine $S_\mu\varphi=\varphi*\mu^\sim$. This is still unsatisfactory since $\varphi\to S_\mu\varphi=\varphi*\mu^\sim$ might not map $\mathfrak{L}_2(G)$ into $\mathfrak{L}_2(G)$ (20.34). In any case, the relation $S_{\alpha\mu}=\alpha S_\mu$ will not hold for all $\mu\in\boldsymbol{M}(G)$ and $\alpha\in K$.

There is, to be sure, another natural $\sim$-representation S of $\boldsymbol{M}(G)$: use right Haar measure ϱ instead of λ and use the representation space $\mathfrak{L}_2(G,\varrho)$. In the notation of (20.32), this representation turns out to be $S_\mu\varphi=\overline{\overline{\varphi}\underset{r}{*}\mu^\sim}$ for $\varphi\in\mathfrak{L}_2(G,\varrho)$. The representation S is equivalent to the representation T given in (22.11). The connecting isometry W between $\mathfrak{L}_2(G,\lambda)$ and $\mathfrak{L}_2(G,\varrho)$ is $W(\varphi)=\varphi^\star$ (20.32.b).

(22.22) Groups with no finite-dimensional unitary representations. Let G be any topological group. We are interested here in those elements of G that can be separated from e by a continuous, finite-dimensional, irreducible, unitary representation of G. We denote by G_0 the set of elements in G that cannot be separated from e by a continuous, finite-dimensional, irreducible, unitary representation of G; *i.e.*, x belongs to G_0 if $V_x = I$ for all such representations V.

(a) If $V_x \neq I$ for some continuous, finite-dimensional, unitary representation V, then x is not in G_0. [By (21.40.a), V is the direct sum of irreducible representations $V^{(1)}, \ldots, V^{(m)}$, which are clearly also continuous. Then for some $j = 1, \ldots, m$, it must be the case that $V_x^{(j)} \neq I$.]

(b) The subset G_0 of G is a closed normal subgroup of G. [This follows from the identity $G_0 = \bigcap \{x \in G : V_x = I$ and V is a continuous, finite-dimensional, unitary representation of $G\}$.]

(c) If H is a closed normal subgroup of G such that G/H is compact, then $G_0 \subset H$. [Let y be in $G \cap H'$. By (22.14), there is a continuous, finite-dimensional, irreducible, unitary representation V' of G/H such that $V'_{yH} \neq V'_H = I$. For $x \in G$, we define $V_x = V'_{xH}$, and obtain a continuous, finite-dimensional, irreducible, unitary representation V of G for which $V_y \neq I$. Thus $y \in G \cap G'_0$ and $G_0 \subset H$.]

(d) For any discrete free group G, we have $G_0 = \{e\}$. [By (4.21.e), the intersection of all normal subgroups having finite index is exactly $\{e\}$. Hence $G_0 = \{e\}$ by (c).]

(e) For our remaining examples, we need the following lemma (v. NEUMANN and WIGNER [1]). Let A belong to $\mathfrak{U}(n)$, the group of unitary $n \times n$ matrices. Suppose that for every $m = 1, 2, \ldots$, there is an integral multiple k_m of m and a nonsingular matrix B_m such that $A^{k_m} = B_m A B_m^{-1}$. Then $A = I$. [Let $x_1, \ldots, x_n$ be the eigenvalues of A. For any m, A^{k_m} has $x_1^{k_m}, \ldots, x_n^{k_m}$ as its eigenvalues and $B_m A B_m^{-1}$ has $x_1, \ldots, x_n$ as its eigenvalues. Thus $x_1^{k_m}, \ldots, x_n^{k_m}$ must be a permutation of $x_1, \ldots, x_n$. Consider now a fixed x_j, $j = 1, \ldots, n$. Since $\{k_m\}_{m=1}^{\infty}$ is an infinite sequence and $\{x_j^{k_1}, x_j^{k_2}, \ldots\} \subset \{x_1, \ldots, x_n\}$, we have $x_j^{k_m} = x_j^{k_l}$ for some k_m and k_l with $k_l < k_m$. That is, $x_j^{k_m - k_l} = 1$ and x_j is a root of unity.

Now choose m_0 so that $x_j^{m_0} = 1$ for $j = 1, \ldots, n$. Then $x_j^{k_{m_0}} = 1$ for $j = 1, \ldots, n$, and as $x_1^{k_{m_0}}, \ldots, x_n^{k_{m_0}}$ is a permutation of $x_1, \ldots, x_n$, we see that all of the eigenvalues of A are equal to 1. Consequently, we have $A = I$.]

(f) (v. NEUMANN and WIGNER [1].) Let G be a topological group. Consider a fixed x in G. If for every $m = 1, 2, \ldots$, there is an integral multiple k_m of m and an element y_m in G such that $x^{k_m} = y_m x y_m^{-1}$, then x belongs to G_0. [Let V be a continuous, finite-dimensional, irreducible, unitary representation of G; we may suppose that the representation space is K^n. Let $A = V_x$ and $B_m = V_{y_m}$ and apply (e) to infer that $V_x = I$.]

(g) Let G be the group of matrices $\begin{pmatrix} x & y \\ 0 & 1 \end{pmatrix}$ where x and y are rational numbers and $x \neq 0$; give G the discrete topology. Then G_0 consists of the matrices having the form $\begin{pmatrix} 1 & y \\ 0 & 1 \end{pmatrix}$. [For typographical convenience, we write (x, y) for $\begin{pmatrix} x & y \\ 0 & 1 \end{pmatrix}$. For a real number α, let $\chi_\alpha((x, y)) = \exp(2\pi i\alpha \cdot \log|x|)$ and $\psi((x, y)) = \frac{x}{|x|} = \operatorname{sgn} x$. Then χ_α and ψ are continuous characters of G, that is, 1-dimensional unitary representations of G. For any $(x, y) \in G$ for which $|x| \neq 1$, there exists an $\alpha \in R$ such that $\chi_\alpha((x, y)) = \exp(2\pi i\alpha \cdot \log|x|) \neq 1$. Also for any $(-1, y) \in G$, we have $\psi((-1, y)) = -1$. Therefore we have $G_0 \subset \{(1, y) \in G : y \in Q\}$. For every matrix $(1, y) \in G$ and positive integer m, the equalities $(m, 0)(1, y)(m, 0)^{-1} = (1, my) = (1, y)^m$ hold. Setting k_m equal to m, we see from (f) that $(1, y) \in G_0$. That is, $\{(1, y) \in G : y \in Q\} \subset G_0$.]

If G is the above group with its relative topology as a subgroup of $\mathfrak{GL}(2, R)$, the same computations show that $G_0 = \{(1, y) : y \in Q\}$. Also, if G is the group of matrices $\begin{pmatrix} x & y \\ 0 & 1 \end{pmatrix}$ where x and y are *real* and $x \neq 0$, then $G_0 = \left\{\begin{pmatrix} 1 & y \\ 0 & 1 \end{pmatrix} : y \in R\right\}$ when G is given its usual topology or the discrete topology.

(h) Let G be the special linear group $\mathfrak{SL}(2, K)$ with its usual topology or with the discrete topology. Then $G_0 = G$; *i.e.*, G admits no nontrivial, finite-dimensional, unitary representations. [Obviously we need only consider the case where G is discrete. Consider the element $\begin{pmatrix} 1 & a \\ 0 & 1 \end{pmatrix}$ in G; a is any complex number. Then for any positive integer m,

$$\begin{pmatrix} m & 0 \\ 0 & \frac{1}{m} \end{pmatrix} \begin{pmatrix} 1 & a \\ 0 & 1 \end{pmatrix} \begin{pmatrix} m & 0 \\ 0 & \frac{1}{m} \end{pmatrix}^{-1} = \begin{pmatrix} 1 & m^2 a \\ 0 & 1 \end{pmatrix} = \begin{pmatrix} 1 & a \\ 0 & 1 \end{pmatrix}^{m^2}.$$

Letting $k_m = m^2$ and applying (f), we find that $\begin{pmatrix} 1 & a \\ 0 & 1 \end{pmatrix} \in G_0$. Since G_0 is a normal subgroup of G and

$$\begin{pmatrix} 1 & 0 \\ a & 1 \end{pmatrix} = \begin{pmatrix} 0 & -1 \\ 1 & 0 \end{pmatrix} \begin{pmatrix} 1 & -a \\ 0 & 1 \end{pmatrix} \begin{pmatrix} 0 & -1 \\ 1 & 0 \end{pmatrix}^{-1},$$

$\begin{pmatrix} 1 & 0 \\ a & 1 \end{pmatrix}$ is in G_0 for all $a \in K$. Now consider an arbitrary matrix $\begin{pmatrix} a & b \\ c & d \end{pmatrix}$ in G. If $c \neq 0$, then

$$\begin{pmatrix} a & b \\ c & d \end{pmatrix} = \begin{pmatrix} 1 & \frac{a-1}{c} \\ 0 & 1 \end{pmatrix} \begin{pmatrix} 1 & 0 \\ c & 1 \end{pmatrix} \begin{pmatrix} 1 & \frac{d-1}{c} \\ 0 & 1 \end{pmatrix} \in G_0,$$

and if $c = 0$, then $d \neq 0$ and

$$\begin{pmatrix} a & b \\ 0 & d \end{pmatrix} = \begin{pmatrix} -b & a \\ -d & 0 \end{pmatrix} \begin{pmatrix} 0 & -1 \\ 1 & 0 \end{pmatrix} \in G_0.$$

Consequently $G_0 = G$.]

(i) Groups with no nontrivial, finite-dimensional, unitary representations can also be found by the following argument, which is strongly analogous to (f) *supra*.

Let G be a group [no topology] and let H be an infinite subgroup of G with the following properties:

(i) there are only finitely many conjugacy classes $\{e\}$, $H_1, \ldots, H_m$ of H with respect to inner automorphisms of G;

(ii) for every H_j, the smallest normal subgroup of H containing H_j is H;

(iii) the smallest normal subgroup of G containing H is G.

Then G admits no nontrivial homomorphism into a compact group.

[Let $\widetilde{G}$ be a compact group and φ a homomorphism of G into $\widetilde{G}$. We prove first: (I) If $\varphi(H_j) \cap \widetilde{W} \neq \emptyset$ for all neighborhoods $\widetilde{W}$ of $\tilde{e}$, then $\varphi(H_j) = \{\tilde{e}\}$. For every neighborhood $\widetilde{U}$ of $\tilde{e}$, there is a neighborhood $\widetilde{W}$ of $\tilde{e}$ such that $\tilde{t}^{-1}\widetilde{W}\tilde{t} \subset \widetilde{U}$ for all $\tilde{t} \in \widetilde{G}$ (4.9). If h_1, h_2 lie in H_j, then $t^{-1} h_1 t = h_2$ for some $t \in G$. Thus if $\varphi(h_1) \in \widetilde{W}$, then $\varphi(h_2) = \varphi(t)^{-1} \varphi(h_1) \varphi(t) \in \widetilde{U}$. Therefore $\varphi(H_j) \subset \widetilde{U}$ and, as $\widetilde{U}$ is arbitrary, (I) follows.

(II) If $\varphi(H_j) = \{\tilde{e}\}$ for some j, then (ii) and (iii) imply that $\varphi(G) = \{\tilde{e}\}$.

(III) If $\varphi(H_j) \neq \{\tilde{e}\}$ for $j = 1, 2, \ldots, m$, (I) implies that $\varphi(H \cap \{e\}') \cap \widetilde{W} = \emptyset$ for some neighborhood $\widetilde{W}$ of $\tilde{e}$. This implies that φ is one-to-one on H and also that $\varphi(H)$ is a discrete subgroup of $\widetilde{G}$. Since H is infinite and $\widetilde{G}$ is compact, this is impossible. Hence this case cannot occur.]

(22.23) Some special representations. (a) Let G be a compact group, and let $x \to V_x$ be a continuous representation of G by bounded operators on a Hilbert space H. Then there is an inner product $\langle\!\langle \zeta, \eta \rangle\!\rangle$ on H with respect to which: the operators V_x are unitary; the mapping $x \to V_x$ is a continuous representation; and H is a Hilbert space. [By virtue of (22.8) and the BANACH-STEINHAUS theorem (B.20), we have $\sup\{\|V_x\| : x \in G\} < \infty$. Let $\langle \zeta, \eta \rangle$ denote the original inner product in H. For $\zeta, \eta \in H$ and $a, x \in G$, we have

$$\begin{aligned} |\langle V_x\zeta, V_x\eta\rangle - \langle V_a\zeta, V_a\eta\rangle| &\leq |\langle V_x\zeta - V_a\zeta, V_x\eta\rangle| + |\langle V_a\zeta, V_x\eta - V_a\eta\rangle| \\ &\leq \|V_x\zeta - V_a\zeta\| \cdot \|V_x\eta\| + \|V_a\zeta\| \cdot \|V_x\eta - V_a\eta\|. \end{aligned}$$

Applying (22.8.i) to the foregoing, we see that the mapping $x \to \langle V_x\zeta, V_x\eta\rangle$ is continuous on G: note that the mapping $x \to \|V_x\eta\|$ is a continuous function on G.

Therefore we can define

$$\langle\!\langle \zeta, \eta \rangle\!\rangle = \int_G \langle V_s\zeta, V_s\eta\rangle \, ds,$$

where $\int_G \cdots ds$ denotes normalized Haar integration on G. We obtain in this fashion a complex number for each pair $(\zeta, \eta) \in H \times H$. It is obvious that (B.39.i)–(B.39.iii) hold for the function $\langle\langle\zeta, \eta\rangle\rangle$. Since $\langle V_s\zeta, V_s\zeta\rangle$ is positive for all $s \in G$ and nonzero $\zeta \in H$, (B.39.iv) also holds. For every $x \in G$, we have

$$\begin{aligned}\langle\langle V_x\zeta, V_x\eta\rangle\rangle &= \int_G \langle V_s V_x\zeta, V_s V_x\eta\rangle\, ds = \int_G \langle V_{sx}\zeta, V_{sx}\eta\rangle\, ds \\ &= \int_G [\langle V_s\zeta, V_s\eta\rangle]_x\, ds = \int_G \langle V_s\zeta, V_s\eta\rangle\, ds = \langle\langle\zeta, \eta\rangle\rangle,\end{aligned}$$

since the Haar integral on G is right invariant. We also have

$$\begin{aligned}|\langle\langle V_x\zeta, \eta\rangle\rangle - \langle\langle\zeta, \eta\rangle\rangle| &= \left|\int_G \langle V_{sx}\zeta - V_s\zeta, V_s\eta\rangle\, ds\right| \\ &\leqq \int_G \|V_{sx}\zeta - V_s\zeta\| \cdot \|V_s\eta\|\, ds;\end{aligned}$$

Theorem (22.8) now shows that the representation $x \to V_x$ is continuous with respect to the inner product $\langle\langle\zeta, \eta\rangle\rangle$.

It remains to show that H is complete in the metric induced by the inner product $\langle\langle\zeta, \eta\rangle\rangle$. Write $|||\zeta|||$ for $\langle\langle\zeta, \zeta\rangle\rangle^{\frac{1}{2}}$. Let $\{\zeta_n\}_{n=1}^{\infty}$ be a Cauchy sequence. Then there is a subsequence $\{\zeta_{n_k}\}_{k=1}^{\infty}$ for which, upon applying (12.4), we have

$$\begin{aligned}\int_G \|V_s\zeta_{n_k} - V_s\zeta_{n_{k+1}}\|\, ds &\leqq \left[\int_G \langle V_s(\zeta_{n_k} - \zeta_{n_{k+1}}), V_s(\zeta_{n_k} - \zeta_{n_{k+1}})\rangle\, ds\right]^{\frac{1}{2}} \\ &= |||\zeta_{n_k} - \zeta_{n_{k+1}}||| < 2^{-k} \qquad (k = 1, 2, \ldots).\end{aligned}$$

Summing this over $k = 1, 2, \ldots$, and applying the monotone convergence theorem, we have

$$\int_G \left[\sum_{k=1}^{\infty} \|V_s\zeta_{n_k} - V_s\zeta_{n_{k+1}}\|\right] ds < 1.$$

Theorem (11.27) implies that the sum $\sum_{k=1}^{\infty} \|V_s\zeta_{n_k} - V_s\zeta_{n_{k+1}}\|$ is finite for λ-almost all $s \in G$. Thus for some $a \in G$, the sequence $\{V_a\zeta_{n_k}\}_{k=1}^{\infty}$ is a Cauchy sequence of vectors in the original norm for H, and hence has a limit, say ζ. Since $V_{a^{-1}}$ is a bounded operator, we infer that ζ_{n_k} has limit $V_{a^{-1}}\zeta$. Write $V_{a^{-1}}\zeta = \zeta'$. We then have

$$\begin{aligned}\lim_{k\to\infty} |||\zeta_{n_k} - \zeta'||| &= \lim_{k\to\infty} \left[\int_G \langle V_s(\zeta_{n_k} - \zeta'), V_s(\zeta_{n_k} - \zeta')\rangle\, ds\right]^{\frac{1}{2}} \\ &\leqq \lim_{k\to\infty} [\sup\{\|V_s\| : s \in G\} \cdot \|\zeta_{n_k} - \zeta'\|] = 0.\end{aligned}$$

It follows as usual that $\lim_{n\to\infty} |||\zeta_n - \zeta'||| = 0$.]

(b) Let n be a positive integer and let $\mathfrak{A}$ be a subgroup of the linear group $\mathfrak{GL}(n, K)$ that is compact in its relative topology as a subspace of K^{n^2}. Then there is a matrix $T \in \mathfrak{GL}(n, K)$ such that all of the matrices $T^{-1}MT$ for $M \in \mathfrak{A}$ are unitary. [Choose any basis $\mathbf{e}_1, \ldots, \mathbf{e}_n$ in K^n and for $M = (m_{jk})_{j,k=1}^{n} \in \mathfrak{A}$, let M' be the linear operator on K^n such that

$M'(\boldsymbol{e}_k) = \sum_{j=1}^{n} m_{jk}\,\boldsymbol{e}_j$ $(k=1, \ldots, n)$. The mapping $M \to M'$ is obviously a representation of $\mathfrak{A}$ by operators on K^n satisfying the hypotheses of part (a). Thus there is an inner product $\langle\!\langle \boldsymbol{x}, \boldsymbol{y} \rangle\!\rangle$ for K^n such that $\langle\!\langle M'\boldsymbol{x}, M'\boldsymbol{y} \rangle\!\rangle = \langle\!\langle \boldsymbol{x}, \boldsymbol{y} \rangle\!\rangle$ for all $M \in \mathfrak{A}$ and all $\boldsymbol{x}, \boldsymbol{y} \in K^n$. Let $\boldsymbol{f}_1, \ldots, \boldsymbol{f}_n$ be an orthonormal basis for K^n with respect to the inner product $\langle\!\langle \boldsymbol{x}, \boldsymbol{y} \rangle\!\rangle$. Let $T = (t_{jk})_{j,k=1}^{n}$ be the matrix such that $\boldsymbol{f}_k = \sum_{j=1}^{n} t_{jk}\,\boldsymbol{e}_j$ $(k=1, \ldots, n)$. Then all of the matrices $T^{-1} M T$ are unitary.]

(c) (WEIL [4], p. 70.) Let G be any group [not necessarily topological], and let τ be a homomorphism of G into $\mathfrak{G}\mathfrak{L}(n, K)$; we write $\tau(x) = (a_{jk}(x))_{j,k=1}^{n}$. Suppose that all of the functions a_{jk} are bounded on G. Then there is a matrix $T \in \mathfrak{G}\mathfrak{L}(n, K)$ such that all of the matrices $T^{-1}\tau(x)T$ are unitary. [It is easy to see that the closure of $\tau(G)$ in $\mathfrak{G}\mathfrak{L}(n, K)$ is a bounded, closed subset of K^{n^2} and hence is a compact subgroup of $\mathfrak{G}\mathfrak{L}(n, K)$. Now apply part (b).]

(d) Let G be any group and let $x \to V_x$ be a finite-dimensional representation of G by operators on K^n such that the functions $x \to \langle V_x \boldsymbol{y}, V_x \boldsymbol{y} \rangle$ are bounded on G for all $\boldsymbol{y} \in K^n$. Then there is an inner product $\langle\!\langle \boldsymbol{y}, \boldsymbol{z} \rangle\!\rangle$ on K^n such that all of the operators V_x are unitary with respect to this inner product. [This is merely a restatement of part (c).]

Notes

The first paper we know of dealing explicitly with *infinite-dimensional* unitary representations of a group is WIGNER [1], in which irreducible, infinite-dimensional, unitary representations of the inhomogeneous Lorentz group are computed. The regular representation V_a $[V_a f = \varepsilon_a * f]$ of G by operators in $\mathfrak{C}(G)$ appears implicitly in WEYL and PETER [1] for compact Lie groups G, and by operators in $\mathfrak{L}_2(G)$ in A. WEIL [4], Ch. V. The first study of infinite-dimensional unitary representations of general locally compact groups is in the fundamental paper of GEL'FAND and RAĬKOV [2]; see also GEL'FAND and RAĬKOV [3]. Representations of G by operators on Banach spaces were introduced by I. E. SEGAL [3]. They have been further studied by LOOMIS [2], § 32, and by SHIGA [1] for compact groups.

Theorem (22.3) for unitary representations of $\boldsymbol{M}(G)$ is due to GODEMENT [1] and for Banach space representations of $\boldsymbol{M}_a(G)$ to LOOMIS [2], Theorem 32B. For $\boldsymbol{A} = \boldsymbol{M}_a(G)$, (22.7) appears in LOOMIS [2], Theorem 32C, and in NAĬMARK [1], § 29, Theorem 1. The fact that weak continuity implies strong continuity for unitary representations seems first to have been remarked by GODEMENT [1], p. 14, footnote 12. A result similar to (22.8) for compact groups appears in SHIGA [1]. So far as we know, (22.8) is new for the groups and representations considered. Theorem (22.10) is based on NAĬMARK [1], §29, Theorem 1.

The basis of (22.11) appears in A. WEIL [4], p. 48; the rest is closely related to observations of GODEMENT [1].

Theorem (22.12) is due, as noted, to GEL'FAND and RAĬKOV. The proof given here is like the proof given by I. E. SEGAL [2]. GODEMENT [1] has published a proof of (22.12) very like the original one, as have several other authors. In the general arrangement of our proof [although not in its details] we have been influenced by NAĬMARK [1], §29. It is worth noting once again that all proofs of (22.12) depend upon the KREĬN-MIL'MAN theorem (B.30) and hence in the end upon *compactness*. In this (22.12) resembles a remarkable number of the most important facts in analysis.

The GEL'FAND-RAĬKOV theorem has provided the basis for the formidable theory of unitary representations of locally compact, noncompact, non-Abelian groups, as it has developed since 1947. This theory is outside of the scope of our work.

Theorem (22.13) was proved for compact groups G with countable open bases by A. GUREVIČ [1]. In fact, she proved more: *every* continuous unitary representation of such a group is the direct sum of irreducible, finite-dimensional, unitary representations. [We will prove this for all compact G in Vol. II, (27.44).] KOOSIS [1] and NACHBIN [1] have published proofs of (22.13); the proof in the text is NACHBIN'S.

Corollary (22.14) is due to WEYL and PETER [1] for compact Lie groups and to VAN KAMPEN [1] for arbitrary compact groups. [VAN KAMPEN pointed out that every continuous function on a compact group is almost periodic [see (18.2)] and then applied VON NEUMANN'S theorem [4] on almost periodic functions and irreducible, finite-dimensional, unitary representations.] PONTRYAGIN [5] also published a proof of (22.14) for compact groups with countable open bases.

Characters of groups have been known and exploited for many years. For some of the history, see the notes to §§23—25. Semicharacters of semigroups were introduced only recently: in independent papers of Š. SCHWARZ [1] and HEWITT and ZUCKERMAN [3]. Theorem (22.17) was proved by VON NEUMANN [4] for locally compact Abelian groups with countable open bases; by J.W. ALEXANDER [1] for discrete Abelian groups; by VAN KAMPEN [1] for arbitrary locally compact Abelian groups. A proof of (22.17) based on the theory of Banach algebras [but postulating the existence of Haar measure] was given by GEL'FAND and RAĬKOV [1].

Theorem (22.18) generalizes (22.19), which is a theorem of A. WEIL [4], pp. 66—67.

À propos of (22.22), it is interesting to remark that locally compact groups G with sufficiently many finite-dimensional, continuous, unitary representations are fairly special. If such a group is *connected*, then it has the form $R^n \times G_0$, where n is a nonnegative integer and G_0 is compact. This startling result was proved by FREUDENTHAL [2] for groups with countable open bases. The general case is due to A. WEIL [4], pp. 126—130. There are many disconnected noncompact groups with sufficiently many finite-dimensional, continuous, unitary representations: see (22.22.d) for example. It would be interesting to have extensions of FREUDENTHAL's theorem to at least some classes of disconnected groups.

Chapter Six

Characters and duality of locally compact Abelian groups

This chapter is devoted to what might be called the "fine structure" of locally compact Abelian groups. We will obtain extremely detailed information about the structure of these groups. The basic tool in this program is the character group, which is defined and studied in §23. The character group is important in large measure because of the PONTRYAGIN-VAN KAMPEN duality theorem, which is stated, proved, and utilized in §24. In §25, we apply the duality theorem to study a number of special locally compact Abelian groups, and in §26 we apply it to certain problems in structure theory and analysis.

§23. The character group of a locally compact Abelian group

(23.1) Definition. Let G be a group, not necessarily Abelian or topological. Let χ_1 and χ_2 be characters of G. The product $\chi_1\chi_2$ is defined as the ordinary pointwise product: $\chi_1\chi_2(x) = \chi_1(x)\,\chi_2(x)$.

(23.2) Theorem. *The set of all characters of G is an Abelian group, with identity element the function 1 and $\chi^{-1} = \bar{\chi}$.*

This theorem is obvious.

(23.3) Definition. Let G be a topological group. The group of all *continuous* characters of G is called the *character group of G*. Character groups will be denoted by letters X, Y, ω, ...

Plainly every topological group has at least one continuous character, namely, 1. Some noncommutative locally compact groups have no others (22.22.h). We shall see in the present section the usefulness of the character group of a locally compact *Abelian* group, and begin with a general theorem.

(23.4) Theorem. *Let G be a locally compact*[1] *group, and let $\boldsymbol{A}$ be a closed subalgebra of $\boldsymbol{M}(G)$ with the following properties:*

[1] Not necessarily Abelian!

(i) *every bounded linear functional on* $\boldsymbol{A}$ *has the form* $\mu \to \int_G h\,d\mu$, *where* h *is a bounded function on* G *such that* $h\xi_E$ *is Borel measurable for all* σ*-compact subsets* E *of* G;

(ii) $a \in G$ *and* $\mu \in \boldsymbol{A}$ *imply* $\varepsilon_a * \mu \in \boldsymbol{A}$ *and* $\mu * \varepsilon_a \in \boldsymbol{A}$.

Let τ *be a multiplicative linear functional on* $\boldsymbol{A}$.[1] *Then there is a continuous character* χ *of* G *such that*

(iii) $\tau(\mu) = \int_G \bar{\chi}\,d\mu$ *for all* $\mu \in \boldsymbol{A}$.

If χ *is any continuous character of* G, *the mapping*

(iv) $\mu \to \int_G \bar{\chi}\,d\mu$[2]

is a multiplicative linear functional on $\boldsymbol{M}(G)$.

Proof. Let μ, ν be measures in $\boldsymbol{A}$ for which $\tau(\mu) \neq 0$, $\tau(\nu) \neq 0$, and let a be any element of G. Then we have

$$\tau(\varepsilon_a * \mu * \nu) = \tau(\varepsilon_a * \mu)\,\tau(\nu) = \tau(\nu)\,\tau(\varepsilon_a * \mu) = \tau(\nu * \varepsilon_a * \mu) = \tau(\nu * \varepsilon_a)\,\tau(\mu).$$

Thus we have

$$\frac{\tau(\varepsilon_a * \mu)}{\tau(\mu)} = \frac{\tau(\nu * \varepsilon_a)}{\tau(\nu)};$$

putting $\mu = \nu$, we have $\tau(\varepsilon_a * \nu) = \tau(\nu * \varepsilon_a)$, and hence

$$\frac{\tau(\varepsilon_a * \mu)}{\tau(\mu)} = \frac{\tau(\varepsilon_a * \nu)}{\tau(\nu)}.$$

Let χ be the function such that $\chi(a) = \dfrac{\tau(\varepsilon_{a^{-1}} * \mu)}{\tau(\mu)}$ for any μ such that $\tau(\mu) \neq 0$, and all $a \in G$. If $\tau(\nu) = 0$, then $\tau(\varepsilon_b * \nu) = 0$ for all $b \in G$. To see this, choose $\mu \in \boldsymbol{A}$ such that $\tau(\mu) = 1$, and write

$$\tau(\varepsilon_b * \nu) = \tau(\mu)\,\tau(\varepsilon_b * \nu) = \tau(\mu * \varepsilon_b * \nu) = \tau(\mu * \varepsilon_b)\,\tau(\nu) = 0.$$

Hence we have $\chi(a)\,\tau(\nu) = \tau(\varepsilon_{a^{-1}} * \nu)$ for all $\nu \in \boldsymbol{A}$ and $a \in G$. Then for a, b in G, we have

$$\chi(ab) = \tau(\varepsilon_{(ab)^{-1}} * \mu) = \tau(\varepsilon_{b^{-1}} * \varepsilon_{a^{-1}} * \mu) = \chi(b)\,\chi(a) = \chi(a)\,\chi(b).$$

Since τ is a bounded linear functional (C.21), we have

$$|\chi(a)| \leqq \|\tau\|\,\|\varepsilon_{a^{-1}} * \mu\| \leqq \|\tau\| \cdot 1 \cdot \|\mu\|,$$

so that χ is bounded. Since $\chi(e) = 1$, it follows that χ is a character of G.

We now prove that χ is continuous. By hypothesis (i), we have $\tau(\mu) = \int_G h\,d\mu$ for all $\mu \in \boldsymbol{A}$, where h is a function as in (i). Choose $\mu \in \boldsymbol{A}$

---

[1] Recall that a multiplicative linear functional is not identically zero (C.12).

[2] Our choice of $\bar{\chi}$ instead of χ in this integral and in (iii) is arbitrary. Inversion formulas to be presented in Vol. II use χ instead of $\bar{\chi}$: we elect to use $\bar{\chi}$ first.

so that $\tau(\mu)=1$, and let E be a σ-compact subset of G such that $|\mu|(E')=0$. For all Borel sets $A\subset G$ and $x\in G$, we have $\varepsilon_{x^{-1}}*|\mu|(A)=|\mu|(xA)$, so that $\varepsilon_{x^{-1}}*|\mu|((x^{-1}E)')=\varepsilon_{x^{-1}}*|\mu|(x^{-1}(E'))=|\mu|(E')=0$. Thus for each $x\in G$, we have

$$\chi(x)=\int_G h(y)\,d\varepsilon_{x^{-1}}*\mu(y)=\int_G \xi_{x^{-1}E}(y)\,h(y)\,d\varepsilon_{x^{-1}}*\mu(y)$$
$$=\int_G \xi_{x^{-1}E}(x^{-1}y)\,h(x^{-1}y)\,d\mu(y)=\int_G \xi_E(y)\,h(x^{-1}y)\,d\mu(y).$$

Now let F be a compact subset of G such that $\lambda(F)>0$. The function $\xi_{F^{-1}E}h$ is Borel measurable on G by hypothesis (i). The function

$$(x,y)\rightarrow \xi_F(x)\,\xi_E(y)\,\xi_{F^{-1}E}(x^{-1}y)\,h(x^{-1}y) \tag{1}$$

is accordingly Borel measurable on $G\times G$ and vanishes outside of the σ-compact set $F\times E$. The estimate

$$\int_G\int_G \xi_F(x)\,\xi_E(y)\,\xi_{F^{-1}E}(x^{-1}y)\,|h(x^{-1}y)|\,d|\mu|(y)\,dx\leq \|h\|_u\,\|\mu\|\,\lambda(F)<\infty$$

is immediate. Theorem (13.10) implies that the function (1) belongs to $\mathfrak{L}_1(G\times G,\lambda\times|\mu|)$, and (13.8) implies that the function

$$x\rightarrow\int_G \xi_F(x)\,\xi_E(y)\,\xi_{F^{-1}E}(x^{-1}y)\,h(x^{-1}y)\,d\mu(y)$$
$$=\xi_F(x)\int_G \xi_E(y)\,h(x^{-1}y)\,d\mu(y)=\xi_F(x)\,\chi(x)$$

is λ-measurable on G [note that $\xi_F(x)\,\xi_E(y)\,\xi_{F^{-1}E}(x^{-1}y)=\xi_F(x)\,\xi_E(y)$ for all $(x,y)\in G\times G$]. That is, the function χ is λ-measurable on F, and so by (22.19), χ is continuous on G.

To establish (iii), let μ_2 be a measure in $\boldsymbol{A}$ such that $\tau(\mu_2)=1$, let μ_1 be any measure in $\boldsymbol{A}$, and let E_1 and E_2 be σ-compact subsets of G such that $|\mu_j|(E_j')=0$ $(j=1,2)$. Then $|\mu_1|*|\mu_2|((E_1E_2)')=0$, E_1E_2 is σ-compact, and so we can write

$$\tau(\mu_1)=\tau(\mu_1)\,\tau(\mu_2)=\tau(\mu_1*\mu_2)=\int_G h(u)\,d\mu_1*\mu_2(u)$$
$$=\int_G \xi_{E_1E_2}(u)\,h(u)\,d\mu_1*\mu_2(u)=\int_G\int_G \xi_{E_1E_2}(xy)\,h(xy)\,d\mu_2(y)\,d\mu_1(x)$$
$$=\int_G\int_G \xi_{x^{-1}E_1E_2}(y)\,h(xy)\,d\mu_2(y)\,d\mu_1(x)$$
$$=\int_{E_1}\int_G \xi_{x^{-1}E_1E_2}(y)\,h(xy)\,d\mu_2(y)\,d\mu_1(x).$$

For $x\in E_1$, we have $x^{-1}E_1E_2\supset x^{-1}xE_2=E_2$. Hence the last integral is equal to

$$\int_{E_1}\int_G \xi_{E_2}(y)\,h(xy)\,d\mu_2(y)\,d\mu_1(x)=\int_{E_1}\chi(x^{-1})\,d\mu_1(x)=\int_G \overline{\chi(x)}\,d\mu_1(x).$$

Since μ_1 is arbitrary, we have established (iii).

Finally, we establish (iv). Since χ is continuous and $|\chi|=1$, χ is in $\mathfrak{L}_1(G,|\mu|*|\nu|)$ for all $\mu,\nu\in\boldsymbol{M}(G)$. Theorem (19.10) then gives

$$\int_G \overline{\chi(u)}\, d\mu * \nu(u) = \int_G \int_G \overline{\chi(xy)}\, d\mu(x)\, d\nu(y)$$
$$= \int_G \int_G \overline{\chi(x)}\, \overline{\chi(y)}\, d\mu(x)\, d\nu(y) = \int_G \overline{\chi(x)}\, d\mu(x) \int_G \overline{\chi(y)}\, d\nu(y).$$

Since $\int_G \overline{\chi(x)}\, d\varepsilon_e(x) = \overline{\chi(e)} = 1$, the mapping (iv) is not identically zero on $\mathbf{M}(G)$: thus it is a multiplicative [obviously linear] functional. □

(23.5) Note. If G is nondiscrete, then $\mathbf{M}(G)$ admits at least one multiplicative linear functional *not* of the form (23.4.iv). See (23.28.a).

(23.6) Theorem. *Let G be a locally compact group. Let $\mathbf{A}$ be a subalgebra of $\mathbf{M}(G)$ such that for every nonvoid open subset U of G, there is a μ in $\mathbf{A}^+$ such that $\mu(U)=1$ and $\mu(U')=0$. Then if χ and ψ are distinct continuous characters of G, the multiplicative linear functionals $\mu \to \int_G \bar{\chi}\, d\mu$ and $\mu \to \int_G \bar{\psi}\, d\mu$ are distinct on the algebra $\mathbf{A}$.*

Proof. Let $a \in G$ be such that $\chi(a) \neq \psi(a)$. Suppose also that $\operatorname{Re}\chi(a) > \operatorname{Re}\psi(a)$. Then there are a neighborhood U of a and a positive number α such that $\operatorname{Re}\chi(x) - \operatorname{Re}\psi(x) \geqq \alpha$ for all $x \in U$. Let μ be a measure in $\mathbf{A}^+$ such that $\mu(U)=1$ and $\mu(U')=0$. Then it is clear that $\operatorname{Re}\int_G \bar{\chi}\, d\mu - \operatorname{Re}\int_G \bar{\psi}\, d\mu \geqq \alpha$. Similar arguments apply to the cases $\operatorname{Re}\chi(a) < \operatorname{Re}\psi(a)$ and $\operatorname{Im}\chi(a) \neq \operatorname{Im}\psi(a)$. □

(23.7) Corollary. *Let G be a locally compact group. Every multiplicative linear functional on $\mathbf{M}_a(G)$ has the form*

(i) $\mu \to \int_G \bar{\chi}\, d\mu,$

where χ is a continuous character of G. Conversely, every such mapping is a multiplicative linear functional on $\mathbf{M}_a(G)$, and distinct continuous characters of G produce distinct multiplicative linear functionals by the mapping (i).

Proof. This follows at once from (12.18) and part (I) of the proof of Theorem (22.10) [page 508], which show that $\mathbf{M}_a(G)$ satisfies the hypotheses of (23.4), together with (23.4) and (23.6). □

The character group of a locally compact group G is useful in studying the structure of G only insofar as the elements of G commute with each other. We state this in precise terms as follows.

(23.8) Theorem. *Let G be a locally compact group. Let C_0 be the group generated by all elements of the form $aba^{-1}b^{-1}$, where $a, b \in G$;*[1] *C_0 is a normal subgroup of G. Let*

$$A = \{x \in G : \chi(x) = 1 \text{ for all continuous characters of } G\}.$$

[1] C_0 is called the *commutator subgroup* of G. See, however, the footnote on page 142 of Volume II.

Then $C_0^- = A$. *The character groups of* G *and* G/A *are isomorphic under the mapping* $\chi \to \chi_1$, *where* χ *is a continuous character of* G *and* $\chi_1(xA) = \chi(x)$ *for all* $x \in G$. *If* H *is a normal subgroup of* G, *then* G/H *is Abelian if and only if* $H \supset C_0$. *If* H *is closed and* G/H *is Abelian, then* $H \supset A$ *and* G/H *is topologically isomorphic with* $(G/A)/(H/A)$.

Proof. If $x \in G$ and $c \in C_0$, then $xcx^{-1} = (xcx^{-1}c^{-1})\,c \in C_0 C_0 = C_0$, so that C_0 is a normal subgroup of G. By (5.3), C_0^- is a closed normal subgroup of G.

Suppose that H is a normal subgroup of G. If G/H is Abelian, then for $a, b \in G$, we have $aHbH = bHaH$ and hence $a^{-1}b^{-1}ab \in H$. It follows that $C_0 \subset H$. Conversely, if $C_0 \subset H$, then for $a, b \in G$, we have $a^{-1}b^{-1}ab \in C_0 \subset H$ and so $aHbH = bHaH$. Thus G/H is Abelian.

If χ is a continuous character of G, then $\chi^{-1}(1)$ is a closed normal subgroup of G. As the intersection of closed normal subgroups, A is a closed normal subgroup. Clearly all elements of the form $aba^{-1}b^{-1}$ belong to A. Therefore we have $C_0^- \subset A$, since C_0^- is the smallest closed subgroup containing these elements. Assume now that $C_0^- \subsetneqq A$ and that $x \in A \cap (C_0^-)'$. Then G/C_0^- is Abelian; it is locally compact and Hausdorff by (5.22) and (5.21). By (22.17), there is a continuous character χ_1 of G/C_0^- such that $\chi_1(xC_0^-) \neq 1$. If φ denotes the natural mapping of G onto G/C_0^-, then $\chi_1 \circ \varphi$ is a continuous character of G and $\chi_1 \circ \varphi(x) \neq 1$. That is, $x \notin A$. This contradiction shows that $C_0^- = A$.

The last statement of the theorem now follows from (5.35).

Finally, for every continuous character χ of G, let χ_1 be the function on G/A such that $\chi_1(xA) = \chi(x)$. If $xA = yA$, then $y^{-1}x \in A$, so that $\chi(y^{-1}x) = \chi(y)^{-1}\chi(x) = 1$. Thus χ_1 is well defined on G/A. It is plain that χ_1 is a character of G/A. If W is an open subset of T, then $\chi_1^{-1}(W) = \{xA \in G/A : \chi_1(xA) \in W\} = \{xA \in G/A : \chi(x) \in W\}$, which is the image of the open set $\chi^{-1}(W)$ under the natural mapping φ of G onto G/A. Since φ is open (5.17), χ_1 is a continuous character of G/A. If ψ is a continuous character of G/A, then $\psi \circ \varphi$ is a continuous character of G and $\psi = (\psi \circ \varphi)_1$. Hence the mapping $\chi \to \chi_1$ is an isomorphism of the character group of G onto the character group of G/A. □

Theorem (23.8) shows that in studying character groups we lose nothing by considering only *Abelian* groups. The ensuing analysis will therefore be couched in terms of Abelian groups. Throughout (23.9) to (23.16), G will denote an arbitrary locally compact Abelian group and X will denote the character group of G.

(23.9) Definition. For every $\mu \in \boldsymbol{M}(G)$, let $\hat{\mu}$ be the complex-valued function on X such that

(i) $\hat{\mu}(\chi) = \int\limits_G \bar{\chi}\, d\mu$.

This function $\hat{\mu}$ is called the *Fourier-Stieltjes transform of* μ. If $\mu \in \boldsymbol{M}_a(G)$, so that $d\mu = f\,d\lambda$ with $f \in \mathfrak{L}_1(G)$, we write $\hat{f}$ for $\hat{\mu}$ and call $\hat{f}$ the *Fourier transform of the function* f. For a subset $\boldsymbol{A}$ of $\boldsymbol{M}(G)$, $\hat{\boldsymbol{A}}$ will denote the set $\{\hat{\mu} : \mu \in \boldsymbol{A}\}$ of functions on X; similarly, $\hat{\mathfrak{D}}$ will denote $\{\hat{f} : f \in \mathfrak{D}\}$ for $\mathfrak{D} \subset \mathfrak{L}_1(G)$.

(23.10) Theorem. *The following relations hold:*

(i) $(\mu + \nu)^\wedge = \hat{\mu} + \hat{\nu}$ *for* $\mu, \nu \in \boldsymbol{M}(G)$;

(ii) $(\alpha\mu)^\wedge = \alpha\hat{\mu}$ *for* $\alpha \in K$ *and* $\mu \in \boldsymbol{M}(G)$;

(iii) $(\mu * \nu)^\wedge = \hat{\mu}\hat{\nu}$ *for* $\mu, \nu \in \boldsymbol{M}(G)$;

(iv) $(\mu^\sim)^\wedge = \bar{\hat{\mu}}$ *for* $\mu \in \boldsymbol{M}(G)$;

(v) $\hat{\varepsilon}_a(\chi) = \overline{\chi(a)}$ *for* $a \in G$ *and* $\chi \in$ X;

(vi) $\sup\{|\hat{\mu}(\chi)| : \chi \in \mathrm{X}\} \leq \|\mu\|$ *for* $\mu \in \boldsymbol{M}(G)$.

All of these relations are simple consequences of the definition. We omit the verification.

(23.11) Theorem. *Let* μ *be a nonzero measure in* $\boldsymbol{M}_a(G)$. *Then there is a* $\chi \in$ X *such that* $\hat{\mu}(\chi) \neq 0$.

Proof. By (22.11) and (21.37), there is an irreducible $\sim$-representation S of $\boldsymbol{M}_a(G)$ for which S_μ is not the zero operator. The algebra $\boldsymbol{M}_a(G)$ is commutative [(19.6), and see also (20.24)]. Hence every operator S_ν $(\nu \in \boldsymbol{M}_a(G))$ commutes with all operators S_μ $(\mu \in \boldsymbol{M}_a(G))$ and so by (21.30), S_ν is a multiple of the identity operator: $S_\nu = \tau(\nu) I$. Clearly τ, defined in this way, is a multiplicative linear functional on $\boldsymbol{M}_a(G)$ and $\tau(\mu) \neq 0$. Referring to (23.7), we see that $\tau(\mu) = \int_G \bar{\chi}\,d\mu$ for some $\chi \in$ X. □

(23.12) Note. The analogue of (23.11) holds also for $\boldsymbol{M}(G)$: if $\mu \in \boldsymbol{M}(G)$ and $\mu \neq 0$, then $\hat{\mu} \neq 0$. The proof is somewhat more complicated, however, and is postponed to Vol. II, (31.5).

We now define a topology in X.

(23.13) Theorem. *For* $f_1, \ldots, f_m \in \mathfrak{L}_1(G)$, $\varepsilon > 0$, *and* $\chi_0 \in$ X, *let* $\Delta(f_1, \ldots, f_m; \varepsilon; \chi_0)$ *be the set* $\{\chi \in \mathrm{X} : |\hat{f}_j(\chi) - \hat{f}_j(\chi_0)| < \varepsilon$ *for* $j = 1, \ldots, m\}$. *Let the sets* $\Delta(f_1, \ldots, f_m; \varepsilon; \chi_0)$ *be taken as a basis for open sets in* X. *Under the topology so defined*[1] *for* X, X *is a locally compact Hausdorff space.*

Proof. By (23.7), X can be identified with the structure space of the commutative Banach algebra $\boldsymbol{M}_a(G)$ (C.23). The Δ-topology is just the Gel'fand topology for this structure space and is therefore locally compact and Hausdorff (C.26). □

(23.14) Theorem. *Let* f *be a function in* $\mathfrak{L}_1(G)$, *and let* ε *be a positive number. There is a neighborhood* U *of* e *depending on* f *and* ε *with the*

---

[1] We will call it provisionally the Δ-topology for X.

following property. If x_0, x in G and χ_0, χ in X are such that $\hat{f}(\chi_0)=1$, $x\in Ux_0$, and $\chi\in\Delta(f, {}_{x_0}f; \varepsilon/3; \chi_0)$, then

(i) $|\chi(x)-\chi_0(x_0)|<\varepsilon$.

In particular, the mapping $(x,\chi)\to\chi(x)$ of $G\times X$ into T is continuous in the given topology for G and the Δ-topology for X.

Proof. Let U be a neighborhood of e such that $\|{}_sf-{}_tf\|_1<\frac{\varepsilon}{3}$ if $st^{-1}\in U$ (20.4.i). For arbitrary $\chi_0, \chi\in X$ and $x_0, x\in G$ such that $\hat{f}(\chi_0)=1$, we have

$$\left.\begin{aligned}|\chi(x)-\chi_0(x_0)| &= |\chi(x)-\chi_0(x_0)\hat{f}(\chi_0)|\leqq|\chi(x)-\chi(x)\hat{f}(\chi)|\\ &\quad+|\chi(x)\hat{f}(\chi)-\chi(x_0)\hat{f}(\chi)|+|\chi(x_0)\hat{f}(\chi)-\chi_0(x_0)\hat{f}(\chi_0)|\\ &=|1-\hat{f}(\chi)|+|({}_xf)^\frown(\chi)-({}_{x_0}f)^\frown(\chi)|+|({}_{x_0}f)^\frown(\chi)-({}_{x_0}f)^\frown(\chi_0)|.\end{aligned}\right\}\qquad(1)$$

The last equality is verified by (23.10.iii) and (23.10.v) when we note that $({}_xf)\,d\lambda=(\varepsilon_{x^{-1}}*f)\,d\lambda$. We also have from (23.10.vi) that

$$|({}_xf)^\frown(\chi)-({}_{x_0}f)^\frown(\chi)|\leqq\|{}_xf-{}_{x_0}f\|_1. \qquad (2)$$

Now if $(x,\chi)\in(Ux_0)\times(\Delta(f, {}_{x_0}f; \varepsilon/3; \chi_0))$, we infer from (1) and (2) that $|\chi(x)-\chi_0(x_0)|<\varepsilon$. □

(23.15) Theorem. *For every compact set $F\subset G$ and every $\varepsilon>0$, let $P(F,\varepsilon)$ be the set $\{\chi\in X: |\chi(x)-1|<\varepsilon$ for all $x\in F\}$. With all sets $P(F,\varepsilon)$ taken as an open basis at the identity 1 in X, X is a topological group, under the construction of (4.5). The P-topology is equal to the Δ-topology of (23.13), and so X under this topology is a locally compact Abelian group.*

Proof. We verify (4.5.i)—(4.5.v), noting first that (4.5.ii) and (4.5.iv) are trivial for the sets $P(F,\varepsilon)$. If $|\chi(x)-1|<\frac{\varepsilon}{2}$ and $|\psi(x)-1|<\frac{\varepsilon}{2}$, then we have $|\chi(x)\psi(x)-1|\leqq|\chi(x)\psi(x)-\psi(x)|+|\psi(x)-1|<\varepsilon$, so that $P(F,\varepsilon/2)\cdot P(F,\varepsilon/2)\subset P(F,\varepsilon)$: this verifies (4.5.i). If $\chi\in P(F,\varepsilon)$, then $\max\{|\chi(x)-1|: x\in F\}=\alpha$ is less than ε, since χ is continuous and F is compact. It is easy to see that $\chi\cdot P(F,\varepsilon-\alpha)\subset P(F,\varepsilon)$. This is (4.5.iii), and thus we have shown that X is a topological group under the P-topology. Given $P(F,\varepsilon)$ and $P(F_0,\varepsilon_0)$, we have

$$P(F\cup F_0, \min(\varepsilon,\varepsilon_0))\subset P(F,\varepsilon)\cap P(F_0,\varepsilon_0),$$

so that (4.5.v) holds and the sets $P(F,\varepsilon)$ form a basis at $1\in X$.

We now prove that the Δ- and P-topologies are equal. Since $\chi_0\cdot\Delta(f_1,\ldots,f_m;\varepsilon;1)=\Delta(\chi_0f_1,\ldots,\chi_0f_m;\varepsilon;\chi_0)$, this will be accomplished when we have shown that every $P(F,\varepsilon)$ contains a $\Delta(f_1,\ldots,f_m;\delta;1)$ and *vice versa*. Suppose that F is a compact subset of G and that ε is positive. Let f be a function in $\mathfrak{L}_1(G)$ such that $\int_G f\,d\lambda=\hat{f}(1)=1$, and let U be a neighborhood of e as in (23.14). Since F is compact, a finite number of neighborhoods Ux cover F: suppose that $x_1,\ldots,x_n\in F$ and

$\bigcup_{k=1}^{n} Ux_k \supset F$. Now suppose that $\chi \in \Delta(f, {}_{x_1}f, {}_{x_2}f, \ldots, {}_{x_n}f; \varepsilon/3; 1)$. For all $x \in F$, we have $x \in Ux_k$ for some $k = 1, \ldots, n$ and $\chi \in \Delta(f, {}_{x_k}f; \varepsilon/3; 1)$. Appealing to (23.14) and taking $\chi_0 = 1$, we infer that $|\chi(x) - 1| < \varepsilon$; that is, $\chi \in \mathsf{P}(F, \varepsilon)$.

Finally, let $\Delta(f_1, \ldots, f_m; \delta; 1)$ be arbitrary; we may suppose that no f_j is zero. Let F be a compact subset of G such that

$$\int_{F'} |f_j| \, d\lambda < \frac{\delta}{4} \quad \text{for} \quad j = 1, \ldots, m.$$

Let $\varepsilon = \frac{\delta}{2} \min\left(\frac{1}{\|f_1\|_1}, \ldots, \frac{1}{\|f_m\|_1}\right)$. Then if $\chi \in \mathsf{P}(F, \varepsilon)$, we can write

$$\begin{aligned} |\hat{f}_j(\chi) - \hat{f}_j(1)| &= \left| \int_G f_j \bar{\chi} \, d\lambda - \int_G f_j \, d\lambda \right| \\ &\leq \int_F |f_j| \, |\chi - 1| \, d\lambda + 2 \int_{F'} |f_j| \, d\lambda < \varepsilon \int_F |f_j| \, d\lambda + \frac{\delta}{2} \leq \delta. \end{aligned}$$

Thus $\mathsf{P}(F, \varepsilon)$ is contained in $\Delta(f_1, \ldots, f_m; \delta; 1)$. □

From this point on, we will always consider X as topologized with the Δ- [or P-] topology.

(23.16) Corollary. *Let F be a compact subset of G such that $\lambda(F) > 0$, and let α be any positive number less than 1. The neighborhood $\mathsf{P}(F, \alpha)$ has compact closure in X.*

Proof. Consider the function $g = \xi_F \in \mathfrak{L}_1(G)$. If $\chi \in \mathsf{P}(F; \alpha)$, then we have

$$|\hat{g}(\chi) - \hat{g}(1)| = \left| \int_G \bar{\chi} g \, d\lambda - \int_G g \, d\lambda \right| \leq \alpha \lambda(F) = \alpha \hat{g}(1),$$

and thus

$$|\hat{g}(\chi)| \geq (1 - \alpha) \, \hat{g}(1).$$

Since $\hat{g}$ is in $\mathfrak{C}_0(\mathsf{X})$ (C.26), the set $\{\chi \in \mathsf{X} : |\hat{g}(\chi)| \geq (1 - \alpha) \, \hat{g}(1)\}$ is compact. Thus the closure of $\mathsf{P}(F, \alpha)$ is also compact. □

(23.17) Theorem. *If G is compact, then X is discrete; and if G is discrete, then X is compact.*

Proof. Let A be a subgroup of the group T, and suppose that $|z - 1| < \sqrt{3}$ for all $z \in A$. Then it is easy to see that $A = \{1\}$. Thus if G is compact, the set $\mathsf{P}(G, \sqrt{3}) = \{1\}$ is a neighborhood of the character 1 in X, so that X is discrete. If G is discrete, then $\boldsymbol{M}_a(G)$ has a unit, so that (C.25) implies that X with the Δ-topology is compact. □

We wish to compute the character groups of a number of locally compact Abelian groups, and we wish also to prove an important duality theorem (24.8). We first establish some simple preliminaries.

(23.18) Theorem. *Let $G_1, \ldots, G_m$ be locally compact Abelian groups, with character groups $\mathsf{X}_1, \ldots, \mathsf{X}_m$ respectively. For every $(\chi_1, \ldots, \chi_m) \in \mathop{\mathsf{P}}_{j=1}^{m} \mathsf{X}_j$,* *let* $[\chi_1, \ldots, \chi_m]$ *denote the function*

$$(x_1, \ldots, x_m) \rightarrow \chi_1(x_1)\,\chi_2(x_2) \cdots \chi_m(x_m)$$

defined on $\mathop{\mathsf{P}}\limits_{j=1}^{m} G_j$. *Then the mapping* $(\chi_1, \ldots, \chi_m) \rightarrow [\chi_1, \ldots, \chi_m]$, *which we denote by* θ, *is a topological isomorphism of* $\mathop{\mathsf{P}}\limits_{j=1}^{m} \mathsf{X}_j$ *onto the character group of* $\mathop{\mathsf{P}}\limits_{j=1}^{m} G_j$.[1]

Proof. It is obvious that every function $[\chi_1, \ldots, \chi_m]$ is a continuous character of $\mathop{\mathsf{P}}\limits_{j=1}^{m} G_j$, and that the mapping θ is an isomorphism of $\mathop{\mathsf{P}}\limits_{j=1}^{m} \mathsf{X}_j$ into the character group of $\mathop{\mathsf{P}}\limits_{j=1}^{m} G_j$. Let ψ be any continuous character of $\mathop{\mathsf{P}}\limits_{j=1}^{m} G_j$. For every $(x_1, \ldots, x_m)$, we have

$$\psi(x_1, \ldots, x_m) = \psi(x_1, e_2, \ldots, e_m)\,\psi(e_1, x_2, e_3, \ldots, e_m) \cdots \psi(e_1, e_2, \ldots, e_{m-1}, x_m)$$

[where e_j is the identity element of G_j]. If we define χ_j on G_j by $\chi_j(x_j) = \psi(e_1, \ldots, e_{j-1}, x_j, e_{j+1}, \ldots, e_m)$, then it is plain that $\chi_j \in \mathsf{X}_j$ and that $\psi = [\chi_1, \ldots, \chi_m]$. Thus the isomorphism θ carries $\mathop{\mathsf{P}}\limits_{j=1}^{m} \mathsf{X}_j$ onto the character group of $\mathop{\mathsf{P}}\limits_{j=1}^{m} G_j$.

If $\mathsf{P}(F, \varepsilon)$ is a neighborhood of the identity of the character group of $\mathop{\mathsf{P}}\limits_{j=1}^{m} G_j$, and if F_j denotes the projection of F onto G_j, then

$$\theta\left(\mathop{\mathsf{P}}\limits_{j=1}^{m} \mathsf{P}\left(F_j, \frac{\varepsilon}{m}\right)\right) \subset \mathsf{P}(F, \varepsilon). \tag{1}$$

If $\mathsf{P}(F_j, \varepsilon_j)$ is a neighborhood of the identity in X_j for each j, if $F = \mathop{\mathsf{P}}\limits_{j=1}^{m} (F_j \cup \{e_j\})$, and if $\varepsilon = \min(\varepsilon_1, \ldots, \varepsilon_m)$, then

$$\mathsf{P}(F, \varepsilon) \subset \theta\left(\mathop{\mathsf{P}}\limits_{j=1}^{m} \mathsf{P}(F_j, \varepsilon_j)\right). \tag{2}$$

Relations (1) and (2) show that θ is a homeomorphism. □

We next point out a trivial but useful fact.

(23.19) Lemma. *Let G be a compact group, not necessarily Abelian, and let χ be a continuous character of G. Then*

(i) $\int_G \chi(x)\,dx = 0$

unless $\chi = 1$, *in which case the integral is* 1.

Proof. For all $a \in G$, we have

$$\int_G \chi(x)\,dx = \int_G \chi(ax)\,dx = \chi(a) \int_G \chi(x)\,dx. \tag{1}$$

If $\chi(a) \neq 1$, (1) can hold only if $\int_G \chi(x)\,dx = 0$. □

---

[1] We abbreviate this assertion by the statement that the character group of $\mathop{\mathsf{P}}\limits_{j=1}^{m} G_j$ is the product $\mathop{\mathsf{P}}\limits_{j=1}^{m} \mathsf{X}_j$ of the character groups.

(23.20) Theorem. *Let G be a compact Abelian group, and let* Y *be a subgroup of* X *such that for every $a \neq e$ in G, there is a $\chi \in \mathsf{Y}$ such that $\chi(a) \neq 1$. Then* Y *is equal to* X.

Proof. Let $\mathfrak{T}$ denote the linear space of all complex-valued functions on G of the form $\sum_{j=1}^{m} \alpha_j \chi_j$, where $\alpha_1, \ldots, \alpha_m$ are in K and $\chi_1, \ldots, \chi_m$ are in Y. Since Y is a subgroup of X, $\mathfrak{T}$ is a subalgebra of $\mathfrak{C}(G)$ closed under the formation of complex conjugates. If a, b are distinct elements of G, then there is a $\chi \in \mathsf{Y}$ such that $\chi(ab^{-1}) \neq 1$, and $\chi(a) \neq \chi(b)$. Thus $\mathfrak{T}$ separates points of G. By the STONE-WEIERSTRASS theorem, $\mathfrak{T}$ is uniformly dense in $\mathfrak{C}(G)$. Now assume that there is a continuous character ψ of G such that $\psi \notin \mathsf{Y}$. Then there is certainly a function $\sum_{j=1}^{m} \alpha_j \chi_j \in \mathfrak{T}$ such that

$$\left\| \psi - \sum_{j=1}^{m} \alpha_j \chi_j \right\|_u < 1;$$

the χ_j's are taken to be distinct. Then, taking note of (23.19), we write

$$1 > \int_G \left| \psi - \sum_{j=1}^{m} \alpha_j \chi_j \right|^2 d\lambda = \int_G \psi \bar{\psi}\, d\lambda - \sum_{j=1}^{m} \alpha_j \int_G \chi_j \bar{\psi}\, d\lambda - \sum_{j=1}^{m} \bar{\alpha}_j \int_G \bar{\chi}_j \psi\, d\lambda$$
$$+ \sum_{j=1}^{m} \sum_{k=1}^{m} \alpha_j \bar{\alpha}_k \int_G \chi_j \bar{\chi}_k\, d\lambda = 1 + \sum_{j=1}^{m} |\alpha_j|^2 \geqq 1.$$

This contradiction proves the theorem. □

Theorem (23.20) is useful in obtaining character groups of compact Abelian groups.

(23.21) Theorem. *Let $\{G_\iota : \iota \in I\}$ be a nonvoid family of compact Abelian groups, and let X_ι be the character group of G_ι for each $\iota \in I$. For an element (χ_ι) in $\mathsf{P}^*_{\iota \in I} \mathsf{X}_\iota$, let $[\chi_\iota]$ be the function on $\mathsf{P}_{\iota \in I} G_\iota$ such that $[\chi_\iota]((x_\iota)) = \prod_{\iota \in I} \chi_\iota(x_\iota)$ [only a finite number of terms in the product are different from 1]. All continuous characters of $\mathsf{P}_{\iota \in I} G_\iota$ have the form $[\chi_\iota]$, and the character group of $\mathsf{P}_{\iota \in I} G_\iota$ is topologically isomorphic with the group $\mathsf{P}^*_{\iota \in I} \mathsf{X}_\iota$ under the discrete topology.*

Proof. The set of all characters $[\chi_\iota]$ is plainly a subgroup of the character group of $\mathsf{P}_{\iota \in I} G_\iota$ that separates points of $\mathsf{P}_{\iota \in I} G_\iota$. Thus we can apply (23.20) to see that the characters $[\chi_\iota]$ are all of the continuous characters of $\mathsf{P}_{\iota \in I} G_\iota$, and (23.17) to see that the character group of $\mathsf{P}_{\iota \in I} G_\iota$ is discrete. It is obvious that the multiplication of characters corresponds to multiplication in the group $\mathsf{P}^*_{\iota \in I} \mathsf{X}_\iota$. □

Theorem (23.21) has an obvious complement, as follows.

(23.22) Theorem. *Let $\{G_\iota : \iota \in I\}$ be an arbitrary nonvoid family of discrete Abelian groups, and let X_ι be the [compact] character group of G_ι*

for each $\iota\in I$. *For every* $(\chi_\iota)\in \mathsf{P}_{\iota\in I}\mathsf{X}_\iota$, *let* $[\chi_\iota]$ *be the function on* $\mathsf{P}^*_{\iota\in I} G_\iota$ *such that* $[\chi_\iota]((x_\iota))=\prod_{\iota\in I}\chi_\iota(x_\iota)$ [*only a finite number of terms in this product are different from* 1]. *Every function* $[\chi_\iota]$ *is a character of* $\mathsf{P}^*_{\iota\in I} G_\iota$, *and all characters of* $\mathsf{P}^*_{\iota\in I} G_\iota$ *are functions* $[\chi_\iota]$. *If* $\mathsf{P}^*_{\iota\in I} G_\iota$ *is given the discrete topology, the mapping* $(\chi_\iota)\rightarrow[\chi_\iota]$ *is a topological isomorphism of* $\mathsf{P}_{\iota\in I}\mathsf{X}_\iota$ *onto the* [*compact*] *character group of* $\mathsf{P}^*_{\iota\in I} G_\iota$.

Proof. It is trivial that all of the functions $[\chi_\iota]$ are characters of $\mathsf{P}^*_{\iota\in I} G_\iota$. For each fixed $\varkappa\in I$, let $G_\varkappa^{(0)}$ be the subgroup of $\mathsf{P}^*_{\iota\in I} G_\iota$ consisting of all (x_ι) such that $x_\iota=e_\iota$ for $\iota\neq\varkappa$. If χ is a character of $\mathsf{P}^*_{\iota\in I} G_\iota$, then χ is a character of the subgroup $G_\varkappa^{(0)}$: write $\chi_\varkappa$ for χ with its domain restricted to $G_\varkappa^{(0)}$. Then it is clear that $\chi=[\chi_\iota]$. The mapping $(\chi_\iota)\rightarrow[\chi_\iota]$ is obviously an isomorphism of $\mathsf{P}_{\iota\in I}\mathsf{X}_\iota$ onto the character group of $\mathsf{P}^*_{\iota\in I} G_\iota$. It is a routine matter to show that this mapping is continuous. Since the groups in question are compact, the mapping is a homeomorphism. □

We next identify the character group of a quotient group.

(23.23) Definition. Let G be a locally compact Abelian group with character group X. For an arbitrary nonvoid subset H of G, let $\mathsf{A}(\mathsf{X}, H)$ be the subset of X consisting of all χ such that $\chi(H)=\{1\}$. The set $\mathsf{A}(\mathsf{X}, H)$ is called the *annihilator of* H *in* X.

(23.24) Remarks. (a) If H is as in (23.23) and H_1 is the smallest closed subgroup of G containing H, then it is clear that $\mathsf{A}(\mathsf{X}, H)=\mathsf{A}(\mathsf{X}, H_1)$.

(b) It is trivial that $\mathsf{A}(\mathsf{X}, \{e\})=\mathsf{X}$. It follows from (22.17) that $\mathsf{A}(\mathsf{X}, H)\subsetneqq\mathsf{X}$ if $H\supsetneqq\{e\}$.

(c) It is obvious that $\mathsf{A}(\mathsf{X}, H)$ is a subgroup of X. To see that $\mathsf{A}(\mathsf{X}, H)$ is closed in X, consider any $\chi_0\notin\mathsf{A}(\mathsf{X}, H)$, and an element $a\in H$ such that $\chi_0(a)\neq1$. The P-neighborhood $\chi_0\cdot\mathsf{P}(\{a\}, |\chi_0(a)-1|)$ of χ_0 is contained in $\mathsf{A}(\mathsf{X}, H)'$, so that $\mathsf{A}(\mathsf{X}, H)$ is P-closed.

(d) Let H be a compact subgroup of G. Then it is easy to see that $\mathsf{P}(H, \sqrt{3})=\mathsf{A}(\mathsf{X}, H)$, so that $\mathsf{A}(\mathsf{X}, H)$ is open.

(e) If H is a closed subgroup of G that is not locally null, then $\mathsf{A}(\mathsf{X}, H)$ is compact. This follows from (23.16) and (c).

(23.25) Theorem. *Let* G *be a locally compact Abelian group with character group* X, *and let* H *be a closed subgroup of* G. *Let* Y *be the character group of the* [*locally compact Abelian*] *group* G/H. *The group* Y *is topologically isomorphic with the group* $\mathsf{A}(\mathsf{X}, H)$. *In fact, if* φ *denotes the natural mapping* $x\rightarrow xH$ *of* G *onto* G/H, *the mapping* $\psi\rightarrow\psi\circ\varphi=\varrho(\psi)$ *is a topological isomorphism of* Y *onto* $\mathsf{A}(\mathsf{X}, H)$.

Proof. Since φ is a continuous homomorphism and $\varphi^{-1}(\{H\})=H$, it is clear that $\psi\circ\varphi\in\mathsf{A}(\mathsf{X}, H)$ for all $\psi\in\mathsf{Y}$. Since $(\psi_1\psi_2)\circ\varphi=(\psi_1\circ\varphi)(\psi_2\circ\varphi)$,

ϱ is a homomorphism. A character $\chi_1 \in \mathsf{A}(\mathsf{X}, H)$ is constant on all cosets xH, and so the function ψ_1 on G/H defined by $\psi_1(xH) = \chi_1(x)$ is well defined; it is obviously a character of G/H. It is also simple to see that ψ_1 is continuous on G/H. For $\varepsilon > 0$, there is a neighborhood U of e in G such that $|\chi_1(x) - 1| < \varepsilon$ for $x \in U$. The set $\{xH : x \in U\}$ is a neighborhood of the identity H in G/H (5.15) in which $|\psi_1(xH) - 1| < \varepsilon$. Thus ψ_1 is continuous at H and therefore continuous throughout G/H (5.40.a). Since $\chi_1 = \psi_1 \circ \varphi$, we have proved that ϱ maps Y onto $\mathsf{A}(\mathsf{X}, H)$. It is obvious that $\psi \circ \varphi = 1$ only if $\psi = 1$, so that ϱ is an isomorphism.

Finally, we show that ϱ is a homeomorphism. If F is a compact subset of G, then $\{xH \in G/H : x \in F\}$ is compact in G/H. Moreover, every compact set in G/H has this form by (5.24.b). For $\varepsilon > 0$, we have

$$\begin{aligned} &\varrho(\{\psi \in \mathsf{Y} : |\psi(xH) - 1| < \varepsilon \text{ for } x \in F\}) \\ &\qquad = \{\chi \in \mathsf{A}(\mathsf{X}, H) : |\chi(x) - 1| < \varepsilon \text{ for } x \in F\}. \end{aligned}$$

Thus ϱ maps an open basis at 1 in Y onto an open basis at 1 in $\mathsf{A}(\mathsf{X}, H)$, and ϱ is a homeomorphism. □

(23.26) Corollary. *Let G be a locally compact Abelian group with character group X, and let H be a closed subgroup of G. If a is in G and not in H, there is a $\chi \in \mathsf{A}(\mathsf{X}, H)$ for which $\chi(a) \neq 1$.*

Proof. Let Y and ϱ be as in (23.25). By (22.17), there is a $\psi \in \mathsf{Y}$ such that $\psi(aH) \neq 1$. Then $\varrho(\psi) \in \mathsf{A}(\mathsf{X}, H)$ and $\varrho(\psi)(a) = \psi(aH) \neq 1$. □

(23.27) Some examples of character groups. We now list six elementary and important examples of character groups. Section 25 contains an extended list of character groups.

(a) Consider the group T. The identity mapping $\exp(it) \to \exp(it)$ $(0 \leq t < 2\pi)$ is obviously a continuous character of T, and by itself it separates points of T. Its integral powers, that is, the functions

$$1, \exp(it), \exp(-it), \exp(2it), \exp(-2it), \ldots, \exp(nit), \exp(-nit), \ldots$$

$(n = 0, 1, 2, 3, \ldots)$, therefore exhaust all continuous characters of T (23.20). As a group under multiplication, these functions are isomorphic with the additive group Z. Thus we may say that the character group of T is the [discrete] group Z.

(b) Consider the group Z. A character χ of Z is plainly determined by the number $\chi(1)$, since $\chi(n) = \chi(1)^n$ $(n \in Z)$, and $\chi(1)$ can be any number in T. Let χ_α denote the character of Z such that $\chi(1) = \alpha$ $(\alpha \in T)$. Then the mapping $\alpha \to \chi_\alpha$ is obviously an isomorphism of T onto the character group of Z, and it is also obvious that this mapping is continuous: if $|\alpha - 1|$ is sufficiently small, then $|\alpha^j - 1|$ is arbitrarily small for a finite number of pre-assigned integers j. Hence the mapping is a homeomorphism, and we say that the character group of Z is T.

(c) Let m be an integer greater than 1, and consider the finite cyclic group $Z(m)$, which we represent as $\{0, 1, \ldots, m-1\}$ with addition modulo m. The mapping $k \to \exp\left(\frac{2\pi i k}{m}\right)$ is a character of $Z(m)$ that separates points. As in (a) we see that every character of $Z(m)$ has the form $k \to \exp\left(\frac{2\pi i k l}{m}\right)$, where l is a fixed integer such that $0 \leqq l \leqq m-1$. Thus the character group of $Z(m)$ is isomorphic with $Z(m)$.

(d) Let G be a finite [discrete] Abelian group; by (A.27), G is isomorphic with $Z(m_1) \times Z(m_2) \times \cdots \times Z(m_s)$ for integers $m_1, m_2, \ldots, m_s$ greater than 1, each of which is a power of a prime. From (c) and (23.18) we see that the character group of G is isomorphic with G itself. Every character of G has the form

$$(k_1, \ldots, k_s) \to \exp\left[2\pi i\left(\frac{k_1 l_1}{m_1} + \cdots + \frac{k_s l_s}{m_s}\right)\right],$$

where $0 \leqq l_j < m_j$ $(j = 1, 2, \ldots, s)$.

(e) Consider the additive group R. For every fixed $y \in R$, the function $x \to \exp(ixy) = \chi_y(x)$ is a continuous character of R. We will show that all continuous characters χ of R have this form. Let $A = \{x \in R : \chi(x) = 1\}$. Then A is a closed subgroup of R, and, as is easy to see, we have $A = R$ or $A = \{0\}$ or $A = \{kb : k \in Z\}$ for some positive real number b. If $A = R$, then $\chi = 1$, so that $\chi(x) = \exp(ix0) = \chi_0(x)$.

Assume that $A = \{0\}$. Then $\chi([0, 1])$ is a connected compact subset of T containing 1: that is, an arc or $\{1\}$. By assumption, $\chi([0, 1])$ is not $\{1\}$, and so there is a number $a \in]0, 1]$ such that $\chi(a) = \exp\left(\frac{2\pi i}{m}\right)$, where m is a nonzero integer. Thus $\chi(ma) = 1$, a contradiction to our assumption[1].

Therefore we may suppose that $A = \{kb : k \in Z\}$, where b is a positive number. Since

$$1 = \chi(b) = \chi\left(2 \cdot \frac{b}{2}\right) = \chi\left(\frac{b}{2}\right)^2$$

and $0 < \frac{b}{2} < b$, we have $\chi\left(\frac{b}{2}\right) = -1$. Since $\chi\left(\frac{b}{4}\right)^2 = \chi\left(\frac{b}{2}\right) = -1$, we must have $\chi\left(\frac{b}{4}\right) = \exp\left(\frac{\pi}{2} i\right)$ or $\chi\left(\frac{b}{4}\right) = \exp\left(-\frac{\pi}{2} i\right)$. If the second equality holds, we look at $\bar{\chi}$, and so may suppose with no loss of generality that $\chi\left(\frac{b}{4}\right) = \exp\left(\frac{\pi}{2} i\right)$. This is the last choice available for a continuous χ. Suppose that we have already proved that

$$\chi\left(\frac{b}{2^k}\right) = \exp\left(\frac{2\pi i}{2^k}\right)$$

[1] We can also reason as follows. The set $\chi(R)$ is a connected subgroup of T, and hence is T or $\{1\}$. If $A = \{0\}$, then χ is one-to-one, and so $\chi(R) = T$. By (5.29), χ is an open mapping and hence a homeomorphism, which is plainly impossible.

for some positive integer k. Then we must have

$$\chi\left(\frac{b}{2^{k+1}}\right) = \exp\left(\frac{2\pi i}{2^{k+1}}\right)$$

or

$$\chi\left(\frac{b}{2^{k+1}}\right) = \exp\left(\frac{2\pi i}{2^{k+1}} + \pi i\right).$$

Assume that the second equality holds. Then $\chi\left(\left[\frac{b}{2^{k+1}}, \frac{b}{2^k}\right]\right)$ is a compact connected set in T and must contain $+1$ or -1. Thus for some $a \in \left]\frac{b}{2^{k+1}}, \frac{b}{2^k}\right[$, we have $\chi(a)=1$ or $\chi(a)=-1$. In either case, $\chi(2a)=1$ and $2a < \frac{b}{2^{k-1}} \leqq b$. This contradicts the fact that b is the smallest positive number x such that $\chi(x)=1$, and so $\chi\left(\frac{b}{2^{k+1}}\right) = \exp\left(\frac{2\pi i}{2^{k+1}}\right)$. It is now obvious that $\chi(rb) = \exp(2\pi i r)$ for all dyadic rational numbers r. By continuity we have $\chi(xb) = \exp(2\pi i x)$ for all real numbers x: equivalently,

$$\chi(x) = \exp\left(ix\left(\frac{2\pi}{b}\right)\right) = \chi_{\frac{2\pi}{b}}(x).$$ [1]

It is easy to see that the mapping $y \rightarrow \chi_y$ is an isomorphism, so that the group R is isomorphic with its own character group. It is also easy to see that the ordinary topology of R agrees with the P-topology on the group of characters χ_y. That is, the mapping $y \rightarrow \chi_y$ is a topological isomorphism of R onto its own character group.

(f) The character group of R^n ($n=2, 3, \ldots$) is topologically isomorphic with R^n. Every continuous character of R^n has the form

$$(x_1, \ldots, x_n) \rightarrow \exp\left(i(x_1 y_1 + \cdots + x_n y_n)\right)$$

for some $(y_1, \ldots, y_n) \in R^n$. This follows at once from (e) and (23.18).

Miscellaneous theorems and examples

(23.28) Comments on Theorem (23.4). (a) Let G be a nondiscrete, locally compact group; it need not be Abelian. For every $\mu \in \boldsymbol{M}(G)$, let $\tau(\mu) = \sum_{x \in G} \mu(\{x\})$. It is easy to see that τ is a multiplicative linear functional on $\boldsymbol{M}(G)$ and that $\tau(\mu^{\sim}) = \overline{\tau(\mu)}$ for all $\mu \in \boldsymbol{M}(G)$. That is, the mapping $\mu \rightarrow \tau(\mu)$ is a 1-dimensional $\sim$-representation of $\boldsymbol{M}(G)$. If $\tau(\mu)$ were representable in the form (23.4.iv), or even in the form (22.3.iii) for some Borel measurable representation of G, then there would be a continuous character χ of G such that

$$\int_G \bar{\chi}\, d\mu = \sum_{x \in G} \mu(\{x\})$$

---

[1] The group R has many [$2^{\mathfrak{c}}$ to be exact] discontinuous characters. We discuss these in (25.6).

for all $\mu \in \boldsymbol{M}(G)$: the continuity of χ follows from (22.19). Thus χ would have to be different from 0 while at the same time $\int_G \bar{\chi}\, d\mu = 0$ for all continuous measures μ. This is clearly impossible, and so τ has no representation as $\int_G \bar{\chi}\, d\mu$.

(b) Let G and $\tau(\mu)$ be as in (a) and let $\sigma(\mu) = \mu(G)$, for all $\mu \in \boldsymbol{M}(G)$. Then σ is also a multiplicative linear functional on $\boldsymbol{M}(G)$ [use (19.11) to show that σ is multiplicative]. The functionals τ and σ are equal on $\boldsymbol{M}_d(G)$ but are different on $\boldsymbol{M}_c(G)$.

(c) Let χ be any character of G, continuous or discontinuous. Let $\tau_\chi(\mu) = \sum_{x \in G} \overline{\chi(x)}\, \mu(\{x\})$. Then τ_χ is a multiplicative linear functional on $\boldsymbol{M}(G)$ that vanishes on $\boldsymbol{M}_c(G)$. The τ of part (a) is τ_1 of the present construction.

(d) Let χ be a continuous character of G. Let $\sigma_\chi(\mu) = \int_G \bar{\chi}\, d\mu$. Then σ_χ is a multiplicative linear functional on $\boldsymbol{M}(G)$. The functional σ of part (b) is σ_1.

(23.29) The annihilator $\mathsf{A}(\mathsf{X}, H)$. (a) Let G be a locally compact Abelian group with character group X, and let H be a closed subgroup of G. If H is also open, then $\mathsf{A}(\mathsf{X}, H)$ is compact. The converse is true but requires the duality theorem (24.8). [The group G/H is discrete by (5.21), and therefore its character group is compact (23.17). Thus (23.25) shows that $\mathsf{A}(\mathsf{X}, H)$ is compact.]

(b) Let G and X be as in (a). If H_1 and H_2 are closed subgroups of G and H_3 is the smallest closed subgroup containing H_1 and H_2, then $\mathsf{A}(\mathsf{X}, H_1) \cap \mathsf{A}(\mathsf{X}, H_2) = \mathsf{A}(\mathsf{X}, H_3)$. Note that H_3 is not necessarily $H_1 H_2$; see the last paragraph of the proof of (4.4).

(23.30) Extending characters of dense subgroups. Let G be a topological group [not necessarily Abelian], and let H be a dense subgroup of G. Every continuous character of H can be extended in one and only one way as a continuous function on G, and this extension is a [continuous] character of G. [To see this, note first that the character of H is uniformly continuous in both the left and right uniform structures of H. Like all uniformly continuous functions, it can be continuously extended over G. It is easy to verify that the extended continuous function is also a character.]

(23.31) Some homomorphisms and automorphisms. (a) The only continuous automorphisms of T onto itself are the identity mapping $\exp(it) \to \exp(it)$ and the conjugate mapping $\exp(it) \to \exp(-it)$. [Clearly a continuous automorphism of T must be a one-to-one continuous character. Now apply (23.27.a).]

(b) Let φ be a continuous homomorphism of T into itself. If $\varphi(T) \neq \{1\}$, then $\varphi(T) = T$, and $T/\varphi^{-1}(\{1\})$ is topologically isomorphic with T. [This follows at once from (23.27.a) and (5.27).]

(c) The continuous homomorphisms τ of R into itself have the form $\tau(x) = \alpha x$ for some $\alpha \in R$. [Let $\alpha = \tau(1)$; then compute $\tau(r)$ for rational r and use continuity.]

(23.32) Groups with no continuous characters except 1. (a) (HEWITT and ZUCKERMAN [1], SMITH [1].) Let E be any topological linear space over R (B.5). Let χ be a continuous character of E, where E is regarded as a topological Abelian group under $+$. Then there is a continuous linear functional f on E (B.2) such that

(i) $\chi(x) = \exp[2\pi i f(x)]$

for all $x \in E$. Conversely, every function (i) is a continuous character of E. [1]

[Choose a balanced neighborhood U of 0 in E (B.5) such that $|\chi(x) - 1| < 2$ for all $x \in U$, *i.e.*, $\chi(x) \neq -1$ if $x \in U$. For each $x \in U$, there is exactly one real number $f_0(x)$ such that $-\frac{1}{2} < f_0(x) < \frac{1}{2}$ and $\chi(x) = \exp[2\pi i f_0(x)]$. We now extend f_0 to all of E: if $x \in E$, $\alpha x \in U$, and $\alpha \neq 0$, let

$$f(x) = \frac{1}{\alpha} f_0(\alpha x). \tag{1}$$

When we have shown that f is well defined, it will follow immediately that f agrees with f_0 on U. Let x be any nonzero element of E, and suppose that $\alpha_0 x$ and $\alpha_1 x$ are in U. The subgroup $\{\alpha x : \alpha \in R\}$ is a closed topological isomorph of R in E. The character χ is continuous on this subgroup, and by (23.27.e), there is a unique $\beta \in R$ such that $\chi(\alpha x) = \exp[2\pi i \alpha \beta]$ for all $\alpha \in R$. Therefore $\alpha_0 \beta = f_0(\alpha_0 x) + n_0$ for some $n_0 \in Z$. If $n_0 \neq 0$, then $|\alpha_0 \beta| > \frac{1}{2}$. Therefore $\frac{1}{2\alpha_0 \beta} \cdot \alpha_0 x = \frac{x}{2\beta}$ belongs to U and $\chi\left(\frac{x}{2\beta}\right) = -1$. This contradiction shows that $n_0 = 0$. Similarly $\alpha_1 \beta = f_0(\alpha_1 x)$ and hence $\beta = \frac{1}{\alpha_0} f_0(\alpha_0 x) = \frac{1}{\alpha_1} f_0(\alpha_1 x)$. Therefore f is well defined and

$$\chi(x) = \exp[2\pi i \beta] = \exp[2\pi i f(x)]. \tag{2}$$

To see that f is homogeneous, let β be a nonzero real number and x any nonzero element of E [the cases $\beta = 0$ and $x = 0$ are trivial]. There is a nonzero $\alpha \in R$ such that $\alpha x \in U$ and $\alpha \beta x \in U$. By (1), we have

$$\beta f(x) = \beta \frac{1}{\alpha\beta} f_0(\alpha\beta x) = \frac{1}{\alpha} f_0(\alpha\beta x) = f(\beta x).$$

It is also simple to show that f is continuous at 0. Let ε be a number such that $0 < \varepsilon < \frac{1}{2}$, and let V be a neighborhood of 0 in E such that

[1] For a theorem dealing with a similar situation on locally compact Abelian groups, see (24.43) *infra*.

$V \subset U$ and $|\chi(x)-1| < 2\sin(\pi\varepsilon)$ for $x \in V$. For $x \in V$, we have

$$|\chi(x) - 1| = 2\sin(\pi|f(x)|) < 2\sin(\pi\varepsilon),$$

which implies that $|f(x)| < \varepsilon$.

Finally we show that f is additive. For $x, y \in E$, relation (2) shows at once that $f(x+y) - f(x) - f(y)$ is an integer. If W is a neighborhood of 0 such that $|f(x)|$, $|f(y)|$, and $|f(x+y)|$ are less than $\frac{1}{3}$ for all $x, y \in W$, then it is clear that $f(x+y) = f(x) + f(y)$ for all $x, y \in W$. For arbitrary $x, y \in E$, let α be a nonzero real number such that αx, αy, and $\alpha(x+y)$ are in W. Then we have $f(x+y) = \frac{1}{\alpha} f(\alpha(x+y)) = \frac{1}{\alpha}[f(\alpha x) + f(\alpha y)] = f(x) + f(y)$.

The converse is trivial: every continuous linear functional f automatically defines a continuous character of E by formula (i).]

(b) Let X be a locally compact Hausdorff space, and let I and ι be as defined in §11. Suppose that $0 < \iota(X) < \infty$ and that $\iota(\{x\}) = 0$ for every $x \in X$. Let $\mathfrak{M}(X, \iota)$ [we write this as $\mathfrak{M}$ throughout the present discussion] be the set of *all* ι-measurable real-valued functions on X, two functions in $\mathfrak{M}$ being identified if they are equal ι-almost everywhere. Define the metric $\varrho(f, g)$ by

(ii) $$\varrho(f, g) = \int_X \frac{|f-g|}{1+|f-g|}\, d\iota.$$

With the topology induced by the metric ϱ, $\mathfrak{M}$ is a topological linear space admitting no nonzero continuous linear functionals[1]. Hence as a topological Abelian group, $\mathfrak{M}$ has no continuous characters save 1. The space $\mathfrak{M}$ is complete under the metric ϱ. [The function $\varphi(t) = \frac{t}{1+t}$ is increasing in $[0, \infty[$, so that for $\alpha, \beta \in R$, we have $\frac{|\alpha+\beta|}{1+|\alpha+\beta|} \leq \frac{|\alpha|+|\beta|}{1+|\alpha|+|\beta|} \leq \frac{|\alpha|}{1+|\alpha|} + \frac{|\beta|}{1+|\beta|}$. From this, the triangle inequality $\varrho(f, h) \leq \varrho(f, g) + \varrho(g, h)$ follows immediately. The other axioms for a metric are obvious. For $f, g, h \in \mathfrak{M}$, it is clear that $\varrho(f+h, g+h) = \varrho(f, g)$. For $\alpha \in R$ and $f \in \mathfrak{M}$, the inequality $\varrho(\alpha f, 0) \leq \max(|\alpha|, 1)\, \varrho(f, 0)$ is easy to check. Thus for $\alpha, \alpha_0 \in R$ and $f, f_0 \in \mathfrak{M}$, we have

$$\begin{aligned} \varrho(\alpha f, \alpha_0 f_0) &\leq \varrho(\alpha f, \alpha f_0) + \varrho(\alpha f_0, \alpha_0 f_0) \\ &\leq \max(|\alpha|, 1)\, \varrho(f - f_0, 0) + \varrho((\alpha - \alpha_0) f_0, 0). \end{aligned}$$

Dominated convergence shows that $\lim_{\alpha \to \alpha_0} \varrho((\alpha - \alpha_0) f_0, 0) = 0$; from this we infer that $(\alpha, f) \to \alpha f$ is a continuous mapping of $R \times \mathfrak{M}$ onto $\mathfrak{M}$. It is also clear that $(f, g) \to f + g$ is a continuous mapping of $\mathfrak{M} \times \mathfrak{M}$ onto $\mathfrak{M}$. Therefore $\mathfrak{M}$ is a topological linear space.

[1] This property of $\mathfrak{M}$ was pointed out to us by Victor L. Klee.

Since ι vanishes for points, we can partition X into ι-measurable sets $A_1, \ldots, A_m$ of arbitrarily small positive measure, say $\iota(A_j) < \varepsilon$ (11.44). Let f be any function in $\mathfrak{M}$, and let $g_j = m f \xi_{A_j}$ ($j = 1, 2, \ldots, m$). Then it is clear that $\varrho(g_j, 0) < \varepsilon$ and that $f = \frac{1}{m} \sum_{j=1}^{m} g_j$. If Φ is a continuous linear functional on $\mathfrak{M}$, then $\Phi(h)$ is arbitrarily small for $\varrho(h, 0)$ sufficiently small. Since $|\Phi(f)| \leqq \max(|\Phi(g_1)|, |\Phi(g_2)|, \ldots, |\Phi(g_m)|)$, it follows that $\Phi(f) = 0$.

The completeness of $\mathfrak{M}$ is proved by an argument similar to that of (12.8), which we omit.]

(c) Let p be a number such that $0 < p < 1$. Let X be a locally compact Hausdorff space and let I and ι be a functional and a measure for X as constructed in §11. Define $\mathfrak{L}_p^r(X, \iota)$ as in (12.1); we abbreviate this as $\mathfrak{L}_p$ in the present discussion. For every $\varepsilon > 0$, let $U_\varepsilon = \{f \in \mathfrak{L}_p : \|f\|_p < \varepsilon\}$. With the U_ε's taken as an open basis at 0, $\mathfrak{L}_p$ is a complete metrizable topological linear space over R. [We first prove that

$$\|f + g\|_p \leqq 2^{\frac{1}{p}-1} (\|f\|_p + \|g\|_p) \quad \text{for} \quad f, g \in \mathfrak{L}_p. \tag{3}$$

If t is a nonnegative number, then

$$1 + t^p \geqq (1 + t)^p.$$

This is easy to see by considering the function $\varphi(t) = 1 + t^p - (1 + t)^p$, for which $\varphi(0) = 0$ and $\varphi'(t) > 0$ for $t > 0$. It follows that

$$|f(x) + g(x)|^p \leqq |f(x)|^p + |g(x)|^p$$

wherever f and g are finite and so

$$\|f + g\|_p^p \leqq \|f\|_p^p + \|g\|_p^p. \tag{4}$$

Also the function $\psi(t) = \left(1 + t^{\frac{1}{p}}\right)(1 + t)^{-\frac{1}{p}}$ has exactly one minimum value in $[0, \infty[$, namely at $t = 1$, and $\psi(1) = 2^{1-\frac{1}{p}}$. This implies that

$$[\|f\|_p^p + \|g\|_p^p]^{\frac{1}{p}} \leqq 2^{\frac{1}{p}-1} [\|f\|_p + \|g\|_p],$$

which together with (4) yields (3).

It is easy to see from (3) that $\mathfrak{L}_p$ is a topological linear space. Since $\mathfrak{L}_p$ has a countable open basis at 0, Theorem (8.3) implies that it admits an invariant metric σ yielding the topology described above: in fact $\|f - g\|_p^p$ is such a metric. The proof that $\mathfrak{L}_p$ is complete in the metric σ is very like the proof of (12.8), and is omitted.]

(d) (Day [1], proof adapted from Robertson [1].) Let p, X, ι, and $\mathfrak{L}_p$ be as in (c), and suppose further that $\iota(\{x\}) = 0$ for all $x \in X$. Then $\mathfrak{L}_p$ admits no nonzero continuous linear functionals, and so $\mathfrak{L}_p$ as an additive group admits no continuous characters save 1. [Let Φ be a nonzero linear functional on $\mathfrak{L}_p$; we will show that it is discontinuous. There is an $f \in \mathfrak{L}_p$ such that $\Phi(f) = 1$. Let I_f be the functional $\varphi \rightarrow \int_X \varphi |f|^p \, d\iota$ on $\mathfrak{C}_{00}(X)$, and let ι_f be the measure corresponding to I_f as in §11. Theorem (14.17) shows that $\iota_f(X) = \|f\|_p^p < \infty$, and it is clear that $\iota_f(\{x\}) = 0$ for all $x \in X$. By (11.44), there is a σ-compact subset E of X such that $\int_E |f|^p \, d\iota = \frac{1}{2} \int_X |f|^p \, d\iota$. Let $g_1 = f \xi_E$ and $h_1 = f \xi_{E'}$. Then $\int_X |g_1|^p \, d\iota = \int_X |h_1|^p \, d\iota = \frac{1}{2} \int_X |f|^p \, d\iota$, and $\Phi(g_1 + h_1) = \Phi(g_1) + \Phi(h_1) = 1$. If $\Phi(g_1) \geqq \frac{1}{2}$, let $f_1 = 2g_1$; otherwise, let $f_1 = 2h_1$. Then we have $\Phi(f_1) \geqq 1$ and $\|f_1\|_p = 2\,\|g_1\|_p = 2^{1-\frac{1}{p}} \|f\|_p$. Continuing this process, we find a sequence $f_1, f_2, \ldots, f_n, \ldots$ of functions in $\mathfrak{L}_p$ such that $\Phi(f_n) \geqq 1$ and $\|f_n\|_p = 2^{(1-\frac{1}{p})n} \|f\|_p$. Since $0 < p < 1$, we have $\lim_{n \to \infty} \|f_n\|_p = 0$, and Φ is necessarily discontinuous.]

(23.33) Character groups of local direct products (BRACONNIER [1]). Consider a family $\{G_\iota\}_{\iota \in I}$ of locally compact Abelian groups, and for each $\iota \in I$, suppose that H_ι is a compact open subgroup of G_ι. Let G denote the local direct product of the groups G_ι relative to the open subgroups H_ι. Write H for the compact open subgroup $\underset{\iota \in I}{\mathsf{P}} H_\iota$ of G. Let X_ι and X be the character groups of G_ι and G, respectively. Identify G_ι with the subgroup of G consisting of all (x_ι) such that $x_\varkappa = e_\varkappa$ for $\varkappa \neq \iota$. For $\chi \in \mathsf{X}$, the restriction $\chi | G_\iota$ is a continuous character of G_ι, which we write as χ_ι. Write $\tau(\chi) = (\chi_\iota) \in \underset{\iota \in I}{\mathsf{P}} \mathsf{X}_\iota$. Then τ is plainly a homomorphism of X into $\underset{\iota \in I}{\mathsf{P}} \mathsf{X}_\iota$. Now look at the restriction $\chi | H$. This function is a continuous character of H, and by (23.21) we infer that $\chi_\iota(H_\iota) = \{1\}$ for all but a finite number of the indices ι. That is, χ_ι is in $\mathsf{A}(\mathsf{X}_\iota, H_\iota)$ for all but a finite number of indices ι, so that $\tau(\chi)$ lies in the local direct product of the groups X_ι relative to the compact open subgroups $\mathsf{A}(\mathsf{X}_\iota, H_\iota)$. Denote this last group by the symbol Y.

Now let (χ_ι) be any element of Y. We will define a $\chi \in \mathsf{X}$ such that $\tau(\chi) = (\chi_\iota)$, as follows. For $(x_\iota) \in G$, let $\chi((x_\iota)) = \prod_{\iota \in I} \chi_\iota(x_\iota)$. Only a finite number of factors in this product are different from 1, since $\chi_\iota(H_\iota) \neq \{1\}$ for only finitely many ι, and $x_\iota \notin H_\iota$ for only finitely many ι. Plainly χ is a character of G, and it is easy to see that χ is continuous. Since $\tau(\chi) = (\chi_\iota)$, we have shown that τ maps X onto Y.

We now show that τ is a topological isomorphism of X onto Y. It is easy to show that τ is an algebraic isomorphism. Hence it suffices to show that τ is a topological isomorphism carrying the compact open subgroup $\mathsf{A}(\mathsf{X}, H)$ of X onto a compact open subgroup of Y. It is clear that $\tau(\mathsf{A}(\mathsf{X}, H)) = \underset{\iota \in I}{\mathsf{P}} \mathsf{A}(\mathsf{X}_\iota, H_\iota)$, and it is a routine matter to show that τ is continuous on $\mathsf{A}(\mathsf{X}, H)$.

The foregoing may be summarized as follows. The character group of G is topologically isomorphic with the local direct product of the character groups of the G_ι relative to the annihilators of the H_ι.

(23.34) Other homomorphism groups. (a) Let G and H be topological groups with identity elements e and f respectively, and suppose that H is Abelian. Let $\mathrm{Hom}(G, H)$ denote the set of all continuous homomorphisms carrying G into H. For $\sigma, \tau \in \mathrm{Hom}(G, H)$, let $\sigma\tau$ be the mapping $x \to \sigma(x)\,\tau(x)$. Under this multiplication, $\mathrm{Hom}(G, H)$ is an Abelian group with identity element the map $x \to f$ and τ^{-1} the mapping $x \to \tau(x)^{-1} = \tau(x^{-1})$. [It is trivial that $\sigma\tau$ is a homomorphism of G into H, and obvious that $\sigma\tau$ is continuous. The group axioms are also easy to verify.]

(b) For a compact subset F of G and a neighborhood U of f in H, let $P(F, U) = \{\sigma \in \mathrm{Hom}(G, H) : \sigma(F) \subset U\}$. The family of all such sets satisfies (4.5.i)—(4.5.v) and accordingly defines a topology under which $\mathrm{Hom}(G, H)$ is a topological Abelian group. [Each of the properties (4.5.i)—(4.5.v) is easy to verify. For example, to establish (4.5.iii), let τ be any element of $P(F, U)$. Then $\tau(F)$ is a compact subset of U. By (4.10), there is a neighborhood V of f in H such that $\tau(F) \cdot V \subset U$. Then if $\sigma \in P(F, V)$ and $x \in F$, we have $(\tau\sigma)(x) = \tau(x)\,\sigma(x) \in \tau(F) \cdot V \subset U$, and hence $\tau\big(P(F, V)\big) \subset P(F, U)$. Since H is a T_1 space and points of G are compact sets, it is easy to see that $\mathrm{Hom}(G, H)$ is a T_1 space.]

(c) Let $G_1, G_2, \ldots, G_m$ be topological groups and H a topological Abelian group. There is a topological isomorphism carrying $\mathrm{Hom}(G_1 \times \cdots \times G_m, H)$ onto $\mathrm{Hom}(G_1, H) \times \cdots \times \mathrm{Hom}(G_m, H)$. [The proof is easily adapted *mutatis mutandis* from the proof of (23.18).]

(d) Let G be a topological group and $H_1, H_2, \ldots, H_m$ topological Abelian groups. There is a topological isomorphism carrying $\mathrm{Hom}(G, H_1 \times \cdots \times H_m)$ onto $\mathrm{Hom}(G, H_1) \times \cdots \times \mathrm{Hom}(G, H_m)$. [For $\tau \in \mathrm{Hom}(G, H_1 \times \cdots \times H_m)$ and $x \in G$, we have $\tau(x) = \big(\tau_1(x), \ldots, \tau_m(x)\big)$, where $\tau_j(x) \in H_j$. It is easy to see that the mapping $x \to \tau_j(x)$ is an element of $\mathrm{Hom}(G, H_j)$. Denote by θ the mapping $\tau \to (\tau_1, \ldots, \tau_m)$ just defined. It is simple to verify that θ is an isomorphism of $\mathrm{Hom}(G, H_1 \times \cdots \times H_m)$ onto $\mathrm{Hom}(G, H_1) \times \cdots \times \mathrm{Hom}(G, H_m)$. Let F be a compact subset of G and V_j neighborhoods of the identity elements in H_j $(j = 1, 2, \ldots, m)$. Then we have $\theta\big(P(F, V_1 \times \cdots \times V_m)\big) = P(F, V_1) \times \cdots \times P(F, V_m)$, so that θ is also a homeomorphism.]

Notes

The word "character" is used by many writers in a sense different from that established in (22.15); we will take up these other characters in Vol. II, §27. For Abelian groups the distinction disappears; and in this sense one can say that characters for finite Abelian groups were invented by FROBENIUS [1]. The idea of making the characters of a group into a group as in (23.2) is due to PONTRYAGIN [1]; the theory was developed much further in PONTRYAGIN [4]. Several other writers in the early 1930's also computed character groups. Explicit computations of characters and tantalizing hints of the modern theory of harmonic analysis on compact Abelian groups appear in WIENER and PALEY [1], [2]. A. HAAR [1] found very sophisticated results about characters of countably infinite discrete Abelian groups G. He essentially defined the character group X of G, identifying it with [0, 1] under a new multiplication, but not describing topology or Haar [!] measure in X. He found the orthogonality relation (23.19.i) and PARSEVAL's equality [see Vol. II, (27.41) and (28.43)]. HAAR [2] also found interesting generalizations of his results to certain non-Abelian countable groups. HAAR's work in [1] was simplified and extended by SZ.-NAGY [2], although still without recognition of the rôle of the character group. FREUDENTHAL [4] pointed out that HAAR's work can be interpreted in terms of PONTRYAGIN's theory.

Theorem (23.4) for $\boldsymbol{M}_a(G)$, where G is Abelian, is due to GEL'FAND and RAĬKOV [1] [although with unneeded hypotheses]. Our proof is very like theirs. Theorem (23.6) for $\boldsymbol{M}_a(G)$, G Abelian, is also due to GEL'FAND and RAĬKOV [1].

All of (23.9)—(23.13) is an application of GEL'FAND's theory of Banach algebras [Appendix C] to $\boldsymbol{M}(G)$ and $\mathfrak{L}_1(G)$. Theorem (23.13) appears explicitly in GEL'FAND and RAĬKOV [1]. The P-topology of (23.15) was invented by PONTRYAGIN: in [4] for countable discrete and compact metric Abelian groups, and in [6], § 30, for all locally compact Abelian groups with countable open bases. VAN KAMPEN [1] used structure theory to topologize the character group; his topology is the same as PONTRYAGIN's. The definition for general locally compact Abelian groups is due to A. WEIL [4], § 27, p. 100. A special case of Theorem (23.15) was known to GEL'FAND and RAĬKOV [1]. The first proof of (23.15) was given by YOSIDA and IWAMURA [1]. Theorems (23.17), (23.18), (23.21), (23.22), and (23.25) are due in one form or another to PONTRYAGIN [4] and [6], VAN KAMPEN [1], and A. WEIL [4]. Lemma (23.19) goes back to WEYL and PETER [1] at least; (23.20) is due to E. HEWITT [2]. The history of (23.27) is impossible to trace. Since 1920 everyone seems to have known these facts, and nearly everyone has proved them.

The result in (23.32.a) is announced without proof by VILENKIN [13]. Necessary and sufficient conditions for the existence of sufficiently many, or no, continuous characters on a topological Abelian group have been given by FØLNER [1], [2]. Analogous conditions appear in COTLAR and RICABARRA [1]. Comparatively little attention has been paid to computing character groups of topological Abelian groups that are not locally compact. Of course the dual spaces of a host of topological linear spaces are known, and so character groups of their additive groups are known (23.32.a). For other computations, see VILENKIN [9]. More comments on this matter appear in the notes to §§24 and 25.

§ 24. The duality theorem

Making use of the structure theory of §9, of Theorem (22.17), and of facts about character groups set forth in §23, we prove here the famous theorem of PONTRYAGIN and VAN KAMPEN asserting that the character group of a character group is the original group. Of course this assertion is vague. Let us make it precise.

Throughout this section, G will denote a locally compact Abelian group and X will denote the character group of G.

(24.1) Definition. For fixed x in G, let x' be the function on X such that $x'(\chi)=\chi(x)$ for all $\chi\in\mathsf{X}$, and let τ be the mapping $\tau(x)=x'$. Let ω be the character group of X. We call ω the *second character group of* G. These notations will be fixed throughout (24.1)—(24.8).

The duality theorem asserts that τ is a topological isomorphism of G onto ω. We will prove this result in a number of steps.

(24.2) Theorem. *The mapping τ is a continuous isomorphism of G into ω.*

Proof. For $\chi,\psi\in\mathsf{X}$ and $x\in G$, we have $x'(\chi\psi)=\chi\psi(x)=\chi(x)\,\psi(x)=x'(\chi)\,x'(\psi)$, by the definition of multiplication in X. Thus each x' is a character of X. The character x' on X is continuous because $|x'(\chi)-x'(\psi)|<\varepsilon$ whenever $\chi\psi^{-1}\in\mathsf{P}(\{x\},\varepsilon)$. If x and y are distinct elements of G, (22.17) shows that there is a $\chi_0\in\mathsf{X}$ for which $\chi_0(x)\neq\chi_0(y)$; hence $x'\neq y'$. For $x,y\in G$ and $\chi\in\mathsf{X}$ we have $(xy)'(\chi)=\chi(x)\,\chi(y)=x'(\chi)\,y'(\chi)$. We have therefore shown that τ is an isomorphism carrying G into ω.

We now show that τ is continuous; by (5.40.a), it suffices to prove that τ is continuous at e. Let Γ be a compact subset of X and let ε be a positive number. Select any neighborhood U of e in G such that U^- is compact, and consider the neighborhood $\mathsf{P}(U^-,\varepsilon/2)$ of 1 in X. As Γ is compact, a finite number of sets $\chi_1\mathsf{P}(U^-,\varepsilon/2),\ldots,\chi_m\mathsf{P}(U^-,\varepsilon/2)$ cover Γ. Let V be a neighborhood of e in G such that $V\subset U$ and $|\chi_j(x)-1|<\frac{\varepsilon}{2}$ for all $x\in V$ and $j=1,2,\ldots,m$. Now, in the P-topology for ω, the set $\{f\in\omega:|f(\chi)-1|<\varepsilon \text{ for } \chi\in\Gamma\}$ is a generic neighborhood of the identity. If $x\in V\subset G$ and $\chi\in\Gamma$, then $|\chi_j(x)-1|<\frac{\varepsilon}{2}$ for all j and $|\chi_j(x)-\chi(x)|<\frac{\varepsilon}{2}$ for some j. Hence $|\chi(x)-1|<\varepsilon$, *i.e.*, $|x'(\chi)-1|<\varepsilon$ for all $\chi\in\Gamma$, and the mapping τ is continuous. $\square$

We now prove the duality theorem in an important special case.

(24.3) Theorem. *Let G be a topological Abelian group that is either compact or discrete. Then the mapping τ of (24.1) is a topological isomorphism of G onto $\mathfrak{Ω}$.*

Proof. Suppose first that G is discrete. Then X is compact (23.17), and the group $\tau(G)$ of continuous characters of X separates points of X. Theorem (23.20) shows that $\tau(G) = \mathfrak{Ω}$. Since G and $\mathfrak{Ω}$ are discrete, τ is trivially a homeomorphism.

Suppose next that G is compact. Then X is discrete and $\mathfrak{Ω}$ is compact (23.17). Since τ is continuous and one-to-one, $\tau(G)$ is homeomorphic with G and as G is compact, $\tau(G)$ is a closed subgroup of $\mathfrak{Ω}$. Assume that $\tau(G) \subsetneqq \mathfrak{Ω}$; then (23.26) implies that there is a continuous character, say ψ, of $\mathfrak{Ω}$ such that $\psi(\tau(G)) = \{1\}$ and $\psi \neq 1$. Since X is discrete, the preceding paragraph shows that every continuous character of $\mathfrak{Ω}$ arises from an element of X. In particular, there is a $\chi_0 \in \mathsf{X}$ such that $\psi(\omega) = \omega(\chi_0)$ for all $\omega \in \mathfrak{Ω}$. We find that $\psi(x') = x'(\chi_0) = \chi_0(x) = 1$ for all $x \in G$, while $\chi_0 \neq 1$. This contradiction proves that $\tau(G) = \mathfrak{Ω}$. ☐

In order to prove the duality theorem for all locally compact Abelian groups, we must first establish (9.5), in order to be able to appeal to (9.8). Three technical lemmas are required for this purpose.

(24.4) Lemma. *Let G be a locally compact Abelian group, let H be a closed subgroup of G, and let ψ be a continuous character of H. Then if H is compact or open, there is a continuous character χ of G such that $\chi(x) = \psi(x)$ for all $x \in H$.*[1]

Proof. Suppose first that H is compact. Then the extensible continuous characters of H are a subgroup of the character group of H that separates points of H (22.17). Hence all continuous characters of H are extensible (23.20).[2]

Suppose next that H is open. The character ψ can be extended to a character of G by (A.7), and since ψ is continuous on H, the extension is necessarily continuous on G (5.40.a). ☐

(24.5) Lemma. *Let G be a locally compact Abelian group and let H be an open subgroup of G. Let X be the character group of G and Y the character group of H. The mapping $\chi \to \chi|H = \sigma(\chi)$ is an open continuous homomorphism of X onto Y with kernel $\mathsf{A}(\mathsf{X}, H)$. Thus Y is topologically isomorphic with $\mathsf{X}/\mathsf{A}(\mathsf{X}, H)$.*

---

[1] We abbreviate this assertion by stating that *all continuous characters of H are extensible over G*. The lemma is true for all H, as we will prove below (24.12).

[2] We *need* the present lemma only for open subgroups H, but the proof for compact subgroups H is so amusing that we include it.

Proof. The mapping σ merely restricts the domain of definition of χ. The definition of multiplication in X makes it obvious that σ is a homomorphism of X into Y, and our Lemma (24.4) implies that $\sigma(\mathsf{X})=\mathsf{Y}$. If E is a compact subset of H, then $\sigma^{-1}\{\chi\in\mathsf{Y}:|\chi(x)-1|<\varepsilon \text{ for } x\in E\}=\mathsf{P}(E,\varepsilon)$. Since E is compact in G, σ is continuous.

Choose a compact subset E of H such that $\lambda(E)>0$, write $\mathsf{P}_{\mathsf{X}}(E,\frac{1}{2})$ for the set $\{\chi\in\mathsf{X}:|\chi(x)-1|<\frac{1}{2} \text{ for } x\in E\}$, and write $\mathsf{P}_{\mathsf{Y}}(E,\frac{1}{2})$ for the set $\{\chi\in\mathsf{Y}:|\chi(x)-1|<\frac{1}{2} \text{ for } x\in E\}$. We know from (23.16) that $\mathsf{P}_{\mathsf{X}}(E,\frac{1}{2})$ has compact closure in X. Let X_1 and Y_1 be the subgroups of X and Y, respectively, generated by $\mathsf{P}_{\mathsf{X}}(E,\frac{1}{2})$ and $\mathsf{P}_{\mathsf{Y}}(E,\frac{1}{2})$. The obvious equality $\sigma(\mathsf{P}_{\mathsf{X}}(E,\frac{1}{2}))=\mathsf{P}_{\mathsf{Y}}(E,\frac{1}{2})$ implies that $\sigma(\mathsf{X}_1)=\mathsf{Y}_1$. As X_1 is σ-compact, Theorem (5.29) shows that σ is an open mapping of X_1 onto Y_1. Since X_1 and Y_1 are open subgroups of X and Y, respectively, we infer that σ is an open homomorphism of X onto Y. □

(24.6) Lemma. *Let G be a discrete Abelian group, and let $\mathsf{P}(F,\varepsilon)$ be an arbitrary neighborhood of the identity in the character group X of G. Then there is a subgroup H of G having a finite number of generators such that $\mathsf{A}(\mathsf{X},H)\subset\mathsf{P}(F,\varepsilon)$.*

Proof. This lemma is trivial. Since G is discrete, F is finite. Let H be the smallest subgroup of G that contains F. □

(24.7) Theorem. *Let G be a compact Abelian group and let U be a neighborhood of e in G. There is a closed subgroup H of G such that $H\subset U$ and G/H is topologically isomorphic with $T^a\times F$, where a is a nonnegative integer and F is a finite group.*

Proof. Consider the [discrete] character group X of G, and let Ξ be any subgroup of X having a finite number of generators. By (A.27), Ξ is isomorphic with $Z^a\times F$, where a is a nonnegative integer and F is a finite group. By (23.18), (23.27.b), and (23.27.d), the character group of Ξ is topologically isomorphic with $T^a\times F$. By (24.3), the mapping τ defined in (24.1) is a topological isomorphism of G onto the character group of X. By (24.5), the character group of Ξ, that is $T^a\times F$, is topologically isomorphic with $\tau(G)/\mathsf{A}(\tau(G),\Xi)$. By (24.6), the neighborhood $\tau(U)$ of $\tau(e)$ in $\tau(G)$ contains a subgroup of the form $\mathsf{A}(\tau(G),\Xi)$. □

Theorem (24.7), which we have just proved, is exactly Theorem (9.5). The reader should note that our argument is not circular; we use neither (9.5) nor any of its consequences to prove (24.7). Since the proof of (24.7) does require some knowledge of characters, we have merely postponed its proof from §9 to this more convenient place.

We can now prove the duality theorem in full generality.

(24.8) Theorem. *Let G be a locally compact Abelian group, let X be the character group of G, and let Ω be the character group of X. The mapping τ of (24.1) is a topological isomorphism carrying G onto Ω.*

Proof. (I) Suppose first that G is compactly generated. We appeal to the structure theorem (9.8), which tells us that G is topologically isomorphic with $R^a \times Z^b \times F$ for nonnegative integers a and b and a compact Abelian group F. The second character group of R can be identified with R. By (23.27.e), every continuous character of R has the form $x \rightarrow \exp(ixy) = \chi_y(x)$ for some $y \in R$, and the mapping $\chi_y \rightarrow y$ is a topological isomorphism of the character group of R onto R. Hence every function in the second character group of R has the form $\chi_y \rightarrow \exp(i\alpha y)$ for some real number α. This is just the function α' operating on the character group of R. Thus for the group R, the function τ of (24.1) maps R *onto* the second character group of R. By (24.3), the groups Z and F also have this property. Theorem (23.18) shows that the function τ defined for the group $R^a \times Z^b \times F$ maps this group *onto* its second character group. Since τ is continuous and G is σ-compact, τ is a topological isomorphism of G onto its second character group (5.29).

(II) Suppose now that G is any locally compact Abelian group, and that τ is defined on G. Let H be a compactly generated open subgroup of G: we may suppose that H contains any fixed compact subset of G (5.14). Let Y denote the character group of H. For $x \in H$, it is plain that $\tau(x) \in \mathsf{A}(\omega, \mathsf{A}(\mathsf{X}, H))$. Suppose that $f \in \mathsf{A}(\omega, \mathsf{A}(\mathsf{X}, H))$. If χ and χ_0 are in X and $\chi|H = \chi_0|H$, then $f(\chi) = f(\chi_0)$. By (24.5), every character ψ in Y has the form $\chi|H$ for some $\chi \in \mathsf{X}$. Hence if we assign to each character ψ in Y the number $f(\chi)$ where $\chi|H = \psi$, we obtain a well-defined character f_0 of Y. The continuity of f_0 follows from the fact that the mapping $\chi \rightarrow \chi|H$ is open (24.5). By (I), we infer that $f(\chi) = \chi(x)$ for some $x \in H$ and all $\chi \in \mathsf{X}$. Thus τ maps H *onto* the subgroup $\mathsf{A}(\omega, \mathsf{A}(\mathsf{X}, H))$ of ω. Since H is open, G/H is discrete (5.21). As the character group of G/H (23.25), $\mathsf{A}(\mathsf{X}, H)$ is thus compact (23.17). The annihilator of a compact subgroup is open (23.24.d), and so $\tau(H) = \mathsf{A}(\omega, \mathsf{A}(\mathsf{X}, H))$ is an open subgroup of ω. By (5.29), τ is a topological isomorphism on H and by (5.40.a), it follows that τ is a topological isomorphism of G onto $\tau(G)$.

(III) To complete the proof, we need only to prove that τ maps G onto ω. We will prove that for every $f \in \omega$, there is an H as in (II) for which $f \in \tau(H)$. Let $\mathsf{P}(F, \varepsilon)$ be a neighborhood of 1 in X such that $|f(\chi) - 1| < \sqrt{3}$ for all $\chi \in \mathsf{P}(F, \varepsilon)$. Let U be any symmetric neighborhood of e in G with compact closure. Let H_1 be the subgroup of G generated by U. Then H_1 is open and compactly generated (5.13). Let Y_1 be the character group of G/H_1, and let ϱ be the topological isomorphism of Y_1 onto $\mathsf{A}(\mathsf{X}, H_1)$ given in (23.25). Since G/H_1 is discrete, it contains a subgroup C with a finite number of generators $x_1H_1, x_2H_1, \ldots, x_mH_1$ such that $\mathsf{A}(\mathsf{Y}_1, C) \subset \varrho^{-1}(\mathsf{P}(F, \varepsilon) \cap \mathsf{A}(\mathsf{X}, H_1))$ (24.6). Now let H be an open, compactly generated subgroup of G that contains U^- and $x_1, \ldots, x_m$ (5.14). If $\chi \in \mathsf{A}(\mathsf{X}, H)$, then $\chi(H_1) = \{1\}$ and $\chi(x_j) = 1$. Therefore we have $\chi \in \mathsf{A}(\mathsf{X}, H_1)$ and $\varrho^{-1}(\chi) \in \mathsf{A}(\mathsf{Y}_1, C)$. It follows that $\chi \in \mathsf{P}(F, \varepsilon)$, and so $|f(\chi) - 1| < \sqrt{3}$. Thus $|f(\chi) - 1| < \sqrt{3}$ for all χ in the subgroup $\mathsf{A}(\mathsf{X}, H)$ of X, which implies that $f(\chi) = 1$ for all $\chi \in \mathsf{A}(\mathsf{X}, H)$. That is, f is in $\mathsf{A}(\omega, \mathsf{A}(\mathsf{X}, H)) = \tau(H)$. □

(24.9) Scholium. Theorem (24.8) permits us to identify G with its own second character group, although conceptually the two objects are of course quite distinct. Given a locally compact Abelian group G with character group X, we consider the mapping of $G \times \mathsf{X}$ into T defined by

(i) $(x, \chi) \to \chi(x)$.

Then every continuous character of G and of X is obtained by fixing χ or x. For a closed subgroup Y of X, we write $A(G, \mathsf{Y})$ for the set $\{x \in G : \chi(x) = 1 \text{ for all } \chi \in \mathsf{Y}\}$.

We now draw some inferences from (24.8).

(24.10) Theorem. *Let H be a closed subgroup of G. Then $H = A(G, \mathsf{A}(\mathsf{X}, H))$.*

Proof. It is obvious that $H \subset A(G, \mathsf{A}(\mathsf{X}, H))$. If $x \notin H$, then by (23.26) there is a $\chi \in \mathsf{A}(\mathsf{X}, H)$ such that $\chi(x) \neq 1$. Thus $x \notin A(G, \mathsf{A}(\mathsf{X}, H))$. □

(24.11) Theorem. *Let H be a closed subgroup of G. The character group of H is topologically isomorphic with $\mathsf{X}/\mathsf{A}(\mathsf{X}, H)$, every continuous character of H having the form $x \to \chi(x)$ for some $\chi \in \mathsf{X}$. Two characters χ_1 and χ_2 in X define the same character of H if and only if $\chi_1\chi_2^{-1} \in \mathsf{A}(\mathsf{X}, H)$.*

Proof. Since G is the character group of X and $H = A(G, \mathsf{A}(\mathsf{X}, H))$, Theorem (23.25) implies that H is the character group of $\mathsf{X}/\mathsf{A}(\mathsf{X}, H)$. Therefore every continuous character of $\mathsf{X}/\mathsf{A}(\mathsf{X}, H)$ has the form $\chi \cdot \mathsf{A}(\mathsf{X}, H) \to \chi(x)$ for some $x \in H$. Hence by (24.8) every continuous character of H has the form $x \to \chi(x)$ for some $\chi \in \mathsf{X}$. The last statement is obvious. □

(24.12) Corollary. *Let H be a closed subgroup of G. Every continuous character ψ of H admits an extension χ over G that is a continuous character of G. If $a \in G \cap H'$, then χ can be determined so that $\chi(a) \neq 1$.*

Proof. The existence of some extension χ follows at once from (24.11). If $\chi(a) \neq 1$ the proof is complete. If $\chi(a) = 1$, let ψ be a continuous character of G/H such that $\psi(aH) \neq 1$ (22.17), let φ be the natural mapping of G onto G/H, and consider the character $\chi \cdot (\psi \circ \varphi)$. □

(24.13) Theorem (24.8) tells us that every topological and algebraic property of a locally compact Abelian group G must be describable in terms of topological or algebraic properties of its character group X, since X determines G both as a group and a topological space. It also reduces the study of compact Abelian groups to the study of the algebraic structure of Abelian groups with no topology: a field that is far from exhausted at the present time. [See for example FUCHS [1].] Let us now examine some of the most important of these dual properties of G and X.

(24.14) Theorem. *For arbitrary G and X, the equality $\mathfrak{w}(G)=\mathfrak{w}(X)$ obtains*[1].

Proof. If G is finite, the theorem is trivial. We may thus suppose that G and X are infinite, and accordingly that $\mathfrak{w}(G)$ is infinite. Let $\mathscr{B}$ be an open basis for G such that $\overline{\overline{\mathscr{B}}}=\mathfrak{w}(G)$. Since $\{U\in\mathscr{B}: U^-$ is compact$\}$ is also an open basis for G, we may suppose that U^- is compact for all $U\in\mathscr{B}$. Let $\mathfrak{B}$ be the set of all functions on G of the form $\max\{\xi_{U_1}, \xi_{U_2}, \ldots, \xi_{U_m}\}$ for $\{U_1, U_2, \ldots, U_m\}\subset\mathscr{B}$; plainly $\overline{\overline{\mathfrak{B}}}=\mathfrak{w}(G)$. Now let F be any compact subset of G and W any open set in G such that $W\supset F$. Since W is the union of sets in $\mathscr{B}$, there are a finite number of sets $U_1, U_2, \ldots, U_m$ in $\mathscr{B}$ such that $F\subset U_1\cup\cdots\cup U_m\subset W$. Thus we have $\xi_F\leqq\max\{\xi_{U_1}, \xi_{U_2}, \ldots, \xi_{U_m}\}\leqq\xi_W$. Since we can find a W for which $\lambda(W\cap F')$ is as small as we like, it follows that $\|\xi_F-g\|_1$ is arbitrarily small for some $g\in\mathfrak{B}$. An argument like that used in proving (12.10) shows that the set of functions of the form

$$(r_1+is_1)\,\xi_{F_1}+\cdots+(r_n+is_n)\,\xi_{F_n} \tag{1}$$

[where $F_1, \ldots, F_n$ are compact and $r_1, \ldots, r_n, s_1, \ldots, s_n$ are rational numbers] is dense in $\mathfrak{L}_1(G)$. It follows that the set $\mathfrak{D}$ of all functions of the form

$$(r_1+is_1)\,g_1+\cdots+(r_n+is_n)\,g_n, \tag{2}$$

where $g_1, \ldots, g_n\in\mathfrak{B}$ and $r_1, \ldots, r_n, s_1, \ldots, s_n$ are rational, is also dense in $\mathfrak{L}_1(G)$. That is, $\mathfrak{L}_1(G)$ contains a dense subset of cardinal number $\mathfrak{w}(G)$.

Now consider the family $\mathscr{A}$ of all subsets of X that are finite intersections of sets of the form

$$\{\chi\in X: |\hat{f}(\chi)-(r+si)|<t\},$$

where r and s are rational numbers, t is a positive rational number, and $f\in\mathfrak{D}$. Note that all sets in $\mathscr{A}$ are open in the Δ-topology. The family $\mathscr{A}$ has cardinal number not exceeding $\overline{\overline{\mathfrak{D}}}=\mathfrak{w}(G)$ and is an open basis for a topology in X in which all of the functions $\hat{f}$ $(f\in\mathfrak{D})$ are continuous. The set of functions $\hat{\mathfrak{D}}$ is uniformly dense in $\widehat{\mathfrak{L}_1(G)}$, since $\mathfrak{D}$ is dense in $\mathfrak{L}_1(G)$ and $\|\hat{\varphi}\|_u\leqq\|\varphi\|_1$ [see (23.10.vi)]. Therefore all $\hat{g}$ $(g\in\mathfrak{L}_1(G))$ are continuous in the topology defined by $\mathscr{A}$. As the Δ-topology is the

[1] For the definition of $\mathfrak{w}$, see the introduction to § 3, p. 9.

weakest topology for X under which all $\hat{g}$ are continuous (23.13), the $\mathscr{A}$- and Δ-topologies are equal and

$$\mathfrak{w}(\mathsf{X}) \leqq \overline{\overline{\mathscr{A}}} \leqq \overline{\overline{\mathfrak{D}}} = \mathfrak{w}(G).$$

Theorem (24.8) now implies that $\mathfrak{w}(G) \leqq \mathfrak{w}(\mathsf{X})$. ☐

(24.15) Theorem. *Let G be a compact Abelian group. Then we have*

(i) $\mathfrak{w}(G) = \dim(\mathfrak{L}_2(G)) = \overline{\overline{\mathsf{X}}}$.

In particular, G is metrizable if and only if X is a countable [discrete] group.

Proof. For the discrete space X, we have $\mathfrak{w}(\mathsf{X}) = \overline{\overline{\mathsf{X}}}$. Thus (24.14) implies that $\mathfrak{w}(G) = \overline{\overline{\mathsf{X}}}$. Let $\mathfrak{T}$ denote the set of functions $\sum_{j=1}^{m} \alpha_j \chi_j$ for $\alpha_j \in K$ and $\chi_j \in \mathsf{X}$. Plainly $\mathfrak{T}$ is a subalgebra of $\mathfrak{C}(G)$ closed under complex conjugation and separating points of G [see (22.17)]. Thus $\mathfrak{T}$ is uniformly dense in $\mathfrak{C}(G)$ and hence dense in $\mathfrak{L}_2(G)$ (12.10). Since X is an orthonormal set in $\mathfrak{L}_2(G)$ (23.19), it follows that X is an orthonormal basis in $\mathfrak{L}_2(G)$ (B.47). That is, $\dim(\mathfrak{L}_2(G)) = \overline{\overline{\mathsf{X}}}$.

The last statement follows from the elementary fact that a compact Hausdorff space is metrizable if and only if it has a countable open basis. ☐

(24.16) Note. Since discrete Abelian groups exist having arbitrary cardinal number, (24.15) and (24.8) show that $\dim(\mathfrak{L}_2(G))$ can be an arbitrary nonzero cardinal number for a compact Abelian group.

We next discuss connectivity properties of G and X.

(24.17) Theorem. *Let B be the closed subgroup of X consisting of all compact elements [see (9.10)], and let C be the component of the identity in G [see (7.1)]. Then $\mathsf{B} = \mathsf{A}(\mathsf{X}, C)$ and $C = A(G, \mathsf{B})$.*

Proof. (I) Suppose that there is a compact subgroup Y of X such that $\{1\} \subsetneqq \mathsf{Y}$. From (23.24.b), (23.24.d), and (24.8), we infer that $A(G, \mathsf{Y}) \subsetneqq G$ and that $A(G, \mathsf{Y})$ is an open subgroup of G. Consequently G is not connected.

(II) Suppose that G is totally disconnected, and let χ be any continuous character of G. Let U be a neighborhood of e in G such that $|\chi(x) - 1| < \sqrt{3}$ for all $x \in U$. By (7.7), U contains a compact open subgroup H. It is clear that $\chi(H) = \{1\}$, *i.e.*, $\chi \in \mathsf{A}(\mathsf{X}, H)$. Since H is open, (23.24.e) implies that $\mathsf{A}(\mathsf{X}, H)$ is compact. Therefore χ is in B, *i.e.*, $\mathsf{B} = \mathsf{X}$.

(III) We now turn to the general case. The [locally compact Abelian] group G/C is totally disconnected (7.3). By part (II), the character group of G/C contains only compact elements. Since the character group of G/C is topologically isomorphic with $\mathsf{A}(\mathsf{X}, C)$ (23.25), we infer that $\mathsf{A}(\mathsf{X}, C) \subset \mathsf{B}$.

To prove the reversed inclusion, we argue as follows. The [locally compact Abelian] group C is connected, and by part (I) its character group contains no compact subgroup different from the identity. The character group of C is topologically isomorphic with $\mathsf{X}/\mathsf{A}(\mathsf{X}, C)$ (24.11). Now if $\chi \in \mathsf{B}$ and $\chi \notin \mathsf{A}(\mathsf{X}, C)$, there is a compact subgroup Y of X whose image under the natural homomorphism of X onto $\mathsf{X}/\mathsf{A}(\mathsf{X}, C)$ is not the identity. This image is a compact subgroup of $\mathsf{X}/\mathsf{A}(\mathsf{X}, C)$; the resulting contradiction shows that $\mathsf{B} \subset \mathsf{A}(\mathsf{X}, C)$, and thus $\mathsf{B} = \mathsf{A}(\mathsf{X}, C)$. Applying (24.10), we have $C = A(G, \mathsf{A}(\mathsf{X}, C)) = A(G, \mathsf{B})$. □

(24.18) Corollary. *The following properties of G are equivalent:*

(i) *G is 0-dimensional;*

(ii) *every element of X lies in a compact subgroup of X;*

(iii) *G contains a compact open 0-dimensional subgroup H.*

Proof. The equivalence of (i) and (ii) follows immediately from (24.17) and (3.5). The equivalence of (i) and (iii) follows from (7.7). □

(24.19) Corollary. *The group G is connected if and only if X contains no compact subgroup different from $\{1\}$.*

Proof. By (24.17), we have $C = A(G, \mathsf{B})$; and $A(G, \mathsf{B}) = G$ if and only if $\mathsf{B} = \{1\}$. □

(24.20) Corollary. *Let G be a compact Abelian group, with component of the identity C. Let Φ be the torsion subgroup*[1] *of X. Then $\Phi = \mathsf{A}(\mathsf{X}, C)$ and $C = A(G, \Phi)$. Also Φ is isomorphic with the [discrete] character group of G/C.*

Proof. The two equalities are (24.17) for compact G, since an element of a discrete group lies in a compact subgroup if and only if it has finite order. The last assertion follows from (23.25). □

A curious consequence of (24.18) is the following.

(24.21) Theorem. *Let G be a torsion group. Then G and X are 0-dimensional*[2].

Proof. Let H be an open compactly generated subgroup of G. It follows from (9.8) that H is compact. Let Y be the character group of H. If $\psi \in \mathsf{Y}$, then $\psi(H)$ is a compact torsion subgroup of T and is hence a finite cyclic group. Thus ψ belongs to a compact subgroup of Y and H is 0-dimensional by (24.18). Consequently, G is 0-dimensional, as H is open in G. Since every element of G obviously lies in a compact subgroup of G, (24.18) implies that X is 0-dimensional. □

---

[1] See (A.4).

[2] See also (25.9) *infra*.

We obtain more relations between G and X by considering the homomorphisms f_n, $f_n(x)=x^n$, and the subgroups $G^{(n)}=f_n(G)$ and $G_{(n)}=f_n^{-1}(e)$ [see (A.2)].

(24.22) Theorem. *Let G be a locally compact Abelian group with character group X, and let n be a positive integer. Then we have*

(i) $\mathsf{A}(\mathsf{X}, G^{(n)}) = \mathsf{X}_{(n)}$

and

(ii) $\mathsf{A}(\mathsf{X}, G_{(n)}) = \mathsf{X}^{(n)-}$.

Proof. Let χ be a character in $\mathsf{A}(\mathsf{X}, G^{(n)})$. For $x\in G$, we have $x^n\in G^{(n)}$, and so $\chi(x^n)=\chi^n(x)=1$. Thus $\chi\in\mathsf{X}_{(n)}$. If $\chi\in\mathsf{X}_{(n)}$, then $\chi^n(x)=\chi(x^n)=1$ for all $x\in G$, so that $\chi\in\mathsf{A}(\mathsf{X}, G^{(n)})$. This proves (i).

To prove (ii), regard G as the character group of X. Then (i) implies that $A(G, \mathsf{X}^{(n)})=G_{(n)}$, and (24.10) and (23.24.a) imply that

$$\mathsf{X}^{(n)-} = \mathsf{A}(\mathsf{X}, A(G, \mathsf{X}^{(n)-})) = \mathsf{A}(\mathsf{X}, A(G, \mathsf{X}^{(n)})) = \mathsf{A}(\mathsf{X}, G_{(n)}).$$

This proves (ii). □

(24.23) Theorem. *If G is divisible, then X is torsion-free. If X is torsion-free, then $G^{(n)}$ is dense in G for $n=1, 2, 3, \ldots$. If G is discrete or compact, then G is divisible if and only if X is torsion-free*[1].

Proof. If G is divisible, then $G^{(n)}=G$ for all n, and (24.22.i) implies that $\mathsf{X}_{(n)}=\{1\}$. Thus X is torsion-free.

Suppose that X is torsion-free. Then (24.22.ii) shows that $A(G, \mathsf{X}_{(n)})= G=G^{(n)-}$. If G is discrete, then $G^{(n)}$ is trivially closed. If G is compact, $G^{(n)}$ is a continuous image of the compact space G and is hence compact and closed. Thus in these cases $G=G^{(n)}$ for all n and G is divisible. □

Another consequence of (24.22) follows.

(24.24) Theorem. *Let D be the largest divisible subgroup of G,*[2] *let C be the component of e in G, and let Φ be the torsion subgroup of X. Then we have*

(i) $C\subset D\subset D^-\subset \bigcap_{n=1}^{\infty} G^{(n)-}=A(G,\Phi)$.

If G is compact, $G^{(n)}=G^{(n)-}$ and all inclusions in (i) *are equalities*[3].

Proof. We first prove that

$$\bigcap_{n=1}^{\infty} G^{(n)-}=A(G, \Phi). \tag{1}$$

---

[1] There are nondivisible G's with torsion-free X's: see (24.44).

[2] See (A.6).

[3] If G is noncompact, examples can be found to show that each inclusion in (i) may be proper. See (24.44).

By (24.22.ii), we have $A(G, \mathsf{X}_{(n)}) = G^{(n)-}$, and so

$$A(G, \Phi) = A(G, \bigcup_{n=1}^{\infty} \mathsf{X}_{(n)}) = \bigcap_{n=1}^{\infty} A(G, \mathsf{X}_{(n)}) = \bigcap_{n=1}^{\infty} G^{(n)-}.$$

The inclusion $D \subset \bigcap_{n=1}^{\infty} G^{(n)}$ holds because D is divisible. Taking closures, we have

$$D \subset D^- \subset \Big(\bigcap_{n=1}^{\infty} G^{(n)}\Big)^- \subset \bigcap_{n=1}^{\infty} (G^{(n)-}).$$

This proves all of (i) except for the inclusion $C \subset D$.

To prove that $C \subset D$, it is convenient first to deal with the case in which G, and hence C, are compact. In this case, the character group of C is discrete and torsion-free (24.19), and so C is a divisible group (24.23). Thus we have $C \subset D$. By (24.20), we have $C = A(G, \Phi)$, and so by the foregoing we have

$$C \subset D \subset A(G, \Phi) = C.$$

Thus (i) holds, and all inclusions become equalities, if G is compact.

It is now easy to show that $C \subset D$ for all G. By (9.14), C is topologically isomorphic with a group $R^n \times F$, where F is a compact connected Abelian group and n is a nonnegative integer. The group F is divisible, as proved already, and so C is a divisible group. This implies that $C \subset D$. ☐

For convenience, we collect some of the foregoing results in one statement.

(24.25) Theorem. *Let G be a compact Abelian group. The following properties are equivalent:*

(i) *G is connected;*

(ii) *X is torsion-free;*

(iii) *G is a divisible group.*

Dimension 0 can also be described, as follows.

(24.26) Theorem. *Let G be a compact Abelian group. The following assertions are equivalent:*

(i) *G has dimension 0;*

(ii) *X is a torsion group;*

(iii) *$\chi(G)$ is a finite subgroup of T for every $\chi \in \mathsf{X}$.*

Proof. Corollary (24.18) shows that (i) and (ii) are equivalent. The equivalence of (ii) and (iii) is obvious. ☐

Theorem (24.26) admits an interesting generalization: the dimension of a compact Abelian group is equal to the torsion-free rank of its character group. We first state and prove a preliminary result.

(24.27) Theorem. *Let G be a discrete torsion-free Abelian group of rank $\mathfrak{a}$. Then the character group X of G contains a subset homeomorphic with $]0, 1[^{\mathfrak{a}}$ considered as a topological space.*

Proof. Let $\{x_\iota : \iota \in I\}$ be a maximal independent subset of G [of cardinal number $\mathfrak{a}$] (A.12). Let H be the [free] subgroup of G generated by the x_ι's. Then plainly a character of H can assume arbitrary values at the x_ι's. In particular, for a fixed element $(t_\iota)_{\iota \in I}$ in $]0, 1[^{\mathfrak{a}}$, there is exactly one character χ_0 on H such that $\chi_0(x_\iota) = \exp(2\pi i t_\iota)$ for $\iota \in I$. Let y be any element of G. Then we have $y^n = x_{\iota_1}^{a_1} x_{\iota_2}^{a_2} \cdots x_{\iota_k}^{a_k}$ for nonzero integers n and a_j and $\iota_1, \ldots, \iota_k \in I$. Since G is torsion-free, the element y completely determines the elements $x_{\iota_1}, \ldots, x_{\iota_k}$ and the ratios $\frac{a_1}{n}, \ldots, \frac{a_k}{n}$. Now write $\chi(y) = \exp\left[2\pi i\left(\frac{a_1}{n} t_{\iota_1} + \cdots + \frac{a_k}{n} t_{\iota_k}\right)\right]$. The function χ is well defined on G and is an extension of χ_0. To see that χ is multiplicative, let $y^n = x_{\iota_1}^{a_1} \cdots x_{\iota_k}^{a_k}$ and $z^m = x_{\iota_1}^{b_1} \cdots x_{\iota_k}^{b_k}$ [some a's and b's may be 0]. Then $(yz)^{mn} = x_{\iota_1}^{ma_1 + nb_1} \cdots x_{\iota_k}^{ma_k + nb_k}$, and so $\chi(yz) = \exp\left[2\pi i\left(\frac{ma_1 + nb_1}{mn} t_{\iota_1} + \cdots + \frac{ma_k + nb_k}{mn} t_{\iota_k}\right)\right] = \chi(y)\,\chi(z)$. Each χ defined in this way is therefore a character of G. We thus have a mapping f of $]0, 1[^{\mathfrak{a}}$ into X: the element (t_ι) in $]0, 1[^{\mathfrak{a}}$ goes into the character χ of G constructed above. It is obvious that f is one-to-one. A routine argument establishes the continuity of f and f^{-1}. □

(24.28) Theorem. *Let G be a compact Abelian group with character group X. The dimension of G (3.11) is equal to the torsion-free rank $r_0(X)$.*[1]

Proof. Let n be a nonnegative integer, and suppose that $\mathfrak{U}$ is a finite open covering of G for which every finite closed refinement of $\mathfrak{U}$ has multiplicity at least $n+1$. That is, suppose that $\dim(G) \geqq n$. For every $x \in G$, there is a neighborhood V_x of e such that $x \cdot V_x^2$ is contained in some $U \in \mathfrak{U}$. A finite number of the sets xV_x cover G: say $x_1 V_{x_1} \cup \cdots \cup x_m V_{x_m} = G$. Let $V = V_{x_1} \cap \cdots \cap V_{x_m}$. By (24.7), there is a closed subgroup H of G such that $H \subset V$ and G/H is topologically isomorphic with $T^a \times F$, where a is a nonnegative integer and F is a finite Abelian group. Plainly the character group of G/H is topologically isomorphic with $Z^a \times F$; it is also topologically isomorphic with $A(X, H)$ by (23.25). Therefore we have $r_0(X) \geqq r_0(A(X, H)) \geqq a$. We shall take it as known that $\dim(T^a \times F) = a$;[2] thus we have $\dim(G/H) = a$. The construction of V shows that if $y \in G$, the coset yH is contained in some $U \in \mathfrak{U}$. Since yH is compact, there is a neighborhood W_y of e in G such that yHW_y is contained in some $U \in \mathfrak{U}$ (4.10). Choose $y_1, \ldots, y_p$ in G so that

[1] Here we make the convention that $r_0(X) = \infty$ whenever $r_0(X)$ is infinite.

[2] This is a well-known theorem in dimension theory. It is also well known that its proof is not easy. See for example P. S. Aleksandrov [1], Ch. V.

$\bigcup_{k=1}^{p} y_k W_{y_k} = G$. Let $\mathscr{W}$ be the family of subsets of G/H of the form $\{zH : z \in y_k W_{y_k}\}$; plainly $\mathscr{W}$ is a finite open covering of G/H. Now assume that $\dim(G/H) < n$. Then there is a covering $\{\{xH : x \in F_1\}, \ldots, \{xH : x \in F_q\}\}$ of G/H, where the F_j's are compact subsets of G, that refines $\mathscr{W}$ and has multiplicity less than $n+1$. It is easy to see that $\{HF_1, \ldots, HF_q\}$ is a closed covering of G that refines $\mathscr{U}$ and has multiplicity less than $n+1$. This contradiction shows that $\dim(G/H) \geqq n$. Since we have already proved that $r_0(\mathsf{X}) \geqq a = \dim(G/H)$, we infer that $r_0(\mathsf{X}) \geqq \dim(G)$.

In proving the reversed inequality $r_0(\mathsf{X}) \leqq \dim(G)$, we may suppose that $r_0(\mathsf{X}) > 0$. Then consider any positive integer $k \leqq r_0(\mathsf{X})$, and let $\chi_1, \ldots, \chi_k$ be k independent elements of X having infinite order: $\chi_1^{a_1}\chi_2^{a_2}\cdots\chi_k^{a_k} = 1$ and $a_j \in Z$ imply $a_1 = a_2 = \cdots = a_k = 0$. Let Y be the subgroup of X generated by $\chi_1, \ldots, \chi_k$ and let Ξ be a subgroup of X such that $\Xi \cap \mathsf{Y} = \{1\}$ and Ξ is maximal with respect to the property $\Xi \cap \mathsf{Y} = \{1\}$. Let $\psi \in \mathsf{X}$ and suppose that some power ψ^n of ψ is in Ξ. If $\psi \notin \Xi$, there are a power ψ^m and a $\xi \in \Xi$ such that $\psi^m \xi = \chi_1^{a_1} \cdots \chi_k^{a_k}$, where not all of the integers a_j are zero. Then $\psi^{nm}\xi^n = \chi_1^{na_1} \cdots \chi_k^{na_k}$, and $\psi^{nm}\xi^n$ is in Ξ. Hence $na_1 = \cdots = na_k = 0$, a contradiction. Therefore ψ is in Ξ if ψ^n is in Ξ, *i.e.*, X/Ξ is torsion-free. The cosets $\chi_1\Xi, \ldots, \chi_k\Xi$ are plainly a maximal independent set in the group X/Ξ. From (24.27), we infer that the character group of X/Ξ contains a homeomorph of $]0, 1[^k$; clearly $]0, 1[^k$ contains a homeomorph of $[0, 1]^k$. Since the character group of X/Ξ is topologically isomorphic with $A(G, \Xi)$, we have proved that G contains a closed subspace homeomorphic with $[0, 1]^k$. Since $\dim([0, 1]^k) = k$, we have proved that $r_0(\mathsf{X}) \leqq \dim(G)$. □

The assertions (24.18)—(24.26) depend upon the basic result (24.17) [along with (24.8), of course]. We next give a quite different application of (24.17), to the effect that every locally compact Abelian group admits a group of the form R^a ($a = 0, 1, 2, \ldots$) as a direct factor, and that there is a largest value of a, described simply in terms of the structure of G.

(24.29) Theorem. *Let G be an arbitrary locally compact Abelian group, with character group X, and let C and K, respectively, be the components of the identity in G and X. Let n be the nonnegative integer and E a compact connected Abelian group such that there is a topological isomorphism σ carrying $R^n \times E$ onto C, as in* (9.14). *Let m be the nonnegative integer and F a compact connected Abelian group such that there is a topological isomorphism τ carrying $R^m \times F$ onto K. Then we have:*

(i) *G is topologically isomorphic with $R^n \times A(G, \tau(R^m))$;*

(ii) *X is topologically isomorphic with $R^m \times \mathsf{A}(\mathsf{X}, \sigma(R^n))$;*

(iii) *$m = n$;*

(iv) *n is the largest integer a such that G or X contains a topological isomorph of the group R^a.*

Proof. Let B and B be the closed subgroups of G and X, respectively, consisting of compact elements (9.10). By (24.17) and (24.8), we have

$$\mathsf{B}=\mathsf{A}(\mathsf{X}, C), \quad \mathsf{K}=\mathsf{A}(\mathsf{X}, B), \quad B=A(G, \mathsf{K}), \quad C=A(G, \mathsf{B}). \quad (1)$$

Since τ is a topological isomorphism, $\tau(R^m)$ is locally compact in its relative topology as a subgroup of X and is therefore closed (5.11). It follows that $\mathsf{B}\cap\tau(R^m)=\{1\}$. By (9.26.a), the subgroup BC of G is open [and closed]. By (1), we have

$$BC=A(G, \mathsf{K})\cdot C\subset A(G, \tau(R^m))\cdot C.$$

Hence $A(G, \tau(R^m))\cdot C$ is an open and closed subgroup of G. If $A(G, \tau(R^m))\cdot C$ were not equal to G, there would be a character $\chi\in\mathsf{X}$ such that $\chi=1$ on $A(G, \tau(R^m))$ and on C while χ is not identically 1 (23.26). By (24.10), we would have

$$\chi\in\mathsf{A}(\mathsf{X}, A(G, \tau(R^m)))\cap\mathsf{A}(\mathsf{X}, C)=\tau(R^m)\cap\mathsf{B}=\{1\},$$

which is a contradiction. We infer that

$$A(G, \tau(R^m))\cdot C=G. \quad (2)$$

We can write $C=\sigma(E)\,\sigma(R^n)$; since E is a compact group, $\sigma(E)$ is a compact subgroup of G, and hence $\sigma(E)\subset B$. Thus $\sigma(E)\subset A(G, \mathsf{K})\subset A(G, \tau(R^m))$, and so (2) implies that

$$A(G, \tau(R^m))\cdot\sigma(R^n)=G. \quad (3)$$

The same argument shows that

$$\mathsf{A}(\mathsf{X}, \sigma(R^n))\cdot\tau(R^m)=\mathsf{X}.$$

Thus if $x\in A(G, \mathsf{A}(\mathsf{X}, \sigma(R^n)))=\sigma(R^n)$ and $x\in A(G, \tau(R^m))$, we have $x=e$. That is,

$$A(G, \tau(R^m))\cap\sigma(R^n)=\{e\}. \quad (4)$$

Since $\sigma(R^n)$ is compactly generated, (6.21) shows that G is topologically isomorphic with $\sigma(R^n)\times A(G, \tau(R^m))$. We have thus proved (i); (i) and (24.8) yield (ii).

By (24.11), the character group of $\sigma(R^n)$ is topologically isomorphic with $\mathsf{X}/\mathsf{A}(\mathsf{X}, \sigma(R^n))$, and by (ii) this group is topologically isomorphic with R^m. This implies at once that $m=n$, since the character group of R^n is topologically isomorphic with R^n (23.27.f).

To prove (iv), suppose that G contains a topological isomorph of R^a. Theorem (9.14) shows that $a\leq n$. □

(24.30) Theorem. *A locally compact Abelian group G is topologically isomorphic with $R^n \times G_0$, where G_0 is a locally compact Abelian group containing a compact open subgroup. The group $A(G, \tau(R^n))$ of (24.29) may be taken as G_0. If G is also topologically isomorphic with $R^m \times G_1$ and G_1 contains a compact open subgroup, then $m = n$.*

Proof. Let G_0 be the group $A(G, \tau(R^n))$ of (24.29). Let H be a compactly generated open subgroup of G_0. By (9.8), H is topologically isomorphic with $R^a \times Z^b \times F$, where F is a compact group. If $a \geqq 1$, then $R^n \times G_0$ would contain a topological isomorph of R^{n+a}, contrary to (24.29.iv). Thus we have $a = 0$. Since Z^b is discrete, F is open in $Z^b \times F$, and its topological isomorph is open in H and hence in G_0.

We now prove the second statement. Let C_0 and C_1 be the components of the identities of G_0 and G_1, respectively. Then plainly C_0 and C_1 are compact and the component of e in G is topologically isomorphic with both $R^n \times C_0$ and $R^m \times C_1$. Corollary (9.13) now implies that $n = m$. ☐

Another application of (24.8) follows. We will use it in classifying compact monothetic groups (25.11)—(25.17).

(24.31) Theorem. *Let G be a compact Abelian group with character group X. Let $\mathfrak{m}$ be a nonzero cardinal number. The following conditions are equivalent:*

(i) *$G \cap \{e\}'$ contains a subset A of cardinal number $\mathfrak{m}$ such that the smallest subgroup H containing A is dense in G;*

(ii) *there is an index class I of cardinal number $\mathfrak{m}$ and an isomorphism τ of X into $\underset{\iota \in I}{\mathsf{P}} T_{(\iota)}$ such that $\pi_\iota(\tau(\mathsf{X})) \neq \{1\}$ for each projection π_ι, and the functions $\pi_\iota \circ \tau$ are distinct for distinct $\iota \in I$.*

Proof. Suppose that (i) holds. A continuous character on G is determined by its values on H, and these in turn are determined by its values on A. For $\chi \in \mathsf{X}$, let $\tau(\chi)$ be the element $(\chi(a))_{a \in A}$ of $\underset{a \in A}{\mathsf{P}} T_{(a)}$. Plainly τ is an isomorphism of X onto $\tau(\mathsf{X})$. Since $a \neq e$ for $a \in A$, there is for each $a \in A$ a character $\chi \in \mathsf{X}$ such that $\chi(a) \neq 1$ (22.17). Thus the projection of $\tau(\mathsf{X})$ onto the a-th axis contains a number different from 1 for each $a \in A$. If $a \neq b$, there is a $\chi \in \mathsf{X}$ such that $\chi(a) \neq \chi(b)$. Thus (ii) holds.

Suppose conversely that (ii) holds. For every $\iota \in I$, the mapping $\chi \rightarrow \pi_\iota \circ \tau(\chi)$ is a character of X different from the identity, and so by (24.8) there is a point $a_\iota \in G$ such that $\chi(a_\iota) = \pi_\iota \circ \tau(\chi)$ for all $\chi \in \mathsf{X}$. It is clear from (ii) that $a_\iota \neq a_{\iota'}$ if $\iota \neq \iota'$ and that no a_ι is e. If the subgroup generated by $\{a_\iota\}_{\iota \in I}$ were not dense in G, there would be a $\psi \in \mathsf{X}$ such that $\psi \neq 1$ and $\psi(a_\iota) = 1$ for all ι. That is, $\pi_\iota \circ \tau(\psi) = 1$ for all ι, and $\tau(\psi) = \tau(1)$. This contradicts the hypothesis that τ is an isomorphism. ☐

(24.32) Corollary. *A compact Abelian group G is monothetic if and only if its character group is isomorphic with a subgroup of T_d.*[1]

Proof. Barring the trivial case $G=\{e\}$, this assertion is (24.31) with $\mathfrak{m}=1$. □

In various parts of harmonic analysis, continuous homomorphisms into the additive group R are important. These can be analyzed on the basis of Theorem (24.17).

(24.33) Definition. Let G be a group. A homomorphism of G into the additive group R is called a *real character of* G. For real characters f and g, the *sum* $f+g$ is defined pointwise. The set of all *continuous* real characters of a topological group G is called the *real-character group of* G.

(24.34) Theorem. *Let G be a locally compact Abelian group, and let B be the closed subgroup of compact elements of G. For $x, y \in G$, there is a continuous real character f of G such that $f(x) \neq f(y)$ if and only if $xy^{-1} \notin B$. Thus G admits sufficiently many continuous real characters if and only if $B=\{e\}$.*

Proof. Since R has only one compact subgroup, namely $\{0\}$, it is clear that $f(B)=\{0\}$ if f is a continuous real character. Thus $f(x) \neq f(y)$ implies that $xy^{-1} \notin B$.

Next we examine the quotient group G/B. By (24.17), we have $B=A(G, \mathsf{K})$, where K is the component of 1 in the character group X of G, and so G/B is topologically isomorphic with the character group of the connected group K. By (24.19), G/B contains no compact elements except the identity.

Our proof will thus be complete if we show that there are sufficiently many continuous real characters on every locally compact Abelian group G such that $B=\{e\}$. Let x be an element of G different from e, and let H be a compactly generated open subgroup of G that contains x. By (9.8), H is topologically isomorphic with $R^a \times Z^b$, where a and b are nonnegative integers and a or b is positive. It follows that there is a continuous real character f_0 of H such that $f_0(x) \neq 0$. If we extend f_0 in any manner to a real character f of G [see (A.7)], the real character f will be continuous on G because it is continuous on the open subgroup H. □

(24.35) Corollary. *The following conditions on a locally compact Abelian group G are equivalent:*

(i) *G has sufficiently many continuous real characters;*

(ii) *the character group of G is connected;*

(iii) *G is topologically isomorphic with $R^n \times F$, where n is a nonnegative integer and F is a discrete torsion-free Abelian group.*

---

[1] Recall that T_d denotes the group T with the discrete topology.

Proof. The equivalence of (i) and (ii) follows from (24.19) and (24.34). The equivalence of (i) and (iii) follows from (24.30) and (24.34). □

(24.36) Theorem. *Let G be a locally compact Abelian group, H_1 a closed subgroup of G, and f_1 a continuous real character on H_1. Then there is a continuous real character f of G such that $f|H_1=f_1$.*

Proof. (I) Suppose first that $G=R^c$, where c is a positive integer. By (9.11), H_1 is topologically isomorphic with $R^a\times Z^b$ for nonnegative integers a and b. Let $\boldsymbol{x}_1, \ldots, \boldsymbol{x}_a, \boldsymbol{y}_1, \ldots, \boldsymbol{y}_b$ in R^c be generators of H_1 in the sense that every vector in H_1 can be written uniquely as

$$\alpha_1\boldsymbol{x}_1+\cdots+\alpha_a\boldsymbol{x}_a+m_1\boldsymbol{y}_1+\cdots+m_b\boldsymbol{y}_b,$$

where $\alpha_1, \ldots, \alpha_a\in R$ and $m_1, \ldots, m_b\in Z$. A continuous real character f_1 on H_1 assumes arbitrary real values at $\boldsymbol{x}_1, \ldots, \boldsymbol{x}_a, \boldsymbol{y}_1, \ldots, \boldsymbol{y}_b$, and then is completely determined:

$$\begin{aligned} &f_1(\alpha_1\boldsymbol{x}_1+\cdots+\alpha_a\boldsymbol{x}_a+m_1\boldsymbol{y}_1+\cdots+m_b\boldsymbol{y}_b)\\ &\quad=\alpha_1 f_1(\boldsymbol{x}_1)+\cdots+\alpha_a f_1(\boldsymbol{x}_a)+m_1 f_1(\boldsymbol{y}_1)+\cdots+m_b f_1(\boldsymbol{y}_b).\end{aligned}$$

By (9.11), the vectors $\boldsymbol{x}_1, \ldots, \boldsymbol{x}_a, \boldsymbol{y}_1, \ldots, \boldsymbol{y}_b$ are linearly independent in R^c, and we can extend them to a basis for R^c. It follows that f_1 can be extended to a real linear functional f on R^c, and that f is then also a continuous real character on R^c.

(II) Suppose next that $G=R^c\times F$ where c is a nonnegative integer and F is a compact Abelian group. Let φ denote the projection of $R^c\times F$ onto R^c. By (5.18), the group $\varphi(H_1)$ is closed in R^c. We define f_2 on $\varphi(H_1)$ by the rule

$$f_2(\boldsymbol{x})=f_1((\boldsymbol{x}, z)) \quad \text{where} \quad (\boldsymbol{x}, z)\in H_1.$$

The function f_2 is well defined because $f_1((\{\boldsymbol{0}\}\times F)\cap H_1)=\{0\}$, and f_2 is obviously a real character. Assume that f_2 is not continuous. Then f_2 is not continuous at $\boldsymbol{0}\in R^c$; hence there exists an $\varepsilon>0$ and a sequence $\{(\boldsymbol{x}_n, z_n)\}_{n=1}^{\infty}$ of points in H_1 such that $\lim\limits_{n\to\infty}\boldsymbol{x}_n=\boldsymbol{0}$ and $|f_1((\boldsymbol{x}_n, z_n))|\geqq\varepsilon$ for $n=1, 2, \ldots$. There exists a subnet z_β of the sequence $\{z_n\}_{n=1}^{\infty}$ that converges to an element z in F. We then have $\lim\limits_{\beta}(\boldsymbol{x}_\beta, z_\beta)=(\boldsymbol{0}, z)$, $f_1((\boldsymbol{0}, z))=0$, and $|f_1((\boldsymbol{x}_\beta, z_\beta))|\geqq\varepsilon$ for all β. This contradicts the continuity of f_1, and so f_2 is continuous. By (I), f_2 can be extended to a continuous real character f_3 on R^c. Defining f to be $f_3\circ\varphi$, we obtain a continuous real character on G that extends f_1.

(III) Now suppose that G and H_1 are arbitrary. Let H_2 be a compactly generated open subgroup of G. Then H_2 is topologically isomorphic with $R^a\times Z^b\times F$ (9.8), and since Z^b is discrete, G contains an open subgroup H_3 topologically isomorphic with $R^a\times F$. Part (II) implies that $f_1|(H_1 \cap H_3)$ can be extended to a continuous real character f_3 on H_3. On the open subgroup H_3H_1 of G, let $f_4(h_3h_1) = f_3(h_3) + f_1(h_1)$. It is easy to see that f_4 is well defined, is a continuous real character on H_3H_1, and extends f_1. Since R is divisible, f_4 can be extended to a real character f on G (A.7). Since $f|H_3H_1$ is continuous, f is continuous. □

(24.37) Definition. Let G_1 and G_2 be locally compact Abelian groups with character groups X_1 and X_2 respectively. Let φ be a continuous homomorphism of G_1 into G_2. For each $\chi_2 \in \mathsf{X}_2$, let $\varphi^\sim(\chi_2)$ be the function $\chi_2 \circ \varphi$. Then $\varphi^\sim(\chi_2)$ is obviously in X_1, and $\varphi^\sim$ is a mapping of X_2 into X_1. It is called the *adjoint of* φ.

(24.38) Theorem. *The adjoint mapping $\varphi^\sim$ is a continuous homomorphism of X_2 into X_1, and its kernel is $\mathsf{A}(\mathsf{X}_2, \varphi(G_1))$.*

Proof. For $\chi_2, \psi_2 \in \mathsf{X}_2$ and $x \in G_1$, we have $(\chi_2\psi_2)\circ\varphi(x) = \chi_2(\varphi(x))\,\psi_2(\varphi(x))$ $= \chi_2\circ\varphi(x)\cdot\psi_2\circ\varphi(x)$. Thus $\varphi^\sim$ is a homomorphism. For a compact subset F_1 of G_1, $\varphi(F_1)$ is compact in G_2, and plainly $\varphi^\sim(\mathsf{P}_2(\varphi(F_1), \varepsilon)) \subset \mathsf{P}_1(F_1, \varepsilon)$. Thus $\varphi^\sim$ is continuous. It is trivial that $\varphi^\sim(\chi_2)$ is identically 1 if and only if $\chi_2(y) = 1$ for all $y \in \varphi(G_1)$. □

(24.39) Theorem. *Let H_1 be a closed subgroup of G_1. Then*

$$\varphi^{\sim -1}(\mathsf{A}(\mathsf{X}_1, H_1) \cap \varphi^\sim(\mathsf{X}_2)) = \mathsf{A}(\mathsf{X}_2, \varphi(H_1)).$$

Proof. If $\chi_2(x_2) = 1$ for all $x_2 \in \varphi(H_1)$, then $\chi_2\circ\varphi(H_1) = \{1\}$, so that $\varphi^\sim(\chi_2) \in \mathsf{A}(\mathsf{X}_1, H_1)$. Also, if $\chi_2 \in \varphi^{\sim -1}(\mathsf{A}(\mathsf{X}_1, H_1) \cap \varphi^\sim(\mathsf{X}_2))$, then $\varphi^\sim(\chi_2)(H_1) = \{1\}$, so that $\chi_2(\varphi(H_1)) = \{1\}$. □

(24.40) Theorem. *Suppose that φ is a continuous homomorphism of G_1 into G_2 and that φ is an open mapping of G_1 onto $\varphi(G_1)$. Then $\varphi^\sim$ is an open continuous homomorphism of X_2 onto $\mathsf{A}(\mathsf{X}_1, \varphi^{-1}(e_2))$.*

Proof. By (5.40.d), $\varphi(G_1)$ is a closed subgroup of G_2. The kernel of $\varphi^\sim$ is plainly $\mathsf{A}(\mathsf{X}_2, \varphi(G_1))$. Let ϱ_0 be the mapping of $\mathsf{X}_2/\mathsf{A}(\mathsf{X}_2, \varphi(G_1))$ into X_1 defined by $[\varrho_0(\chi_2\,\mathsf{A}(\mathsf{X}_2, \varphi(G_1)))](x_1) = \chi_2\circ\varphi(x_1)$, for all $x_1 \in G_1$. Since φ is open, $\varphi(G_1)$ is topologically isomorphic with the group $G_1/\varphi^{-1}(e_2)$, and so ϱ_0 is the topological isomorphism ϱ described in (23.25). That is, ϱ_0 maps the character group of $\varphi(G_1)$ [and hence of $G_1/\varphi^{-1}(e_2)$] onto $\mathsf{A}(\mathsf{X}_1, \varphi^{-1}(e_2))$. If σ denotes the natural mapping of X_2 onto $\mathsf{X}_2/\mathsf{A}(\mathsf{X}_2, \varphi(G_1))$, then we have $\varphi^\sim = \varrho_0\circ\sigma$. Since σ is open and ϱ_0 is a homeomorphism, $\varphi^\sim$ is open. □

(24.41) We now list without proofs some additional properties of adjoint homomorphisms.

(a) The homomorphism φ satisfies $\varphi^{\sim\sim} = \varphi$. This follows from (24.8).

(b) The homomorphism $\varphi^\sim$ is one-to-one if and only if $\varphi(G_1)$ is dense in G_2, and $\varphi^\sim(\mathsf{X}_2)$ is dense in X_1 if and only if φ is one-to-one.

(c) The homomorphism $\varphi^{\sim}$ is a topological isomorphism of X_2 onto X_1 if and only if φ is a topological isomorphism of G_1 onto G_2.

(d) Let f_n be the mapping $x \to x^n$ of G into G. Then the adjoint homomorphism $f_n^{\sim}$ is the mapping $\chi \to \chi^n$ of X into X.

Miscellaneous theorems and examples

(24.42) Let H and H_0 be closed subgroups of G. Then $H \cap H_0 = \{e\}$ if and only if $(\mathsf{A}(\mathsf{X}, H) \cdot \mathsf{A}(\mathsf{X}, H_0))^- = \mathsf{X}$. [Using (23.29.b) and (24.10), we see that

$$A\left(G, (\mathsf{A}(\mathsf{X}, H) \cdot \mathsf{A}(\mathsf{X}, H_0))^-\right) = A\left(G, \mathsf{A}(\mathsf{X}, H)\right) \cap A\left(G, \mathsf{A}(\mathsf{X}, H_0)\right)$$
$$= H \cap H_0;$$

the assertion follows.]

(24.43) Real characters (DIXMIER [2]). Let G be a locally compact Abelian group with character group X. Let f be a continuous real character of G. Then the mapping

(i) $x \to \exp[i f(x)]$

is plainly an element of X. Every element of X has the form (i) if and only if X is the union of one-parameter subgroups. [Let φ be a continuous homomorphism of R into X, and let $\varphi^{\sim}$ be the adjoint homomorphism of G into the character group R of R. For $x \in G$, $\varphi^{\sim}(x)$ is the real number such that $\varphi(t)(x) = \exp[i\varphi^{\sim}(x) \cdot t]$ for all real numbers t [note that $\varphi(t)$ is an element of X]. Thus if $\chi \in \mathsf{X}$ and χ lies in a one-parameter subgroup of X, then χ has the form (i). On the other hand, if f is a continuous real character of G, the characters $x \to \exp[itf(x)]$ $(t \in R)$ form a one-parameter subgroup of X.]

(24.44) Examples relating to (24.23) and (24.24). (a) (H. FREUDENTHAL; see HARTMAN and RYLL-NARDZEWSKI [1].) There are locally compact Abelian groups G that are not divisible, whose character groups are torsion-free. [Let $\mathfrak{m}$ be any infinite cardinal number, and let G be the subgroup of $T^{\mathfrak{m}}$ consisting of all (x_ι) such that $x_\iota = \pm 1$ for all but a finite number of indices ι. Let H be the subgroup of G consisting of all (x_ι) such that $x_\iota = 1$ for all but a finite number of indices ι. It is easy to see that $G^{(n)} = H$ if n is even and $G^{(n)} = G$ if n is odd. For a finite set Λ of indices ι, let U_Λ be the set of all (x_ι) such that $x_\iota = 1$ for $\iota \in \Lambda$ and $x_\iota = \pm 1$ for $\iota \notin \Lambda$. Let all U_Λ be taken as an open basis at the identity (1_ι). The subgroup U_Λ is clearly topologically isomorphic with $\{-1, 1\}^{\mathfrak{m}}$, so that U_Λ is compact. Thus G is locally compact, and H is a dense subgroup of G. If χ is a continuous character on G of finite order k, then $\{1\} = \chi(G^{(2k)}) = \chi(H) = \chi(G)$. Thus X is torsion-free, and G is plainly nondivisible.]

(b) The group G described in part (a) is obviously 0-dimensional, so that the component of (1_ι) is $\{(1_\iota)\}$. The largest divisible subgroup of G is H. Thus we have $C \subsetneqq D = H \subsetneqq H^- = G$ for this group.

(c) The proper inclusion $D^- \subsetneqq \bigcap_{n=1}^{\infty} G^{(n)}$ occurs in a number of *discrete* Abelian groups. [Let G be the set of all sequences $(n; m_2, m_3, m_4, \ldots, m_k, \ldots)$, where $n \in Z$, $m_k \in \{0, 1, 2, \ldots, k-1\}$, and only a finite number of entries are different from 0. We define the product of two of these sequences by the formula

$$(n; m_2, m_3, \ldots)(n'; m_2', m_3', \ldots)$$
$$= \left(n + n' + \sum_{k=2}^{\infty} \left[\frac{m_k + m_k'}{k}\right]; m_2 + m_2' \ (\mathrm{mod}\ 2), \ldots, m_k + m_k' \ (\mathrm{mod}\ k), \ldots\right).$$

It is easy to see that G is an Abelian group. It is also clear that if $(n; m_2, m_3, \ldots, m_k, \ldots)$ is in $G^{(k)}$, then $m_k = 0$. Hence $\bigcap_{k=2}^{\infty} G^{(k)}$ contains only elements of the form $(n; 0, 0, 0, \ldots)$. Since

$$(0; 0, \ldots, 0_{(k-1)}, 1_{(k)}, 0_{(k+1)}, \ldots)^{nk} = (n; 0, 0, \ldots),$$

we see that $\bigcap_{k=1}^{\infty} G^{(k)}$ consists exactly of the subgroup $\{(n; 0, 0, \ldots) : n \in Z\}$. This group is reduced (A.5), and so $D = D^- = \{(0; 0, 0, \ldots)\} \subsetneqq \bigcap_{k=1}^{\infty} G^{(k)}$.]

(d) The component C of a locally compact Abelian group is always divisible, as shown in (24.24). The component C may fail to be divisible for some non-Abelian, noncompact, locally compact groups[1]. [The group $\mathfrak{S}\mathfrak{L}(2, R)$ is connected. First, the subgroups $\left\{\begin{pmatrix} 1 & x \\ 0 & 1 \end{pmatrix} : x \in R\right\}$ and $\left\{\begin{pmatrix} 1 & 0 \\ y & 1 \end{pmatrix} : y \in R\right\}$ are obviously connected. For any $\begin{pmatrix} a & b \\ c & d \end{pmatrix}$ in $\mathfrak{S}\mathfrak{L}(2, R)$ such that $b \neq 0$, we have

$$\begin{pmatrix} a & b \\ c & d \end{pmatrix} = \begin{pmatrix} 1 & 0 \\ \frac{d-1}{b} & 1 \end{pmatrix} \begin{pmatrix} 1 & b \\ 0 & 1 \end{pmatrix} \begin{pmatrix} 1 & 0 \\ \frac{a-1}{b} & 1 \end{pmatrix}.$$

Since the component of E is a closed normal subgroup, $\mathfrak{S}\mathfrak{L}(2, R)$ is connected. If α is any negative number different from -1, then there is no real matrix $\begin{pmatrix} x & y \\ z & w \end{pmatrix}$ such that $\begin{pmatrix} x & y \\ z & w \end{pmatrix}^2 = \begin{pmatrix} \alpha & 0 \\ 0 & \frac{1}{\alpha} \end{pmatrix}$.]

(e) Another example illustrating (d) is given by $\mathfrak{G}\mathfrak{L}(2, R)$. [The subgroups $\left\{\begin{pmatrix} x & 0 \\ y & z \end{pmatrix} : x > 0,\ z > 0,\ y \in R\right\}$ and $\left\{\begin{pmatrix} x & y \\ 0 & z \end{pmatrix} : x > 0,\ z > 0,\ y \in R\right\}$ are

[1] The definition of a divisible non-Abelian group is exactly the same as the definition of a divisible Abelian group (A.5).

obviously connected. If $\begin{pmatrix} a & b \\ c & d \end{pmatrix} \in \mathfrak{GL}(2, R)$ and $b \neq 0$, then

$$\begin{pmatrix} a & b \\ c & d \end{pmatrix} = \begin{pmatrix} 1 & 0 \\ \frac{d-1}{b} & 1 \end{pmatrix} \begin{pmatrix} ad-bc & b \\ 0 & 1 \end{pmatrix} \begin{pmatrix} 1 & 0 \\ \frac{bc-ad+a}{b} & 1 \end{pmatrix}.$$

Thus if $ad-bc>0$, the matrix $\begin{pmatrix} a & b \\ c & d \end{pmatrix}$ must belong to the component $\mathfrak{C}$ of E; that is, $\mathfrak{C} \supset \{A \in \mathfrak{GL}(2, R) : \det A > 0\}$. This inclusion is actually an equality, since otherwise $\mathfrak{C}$ would equal $\mathfrak{GL}(2, R)$, contrary to the fact that the continuous function $A \to \det A$ maps $\mathfrak{GL}(2, R)$ onto $]-\infty, 0[\cup]0, \infty[$. If α and β are distinct nonzero real numbers and at least one is negative, then there is no real matrix $\begin{pmatrix} x & y \\ z & w \end{pmatrix}$ such that $\begin{pmatrix} x & y \\ z & w \end{pmatrix}^2 = \begin{pmatrix} \alpha & 0 \\ 0 & \beta \end{pmatrix}$.]

(24.45) The component of e as a direct factor (BRACONNIER [1]). (a) Let G be a locally compact Abelian group. If the component C of e is open, then G is topologically isomorphic with $C \times (G/C)$. [By (24.24), C is divisible. Now apply (6.22.b).]

(b) If G is locally connected, then G is topologically isomorphic with $C \times (G/C)$. [For, C contains a neighborhood of e, and (5.5) implies that C is open. Now apply (a).]

(24.46) Pure subgroups. (a) (HARTMAN and HULANICKI [1].) Let G be a locally compact Abelian group, H a closed subgroup of G, and X and Y the character groups of G and H respectively. If every character $\chi_1 \in \mathsf{Y}$ admits an extension $\chi_2 \in \mathsf{X}$ having the same order as χ_1, then H is said to have *property* (H). The subgroup H has property (H) if and only if the annihilator $\mathsf{A}(\mathsf{X}, H)$ is a pure subgroup of X. [Suppose H has property (H), that $\chi_1 \in \mathsf{X}$, and that $\chi_1^n \in \mathsf{A}(\mathsf{X}, H)$. Then χ_1 *as an element of* Y has order a, where a divides n. Let χ_2 be an extension of χ_1 in X having order a. Then $(\bar{\chi}_2\chi_1)^n = \chi_1^n$ and $\bar{\chi}_2\chi_1 \in \mathsf{A}(\mathsf{X}, H)$. Thus $\mathsf{A}(\mathsf{X}, H)$ is pure in X.

If $\mathsf{A}(\mathsf{X}, H)$ is pure, χ_0 is in Y, and the order of χ_0 is n, let χ_1 be an arbitrary extension of χ_0 such that $\chi_1 \in \mathsf{X}$. The character χ_1^n is in $\mathsf{A}(\mathsf{X}, H)$. Since $\mathsf{A}(\mathsf{X}, H)$ is pure, there is a $\chi_2 \in \mathsf{A}(\mathsf{X}, H)$ such that $\chi_2^n = \chi_1^n$. Then $\bar{\chi}_2\chi_1$ is an extension of χ_0 such that $(\bar{\chi}_2\chi_1)^n = 1$.]

(b) Let G be a discrete Abelian group, and let H be a pure subgroup of G such that G/H is divisible. Then $\mathsf{A}(\mathsf{X}, H)$ is a pure subgroup of X. [Let χ_0 be a character of H, of order $n < \infty$. We will show that χ_0 can be extended to a character of G also of order n. This will show that $\mathsf{A}(\mathsf{X}, H)$ is pure, by part (a). Since G/H is divisible, every $x \in G$ can be written in the form $x = hy^n$ for some $h \in H$ and $y \in G$. Define $\chi(x) = \chi_0(h)$. To see that χ is well defined, suppose that $h_1 y_1^n = h_2 y_2^n$. Then $h_1 h_2^{-1} =$

$(y_2 y_1^{-1})^n$, and since H is pure, there is an element $h_3 \in H$ such that $h_3^n = h_1 h_2^{-1}$. Thus we have $1 = \chi_0(h_3)^n = \chi_0(h_3^n) = \chi_0(h_1)\,\chi_0(h_2)^{-1}$. That is, $\chi_0(h_1) = \chi_0(h_2)$, and χ is well defined. To see that χ is a character of G, let x_1 and x_2 be any elements of G, and let $x_j = h_j y_j^n$ $(j=1, 2)$. Then $x_1 x_2 = h_1 h_2 (y_1 y_2)^n$, and as above we have $\chi(x_1 x_2) = \chi_0(h_1 h_2) = \chi_0(h_1)\,\chi_0(h_2) = \chi(x_1)\,\chi(x_2)$.]

(c) (BRACONNIER [1].) The closure of a pure subgroup need not be pure. [Let I be any infinite index set. For each $\iota \in I$, let $G_\iota = Z(4)$, which we represent as $\{0, 1, 2, 3\}$ with addition modulo 4. For $\iota \in I$, let $H_\iota = \{0, 2\}$. Topologize $G = \underset{\iota\in I}{\mathsf{P}} G_\iota$ as in (6.15.c). Thus a generic neighborhood of the identity has the form

$$U_\Lambda = \{x \in G : x_\iota \in \{0, 2\} \text{ for all } \iota \text{ and } x_\iota = 0 \text{ for } \iota \in \Lambda\},$$

where Λ is a finite subset of I. It is easy to see that G is a locally compact, Abelian, 0-dimensional group. Let G_0 be the subgroup consisting of all $x \in G$ such that $\{\iota \in I : x_\iota \neq 0\}$ is finite. It is evident that G_0 is pure in G: if $y \in G$ and $y^n \in G_0$, then there is a $z \in G_0$ such that $z^n = y^n$. The closure G_0^- of G_0 consists of all $x \in G$ such that $\{\iota \in I : x_\iota = 1 \text{ or } 3\}$ is finite. This subgroup is not pure: choose x such that $x_\iota \in \{0, 2\}$ for all $\iota \in I$ and $\{\iota \in I : x_\iota = 2\}$ is infinite. If $y_\iota = 1$ for all ι such that $x_\iota = 2$ and $y_\iota = 0$ for all other ι, then $y^2 = x$ but there is no $z \in G_0^-$ such that $z^2 = x$.]

(24.47) (KAKUTANI [4].) Let G be an infinite discrete Abelian group with [compact] character group X. Theorem (24.15) shows that $\mathfrak{w}(\mathsf{X}) = \overline{\overline{G}}$. The cardinal number of X itself is $2^{\overline{\overline{G}}}$. [Since elements of X are complex-valued functions on G, it follows that $\overline{\overline{\mathsf{X}}} \leq (\overline{\overline{K}})^{\overline{\overline{G}}} = \mathfrak{c}^{\overline{\overline{G}}} = 2^{\overline{\overline{G}}}$. To verify the reversed inequality, we first suppose that G is countable. Since X is compact, (4.26) implies that $\overline{\overline{\mathsf{X}}} \geq \mathfrak{c} = 2^{\overline{\overline{G}}}$. Finally, suppose that $\overline{\overline{G}} > \aleph_0$. Let $G = \{x_1, x_2, \ldots, x_\alpha, \ldots\}$ be a well ordering of G, where α runs through all ordinals less than the smallest ordinal having cardinal number $\overline{\overline{G}}$ and where $x_1 = e$. For each $\alpha > 1$, let H_α denote the group generated by $\{x_\beta \in G : \beta < \alpha\}$. Let A consist of those elements x_α of G for which $x_\alpha \notin H_\alpha$. The set A generates the entire group G. For, if A_0 denotes the group generated by A and $A_0 \neq G$, there is a least element x_{α_0} in $G \cap A_0'$. It is clear that $\alpha_0 > 1$ and $x_{\alpha_0} \notin A$. Therefore $x_{\alpha_0} = x_{\beta_1}^{n_1} \cdots x_{\beta_l}^{n_l}$ where $\beta_j < \alpha_0$. Since all of the x_{β_j} are in A_0, x_{α_0} is also in A_0, a contradiction. Since $\overline{\overline{G}} > \aleph_0$, it follows that $\overline{\overline{A}} = \overline{\overline{G}}$.

We now show how to construct a character χ of G by transfinite induction. Let $\chi(x_1) = 1$ and suppose that $\chi(x_\beta)$ has been defined for all $\beta < \alpha$. If $x_\alpha \notin A$, *i.e.*, $x_\alpha \in H_\alpha$, then $\chi(x_\alpha)$ is determined by the values $\{\chi(x_\beta) : \beta < \alpha\}$. Suppose that $x_\alpha \in A$. If no power of x_α lies in H_α, then we may plainly define $\chi(x_\alpha)$ to be any number in T. If some power of

x_α lies in H_α and m is the least positive integer such that $x_\alpha^m \in H_\alpha$, then we may define $\chi(x_\alpha)$ to be any m-th root of $\chi(x_\alpha^m)$. This procedure always defines a character of G and moreover, for each $x_\alpha \in A$, at least two different choices of $\chi(x_\alpha)$ are possible. It follows that G admits at least $2^{\overline{\overline{A}}}$ distinct characters. That is, $\overline{\overline{\mathsf{X}}} \geqq 2^{\overline{\overline{A}}} = 2^{\overline{\overline{G}}}$.]

(24.48) (HEWITT and STROMBERG [1].) Let G be a locally compact Abelian group with character group X. Let $\mathfrak{a}$ be the least cardinal number of an open basis at e in G and $\mathfrak{b}$ the least cardinal number of a family of compact subsets of X whose union is X. Then $\mathfrak{a}$ and $\mathfrak{b}$ are equal. [Let $\mathscr{V}$ be an open basis at e such that each V in $\mathscr{V}$ has compact closure. The sets $\mathsf{P}(V^-, \frac{1}{2})^-$, where $V \in \mathscr{V}$, are compact subsets of X (23.16), and it is obvious that their union is X. Hence $\mathfrak{a} \geqq \mathfrak{b}$. If $\mathfrak{b} = 1$, then also $\mathfrak{a} = 1$ (23.17). If $\mathfrak{b} \neq 1$, then $\mathfrak{b}$ is infinite. Thus suppose that $\mathfrak{b} \geqq \aleph_0$ and that $\mathscr{A} = \{\mathsf{A}_\lambda : \lambda \in \Lambda\}$ is any family of compact subsets of X such that $\bigcup_{\lambda \in \Lambda} \mathsf{A}_\lambda = \mathsf{X}$ and $\overline{\overline{\mathscr{A}}} = \mathfrak{b}$. Plainly there are open sets $\mathsf{Y}_\lambda \supset \mathsf{A}_\lambda$ such that the sets Y_λ^- are compact. Let $\mathscr{B}$ be the family of all finite unions of sets $\mathsf{Y}_{\lambda_1}^- \cup \cdots \cup \mathsf{Y}_{\lambda_k}^-$. The sets $\left\{x \in G : |\chi(x) - 1| < \frac{1}{n} \text{ for all } \chi \in \mathsf{B}\right\}$, where $\mathsf{B} \in \mathscr{B}$, form a basis of neighborhoods at e in G of cardinal number $\mathfrak{b}$. In fact, for an arbitrary neighborhood V of e in G, (24.8) shows that there are a compact subset Φ of X and a positive integer n such that $\left\{x \in G : |\chi(x) - 1| < \frac{1}{n} \text{ for all } \chi \in \Phi\right\} \subset V$. There is some B in $\mathscr{B}$ containing Φ; this proves that $\mathfrak{a} \leqq \mathfrak{b}$.]

(24.49) Remark on Theorem (24.30). We can obtain part of (24.30) directly from (9.8) without appealing to the duality theorem. Let H be a compactly generated open subgroup of G. By (9.8), H can be regarded as $R^m \times Z^p \times F$, where F is a compact Abelian group and m and p are nonnegative integers. The subgroup $R^m \times F$ of H is plainly open in G. Since R^m is a divisible group, the projection mapping of $R^m \times F$ onto R^m can be extended to a homomorphism of G onto R^m (A.7). This homomorphism is continuous on G since it is continuous on $R^m \times F$ (5.40.a). By (6.22.a), G is topologically isomorphic with $R^m \times (G/R^m)$. Since the natural mapping of G onto G/R^m is an open mapping (5.17), the image of $R^m \times F$ is open in G/R^m. That is, $(R^m \times F)/R^m$ is open in G/R^m. Since $(R^m \times F)/R^m$ is topologically isomorphic with F, we have proved that G/R^m contains a compact open subgroup.

Notes

The duality theorem (24.8) is the creation of L. S. PONTRYAGIN and E. R. VAN KAMPEN; other writers have made contributions, as noted below. Theorem (24.7) is due to VAN KAMPEN [1], p. 458. Theorem (24.3) for G having a countable open basis was hinted at in PONTRYAGIN [1], announced in PONTRYAGIN [3], and proved in PONTRYAGIN [4]. ALEXANDER [1] proved (24.3) for arbitrary discrete Abelian groups; ALEXANDER and ZIPPIN [1] simplified the argument. Theorem (24.8) was first proved by VAN KAMPEN [1]. VAN KAMPEN'S topology for X is formally different from PONTRYAGIN'S, but is actually the same. Our proof of (24.8) is basically the same as VAN KAMPEN'S; we have also been guided by PONTRYAGIN [7], Ch. 6. A. WEIL [4], §28, pp. 102–109, has given a different proof of (24.8). RAĬKOV [announced in [1], full details in [3]] and CARTAN and GODEMENT [1] have given a short and totally different proof of (24.8), based on Fourier transform theory. We have chosen to give the "classical" structure-theoretic proof of (24.8) for several reasons. First, we needed Theorem (9.8) for other purposes; (9.8) is not a consequence of (24.8), while (9.8) is the key to our proof of (24.8). Second, (24.10) and (23.25), which are needed for various purposes, are not consequences of (24.8). Third, we feel that the classical proof of (24.8) is more illuminating than the proof based on Fourier transforms.

Theorems (23.25), (24.10), and (24.11) are due to PONTRYAGIN [4] for countable discrete groups; to VAN KAMPEN [1] for general locally compact Abelian groups.

Theorem (24.14) is due to KAKUTANI [4] for compact G, and in the general case to PONTRYAGIN [7], §40, Theorem 57. Our proof is new. Theorem (24.17) and the proof given here are also PONTRYAGIN'S [7], §40, Example 73, p. 282. Parts of (24.18), (24.25), and (24.26) are due, in various forms, to PONTRYAGIN [4] and VAN KAMPEN [1]. Theorem (24.21) [but not our proof] is due to BRACONNIER [1], p. 51. Various parts of (24.23) and (24.24) are due to KAPLANSKY [1], p. 55 [without proofs], to HARTMAN and RYLL-NARDZEWSKI [1], and to BRACONNIER [1]. In (24.23), we correct an error in HALMOS [1].

Theorem (24.28) is due to PONTRYAGIN [4] for G with a countable open basis and to VAN KAMPEN [1] for arbitrary G. Our proof of (24.28), and (24.27), are from PONTRYAGIN [7], §38, Theorem 47, p. 263. It is interesting to note that ARHANGEL'SKIĬ [1] and PASYNKOV [1] have proved that all standard definitions of dimension are the same for locally compact groups.

Theorem (24.29) is due to PONTRYAGIN [7], §40, Example 74, p. 283. Its immediate consequence (24.30) is due to VAN KAMPEN [1], p. 461; a proof also appears in A. WEIL [4], §29, p. 110.

Real characters as defined in (24.33) have been studied by a number of writers. Corollary (24.35) is due to G.W. MACKEY [2], and (24.36) to J. DIXMIER [2] [our proof of (24.36) is different from DIXMIER'S]. RISS [1], [2] has two notes on real characters and has utilized real characters heavily in his theory of distributions on locally compact Abelian groups [3]. Real characters of certain semigroups have been considered by DEVINATZ and NUSSBAUM [1]. DIXMIER [2] has also characterized in a manner different from (24.43) the locally compact Abelian groups such that every continuous character has the form $\exp(if)$ where f is a continuous real character.

Definition (24.37) is due to FREUDENTHAL [3]. The adjoint mapping has been used by many writers, for example: A. WEIL [4], §28, p. 103; ANZAI and KAKUTANI [1]; BRACONNIER [1]. Theorem (24.40) is due to BRACONNIER [1].

Thousands of pages in research journals have been devoted to studies of the structure of locally compact Abelian groups, of characters of all sorts of groups, and of extensions of the duality theorem. We now mention a few of the most interesting results and papers not dealt with in the main text.

PONTRYAGIN [7], §38, Theorem 48, p. 264, has characterized compact, locally connected, Abelian groups G in terms of algebraic properties of the character group X. A special case of his theorem is the following. A compact Abelian group G is connected and locally connected if and only if every finite subset of X is contained in a pure subgroup of X having a finite number of generators. If X is second countable and G is connected and locally connected, then G is also arcwise connected and has the form $R^n \times T^{\mathfrak{m}}$, where n is a nonnegative integer and $\mathfrak{m} \leqq \aleph_0$: this is due also to PONTRYAGIN [4], p. 380. DIXMIER [2] has constructed an uncountable group X for which G is connected and locally connected but not arcwise connected.

N. YA. VILENKIN [4], [7], [8], [13] has made elaborate computations of character groups, for both locally compact and non locally compact Abelian groups. G. C. PRESTON [1] has shown that continuous homomorphisms of 0-dimensional locally compact Abelian groups into certain groups other than T can serve to produce theorems like (24.8), although the "character" groups may be quite different from the character groups treated in the main text.

Some attention has been given to extensions of (24.8) to topological Abelian groups that are not locally compact. See the following: T. ISIWATA [1]; S. KAPLAN [1], [2]; H. LEPTIN [1], [2], [3], [4]; H. SCHÖNEBORN [1], [2], [3]; M. F. SMITH [1].

§25. Special structure theorems

We first construct the character groups of a number of specific groups [(25.1) to (25.7)]. In doing this, the duality theorem (24.8) and also Theorems (24.10) and (24.11) are very useful.

(25.1) The $\boldsymbol{a}$-adic numbers. Consider the additive group $\Omega_{\boldsymbol{a}}$ of all $\boldsymbol{a}$-adic numbers, which is discussed in (10.2)—(10.5). We will show that the character group of $\Omega_{\boldsymbol{a}}$ is topologically isomorphic with $\Omega_{\boldsymbol{a}^\star}$, where $a_n^\star = a_{-n}$ for all $n \in Z$. In particular, for any integer $r > 1$, the character group of Ω_r is Ω_r itself.

Consider first a continuous character χ of $\Omega_{\boldsymbol{a}}$ different from the character 1. Then for some integer l, we have $|\chi(\boldsymbol{x}) - 1| < \sqrt{3}$ for all $\boldsymbol{x} \in \Lambda_l$. Since Λ_l is a subgroup of $\Omega_{\boldsymbol{a}}$, it follows as in (23.17) that $\chi(\Lambda_l) = \{1\}$. Now consider all integers l, negative or nonnegative, for which $\chi(\Lambda_l) = \{1\}$. Since $\chi \neq 1$, there is a least integer with this property; it may or may not be negative, but in any case we denote its negative by k. In other words,

$$\chi(\Lambda_m) = \{1\} \quad \text{if and only if} \quad m \geqq -k. \tag{1}$$

For $n \in Z$, let $\boldsymbol{u}^{(n)}$ be the element of $\Omega_{\boldsymbol{a}}$ such that $u_p^{(n)} = \delta_{np}$ $(p \in Z)$. By (1), we have $\chi(\boldsymbol{u}^{(n)}) = 1$ for $n \geqq -k$. Note that χ on Λ_{-k-1} is determined by its value at $\boldsymbol{u}^{(-k-1)}$ because the set of multiples $\{\boldsymbol{u}^{(-k-1)}, 2\boldsymbol{u}^{(-k-1)}, \ldots\}$ is dense in Λ_{-k-1}. It follows that $\chi(\boldsymbol{u}^{(-k-1)}) \neq 1$. For each $n \in Z$, let λ_n be the unique number in $[0, 1[$ for which $\chi(\boldsymbol{u}^{(-n)}) = \exp[2\pi i \lambda_n]$. Thus we have $\lambda_n = 0$ for $n \leqq k$. Since $a_n^\star \boldsymbol{u}^{(-n)} = a_{-n} \boldsymbol{u}^{(-n)} = \boldsymbol{u}^{(-n+1)}$, we infer that $a_n^\star \lambda_n = \lambda_{n-1}$ (mod 1), that is,

$$a_n^\star \lambda_n = \lambda_{n-1} + y_n \tag{2}$$

for some integer y_n. Moreover, we have $y_n = a_n^\star \lambda_n - \lambda_{n-1} \leqq a_n^\star \lambda_n < a_n^\star$ and $y_n = a_n^\star \lambda_n - \lambda_{n-1} \geqq -\lambda_{n-1} > -1$, so that $0 \leqq y_n \leqq a_n^\star - 1$. Thus we have defined an element $\boldsymbol{y} = (y_n)$ of $\Omega_{\boldsymbol{a}^\star}$ such that $y_n = 0$ for $n \leqq k$ and $y_{k+1} \neq 0$: since $0 < \lambda_{k+1} < 1$ and $a_{k+1}^\star \geqq 2$, (2) shows that $0 \neq a_{k+1}^\star \lambda_{k+1} = \lambda_k + y_{k+1} = y_{k+1}$. An elementary computation using (2) shows that

$$\lambda_n = \sum_{j=k+1}^{n} \frac{y_j}{a_j^\star a_{j+1}^\star \cdots a_n^\star}, \tag{3}$$

for $n = k+1, k+2, \ldots$. Given $\boldsymbol{x}$ in $\Omega_{\boldsymbol{a}}$, let us compute $\chi(\boldsymbol{x})$. If $\boldsymbol{x} \in \Lambda_{-k}$, then $\chi(\boldsymbol{x}) = 1$. Otherwise there is an integer $m < -k$ such that $\boldsymbol{x} = x_m \boldsymbol{u}^{(m)} + x_{m+1} \boldsymbol{u}^{(m+1)} + \cdots + x_{-k-1} \boldsymbol{u}^{(-k-1)} + \boldsymbol{z}$, where the x_j's are integers and $\boldsymbol{z} \in \Lambda_{-k}$. Therefore we have

$$\chi(\boldsymbol{x}) = \exp[2\pi i (x_m \lambda_{-m} + x_{m+1} \lambda_{-m-1} + \cdots + x_{-k-1} \lambda_{k+1})]. \tag{4}$$

Relations (3) and (4) show that χ is determined by the sequence $(y_n) \in \Omega_{\boldsymbol{a}^\star}$. If χ is the character 1, let $\boldsymbol{y} = \boldsymbol{0}$ and $\lambda_n = 0$ for $n \in Z$. Then $1 = 1(\boldsymbol{u}^{(-n)}) = \exp[2\pi i \lambda_n]$ and the relation (2) holds for all $n \in Z$.

We now show how to obtain a continuous character of $\Omega_{\boldsymbol{a}}$ from an element $\boldsymbol{y} \in \Omega_{\boldsymbol{a}^\star}$. First consider the zero $\boldsymbol{y} = \boldsymbol{0}$ of $\Omega_{\boldsymbol{a}^\star}$. Set $\lambda_n = 0$ for all

$n \in Z$, and notice that (2) holds. We then require that $\chi(\boldsymbol{u}^{(-n)}) = \exp[2\pi i \lambda_n]$ for all $n \in Z$; in other words, $\chi = 1$. Now consider any element $\boldsymbol{y}$ in $\Omega_{\boldsymbol{a}^\star}$ different from $\boldsymbol{0}$. Let k be the least integer such that $y_{k+1} \neq 0$. Define the sequence $\{\lambda_n\}_{n=k+1}^{\infty}$ by (3), and for $n \leqq k$, let $\lambda_n = 0$. It is easy to see that the relation (2) holds for all integers n when the λ_n's are so defined. For $\boldsymbol{x} \in \Lambda_{-k}$, let $\chi(\boldsymbol{x}) = 1$ and for $\boldsymbol{x} \in \Omega_{\boldsymbol{a}} \cap (\Lambda_{-k})'$, define $\chi(\boldsymbol{x})$ by relation (4). To show that this defines a continuous character χ of $\Omega_{\boldsymbol{a}}$, it suffices to prove that χ is multiplicative: the continuity of χ will then follow from the fact that χ is identically 1 on the neighborhood Λ_{-k} of $\boldsymbol{0}$. Suppose that $\boldsymbol{x}$ and $\boldsymbol{z}$ belong to $\Omega_{\boldsymbol{a}}$. If either $\boldsymbol{x}$ or $\boldsymbol{z}$ belongs to Λ_{-k}, it is easy to verify that

$$\chi(\boldsymbol{x} + \boldsymbol{z}) = \chi(\boldsymbol{x})\,\chi(\boldsymbol{z}); \tag{5}$$

we omit the details. Suppose then that $x_n = z_n = 0$ for $n < l$, that $x_l + z_l \neq 0$, and that $l < -k$. Let $\boldsymbol{x} + \boldsymbol{z} = \boldsymbol{w} = (w_n)$ and $t_{l-1} = 0$. Obviously $w_n = 0$ for $n < l$, and for $n \geqq l$, we have $x_n + z_n + t_{n-1} = t_n a_n + w_n$; each t_n is either 0 or 1. From this and (2) it follows that

$$\begin{aligned}
&(x_l \lambda_{-l} + x_{l+1} \lambda_{-l-1} + \cdots + x_{-k-1} \lambda_{k+1}) \\
&\quad + (z_l \lambda_{-l} + z_{l+1} \lambda_{-l-1} + \cdots + z_{-k-1} \lambda_{k+1}) \\
&\quad = (w_l \lambda_{-l} + w_{l+1} \lambda_{-l-1} + \cdots + w_{-k-1} \lambda_{k+1}) \\
&\quad + t_l y_{-l} + t_{l+1} y_{-l-1} + \cdots + t_{-k-1} y_{k+1},
\end{aligned}$$

and since $t_l y_{-l} + t_{l+1} y_{-l-1} + \cdots + t_{-k-1} y_{k+1}$ is an integer, (4) implies that (5) holds.

Thus we have associated with every $\boldsymbol{y}$ in $\Omega_{\boldsymbol{a}^\star}$ a continuous character $\chi_{\boldsymbol{y}}$ of $\Omega_{\boldsymbol{a}}$. We will show that the mapping $\boldsymbol{y} \to \chi_{\boldsymbol{y}}$ is a topological isomorphism of $\Omega_{\boldsymbol{a}^\star}$ onto the character group of $\Omega_{\boldsymbol{a}}$. Given χ in the character group of $\Omega_{\boldsymbol{a}}$, we construct the $\boldsymbol{y}$ in $\Omega_{\boldsymbol{a}^\star}$ and the sequence $\{\lambda_n\}_{n=-\infty}^{\infty}$ of numbers that satisfy (2). The character $\chi_{\boldsymbol{y}}$ constructed for this $\boldsymbol{y}$ agrees with χ on all $\boldsymbol{u}^{(n)}$, and hence $\chi = \chi_{\boldsymbol{y}}$. The mapping $\boldsymbol{y} \to \chi_{\boldsymbol{y}}$ therefore maps $\Omega_{\boldsymbol{a}^\star}$ *onto* the character group of $\Omega_{\boldsymbol{a}}$. Let $\boldsymbol{y}$ and $\boldsymbol{z}$ be nonzero elements of $\Omega_{\boldsymbol{a}^\star}$, and let $\boldsymbol{w} = \boldsymbol{y} + \boldsymbol{z}$. Let l be the least integer such that $y_{l+1} + z_{l+1} \neq 0$. Let $\{\lambda_n\}_{n=-\infty}^{\infty}$, $\{\mu_n\}_{n=-\infty}^{\infty}$, and $\{\nu_n\}_{n=-\infty}^{\infty}$ be the sequences of numbers defined as above that correspond to $\boldsymbol{y}$, $\boldsymbol{z}$, and $\boldsymbol{w}$, respectively. To show that $\chi_{\boldsymbol{y}} \chi_{\boldsymbol{z}} = \chi_{\boldsymbol{w}}$, we need only show that $\chi_{\boldsymbol{y}}(\boldsymbol{u}^{(-n)})\,\chi_{\boldsymbol{z}}(\boldsymbol{u}^{(-n)}) = \chi_{\boldsymbol{w}}(\boldsymbol{u}^{(-n)})$ for all $n \in Z$. This is obvious for $n \leqq l$. For $n > l$, we have $y_n + z_n + t_{n-1} = t_n a_n^\star + w_n$; t_l is 0 and each t_n is either 0 or 1. We have

$$\nu_n = \lambda_n + \mu_n - t_n \quad \text{for} \quad n = l, l+1, \ldots. \tag{6}$$

The relation (6) is trivial for $n = l$; and in general we have

$$\begin{aligned}
\nu_{n+1} &= \frac{\nu_n + w_{n+1}}{a_{n+1}^\star} = \frac{(\lambda_n + \mu_n - t_n) + (y_{n+1} + z_{n+1} + t_n - t_{n+1} a_{n+1}^\star)}{a_{n+1}^\star} \\
&= \frac{\lambda_n + y_{n+1}}{a_{n+1}^\star} + \frac{\mu_n + z_{n+1}}{a_{n+1}^\star} - t_{n+1} = \lambda_{n+1} + \mu_{n+1} - t_{n+1}.
\end{aligned}$$

For $n \geqq l$, (6) shows that

$$\begin{aligned} \chi_{\boldsymbol{w}}(\boldsymbol{u}^{(-n)}) &= \exp[2\pi i(\lambda_n + \mu_n - t_n)] \\ &= \exp[2\pi i \lambda_n] \exp[2\pi i \mu_n] \cdot 1 = \chi_{\boldsymbol{y}}(\boldsymbol{u}^{(-n)}) \, \chi_{\boldsymbol{z}}(\boldsymbol{u}^{(-n)}). \end{aligned}$$

It is easy to see that the mapping $\boldsymbol{y} \rightarrow \chi_{\boldsymbol{y}}$ is one-to-one.

To complete the argument, we must show that the mapping $\boldsymbol{y} \rightarrow \chi_{\boldsymbol{y}}$ is a homeomorphism. To avoid confusion, the neighborhoods of $\mathbf{0}$ in $\Omega_{\boldsymbol{a}^\star}$ that have the form Λ_m (10.4) will be written as $\Lambda_m^{(\star)}$. We first observe that

$$\{\chi_{\boldsymbol{y}} : \boldsymbol{y} \in \Lambda_{-m+1}^{(\star)}\} = \{\chi_{\boldsymbol{y}} : \chi_{\boldsymbol{y}}(\Lambda_m) = \{1\}\}, \tag{7}$$

for $m \in Z$. Let $\mathsf{P}(F, \varepsilon)$ be a neighborhood of 1 in the character group of $\Omega_{\boldsymbol{a}}$. Since F is a compact subset of $\Omega_{\boldsymbol{a}}$, F is a subset of some set Λ_m. Hence by (7) we have

$$\begin{aligned} \{\chi_{\boldsymbol{y}} : \boldsymbol{y} \in \Lambda_{-m+1}^{(\star)}\} &\subset \{\chi_{\boldsymbol{y}} : |\chi_{\boldsymbol{y}}(\boldsymbol{x}) - 1| < \varepsilon \text{ for all } \boldsymbol{x} \in \Lambda_m\} \\ &= \mathsf{P}(\Lambda_m, \varepsilon) \subset \mathsf{P}(F, \varepsilon). \end{aligned}$$

Thus the mapping $\boldsymbol{y} \rightarrow \chi_{\boldsymbol{y}}$ is continuous; (5.29) now implies that $\boldsymbol{y} \rightarrow \chi_{\boldsymbol{y}}$ is a homeomorphism.

It is useful to have an explicit formula for characters of $\Omega_{\boldsymbol{a}}$. If $\boldsymbol{y} \in \Omega_{\boldsymbol{a}^\star}$ and $\boldsymbol{x} \in \Omega_{\boldsymbol{a}}$, then

$$\chi_{\boldsymbol{y}}(\boldsymbol{x}) = \exp\left[2\pi i \left(\sum_{n=-\infty}^{\infty} x_n \left(\sum_{s=n}^{\infty} \frac{y_{-s}}{a_n a_{n+1} \cdots a_s}\right)\right)\right]. \tag{8}$$

This is verified by combining formulas (4) and (3). For each given $\boldsymbol{x}$ and $\boldsymbol{y}$, the sums in (8) are actually finite. In fact, if $x_j = 0$ for $j < m$ and $y_l = 0$ for $l \leqq k$, then

$$\chi_{\boldsymbol{y}}(\boldsymbol{x}) = \exp\left[2\pi i \left(\sum_{n=m}^{-k-1} x_n \left(\sum_{s=n}^{-k-1} \frac{y_{-s}}{a_n a_{n+1} \cdots a_s}\right)\right)\right], \tag{9}$$

where void sums are taken to be 0.

(25.2) The $\boldsymbol{a}$-adic integers. Consider the additive group $\Delta_{\boldsymbol{a}}$ of all $\boldsymbol{a}$-adic integers, where $\boldsymbol{a} = (a_0, a_1, \ldots)$. We will find the character group of $\Delta_{\boldsymbol{a}}$ with the aid of Theorem (24.11). We extend $\boldsymbol{a}$ to an arbitrary doubly infinite sequence $(\ldots, a_{-1}, a_0, a_1, \ldots)$ such that $a_n \geqq 2$ for all $n \in Z$ [for example, we might let $a_n = 2$ for $n < 0$]. By (24.11) and (25.1), every continuous character χ of $\Delta_{\boldsymbol{a}}$ has the form $\chi_{\boldsymbol{y}} | \Delta_{\boldsymbol{a}}$ for some $\boldsymbol{y} \in \Omega_{\boldsymbol{a}^\star}$.

Define $\boldsymbol{u}$ in $\Delta_{\boldsymbol{a}}$ by $u_n = 1$ for $n = 0$ and $u_n = 0$ for $n \in Z$, $n \neq 0$. Then a continuous character χ on $\Delta_{\boldsymbol{a}}$ is determined by the value $\chi(\boldsymbol{u})$, since $\boldsymbol{u}$ generates a dense subgroup of $\Delta_{\boldsymbol{a}}$. It follows that the [discrete!] character group of $\Delta_{\boldsymbol{a}}$ is isomorphic with the subgroup $\{\chi_{\boldsymbol{y}}(\boldsymbol{u}) : \boldsymbol{y} \in \Omega_{\boldsymbol{a}^\star}\}$ of T. For $\boldsymbol{y} \in \Omega_{\boldsymbol{a}^\star}$, relation (8) of (25.1) shows that

$$\chi_{\boldsymbol{y}}(\boldsymbol{u}) = \exp\left[2\pi i \left(\sum_{n=0}^{\infty} \frac{y_{-n}}{a_0 a_1 \cdots a_n}\right)\right]; \tag{1}$$

the sums in (1) are finite for each $\boldsymbol{y}$. Thus $\{\chi_{\boldsymbol{y}}(\boldsymbol{u}) : \boldsymbol{y} \in \Omega_{\boldsymbol{a}^*}\}$ consists of all numbers of the form $\exp\left[2\pi i\left(\frac{l}{a_0 a_1 \cdots a_r}\right)\right]$. We call this group $Z(\boldsymbol{a}^\infty)$; it is an obvious generalization of the group $Z(p^\infty)$ defined in §1.

Consider a number $t = \exp\left[2\pi i\left(\frac{l}{a_0 a_1 \cdots a_r}\right)\right]$ in $Z(\boldsymbol{a}^\infty)$. We may suppose that $0 \leq l < a_0 a_1 \cdots a_r$. Then l can be written uniquely as

$$l = y_0 a_1 a_2 \cdots a_r + y_{-1} a_2 \cdots a_r + \cdots + y_{-r+1} a_r + y_{-r},$$

where each y_j is an integer satisfying $0 \leq y_j < a_{-j}$. If we take y_j to be 0 for $j \notin \{-r, -r+1, \ldots, 0\}$, then (1) becomes

$$\chi_{\boldsymbol{y}}(\boldsymbol{u}) = \exp\left[2\pi i\left(\sum_{n=0}^{r} \frac{y_{-n}}{a_0 a_1 \cdots a_n}\right)\right] = \exp\left[2\pi i\left(\sum_{n=0}^{r} \frac{y_{-n} a_{n+1} a_{n+2} \cdots a_r}{a_0 a_1 \cdots a_r}\right)\right] = t.$$

Using this and (25.1.9), we see that the character χ_t of $\Delta_{\boldsymbol{a}}$ that corresponds to t in $Z(\boldsymbol{a}^\infty)$ is defined by

$$\left.\begin{aligned} \chi_t(\boldsymbol{x}) &= \exp\left[2\pi i\left(\sum_{n=0}^{r} x_n \left(\sum_{s=n}^{r} \frac{y_{-s}}{a_n a_{n+1} \cdots a_s}\right)\right)\right] \\ &= \exp\left[2\pi i\left(\sum_{n=0}^{r} x_n \left(\sum_{s=0}^{n-2} y_{-s} a_{s+1} \cdots a_{n-1} + y_{-n+1}\right.\right.\right. \\ &\qquad\qquad \left.\left.\left. + \sum_{s=n}^{r} \frac{y_{-s}}{a_n a_{n+1} \cdots a_s}\right)\right)\right] \\ &= \exp\left[2\pi i\left(\sum_{n=0}^{r} x_n a_0 a_1 \cdots a_{n-1} \left(\sum_{s=0}^{r} \frac{y_{-s}}{a_0 a_1 \cdots a_s}\right)\right)\right] \\ &= \exp\left[\frac{2\pi i l}{a_0 a_1 \cdots a_r}(x_0 + a_0 x_1 + \cdots + a_0 a_1 \cdots a_{r-1} x_r)\right] \\ &= t^{x_0 + a_0 x_1 + \cdots + a_0 a_1 \cdots a_{r-1} x_r}, \end{aligned}\right\} \quad (2)$$

for $\boldsymbol{x} \in \Delta_{\boldsymbol{a}}$.

(25.3) The $\boldsymbol{a}$-adic solenoids. Let $\Sigma_{\boldsymbol{a}}$ denote the $\boldsymbol{a}$-adic solenoid described in (10.12). By (10.13) $\Sigma_{\boldsymbol{a}}$ is a compact group. The character group of $R \times \Delta_{\boldsymbol{a}}$ is $R \times Z(\boldsymbol{a}^\infty)$, where $Z(\boldsymbol{a}^\infty)$ is as defined in (25.2). The continuous characters of $R \times \Delta_{\boldsymbol{a}}$ are the functions

$$(\xi, \boldsymbol{x}) \rightarrow \exp\left[2\pi i\left(\alpha\xi + \frac{l}{a_0 a_1 \cdots a_n}(x_0 + a_0 x_1 + \cdots + a_0 a_1 \cdots a_{n-1} x_n)\right)\right], \quad (1)$$

where $\alpha \in R$ and l is an integer, $0 \leq l < a_0 a_1 \cdots a_n$. By the definition of $\Sigma_{\boldsymbol{a}}$ in (10.12) and Theorem (23.25), the continuous characters of $\Sigma_{\boldsymbol{a}}$ can be identified with the functions (1) that are equal to 1 at $(1, \boldsymbol{u})$. It is easy to see that the admissible values of α are those such that $\alpha + \frac{l}{a_0 a_1 \cdots a_n}$ is an integer for some choice of n and l. Thus α has to be a rational number of the form $\frac{m}{a_0 a_1 \cdots a_n}$. Moreover, if α has the form $\frac{m}{a_0 a_1 \cdots a_n}$, then for some integer l, $0 \leqq l < a_0 a_1 \cdots a_n$, $\alpha + \frac{l}{a_0 a_1 \cdots a_n}$ is an integer, and (1) can be written as

$$\left.\begin{aligned}(\xi, \boldsymbol{x}) \rightarrow &\exp\left[2\pi i(\alpha\xi - \alpha(x_0 + a_0 x_1 + \cdots + a_0 a_1 \cdots a_{n-1} x_n))\right] \\ &= \exp\left[2\pi i \alpha(\xi - (x_0 + a_0 x_1 + \cdots + a_0 a_1 \cdots a_{n-1} x_n))\right].\end{aligned}\right\} \quad (2)$$

Thus, given a rational number $\alpha = \frac{m}{a_0 a_1 \cdots a_n}$, (2) defines a continuous character of $\Sigma_{\boldsymbol{a}}$, and distinct values of α yield distinct characters. Therefore we have a topological isomorphism of the character group of $\Sigma_{\boldsymbol{a}}$ onto the [discrete] additive group of all rational numbers of the form $\frac{m}{a_0 a_1 \cdots a_n}$.

Theorem (24.8) implies that two groups $\Sigma_{\boldsymbol{a}}$ and $\Sigma_{\boldsymbol{b}}$ are topologically isomorphic if and only if the additive groups $\left\{\frac{m}{a_0 a_1 \cdots a_n} : n = 0, 1, \ldots; m \in Z\right\}$ and $\left\{\frac{m}{b_0 b_1 \cdots b_n} : n = 0, 1, \ldots; m \in Z\right\}$ are isomorphic.

(25.4) The discrete additive group $\boldsymbol{Q}$. The character group of the discrete additive group Q of rationals is topologically isomorphic with $\Sigma_{\boldsymbol{a}}$ for any $\boldsymbol{a}$ such that Q is equal to [or isomorphic with] the set $\left\{\frac{m}{a_0 a_1 \cdots a_n} : n = 0, 1, \ldots; m \in Z\right\}$. This follows from (24.8) and (25.3).

In particular, if $\boldsymbol{a} = (2, 3, 4, 5, \ldots)$, then $\Sigma_{\boldsymbol{a}}$ is the character group of Q, since $Q = \left\{\frac{m}{n!} : n = 0, 1, \ldots; m \in Z\right\}$. In this case, (25.3.2) [with an inconsequential change of notation] becomes

$$(\xi, \boldsymbol{x}) \rightarrow \exp\left[2\pi i \frac{m}{n!}\left(\xi - (x_0 + 2!\, x_1 + \cdots + (n-1)!\, x_{n-2})\right)\right]. \quad (1)$$

Consider a fixed element $(\xi, \boldsymbol{x}) + B$ in $\Sigma_{\boldsymbol{a}}$. By (24.9), we see that the character χ of Q corresponding to this element is defined by

$$\chi\left(\frac{m}{n!}\right) = \exp\left[2\pi i \frac{m}{n!}\left(\xi - (x_0 + 2!\, x_1 + \cdots + (n-1)!\, x_{n-2})\right)\right]. \quad (2)$$

Note, in particular, that $\chi(1) = \exp[2\pi i \xi]$.

(25.5) The following characterization of the character group of Q may also be illuminating. Let χ be a character of Q. For $n = 1, 2, \ldots$, let $\alpha_n = \chi\left(\frac{1}{n!}\right)$. We have

$$\alpha_n = \chi\left(\frac{1}{n!}\right) = \chi\left(\frac{n+1}{(n+1)!}\right) = \alpha_{n+1}^{n+1}, \quad (1)$$

for $n = 1, 2, \ldots$. The relation (1) is also sufficient to produce a character of Q: if $\{\alpha_n\}_{n=1}^{\infty}$ is a sequence of numbers in T such that (1) holds, then the function χ on Q such that

$$\chi\left(\frac{m}{n!}\right)=\alpha_n^m \tag{2}$$

is well defined and is a character of Q; we omit the details. For $n=2, 3, \ldots$, let f_n be the mapping of T onto itself such that $f_n(\alpha)=\alpha^n$. Then a character of Q can be identified with a sequence $(\alpha_1, \alpha_2, \ldots, \alpha_n, \ldots)\in T^{\aleph_0}$ such that $f_n(\alpha_n)=\alpha_{n-1}$ for $n=2, 3, \ldots$. Thus the character group of Q is isomorphic to a projective limit as follows [see (6.13)]. Consider the directed set $\{1, 2, \ldots\}$ and for $m<n$, let g_{nm} be the continuous open homomorphism of T onto itself defined by $g_{nm}(\alpha)=\alpha^{\frac{n!}{m!}}$ [g_{nm} is open by (5.29)]. All that we have done is require $g_{n(n-1)}$ to be f_n; g_{nm} was computed by iteration. The character group of Q is isomorphic to the projective limit of the inverse mapping system $\{1, 2, \ldots\}$, $T_{(n)}$, g_{nm}, and the P-topology of the character group of Q obviously agrees with the Cartesian product topology of the projective limit.

(25.6) The discrete group of real numbers. To obtain the character group of R_d, consider any Hamel basis H for R over the rational number field Q, so that every $x\in R$ has one and only one representation as a [finite] sum, $x=\sum_{k=1}^{n}\alpha_k h_k$ $(\alpha_k\in Q, h_k\in H)$. Thus R is algebraically isomorphic to the group $Q^{\mathfrak{c}*}$. Regarding Q and $Q^{\mathfrak{c}*}$ as discrete, we compute the character group of $Q^{\mathfrak{c}*}$ from Theorem (23.22). Transferring this computation back to R_d, we see that the character group of R_d is topologically isomorphic with $(\Sigma_{\boldsymbol{a}})^{\mathfrak{c}}$, where $\boldsymbol{a}=(2, 3, 4, \ldots)$.

We can describe a character χ of R_d somewhat more explicitly as follows. The values of χ on H are quite arbitrary numbers of absolute value 1. For $h\in H$, the numbers $\chi\left(\frac{1}{2}h\right)$, $\chi\left(\frac{1}{6}h\right)$, $\ldots$, $\chi\left(\frac{1}{n!}h\right)$, $\ldots$ are determined in accordance with (25.4) or (25.5). There are $\mathfrak{c}^{\mathfrak{c}}=2^{\mathfrak{c}}$ *discontinuous* characters of R.

(25.7) The discrete group T_d. To find the character group of T_d, we regard T_d as R_d/Z, and compute all characters χ of R_d such that $\chi(Z)=\{1\}$ (23.25). This can be done by choosing a Hamel basis H for R over Q such that $1\in H$, and then constructing all characters χ of R_d for which $\chi(1)=1$, by the process of (25.6). Thus $\chi(h)$ is an arbitrary number of absolute value 1 for all $h\neq 1$ in H, and χ at rh $(r\in Q)$ is described as in (25.6). The numbers $\chi\left(\frac{1}{n!}\right)$ $(n=2, 3, \ldots)$, are determined by the conditions $\chi\left(\frac{1}{2!}\right)^2=1$ and $\chi\left(\frac{1}{(n+1)!}\right)^{n+1}=\chi\left(\frac{1}{n!}\right)$ $(n=2, 3, \ldots)$. For $\boldsymbol{a}=(2, 3, 4, 5, \ldots)$, therefore, we see that the character group of T_d is topologically isomorphic with $(\Sigma_{\boldsymbol{a}})^{\mathfrak{c}}\times G_1$, where G_1 is a compact group that is the projective limit of a sequence of copies of T, with "first" element equal to $\{1\}$. Note that G_1 is the character group of the discrete group Q/Z.

The group G_1 can be identified in various ways. The group Q/Z is isomorphic with $\mathsf{P}^*_{p\in P} Z(p^\infty)$, where P denotes the set of all prime numbers (A.14). By (25.2) and (24.8), the character group of $Z(p^\infty)$ is Δ_p. By (23.22), the character group of Q/Z is topologically isomorphic with $\mathsf{P}_{p\in P} \Delta_p$, so that $G_1 = \mathsf{P}_{p\in P} \Delta_p$.

Finally, the group Q/Z is isomorphic with $Z(\boldsymbol{a}^\infty)$, where $\boldsymbol{a} = (2, 3, 4, 5, \ldots)$. Therefore G_1 is topologically isomorphic with $\Delta_{\boldsymbol{a}}$.

We now classify all compact Abelian torsion-free and torsion groups, as follows.

(25.8) Theorem. *A compact Abelian group G is torsion-free if and only if it is topologically isomorphic with $(\Sigma_{\boldsymbol{a}})^{\mathfrak{m}} \times \mathsf{P}_{p\in P} \Delta_p^{\mathfrak{n}_p}$, where P is the set of all primes, $\mathfrak{m}$ and $\mathfrak{n}_2, \mathfrak{n}_3, \ldots, \mathfrak{n}_p, \ldots$ are arbitrary cardinal numbers, and $\boldsymbol{a} = (2, 3, 4, \ldots)$.*

Proof. By (24.23), G is torsion-free if and only if its character group X is divisible. A divisible group has the form $\mathsf{P}^*_{\iota\in I} H_\iota$, where each H_ι is either Q or $Z(p^\infty)$ for some prime p, and of course all such groups are divisible. Now apply (25.4), (25.2), and (23.22). □

(25.9) Theorem. *A compact Abelian torsion group G is topologically isomorphic with*

(i) $\mathsf{P}_{\iota\in I} Z(b_\iota)$,

where there are only finitely many distinct positive integers b_ι and I is an arbitrary nonvoid index set.

Proof. Let $G_{(n)} = \{x \in G : x^n = e\}$ $(n = 2, 3, 4, \ldots)$. Each $G_{(n)}$ is a closed subgroup of G, and $G = \bigcup_{n=1}^{\infty} G_{(n)}$. As pointed out in (5.28), some $G_{(n)}$ has nonvoid interior, and so is open. Hence $G/G_{(n)}$ is discrete and compact, and therefore finite. Let $a = \overline{\overline{G/G_{(n)}}}$. Then $x^{an} = e$ for all $x \in G$. That is, the orders of all elements of G are bounded. The same is true of the discrete character group X of G. By (A.25), X is isomorphic with $\mathsf{P}^*_{\iota\in I} Z(b_\iota)$, where the b_ι's are bounded. An appeal to (23.27.c) and (23.22) completes the proof. □

(25.10) Corollary. *Let G be a compact Abelian group not of the form (25.9.i). Then every neighborhood of e in G contains an element of infinite order.*

Proof. Let V be a neighborhood of e. Then V^- is compact, every subgroup $G_{(n)}$ has void interior, and by (5.28), we see that $\bigcup_{n=1}^{\infty} (G_{(n)} \cap V^-) \subsetneqq V^-$. □

We next make a detailed analysis of the structure of monothetic groups.

(25.11) Theorem. *Let G be a locally compact Abelian group with character group X. Let S be the set of elements $a \in G$ such that $M_a = \{a^n : n \in Z\}^- = G$.*[1] *Then*

$$S = \{x \in G : \chi(x) \neq 1 \text{ for all } \chi \in \mathsf{X}, \chi \neq 1\}.$$

The group G is monothetic if and only if $S \neq \varnothing$.

Proof. Suppose that $M_a \subsetneqq G$. Then by (24.12), there is a $\chi_0 \in \mathsf{X}$ such that $\chi_0(a) = 1$ and $\chi_0 \neq 1$. Conversely, if $M_a = G$, then plainly a character χ such that $\chi(a) = 1$ must be 1. The last statement is obvious. □

(25.12) Theorem. *There is a largest compact monothetic group G_0, in the sense that every compact monothetic group is a continuous homomorphic image of G_0. Furthermore, every continuous homomorphic image of G_0 is a compact monothetic group.*

Proof. Let G_0 be the character group of T_d described in (25.7). If G is a compact monothetic group with character group X, then (24.32) shows that X may be regarded as a subgroup of T_d. By (24.11), G is topologically isomorphic with $G_0/A(G_0, \mathsf{X})$. □

(25.13) Theorem. *A discrete Abelian group G is isomorphic with a subgroup of T_d if and only if the p-ranks $r_p(G)$ of G are 0 or 1 for all primes p and the torsion-free rank $r_0(G)$ of G is less than or equal to $\mathfrak{c}$* (A.12).

Proof. If the ranks of G are as stated, imbed G in a divisible group E such that $r_p(E) = r_p(G)$ for $p = 0, 2, 3, 5, 7, 11, \ldots$ (A.16). By (A.14), E is isomorphic with

$$Q^{r_0(E)*} \times \underset{p \in P}{\mathsf{P}^*}\left(Z(p^\infty)^{r_p(E)}\right).$$

We also know that T_d is isomorphic with

$$Q^{\mathfrak{c}*} \times \underset{p \in P}{\mathsf{P}^*} Z(p^\infty).$$

Since $r_0(E) \leqq \mathfrak{c}$ and $r_p(E) \leqq 1$ for all primes p, it follows that E and hence G are isomorphic with subgroups of T_d.

It is obvious that $r_p(G) = 0$ or 1 and $r_0(G) \leqq \mathfrak{c}$ for every subgroup G of T_d. □

(25.14) Theorem. *A compact connected Abelian group G is monothetic if and only if $\mathfrak{w}(G) \leqq \mathfrak{c}$.*

Proof. By (24.15), we have $\mathfrak{w}(G) = \overline{\overline{\mathsf{X}}}$. Since G is connected, X is torsion-free (24.25). Thus $r_p(\mathsf{X}) = 0$ for all primes p. If $\overline{\overline{\mathsf{X}}} \leqq \mathfrak{c}$, then obviously $r_0(\mathsf{X}) \leqq \mathfrak{c}$. If $r_0(\mathsf{X}) \leqq \mathfrak{c}$, then (25.13) shows that X is isomorphic

[1] We call the elements *of S generators of G.*

with a subgroup of T_d, and hence $\overline{\overline{\mathsf{X}}} \leqq \mathfrak{c}$. Now (25.13) and (24.32) imply that G is monothetic if and only if $r_0(\mathsf{X}) \leqq \mathfrak{c}$. □

(25.15) Corollary. *The group $T^{\mathfrak{m}}$ is monothetic if and only if $\mathfrak{m} \leqq \mathfrak{c}$.*

Proof. This follows immediately from (25.14) and the fact that $\mathfrak{w}(T^{\mathfrak{m}}) = \max(\mathfrak{m}, \aleph_0)$.

Alternatively, if $\mathfrak{m} \leqq \mathfrak{c}$, one can find directly an element of $T^{\mathfrak{m}}$ whose powers are dense in $T^{\mathfrak{m}}$. Represent $T^{\mathfrak{m}}$ as the group of all functions (x_ι) on a set $I = \{\iota\}$ such that $\overline{\overline{I}} = \mathfrak{m}$, with values in T. Consider a Hamel basis H for R over Q such that $1 \in H$. Let ψ be a one-to-one mapping of I into $H \cap \{1\}'$, and let $a_\iota = \exp[2\pi i \psi(\iota)]$ for all $\iota \in I$. Since every continuous character of $T^{\mathfrak{m}}$ has the form

$$(x_\iota) \rightarrow x_{\iota_1}^{n_1} x_{\iota_2}^{n_2} \cdots x_{\iota_k}^{n_k}$$

for $\{\iota_1, \ldots, \iota_k\} \subset I$ and $n_1, \ldots, n_k \in Z$, we see that only the character 1 maps (a_ι) onto 1. Thus (25.11) implies that (a_ι) is a generator of $T^{\mathfrak{m}}$. □

For a sequence $\boldsymbol{a} = (a_0, a_1, a_2, \ldots)$ of integers greater than 1, $\Delta_{\boldsymbol{a}}$ is an infinite 0-dimensional compact monothetic group [see (10.5) and (10.6)]. We will now show that these are the only such groups.

(25.16) Theorem. *Let G be an infinite topological group. The following statements are equivalent:*

(i) *G is a 0-dimensional compact monothetic group;*

(ii) *G is topologically isomorphic with some group $\Delta_{\boldsymbol{a}}$;*

(iii) *G is topologically isomorphic with a direct product $\underset{p \in P}{\mathsf{P}} A_p$, where P is the set of all prime integers, and each A_p is $\{e\}$, $Z(p^{r_p})$ [r_p is a positive integer], or Δ_p.*

Proof. By (10.5) and (10.6), (ii) implies (i).

Suppose that G is a 0-dimensional compact monothetic group; we will prove (iii). By (24.26) and (24.32) the character group X of G may be regarded as a torsion subgroup of T_d. It follows that the p-primary direct factor X_p of X (A.3) is a subgroup of $Z(p^\infty)$. Since all proper subgroups of $Z(p^\infty)$ are finite cyclic groups with order a power of p, we have $\mathsf{X} = \underset{p \in P}{\mathsf{P}^*} \mathsf{B}_p$, where B_p is $\{e\}$, $Z(p^{r_p})$, or $Z(p^\infty)$. Theorem (23.22) now shows that $G = \underset{p \in P}{\mathsf{P}} A_p$, where each A_p is $\{e\}$, $Z(p^{r_p})$, or Δ_p.

Finally, consider the direct product $\underset{p \in P}{\mathsf{P}} A_p$ as described in (iii). Let $P_1 = \{p \in P : A_p = \Delta_p\}$ and $P_2 = \{p \in P : A_p = Z(p^{r_p})$ for some $r_p \geqq 1\}$. It is easy to see that there is a sequence $\boldsymbol{a} = (a_0, a_1, a_2, \ldots)$ of prime integers consisting of infinitely many p's for each $p \in P_1$, exactly r_p p's for each $p \in P_2$, and no other p's. In order to conclude that $\Delta_{\boldsymbol{a}}$ is topologically isomorphic with $\underset{p \in P}{\mathsf{P}} A_p$, it suffices to show that $Z(\boldsymbol{a}^\infty)$ is isomorphic with $\underset{p \in P}{\mathsf{P}^*} \mathsf{B}_p$ where B_p is $\{e\}$, $Z(p^{r_p})$, or $Z(p^\infty)$ according as A_p is $\{e\}$,

$Z(p^{r_p})$, or Δ_p. Since $Z(\mathfrak{a}^\infty)$ is a torsion group, we can apply (A.3). If $p \in P_1$, then the elements in $Z(\mathfrak{a}^\infty)$ having order a power of p are exactly the numbers of the form $\exp\left[2\pi i \frac{m}{p^n}\right]$, *i.e.*, the group $Z(p^\infty)$. If $p \in P_2$, then the elements having order a power of p are the numbers $\exp\left[2\pi i \frac{m}{p^n}\right]$ where $n \leqq r_p$, *i.e.*, the group $Z(p^{r_p})$. If $p \notin P_1 \cup P_2$, then no element of $Z(\mathfrak{a}^\infty)$ has order a power of p. Hence $Z(\mathfrak{a}^\infty)$ is isomorphic with $\underset{p\in P}{\mathsf{P}^*} \mathsf{B}_p$. □

(25.17) Theorem. *Let G be a compact Abelian group with component of the identity C. The group G is monothetic if and only if $\mathfrak{w}(G) \leqq \mathfrak{c}$ and G/C is a [compact Abelian] group of the form $\Delta_{\mathfrak{a}}$.*

Proof. If G is monothetic, then clearly $\mathfrak{w}(G) = \overline{\overline{\mathsf{X}}} \leqq \mathfrak{c}$ and the quotient group G/C is monothetic. By (7.3) and (3.5), G/C is 0-dimensional. Thus G/C has the form $\Delta_{\mathfrak{a}}$ by (25.16).

Conversely, suppose that the conditions on G and G/C hold. Clearly the torsion-free rank $r_0(\mathsf{X})$ of X does not exceed $\overline{\overline{\mathsf{X}}} = \mathfrak{w}(G) \leqq \mathfrak{c}$. The character group of G/C is the torsion subgroup Φ of X [see (24.20)], and since G/C has the form $\underset{p\in P}{\mathsf{P}} \Delta_p$ (25.16), it follows that $r_p(\Phi) = 0$ or 1 for all p. Since $r_p(\Phi) = r_p(\mathsf{X})$ for every prime p, (25.13) and (24.32) show that G is monothetic. □

We now describe the compact solenoidal groups, defined in (9.2).

(25.18) Theorem. *Let G be a compact Abelian group with character group X. The following conditions are equivalent:*

(i) *G is solenoidal;*

(ii) *G is connected and $\mathfrak{w}(G) \leqq \mathfrak{c}$;*

(iii) *X is isomorphic with a subgroup of R_d;*

(iv) *X is torsion-free and $\overline{\overline{\mathsf{X}}} \leqq \mathfrak{c}$;*

(v) *X is torsion-free and $r_0(\mathsf{X}) \leqq \mathfrak{c}$.*[1]

Proof. Suppose that (i) holds, and let φ denote a continuous homomorphism of R onto a dense subgroup of G. A character $\chi \in \mathsf{X}$ is completely determined by its behavior on $\varphi(R)$, and for $\chi \in \mathsf{X}$, $\chi \circ \varphi$ is a continuous character of R. If χ_1 and χ_2 are distinct characters on $\varphi(R)$, then obviously $\chi_1 \circ \varphi$ and $\chi_2 \circ \varphi$ are distinct characters on R. The mapping $\chi \to \chi \circ \varphi$ is therefore an isomorphism of X into the character group of R, namely R (23.27.e), so that (iii) holds.

Suppose next that (iii) holds, and that τ is an isomorphism of X into R_d. For every $\alpha \in R$, the mapping $\chi \to \exp[i\alpha\tau(\chi)]$ is a character of X, and so (24.8) shows that there is a unique point $\varphi(\alpha)$ in G for which $\exp[i\alpha\tau(\chi)] = \chi(\varphi(\alpha))$ for all $\chi \in \mathsf{X}$. It is easy to see that φ is a continuous homomorphism of R into G; φ is just the adjoint homomorphism

[1] As in (25.13), $r_0(\mathsf{X})$ denotes the torsion-free rank of X.

of the homomorphism τ. If $\varphi(R)$ were not dense in G, there would be a character $\chi_0 \in \mathsf{X}$ such that $\chi_0(\varphi(R)) = \{1\}$ and $\chi_0 \neq 1$. Since τ is an isomorphism, we have $\tau(\chi_0) \neq 0$ and hence $\exp[i\alpha_0\tau(\chi_0)] \neq 1$ for some $\alpha_0 \in R$. This contradiction proves (i).

Since R_d is isomorphic with Q^{c*}, the equivalence of (iii), (iv), and (v) follows readily from (A.16) and (A.14). The equivalence of (ii) and (iv) follows from (24.15) and (24.25). □

Theorem (10.13) shows that the $\boldsymbol{a}$-adic solenoids defined in (10.12) are solenoidal groups.

The following result is analogous to (25.12).

(25.19) Theorem. *The group* $(\Sigma_{\boldsymbol{a}})^c$, *where* $\boldsymbol{a} = (2, 3, 4, \ldots)$, *is the largest compact solenoidal group: a topological group* G *is a compact solenoid if and only if* G *is a continuous homomorphic image of* $(\Sigma_{\boldsymbol{a}})^c$.

Proof. This follows at once from (25.18) and (24.11), since the character group of R_d is topologically isomorphic with $(\Sigma_{\boldsymbol{a}})^c$ (25.6). □

(25.20) Theorem. *A locally compact Abelian group* G *is connected if and only if the smallest closed subgroup of* G *that contains all one-parameter subgroups, say* H, *is* G *itself.*

Proof. Let C be the component of e in G. Then C contains all one-parameter subgroups of G and hence $C \supset H$. Thus $C = G$ if $H = G$.

Conversely, suppose that G is connected. By (9.14) G is topologically isomorphic with $R^n \times F$, where F is compact, connected, and Abelian. If the subgroup generated by one-parameter subgroups of F is dense in F, then the same is obviously true of $R^n \times F$. Thus we may suppose that G is compact as well as connected. Let X be the character group of G. Assume that $H \neq G$. Then there is a character $\chi_0 \in \mathsf{X}$ such that $\chi_0(H) = \{1\}$ and $\chi_0 \neq 1$. Since G is connected, χ_0 has infinite order (24.25), and there is a real character τ of X such that $\tau(\chi_0) \neq 0$ (24.34). Now for every $\alpha \in R$, the mapping $\chi \to \exp[i\alpha\tau(\chi)]$ is a character of X and so $\exp[i\alpha\tau(\chi)] = \chi(\varphi(\alpha))$, where φ is a continuous homomorphism of R into G. For $\chi = \chi_0$, we get $\chi_0(\varphi(\alpha)) = \exp[i\alpha\tau(\chi_0)]$, and since $\tau(\chi_0) \neq 0$, there is an $\alpha_0 \in R$ such that $\chi_0(\varphi(\alpha_0)) \neq 1$. Since $\varphi(R)$ is a one-parameter subgroup of G, we have a contradiction. □

We will now determine completely the algebraic structure of compact Abelian groups. We require a lemma, which is of interest on its own account.

(25.21) Theorem. *Let* H *be a compact Abelian group that is a pure subgroup of some Abelian group* G. *Then* H *is algebraically a direct factor of* G: G *is isomorphic with* $H \times (G/H)$.

Proof. We emphasize that the direct product decomposition $H \times (G/H)$ is *not* topological: G need not have any topology at all. Let A be a subset of G containing just one element in each coset xH. It is easy to see that G admits H as a direct factor if and only if A can be chosen to be a subgroup of H. For, if A is a subgroup of G, then $HA = G$ and $H \cap A = \{e\}$, so that (2.4) applies. The converse is obvious.

For an A as just described, let $\mathfrak{a} = \overline{\overline{A}}$, and consider the compact group $H^{\mathfrak{a}}$, which we will write as $\underset{a \in A}{\mathsf{P}} H_{(a)}$, where each $H_{(a)}$ is equal to H. For $(x_a) \in H^{\mathfrak{a}}$, let $\Phi((x_a))$ be the set $\{a x_a^{-1} : a \in A\}$. It is plain that $\Phi((x_a))$ contains exactly one element of each coset xH and that each set selecting just one element from each coset xH can be obtained from a fixed A as some $\Phi((x_a))$. Now, for a subgroup G_0 of G such that $H \subset G_0 \subset G$, let $\Psi(G_0)$ be the set of all $(x_a) = H^{\mathfrak{a}}$ such that $\Phi((x_a)) \cap G_0$ is a *subgroup* of G_0. As pointed out above, $\Psi(G_0)$ is nonvoid if and only if H is a direct factor of G_0.

Given $a, b \in A$ and $(x_a) \in H^{\mathfrak{a}}$, consider $(a x_a^{-1})(b x_b^{-1})$. This element lies in a coset cH, where $c \in A$ is determined by a and b alone. Thus (x_a) is in $\Psi(G_0)$ if and only if $(a x_a^{-1})(b x_b^{-1}) = c x_c^{-1}$ for all $a, b \in A \cap G_0$. This condition is equivalent to $x_a x_b x_c^{-1} = a b c^{-1}$. Hence we infer at once that $\Psi(G_0)$ is a closed subset of $H^{\mathfrak{a}}$.

Let $\{G_\iota\}_{\iota \in I}$ be the family of all subgroups of G such that $H \subset G_\iota \subset G$ and G_ι/H is finitely generated. It is obvious that $G = \bigcup_{\iota \in I} G_\iota$, and that for every $\{\iota_1, \ldots, \iota_n\} \subset I$, there is an $\iota_0 \in I$ such that $G_{\iota_0} \supset G_{\iota_1} \cup \cdots \cup G_{\iota_n}$. It is also apparent that $\Psi(G) = \bigcap_{\iota \in I} \Psi(G_\iota)$. By (A.27) and (A.22), H is a direct factor of each G_ι; that is, $\Psi(G_\iota) \neq \varnothing$ for each $\iota \in I$. We infer at once that $\bigcap_{\iota \in I} \Psi(G_\iota) \neq \varnothing$, since $H^{\mathfrak{a}}$ is compact. Therefore $\Psi(G)$ is nonvoid, and H is a direct factor of G. □

(25.22) Theorem. *Let G be a compact 0-dimensional Abelian group. Algebraically, G is isomorphic with a group*

(i) $\underset{p \in P}{\mathsf{P}} \left[\Delta_p^{\mathfrak{a}_p} \times \underset{\iota \in I_p}{\mathsf{P}} Z(p^{r_\iota})\right]$

where P is the set of all primes, $\mathfrak{a}_p$ is an arbitrary cardinal number [possibly 0], I_p is an arbitrary index class [possibly void], and each r_ι is a positive integer. Also every group of the form (i) *is [topologically] isomorphic to a compact Abelian group.*

Proof. Let X be as usual the character group of G. Then X is a torsion group, and hence is the weak direct product of its p-primary components X_p (A.3). By (23.21) and (24.8), G is topologically isomorphic with the complete direct product $\underset{p \in P}{\mathsf{P}} H_p$, where H_p is the character group of X_p. We may therefore restrict ourselves to the case in which X itself is p-primary for some prime p. By (A.24), X contains a subgroup B such that:

$$\mathsf{B} \text{ is isomorphic with a weak direct product } \underset{\iota \in I_p}{\mathsf{P}^*} Z(p^{r_\iota}); \tag{1}$$

$$\mathsf{B} \text{ is a pure subgroup of } \mathsf{X}; \tag{2}$$

$$\mathsf{X}/\mathsf{B} \text{ is divisible.} \tag{3}$$

Now consider the subgroup $A(G, \mathsf{B})$ of G. Since (2) and (3) hold, (24.46.b) implies that $A(G, \mathsf{B})$ is a pure subgroup of G. Since $A(G, \mathsf{B})$ is compact, (25.21) shows that $A(G, \mathsf{B})$ is a direct factor of G. Thus G is algebraically, although not in general topologically, isomorphic with $A(G, \mathsf{B}) \times (G/A(G, \mathsf{B}))$. The group $A(G, \mathsf{B})$ is topologically isomorphic with the character group of X/B. In view of (3) and the hypothesis that X is a p-primary group, X/B is a weak direct product $Z(p^\infty)^{\mathfrak{a}_p *}$ (A.14). Its character group is thus the full direct product $\Delta_p^{\mathfrak{a}_p}$. The group $G/A(G, \mathsf{B})$ is topologically isomorphic with the character group of $\mathsf{A}(\mathsf{X}, A(G, \mathsf{B})) = \mathsf{B}$. In view of (1), $G/A(G, \mathsf{B})$ is topologically isomorphic with $\underset{\iota \in I_p}{\mathsf{P}} Z(p^{r_\iota})$. Hence G has algebraically the form (i).

The last statement is trivial: giving Δ_p its usual topology and $Z(p^{r_\iota})$ the discrete topology, we see that each group (i) is a compact group. □

(25.23) Theorem. *Let G be a connected compact Abelian group, with $\mathfrak{w}(G) = \mathfrak{m}$. Then G is algebraically isomorphic with*

(i) $Q^{2^{\mathfrak{m}} *} \times \underset{p \in P}{\mathsf{P}^*} Z(p^\infty)^{\mathfrak{b}_p *}$,

where P is the set of all primes, and each cardinal number $\mathfrak{b}_p$ is finite or is $2^{\mathfrak{e}_p}$ for an infinite cardinal $\mathfrak{e}_p \leqq \mathfrak{m}$.

Proof. Since G is connected and compact, its character group X is torsion-free and G itself is divisible (24.25). Theorem (A.14) implies that G is isomorphic with a weak direct product

$$Q^{\mathfrak{n} *} \times \underset{p \in P}{\mathsf{P}^*} Z(p^\infty)^{\mathfrak{b}_p *}. \tag{1}$$

We must show that $\mathfrak{n}$ and $\mathfrak{b}_p$ are as described.

We begin by identifying $\mathfrak{b}_p$ for a fixed prime number p. Consider the group $\mathsf{X}/\mathsf{X}^{(p)}$, where $\mathsf{X}^{(p)} = \{\chi^p : \chi \in \mathsf{X}\}$ (A.2). All elements of this group save the identity have order exactly p [note that p is a prime]. Hence $\mathsf{X}/\mathsf{X}^{(p)}$ is a weak direct product $Z(p)^{\mathfrak{e}_p *}$, where $\mathfrak{e}_p$ is a cardinal number, possibly 0 (A.25) [we set $\mathfrak{e}_p = 0$ if and only if $\mathsf{X} = \mathsf{X}^{(p)}$]. Since $Z(p)^{\mathfrak{e}_p *}$ has cardinal number $p^{\mathfrak{e}_p}$ if $\mathfrak{e}_p < \aleph_0$ and $\mathfrak{e}_p$ if $\mathfrak{e}_p \geqq \aleph_0$, we see that $\mathfrak{e}_p \leqq \overline{\overline{\mathsf{X}}} = \mathfrak{m}$. By (24.22), we have $G_{(p)} = A(G, \mathsf{X}^{(p)})$, and so $G_{(p)}$ is topologically isomorphic with the character group of $\mathsf{X}/\mathsf{X}^{(p)}$. Since p is a prime, $G_{(p)}$ is exactly the subgroup of G of elements of order 1 or p. The character group of $Z(p)^{\mathfrak{e}_p *}$ is $Z(p)^{\mathfrak{e}_p}$, and so there are in G exactly $p^{\mathfrak{e}_p}$ elements of order 1 or p. In the group (1), it is clear that there are $p^{\mathfrak{b}_p}$ elements of order 1 or p if $\mathfrak{b}_p$ is finite and $\mathfrak{b}_p$ such elements if $\mathfrak{b}_p$ is infinite. It follows that $\mathfrak{b}_p = \mathfrak{e}_p$ if $\mathfrak{b}_p$ is finite and $\mathfrak{b}_p = p^{\mathfrak{e}_p} = 2^{\mathfrak{e}_p}$ if $\mathfrak{b}_p$ is infinite.

We next show that the cardinal number $\mathfrak{n}$ in (1) must be $2^{\mathfrak{m}}$. Let B_0 be a maximal independent subset of X (A.10); thus $\overline{\overline{\mathsf{B}}}_0 = r_0(\mathsf{X})$. Let B_1 be the smallest subgroup of X containing B_0. Then B_1 is a free Abelian group, isomorphic with $Z^{r_0(\mathsf{X})*}$. For every $\chi \in \mathsf{X}$, $\chi \neq 1$, there are nonzero integers $n, m_1, m_2, \ldots, m_s$ and elements $\psi_1, \psi_2, \ldots, \psi_s \in \mathsf{B}_0$ such that $\chi^n = \psi_1^{m_1}\psi_2^{m_2}\cdots\psi_s^{m_s}$. If we assign to χ the s-tuple $(\psi_1, \ldots, \psi_s)$ and the s-tuple $\left(\frac{m_1}{n}, \ldots, \frac{m_s}{n}\right)$ of nonzero rational numbers, then χ is uniquely determined by them; this is true because X is torsion-free. It follows that $\mathfrak{w}(G) = \mathfrak{m} = \overline{\overline{\mathsf{X}}} = \max(\aleph_0, r_0(\mathsf{X}))$. The character group of B_1 is $T^{r_0(\mathsf{X})}$. A routine calculation shows that the torsion-free rank of $T^{r_0(\mathsf{X})}$ is equal to $(\mathfrak{c})^{r_0(\mathsf{X})} = (2^{\aleph_0})^{r_0(\mathsf{X})} = 2^{\aleph_0 \cdot r_0(\mathsf{X})} = 2^{\max(\aleph_0, r_0(\mathsf{X}))} = 2^{\mathfrak{m}}$. Since B_1 is a subgroup of X, the character group $T^{r_0(\mathsf{X})}$ of B_1 is topologically isomorphic with $G/A(G, \mathsf{B}_1)$. It follows that $\mathfrak{n} = r_0(G) \geqq r_0(G/A(G, \mathsf{B}_1)) = 2^{\mathfrak{m}}$. Since $\overline{\overline{G}} = 2^{\mathfrak{m}}$ (24.47), we must have $\mathfrak{n} \leqq 2^{\mathfrak{m}}$. □

(25.24) Theorem. *Let* $\mathfrak{m}$ *be an infinite cardinal number, and let* $\mathfrak{b}_2, \mathfrak{b}_3, \mathfrak{b}_5, \mathfrak{b}_7, \mathfrak{b}_{11}, \ldots$ *be cardinal numbers satisfying the restrictions of* (25.23). *Then there is a compact connected Abelian group* G *with weight* $\mathfrak{m}$ *that is algebraically isomorphic with the group* (25.23.i).

Proof. Let p be an arbitrary prime number, and let L_p be the additive group of all rational numbers m/n, where m is an arbitrary integer and n is a nonzero integer prime to p.[1] Let $\mathfrak{c}_p = \mathfrak{b}_p$ if $\mathfrak{b}_p$ is finite and $\mathfrak{c}_p = \mathfrak{e}_p$ if $\mathfrak{b}_p$ is infinite and so equal to $2^{\mathfrak{e}_p}$. Let X be the discrete group $\left(\mathop{\mathsf{P}^*}_{p\in P} L_p^{\mathfrak{c}_p*}\right) \times Q^{\mathfrak{m}*}$. Let $\boldsymbol{b}_p$ be the [increasing] sequence of all integers greater than 1 and prime to p, and let $\boldsymbol{a} = (2, 3, 4, \ldots)$. The character group, say G, of X is topologically isomorphic with $\left(\mathop{\mathsf{P}}_{p\in P} \Sigma_{\boldsymbol{b}_p}^{\mathfrak{c}_p}\right) \times \Sigma_{\boldsymbol{a}}^{\mathfrak{m}}$. Since $\overline{\overline{\mathsf{X}}} = \mathfrak{m}$, we have $\mathfrak{w}(G) = \overline{\overline{\mathsf{X}}} = \mathfrak{m}$, and Theorem (24.25) implies that G is connected. Obviously we have $r_0(G) \leqq \overline{\overline{G}} = 2^{\mathfrak{m}}$. Since Q is divisible, its character group $\Sigma_{\boldsymbol{a}}$ is torsion-free (24.23) and hence $2^{\mathfrak{m}} = \mathfrak{c}^{\mathfrak{m}} = \overline{\overline{(\Sigma_{\boldsymbol{a}}^{\mathfrak{m}})}} = \max(\aleph_0, r_0(\Sigma_{\boldsymbol{a}}^{\mathfrak{m}})) = r_0(\Sigma_{\boldsymbol{a}}^{\mathfrak{m}}) \leqq r_0(G)$. Since G is divisible (24.25), it is the weak direct product of copies of Q and $Z(p^\infty)$ for primes p, and the number of copies of Q is exactly $r_0(G) = 2^{\mathfrak{m}}$. As in the proof of (25.23), in order to infer that the decomposition of G contains $\mathfrak{b}_p$ copies of $Z(p^\infty)$, we need only to show that $\mathsf{X}/\mathsf{X}^{(p)}$ is isomorphic with $Z(p)^{\mathfrak{c}_p*}$. If q is a prime distinct from p, then $L_q^{(p)} = L_q$, as is easy to see, and of course $Q^{(p)} = Q$. It follows that $\mathsf{X}/\mathsf{X}^{(p)}$ is isomorphic with $(L_p/L_p^{(p)})^{\mathfrak{c}_p*}$. For n prime to

[1] The construction used here is due to R. S. PIERCE [oral communication].

p, and m any integer, there is exactly one $k \in \{0, 1, \ldots, p-1\}$ such that $m - nk$ is a multiple of p. Thus each coset of $L_p^{(p)}$ in L_p contains exactly one k. It follows that $L_p/L_p^{(p)}$ is isomorphic with $Z(p)$, so that $\mathsf{X}/\mathsf{X}^{(p)}$ is isomorphic with $Z(p)^{\mathfrak{c}_p *}$. $\square$

Theorems (25.22)—(25.24) can be combined to give a complete characterization of the algebraic structure of compact Abelian groups.

(25.25) Theorem. *Let P be the set of all primes. For all $p \in P$, let $\mathfrak{a}_p$ be an arbitrary cardinal number, possibly 0, let I_p be an arbitrary index class, possibly void, and let r_ι be an arbitrary positive integer for each $\iota \in I_p$. Let $\mathfrak{n}$ be a cardinal number that is 0 or $2^{\mathfrak{m}}$ for an infinite cardinal $\mathfrak{m}$. For all $p \in P$, let $\mathfrak{b}_p$ be a cardinal not exceeding $\mathfrak{n}$ such that $\mathfrak{b}_p$ is finite or has the form $2^{\mathfrak{e}_p}$ for an infinite cardinal $\mathfrak{e}_p \leqq \mathfrak{m}$. Every compact Abelian group is algebraically isomorphic with a group*

(i) $\underset{p \in P}{\mathsf{P}} \left[\Delta_p^{\mathfrak{a}_p} \times \underset{\iota \in I_p}{\mathsf{P}} Z(p^{r_\iota})\right] \times \underset{p \in P}{\mathsf{P}^*} Z(p^\infty)^{\mathfrak{b}_p *} \times Q^{\mathfrak{n}*}$.

Also every group (i) *is algebraically isomorphic with a compact Abelian group.*

Proof. Let G be a compact Abelian group. Then the component C of e in G is the largest divisible subgroup of G (24.24), and so is an algebraic direct factor of G (A.8). Thus G is isomorphic with $(G/C) \times C$. Since C is a connected compact group, it is isomorphic with a group (25.23.i). Since G/C is 0-dimensional [(7.3), (5.22), and (3.5)], it is isomorphic with a group (25.22.i). Hence G is isomorphic with a group (i). Given a group of the form (i), it is isomorphic with the direct product of two compact groups, as (25.22) and (25.24) show. $\square$

Miscellaneous theorems and examples

(25.26) Characters of Q. (a) If Q has its relative topology as a subspace of R, then every continuous character of Q has the form $r \to \exp(iry)$ for some $y \in R$. [By (23.30), every continuous character of Q has a continuous extension over R.]

(b) Let $\Sigma_{\boldsymbol{a}}$ denote the character group of Q as in (25.4), $\boldsymbol{a} = (2, 3, 4, \ldots)$. The group $\Sigma_{\boldsymbol{a}}$ is topologically isomorphic with $[0, 1[\times \Delta_{\boldsymbol{a}}$ when it is made into a topological group as in (10.15). Let H denote the subgroup of $[0, 1[\times \Delta_{\boldsymbol{a}}$ of all elements $(\xi, \boldsymbol{x})$ such that either $x_n = 0$ for all but finitely many n or else $x_n = n+1$ for all but finitely many n. [This subgroup is the dense subgroup $\varphi(R \times \{\mathbf{0}\})$ described in the proof of (10.13).] Then H consists of exactly those elements of $\Sigma_{\boldsymbol{a}}$ whose corresponding characters are continuous when Q is regarded as a subspace of R.

[The proof of the bracketed remark is elementary and is omitted. Suppose that $(\xi, \boldsymbol{x}) \in [0, 1[\times \Delta_{\boldsymbol{a}}$ and that $x_n = 0$ for $n \geqq n_0$. If χ is the character of Q corresponding to $(\xi, \boldsymbol{x})$ and t is the number

$\xi-(x_0+2!x_1+\cdots+n_0!x_{n_0-1})$, then using (25.4.2) we find that $\chi(r)=\exp[2\pi i t r]$ for all $r\in Q$, and hence χ is continuous. Since the continuous characters form a group, every element in H corresponds to a continuous character.

Now consider a continuous character χ of Q such that $\chi(r)=\exp[2\pi i t r]$ for all $r\in Q$, where t is a negative real number. Then for some value of ξ in $[0,1[$ and some sequence $\{x_0, x_1, \ldots, x_{n_0-1}\}$ of integers where $0\leqq x_j\leqq j+1$, we have $t=\xi-(x_0+2!x_1+\cdots+n_0!x_{n_0-1})$. If we set $x_j=0$ for $j\geqq n_0$, then the character χ corresponding to $(\xi, \boldsymbol{x})$ satisfies the identity $\chi(r)=\exp[2\pi i t r]$ for $r\in Q$. Since H is a group, every continuous character of Q corresponds to an element in H.]

(c) (HALMOS [1].) The character group $\Sigma_{\boldsymbol{a}}$ of Q is algebraically isomorphic with the additive group R $[\boldsymbol{a}=(2,3,4,\ldots)]$. [By (24.23), $\Sigma_{\boldsymbol{a}}$ is a divisible torsion-free group. By Theorem (A.14), $\Sigma_{\boldsymbol{a}}$ is algebraically the weak direct product of a certain number of copies of Q. Since the cardinal number of $\Sigma_{\boldsymbol{a}}$ is $\mathfrak{c}$, exactly $\mathfrak{c}$ copies of Q are needed. The group R is also of this form; thus the two groups are isomorphic. Hence we can impose a topology on R making it into a compact Abelian group.]

(d) The character group $\Sigma_{\boldsymbol{a}}$ of Q is connected but is not the union of its one-parameter subgroups. [Every real character of Q has the form $r\to tr$ for some $t\in R$. If $\Sigma_{\boldsymbol{a}}$ were the union of one-parameter subgroups, then (24.43) would imply that every character of Q has the form $r\to\exp[itr]$, $t\in R$. This is plainly false; see for example (b).]

(e) There is a character χ of R such that $\chi(x)$ has infinite order in T for all $x\neq 0$. Such a χ is necessarily discontinuous; see (23.27.e). [Let H be a Hamel basis for R over Q such that $\pi\in H$. Let τ be a one-to-one mapping of H onto $H\cap\{\pi\}'$. Let χ be the character of R such that $\chi(rh)=\exp(ir\tau(h))$ if $h\in H$ and $r\in Q$. A number α in T has finite order if and only if $\alpha=\exp(2\pi i s)$ for some $s\in Q$. Suppose that $x\in R$ and $\chi(x)=\exp(2\pi i s)$. Writing $x=r_1h_1+\cdots+r_mh_m$ where $h_j\in H$ and $r_j\in Q$, we have $\exp[i(r_1\tau(h_1)+\cdots+r_m\tau(h_m)-2\pi s)]=1$. Thus for some integer k, we have $r_1\tau(h_1)+\cdots+r_m\tau(h_m)-\pi(2s+2k)=0$. It follows that $r_1=r_2=\cdots=r_m=0$, and hence $x=0$.]

(25.27) Generators of monothetic groups (HALMOS and SAMELSON [1]). (a) Let G be a compact, connected Abelian group such that $\mathfrak{w}(G)=\aleph_0$, and let S be the set of all generators of G (25.11). Then S is λ-measurable and $\lambda(S)=1$. [Since $\mathfrak{w}(G)=\aleph_0$, X is countably infinite: $X=\{1,\chi_1,\chi_2,\chi_3,\ldots,\chi_n,\ldots\}$. Let $H_n=\{x\in G:\chi_n(x)=1\}$. Then H_n is a closed subgroup of G and G/H_n is infinite because G is connected. Therefore $\lambda(H_n)=0$, and so $\lambda(\bigcap_{n=1}^{\infty}H_n')=1$. It follows from (25.11) that $S=\bigcap_{n=1}^{\infty}H_n'$.]

(b) The set S fails to be λ-measurable for some compact connected monothetic groups. [Consider any group $T^{\mathfrak{m}}$, where $\aleph_0<\mathfrak{m}\leqq\mathfrak{c}$, represented as in (25.15). It follows easily from (25.11) that $(a_\iota)\in T^{\mathfrak{m}}$ belongs to S if and only if $a_\iota=\exp[2\pi i t_\iota]$ where the numbers t_ι are distinct and the set $\{t_\iota:\iota\in I\}\cup\{1\}$ is rationally independent. As in (16.13.f) we see that $\lambda(F)=0$ if F is any compact subset of S, since $S\subset\{(x_\iota):x_\iota\neq-1$ for all $\iota\in I\}$. Now assume that F is a compact subset of S' such that $\lambda(F)>0$. Then by (19.30.b) and the argument used in (16.13.f), there is a subset D of F such that D is λ-measurable, $\lambda(D)>0$, and D has the following property: there is a countable subset Σ of I such that $(x_\iota)\in D$, $(y_\iota)\in T^{\mathfrak{m}}$, and $x_\iota=y_\iota$ for $\iota\in\Sigma$ imply $(y_\iota)\in D$. Thus the set D is "restricted" only on the countable set of coordinates Σ. The set $\pi_\Sigma(D)$ [π_Σ is the projection map defined as in (6.8)] has measure in $\underset{\iota\in\Sigma}{\mathsf{P}} T_{(\iota)}$ equal to $\lambda(D)$. Hence $\pi_\Sigma(D)$ contains a generator $(a_\iota)_{\iota\in\Sigma}$ of $\underset{\iota\in\Sigma}{\mathsf{P}} T_{(\iota)}$ [see (a)]. It is easy to extend $(a_\iota)_{\iota\in\Sigma}$ to a generator $(a_\iota)_{\iota\in I}$ of $T^{\mathfrak{m}}$. Hence we have $D\not\subset S'$, a contradiction.]

(c) The group $Z(p^r)$ has p^r-p^{r-1} generators, since the congruence $ax=b \pmod{p^r}$ has a solution x for all b if and only if a and p^r, hence a and p, are relatively prime. Thus the measure of the set of generators of $Z(p^r)$ is $1-\frac{1}{p}$. Consider next the group Δ_p. It follows from (10.16.a) that the set S of generators of Δ_p is Λ_1'. The measure of Λ_1 in Δ_p is $1/p$ [see (15.17.k)], so that $\lambda(S)=1-\frac{1}{p}$. Now consider an arbitrary 0-dimensional compact monothetic group, written as $\underset{p\in P}{\mathsf{P}} A_p$, where each A_p is as in (25.16). By (23.21), every continuous character of $\underset{p\in P}{\mathsf{P}} A_p$ different from 1 has the form $(x_p)\rightarrow\chi_{p_1}(x_{p_1})\cdots\chi_{p_l}(x_{p_l})$, where $\{p_1,p_2,\ldots,p_l\}$ is a nonvoid subset of the set of primes for which $A_p\neq\{e\}$, and each χ_{p_j} is a continuous character of A_{p_j} different from 1. Let S_p denote the set of generators of A_p, and S_0 the set of generators of $\underset{p\in P}{\mathsf{P}} A_p$. It is clear from (25.11) and the preceding two sentences that $S_0\subset\underset{p\in P}{\mathsf{P}} S_p$. On the other hand, if $(x_p)\in\underset{p\in P}{\mathsf{P}} S_p$, and χ is a continuous character of $\underset{p\in P}{\mathsf{P}} A_p$ different from 1, then $\chi((x_p))=\exp\left[2\pi i\left(\frac{a_1}{p_1^{n_1}}+\frac{a_2}{p_2^{n_2}}+\cdots+\frac{a_l}{p_l^{n_l}}\right)\right]$, where a_j is prime to p_j [see (23.27.c) and (25.2.2)]. Since the primes p_j are distinct, it is easy to see that $\chi((x_p))\neq1$. Thus $S_0=\underset{p\in P}{\mathsf{P}} S_p$, and so $\lambda(S_0)=\prod_{p\in P}\left(1-\frac{1}{p}\right)$, the product being taken over all the primes p for which $A_p\neq\{e\}$. Since $\sum_{p\in P}\frac{1}{p}=\infty$, we can select the A_p's so that $\lambda(S_0)$ is any number in the interval $[0, 1[$.

(25.28) More about $\Delta_{\boldsymbol{a}}$ and $\boldsymbol{Z(a^\infty)}$. (a) Let $\boldsymbol{a}=(a_0, a_1, a_2, \ldots)$ be an arbitrary sequence of integers greater than 1. Let P_1 be the [perhaps void] set of all prime numbers p such that for every power p^n there is a k such that $p^n | a_0 a_1 \cdots a_k$. Let P_2 be the [perhaps void] set of all primes not in P_1 that divide some number $a_0 a_1 \cdots a_k$. For $p \in P_2$, let r_p be the largest power of p such that $p^{r_p} | a_0 a_1 \cdots a_k$ for some k. Then $\Delta_{\boldsymbol{a}}$ is topologically isomorphic with $\underset{p \in P_1}{\mathsf{P}} \Delta_p \times \underset{p \in P_2}{\mathsf{P}} Z(p^{r_p})$. Also $Z(\boldsymbol{a}^\infty)$ is isomorphic with $\underset{p \in P_1}{\mathsf{P}^*} Z(p^\infty) \times \underset{p \in P_2}{\mathsf{P}^*} Z(p^{r_p})$. [Argue as in the proof of (25.16).]

(b) Let $\boldsymbol{a} = (a_0, a_1, a_2, \ldots)$ and $\boldsymbol{b} = (b_0, b_1, b_2, \ldots)$ be arbitrary sequences of integers greater than 1. The following statements are equivalent:

(i) $\Delta_{\boldsymbol{a}}$ and $\Delta_{\boldsymbol{b}}$ are topologically isomorphic;

(ii) $\Delta_{\boldsymbol{a}}$ and $\Delta_{\boldsymbol{b}}$ are algebraically isomorphic;

(iii) $Z(\boldsymbol{a}^\infty)$ and $Z(\boldsymbol{b}^\infty)$ are isomorphic;

(iv) $Z(\boldsymbol{a}^\infty) = Z(\boldsymbol{b}^\infty)$;

(v) for every prime power p^r, p^r divides some $a_0 a_1 \cdots a_k$ if and only if it divides some $b_0 b_1 \cdots b_l$.

[By the duality theorem (24.8), (i) and (iii) are equivalent. It is obvious that (i) implies (ii) and that (iv) implies (iii). It is easy to verify that (iii) implies (v) and that (iv) and (v) are equivalent. It suffices then to show that (ii) fails whenever (iii) fails. If $Z(\boldsymbol{a}^\infty)$ and $Z(\boldsymbol{b}^\infty)$ are not isomorphic, then for some prime p, the p-primary factors of $Z(\boldsymbol{a}^\infty)$ and $Z(\boldsymbol{b}^\infty)$ are different. Hence the corresponding factors of $\Delta_{\boldsymbol{a}}$ and $\Delta_{\boldsymbol{b}}$ are different. If $Z(p^r)$ is a factor of $\Delta_{\boldsymbol{a}}$, and Δ_p, $Z(p^s)$, or $\{e\}$ is the corresponding factor of $\Delta_{\boldsymbol{b}}$ where $s < r$, then $\Delta_{\boldsymbol{a}}$ has an element of order p^r and $\Delta_{\boldsymbol{b}}$ does not. If Δ_p is a factor of $\Delta_{\boldsymbol{a}}$ and $\{e\}$ is the corresponding factor of $\Delta_{\boldsymbol{b}}$, then $\Delta_{\boldsymbol{a}}^{(p)} \subsetneqq \Delta_{\boldsymbol{a}}$ and $\Delta_{\boldsymbol{b}}^{(p)} = \Delta_{\boldsymbol{b}}$.]

(25.29) (Braconnier [1], p. 41.) Let G be a locally compact Abelian group such that every element of G different from e has order p, where p is a fixed prime. Then G is topologically isomorphic with

(i) $Z(p)^{\mathfrak{m}} \times Z(p)^{\mathfrak{n}*}$,

where $\mathfrak{m}$ and $\mathfrak{n}$ are arbitrary cardinal numbers, $Z(p)^{\mathfrak{m}}$ has its usual compact topology, and $Z(p)^{\mathfrak{n}*}$ is taken discrete. Conversely, every group of the form (i) is locally compact and Abelian, and every element except e has order p. [By (24.30), G contains a compact open subgroup H. By (25.9), H is topologically isomorphic with $Z(p)^{\mathfrak{m}}$ for some cardinal number $\mathfrak{m}$. It is clear that if $x \in H$ and k is an integer prime to p, then there is an element $y \in H$ such that $y^k = x$. Consequently H is a pure subgroup of G and is *algebraically* a direct factor of G (25.21). It is easy to see that H, being open, is also a topological direct factor of G (6.22.a). Since G/H is a discrete Abelian group all of whose non-identity elements have order p, G/H is isomorphic with $Z(p)^{\mathfrak{n}*}$ for some cardinal number $\mathfrak{n}$ (A.25).]

(25.30) Remarks on direct factorization. (a) Let X be a discrete Abelian group and let Y be a subgroup of X such that X/Y is isomorphic with $Z^{\mathfrak{a}*}$ for some nonzero cardinal number $\mathfrak{a}$. Then Y is a direct factor of X: X is isomorphic with $Y \times (X/Y)$. [Since X/Y is torsion-free, Y is evidently a pure subgroup of X. Now apply (A.22).]

(b) Let G be a compact Abelian group and H a closed subgroup of G such that G/H is torsion-free. Then H is a topological direct factor of G: G is topologically isomorphic with $H \times (G/H)$. [Let X denote the character group of G. Then $\mathsf{A}(\mathsf{X}, H)$, being isomorphic with the character group of G/H, is divisible (24.23) and hence is a direct factor of X: X is isomorphic with $\mathsf{A}(\mathsf{X}, H) \times \mathsf{B}$, where B is a subgroup of X (A.8). It follows that B is isomorphic with $\mathsf{X}/\mathsf{A}(\mathsf{X}, H)$, so that B is isomorphic with the character group of H. Hence G is topologically isomorphic with $(G/H) \times H$.]

(c) Let G be a locally compact torsion-free Abelian group, and let C be the component of e in G. Then C is a topological direct factor of G. [Suppose first that G contains a compact open subgroup H. By (7.8), C is a subgroup of H, and so is compact. By (25.8), H is topologically isomorphic with a group $\Sigma_{\boldsymbol{a}}^{\mathfrak{m}} \times \underset{p \in P}{\mathsf{P}} \Delta_p^{\mathfrak{n}_p}$, and C is topologically isomorphic with $\Sigma_{\boldsymbol{a}}^{\mathfrak{m}}$. Thus C is a topological direct factor of H, and there is a continuous homomorphism π of H onto C whose restriction to C is the identity map. As C is divisible (24.25), π admits an extension π' over G that is a homomorphism of G onto C and is of course the identity on C (A.7). Since π is continuous on the open subgroup H, π' is continuous on G. By (6.22.a), C is a topological direct factor of G.

For an arbitrary G, use (24.30) to write G as $R^n \times G_1$, where G_1 contains a compact open subgroup [and is plainly torsion-free]. Let C_1 be the component of the identity in G_1. Then G is topologically isomorphic with $R^n \times C_1 \times (G_1/C_1)$, and it is easy to see that $R^n \times C_1$ is the component of the identity in $R^n \times C_1 \times (G_1/C_1)$.]

(25.31) Subgroups $T^{\mathfrak{a}}$ as direct factors. An idea used in proving (25.21) can be used to show that subgroups $T^{\mathfrak{a}}$ are topological direct factors of locally compact Abelian groups.

(a) Let G be a compact Abelian group, $\mathfrak{a}$ a nonzero cardinal, and suppose that G contains a subgroup H topologically isomorphic with $T^{\mathfrak{a}}$. Then H is closed, and G is topologically isomorphic with $H \times (G/H)$. Informally, we write: subgroups $T^{\mathfrak{a}}$ are direct factors. [Since compact subsets are closed in Hausdorff spaces, H is a closed subgroup of G. Let X and Y be the character groups of G and H respectively. Then Y is isomorphic with $\mathsf{X}/\mathsf{A}(\mathsf{X}, H)$ and also with $Z^{\mathfrak{a}*}$. By (25.30.a), $\mathsf{A}(\mathsf{X}, H)$ is a direct factor of X: X is isomorphic with $\mathsf{A}(\mathsf{X}, H) \times \mathsf{X}/\mathsf{A}(\mathsf{X}, H)$ and hence with $\mathsf{A}(\mathsf{X}, H) \times \mathsf{Y}$. If we regard G as the character group of X

and use (23.18) and (23.25), we see that G is topologically isomorphic with $(G/H)\times H$.]

(b) Let G be a locally compact Abelian group containing a subgroup H topologically isomorphic with $T^{\mathfrak{a}}$ for a nonzero cardinal number $\mathfrak{a}$. Then H is a topological direct factor of G. [By (24.30), G is topologically isomorphic with $R^n\times G_0$, where G_0 is a locally compact Abelian group containing a compact open subgroup F. By part (a), H is a direct factor of F; thus there is a continuous homomorphism π of F onto H whose restriction to H is the identity map. Since H is divisible, π admits an extension π' of G_0 onto H that is a homomorphism and is the identity map on H. Since π' is continuous on the open subgroup F, it is continuous on G_0. By (6.22.a), G_0 is topologically isomorphic with $H\times(G_0/H)$, and hence G is topologically isomorphic with $H\times(G_0/H)\times R^n$.]

(c) Theorem (25.21) gives rise to the conjecture that a locally compact Abelian group with a compact open subgroup *might* be the direct product of a compact group and a discrete group. This is disproved by the group Ω_p for an arbitrary prime p. Every compact subgroup of Ω_p is a group Λ_k for some integer k (10.16.a), and no Λ_k is pure in Ω_p, let alone a direct factor.

(25.32) Minimal divisible extensions. It is proved in (A.15) and (A.16) that every Abelian group G can be imbedded in a minimal divisible group E, and that E is unique up to an isomorphism leaving the elements of G fixed. That is, E is as unique as it can possibly be. We examine here a few of these minimal divisible extensions and a natural topology for them.

(a) Let G be a topological Abelian group and E the minimal divisible extension of G. Let $\mathfrak{U}$ be an open basis at e *in the group* G. Let $\mathfrak{U}$ be taken as an open basis at e for the larger group E. As noted in (4.18.h), E is a topological group containing G as an open subgroup.

(b) The minimal divisible extension of Δ_p is Ω_p, and for a finite cardinal number n, the minimal divisible extension of Δ_p^n is Ω_p^n. The topology described in part (a) coincides with the usual topology of Ω_p^n, when Δ_p and Δ_p^n are given their usual topologies. [The set Ω_p is a field (10.11) of characteristic 0 and hence is divisible. Therefore Ω_p^n is a divisible group containing Δ_p^n. For every $(x_1, x_2, \ldots, x_n)=\boldsymbol{x}\in\Omega_p^n$, there is a nonnegative integer k such that $p^k\boldsymbol{x}$ is in Δ_p^n.[1] It follows that Ω_p^n/Δ_p^n is a torsion group. Since Ω_p^n is torsion-free, (A.17) shows that Ω_p^n is the minimal divisible extension of Δ_p^n. The remarks about the topology of Ω_p^n are obvious.]

---

[1] We lapse into additive notation in dealing with Ω_p^n, since the group operation we are concerned with in Ω_p is addition.

(c) The strict analogue of (b) fails for *infinite* cardinal numbers $\mathfrak{n}$. The group $\Omega_p^{\mathfrak{n}}$ in this case is of course a divisible supergroup of $\Delta_p^{\mathfrak{n}}$, but it is too large. Consider $\Omega_p^{\mathfrak{n}}$ as the group of all Ω_p-valued functions (x_ι) on a set I such that $\overline{\overline{I}}=\mathfrak{n}$. Let σ_p be the function σ on Ω_p defined in (10.4). For $(x_\iota)\in\Omega_p^{\mathfrak{n}}$, let $\nu_p((x_\iota))=\sup\{\sigma_p(x_\iota, 0):\iota\in I\}$. Then the minimal divisible extension of $\Delta_p^{\mathfrak{n}}$ is the subgroup of $\Omega_p^{\mathfrak{n}}$ consisting of all (x_ι) for which $\nu_p((x_\iota))$ is finite. We denote this group by the symbol $\Omega_p^{\mathfrak{n}}{}'$. Note that $\Omega_p^{\mathfrak{n}}{}'=\bigcup_{k=-\infty}^{\infty}\Lambda_k^{\mathfrak{n}}$ where Λ_k is as defined in (10.4). [It is easy to see that $\Omega_p^{\mathfrak{n}}{}'$ is a divisible extension of $\Delta_p^{\mathfrak{n}}$ and also that $p^k(x_\iota)\in\Delta_p^{\mathfrak{n}}$ if $(x_\iota)\in\Omega_p^{\mathfrak{n}}{}'$ and k is a suitably selected nonnegative integer. Then argue as in part (b).]

(d) Consider any group $\underset{p\in P}{\mathsf{P}}\Delta_p^{\mathfrak{n}_p}$, where P is the set of all prime integers, and $\mathfrak{n}_2, \mathfrak{n}_3, \mathfrak{n}_5, \mathfrak{n}_7, \ldots$ are arbitrary cardinal numbers.[1] The minimal divisible extension E of $\underset{p\in P}{\mathsf{P}}\Delta_p^{\mathfrak{n}_p}$ is a certain subgroup of $\underset{p\in P}{\mathsf{P}}\Omega_p^{\mathfrak{n}_p}{}'$, identified as follows. Write a generic element of this group as $(x_2, x_3, x_5, x_7, \ldots)$ where $x_p\in\Omega_p^{\mathfrak{n}_p}{}'$. Our group E is the set of all $(x_2, x_3, x_5, \ldots)$ such that all but a finite number of the x_p's lie in $\Delta_p^{\mathfrak{n}_p}$. Suppose that $\Omega_p^{\mathfrak{n}_p}{}'$ is topologized as in part (a), where $\Delta_p^{\mathfrak{n}_p}$ is furnished with its usual topology, and that E is topologized as in part (a). Then E is the local direct product of the groups $\Omega_p^{\mathfrak{n}_p}{}'$ relative to the open subgroups $\Delta_p^{\mathfrak{n}_p}$ (6.16). [It is easy to see that $E/\underset{p\in P}{\mathsf{P}}\Delta_p^{\mathfrak{n}_p}$ is a torsion group. In fact, if $(x_2, x_3, x_5, \ldots)\in E$, $x_p\in\Delta_p^{\mathfrak{n}_p}$ for $p>p_0$, and $\nu_p(x_p)\leqq 2^k$ for $p=2, 3, 5, \ldots, p_0$, then $(2\cdot 3\cdot 5\cdots p_0)^k(x_2, x_3, x_5, \ldots)$ is in $\underset{p\in P}{\mathsf{P}}\Delta_p^{\mathfrak{n}_p}$. We must also show that E is divisible. This fact rests upon a simple property of the fields Ω_p. Let $\boldsymbol{y}=(y_l)_{l=-\infty}^{\infty}$ be a nonzero element of Ω_p, let $y_m\neq 0$, and $y_l=0$ for $l<m$. That is, $\sigma_p(\boldsymbol{y}, \boldsymbol{0})=2^{-m}$. Then for every prime $q\neq p$, we have $\sigma_p\left(\frac{1}{q}\boldsymbol{y}, \boldsymbol{0}\right)=\sigma_p(\boldsymbol{y}, \boldsymbol{0})$, and $\sigma_p\left(\frac{1}{p}\boldsymbol{y}, \boldsymbol{0}\right)=2^{-m+1}$. Let $(x_2, x_3, x_5, \ldots)$ be an element of E, and suppose that $x_p\in\Delta_p^{\mathfrak{n}_p}$ for $p>p_0$. For every prime q, the equation $(x_2, x_3, x_5, \ldots)=(qy_2, qy_3, qy_5, \ldots)$ has a solution for which $\nu_p(y_p)=\nu_p(x_p)$ if $p\neq q$ and $\nu_q(y_q)=2\nu_q(x_q)$. Thus we have $y_p\in\Delta_p^{\mathfrak{n}_p}$ for all $p>\max(p_0, q)$. Since q is an arbitrary prime, E is divisible.]

(e) Consider two sequences of cardinal numbers $\mathfrak{n}_2, \mathfrak{n}_3, \mathfrak{n}_5, \mathfrak{n}_7, \ldots$ and $\bar{\mathfrak{n}}_2, \bar{\mathfrak{n}}_3, \bar{\mathfrak{n}}_5, \bar{\mathfrak{n}}_7, \ldots$ as in part (d). Form the groups $\underset{p\in P}{\mathsf{P}}\Delta_p^{\mathfrak{n}_p}$ and $\underset{p\in P}{\mathsf{P}}\Delta_p^{\bar{\mathfrak{n}}_p}$, and the minimal divisible extensions E and $\overline{E}$, respectively, of these groups. Suppose that E and $\overline{E}$ are topologically isomorphic. Then $\mathfrak{n}_p=\bar{\mathfrak{n}}_p$ for all primes p. [The subgroup $\{(0, \ldots, 0, x_p, 0, \ldots): x_p\in\Omega_p^{\mathfrak{n}_p}{}'\}$ of E is exactly the set of all x in E such that $\lim_{k\to\infty} p^k x=0$. Hence if E and $\overline{E}$ are topologically isomorphic, so are $\Omega_p^{\mathfrak{n}_p}{}'$ and $\Omega_p^{\bar{\mathfrak{n}}_p}{}'$. Let τ be a topological isomorphism of $\Omega_p^{\mathfrak{n}_p}{}'$ onto $\Omega_p^{\bar{\mathfrak{n}}_p}{}'$. Regard these groups as linear

---

[1] We make the convenient convention that Δ_p^0 is the one-element group.

spaces over the field Ω_p. Then it is easy to see that τ is a linear mapping. If $\mathfrak{n}_p$ is finite, then $\mathfrak{n}_p$ is the dimension of $\Omega_p^{\mathfrak{n}_p\prime}$ as a linear space over Ω_p, and so $\mathfrak{n}_p=\overline{\mathfrak{n}}_p$ if either $\mathfrak{n}_p$ or $\overline{\mathfrak{n}}_p$ is finite. If $\mathfrak{n}_p$ and $\overline{\mathfrak{n}}_p$ are infinite, consider any compact open subgroup F of $\Omega_p^{\mathfrak{n}_p\prime}$. This subgroup is a finite union of cosets of a group topologically isomorphic with $\Delta_p^{\mathfrak{n}_p}$. The character group of $\Delta_p^{\mathfrak{n}_p}$ is isomorphic with $Z(p^\infty)^{\mathfrak{n}_p*}$, and (24.15) implies that $\mathfrak{w}(\Delta_p^{\mathfrak{n}_p})=\overline{\overline{Z(p^\infty)^{\mathfrak{n}_p*}}}=\mathfrak{n}_p$. It follows that $\mathfrak{w}(F)=\mathfrak{n}_p$, and we infer with no hesitation that $\mathfrak{n}_p=\overline{\mathfrak{n}}_p$.]

(25.33) The structure of torsion-free divisible groups (adapted from MACKEY [1]). (a) Let G be a locally compact, divisible, torsion-free Abelian group. Then G is topologically isomorphic with a group

(i) $R^{\mathfrak{p}}\times\Sigma_{\boldsymbol{a}}^{\mathfrak{q}}\times Q^{\mathfrak{r}*}\times E$,

where: $\mathfrak{p}$ is a finite cardinal; $\mathfrak{q}$ and $\mathfrak{r}$ are arbitrary cardinals[1]; $\boldsymbol{a}=(2,3,4,5,\ldots)$; $Q^{\mathfrak{r}*}$ is topologized discretely; and E is the minimal divisible extension of a certain group $\underset{p\in P}{\mathsf{P}}\Delta_p^{\mathfrak{n}_p}$, constructed in (25.32.d). Conversely, every group of the form (i) is locally compact, divisible, torsion-free, and Abelian. [By (25.30.c) G is topologically isomorphic with a group

$$R^{\mathfrak{p}}\times C_1\times G_2, \tag{1}$$

where C_1 is compact, connected and torsion-free and G_2 is the quotient group of G by its component of e. By (25.8), C_1 is topologically isomorphic with $\Sigma_{\boldsymbol{a}}^{\mathfrak{q}}$ for some cardinal $\mathfrak{q}$. We are left with the 0-dimensional [see (7.3) and (3.5)] group G_2. It is plain that G_2 is also divisible and torsion-free. Let H be a compact open subgroup of G_2 (7.7). If H is the identity, then G_2 is discrete and is topologically isomorphic with $Q^{\mathfrak{r}*}$ for some cardinal $\mathfrak{r}$ (A.14). If H is not the identity, then it is an infinite, 0-dimensional, compact, torsion-free group and necessarily has the form $\underset{p\in P}{\mathsf{P}}\Delta_p^{\mathfrak{n}_p}$ (25.8). The intersection E_0 of all of the divisible subgroups of G_2 that contain H is again a divisible subgroup of G_2. This follows from the fact that G_2 is torsion-free. It is trivial that there is no divisible group D for which $H\subset D\subsetneqq E_0$. Thus E_0 is a minimal divisible extension of H and therefore is isomorphic as a group with the group E constructed in (25.32.d). Since H is open in G_2, it is open in E_0 and so E_0 is topologically isomorphic with the group E of (25.32.d). As E_0 is open in G_2 and divisible, it is a topological direct factor of G_2 (6.22.b): G_2 is topologically isomorphic with

$$(G_2/E_0)\times E_0; \tag{2}$$

and G_2/E_0 is discrete, divisible, and torsion-free. By (A.14), G_2/E_0 is isomorphic with $Q^{\mathfrak{r}*}$ for some cardinal $\mathfrak{r}$. Combining (1) and (2), we obtain (i). The converse is trivial.]

---

[1] We again agree that a group to the zero power is the one-element group.

(b) Two groups $G=R^{\mathfrak{p}}\times\Sigma_{\boldsymbol{a}}^{\mathfrak{q}}\times Q^{\mathfrak{r}*}\times E$ and $\overline{G}=R^{\bar{\mathfrak{p}}}\times\Sigma_{\boldsymbol{a}}^{\bar{\mathfrak{q}}}\times Q^{\bar{\mathfrak{r}}*}\times\overline{E}$ of the form (i) are topologically isomorphic if and only if $\mathfrak{p}=\bar{\mathfrak{p}}$, $\mathfrak{q}=\bar{\mathfrak{q}}$, $\mathfrak{r}=\bar{\mathfrak{r}}$, and $\mathfrak{n}_p=\bar{\mathfrak{n}}_p$ [1] for $p=2, 3, 5, 7, \ldots$. [It is convenient to examine only one group and to show that the numbers $\mathfrak{p}$, $\mathfrak{q}$, $\mathfrak{r}$, and $\mathfrak{n}_p$ are defined in terms of the structure of G; thus they must be preserved under a topological isomorphism. The number $\mathfrak{p}$ is the largest integer n such that G contains a topological isomorph of R^n (9.14). The subgroup $R^{\mathfrak{p}}\times\Sigma_{\boldsymbol{a}}^{\mathfrak{q}}$ is the component C of e in G, and $\Sigma_{\boldsymbol{a}}^{\mathfrak{q}}$ is the subgroup B_0 of compact elements of C. Since $\mathfrak{q}$ is the torsion-free rank of the character group $Q^{\mathfrak{q}*}$ of $\Sigma_{\boldsymbol{a}}^{\mathfrak{q}}$, $\mathfrak{q}$ is well defined.

It is evident that every element of E lies in a compact subgroup of E. Hence the character group Y of E is 0-dimensional (24.18). The character group X of G is topologically isomorphic with $R^{\mathfrak{p}}\times Q^{\mathfrak{q}*}\times\Sigma_{\boldsymbol{a}}^{\mathfrak{r}}\times\mathsf{Y}$.

Arguing as in the case of the cardinal $\mathfrak{q}$ above, we see that $\mathfrak{r}$ is well defined. Obviously the group B of compact elements in G is $\Sigma_{\boldsymbol{a}}^{\mathfrak{q}}\times E$. Thus E is topologically isomorphic with B/B_0, and hence is well defined up to topological isomorphism. The uniqueness of the cardinal numbers $\mathfrak{n}_2, \mathfrak{n}_3, \mathfrak{n}_5, \ldots$ that define E was proved in (25.32.e).]

(25.34) Groups topologically isomorphic with their character groups. If a locally compact Abelian group G is topologically isomorphic with its character group X, we will say that G is *self-dual*. The group $G\times\mathsf{X}$, for arbitrary G, is a trivial example. If G_1 and G_2 are self-dual, then so is $G_1\times G_2$.

(a) Examples of self-dual groups already met with in the text are the following:

(i) a finite Abelian group G (23.27.d);

(ii) R (23.27.e);

(iii) $\Omega_{\boldsymbol{a}}$ where $\boldsymbol{a}=\{a_k\}_{k=-\infty}^{\infty}$ is such that $a_k=a_{-k}$ for all $k\in Z$ (25.1).

(b) More examples of self-dual groups are found from (23.33). Let I be an index class. For each $\iota\in I$, let G_ι be a locally compact self-dual Abelian group with character group X_ι. Suppose also that there is a topological isomorphism τ_ι of G_ι onto X_ι and a compact open subgroup H_ι of G_ι such that $\tau_\iota(H_\iota)=\mathsf{A}(\mathsf{X}_\iota, H_\iota)$. Then H_ι is topologically isomorphic with the character group of G_ι/H_ι (23.25). Now form the local direct product G of the G_ι relative to the compact open subgroups H_ι. It follows immediately from (23.33) that G is self-dual.

(c) For a compact Abelian group H with character group Y, the group $H\times\mathsf{Y}$ is a trivial example of a group satisfying the requirements of part (b). Another example is the group $Z(p^{2r})$, written as $\{0, 1, 2, \ldots, p^{2r}-1\}$, with the subgroup $\{0, p^r, 2p^r, \ldots, (p^r-1)\,p^r\}$ [p a prime, r a

[1] We write $\overline{E}$ for the minimal divisible extension of $\underset{p\in P}{\mathrm{P}}\Delta_p^{\bar{\mathfrak{n}}_p}$.

positive integer]. Still another example is provided by any group Ω_r and its compact open subgroup Δ_r, where r is an integer greater than 1.

(d) The foregoing examples illustrate the wealth of self-dual groups. No complete classification of these groups is known to the writers.

(25.35) Products of {0, 1} again (adapted from VILENKIN [17]). The following result is related to Theorem (9.15). Let G be a compact Abelian group. For some cardinal number $\mathfrak{m}$, there is a continuous mapping of $\{0, 1\}^{\mathfrak{m}}$ onto G; $\mathfrak{m}$ can be taken to be $\max[\aleph_0, r(\mathsf{X})]$, where X is the character group of G. [We break the proof up into two steps. (I) If Y is a compact metric space, then there is a continuous mapping of $\{0, 1\}^{\aleph_0}$ onto Y. This result and its proof are similar to the contents of the footnote on page 98. The space Y is the union of closed sets $B_1, \ldots, B_{a_1}$ with nonvoid interior such that the diameter of each B_i is less than 1. Each B_i is the union of a finite number of nonvoid closed sets $B_{i1}, B_{i2}, \ldots, B_{ia_2}$ [a_2 is the same positive integer for each i], where each B_{ij} has diameter less than $\frac{1}{2}$. Continuing this construction in the indicated manner, we find a sequence $\{a_n\}_{n=1}^{\infty}$ of positive integers and nonvoid closed subsets $B_{i_1 i_2 \cdots i_m}$ of Y such that $B_{i_1 i_2 \cdots i_m}$ has diameter less than 2^{-m+1}, $B_{i_1 i_2 \cdots i_{m+1}}$ is a subset of $B_{i_1 i_2 \cdots i_m}$, and $\bigcup \{B_{i_1 i_2 \cdots i_m} : 1 \leqq i_k \leqq a_k, k = 1, \ldots, m\} = Y$ for each m. For each sequence $\boldsymbol{i} = (i_1, i_2, \ldots)$, where $1 \leqq i_k \leqq a_k$ for $k = 1, 2, \ldots$, define $\Phi(\boldsymbol{i})$ to be the unique point in $\bigcap_{m=1}^{\infty} B_{i_1 i_2 \cdots i_m}$. Then Φ is a continuous mapping of $\mathsf{P}_{m=1}^{\infty} \{1, \ldots, a_m\}$ onto Y. The last two paragraphs of the proof of (9.15) prove that $\mathsf{P}_{m=1}^{\infty} \{1, \ldots, a_m\}$ is homeomorphic with $\{0, 1\}^{\aleph_0}$.

(II) Let G be a compact Abelian group. The discrete group X can be imbedded in a divisible group Y having the same rank as X [see (A.15) and (A.16)]. By (A.14), Y is isomorphic with

$$Q^{r_0(\mathsf{X})*} \times \mathop{\mathsf{P}^*}_{p \in P} \left(Z(p^{\infty})^{r_p(\mathsf{X})*}\right),$$

where P denotes the set of all primes. The character group H of Y is topologically isomorphic with $(\Sigma_{\boldsymbol{a}})^{r_0(\mathsf{X})} \times \mathop{\mathsf{P}}_{p \in P} (\Delta_p)^{r_p(\mathsf{X})}$, as (23.22), (25.4), and (25.2) show. By (24.11), the character group G of X is topologically isomorphic with $H/A(H, \mathsf{X})$; that is, G is a continuous image of H. It suffices then to prove that H is a continuous image of $\{0, 1\}^{\mathfrak{m}}$ where $\mathfrak{m} = \max[\aleph_0, r(\mathsf{X})]$.

The groups $\Sigma_{\boldsymbol{a}}$ and Δ_p are compact and metrizable by (24.48). Thus by (I), $\Sigma_{\boldsymbol{a}}$ and Δ_p are continuous images of $\{0, 1\}^{\aleph_0}$. It follows that H is a continuous image of $\{0, 1\}^{\mathfrak{m}}$ where $\mathfrak{m} = \max[\aleph_0, r_0(\mathsf{X}) + r_2(\mathsf{X}) + r_3(\mathsf{X}) + \cdots] = \max[\aleph_0, r(\mathsf{X})]$; see (A.12) and (A.13).]

(25.36) The uniqueness of T in the duality theorem (adapted from PONTRYAGIN [7], Example 72). The choice of T as the value group for characters of locally compact Abelian groups is natural enough on various grounds: characters are 1-dimensional unitary representations; characters are generalizations of the famous functions $\exp(ix) \to \exp(inx)$ on T and $x \to \exp(ixy)$ on R; complex-valued functions are well-known objects and hence presumably amenable to the techniques and concepts of ordinary analysis. But even more, *no other locally compact Abelian group* would have given us Theorem (24.8). In fact, let H be a locally compact Abelian group, and suppose that $\mathrm{Hom}\big(\mathrm{Hom}(T, H), H\big)$ is topologically isomorphic with T. [Notation is as in (23.34).] Thus we suppose much less than the full duality theorem. It then follows that $H = T$.

[Let τ be any element of $\mathrm{Hom}(T, H)$ not carrying T onto the identity f of H; clearly such a τ exists. The kernel of τ is a proper closed subgroup of T and is hence finite. It follows that $\tau(T)$ is topologically isomorphic with T (5.39.j). Let H_1 be the connected component of f in H, and let E be the largest compact subgroup of H_1 [see (9.14)]. Since $\tau(T)$ is compact and connected, it is clear that $\tau(T) \subset E$. By (25.31.a), $\tau(T)$ is a direct factor of E: E is topologically isomorphic with $\tau(T) \times E_1$, where E_1 is a closed subgroup of E. If $\sigma \in \mathrm{Hom}(T, H)$, it is clear that $\sigma(T) \subset E$. Thus replacing topological isomorphisms by equalities and using (23.34.d), we have

$$\mathrm{Hom}(T, H) = \mathrm{Hom}\big(T, \tau(T) \times E_1\big) = \mathrm{Hom}\big(T, \tau(T)\big) \times \mathrm{Hom}(T, E_1). \quad (1)$$

Applying (23.34.c) to (1), we obtain

$$\left.\begin{aligned} \mathrm{Hom}\big(\mathrm{Hom}(T, H), H\big) &= \mathrm{Hom}\big(\mathrm{Hom}(T, \tau(T)) \times \mathrm{Hom}(T, E_1), H\big) \\ &= \mathrm{Hom}\big(\mathrm{Hom}(T, \tau(T)), H\big) \times \mathrm{Hom}\big(\mathrm{Hom}(T, E_1), H\big). \end{aligned}\right\} \quad (2)$$

Since $\tau(T)$ is topologically isomorphic with T, $\mathrm{Hom}\big(T, \tau(T)\big)$ is topologically isomorphic with the [discrete] character group Z of T. It is an elementary matter to verify that $\mathrm{Hom}(Z, H)$ is topologically isomorphic with H. From (2) and our hypothesis, therefore, we have

$$T = H \times \mathrm{Hom}\big(\mathrm{Hom}(T, E_1), H\big). \quad (3)$$

Obviously the factor H in (3) is a connected subgroup of T. It follows either that $H = \{f\}$, a palpable absurdity, or that $H = T$, as we wished to show.]

Notes

So far as we know, (25.1) is new. The result of (25.2) is stated by ABE [1]; it is of course only a slight generalization of the fact that the character group of Δ_p is $Z(p^\infty)$. This fact was published by W. KRULL [1], and has been published many times since.

As remarked in the Notes to §10, the group $\Sigma_{\mathbf{a}}$ for a general $\mathbf{a} = (a_0, a_1, a_2, \ldots)$ was constructed by VAN DANTZIG [6]. The fact that $\Sigma_{\mathbf{a}}$ is the character group of Q_d if $\mathbf{a} = (2, 3, 4, \ldots)$ was pointed out by ABE [1] and by ANZAI and KAKUTANI [1]. Our computation in (25.4) seems more explicit than theirs. BRACONNIER [1], p. 19, also mentions the character group of Q_d, but in describing it as a solenoid he refers to VAN DANTZIG [1], where only the group $\Sigma_{(n,n,n,\ldots)}$ [which is *not* the character group of Q_d] is constructed. Other writers who at this period studied the character group of Q_d made no mention that it is $\Sigma_{(2,3,4,\ldots)}$ [see for example HALMOS [1] and MACKEY [1]]. The construction in (25.5) appears in part in VON NEUMANN [4], p. 477.

The construction of a Hamel basis for R over Q is due, unsurprisingly, to G. HAMEL [1]. The computation of *all* characters of R_d given in (25.6) appears in MAAK [1], §23, Satz 4, pp. 89—90. We have no reference for (25.7), although its content must be widely known.

Theorem (25.8) is a strengthened form of BRACONNIER [1], p. 17, Corollaire; (25.9) and (25.10) were suggested by a remark of W. RUDIN [1], Lemma 3.

Monothetic groups were introduced by VAN DANTZIG [1], and have been studied by HALMOS and SAMELSON [1], ECKMANN [1], and ANZAI and KAKUTANI [1]. Much of (25.11)—(25.17) is drawn from these sources. Theorem (25.13) was kindly communicated to us orally by R. A. BEAUMONT and R. S. PIERCE. The implication (i) implies (ii) in (25.16) is due to VAN DANTZIG [6]; the rest of (25.16) appears to be new. Essentially all of (25.18) and (25.19) are due to ANZAI and KAKUTANI [1]. Theorem (25.20) is stated in MACKEY [2].

A large number of writers have contributed to (25.21)—(25.25), and as frequently happens some duplication has occurred. The problem of classifying the algebraic structure of compact Abelian groups was raised by HALMOS [1]. Incomplete results were obtained by KAPLANSKY [1], pp. 55—56; see also BALCERZYK [1]. Theorem (25.21) [which is the key to (25.25)] is due to ŁOŚ [1] and [2]. Theorems (25.22), (25.23), and all but the last sentence of (25.25) are due to HARRISON [1] and HULANICKI [2] and [3]. FUCHS [2] has given a simplified version of HULANICKI'S construction. Finally, (25.23) was communicated to us orally in May 1959 by S. KAKUTANI as an unpublished result of S. KAKUTANI and T. NAKAYAMA. So far as we know, (25.24) is new.

The material presented in the text is only a part [although we trust an important part] of what is known of the structure of various classes of locally compact Abelian groups. The reader interested in pursuing the subject of the present section further should first consult BRACONNIER [1], Ch. II. There is also a formidable sequence of papers by N. YA. VILENKIN, of somewhat uneven merit, which should be examined. These are: VILENKIN [1], [3], [5], [7], [10], [11], [12], [15], [16].

§ 26. Miscellaneous consequences of the duality theorem

In this section we present a number of constructions and theorems which depend in one way or another upon (24.8). We begin with a study of automorphism groups, which do not *ab initio* depend on (24.8) but whose most interesting application [among those given here] does: see (26.10).

(26.1) Definition. Let G be a topological group, and let $\mathfrak{S}(G)$ be the set of all *topological automorphisms* of G onto itself, that is, the set of all mappings of G onto G that are simultaneously automorphisms and homeomorphisms. For $\sigma, \tau \in \mathfrak{S}(G)$, let $\sigma \circ \tau$ be as usual the composition of σ and τ: $\sigma \circ \tau(x) = \sigma(\tau(x))$ for all $x \in G$. Let ι be the identity mapping on G: $\iota(x) = x$ for all $x \in G$. Let τ^{-1} be the usual inverse of τ: $\tau^{-1}(x)$ is the [unique!] element $y \in G$ such that $\tau(y) = x$. [1]

(26.2) It is trivial that $\mathfrak{S}(G)$ is a group under the operations described in (26.1). In fact, $\mathfrak{S}(G)$ is a subgroup of the group of all one-to-one mappings of G onto itself, referred to in (2.8.c).

(26.3) Definition. Let G and $\mathfrak{S}(G)$ be as in (26.1). For a compact subset F of G and a neighborhood U of e in G, let $\mathfrak{B}(F, U)$ be the set of all $\tau \in \mathfrak{S}(G)$ such that $\tau(x) \in Ux$ and $\tau^{-1}(x) \in Ux$ for all $x \in F$.

We will show that for a locally compact group G, the sets $\mathfrak{B}(F, U)$ satisfy conditions (4.5.i)—(4.5.v) and hence define a topology for $\mathfrak{S}(G)$ under which $\mathfrak{S}(G)$ is a topological group. A lemma is needed.

(26.4) Lemma. *Let F be a compact subset of G and U a neighborhood of e in G. For each $\tau \in \mathfrak{B}(F, U)$, there is a neighborhood V of e [depending upon F, U, and τ] such that $V\tau(x) \subset Ux$ for all $x \in F$.*

Proof. For $x \in G$, let $\psi(x) = \tau(x)\,x^{-1}$. Then ψ is a continuous mapping of G into G. Since $\tau(x)\,x^{-1} \in U$ for all $x \in F$, it follows that $\psi(F)$ is a compact subset of U. By (4.10), there is a neighborhood V of e in G such that $V\psi(F) \subset U$. Now if $x \in F$, then $V\psi(x) \subset U$ so that $V\tau(x)\,x^{-1} \subset U$ and $V\tau(x) \subset Ux$. □

---

[1] The reader should note that $\tau^{-1}(x)$ is *not* necessarily the same as $\tau(x^{-1})$. The symbols x^{-1} and τ^{-1} have different, and unrelated, meanings.

(26.5) Theorem. *Let G be a locally compact group. As F runs through all compact subsets of G and U through all neighborhoods of e in G, the sets $\mathfrak{B}(F, U)$ of* (26.3) *describe a family of sets satisfying axioms* (4.5.i)—(4.5.v) *and so in accordance with* (4.5) *define a topology on $\mathfrak{S}(G)$ under which it is a topological group.*

Proof. To verify (4.5.i), let $\mathfrak{B}(F, U)$ be an arbitrary neighborhood of ι; let V be a neighborhood of e in G such that $V^2 \subset U$ and V^- is compact. Suppose that $\sigma, \tau \in \mathfrak{B}(V^- \cdot F, V)$. Then for all $x \in F$, we have

$$\sigma(\tau(x))\, x^{-1} = \sigma(\tau(x))\, \tau(x)^{-1} \tau(x)\, x^{-1} \in \sigma(\tau(x))\, \tau(x)^{-1} \cdot V$$

and $\tau(x) \in V \cdot F \subset V^- \cdot F$. Hence $\sigma(\tau(x))\, \tau(x)^{-1} \in V$, and so

$$\sigma(\tau(x))\, x^{-1} \in V^2 \subset U.$$

Similarly we see that

$$(\sigma \circ \tau)^{-1}(x) \cdot x^{-1} = \tau^{-1}(\sigma^{-1}(x))\, (\sigma^{-1}(x))^{-1} \sigma^{-1}(x)\, x^{-1}$$

is in U for all $x \in F$. Since $V^- \cdot F$ is compact (4.4), axiom (4.5.i) is established.

Since $\mathfrak{B}(F, U) = (\mathfrak{B}(F, U))^{-1}$, axiom (4.5.ii) is trivially satisfied.

We now verify (4.5.iii). Given a compact set $F \subset G$, a neighborhood U of e in G, and a fixed $\tau \in \mathfrak{B}(F, U)$, let V be a neighborhood of e in G such that $V\tau(x) \subset Ux$ and $V\tau^{-1}(x) \subset Ux$ for all $x \in F$ [see (26.4) above]. Let W be a neighborhood of e such that $\tau(W) \subset V$. Then the inclusion

$$\tau \circ \mathfrak{B}(F \cup \tau^{-1}(F), W \cap V) \subset \mathfrak{B}(F, U) \tag{1}$$

obtains. In fact, let σ be any automorphism in $\mathfrak{B}(F \cup \tau^{-1}(F), W \cap V)$, and let x be any element of F. Then $\sigma(x) = wx$ where $w \in W$, and so we have

$$\tau \circ \sigma(x) = \tau(w)\, \tau(x) \in V\tau(x) \subset Ux. \tag{2}$$

Since $\tau^{-1}(x) \in \tau^{-1}(F)$, we also have

$$\sigma^{-1} \circ \tau^{-1}(x) \in V\tau^{-1}(x) \subset Ux. \tag{3}$$

The relations (2) and (3) imply (1).

Axiom (4.5.iv) is simple: the inclusion

$$\tau \circ \mathfrak{B}(\tau^{-1}(F), \tau^{-1}(U)) \circ \tau^{-1} \subset \mathfrak{B}(F, U)$$

is immediate. Axiom (4.5.v) is also obvious.

It remains to show that $\mathfrak{S}(G)$ is a T_0 group. If τ in $\mathfrak{S}(G)$ is not the identity automorphism ι, then there is an $a \in G$ such that $\tau(a) \neq a$. Since G is a T_1 group (4.8), there is a neighborhood U of e in G such that $\tau(a) \notin Ua$. Then $\mathfrak{B}(\{a\}, U)$ is a neighborhood of ι not containing τ. It follows that $\mathfrak{S}(G)$ is a T_0 group. □

(26.6) The group $\mathfrak{S}(G)$ need not be locally compact for locally compact G. Even if G is compact and Abelian, $\mathfrak{S}(G)$ can fail to be locally compact: see (26.18.k).

The two following theorems are of some independent interest but for our purposes are mainly lemmas leading up to (26.10).

(26.7) Theorem. *Let G be a topological group and let H be a locally compact normal subgroup of G. For $s \in G$, let ϱ_s denote the inner automorphism $x \rightarrow s x s^{-1}$ of G onto G. The mapping*

(i) $s \rightarrow \varrho_s | H$

is a continuous homomorphism of G into $\mathfrak{S}(H)$.

Proof. It is obvious that (i) is a homomorphism of G into $\mathfrak{S}(H)$. By (5.40.a), it suffices to prove that (i) is continuous at e. Thus let F be any compact subset of H and U any neighborhood of e in G. Let V be a neighborhood of e in G such that $V^2 \subset U$. Since F is compact in H, it is obviously compact in G. Use (4.9) to choose a symmetric neighborhood W of e in G such that $W \subset V$ and $xWx^{-1} \subset V$ for all $x \in F$. We claim that $\varrho_s | H$ is in $\mathfrak{B}(F, U \cap H)$ for all $s \in W$. In fact, if $s \in W$ and $x \in F$, we have

$$\varrho_s(x) \cdot x^{-1} = (s x s^{-1}) x^{-1} = s (x s^{-1} x^{-1}) \in W V \subset V^2 \subset U, \tag{1}$$

and plainly $(s x s^{-1}) x^{-1}$ is in H. Since $s^{-1} \in W$, we also have

$$\varrho_s^{-1}(x) \cdot x^{-1} = \varrho_{s^{-1}}(x) \cdot x^{-1} \in U \cap H. \tag{2}$$

The computations (1) and (2) show that $\varrho_s | H$ is in $\mathfrak{B}(F, U \cap H)$. □

(26.8) Theorem. *Let G be a 0-dimensional locally compact group. Then $\mathfrak{S}(G)$ is also 0-dimensional; in fact, open subgroups form an open basis at ι.*[1]

Proof. Let H be any compact open subgroup of G, and let F be a compact subset of G such that $F \supset H$. If $\sigma, \tau \in \mathfrak{B}(F, H)$ and $x \in F$, then for some $h \in H$ we have $\tau^{-1}(x) = hx$ and hence $\sigma \circ \tau^{-1}(x) = \sigma(hx) = \sigma(h)\,\sigma(x) \in H h H x = Hx$. Similarly we have $(\sigma \circ \tau^{-1})^{-1}(x) = \tau \circ \sigma^{-1}(x) \in Hx$. Therefore $\sigma \circ \tau^{-1} \in \mathfrak{B}(F, H)$ and $\mathfrak{B}(F, H)$ is an open subgroup of $\mathfrak{S}(G)$.

Now if G is 0-dimensional, F is compact in G, and U is a neighborhood of e in G, there is a compact open subgroup H of G such that $H \subset U$ (7.7). Thus we have $\mathfrak{B}(F \cup H, H) \subset \mathfrak{B}(F, U)$, so that open subgroups are a basis at ι. □

The next result, also useful for (26.10), is of considerable importance on its own account.

---

[1] Since $\mathfrak{S}(G)$ need not be locally compact, we cannot use Theorem (7.7): our assertion about open subgroups is stronger than the assertion that $\mathfrak{S}(G)$ is 0-dimensional.

(26.9) Theorem. *Let G be a locally compact Abelian group with character group X. For every $\tau \in \mathfrak{S}(G)$, let $\tau^{\sim}$ be the adjoint homomorphism defined in* (24.37), i.e., *let $\tau^{\sim}(\chi)(x) = \chi(\tau(x))$ for all $\chi \in X$ and $x \in G$. The mapping*

(i) $\tau \to \tau^{\sim}$

is a topological anti-isomorphism of $\mathfrak{S}(G)$ onto $\mathfrak{S}(X)$.

Proof. As noted in (24.41.c), every $\tau^{\sim}$ is an element of $\mathfrak{S}(X)$. Since continuous characters of G separate points of G, it is clear that $\tau_1 \neq \tau_2$ implies $\tau_1^{\sim} \neq \tau_2^{\sim}$. If σ is an element of $\mathfrak{S}(X)$, then, regarding G as the character group of X, we see that $\sigma^{\sim}$ is an element of $\mathfrak{S}(G)$, and by (24.41.a), $(\sigma^{\sim})^{\sim} = \sigma$. Thus (i) maps $\mathfrak{S}(G)$ onto $\mathfrak{S}(X)$ and is one-to-one.

For $\sigma, \tau \in \mathfrak{S}(G)$, for $\chi \in X$, and $x \in G$, we have

$$(\sigma \circ \tau)^{\sim}(\chi)(x) = \chi(\sigma \circ \tau(x)) = \chi(\sigma(\tau(x))) = \sigma^{\sim}(\chi)(\tau(x)) = (\tau^{\sim} \circ \sigma^{\sim})(\chi)(x).$$

Thus (i) is an anti-isomorphism.

It remains to show that (i) is a homeomorphism. According to (24.8), the sets $\{x \in G : |\chi(x) - 1| < \varepsilon \text{ for all } \chi \in \Phi\}$ form an open basis at e in G as Φ runs through all compact subsets of X and ε runs through all positive real numbers. Thus a generic neighborhood of ι in $\mathfrak{S}(G)$ can be described as

$$\left.\begin{aligned} \mathfrak{A}(F, \Phi, \varepsilon) = \{\tau \in \mathfrak{S}(G) : |\chi(\tau(x)\, x^{-1}) - 1| < \varepsilon \text{ and} \\ |\chi(\tau^{-1}(x)\, x^{-1}) - 1| < \varepsilon \text{ for all } x \in F \text{ and } \chi \in \Phi\}. \end{aligned}\right\} \quad (1)$$

Here F is compact in G, Φ is compact in X, and ε is a positive real number. We have also

$$\chi(\tau(x)\, x^{-1}) = \chi(\tau(x))\, \chi(x^{-1}) = \tau^{\sim}(\chi)(x)\, \chi^{-1}(x) = (\tau^{\sim}(\chi)\, \chi^{-1})(x)$$

and similarly

$$\chi(\tau^{-1}(x)\, x^{-1}) = ((\tau^{\sim})^{-1}(\chi)\, \chi^{-1})(x).$$

It follows that $\mathfrak{A}(F, \Phi, \varepsilon)^{\sim}$ is a set of exactly the same sort as $\mathfrak{A}(F, \Phi, \varepsilon)$ but in the group $\mathfrak{S}(X)$. Thus (i) is a homeomorphism. □

With the foregoing results we can prove an interesting fact about the centers of connected groups.

(26.10) Theorem. *Let G be a connected group and H a locally compact normal subgroup of G. Suppose either that H is* 0*-dimensional or that H is Abelian and has the property that every element of H lies in a compact subgroup of H. Then H is contained in the center of G.*[1]

[1] For analogous but much simpler facts about the center of a connected group, see (7.17), p. 65.

Proof. If H is 0-dimensional, then $\mathfrak{S}(H)$ is 0-dimensional (26.8). If H is Abelian and every element of H lies in a compact subgroup of H, then the character group Y of H is 0-dimensional (24.18), so that $\mathfrak{S}(\mathsf{Y})$ is 0-dimensional. By (26.9), $\mathfrak{S}(H)$ is also 0-dimensional. By (26.7), the mapping $s \to \varrho_s | H$ of G into $\mathfrak{S}(H)$ is continuous. Since G is connected, each ϱ_s must be the identity ι on H. That is, $sxs^{-1} = x$ for all $x \in H$ and $s \in G$, and H is contained in the center of G. □

We now turn to another construction, using subsets of the character group of a locally compact Abelian group G to imbed homomorphic images of G in compact Abelian groups.

(26.11) Definition. Let G be a locally compact Abelian group with character group X, and let Γ be any nonvoid subset of X. We define a mapping Φ_Γ of G into $\underset{\chi\in\Gamma}{\mathsf{P}} T_{(\chi)}$ [each $T_{(\chi)}$ is a replica of T] as follows: for each $x \in G$, $\Phi_\Gamma(x) = (\chi(x)) \in \underset{\chi\in\Gamma}{\mathsf{P}} T_{(\chi)}$. It is very easy to see that Φ_Γ is a continuous homomorphism, and that Φ_Γ is one-to-one if and only if Γ separates points of G. Now let $b_\Gamma G = (\Phi_\Gamma(G))^-$ [closure in $\underset{\chi\in\Gamma}{\mathsf{P}} T_{(\chi)}$]. Such groups $b_\Gamma G$ are called *Bohr compactifications of* G. The group $b_{\mathsf{X}} G$ will be written bG.

(26.12) Theorem. *The character group of $b_\Gamma G$ is topologically isomorphic with the subgroup Σ_d of X_d generated by Γ; in particular the character group of bG is topologically isomorphic with X_d. Thus $b_\Gamma G$ is topologically isomorphic with the [compact] character group of Σ_d, and bG is topologically isomorphic with the character group of X_d.*

Proof. For $\chi \in \Gamma$, let π_χ denote the projection of $\underset{\chi\in\Gamma}{\mathsf{P}} T_{(\chi)}$ onto $T_{(\chi)}$. For $x \in G$, we have $\pi_\chi(\Phi_\Gamma(x)) = \chi(x)$, an equality which is tautological but nonetheless useful. The projections π_χ are continuous characters of the entire group $\underset{\chi\in\Gamma}{\mathsf{P}} T_{(\chi)}$ and when restricted to $b_\Gamma G$ are of course continuous characters of $b_\Gamma G$. They also separate points of $b_\Gamma G$ and so by (23.20), the character group of $b_\Gamma G$ consists of all functions $\pi_{\chi_1}^{\alpha_1}\pi_{\chi_2}^{\alpha_2}\cdots\pi_{\chi_m}^{\alpha_m}$, where $\alpha_1, \ldots, \alpha_m$ are integers. It is clear that $\pi_{\chi_1}^{\alpha_1}\pi_{\chi_2}^{\alpha_2}\cdots\pi_{\chi_m}^{\alpha_m} = 1$ on $b_\Gamma G$ if and only if $\chi_1^{\alpha_1}\chi_2^{\alpha_2}\cdots\chi_m^{\alpha_m} = 1$ on G. Thus the character group of $b_\Gamma G$ is topologically isomorphic with Σ_d. The remaining statements about $b_\Gamma G$ and bG follow at once from the foregoing and (24.8). □

Another characterization of Bohr compactifications follows.

(26.13) Theorem. *Let G and X be as in* (26.11). *Let H be a compact Abelian group with [discrete] character group Y. The following assertions are equivalent:*

(i) *there is a continuous homomorphism φ of G onto a dense subgroup of H;*

(ii) *H is topologically isomorphic with a Bohr compactification of G;*
(iii) *Y is topologically isomorphic with a subgroup of X_d.*

Proof. If (i) holds, then the adjoint homomorphism $\varphi^{\sim}$ of (24.37) is an isomorphism carrying Y into X (24.41.b); that is, (iii) holds. If (iii) holds, and τ is an isomorphism of Y into X_d, then τ is also a continuous isomorphism into X. Hence $\tau^{\sim}$ is a continuous homomorphism of the character group of X onto a dense subgroup of the character group of Y. By (24.8), (i) thus holds.

It is trivial that (ii) implies (i). If (i) holds, take Γ to be the group of all characters $\chi \circ \varphi$ of G, with $\chi \in \mathsf{Y}$. Since H and $b_\Gamma G$ have isomorphic character groups, they are topologically isomorphic. □

With the construction of bG we can prove the following three approximation theorems.

(26.14) Theorem. *Let G be an infinite locally compact Abelian group; let $\chi_1, \chi_2, \ldots, \chi_m$ be elements of the character group X of G; and let ε be a positive real number. Then there is an $x \in G$ such that $x \neq e$ and $|\chi_j(x)-1| < \varepsilon$ for $j=1, 2, \ldots, m$. Dually, let $x_1, x_2, \ldots, x_m$ be elements of G and ε a positive number. There is a $\chi \in \mathsf{X}$ such that $\chi \neq 1$ and $|\chi(x_j)-1| < \varepsilon$ for $j=1, 2, \ldots, m$.*

Proof. The first statement is obvious if G is nondiscrete: there is a neighborhood U of e such that $|\chi_j(x)-1| < \varepsilon$ for $j=1, \ldots, m$ and $x \in U$; and clearly U is not equal to $\{e\}$.

Assume now that G is discrete and that the first assertion fails. Write Φ for the isomorphism Φ_{X} of (26.11). Then $\Phi(G)$ is a discrete subgroup of bG, since the topology of $\Phi(G)$ is simply that in which neighborhoods of $\Phi(e)$ are sets $\{\Phi(x): x \in G$ and $|\chi_j(x)-1| < \varepsilon$ for $j=1, 2, \ldots, m\}$. Since discrete subgroups of bG are closed (5.10), it would follow that $\Phi(G)=bG$, and that bG is an infinite discrete compact group. This is an obvious impossibility. The second statement follows from the first and (24.8). □

(26.15) Theorem. *Let G and X be as in (26.11), and let $\Gamma=\Sigma$ be any subgroup of X. Let f be any character of the group Σ_d, let $\chi_1, \chi_2, \ldots, \chi_m$ be arbitrary elements of Σ, and let ε be a positive number. Then there is a point $x \in G$ such that*

(i) $|f(\chi_j)-\chi_j(x)| < \varepsilon \quad (j=1, 2, \ldots, m)$.[1]

Proof. Consider the compact group $b_\Sigma G$. As shown in the proof of (26.12), every continuous character of $b_\Sigma G$ has the form $(t_\chi)_{\chi\in\Sigma} \rightarrow t_{\chi_0}$ for some fixed $\chi_0 \in \Sigma$. Theorem (24.8) shows that every character f of Σ_d

[1] This assertion is an abstract form of KRONECKER's approximation theorem. For various applications, see (26.19).

has the form $f(\chi)=t_\chi$ for some fixed element (t_χ) of $b_\Sigma G$. Since $b_\Sigma G$ is the closure in the Cartesian product topology of $\Phi_\Sigma(G)$, (i) follows. ☐

(26.16) Corollary. *Let G be a locally compact Abelian group. Then every character ψ of G is the pointwise limit of continuous characters, in the sense that for every $\varepsilon>0$ and every finite subset $\{x_1, x_2, \ldots, x_m\}$ of G, there is a continuous character χ of G such that*

$$|\chi(x_j)-\psi(x_j)|<\varepsilon \qquad (j=1, 2, \ldots, m).$$

Proof. Consider X as the group G of (26.15), consider G as the group Σ, and apply (26.15). ☐

Many classical theorems in the theory of equidistributed sequences can be subsumed under a general theorem concerning Haar measure on certain compact groups. It seems worthwhile to present this general theorem; some of its applications appear in (26.20).

(26.17) Theorem. *Let G be a locally compact, σ-compact, Abelian group, with Haar measure λ, and let $\{H_n\}_{n=1}^\infty$ be a sequence of sets as in* (18.14) *such that $\lim\limits_{n\to\infty} \frac{1}{\lambda(H_n)} \int\limits_{H_n} f\,d\lambda$ is the mean value of f for every continuous almost periodic function f on G. Let H be any compact Abelian group, and let φ be a continuous homomorphism of G onto a dense subgroup of H. Let μ denote normalized Haar measure on H. Then for every $g\in\mathfrak{C}(H)$, $g\circ\varphi$ is almost periodic and*

$$\text{(i)}\quad \int\limits_H g\,d\mu = \lim_{n\to\infty} \frac{1}{\lambda(H_n)} \int\limits_{H_n} (g\circ\varphi)\,d\lambda.$$

Furthermore, if U is an open subset of H such that $\mu(U)=\mu(U^-)$, then

$$\text{(ii)}\quad \mu(U) = \lim_{n\to\infty} \frac{1}{\lambda(H_n)} \int\limits_{H_n} (\xi_U\circ\varphi)\,d\lambda.$$

Proof. For a real-valued, bounded, λ-measurable function h on G, let us write $\underline{\omega}(h) = \varliminf\limits_{n\to\infty} \frac{1}{\lambda(H_n)} \int\limits_{H_n} h\,d\lambda$ and $\overline{\omega}(h) = \varlimsup\limits_{n\to\infty} \frac{1}{\lambda(H_n)} \int\limits_{H_n} h\,d\lambda$; for a complex-valued, bounded, λ-measurable h, write $\omega(h) = \lim\limits_{n\to\infty} \frac{1}{\lambda(H_n)} \int\limits_{H_n} h\,d\lambda$ if the limit exists. Let χ be a continuous character of H different from 1. Then $\chi\circ\varphi$ is a continuous character of G and, since $\varphi(G)$ is dense in H, $\chi\circ\varphi$ is different from 1. As (18.2) immediately shows, $\chi\circ\varphi$ is almost periodic. Let M be the mean value for $\mathfrak{A}_c(G)$ constructed in (18.8). It is simple to prove that $M(\chi\circ\varphi)=0$. From this and (18.14) we see that $\omega(\chi\circ\varphi)=0$. By Lemma (23.19), we have $\int\limits_H \chi\,d\mu=0$. In other words,

$$\int\limits_H \chi\,d\mu = \omega(\chi\circ\varphi)$$

for continuous characters χ on H different from 1; this equality also holds for $\chi=1$ since $\int_H 1\,d\mu=1=\omega(1\circ\varphi)$. Since $\int_H$ and ω are plainly linear and can be interchanged with uniform limits of sequences, we infer that

$$\int_H g\,d\mu=\omega(g\circ\varphi)$$

for every complex-valued function g on H that is the uniform limit of a sequence of linear combinations of continuous characters of H. The STONE-WEIERSTRASS theorem shows that these functions g are all of the continuous functions on H. This proves relation (i). Since every g in $\mathfrak{C}(H)$ is a uniform limit of a sequence of linear combinations of continuous characters, the same is true for $g\circ\varphi$, so that $g\circ\varphi$ is almost periodic (18.3).

To prove (ii), let $\{f_n\}_{n=1}^{\infty}$ and $\{g_n\}_{n=1}^{\infty}$ be sequences of functions in $\mathfrak{C}^+(H)$ such that $f_n\leqq f_{n+1}$, $g_{n+1}\leqq g_n$, $0\leqq f_n\leqq\xi_U\leqq\xi_{U^-}\leqq g_n$, and $\lim\limits_{n\to\infty}\int_H f_n\,d\mu=\lim\limits_{n\to\infty}\int_H g_n\,d\mu=\mu(U)=\mu(U^-)$. The regularity of Haar measure ensures the existence of such sequences of functions. It is obvious that

$$\omega(f_n\circ\varphi)\leqq\underline{\omega}(\xi_U\circ\varphi)\leqq\overline{\omega}(\xi_U\circ\varphi)\leqq\overline{\omega}(\xi_{U^-}\circ\varphi)\leqq\omega(g_n\circ\varphi).$$

Taking limits as $n\to\infty$ in these inequalities, and using (i), we find

$$\mu(U)\leqq\underline{\omega}(\xi_U\circ\varphi)\leqq\overline{\omega}(\xi_U\circ\varphi)\leqq\overline{\omega}(\xi_{U^-}\circ\varphi)\leqq\mu(U^-).$$

Since $\mu(U)=\mu(U^-)$ by hypothesis, (ii) follows. □

Miscellaneous theorems and examples

(26.18) Examples of automorphism groups. (a) The group Z admits just two automorphisms: ι and the mapping $x\to -x$. The group $\mathfrak{S}(Z)$ is obviously the discrete two-element group.

(b) The group $\mathfrak{S}(T)$ contains just two elements: ι and $z\to\bar{z}$ (23.31.a). The group $\mathfrak{S}(T)$ is the discrete two-element group. [This also follows from (a) and (26.9).]

(c) It is elementary to show that every automorphism $\tau\in\mathfrak{S}(R)$ has the form $\tau(x)=\alpha x$ for a nonzero real number α. It is also easy to see that the topology of $\mathfrak{S}(R)$ is the topology of $]-\infty,0[\cup]0,\infty[$ as a subspace of R. Thus $\mathfrak{S}(R)$ is topologically isomorphic with the multiplicative group of all nonzero real numbers.

(d) Let $r>1$ be an integer that is a prime power. The automorphism group $\mathfrak{S}(\Omega_r)$ is topologically isomorphic with the multiplicative group of all nonzero r-adic numbers, with its relative topology as a subspace of Ω_r. This is proved by an argument like that used to prove (c).

(e) The automorphism group $\mathfrak{S}(\Delta_p)$, where p is an arbitrary prime number, is topologically isomorphic with the multiplicative subgroup of the *ring* Δ_p consisting of all $\boldsymbol{x}=(x_0, x_1, x_2, \ldots)\in\Delta_p$ for which $x_0\neq 0$; this subgroup has its relative topology as a subspace of Δ_p. [Consider any $\boldsymbol{x}\in\Delta_p$ such that $x_0\neq 0$, and define $\tau_{\boldsymbol{x}}$ by $\tau_{\boldsymbol{x}}(\boldsymbol{v})=\boldsymbol{x}\boldsymbol{v}$ for all $\boldsymbol{v}\in\Delta_p$. It is easy to see that $\tau_{\boldsymbol{x}}\in\mathfrak{S}(\Delta_p)$. Now let τ be any element of $\mathfrak{S}(\Delta_p)$. Write $\boldsymbol{u}$ for the element $(1, 0, 0, \ldots)\in\Delta_p$. Since multiples of $\boldsymbol{u}$ are dense in Δ_p, the same is true of $\tau(\boldsymbol{u})$. Thus $\tau(\boldsymbol{u})=\boldsymbol{x}$ where $x_0\neq 0$, from which it follows that $\tau=\tau_{\boldsymbol{x}}$. Hence the mapping $\boldsymbol{x}\to\tau_{\boldsymbol{x}}$ is an isomorphism onto $\mathfrak{S}(\Delta_p)$; it is also easy to see that $\mathfrak{B}(\Delta_p, \Lambda_k)=\{\tau_{\boldsymbol{x}}:\boldsymbol{x}\in\boldsymbol{u}+\Lambda_k\}$ for $k\geqq 1$.]

(f) If G is a finitely generated discrete group, then $\mathfrak{S}(G)$ is discrete. [If F generates G, then $\mathfrak{B}(F, \{e\})=\{\iota\}$.]

(g) Let m be a positive integer. Then the automorphism group $\mathfrak{S}(Z^m)$ is isomorphic with the discrete group of all $m\times m$ matrices A having integer entries and for which $\det A=1$ or -1. For such a matrix $A=(a_{jk})_{j,k=1}^m$, the corresponding automorphism τ of Z^m is given by

$$\tau(k_1, \ldots, k_m)=\left(\sum_{j=1}^m k_j a_{1j}, \ldots, \sum_{j=1}^m k_j a_{mj}\right). \tag{1}$$

[The only nonobvious point is that $\det A=\pm 1$. The matrix A^{-1} is the matrix corresponding to τ^{-1}. Both $\det A$ and $\det A^{-1}$ are nonzero integers, and $(\det A)(\det A^{-1})=\det AA^{-1}=\det I=1$. The group $\mathfrak{S}(Z^m)$ is discrete by part (f).]

(h) Theorem (26.9) shows that the automorphism group of T^m is topologically isomorphic with $\mathfrak{S}(Z^m)$, which is described in part (g). Specifically, if $\tau\in\mathfrak{S}(Z^m)$ and $A=(a_{jk})_{j,k=1}^m$ is the corresponding matrix, then

$$\tau^{\sim}(z_1, \ldots, z_m)=(z_1^{a_{11}}z_2^{a_{21}}\cdots z_m^{a_{m1}}, \ldots, z_1^{a_{1m}}z_2^{a_{2m}}\cdots z_m^{a_{mm}}), \tag{2}$$

for $(z_1, \ldots, z_m)\in T^m$. By Theorem (26.9), all elements of $\mathfrak{S}(T^m)$ have the form (2). [Write $\tau^{\sim}(z_1, \ldots, z_m)=(w_1, \ldots, w_m)$. Then w_j is the value at $(\delta_{1j}, \delta_{2j}, \ldots, \delta_{mj})\in Z^m$ of the character of Z^m defined by $\tau^{\sim}(z_1, \ldots, z_m)$; equivalently, w_j is the value at $\tau(\delta_{1j}, \delta_{2j}, \ldots, \delta_{mj})=(a_{1j}, a_{2j}, \ldots, a_{mj})$ of the character of Z^m defined by $(z_1, \ldots, z_m)$. It follows that $w_j=z_1^{a_{1j}}\cdots z_m^{a_{mj}}$.]

(i) The automorphism group of R^m is topologically isomorphic with $\mathfrak{GL}(R, m)$ for every positive integer m. This is shown as in (g); the topology of $\mathfrak{S}(R^m)$ is easily shown to be that of $\mathfrak{GL}(R, m)$.

(j) The situation alters radically when we consider the discrete group $Z^{\mathfrak{m}*}$ for an infinite cardinal number $\mathfrak{m}$. It is trivial that $Z^{\mathfrak{m}*}$ contains a maximal independent set B of cardinal number $\mathfrak{m}$, and that every element of $Z^{\mathfrak{m}*}$ can be written in exactly one way as $k_1b_1+\cdots+k_sb_s$ for $k_j\in Z$ and

$b_j \in B$ [for an obvious reason we write $Z^{\mathfrak{m}*}$ additively]. An element $\tau \in \mathfrak{S}(Z^{\mathfrak{m}*})$ is thus completely determined by its values on B, and of these all that can be said is that $\tau(B)$ is another maximal independent set in $Z^{\mathfrak{m}*}$, in terms of which every element of $Z^{\mathfrak{m}*}$ can be uniquely expressed. Clearly neighborhoods $\mathfrak{V}(\{b_1, \ldots, b_m\}, 0)$ are an open basis at ι in $\mathfrak{S}(Z^{\mathfrak{m}*})$, and each $\mathfrak{V}(\{b_1, \ldots, b_m\}, 0)$ is an open [and hence closed] subgroup of $\mathfrak{S}(Z^{\mathfrak{m}*})$. If $\mathfrak{S}(Z^{\mathfrak{m}*})$ were locally compact, some $\mathfrak{V}(\{b_1, \ldots, b_m\}, 0)$ would be compact. Let b_{m+1} be any element of B different from $b_1, \ldots, b_m$. If $\mathfrak{V}(\{b_1, \ldots, b_m\}, 0)$ were compact, we could choose $\tau^{(1)}, \ldots, \tau^{(n)}$ in $\mathfrak{V}(\{b_1, \ldots, b_m\}, 0)$ such that $\bigcup_{k=1}^{n} \tau^{(k)} \mathfrak{V}(\{b_{m+1}\}, 0) \supset \mathfrak{V}(\{b_1, \ldots, b_m\}, 0)$. For an automorphism σ in $\bigcup_{k=1}^{n} \tau^{(k)} \mathfrak{V}(\{b_{m+1}\}, 0)$, the element $\sigma(b_{m+1})$ is one of the elements $\tau^{(1)}(b_{m+1}), \ldots, \tau^{(n)}(b_{m+1})$. There is evidently an element $\sigma \in \mathfrak{V}(\{b_1, \ldots, b_m\}, 0)$ such that $\sigma(b_{m+1})$ lies outside of any preassigned set of cardinal number less than $\mathfrak{m}$. Hence $\mathfrak{V}(\{b_1, \ldots, b_m\}, 0)$ is not compact, and $\mathfrak{S}(Z^{\mathfrak{m}*})$ is not locally compact.

(k) For an infinite cardinal $\mathfrak{m}$, the compact Abelian group $T^{\mathfrak{m}}$ has an automorphism group that is not locally compact. [This follows from part (j) and (26.9).]

(26.19) Approximation theorems (Hewitt and Zuckerman [1]). By using (26.15) and (26.16) together with the computations of characters given in §25, we obtain sharpened forms of Kronecker's approximation theorems as well as new approximation theorems. The following examples, while obviously not exhaustive, illustrate the method adequately.

(a) Let a and b be positive integers; let $h_1, h_2, \ldots, h_b$ be real numbers that are linearly independent over Q; let $\eta_j = \sum_{k=1}^{b} \frac{l_{j,k}}{m_{j,k}!} h_k$ $(j = 1, 2, \ldots, a)$, where the l's are integers and the m's are positive integers. Let $\xi_1, \ldots, \xi_b$ be numbers in $[0, 1[$ and $\boldsymbol{x}^{(1)}, \ldots, \boldsymbol{x}^{(b)}$ be elements of $\Delta_{\boldsymbol{a}}$ $[\boldsymbol{a} = (2, 3, 4, 5, \ldots)]$. Let ε be a positive number. There is a real number t such that

$$\text{(i)} \quad \left| \eta_j t - \left\{ \sum_{k=1}^{b} \frac{l_{j,k}}{m_{j,k}!} \left(\xi_k - x_0^{(k)} - 2!\, x_1^{(k)} - \cdots \right.\right.\right. \\ \left.\left.\left. - (m_{j,k} - 1)!\, x_{m_{j,k}-2}^{(k)} \right) \right\} \right| < \varepsilon \pmod 1$$

for $j = 1, 2, \ldots, a$.[1] [Here we use (26.16). For the group G of (26.16), take R. Computing an arbitrary character of R by (25.6) and (25.4), we obtain (i).]

(b) One form of the classical approximation theorem of Kronecker is obtained by putting $\eta_j = h_j$ [$j = 1, \ldots, a$ and $a = b$] and $\boldsymbol{x}^{(1)} = \cdots = \boldsymbol{x}^{(b)} = \boldsymbol{0}$ in (a). Thus we have

[1] For a real number α, the expression $|\alpha| < \varepsilon \pmod 1$ means that $|\alpha - k| < \varepsilon$ for some integer k.

(ii) $|h_j t - \xi_j| < \varepsilon \pmod 1 \quad (j = 1, 2, \ldots, a)$,

where $\xi_1, \ldots, \xi_a$ are arbitrary real numbers and $h_1, \ldots, h_a$ are rationally independent real numbers.

(c) Let $h_1, h_2, \ldots, h_b$ be real numbers such that $1, h_1, h_2, \ldots, h_b$ are linearly independent over Q, and let $h_0 = 1$. Let $\eta_j = \sum_{k=0}^{b} \frac{l_{j,k}}{m_{j,k}!} h_k$ $(j = 1, 2, \ldots, a)$, where the l's are integers and the m's are positive integers. Let $\boldsymbol{a} = (2, 3, 4, 5, \ldots)$. Let $\boldsymbol{x}^{(0)}, \boldsymbol{x}^{(1)}, \ldots, \boldsymbol{x}^{(b)}$ be elements of $\Delta_{\boldsymbol{a}}$; let $\xi_1, \xi_2, \ldots, \xi_b$ be numbers in $[0, 1[$; let $\xi_0 = 0$; and let ε be a positive real number. There is then an integer n such that

$$\text{(iii)} \quad \Bigg| \eta_j n - \Bigg\{ \sum_{k=0}^{b} \frac{l_{j,k}}{m_{j,k}!} \big(\xi_k - x_0^{(k)} - 2!\, x_1^{(k)} - \cdots \\ - (m_{j,k} - 1)!\, x_{m_{j,k}-2}^{(k)} \big) \Bigg\} \Bigg| < \varepsilon \pmod 1$$

for $j = 1, 2, \ldots, a$. [Here we use (26.16) and (25.7). The expression $\{\cdots\}$ in (iii) is obtained from the value of an arbitrary character of the group T computed at $\exp[2\pi i \eta_j]$, and $\eta_j n$ from the value of a *continuous* character of T computed at $\exp[2\pi i \eta_j]$.]

(d) We obtain another version of KRONECKER's approximation theorem by putting $\eta_j = h_j$ [$j = 1, \ldots, a$ and $a = b$] and $\boldsymbol{x}^{(0)} = \cdots = \boldsymbol{x}^{(b)} = \boldsymbol{0}$ in (c). Thus we have

(iv) $|h_j n - \xi_j| < \varepsilon \pmod 1 \quad (j = 1, 2, \ldots, a)$,

for some integer n, where $\xi_1, \ldots, \xi_a$ are arbitrary real numbers and $h_1, \ldots, h_a$ are real numbers such that $1, h_1, \ldots, h_a$ are linearly independent over Q.

(e) Let $\boldsymbol{b}^{(j)} = (\beta_1^{(j)}, \ldots, \beta_n^{(j)})$ $(j = 1, 2, \ldots, m)$ be elements of R^n such that $\boldsymbol{b}^{(1)}, \ldots, \boldsymbol{b}^{(m)}$, $(1, 0, \ldots, 0_{(n)})$, $(0, 1, 0, \ldots, 0_{(n)})$, $\ldots$, $(0, 0, \ldots, 0, 1_{(n)})$ are linearly independent over Q in the vector space R^n. Let $\alpha_1, \ldots, \alpha_m$ be any real numbers and ε a positive number. There is an element $(k_1, k_2, \ldots, k_n) \in Z^n$ such that

(v) $\left| \alpha_j - \sum_{l=1}^{n} k_l \beta_l^{(j)} \right| < \varepsilon \pmod 1 \quad (j = 1, 2, \ldots, m)$.

[Apply (26.16) to T^n and $(T_d)^n$. To obtain the character on $(T_d)^n$, regard $(T_d)^n$ as $(R_d)^n/Z^n$ and find a character ψ of $(R_d)^n$ such that $\psi(\boldsymbol{b}^{(j)}) = \exp[2\pi i \alpha_j]$ for $j = 1, \ldots, m$, and $\psi(Z^n) = 1$. To obtain characters on T^n, use (23.18).]

(f) Let E be a topological linear space over R, let $f_1, f_2, \ldots, f_m$ be continuous linear functionals on E that are linearly independent over Q, let $\alpha_1, \ldots, \alpha_m$ be any real numbers, and let ε be a positive number. There is an $x \in E$ such that

(vi) $|\alpha_j - f_j(x)| < \varepsilon \pmod 1 \quad (j = 1, 2, \ldots, m)$.

[Use (26.15) with E as the Abelian group G. As the group Σ, take the group of all characters $x \to \exp[2\pi i f(x)]$ for $f \in E^*$. If $f_1, f_2, \ldots, f_m$ are elements of E^* independent over Q, then some linear functional F on E^* has the property that $F(f_j) = \alpha_j$. Thus $\exp[2\pi i F(f_j)] = \exp[2\pi i \alpha_j]$; (vi) follows at once from (26.15).]

(g) Let $\boldsymbol{x}^{(1)}, \boldsymbol{x}^{(2)}, \ldots, \boldsymbol{x}^{(a)}$ be elements of the group Δ_p [p is any prime] that are independent:

$$\sum_{j=1}^{a} m_j \boldsymbol{x}^{(j)} = \boldsymbol{0} \quad \text{and} \quad m_j \in Z \quad \text{imply} \quad m_1 = \cdots = m_a = 0.$$

Let $\alpha_1, \alpha_2, \ldots, \alpha_a$ be any real numbers, and let ε be any positive real number. There are positive integers l and n such that

$$\text{(vii)} \quad \left| \alpha_j - l \sum_{k=0}^{n-1} \frac{x_k^{(j)}}{p^{n-k}} \right| < \varepsilon \pmod 1 \qquad (j = 1, 2, \ldots, a).$$

[The p-adic integers $\boldsymbol{x}^{(1)}, \boldsymbol{x}^{(2)}, \ldots, \boldsymbol{x}^{(a)}$ generate a free subgroup of Δ_p, and hence a character of Δ_p can assume arbitrary values of absolute value 1 at $\boldsymbol{x}^{(1)}, \ldots, \boldsymbol{x}^{(a)}$. Continuous characters of Δ_p are given by (25.2.2). Combining these observations with (26.16), we obtain (vii).]

(h) If ψ is any character of Q and $r_1, r_2, \ldots, r_m$ are any elements in Q, there is a continuous character χ of Q such that $\chi(r_j) = \psi(r_j)$ for $j = 1, \ldots, m$. [Although this is a strong approximation theorem, it is really quite elementary and does not depend upon (26.15) or (26.16). Let $\boldsymbol{a} = (2, 3, 4, \ldots)$ and let $(\xi, \boldsymbol{x})$ be the element of $\Sigma_{\boldsymbol{a}} = [0, 1[\times \Delta_{\boldsymbol{a}}$ that corresponds to ψ as in (25.4). Let n be a positive integer such that $n! r_j$ is an integer for $j = 1, \ldots, m$. Let $t = \xi - (x_0 + 2! x_1 + \cdots + (n-1)! x_{n-2})$; by (25.4.2) we have $\psi(r_j) = \exp[2\pi i r_j t]$. Define χ by $\chi(r) = \exp[2\pi i r t]$ for all $r \in Q$.]

(26.20) Examples on equidistributions. Theorem (26.17) has many applications. The following are a small sample.

(a) Let H be a compact monothetic group, with $\{x^n\}_{n=-\infty}^{\infty}$ dense in H. Let U be any open subset of H such that $\mu(U) = \mu(U^-)$ [μ is normalized Haar measure on H]. Then

$$\text{(i)} \quad \mu(U) = \lim_{m \to \infty} \frac{1}{2m+1} \sum_{j=-m}^{m} \xi_U(x^j) = \lim_{m \to \infty} \frac{1}{m-a+1} \sum_{j=a}^{m} \xi_U(x^j)$$

$$= \lim_{m \to \infty} \frac{1}{m+b+1} \sum_{j=-m}^{b} \xi_U(x^j).$$

[The first equality follows from (26.17) by taking $G = Z$, $\varphi(n) = x^n$, and $H_m = \{-m, -m+1, \ldots, 0, \ldots, m-1, m\}$. The proofs of the remaining equalities are similar; see (18.15.b).]

(b) Let H be a compact solenoidal group, and let φ be a continuous homomorphism of R onto a dense subgroup of H. Let μ and U be as in (a).

Then

(ii) $$\mu(U) = \lim_{T\to\infty} \frac{1}{2T}\int_{-T}^{T} \xi_U(\varphi(t))\,dt = \lim_{T\to\infty} \frac{1}{T-a}\int_{a}^{T} \xi_U(\varphi(t))\,dt = \lim_{T\to\infty} \frac{1}{T+b}\int_{-T}^{b} \xi_U(\varphi(t))\,dt.$$

[The proof is similar to the proof of (a); use (18.15.a).]

(c) Let $\alpha_1, \dots, \alpha_m$ be irrational real numbers. Let U be an open subset of T^m such that $\mu(U) = \mu(U^-)$ [μ denotes normalized Haar measure on T^m]. For every positive integer n, let $A(n)$ be the number of distinct m-tuples $(k_1, \dots, k_m)$ of integers such that $|k_j| \leqq n$ for $j = 1, \dots, m$ and $(\exp[2\pi i \alpha_1 k_1], \dots, \exp[2\pi i \alpha_m k_m])$ lies in U. Then

(iii) $$\lim_{n\to\infty} \frac{A(n)}{(2n+1)^m} = \mu(U).$$

[The mapping $(k_1, \dots, k_m) \to (\exp[2\pi i \alpha_1 k_1], \dots, \exp[2\pi i \alpha_m k_m])$ is a [continuous] homomorphism of Z^m onto a dense subgroup of T^m. Let H_n be the set of all m-tuples $(k_1, \dots, k_m)$ of integers such that $|k_j| \leqq n$ for $j = 1, \dots, m$. Since $\{H_n\}_{n=1}^{\infty}$ satisfies (18.10.i), (26.17) implies (iii).]

(d) Let $\alpha_1, \dots, \alpha_m$, U, and μ be as in (c). Suppose further that the set $\{\alpha_1, \dots, \alpha_m, 1\}$ is linearly independent over Q. For positive integers n, let $B(n)$ denote the number of integers k such that $|k| \leqq n$ and $(\exp[2\pi i \alpha_1 k], \dots, \exp[2\pi i \alpha_m k])$ lies in U. Then

(iv) $$\lim_{n\to\infty} \frac{B(n)}{2n+1} = \mu(U).$$

[The set $\{(\exp[2\pi i \alpha_1 k], \dots, \exp[2\pi i \alpha_m k]) : k \in Z\}$ is a dense subgroup of T^m by virtue of (26.19.d). Now apply part (a).]

(26.21) The modular function for $\mathfrak{S}(G)$ (BRACONNIER [1], pp. 75 and 76). Let G be any locally compact group with Haar measure λ. The modular function $\tau \to \Delta(\tau)$ defined in (15.26) is a *continuous* homomorphism of $\mathfrak{S}(G)$ into the multiplicative group $]0, \infty[$. [In view of (15.26), we need to prove only continuity. Let $f \in \mathfrak{C}_{00}^{+}(G)$ be such that $\int_G f\,d\lambda = 1$, and let F be a compact subset of G such that $f(F') = 0$. Let V be a fixed neighborhood of e with compact closure. For $\varepsilon > 0$, let U be a neighborhood of e such that $U \subset V$ and $|f(y) - f(z)| < \varepsilon$ if $yz^{-1} \in U$ (15.4). Now let τ be any automorphism in $\mathfrak{B}(U^- F, U)$. Then $\tau^{-1}(F) \subset U^- F$, and $|f(\tau(x)) - f(x)| < \varepsilon$ if $x \in U^- F$. Therefore we have

$$|\Delta(\tau) - 1| = |\Delta(\tau) - 1| \int_G f\,d\lambda = \left|\int_{\tau^{-1}(F)} f(\tau(x))\,dx - \int_F f(x)\,dx\right|$$
$$\leqq \int_{U^- F} |f(\tau(x)) - f(x)|\,dx \leqq \int_{U^- F} \varepsilon\,dx = \varepsilon\,\lambda(U^- F) \leqq \varepsilon\,\lambda(V^- F).]$$

Notes

Definition (26.3), (26.5) [barring changes needed to show that $\mathfrak{B}(F, U)$ is open], (26.8), and (26.9) are taken from BRACONNIER [1], Ch. IV.

Theorem (26.10) generalizes IWASAWA [2], Theorem 4, p. 515. Automorphism groups were studied earlier by ABE [1], but with a topology different from that defined in (26.3). A detailed investigation of homeomorphism groups of locally compact Hausdorff spaces has been made by ARENS [1]. For results related to (26.10), see HOFMANN [1].

We note also the following interesting and apparently difficult theorem of IWASAWA [2]. Let G be a compact group and $\mathfrak{J}(G)$ the group of all inner automorphisms of G. Then $\mathfrak{J}(G)$ is a compact normal subgroup of $\mathfrak{S}(G)$, and $\mathfrak{S}(G)/\mathfrak{J}(G)$ is a 0-dimensional group.

The constructions and results of (26.11)—(26.13) are due to ANZAI and KAKUTANI [1]. Theorem (26.14) seems to be new; (26.15) and (26.16) are from HEWITT and ZUCKERMAN [1]. Theorem (26.17) appears to be new.

Appendix A

Abelian groups

In this appendix, we set forth part of the algebraic theory of Abelian groups. We limit ourselves to those parts of the theory that are needed for our study of the structure of locally compact Abelian groups [§9] and the duality theory of locally compact Abelian groups [§§23—25]. A more complete treatment appears in KAPLANSKY [1] and a far more complete treatment in FUCHS [1].

We will use multiplicative notation for the group operation except in dealing with Z, Q, R, $Z(n)$, and a few linear spaces. The exceptions will be specifically pointed out as they occur. Throughout this appendix, "group" means "Abelian group" and the symbol "G" will be used to denote an arbitrary Abelian group.

(A.1) Definition. If every element of a group G has finite order, then G is called a *torsion group*. If every element of G except e has infinite order, then G is called a *torsion-free group*. A torsion group G is said to be *p-primary* [for a prime p] if the order of every element is a power of p. If G is a torsion group and if the order of every element is less than some fixed positive integer, then G is said to be of *bounded order*.

(A.2) Definition. Let G be a group. For $n=2, 3, \ldots$, let f_n be the mapping of G into G defined by $f_n(x)=x^n$. We define $G^{(n)}=f_n(G)$ and $G_{(n)}=f_n^{-1}(e)$.

We first prove an elementary but very useful theorem on the decomposition of torsion groups.

(A.3) Theorem. *Let G be a torsion group. For each prime p, let G_p be the subset of G consisting of elements whose order is a power of p. Then* *each G_p is a p-primary group and G is isomorphic with* $\mathsf{P}^*_{p\in P} G_p$ *[P is the set of all primes].*

Proof. Obviously each G_p is a p-primary group. To show that G is isomorphic with $\mathsf{P}^*_{p\in P} G_p$, we apply (2.5). Consider any element $x\in G$ having order n. Then $n=p_1^{r_1}p_2^{r_2}\cdots p_k^{r_k}$, where $p_1, p_2, \ldots, p_k$ are distinct primes. If $n_j=n/p_j^{r_j}$, then x^{n_j} has order $p_j^{r_j}$ and belongs to G_{p_j}. Since $n_1, n_2, \ldots, n_k$ are relatively prime, there are integers $a_1, a_2, \ldots, a_k$ such that $a_1n_1+a_2n_2+\cdots+a_kn_k=1$. Then

$$x=(x^{n_1})^{a_1}\cdot(x^{n_2})^{a_2}\cdots(x^{n_k})^{a_k}\in G_{p_1}\cdot G_{p_2}\cdots G_{p_k},$$

and (2.5.i) is proved.

Suppose now that $x\in G_p$ and that $x=x_1x_2\cdots x_n$ where $x_j\in G_{p_j}$ and the p_j are all different from p. Then x has order a power of p and $x_1x_2\cdots x_n$ has order $p_1^{r_1}p_2^{r_2}\cdots p_n^{r_n}$ for some integers $r_1, r_2, \ldots, r_n$. It follows that x must be e; so that (2.5.ii) is established. □

(A.4) Theorem. *Every group G contains a largest torsion subgroup F; the equality $F=\bigcup_{n=1}^{\infty} G_{(n)}$ obtains; and G/F is torsion-free.*

Proof. Let F be the set of elements in G having finite order; then F is plainly the largest torsion subgroup of G and $F=\bigcup_{n=1}^{\infty} G_{(n)}$. Suppose that $(xF)^n=x^nF=F$ for some n. Then $x^n\in F$ and $x^{nm}=e$ for some m; *i.e.*, $x\in F$ and $xF=F$. □

We next introduce a very important if very special class of groups.

(A.5) Definition. A group G is said to be *divisible* if $G^{(n)}=G$ for $n=2, 3, \ldots$. In other words, given $x\in G$ and n, there is a $y\in G$ such that $x=y^n$. If $\{e\}$ is the only divisible subgroup of a group G, then G is said to be *reduced.*

A direct product or weak direct product of groups is divisible if and only if each factor is divisible. A homomorphic image of a divisible group is again divisible. The group Q and the groups $Z(p^\infty)$ are divisible. We will prove in (A.14) that every divisible group is a weak direct product of these groups.

(A.6) Theorem. *Every group G contains a largest divisible subgroup*[1] *D.*

Proof. Let D be the smallest group containing all divisible subgroups of G. If x is in D, then $x=x_1\cdots x_m$ where x_j belongs to a divisible group D_j for $j=1, \ldots, m$. For $n>1$, we have $x_j=y_j^n$ for some $y_j\in D_j$. Then $x_1\cdots x_m=(y_1\cdots y_m)^n$ and $y_1\cdots y_m\in D$; that is, D is divisible. □

---

[1] By *divisible subgroup* we of course mean a subgroup that is divisible when looked at by itself.

The next theorem has several useful applications in the theory of topological groups.

(A.7) Theorem. *Let H be a subgroup of G. Every homomorphism φ of H into a divisible group D can be extended to a homomorphism of G into D.*

Proof. By ZORN'S lemma [or by a well-ordering argument], it suffices to choose any $x \notin H$ and extend φ to the group

$$H_0 = \{x^n h : h \in H \text{ and } n \in Z\}.$$

If $x^n \notin H$ for all $n \geqq 2$, let $\psi(x^n h) = \varphi(h)$ for $x^n h \in H_0$. Then ψ is a well-defined homomorphism that extends φ. If $x^n \in H$ for some $n \geqq 2$, we take k to be the least such integer. There is an element $z \in D$ such that $z^k = \varphi(x^k)$. Defining $\psi(x^n h)$ to be $z^n \varphi(h)$ for $x^n h \in H_0$, we again obtain a well-defined homomorphism that extends φ. □

(A.8) Theorem. *Let H be a divisible subgroup of G and let A be a subgroup of G such that $H \cap A = \{e\}$. Then there is a subgroup A_0 of G such that $H \cap A_0 = \{e\}$, $A \subset A_0$, and $HA_0 = G$;* i.e., *G is isomorphic with $H \times A_0$. In particular, every divisible subgroup of a group is a direct factor of the group.*

Proof. Since $H \cap A = \{e\}$, HA is isomorphic with $H \times A$. Let φ denote the projection of HA onto H; then $\varphi(x) = x$ for $x \in H$ and $A = \varphi^{-1}(e)$. By (A.7), φ can be extended to a homomorphism ψ of G onto H; note that $\psi \circ \psi = \psi$. Let $A_0 = \psi^{-1}(e)$. Then $A_0 \cap H = \{e\}$ and $A \subset A_0$. Moreover, if $x \in G$, then $x = [x\psi(x^{-1})]\,\psi(x)$; and we have $x\psi(x^{-1}) \in A_0$ and $\psi(x) \in H$. □

Before completing our study of divisible groups, we will define and study the notion of *rank*. The following simple fact is useful for this purpose.

(A.9) Theorem. *Let p be a prime integer and let $\{0, 1, \ldots, p-1\}$ be a realization of the cyclic group $Z(p)$. Define multiplication in $Z(p)$ as ordinary multiplication modulo p. Then $Z(p)$ is a field. Let G be a group such that every element except e has order p, and regard G as an* additive *group. For $a \in \{0, 1, \ldots, p-1\}$ and $x \in G$, we define ax to be, as usual, the sum $x + x + \cdots + x$ [a times]. Then G is a linear space over the field $Z(p)$.*

The proof of this theorem is elementary, and we omit it. The order restriction on G is needed in order to prove that

$$[(a+b) \bmod p] \cdot x = ax + bx$$

and

$$[(ab) \bmod p] \cdot x = a(bx)$$

for $a, b \in \{0, 1, \ldots, p-1\}$ and $x \in G$.

(A.10) Definition. A finite subset $\{x_1, \ldots, x_k\}$ of a group G is said to be *independent* if it does not contain e and if $x_1^{n_1} \cdots x_k^{n_k} = e$ implies that

$x_1^{n_1} = \cdots = x_k^{n_k} = e$ [each n_j is an integer]. An infinite subset L of G is said to be *independent* if every finite subset of L is independent[1]. If an independent set L generates the group G, then L is called a *basis of* G.

The following rather cumbersome theorem will allow us to define various notions of rank.

(A.11) Theorem. *Let G be any group. Let $\mathscr{A}$ be the family of all independent sets L in G consisting only of elements whose order is infinite or a power of a prime and such that L is maximal with respect to these properties. Similarly, let $\mathscr{A}_0$ [$\mathscr{A}_p$] be the family of independent sets L_0 [L_p] in G consisting of elements whose order is infinite [a power of p] and such that L_0 [L_p] is maximal with respect to these properties; p denotes any prime integer. Then $\mathscr{A}$, $\mathscr{A}_0$, and $\mathscr{A}_p$ are nonvoid. Every set L in $\mathscr{A}$ is the [disjoint] union $L_0 \cup \left(\bigcup_{p \in P} L_p\right)$ of sets in $\mathscr{A}_0$ and $\mathscr{A}_p$ [P denotes the set of all primes]. All the sets in $\mathscr{A}$ have the same cardinal number; all the sets in $\mathscr{A}_0$ [$\mathscr{A}_p$] have the same cardinal number.*

Proof. (I) An elementary ZORN's lemma argument shows that $\mathscr{A}$, $\mathscr{A}_0$, and $\mathscr{A}_p$ are all nonvoid, since the void set is independent.

(II) Consider a set L in $\mathscr{A}$, and let L_0 [L_p] consist of the elements x in L having infinite order [order a power of p]. If L_0 does not belong to $\mathscr{A}_0$, there must exist an element $a \in G \cap L_0'$ having infinite order and for which $L_0 \cup \{a\}$ is independent. Then $L \cup \{a\}$ is easily seen to be independent, contradicting the hypothesis that $L \in \mathscr{A}$. It is a little more delicate to show that L_p belongs to $\mathscr{A}_p$. If not, there is an $a \in G$ having order p^r such that $L_p \cup \{a\}$ is independent and $L \cup \{a\}$ is dependent. Then $a^n s_0 s_p t_p = e$ where $0 < n < p^r$, s_0 is a product of powers of elements from L_0, s_p is a product of powers of elements from L_p, and t_p is a product of powers of elements from $\bigcup_{q \neq p} L_q$. There is a positive integer m relatively prime to p such that $t_p^m = e$. Thus $a^{nm} s_0^m s_p^m = e$ and $a^{nm} \neq e$. Since $L_p \cup \{a\}$ is independent, s_0 cannot be equal to e. Now for some power p^k, we have $s_0^{mp^k} = (a^{nm} s_0^m s_p^m)^{p^k} = e$ so that L_0 is not independent; this is a contradiction.

(III) In view of part (II), sets in $\mathscr{A}$ have the same cardinal number, if the corresponding assertions for $\mathscr{A}_0$ and $\mathscr{A}_p$ are true. Let us prove these facts; we deal first with $\mathscr{A}_p$. For this purpose, we may suppose with no loss of generality that G itself is p-primary; then we want to show that all maximal independent sets in G have the same cardinal number. We will also write G as an *additive* group. By (A.9), the subgroup $G_{(p)} = \{x \in G : px = 0\}$ is a linear space over $Z(p)$. Since all bases of $G_{(p)}$ have the same cardinal number [this is a well-known fact in the theory of linear spaces], it suffices to prove the following.

[1] It is also convenient to consider the void set as an independent subset of G.

Let L be a maximal independent set in G. For $x \in L$, let p^{l_x} be the order of x. Then $\widetilde{L} = \{p^{l_x - 1}x : x \in L\}$ is a basis for $G_{(p)}$. It is obvious that $\widetilde{L}$ is linearly independent in $G_{(p)}$. If $\widetilde{L}$ is not maximal, there is an element $a \in G_{(p)} \cap (\widetilde{L})'$ such that $\widetilde{L} \cup \{a\}$ is linearly independent. Since $L \cup \{a\}$ is dependent in G, we have

$$n a = n_1 x_1 + \cdots + n_k x_k \tag{1}$$

for some integers n_j and n where $0 < n < p$. Then

$$0 = p n a = p n_1 x_1 + \cdots + p n_k x_k;$$

hence $p n_j x_j = 0$ for each j and $p n_j = m_j p^{l_{x_j}}$ for certain integers m_j. In view of this, relation (1) becomes

$$n a = m_1 p^{l_{x_1} - 1} x_1 + \cdots + m_k p^{l_{x_k} - 1} x_k;$$

since $n a \neq 0$, this violates the linear independence of $\widetilde{L} \cup \{a\}$ in $G_{(p)}$.

(IV) We now show that all sets in $\mathscr{A}_0$ have the same cardinal number in the case that G is torsion-free. We return to multiplicative notation. Suppose first that G contains a finite maximal independent set $L = \{x_1, x_2, \ldots, x_k\}$. Let $\{y_1, y_2, \ldots, y_r\}$ be any finite independent subset of G. We will prove that $r \leqq k$. The maximality of L implies that $y_1^n = x_1^{m_1} \cdots x_k^{m_k}$, where $n \neq 0$. Since G is torsion-free, not all m_j are 0; by relabelling if necessary, we may suppose that $m_1 \neq 0$. It is then easy to see that $\{y_1, x_2, \ldots, x_k\}$ is a maximal independent set. Thus if $r \geqq 2$, we have $y_2^t = y_1^{s_1} x_2^{s_2} \cdots x_k^{s_k}$, where at least one of $s_2, \ldots, s_k$ is not zero. Let us say that $s_2 \neq 0$; then $\{y_1, y_2, x_3, \ldots, x_k\}$ is a maximal independent set. The process continues until we run out of y's or x's. If $r > k$, we find that $\{y_1, \ldots, y_k\}$ is a maximal independent set properly contained in the independent set $\{y_1, \ldots, y_r\}$. This contradiction proves (IV) if G contains a finite maximal independent set.

We complete the proof of (IV) by showing that any infinite set L in $\mathscr{A}_0$ has cardinal number $\overline{\overline{G}}$. It is trivial that $\overline{\overline{L}} \leqq \overline{\overline{G}}$. If $z \in G$, then $z^n = x_1^{n_1} \cdots x_k^{n_k}$ for some integers $n, n_1, \ldots, n_k$ where $n \neq 0$ and $x_1, \ldots, x_k \in L$. We associate with z the element $\varphi(z) = (s_x)_{x \in L}$ in $Q^{\overline{\overline{L}}*}$ such that $s_{x_j} = \frac{n_j}{n}$ for $j = 1, \ldots, k$ and $s_x = 0$ for $x \notin \{x_1, \ldots, x_k\}$. If $z \neq w$, then we have $\varphi(z) \neq \varphi(w)$. For if $z^n = x_1^{n_1} \cdots x_k^{n_k}$ and $w^m = x_1^{m_1} \cdots x_k^{m_k}$, where $m_j n = m n_j$ for $j = 1, \ldots, k$, then $z^{mn} = w^{mn}$ and so $z = w$, since G is torsion-free. Similarly one sees that $\varphi(z)$ is uniquely determined by z [although this fact is not essential for the proof]. Consequently, φ is a one-to-one mapping of G into $Q^{\overline{\overline{L}}*}$; and so we find that $\overline{\overline{G}} \leqq \overline{\overline{L}} \cdot \aleph_0 = \overline{\overline{L}}$.

(V) Now let G be any group. Let F be the torsion subgroup of G (A.4) and let L be in $\mathscr{A}_0$. We will show that $\{xF : x \in L\}$ is a maximal independent set in the torsion-free group G/F; part (IV) then implies that all elements in $\mathscr{A}_0$ have the same cardinal number. Suppose that

$$(x_1F)^{n_1}\cdots(x_kF)^{n_k}=F$$

for $x_1, \ldots, x_k \in L$. Then $x_1^{n_1}\cdots x_k^{n_k}$ belongs to F and hence for some integer $m \geqq 1$, we have $(x_1^{n_1}\cdots x_k^{n_k})^m=e$. This implies that $x_j^{n_j m}=e$ and thus $n_j=0$ for $j=1, \ldots, k$. Hence $\{xF: x\in L\}$ is independent in G/F. If it is not maximal, then there is an a in $G\cap L'$ having infinite order such that $\{xF: x\in L\}\cup\{aF\}$ is independent. However, $a^n=x_1^{m_1}\cdots x_s^{m_s}$ for some $x_1, \ldots, x_s\in L$ and $n\neq 0$, so that $(aF)^n=(x_1F)^{m_1}\cdots(x_sF)^{m_s}$, contrary to the choice of a. □

(A.12) Definition. Let G, $\mathscr{A}$, $\mathscr{A}_0$, and $\mathscr{A}_p$ be as in (A.11). The cardinal number of any set in $\mathscr{A}$ is called the *rank of* G and is denoted by $r(G)$. Similarly, the cardinal number of any set in $\mathscr{A}_0$ [$\mathscr{A}_p$] is called the *torsion-free rank* [the *p-rank*] *of* G and is denoted by $r_0(G)$ [$r_p(G)$].

(A.13) Theorem. *Let F be the torsion subgroup of G* (A.4). *Then $r_0(G)=r_0(G/F)=r(G/F)$ and $r_p(G)=r_p(F)$ for all primes p.*

Proof. The equalities $r_p(G)=r_p(F)$ and $r_0(G/F)=r(G/F)$ follow immediately from the definitions. The equality $r_0(G)=r(G/F)$ is proved in part (V) of the proof of (A.11). □

We now identify all divisible groups.

(A.14) Theorem. *Let G be a divisible group. Then G is isomorphic with*

$$Q^{r_0(G)*}\times \underset{p\in P}{\mathsf{P}^*}\left(Z(p^\infty)^{r_p(G)*}\right),$$

where P denotes the set of all primes.

Proof. Let F be the torsion subgroup of G. It is obvious that F is divisible. Theorem (A.8) shows that G is isomorphic with $F\times H$, where H is isomorphic with G/F.

In order to apply linear space theory, we write H as an *additive* group. Theorem (A.4) states that G/F, and hence H, is torsion-free and, being a homomorphic image of G, it is also divisible. Let x be in H and n a positive integer. Then $x=ny$ for some $y\in H$. The element y is unique since $ny=ny_0$ implies that $n(y-y_0)=0$ and hence $y=y_0$. In other words, $\frac{1}{n}x$ is uniquely defined, and it follows that rx is uniquely defined for all $r\in Q$ and $x\in H$. It is now easy to show that H is a linear space over the field Q. As such, H contains a basis having cardinal number $\mathfrak{n}_0$, say. Thus H is isomorphic with $Q^{\mathfrak{n}_0 *}$.

The divisible torsion group F is isomorphic with $\underset{p\in P}{\mathsf{P}^*} G_p$, where each G_p is a divisible p-primary group (A.3). Fix p and consider a family of subgroups $\{A_\iota\}_{\iota\in I}$ of G_p such that each A_ι is isomorphic with $Z(p^\infty)$ and

$$\left.\begin{array}{l}\text{if } x_1 x_2 \cdots x_k = e \text{ where } x_j \in A_{\iota_j} \text{ and the } \iota_j \text{ are distinct,} \\ \text{then } x_j = e \text{ for } j = 1, \ldots, k.\end{array}\right\} \quad (1)$$

By virtue of (2.5), the smallest subgroup of G_p that contains all of the A_ι is isomorphic with a weak direct product of groups $Z(p^\infty)$. Let $\mathscr{A}$ denote the family of all such families of subgroups of G_p. It is easy to see that ZORN's lemma applies: that is, there is a maximal family $\{B_\iota\}_{\iota\in I}$ in $\mathscr{A}$. Let B be the smallest subgroup of G_p that contains all of the B_ι's. Then B is the weak direct product of $Z(p^\infty)$'s and is divisible. If $B \subsetneqq G_p$, (A.8) shows that G_p contains a nontrivial divisible subgroup C such that $C \cap B = \{e\}$. Using the divisibility of C, we find a sequence $x_1, x_2, \ldots$ of elements of C such that $x_1^p = e$, $x_2^p = x_1$, and so on. Consider the group B_0 generated by $\{x_n\}_{n=1}^\infty$ and map each x_j onto the complex number $\exp\left[2\pi i \frac{1}{p^j}\right]$. This mapping can obviously be extended to an isomorphism of B_0 onto $Z(p^\infty)$. Thus $\{B_\iota\}_{\iota\in I} \cup \{B_0\}$ belongs to $\mathscr{A}$. This contradiction proves that $B = G_p$.

Summarizing, we see that G is isomorphic with

$$Q^{\mathfrak{n}_0 *} \times \underset{p\in P}{\mathsf{P}^*} \left(Z(p^\infty)^{\mathfrak{n}_p *}\right),$$

for some cardinals $\mathfrak{n}_0$ and $\mathfrak{n}_p$. The torsion-free and p-ranks of this product are $\mathfrak{n}_0$ and $\mathfrak{n}_p$, respectively. Hence $\mathfrak{n}_0 = r_0(G)$ and $\mathfrak{n}_p = r_p(G)$ for $p \in P$. $\square$

For some of the refined analysis of locally compact Abelian groups, we need to imbed an arbitrary group in a divisible group, and in fact in a particular way.

(A.15) Theorem. *A group G can be imbedded in a divisible group E that is minimal in the sense that if $G \subset D \subset E$ and D is a divisible group, then $D = E$.*

Proof. First we show that G can be imbedded in some divisible group E_0. Let $\mathfrak{m} = \overline{\overline{G}}$ and for each $x \in G$, let Z_x be the group Z. Consider $Z^{\mathfrak{m}*} = \underset{x\in G}{\mathsf{P}^*} Z_x$, and for $(k_x) \in Z^{\mathfrak{m}*}$, let

$$\varphi((k_x)) = \prod_{x\in G} x^{k_x};$$

for each (k_x), this is a finite product. Then φ is a homomorphism of $Z^{\mathfrak{m}*}$ onto G, and G is isomorphic with $Z^{\mathfrak{m}*}/N$ where $N = \varphi^{-1}(e)$. Now imbed $Z^{\mathfrak{m}*}/N$ in $Q^{\mathfrak{m}*}/N$. This group is divisible, since it is a homomorphic image of $Q^{\mathfrak{m}*}$. Let $E_0 = Q^{\mathfrak{m}*}/N$, and identify G with its isomorphic image contained in E_0.

Let H be a divisible subgroup of E_0 maximal with respect to the relation $G \cap H = \{e\}$. By (A.8), there is a subgroup E of E_0 such that $G \subset E$, $E \cap H = \{e\}$, and $EH = E_0$. Thus E is a direct factor of E_0 and hence is divisible. Suppose that $G \subset D \subsetneqq E$ and that D is divisible. By (A.8), there is a divisible subgroup H_0 of E such that $H_0 \neq \{e\}$, $DH_0 = E$, and $D \cap H_0 = \{e\}$. Now HH_0 is a proper supergroup of H, is isomorphic with $H \times H_0$, and satisfies the relation $(HH_0) \cap G = \{e\}$. Thus HH_0 is a divisible supergroup of H that is disjoint from G. This contradicts the maximality of H. □

(A.16) Theorem. *Let G be a group and E a divisible extension of G that is minimal in the sense of* (A.15). *Then $r_0(G) = r_0(E)$ and $r_p(G) = r_p(E)$ for all primes p. If E_1 is another minimal divisible extension of G, there is an isomorphism φ carrying E onto E_1 that is the identity mapping on G.*

Proof. It is obvious that $r_0(G) \leqq r_0(E)$ and $r_p(G) \leqq r_p(E)$ for all primes p. Consider now any L_0 in $\mathscr{A}_0$ for the group G. Then L_0 is also in $\mathscr{A}_0$ for the group E. To see this, let H be the group generated by L_0, and assume that a in E has infinite order and $a^n \notin H$ for $n = 1, 2, \ldots$. There is a subgroup Q_0 of E containing a and isomorphic with the rational number field Q: to see this, use the decomposition of E given in (A.14). Now $Q_0 \cap G = \{e\}$, since otherwise some integral power a^r of a would lie in G and $L_0 \cup \{a^r\}$ would be an independent set in G containing only elements of infinite order. By (A.8), there is a divisible subgroup D of E such that $G \subset D$, $D \cap Q_0 = \{e\}$, and $DQ_0 = E$. This contradicts the minimality of E. Hence L_0 is in $\mathscr{A}_0$ for the group E; this proves that $r_0(E) \leqq r_0(G)$. Similarly, we have $r_p(E) \leqq r_p(G)$ for all primes p.

Let E_1 be any minimal divisible group containing G. By (A.7), the identity mapping of G onto itself can be extended to a homomorphism φ of E into E_1. Since $\varphi(E)$ is a divisible group containing G, we have $\varphi(E) = E_1$. Assume that φ is not one-to-one, *i.e.*, $\varphi^{-1}(e) \neq \{e\}$. Since $\varphi^{-1}(e) \cap G = \{e\}$, there is an element a in $\varphi^{-1}(e)$ such that $a \neq e$ and such that $a^n \in G$ implies $a^n = e$. As in the previous paragraph, we see that there is a divisible group D such that $G \subset D$ and $a \notin D$. This contradicts the minimality of E. □

(A.17) Theorem. *Let G be a group and E a minimal divisible extension of G. Then E/G is a torsion group. If G is imbedded in a divisible torsion-free group D and D/G is a torsion group, then D is a minimal divisible extension of G.*

Proof. Let

$$E_0 = \{x \in E : x^n \in G \text{ for some positive integer } n\}.$$

Given $x \in E_0$ and an integer $k \geqq 1$, we have $x^n \in G$ for some n and $z^k = x$ for some $z \in E$. Since z^{kn} is in G, z belongs to E_0. Therefore E_0 is divisible and $E_0 = E$. Hence E/G is a torsion group.

Suppose that D is as in the second statement of the theorem and that $G \subset D_0 \subset D$, where D_0 is divisible. If $x \in D$, then $x^n \in G$ for some integer n. Since D_0 is divisible, $x^n = y^n$ where $y \in D_0$. Since D is torsion-free and $(xy^{-1})^n = e$, it follows that $x = y \in D_0$. That is, D_0 is equal to D. □

(A.18) Note. A subgroup H of a group G that is generated by an independent set is the weak direct product of cyclic groups, the generators of these cyclic groups being the elements of the independent set. We omit the proof of this elementary fact.

For our study of the structure of compact Abelian groups, we require another group-theoretic concept, which we will now introduce and discuss.

(A.19) Definition. Let G be a group. A subgroup H of G is said to be *pure* [*in* G] if whenever $x \in G$ and $x^n \in H$, there is an element y in H such that $y^n = x^n$.

(A.20) Theorem. *Let G be a group and H a subgroup of G. The following statements are equivalent:*

(i) *H is pure in G;*

(ii) *$H^{(n)} = H \cap G^{(n)}$ for $n = 2, 3, \ldots$;*

(iii) *$f_n^{-1}(H) = H \cdot G_{(n)}$ for $n = 2, 3, \ldots$, where f_n is the mapping defined in* (A.2).

Proof. The inclusion $H^{(n)} \subset H \cap G^{(n)}$ always holds. It is easy to see that the purity of H in G is equivalent to the inclusion $H^{(n)} \supset H \cap G^{(n)}$.

The inclusion $f_n^{-1}(H) \supset H \cdot G_{(n)}$ always holds. Suppose that H is pure in G and that $x \in f_n^{-1}(H)$. Then $x^n \in H$ and $x^n = y^n$ for some $y \in H$. It follows that $y^{-1}x \in G_{(n)}$ so that $x = y(y^{-1}x) \in H \cdot G_{(n)}$. Thus $f_n^{-1}(H) \subset H \cdot G_{(n)}$. Finally, suppose that $f_n^{-1}(H) \subset H \cdot G_{(n)}$ and that $x^n \in H$. Then $x = hy$ where $h \in H$ and $y^n = e$. Thus $h^n = x^n$ and H is pure in G. □

(A.21) Theorem. *Let H be a pure subgroup of G. Then every coset in G/H contains an element y such that the order of y in G is equal to the order of yH in G/H.*

Proof. If a coset has infinite order in G/H, then all of its elements have infinite order. Suppose then that the coset, say xH, has order n. Then $x^n \in H$ and H contains an element z such that $z^n = x^n$. Let $y = xz^{-1}$; obviously $y \in xH$ and $y^n = e$. If the order of y were less than n, then the order of yH would also be less than n. Hence the order of y is exactly n. □

(A.22) Theorem. *If H is a pure subgroup of G, and G/H is isomorphic with a weak direct product of cyclic groups, then G is isomorphic with $H \times (G/H)$.*

Proof. For each cyclic factor of G/H, choose a generator $x_\alpha H$. By virtue of (A.21), we may suppose that the order of each x_α in G is equal to the order of $x_\alpha H$ in G/H. Let N be the subgroup of G generated by all the x_α. Let us show that $HN=G$ and $H\cap N=\{e\}$. This will imply that G is isomorphic with $H\times N$ and that N is isomorphic with G/H (2.4).

Consider any y in G. Then $yH=(x_{\alpha_1}H)^{n_1}\cdots(x_{\alpha_k}H)^{n_k}$ for some integers $n_1, \ldots, n_k$ and indices $\alpha_1, \ldots, \alpha_k$. Consequently, $y=x_{\alpha_1}^{n_1}\cdots x_{\alpha_k}^{n_k}h$ for some $h\in H$; *i.e.*, y belongs to NH and hence $G=NH$. Suppose that h belongs to both H and N; then $h=x_{\alpha_1}^{m_1}\cdots x_{\alpha_s}^{m_s}$ and $(x_{\alpha_1}H)^{m_1}\cdots(x_{\alpha_s}H)^{m_s}=H$. Thus each m_j is a multiple of the order of $x_{\alpha_j}H$ and hence of the order of x_{α_j}. This implies that $x_{\alpha_j}^{m_j}=e$ for all j, so that $h=e$. This proves the relation $H\cap N=\{e\}$. □

Theorem (A.24) *infra* is another group-theoretic fact needed in the detailed study of compact Abelian groups. We first state and prove a lemma.

(A.23) Lemma. *A nondivisible p-primary group G contains a nontrivial cyclic subgroup that is pure in G.*

Proof. (I) We first assume that for every x in G having order p and for every positive integer k, there is a $y\in G$ such that $y^{p^k}=x$. We will prove that in this case G must be divisible, which is contrary to our hypothesis. Suppose that x is in G, q is a prime different from p, and that p^l is the order of x. Then $1=ap^l+bq$ for some integers a and b, and hence $x=(x^{p^l})^a(x^b)^q=(x^b)^q$. To show that every element x in G is divisible by p, we apply finite induction to k, where p^k is the order of x. The assumption yields this for $k=1$. Suppose that all elements of order less than or equal to p^{k-1} are divisible by p. Let x have order p^k; by the assumption $x^{p^{k-1}}=y^{p^k}$ for some $y\in G$. Since $(y^px^{-1})^{p^{k-1}}=e$, y^px^{-1} is divisible by p: $y^px^{-1}=z^p$ for some $z\in G$. Thus $x=(yz^{-1})^p$, and G is divisible.

(II) Choose an element x in G and a positive integer k such that x has order p and $x=y^{p^{k-1}}$ for some y in G, but for which the equation $x=z^{p^k}$ has no solution z. Let H be the subgroup of G generated by y. Suppose that $w^m=y^t$ for integers t and m, where $0<t<p^k$. Then $t=up^l$ where $l\leqq k-1$ and u is prime to p. We show that p^{l+1} does not divide m. If p^{l+1} divides m, then we have $y^{up^l}=w^{p^{l+1}r}$ for some integer r. Now $1=au+bp$ for some integers a and b, so that

$$\begin{aligned}x=y^{p^{k-1}}&=y^{aup^{k-1}}y^{bp^k}\\&=y^{aup^{k-1}}=(y^{up^l})^{ap^{k-l-1}}=(w^{p^{l+1}r})^{ap^{k-l-1}}=(w^{ar})^{p^k},\end{aligned}$$

which is a contradiction. Since the greatest common divisor of m and p^k is $p^{l'}$ where $l'\leqq l$, there are integers c and d for which $t=up^l=cm+dp^k$. Hence $w^m=y^t=(y^c)^m(y^{p^k})^d=(y^c)^m$. □

To prove Theorem (A.24), we need another concept. A subset L of a group G is called *pure independent* if it is independent and the smallest subgroup of G containing L is pure in G. The usual ZORN's lemma argument shows that every pure independent subset of a group is contained in a maximal pure independent subset.

(A.24) Theorem. *Let G be a p-primary group. Then G contains a subgroup B such that:*

(i) *B is isomorphic with a weak direct product of cyclic groups;*

(ii) *B is a pure subgroup of G;*

(iii) *the quotient group G/B is divisible.*

Proof. Let L be a maximal pure independent subset of G. If L is void, (A.23) shows that G is divisible, and in this case we take $B=\{e\}$. If L is nonvoid, let B be the subgroup of G generated by L. Then assertions (i) and (ii) are immediate. If G/B is not divisible, then again by (A.23), G/B contains a nontrivial pure cyclic subgroup; this subgroup is generated by $x_0 B$, say. Since B is pure in G, we may suppose that x_0 has the same order in G as $x_0 B$ has in G/B (A.21). It follows that $L \cup \{x_0\}$ is an independent subset of G. To arrive at a contradiction, we will show that the group B_0 generated by $L \cup \{x_0\}$ is pure in G. Suppose that $bx_0^n = y^m$ for some $y \in G$ and $b \in B$. We then have $x_0^n B = y^m B$. Since the group generated by $x_0 B$ is pure in G/B, there is an integer k such that $x_0^n B = (x_0^k B)^m$. Since also $x_0^n B = y^m B$, we find that $(x_0^k y^{-1})^m = x_0^{km} y^{-m}$ belongs to B. The purity of B yields a $b_0 \in B$ for which $(x_0^k y^{-1})^m = b_0^m$. Thus $y^m = (x_0^k b_0^{-1})^m$, and the group B_0 is pure in G. □

We now classify groups of bounded order.

(A.25) Theorem. *Every group G of bounded order is isomorphic with a weak direct product of cyclic groups $\mathop{\mathrm{P}^*}_{\iota \in I} Z(p_\iota^{r_\iota})$, where only finitely many distinct primes p_ι and positive integers r_ι occur. Conversely, every such group has bounded order.*

Proof. By (A.3), we may suppose that G is p-primary. Let B be a subgroup of G as in (A.24). Since G/B is divisible and of bounded order, it follows that $G/B = \{B\}$. Therefore $G = B$ and G is isomorphic with a weak direct product of cyclic groups. Since each factor is p-primary and of bounded order, it has the form $Z(p^r)$. The remainder of the theorem is obvious[1]. □

If a group G is generated by some finite subset, then G is said to be *finitely generated*. Using the next theorem, we will prove the fundamental theorem (A.27) which characterizes all finitely generated [Abelian] groups.

[1] This deduction of (A.25) from (A.24) was pointed out to us by R. S. PIERCE.

(A.26) Theorem. *Let n be a positive integer, and let H be a subgroup of Z^n different from $\{0\}$. Then there are a basis $\{\boldsymbol{x}_1, \ldots, \boldsymbol{x}_n\}$ for Z^n [see* (A.10)] *and a sequence $\{k_1, \ldots, k_r\}$ of positive integers such that $r \leqq n$,*

(i) *$\{k_1\boldsymbol{x}_1, \ldots, k_r\boldsymbol{x}_r\}$ is a basis for H,*

and

(ii) *k_j divides k_{j+1} for $j=1, \ldots, r-1$.*

Proof. The number n is the cardinal number of all bases of Z^n. If $n=1$, then Z^n is cyclic and the theorem is obvious. Suppose that the theorem is true for $n-1$. For every $\boldsymbol{h} \in H$ and every basis $\{\boldsymbol{y}_1, \ldots, \boldsymbol{y}_n\}$ of Z^n, we can write $\boldsymbol{h}=l_1\boldsymbol{y}_1+\cdots+l_n\boldsymbol{y}_n$. There is a least possible positive value k_1 for the integers l_j as $\boldsymbol{h}$ runs through all nonzero elements of H and $\{\boldsymbol{y}_1, \ldots, \boldsymbol{y}_n\}$ runs through all bases for Z^n. Let $\boldsymbol{h}_0$ be an element of H such that

$$\boldsymbol{h}_0=k_1\boldsymbol{y}_1+m_2\boldsymbol{y}_2+\cdots+m_n\boldsymbol{y}_n, \tag{1}$$

where $\{\boldsymbol{y}_1, \ldots, \boldsymbol{y}_n\}$ is a basis for Z^n and $m_2, \ldots, m_n$ are integers.

Let us show that k_1 divides m_j for $j=2, \ldots, n$. If $m_j=a_jk_1+b_j$ where $0 \leqq b_j < k_1$, then we have

$$\boldsymbol{h}_0=k_1\boldsymbol{x}_1+b_2\boldsymbol{y}_2+\cdots+b_n\boldsymbol{y}_n,$$

where $\boldsymbol{x}_1=\boldsymbol{y}_1+a_2\boldsymbol{y}_2+\cdots+a_n\boldsymbol{y}_n$. It is easy to see that $\{\boldsymbol{x}_1, \boldsymbol{y}_2, \ldots, \boldsymbol{y}_n\}$ is a basis for Z^n. The minimality of k_1 implies that each b_j is zero. Note that $\boldsymbol{h}_0=k_1\boldsymbol{x}_1$.

Now let G be the subgroup of Z^n generated by $\{\boldsymbol{y}_2, \ldots, \boldsymbol{y}_n\}$. Then G is isomorphic with Z^{n-1}. If $G \cap H=\{0\}$, then we have $H=\{lk_1\boldsymbol{x}_1 : l \in Z\}$, and the theorem is evidently true for H. Otherwise, our inductive hypothesis implies that there are a basis $\{\boldsymbol{x}_2, \ldots, \boldsymbol{x}_n\}$ for G and a sequence $\{k_2, \ldots, k_r\}$ of positive integers such that

$$\{k_2\boldsymbol{x}_2, \ldots, k_r\boldsymbol{x}_r\} \quad \text{is a basis for} \quad G \cap H$$

and

$$k_j \quad \text{divides} \quad k_{j+1} \quad \text{for} \quad j=2, \ldots, r-1.$$

It is easy to see that $\{\boldsymbol{x}_1, \ldots, \boldsymbol{x}_n\}$ is a basis for Z^n.

Suppose now that $\boldsymbol{h}$ is any element in H; we write $\boldsymbol{h}=l_1\boldsymbol{x}_1+\cdots+l_n\boldsymbol{x}_n$. Write $l_1=s_1k_1+c$, where $0 \leqq c < k_1$. Then we have

$$\boldsymbol{h}-s_1\boldsymbol{h}_0=c\boldsymbol{x}_1+l_2\boldsymbol{x}_2+\cdots+l_n\boldsymbol{x}_n,$$

and the minimality of k_1 implies that $c=0$. Thus $\boldsymbol{h}-s_1\boldsymbol{h}_0$ belongs to $G \cap H$, so that $\boldsymbol{h}-s_1\boldsymbol{h}_0=s_2k_2\boldsymbol{x}_2+\cdots+s_rk_r\boldsymbol{x}_r$ and $\boldsymbol{h}=s_1k_1\boldsymbol{x}_1+s_2k_2\boldsymbol{x}_2+\cdots+s_rk_r\boldsymbol{x}_r$. Since $\{k_1\boldsymbol{x}_1, \ldots, k_r\boldsymbol{x}_r\}$ is an independent set, we have proved (i).

It remains only to verify that k_1 divides k_2. Let $k_2 = a k_1 + b$ where $0 \leqq b < k_1$. Then $\{\boldsymbol{x}_1 + a\boldsymbol{x}_2, \boldsymbol{x}_2, \ldots, \boldsymbol{x}_n\}$ is a basis for Z^n and

$$k_1 \boldsymbol{x}_1 + k_2 \boldsymbol{x}_2 = k_1 (\boldsymbol{x}_1 + a\boldsymbol{x}_2) + b\boldsymbol{x}_2.$$

The minimality of k_1 shows that $b = 0$. □

(A.27) Theorem. *A finitely generated group G is isomorphic with the direct product of a finite number of cyclic groups, where each factor is infinite or has order a power of a prime.*

Proof. Let $\{y_1, \ldots, y_n\}$ be a set that generates G. For $(m_1, \ldots, m_n)$ in Z^n, let

$$\varphi((m_1, \ldots, m_n)) = \prod_{j=1}^{n} y_j^{m_j}.$$

Then φ is a homomorphism of Z^n onto G, and G is isomorphic with Z^n/H, where $H = \varphi^{-1}(e)$. If $H = \{\boldsymbol{0}\}$, then G is isomorphic with Z^n. Otherwise, (A.26) shows that there are a basis $\{\boldsymbol{x}_1, \ldots, \boldsymbol{x}_n\}$ for Z^n and a sequence $\{k_1, \ldots, k_r\}$ of positive integers such that $\{k_1\boldsymbol{x}_1, \ldots, k_r\boldsymbol{x}_r\}$ is a basis for H. The mapping ψ of Z^n onto $Z(k_1) \times \cdots \times Z(k_r) \times Z^{n-r}$ defined by

$$\psi(m_1\boldsymbol{x}_1 + \cdots + m_n\boldsymbol{x}_n) = \big(m_1 (\text{mod } k_1), \ldots, m_r (\text{mod } k_r), m_{r+1}, \ldots, m_n\big),$$

is a homomorphism of Z^n onto $Z(k_1) \times \cdots \times Z(k_r) \times Z^{n-r}$ whose kernel is H. It follows that G is isomorphic with $Z(k_1) \times \cdots \times Z(k_r) \times Z^{n-r}$. Finally, by (A.25) each $Z(k_r)$ is a product of cyclic groups, each having order a power of a prime. □

Appendix B

Topological linear spaces

In this appendix, we sketch the theory of topological linear spaces. Because of the plethora of lucid textbooks on various parts of the subject, we have merely given references for many classical theorems. We thought it worthwhile to prove the KREĬN-MIL'MAN theorem, however, because of its vital rôle in the theory of representations [§22] and the fact that no simple proof seemed readily citable.

(B.1) Definition. Let F be a field and E an Abelian group, written additively. Suppose that for each $\alpha \in F$ and $x \in E$, an element αx in E, called the *product of* α *and* x, is defined, for which the following axioms hold:

(i) $\alpha(x+y) = \alpha x + \alpha y$;

(ii) $(\alpha+\beta)x = \alpha x + \beta x$;

(iii) $(\alpha\beta)x = \alpha(\beta x)$;

(iv) $1x = x$;

where $x, y \in E$ and $\alpha, \beta \in F$. Then E is said to be a *linear space* [or *vector space*] *over* F, and F is called the *scalar field for* E. The symbol 0 denotes the identity element of the additive group E and also zero in F. The linear space E is called a *real* or *complex linear space* if F is R or K, respectively. For a subset A of E and $\alpha \in F$, αA denotes the set $\{\alpha x : x \in A\}$.

Whenever an expression such as "linear space", "Banach space", "Hilbert space", **or** "linear functional" **appears in this monograph and the scalar field is not specified, the scalar field is the complex number field** K.

(B.2) Definitions. Let E be a linear space over a field F. A subset E_0 of E that is itself a linear space over F is called a *linear subspace of* E. Let E_0 be a linear subspace of E and consider the quotient *group* E/E_0. If x and x' belong to the same coset of E_0 and α belongs to F, then αx and $\alpha x'$ belong to the same coset. Therefore the group E/E_0 admits a well-defined scalar multiplication, namely $\alpha(xE_0) = (\alpha x)E_0$. Then E/E_0 is a linear space over F; it is called the *quotient space* or *difference space of* E *by* E_0.

Let E be a real linear space. For $x_1, \ldots, x_m \in E$, an element $\alpha_1 x_1 + \cdots + \alpha_m x_m$, where $\alpha_1, \ldots, \alpha_m \in R$, each α_j is nonnegative, and $\alpha_1 + \cdots + \alpha_m = 1$, is called a *convex combination of* $x_1, \ldots, x_m$. A subset A of E is said to be *convex* if each convex combination of elements in A again belongs to A.

The *direct sum* of linear spaces E_ι, $\iota \in I$, is the usual direct product $\underset{\iota \in I}{\mathsf{P}} E_\iota$ of the additive groups E_ι. Scalar multiplication is defined coordinatewise. We frequently write the direct sum of a finite family of linear spaces as $E_1 \oplus \cdots \oplus E_m$.

Let E and E' be linear spaces over the same field F. A mapping T of E into E' is said to be *additive* if $T(x+y) = T(x) + T(y)$ for all $x, y \in E$, and T is said to be *homogeneous* if $T(\alpha x) = \alpha T(x)$ for all $x \in E$ and $\alpha \in F$. A function that is both additive and homogeneous is said to be a *linear function*. Thus a function T is linear if and only if $T(\alpha x + y) = \alpha T(x) + T(y)$ for $x, y \in E$ and $\alpha \in F$. Suppose that E and E' are linear spaces over K. An additive function T such that $T(\alpha x) = \bar{\alpha} T(x)$ for $x \in E$ and $\alpha \in K$ is said to be *conjugate-linear*. The term "linear transformation" is synonymous with "linear function". An *operator* is a linear transformation of a linear space into itself.

A linear or conjugate-linear function on a linear space E into its scalar field is called a *linear* [or *conjugate-linear*] *functional*. We emphasize that linear functionals always assume values in the scalar field of E.

(B.3) Theorem. *Let* E *be a complex linear space, and let* L *be a linear functional on* E. *For* $x \in E$, *write* $L(x) = L_1(x) + iL_2(x)$, *where* L_1 *and* L_2 *are real-valued. Then we have:*

(i) $L_j(x+y)=L_j(x)+L_j(y)$ *for* $x, y\in E$ *and* $j=1, 2$;

(ii) $L_j(\alpha x)=\alpha L_j(x)$ *for* $x\in E$, $\alpha\in R$, *and* $j=1, 2$;

(iii) $L_2(x)=-L_1(ix)$ *for* $x\in E$.

Proof. Properties (i) and (ii) are obvious from the additivity and homogeneity of L. Property (iii) is proved from the identities $L_1(ix)+iL_2(ix)=iL(x)=iL_1(x)-L_2(x)$. □

(B.4) Definition. Consider the Cartesian product $E\times E'$ of complex linear spaces E and E'. A *bilinear functional* L is a functional on $E\times E'$ that is linear in the first variable and conjugate-linear in the second variable. In other words,

$$L(\alpha x+y, x')=\alpha L(x, x')+L(y, x')$$

and

$$L(x, \alpha x'+y')=\bar{\alpha}L(x, x')+L(x, y')$$

for $x, y\in E$, $x', y'\in E'$, and $\alpha\in K$.

(B.5) Definition. Let E be a real or complex linear space that is also a topological T_0 space. Suppose that the mapping $(x, y)\to x+y$ of $E\times E$ onto E is continuous and that the mapping $(\alpha, x)\to\alpha x$ of $F\times E$ onto E is continuous. Note that F is either R or K. Then E is said to be a *topological linear space*. A topological linear space over R is said to be *locally convex* if there is an open basis at 0 consisting of convex sets.

A neighborhood U of 0 in a topological linear space E is said to be *balanced* if $x\in U$, $\alpha\in F$, and $|\alpha|\leqq 1$ imply $\alpha x\in U$.

Theorem (8.4) applies to every topological linear space E, since E is a topological group under addition. Hence a topological linear space is completely regular.

(B.6) Theorem. *Let E be a topological linear space. The family of all balanced neighborhoods of* 0 *is an open basis at* 0.

Proof. Let U be any neighborhood of 0. By the continuity of scalar multiplication at 0, there is a positive number ε and a neighborhood W of 0 such that $|\alpha|<\varepsilon$ and $x\in W$ imply $\alpha x\in U$. Let $V=\bigcup\{\alpha W:|\alpha|<\varepsilon\}$. Then V is a balanced neighborhood of 0 and $V\subset U$. □

(B.7) Definition. Let E be a linear space over F, where F is R or K. A *norm* on E is a real-valued function [whose value at $x\in E$ is denoted by $\|x\|$] having the following properties:

(i) $\|x\|>0$ whenever $x\neq 0$;

(ii) $\|\alpha x\|=|\alpha|\cdot\|x\|$ for all $x\in E$ and $\alpha\in F$;

(iii) $\|x+y\|\leqq\|x\|+\|y\|$ for all $x, y\in E$.

Property (iii) is known as the *triangle inequality*. We call E a *normed linear space*. The set $\{x\in E:\|x\|\leqq 1\}$ is called the *unit ball of* E.

Let E be a normed linear space, and for $x, y \in E$, let $d(x, y) = \|x-y\|$. Then d is a metric on E. [The equality $\|0\|=0$ follows from (B.7.ii).] With the topology induced by d, E is a locally convex topological linear space; this topology is called the *norm topology of* E.

The fields K and R are always normed by $\|\alpha\| = |\alpha|$; this amounts to defining $\|1\|=1$.

(B.8) Definition. Let T be a linear transformation of a normed linear space E into a normed linear space E', where E and E' are both real or both complex. If there is a positive number A such that $\|T(x)\| \leqq A\|x\|$ for all $x \in E$, then T is said to be *bounded*. The infimum of all such values of A is defined to be the *norm of* T and is denoted by $\|T\|$. If E' is the scalar field [*i.e.*, R or K], then T is called a *bounded linear functional.*

(B.9) Theorem. *If T is a bounded linear transformation of a normed linear space E into another normed linear space E', then*

$$\|T\| = \sup\{\|T(x)\| : x \in E, \|x\| \leqq 1\} = \sup\left\{\frac{\|T(x)\|}{\|x\|} : x \in E, x \neq 0\right\}.$$

(B.10) Theorem. *A linear transformation T mapping a real or complex normed linear space into another such space is bounded if and only if it is continuous.*

Theorems (B.9) and (B.10) are very simple; we omit their proofs.

(B.11) Theorem. *Let E be a real or complex normed linear space and D a dense linear subspace of E. A bounded linear functional L defined on D can be extended to a bounded linear functional L_0 on E. Moreover, $\|L_0\| = \|L\|$, and the extension is unique.*

Proof. If $x \in E$ and $\lim\limits_{n\to\infty} \|x - x_n\| = 0$, where each x_n is in D, then $\{L(x_n)\}_{n=1}^{\infty}$ is a Cauchy sequence in K or R and so has a limit. Define $L_0(x)$ to be this limit. This defines L_0 uniquely on E, L_0 is a linear functional on E, and $\|L_0\| = \|L\|$. □

We next state the essential part of the HAHN-BANACH theorem for real linear spaces and some of its consequences.

(B.12) Theorem. *Let H be a linear subspace of the real linear space E. Let p be a real-valued function on E such that*

(i) $p(\alpha x) = \alpha p(x)$ *if* $\alpha \geqq 0$ *and* $x \in E$;

(ii) $p(x+y) \leqq p(x) + p(y)$ *for all* $x, y \in E$.

Let L be a linear functional on H such that $L(x) \leqq p(x)$ for $x \in H$. Let x_0 belong to $E \cap H'$, and let H_0 be the linear subspace of E generated by x_0 and H. Then we have

$$\sup\{-p(-h-x_0) - L(h) : h \in H\} \leqq \inf\{p(h+x_0) - L(h) : h \in H\}.$$

Let α_0 *be any real number such that* $\sup\{-p(-h-x_0)-L(h): h\in H\}\leqq \alpha_0 \leqq \inf\{p(h+x_0)-L(h): h\in H\}$. *Then* L *can be extended to a linear functional* L_0 *on* H_0 *such that* $L_0(x_0)=\alpha_0$ *and* $L_0(x)\leqq p(x)$ *for all* $x\in H_0$.

(B.13) Theorem. *Let* E, H, p, *and* L *be as in* (B.12). *Then* L *can be extended to a linear functional* $\widetilde{L}$ *on* E *such that* $\widetilde{L}(x)\leqq p(x)$ *for all* $x\in E$.

(B.14) Theorem. *Let* E *be a real or complex normed linear space and* H *any linear subspace of* E. *Then any bounded linear functional* L *defined on* H *can be extended to a linear functional* $\widetilde{L}$ *on* E *such that the norm of* $\widetilde{L}$ *on* E *is equal to the norm of* L *on* H.

(B.15) Corollary. *Let* H *be a linear subspace of a real or complex normed linear space* E. *For every* $x\in E\cap(H^-)'$, *there is a linear functional* L *on* E *such that* $L(H)=\{0\}$, $L(x)=1$, *and* $\|L\|=1/d$, *where* $d=\inf\{\|x-h\|: h\in H\}$.

All of Theorems (B.12)—(B.15) have been referred to as "the HAHN-BANACH theorem".

Proofs of Theorem (B.13) appear in: DAY [6], p. 9; DUNFORD and SCHWARTZ [1], p. 62; HILLE and PHILLIPS [1], p. 29; KÖTHE [1], p. 193; MUNROE [1], p. 56; and NAĬMARK [1], §(I.1.9). Theorem (B.14) and Corollary (B.15) are stated and proved in DUNFORD and SCHWARTZ [1] and HILLE and PHILLIPS [1].

(B.16) Definition. A [real or complex] normed linear space E which is complete in the metric induced by its norm is said to be a [real or complex] *Banach space*.

(B.17) Theorem. *Suppose that* E *is a real or complex Banach space and that* H *is a closed linear subspace of* E. *For* $x+H\in E/H$, *let*

$$\|x+H\|=\inf\{\|y\|: y\in x+H\}.$$

With this norm and with the linear operations of (B.2), *the difference space* E/H *is a Banach space.*

Proof. If $\|x+H\|=0$, there is a sequence $\{h_n\}_{n=1}^{\infty}$ in H such that $\lim_{n\to\infty}\|x+h_n\|=0$. Since H is closed and $\lim_{n\to\infty}\|x-(-h_n)\|=0$, x belongs to H and $x+H=H$. Thus (B.7.i) holds. It is trivial to verify that $\|\alpha x+H\|=|\alpha|\cdot\|x+H\|$ for α in the scalar field and $x\in E$, and that $\|x+y+H\|\leqq\|x+H\|+\|y+H\|$ for $x, y\in E$.

To show that E/H is complete, let $\{x_n+H\}_{n=1}^{\infty}$ be a sequence of cosets in E/H such that $\lim_{n,m\to\infty}\|x_m-x_n+H\|=0$. We can now choose a subsequence [which for convenience we again write as $\{x_n+H\}_{n=1}^{\infty}$] such that $\|x_m-x_n+H\|<2^{-n}$ for $m\geqq n$. In particular, for each n, there is an $h_n\in H$ such that $\|x_{n+1}-x_n+h_n\|<2^{-n}$. Let $y_n=x_n+h_{n-1}+\cdots+h_1$ for $n=1, 2, \ldots$. Then $\{y_n\}_{n=1}^{\infty}$ is a Cauchy sequence in E, since for

$m \geqq n$, we have

$$\|y_m - y_n\| = \left\|\sum_{k=n}^{m-1} [x_{k+1} - x_k + h_k]\right\| \leqq \sum_{k=n}^{m-1} \|x_{k+1} - x_k + h_k\| < 2^{-n+1}.$$

Since E is complete, there is a $y \in E$ such that $\lim_{n\to\infty} \|y - y_n\| = 0$. Therefore

$$\lim_{n\to\infty} \|(y+H) - (x_n + H)\| = \lim_{n\to\infty} \|(y+H) - (y_n + H)\|$$
$$\leqq \lim_{n\to\infty} \|y - y_n\| = 0,$$

because $x_n + H = y_n + H$ for all n. That is, $y + H = \lim_{n\to\infty} (x_n + H)$, and so E/H is complete. □

We next take up the BANACH-STEINHAUS theorem. Some preliminary results on category are needed.

(B.18) Definition. Let X be a topological space. A subset A of X is *nowhere dense in* X if the interior of A^- is void. A subset A of X is said to be of the *first category in* X if it is the union of a countable family of nowhere dense sets. If a subset A of X is not of the first category, then it is said to be of the *second category in* X.

(B.19) Theorem. *Let X be a complete metric space, a locally countably compact regular space, or a locally compact Hausdorff space. Then X is of the second category in itself.*

The first assertion is proved in KELLEY [2], p. 200. For the others, see (5.28) *supra*.

(B.20) Theorem [BANACH-STEINHAUS]. *Let E and E' be Banach spaces over the same scalar field. Let $\mathfrak{A}$ be a set of bounded linear transformations of E into E'. If the set A of points x for which* $\sup\{\|T(x)\| : T \in \mathfrak{A}\} < \infty$ *is of the second category in E, then the set $\{\|T\| : T \in \mathfrak{A}\}$ of numbers is bounded. In other words, there is a constant β such that $\|T(x)\| \leqq \beta \|x\|$ for all $x \in E$ and all $T \in \mathfrak{A}$.*

Proof. For $m = 1, 2, \ldots$, let A_m denote the set of elements x in A such that $\sup\{\|T(x)\| : T \in \mathfrak{A}\} \leqq m$. For each T the set $\{x \in E : \|T(x)\| \leqq m\}$ is closed because the mapping $x \to \|T(x)\|$ is continuous. Therefore $A_m = \bigcap_{T\in\mathfrak{A}} \{x \in E : \|T(x)\| \leqq m\}$ is closed. Since $A = \bigcup_{m=1}^{\infty} A_m$ is of the second category in E, some A_m, say A_{m_0}, must have nonvoid interior. That is, there is an x_0 in E and an $\varepsilon > 0$ such that

$$\{y \in E : \|y - x_0\| < \varepsilon\} \subset A_{m_0}.$$

For any nonzero element x in E, the element $\frac{\varepsilon}{2\|x\|} x + x_0$ belongs to the set $\{y \in E : \|y - x_0\| < \varepsilon\}$ and therefore we have $\left\|T\left(\frac{\varepsilon}{2\|x\|} x + x_0\right)\right\| \leqq m_0$

for all $T \in \mathfrak{A}$. Consequently, we have

$$\|T(x)\| = \frac{2\|x\|}{\varepsilon} \left\| T\left(\frac{\varepsilon}{2\|x\|} x\right) \right\|$$
$$\leqq \frac{2\|x\|}{\varepsilon} \left[\left\| T\left(\frac{\varepsilon}{2\|x\|} x + x_0\right) \right\| + \|T(-x_0)\| \right] \leqq \frac{2\|x\|}{\varepsilon} [2m_0],$$

and hence $\|T\| \leqq \frac{4m_0}{\varepsilon}$ for all $T \in \mathfrak{A}$. □

(B.21) Definition. Let E be a real or complex topological linear space. The set of all continuous linear functionals on E is denoted by E^*; E^* is called the *conjugate space of* E. For $L, M \in E^*$ and $x \in E$, we define $(L+M)(x) = L(x) + M(x)$. Also for $L \in E^*$ and a scalar α, we define $(\alpha L)(x) = \alpha \cdot L(x)$.

(B.22) Theorem. *Let E be a real or complex normed linear space. With the linear operations defined in* (B.21) *and the norm defined in* (B.8), *E^* is a Banach space.*

Proof. We omit the proof that E^* is a normed linear space. Let $\{L_n\}_{n=1}^{\infty}$ be a Cauchy sequence in E^*. For $x \in E$, we have $|L_n(x) - L_m(x)| \leqq \|x\| \cdot \|L_n - L_m\|$ so that $\{L_n(x)\}_{n=1}^{\infty}$ is a Cauchy sequence of numbers. We define $L(x)$ to be $\lim\limits_{n\to\infty} L_n(x)$. Since

$$|\alpha L(x) + L(y) - L(\alpha x + y)|$$
$$\leqq |\alpha L(x) - \alpha L_n(x)| + |L(y) - L_n(y)| + |L_n(\alpha x + y) - L(\alpha x + y)|,$$

it follows that $L(\alpha x + y) = \alpha L(x) + L(y)$ for $x, y \in E$ and α in R or K.

Suppose that $\|L_n - L_m\| < \varepsilon$ for $n, m \geqq N$. Then for $x \in E$, we have

$$|L(x) - L_n(x)| \leqq |L(x) - L_m(x)| + |L_m(x) - L_n(x)|$$
$$\leqq |L(x) - L_m(x)| + \varepsilon \|x\|.$$

Taking the limit as $m \to \infty$, we find that $|L(x) - L_n(x)| \leqq \varepsilon \|x\|$ for $x \in E$. This shows that L is bounded and that $\lim\limits_{n\to\infty} \|L - L_n\| = 0$. □

(B.23) Let E be a real or complex normed linear space. For each x in E, let x' denote the bounded linear functional on E^* defined by

(i) $x'(L) = L(x)$ for $L \in E^*$.

Plainly x' belongs to the conjugate space E^{**} of E^*, and $\|x'\| = \|x\|$. The mapping $x \to x'$ is thus a linear norm-preserving mapping of E into E^{**}. The space E is said to be *reflexive* if this mapping carries E *onto* E^{**} [E must be complete for this to occur].

In addition to the norm topology, there are other important topologies that may be defined on normed linear spaces. We define two such topologies.

(B.24) Definitions. Let E be a real or complex normed linear space. For each $x \in E$, each finite subset $\{L_1, \ldots, L_n\}$ of E^*, and each $\varepsilon > 0$, define

(i) $U(x; L_1, \ldots, L_n; \varepsilon) = \{y \in E : |L_k(y) - L_k(x)| < \varepsilon \text{ for } k = 1, \ldots, n\}$.
The sets (i) are an open basis for a topology for E; this topology is called the *weak topology for* E. It is the weakest topology on E for which each L in E^* is continuous.

Consider the conjugate space E^*. For each $L \in E^*$, each finite subset $\{x_1, \ldots, x_n\}$ of E, and each $\varepsilon > 0$, define

(ii) $U(L; x_1, \ldots, x_n; \varepsilon) = \{M \in E^* : |M(x_k) - L(x_k)| < \varepsilon \text{ for } k = 1, \ldots, n\}$.
The sets (ii) are an open basis for a topology for E^*; this topology is called the *weak-$*$ topology for* E^*. It is the weakest topology on E^* for which each x' in E^{**} is continuous on E^*.

The weak and weak-$*$ topologies on E^* coincide if E is reflexive.

(B.25) Theorem [Alaoglu]. *Let E be a real or complex normed linear space. Then the unit ball $S^* = \{L \in E^* : \|L\| \leq 1\}$ in E^* is compact in the weak-$*$ topology of E^*.*

Proof. For each $x \in E$, let I_x denote the set of scalars α such that $|\alpha| \leq \|x\|$.[1] For $L \in S^*$, let $\tau(L)$ be the element in $\underset{x \in E}{\mathsf{P}} I_x$ such that $(\tau(L))_x = L(x)$. It is elementary to verify that τ is a homeomorphism of S^* into $\underset{x \in E}{\mathsf{P}} I_x$. By the Tihonov product theorem, $\underset{x \in E}{\mathsf{P}} I_x$ is compact; hence it suffices to show that $\tau(S^*)$ is closed in $\underset{x \in E}{\mathsf{P}} I_x$.

Let (t_x) be in $\tau(S^*)^- \subset \underset{x \in E}{\mathsf{P}} I_x$. Let x and y be fixed in E. For $\varepsilon > 0$, there is an L in S^* such that $|t_x - L(x)| < \varepsilon$, $|t_y - L(y)| < \varepsilon$, and $|t_{x+y} - L(x+y)| < \varepsilon$. It follows that $|t_x + t_y - t_{x+y}| < 3\varepsilon$. Thus $x \to t_x$ is additive; homogeneity is proved in like manner. Since $|t_x| \leq \|x\|$, the mapping $x \to t_x$ is a bounded linear functional on E whose norm does not exceed 1; hence (t_x) belongs to $\tau(S^*)$. □

We now take up the Kreĭn-Mil'man theorem. Some preliminaries are needed.

(B.26) Lemma. *Let E be a real, locally convex, topological linear space, C a closed convex subset of E, and z a point in $E \cap C'$. Then there is a continuous [real] linear functional L_0 on E such that*

(i) $\inf\{L_0(x) : x \in C\} > L_0(z)$.

Proof. Since (i) is equivalent to the inequality

$$\inf\{L_0(x) : x \in C - z\} > L_0(0) = 0,$$

[1] For a real space E, I_x is a closed interval in R. For complex space E, I_x is a closed disk in K. In both cases I_x is *compact*.

we may suppose that $z=0$. Let U be a convex neighborhood of 0 such that U^- is disjoint from C, and let $V=U\cap(-U)$. Let c be any point in C. Then $C_0=V^-+(-C)+c$ is easily seen to be a convex set such that $0\in V\subset C_0$ and $c\notin C_0$.[1] If $x\in E$, then $\frac{1}{\alpha}x$ belongs to C_0 if α is a sufficiently large real number; we define

$$p(x)=\inf\left\{\alpha\in R:\alpha>0,\ \frac{1}{\alpha}x\in C_0\right\}. \tag{1}$$

It is trivial to verify that p satisfies (B.12.i). Suppose that $\frac{1}{\alpha}x$ and $\frac{1}{\beta}y$ belong to C_0. Then

$$\frac{1}{\alpha+\beta}(x+y)=\frac{\alpha\left(\frac{1}{\alpha}x\right)+\beta\left(\frac{1}{\beta}y\right)}{\alpha+\beta};$$

as this is a convex combination of $\frac{1}{\alpha}x$ and $\frac{1}{\beta}y$, $\frac{1}{\alpha+\beta}(x+y)$ belongs to C_0. This shows that $p(x+y)\leqq p(x)+p(y)$; *i.e.*, (B.12.ii) holds.

Note that $p(c)\geqq 1$. For $\alpha\in R$, let $L(\alpha c)=\alpha$. Then L is a linear functional on $\{\alpha c:\alpha\in R\}$. If $\alpha\geqq 0$, then $L(\alpha c)=\alpha\leqq\alpha p(c)=p(\alpha c)$; if $\alpha<0$, then $L(\alpha c)=\alpha<0\leqq p(\alpha c)$. Thus the HAHN-BANACH theorem (B.13) shows that L can be extended to a linear functional L_0 on E such that $L_0(x)\leqq p(x)$ for all $x\in E$.

For every $x\in V^-$ and $y\in C$, the element $x-y+c$ is in C_0, so that $p(x-y+c)\leqq 1$. Therefore $L_0(x-y+c)\leqq 1$. Since $L_0(c)=1$ and L_0 is linear, we infer that $L_0(x)\leqq L_0(y)$. Therefore we have

$$\sup\{L_0(x):x\in V^-\}\leqq\inf\{L_0(y):y\in C\}\leqq L_0(c)=1. \tag{2}$$

There is a positive real number δ such that $\delta c\in V$. Thus $L_0(\delta c)=\delta$, and so $\sup\{L_0(x):x\in V^-\}$ is positive. In other words, we have $\inf\{L_0(y):y\in C\}>0=L_0(0)$.

To see that L_0 is continuous, consider any $\varepsilon>0$. If x belongs to εV, then $\frac{1}{\varepsilon}x$ belongs to V and (2) implies that $L_0(x)=\varepsilon L_0\left(\frac{1}{\varepsilon}x\right)\leqq\varepsilon$. Since $V=-V$, we also have $-L_0(x)\leqq\varepsilon$ so that $|L_0(x)|\leqq\varepsilon$. Thus L_0 is continuous at 0, from which it follows that L_0 is continuous everywhere on E (5.40.a). □

(B.27) Definition. Let C be a convex subset of a real linear space E. A convex subset B of C is said to be an *extreme subset of* C if the relations $x, y\in C$, $\alpha x+(1-\alpha)y\in B$, and $0<\alpha<1$ imply that x and y are in B. If $\{x\}$ is an extreme subset of C, then x is said to be an *extreme point of* C.

(B.28) Remarks. Let C be a convex subset of a real linear space. The intersection of a family of extreme subsets of C is an extreme subset

[1] Note that the closure in E of a convex set is convex.

of C. If B is an extreme subset of C, and A is an extreme subset of B, then A is an extreme subset of C. These statements are easy to verify; we omit the details.

(B.29) Lemma. *Let C be a compact convex subset of a real topological linear space E, let L be a continuous linear functional on E, and let $\beta = \min\{L(x) : x \in C\}$. Then the set $S = \{x \in C : L(x) = \beta\}$ is a nonvoid compact extreme subset of C.*

Proof. Plainly S is compact, convex, and nonvoid. For $x \in C$, $y \in C$, and $\alpha x + (1-\alpha) y \in S$, assume that x, say, belongs to $C \cap S'$. Then $L(x) > \beta$ and hence $L(\alpha x + (1-\alpha) y) = \alpha L(x) + (1-\alpha) L(y) > \alpha\beta + (1-\alpha) L(y) \geqq \alpha\beta + (1-\alpha)\beta = \beta$, which is a palpable contradiction. □

(B.30) Theorem [KREĬN-MIL'MAN]. *Let C be a compact convex subset of a real, locally convex, topological linear space E. Let B denote the set of all convex combinations of extreme points of C. Then $B^- = C$.*

Proof[1]. The fact that C has any extreme points at all is itself nontrivial. This will follow from step (I) of the proof, since C is a compact extreme subset of itself.

(I) Every nonvoid, closed, extreme subset A of C contains an extreme point of C. Let $\mathscr{A}$ be the family of all nonvoid closed subsets of A that are extreme subsets of C. If $\mathscr{L}$ is a subfamily of $\mathscr{A}$ linearly ordered by inclusion, then $\bigcap\{B : B \in \mathscr{L}\}$ is nonvoid since C is compact. Moreover, by (B.28), $\bigcap\{B : B \in \mathscr{L}\}$ is an extreme subset of C and hence belongs to $\mathscr{A}$. ZORN's lemma implies the existence of a minimal set B_0 in $\mathscr{A}$. Assume that B_0 contains two distinct points y and z. A special case of (B.26) shows that there is a continuous linear functional L on E such that $L(z) < L(y)$. If $\beta = \min\{L(x) : x \in B_0\}$, then $\{x \in B_0 : L(x) = \beta\}$ is a nonvoid extreme subset of B_0 (B.29), and it does not contain y. But by (B.28), $\{x \in B_0 : L(x) = \beta\}$ must also be an extreme subset of C, and this contradicts the minimality of B_0.

(II) Let B denote the set of all convex combinations of extreme points of C. Then B^- is a compact convex subset of C. Assume that $C \cap (B^-)'$ is nonvoid and contains, say, x_0. By (B.26), there is a continuous linear functional L on E such that $\min\{L(x) : x \in B^-\} > L(x_0)$. Let $\beta = \min\{L(x) : x \in C\}$; then $\{x \in C : L(x) = \beta\}$ is a nonvoid closed extreme subset of C disjoint from B^-. By step (I), the set $\{x \in C : L(x) = \beta\}$ contains an extreme point of C. This is plainly impossible. □

We now take up the problem of representing linear functionals on certain spaces of functions as differences of nonnegative linear functionals. The matter can be treated in more generality than is done here,

---

[1] This proof is taken from KELLEY [1].

but at the cost of technical complications. See for example BOURBAKI [2], pp. 17—40, or BIRKHOFF [2], pp. 238—248.

(B.31) Definitions. Let Y be a nonvoid set, and let $\mathfrak{F}$ be a *real* linear space of real-valued functions on Y [linear operations are pointwise] such that $f \in \mathfrak{F}$ implies $|f| \in \mathfrak{F}$. Then if $f, g \in \mathfrak{F}$, we have $\max(f, g) = \frac{1}{2}(f+g+|f-g|) \in \mathfrak{F}$ and $\min(f, g) = \frac{1}{2}(f+g-|f-g|) \in \mathfrak{F}$. The symbol $\mathfrak{F}^+$ will denote the set of all nonnegative functions in $\mathfrak{F}$. A linear functional L on $\mathfrak{F}$ is said to be *nonnegative* if $L(f) \geqq 0$ for all $f \in \mathfrak{F}^+$. A linear functional L on $\mathfrak{F}$ is said to be *strictly positive* if $L(f) > 0$ for all f in $\mathfrak{F}^+$ except $f=0$. A linear functional L on $\mathfrak{F}$ will be called *relatively bounded* if for every $f \in \mathfrak{F}^+$, the set $\{L(g) : g \in \mathfrak{F}, |g| \leqq f\}$ is a bounded set of real numbers[1].

Throughout (B.32)—(B.37), Y and $\mathfrak{F}$ will be as in (B.31).

(B.32) Lemma. *Let L be a real-valued function on $\mathfrak{F}^+$ such that:*

(i) $L(f+g) = L(f) + L(g)$ *for* $f, g \in \mathfrak{F}^+$ [*L is additive*];

(ii) $L(\alpha f) = \alpha L(f)$ *for* $f \in \mathfrak{F}^+$ *and* $\alpha \geqq 0$ [*L is nonnegative homogeneous*].

Then L can be extended in exactly one way so as to be a linear functional on $\mathfrak{F}$.

Proof. Since $f = \frac{|f|+f}{2} - \frac{|f|-f}{2}$, every f in $\mathfrak{F}$ is the difference of two functions in $\mathfrak{F}^+$. If $f = f_1 - f_2$ where $f_1, f_2 \in \mathfrak{F}^+$, then define $L_r(f) = L(f_1) - L(f_2)$. If $f = f_1 - f_2 = f_3 - f_4$ with all $f_j \in \mathfrak{F}^+$, then $f_1 + f_4 = f_3 + f_2$, and so by (i), $L(f_1) - L(f_2) = L(f_3) - L(f_4)$. That is, L_r is defined uniquely on $\mathfrak{F}$ and is an extension of L. It is easy to verify that L_r is a linear functional on $\mathfrak{F}$. □

(B.33) Lemma. *Every nonnegative linear functional L on $\mathfrak{F}$ is relatively bounded.*

Proof. If $f \in \mathfrak{F}^+$, $g \in \mathfrak{F}$, and $|g| \leqq f$, then $f - g \geqq 0$ and $f + g \geqq 0$. Therefore $L(f) - L(g) \geqq 0$ and $L(f) + L(g) \geqq 0$, so that $|L(g)| \leqq L(f)$. □

The linear space $\mathfrak{F}^*$ of all relatively bounded [real] linear functionals on $\mathfrak{F}$ admits a partial ordering: for $L, M \in \mathfrak{F}^*$, we write $L \geqq M$ if $L - M$ is a nonnegative linear functional.

(B.34) Theorem. *Under the partial ordering $\geqq$, $\mathfrak{F}^*$ is a lattice. That is, given $L, M \in \mathfrak{F}^*$, there is a least element* $\max(L, M)$ *of $\mathfrak{F}^*$ greater than or equal to L and M, and there is a greatest element* $\min(L, M)$ *of $\mathfrak{F}^*$ less than or equal to L and M. For $f \in \mathfrak{F}^+$, we have*

(i) $\max(L, M)(f) = \sup\{L(g) + M(f-g) : g \in \mathfrak{F}^+, g \leqq f\}$

and

(ii) $\min(L, M)(f) = \inf\{L(g) + M(f-g) : g \in \mathfrak{F}^+, g \leqq f\}$.

---

[1] If $\mathfrak{F}$ is a normed linear space such that $\|g\| \leqq \|f\|$ whenever $|g| \leqq |f|$ and $f, g \in \mathfrak{F}$, then a bounded linear functional on $\mathfrak{F}$ is relatively bounded.

Proof. If $I\in\mathfrak{F}^*$, $I\geqq L$, and $I\geqq M$, it is clear that $I(f)$ must be greater than or equal to the right side of (i) for $f\in\mathfrak{F}^+$. Let $I_1(f)$ denote the right side of (i); note that $I_1(f)$ is a real number for all $f\in\mathfrak{F}^+$ since L and M are relatively bounded. We next show that I_1 is additive and nonnegative homogeneous on $\mathfrak{F}^+$. Since $L(\alpha g)=\alpha L(g)$ and $M(\alpha g)=\alpha M(g)$ for $\alpha\in R$ and $g\in\mathfrak{F}$, it follows at once that $I_1(\alpha f)=\alpha I_1(f)$ for $f\in\mathfrak{F}^+$ and $\alpha\geqq 0$.

To prove that I_1 is additive on $\mathfrak{F}^+$, take f_1 and f_2 in $\mathfrak{F}^+$, and $0\leqq g_j\leqq f_j$ $(j=1, 2)$ such that $L(g_j)+M(f_j-g_j)+\frac{\varepsilon}{2}>I_1(f_j)$. Then

$$I_1(f_1+f_2)\geqq L(g_1+g_2)+M(f_1+f_2-g_1-g_2)>I_1(f_1)+I_1(f_2)-\varepsilon.$$

Hence we have

$$I_1(f_1+f_2)\geqq I_1(f_1)+I_1(f_2).$$

To prove the reversed inequality, choose $g\in\mathfrak{F}^+$ such that $g\leqq f_1+f_2$ and $L(g)+M(f_1+f_2-g)+\varepsilon>I_1(f_1+f_2)$. Write $g_1=\min(g, f_1)$ and $g_2=g-g_1$. Then we have $0\leqq g_1\leqq f_1$, $0\leqq g_2\leqq f_2$, $g_1+g_2=g$, and so

$$\begin{aligned}I_1(f_1+f_2)&<L(g_1+g_2)+M(f_1+f_2-g_1-g_2)+\varepsilon\\&=L(g_1)+M(f_1-g_1)+L(g_2)+M(f_2-g_2)+\varepsilon\leqq I_1(f_1)+I_1(f_2)+\varepsilon.\end{aligned}$$

Therefore I_1 is additive on $\mathfrak{F}^+$. By (B.32), I_1 can be extended uniquely to a linear functional on $\mathfrak{F}$; clearly this extension is the least element $\max(L, M)$ greater than or equal to L and M. Obviously L is relatively bounded, and by (B.33), $\max(L, M)-L$ is relatively bounded; hence $\max(L, M)$ is relatively bounded.

Relation (ii) follows from the identity $\min(L, M)=-\max(-L, -M)$. □

(B.35) Theorem. *The space $\mathfrak{F}^*$ is a complete lattice. That is, if $\{L_\gamma\}_{\gamma\in\Gamma}$ is any subset of $\mathfrak{F}^*$ that is bounded above [below], then this set has a least upper [greatest lower] bound.*

Proof. Suppose that $L_\gamma\leqq M$ for all $\gamma\in\Gamma$. For $f\in\mathfrak{F}^+$, let

$$I(f)=\sup\{\max(L_{\gamma_1}, \ldots, L_{\gamma_m})(f):\{\gamma_1, \ldots, \gamma_m\}\subset\Gamma\}.$$

It is easy to show that I is additive and nonnegative homogeneous on $\mathfrak{F}^+$, so that I can be extended to a linear functional I_0 on $\mathfrak{F}$ (B.32). It is also easy to see that I_0 is the least linear majorant of all of the L_γ's. For any $\gamma_0\in\Gamma$, $I_0-L_{\gamma_0}$ is nonnegative and hence relatively bounded (B.33). Therefore $I_0=(I_0-L_{\gamma_0})+L_{\gamma_0}$ is relatively bounded.

Sets $\{L_\gamma\}_{\gamma\in\Gamma}$ bounded below are treated by examining $\{-L_\gamma\}_{\gamma\in\Gamma}$. □

(B.36) Theorem. *A linear functional on $\mathfrak{F}$ can be expressed as the difference of two nonnegative linear functionals if and only if it is relatively bounded.*

Proof. If a linear functional is the difference of nonnegative linear functionals, then it is relatively bounded by (B.33). To prove the converse, consider a relatively bounded linear functional L on $\mathfrak{F}$. Define $I_1 = \max(L, 0)$ and $I_2 = I_1 - L$. Then I_1 and I_2 are nonnegative linear functionals on $\mathfrak{F}$ and $L = I_1 - I_2$. □

(B.37) Theorem. *Let L be a relatively bounded linear functional on $\mathfrak{F}$. Then there are unique nonnegative linear functionals I_1 and I_2 such that $L = I_1 - I_2$ and $\min(I_1, I_2) = 0$. Furthermore, we have $I_1 = \max(L, 0)$ and $I_2 = -\min(L, 0)$.*

Proof. Suppose that $L = J_1 - J_2$ where J_1 and J_2 are nonnegative linear functionals on $\mathfrak{F}$. Then for $f \in \mathfrak{F}^+$, we have

$$\begin{aligned} 0 \leqq \min(J_1, J_2)(f) &= \inf\{J_1(f-g) + J_2(g) : g \in \mathfrak{F}^+, g \leqq f\} \\ &= \inf\{J_1(f) - J_1(g) + J_1(g) - L(g) : g \in \mathfrak{F}^+, g \leqq f\} \\ &= \inf\{J_1(f) - L(g) : g \in \mathfrak{F}^+, g \leqq f\} \\ &= J_1(f) - \sup\{L(g) : g \in \mathfrak{F}^+, g \leqq f\} = J_1(f) - \max(L, 0)(f). \end{aligned}$$

Therefore we have $\min(J_1, J_2) = J_1 - \max(L, 0) \geqq 0$. It follows that $\min(J_1, J_2) = 0$ if and only if $J_1 = \max(L, 0)$.

The foregoing together with (B.36) shows that $L = I_1 - I_2$, where $I_1 = \max(L, 0)$ and $\min(I_1, I_2) = 0$. Since $-L = I_2 - I_1$, we must also have $I_2 = \max(-L, 0) = -\min(L, 0)$. □

Our final result along these lines concerns complex functionals on complex linear spaces.

(B.38) Theorem. *Let Y be a nonvoid set and $\mathfrak{F}$ a complex linear space of complex-valued functions on Y. Let $\mathfrak{F}^r$ and $\mathfrak{F}^+$ denote the sets of real-valued and nonnegative real-valued functions in $\mathfrak{F}$, respectively. Suppose that $f \in \mathfrak{F}$ implies $\bar{f} \in \mathfrak{F}$ and that every $f \in \mathfrak{F}^r$ can be written in some fashion as the difference of two functions in $\mathfrak{F}^+$. Let Φ be any complex-valued function defined on $\mathfrak{F}^+$ such that:*

(i) *$\Phi(f_1 + f_2) = \Phi(f_1) + \Phi(f_2)$ for $f_1, f_2 \in \mathfrak{F}^+$;*

(ii) *$\Phi(\alpha f) = \alpha \Phi(f)$ for $f \in \mathfrak{F}^+$ and $\alpha \geqq 0$ [Φ is nonnegative homogeneous].*

Then Φ can be extended in exactly one way so as to be a complex linear functional on $\mathfrak{F}$.

Proof. Just as in Lemma (B.32), Φ can be extended uniquely to a function Φ_r on $\mathfrak{F}^r$ that is additive and real homogeneous. If $f \in \mathfrak{F}$, then $\operatorname{Re} f = \frac{1}{2}(f + \bar{f})$ and $\operatorname{Im} f = \frac{1}{2i}(f - \bar{f})$ are also in $\mathfrak{F}$. Hence f can be *uniquely* written as $f = f_1 + i f_2$ where $f_1, f_2 \in \mathfrak{F}^r$. The only possible way to extend Φ_r linearly over $\mathfrak{F}$ is to write $\Phi_c(f) = \Phi_r(f_1) + i\Phi_r(f_2)$. An obvious calculation shows that Φ_c is additive and complex homogeneous on $\mathfrak{F}$. □

We now sketch the theory of inner product and Hilbert spaces. We give definitions largely to establish our notation. Theorems proved in HALMOS [3] are stated without proof.

(B.39) Definition. Let E be a complex linear space. Suppose that there is a complex-valued function on $E \times E$, whose value at (x, y) is written as $\langle x, y\rangle$, having the following properties:

(i) $\langle x+y, z\rangle = \langle x, z\rangle + \langle y, z\rangle$ for $x, y, z \in E$;

(ii) $\langle \alpha x, y\rangle = \alpha \langle x, y\rangle$ for $x, y \in E$ and $\alpha \in K$;

(iii) $\langle x, y\rangle = \overline{\langle y, x\rangle}$ for $x, y \in E$;

(iv) $\langle x, x\rangle > 0$ if $x \neq 0$.

Then the function $(x, y) \to \langle x, y\rangle$ is called an *inner product* and E is said to be an *inner product space*.

(B.40) Note. If E is an inner product space, then there is a natural norm for E; for $x \in E$, define $\|x\| = \sqrt{\langle x, x\rangle}$. This is a legitimate norm for E. This is not trivial; in fact, the triangle inequality depends upon the CAUCHY-BUNYAKOVSKIĬ-SCHWARZ inequality: $|\langle x, y\rangle| \leqq \|x\| \cdot \|y\|$ [see HALMOS [3], p. 16].

(B.41) Theorem. *If E is an inner product space, then*

(i) $\langle x, y\rangle = \frac{1}{4}\|x+y\|^2 - \frac{1}{4}\|x-y\|^2 + \frac{i}{4}\|x+iy\|^2 - \frac{i}{4}\|x-iy\|^2$,

for $x, y \in E$.

This is proved by a direct computation.

(B.42) Definition. A *linear isometry* is a linear mapping T of a normed linear space E onto another normed linear space E' that preserves the norm; *i.e.* $\|T(x)\| = \|x\|$ for all $x \in E$.

(B.43) Theorem. *A linear isometry T between two inner product spaces E and E' also preserves the inner product*; i.e., $\langle T(x), T(y)\rangle = \langle x, y\rangle$ *for all* $x, y \in E$.

This follows from (B.41).

The most important class of inner product spaces for our purposes are those for which the associated norm induces a complete metric.

(B.44) Definition. An inner product space that is a Banach space under the norm $\|x\| = \langle x, x\rangle^{\frac{1}{2}}$ is called a *Hilbert space*.

In accordance with a common convention, we will denote Hilbert spaces by H, H', etc. rather than by E, E', etc. In consonance with the notation employed in the main text [Chapter Five, *passim*], we write elements of an abstract Hilbert space as $\xi, \eta, \omega, \ldots$.

(B.45) Theorem [F. RIESZ]. *Let H be a Hilbert space. For every bounded linear functional L on H, there is a unique $\eta \in H$ such that*

$L(\xi)=\langle\xi,\eta\rangle$ for all $\xi\in H$. For every conjugate-linear bounded functional M on H, there is a unique $\omega\in H$ such that $M(\xi)=\langle\omega,\xi\rangle$ for all $\xi\in H$.

(B.46) Corollary. *The mapping $L\rightarrow\eta$ described in* (B.45) *is a conjugate-linear isometry carrying the space H^* onto H.*

A proof of (B.45) is given in HALMOS [3], p. 31.

(B.47) Definition. Let H be a Hilbert space. Two elements ξ and η in H are said to be *orthogonal* if $\langle\xi,\eta\rangle=0$. More generally, two subsets A and B of H are *orthogonal* if $\langle\xi,\eta\rangle=0$ for all $\xi\in A$ and $\eta\in B$. If ξ and η are orthogonal, we write $\xi\perp\eta$; similarly $A\perp B$ means that A and B are orthogonal. For a subset A of H, $A^{\perp}$ will denote the set $\{\xi\in H:\xi\perp\eta \text{ for all } \eta\in A\}$.

A single subset A of H is said to be *orthogonal* if $\langle\xi,\eta\rangle=0$ for all $\xi,\eta\in A$ such that $\xi\neq\eta$. If we also have $\langle\xi,\xi\rangle=1$ for all $\xi\in A$, then A is said to be *orthonormal*. An *orthonormal basis of H* is a maximal orthonormal subset of H.

(B.48) Theorem. *Let M be a closed subspace of a Hilbert space H. Then $M^{\perp}$ is a closed subspace of H, $M^{\perp}$ is orthogonal to M, $M^{\perp\perp}=M$, and $H=M\oplus M^{\perp}$.*

See HALMOS [3], §12, for the proof.

The next theorem justifies the definition following it. It is proved in HALMOS [3], p. 29.

(B.49) Theorem. *Let H be a Hilbert space. Then H has at least one orthonormal basis, and any two orthonormal bases of H have the same cardinal number.*

(B.50) Definition. The *dimension* of a Hilbert space H is the cardinal number of any of its orthonormal bases.

We now consider adjoint operators. For the generality required in §22, we must use an arbitrary complex normed linear space E, and define a space $E^{\sim}$ of functionals on E that imitates so far as possible the formation of inner products in an inner product space[1]. Again in consonance with the notation of Chapter Five, we write elements of linear spaces as $\xi,\eta,\omega,\ldots$. In handling operators T on E, it is standard and convenient to write $T\xi$ for $T(\xi)$; we shall frequently do this in what follows.

(B.51) Definitions. Let E be a complex normed linear space. Let $E^{\sim}$ denote the set of all bounded *conjugate-linear* functionals on E. These functionals are added and multiplied by complex numbers pointwise on E, and thus form a normed linear space. For $\omega\in E^{\sim}$ and $\xi\in E$, we write the value of ω at ξ as $\langle\omega,\xi\rangle$. [2]

[1] The following treatment of adjoint operators was suggested to us in conversation by N. ARONSZAJN.

[2] A handy mnemonic is the rule "functionals come first".

(B.52) Remarks. The mapping $(\omega, \xi) \to \langle \omega, \xi \rangle$ has some of the formal properties of the inner product in an inner product space. Thus, for ω's in $E^\sim$, ξ's in E, and α in K, we have:

$$\langle \omega_1 + \omega_2, \xi \rangle = \langle \omega_1, \xi \rangle + \langle \omega_2, \xi \rangle;$$
$$\langle \alpha\omega, \xi \rangle = \alpha \langle \omega, \xi \rangle;$$
$$\langle \omega, \xi_1 + \xi_2 \rangle = \langle \omega, \xi_1 \rangle + \langle \omega, \xi_2 \rangle;$$
$$\langle \omega, \alpha\xi \rangle = \bar{\alpha} \langle \omega, \xi \rangle;$$
$$|\langle \omega, \xi \rangle| \leqq \|\omega\| \cdot \|\xi\|.$$

Since E and $E^\sim$ are in general different entities, we can no longer write $\langle \xi, \xi \rangle$.

There exists a one-to-one correspondence between $E^\sim$ and E^*, as follows. If $\omega \in E^\sim$ and $L(\xi) = \overline{\omega(\xi)} = \overline{\langle \omega, \xi \rangle}$, then L belongs to E^*. Conversely, if $L \in E^*$ and $\omega(\xi) = \overline{L(\xi)}$, then ω belongs to $E^\sim$. Using this correspondence, we make $E^\sim$ into a complex Banach space. If E is reflexive, every element in $E^{\sim\sim}$ is determined by an element η in E. That is, for each $\eta \in E$, the functional $\omega \to \overline{\langle \omega, \eta \rangle}$ on $E^\sim$ belongs to $E^{\sim\sim}$ and all elements of $E^{\sim\sim}$ have this form. Moreover, this correspondence is a one-to-one, linear, norm-preserving map of E onto $E^{\sim\sim}$.

Consider a Hilbert space H. Then H and H^* can be identified, as (B.46) shows. It is plain that H and $H^\sim$ can also be identified: every bounded conjugate-linear functional on H has the form $\xi \to \langle \eta, \xi \rangle$ for some $\eta \in H$, as (B.45) implies. [Here $\langle \eta, \xi \rangle$ denotes the inner product in H.]

(B.53) Definition. Let E be a complex normed linear space and let T be a bounded operator[1] on E. For each ω in $E^\sim$, the mapping $\xi \to \langle \omega, T\xi \rangle$ belongs to $E^\sim$; we define this mapping to be $T^\sim \omega$. Thus

(i) $\langle \omega, T\xi \rangle = \langle T^\sim \omega, \xi \rangle$

for $\omega \in E^\sim$ and $\xi \in E$. The function $T^\sim$ carrying $E^\sim$ into $E^\sim$ is called the *adjoint of the operator* T.

It is easy to show that $T^\sim$ is a bounded linear transformation, *i.e.*, $T^\sim$ is a bounded operator on $E^\sim$.

(B.54) Discussion. For an operator T on $E^\sim$, there is of course an adjoint operator $T^\sim$ on $E^{\sim\sim}$. If E is a *reflexive* Banach space, then $E^{\sim\sim}$ can be identified with E, as noted in (B.52), and $T^\sim$ may be regarded as defined on E. In fact, if the element in $E^{\sim\sim}$ that corresponds to ξ in E is denoted by ξ^0, then the equality (B.53.i) takes on the form

$$\langle \xi^0, T\omega \rangle = \langle T^\sim \xi^0, \omega \rangle$$

---

[1] Recall that "operator on E" means "linear transformation carrying E into itself" (B.2).

for $\xi^0 \in E^{\sim\sim}$ and $\omega \in E^{\sim}$. Identifying $E^{\sim\sim}$ with E, we obtain

(i) $\langle T\omega, \xi\rangle = \langle \omega, T^{\sim}\xi\rangle$

for $\xi \in E$ and $\omega \in E^{\sim}$.

For the case of Hilbert spaces, the above definition of adjoint operator reduces to the ordinary one.

(B.55) Theorem. *Let H be a Hilbert space. If T and U are bounded operators on H and α is in K, then:*

(i) $\|T^{\sim}\| = \|T\|$;

(ii) $T^{\sim\sim} = T$;

(iii) $(\alpha T)^{\sim} = \bar{\alpha} T^{\sim}$;

(iv) $(T+U)^{\sim} = T^{\sim} + U^{\sim}$;

(v) $(TU)^{\sim} = U^{\sim} T^{\sim}$.

For the proofs, see Halmos [3], p. 39.

(B.56) Definitions. Let H be a Hilbert space and T a bounded operator on H. If $T = T^{\sim}$, then T is said to be *Hermitian.* It is easy to show that a bounded operator T is Hermitian if and only if $\langle T\xi, \xi\rangle$ is real for all $\xi \in H$. If $TT^{\sim} = T^{\sim}T$, then T is said to be *normal.* If T is a linear isometry of H onto H, then T is said to be *unitary*. A bounded operator T on H is said to be *positive-definite* if $\langle T\xi, \xi\rangle \geqq 0$ for all $\xi \in H$. In particular, every positive-definite operator is Hermitian.

Let M be a closed linear subspace of H. Every element ξ in H can be written in exactly one way as $\xi_1 + \xi_2$, where $\xi_1 \in M$ and $\xi_2 \in M^{\perp}$ (B.48). The mapping P defined by $P\xi = \xi_1$ is called the *projection of H onto M.*

For a Hilbert space H, $\mathfrak{B}(H)$ will denote the linear space of all bounded operators on H.

(B.57) Theorem. *A function P mapping a Hilbert space H into itself is a projection if and only if it is an Hermitian idempotent operator. Whenever this is the case, P is the projection of H onto $M = \{\xi \in H : P\xi = \xi\}$, and if $M \neq \{0\}$, then $\|P\| = 1$.*

This theorem is treated in Halmos [3], §26.

(B.58) Theorem. *For a bounded operator T on a Hilbert space H, the following statements are equivalent:*

(i) $T = 0$;

(ii) $\langle T\xi, \eta\rangle = 0$ *for all* $\xi, \eta \in H$;

(iii) $\langle T\xi, \xi\rangle = 0$ *for all* $\xi \in H$.

For the proof, see Halmos [3], §22.

(B.59) Theorem. *Let H' be a dense subspace of a Hilbert space H. If T' is a linear isometry of H' onto a dense subspace of H, then T' can be extended uniquely to a unitary operator T on H.*

This theorem is very like (B.11). We omit the proof.

(B.60) Theorem. *Let H be a Hilbert space. Suppose that L is a bilinear functional on $H \times H$ that is bounded*: i.e., *there is a positive real number A such that $|L(\xi, \eta)| \leqq A \|\xi\| \cdot \|\eta\|$ for all $\xi, \eta \in H$. Then there exists a unique bounded operator T on H such that*

(i) $L(\xi, \eta) = \langle T\xi, \eta \rangle$

for all $\xi, \eta \in H$.

For a proof, see HALMOS [3], §22.

We will have occasion to consider direct sums of arbitrarily many Hilbert spaces and also direct sums of operators.

(B.61) Definition. For an arbitrary family $\{H_\gamma\}_{\gamma \in \Gamma}$ of Hilbert spaces, the direct sum $H = \bigoplus_{\gamma \in \Gamma} H_\gamma$ is defined as follows. Let H be the additive subgroup of the additive group $\mathsf{P}_{\gamma \in \Gamma} H_\gamma$ consisting of all $(\xi_\gamma) \in \mathsf{P}_{\gamma \in \Gamma} H_\gamma$ for which $\sum_{\gamma \in \Gamma} \|\xi_\gamma\|^2 < \infty$. Define scalar multiplication in H coordinatewise, and an inner product in H by

(i) $\langle (\xi_\gamma), (\eta_\gamma) \rangle = \sum_{\gamma \in \Gamma} \langle \xi_\gamma, \eta_\gamma \rangle$.

With the above definitions, $H = \bigoplus_{\gamma \in \Gamma} H_\gamma$ is a Hilbert space. Elements (ξ_γ) of H are often written as $\sum_{\gamma \in \Gamma} \xi_\gamma$.

(B.62) Definition. Let $\{H_\gamma\}_{\gamma \in \Gamma}$ be a family of Hilbert spaces and for each $\gamma \in \Gamma$, let T_γ be a bounded operator on H_γ. Suppose also that $\sup\{\|T_\gamma\| : \gamma \in \Gamma\} < \infty$. We define the sum $T = \sum_{\gamma \in \Gamma} T_\gamma$ on H by the following relation:

(i) $T((\xi_\gamma)) = (T_\gamma(\xi_\gamma))$.

The function T is a bounded operator on H and $\|T\| = \sup\{\|T_\gamma\| : \gamma \in \Gamma\}$. We omit the proofs of these statements.

(B.63) Theorem. *Let H be a Hilbert space and suppose that $\{H_\gamma\}_{\gamma \in \Gamma}$ is a family of subspaces of H having the property that $H_\gamma \subset H_\delta^\perp$ whenever $\gamma \neq \delta$, and such that $\bigcap_{\gamma \in \Gamma} H_\gamma^\perp = \{0\}$. Then H is isometric with $\bigoplus_{\gamma \in \Gamma} H_\gamma^-$.*

Proof. Let H_0 consist of those elements (ξ_γ) in $\bigoplus_{\gamma \in \Gamma} H_\gamma^-$ such that $\xi_\gamma \in H_\gamma$ for all $\gamma \in \Gamma$ and $\xi_\gamma = 0$ except for finitely many γ. Then H_0 is a dense linear subspace of $\bigoplus_{\gamma \in \Gamma} H_\gamma^-$. For $(\xi_\gamma) \in H_0$, define $\tau((\xi_\gamma)) = \sum_{\gamma \in \Gamma} \xi_\gamma$. It is easy to verify that τ is a linear mapping of H_0 into H. If $(\eta_\gamma) \in H_0$, then $\langle \tau((\xi_\gamma)), \tau((\eta_\gamma)) \rangle = \left\langle \sum_{\gamma \in \Gamma} \xi_\gamma, \sum_{\gamma \in \Gamma} \eta_\gamma \right\rangle = \sum_{\gamma \in \Gamma} \langle \xi_\gamma, \eta_\gamma \rangle = \langle (\xi_\gamma), (\eta_\gamma) \rangle$. Thus τ is a linear isometry of H_0 onto $\tau(H_0)$. To see that $\tau(H_0)$ is dense in H, suppose that $\zeta \in H$ and that $\langle \xi, \zeta \rangle = 0$ for all $\xi \in \tau(H_0)$. Then $\zeta \in H_\gamma^\perp$ for all $\gamma \in \Gamma$, and hence $\zeta = 0$. As ζ is arbitrary, $\tau(H_0)$ is dense in H. A routine argument shows that τ can be extended in exactly one way to be a linear isometry of $\bigoplus_{\gamma\in\Gamma} H_\gamma^-$ onto H. □

Two more results needed in the text rightfully belong in this appendix. However, it is convenient to state and prove them in Appendix C after the elementary theory of Banach algebras has been described. These are the spectral theorem (C.40)—(C.42) and the fact that a positive-definite operator on a Hilbert space has a unique positive-definite square root (C.35).

Appendix C

Introduction to normed algebras

In this appendix we sketch the elementary theory of Banach algebras. We give only that part of the theory needed for our study of $\boldsymbol{M}(G)$ and its subalgebras. A discussion of comparable scope appears in LOOMIS [2]. Far more complete treatments appear in NAĬMARK [1] and RICKART [1]. We include a proof of the spectral theorem, since it will follow simply from machinery needed for other purposes.

(C.1) Definitions. Let A be a linear space over a field F. Suppose that for each pair x, y of elements from A a product xy is defined, that under this operation A is a semigroup, and that this product is related to the linear operations in A as follows:

(i) $x(y+z) = xy + xz$,

(ii) $(x+y)z = xz + yz$,

(iii) $(\alpha x)y = x(\alpha y) = \alpha(xy)$,

for all $x, y, z \in A$ and all $\alpha \in F$. Then A is said to be an *algebra over* F. An *algebra with unit* is an algebra having a multiplicative identity $u \neq 0$; a *commutative algebra* A is one for which $xy = yx$ for all $x, y \in A$. *Real* and *complex algebras* are algebras over R and K, respectively. We recall that an algebra with an unspecified scalar field is a complex algebra (B.1).

A *Banach algebra* [*normed algebra*] A is a Banach space [normed linear space] that is also an algebra and satisfies the following condition:

(iv) $\|xy\| \leqq \|x\| \cdot \|y\|$

for all $x, y \in A$. If A has a unit u, we will often require that

(v) $\|u\| = 1$.

A linear subspace I of A is a *left ideal* if $x \in I$, $z \in A$ imply $zx \in I$. A linear subspace I is a *right ideal* if $x \in I$, $z \in A$ imply $xz \in I$. A *two-sided ideal* is a left ideal that is also a right ideal. An ideal I of A such that $I \neq A$ is a *proper ideal*. An algebra having no proper two-sided ideals except $\{0\}$ is said to be *simple*. In a commutative algebra, the distinctions among left, right, and two-sided ideals vanish. In this case, we write "ideal" for these entities.

An *algebra homomorphism*, as one might expect, is a mapping τ of an algebra A into another algebra B that is linear and also multiplicative: $\tau(xy) = \tau(x)\,\tau(y)$ for $x, y \in A$.

(C.2) Theorem. *Let A be a normed algebra and I a closed two-sided ideal in A. Let A/I denote the normed linear space A/I described in Theorem* (B.17). *For $x+I$, $y+I \in A/I$, define $(x+I)(y+I) = xy+I$. With this multiplication, A/I is a normed algebra, and A/I is a Banach algebra if A is. If A has a unit u and $\|u\| = 1$, then $\|u+I\| = 1$.*

Proof. It is easy to see that the product is well defined and that A/I is an algebra. By (B.17), A/I is a normed linear space which is complete if A is complete.

We next verify that (C.1.iv) holds for A/I. Consider $x+I$, $y+I$ in A/I and $\varepsilon > 0$. For some v and w in I, we have $\|x+v\| \cdot \|y+w\| < \|x+I\| \cdot \|y+I\| + \varepsilon$. Since $vy + xw + vw$ belongs to I, we have

$$\begin{aligned}\|xy+I\| &\leqq \|xy+vy+xw+vw\| = \|(x+v)(y+w)\| \\ &\leqq \|x+v\| \cdot \|y+w\| < \|x+I\| \cdot \|y+I\| + \varepsilon.\end{aligned}$$

Therefore $\|xy+I\| \leqq \|x+I\| \cdot \|y+I\|$; that is, (C.1.iv) holds in A/I.

Suppose finally that A has a unit u and that $\|u\| = 1$. It is obvious that $\|u+I\| \leqq 1$. Moreover, we have $\|u+I\| = \|(u+I)^2\| \leqq \|u+I\|^2$, so that $1 \leqq \|u+I\|$. □

We now show that any algebra can be imbedded in an algebra with unit.

(C.3) Theorem. *Let A be an algebra. Let A_u be the set of all pairs (x, λ) with $x \in A$ and $\lambda \in K$. Define*

(i) $\alpha(x, \lambda) = (\alpha x, \alpha\lambda)$,

(ii) $(x, \lambda) + (y, \mu) = (x+y, \lambda+\mu)$,

(iii) $(x, \lambda)(y, \mu) = (xy + \lambda y + \mu x, \lambda\mu)$,

for (x, λ), $(y, \mu) \in A_u$ and $\alpha \in K$. Then A_u is an algebra with unit $(0, 1)$, *and its subset $\{(x, 0) : x \in A\}$ is isomorphic with A in every way. The algebra A_u is commutative if and only if A is commutative. If A is a normed [Banach] algebra, then A_u with the norm*

(iv) $\|(x, \lambda)\| = \|x\| + |\lambda|$,

is a normed [Banach] algebra.

Proof. All necessary verifications are easy. For example, the inequalities

$$\begin{aligned}\|(x, \lambda)(y, \mu)\| &= \|(xy + \lambda y + \mu x, \lambda\mu)\| = \|xy + \lambda y + \mu x\| + |\lambda\mu| \\ &\leqq \|x\| \cdot \|y\| + |\lambda| \cdot \|y\| + |\mu| \cdot \|x\| + |\lambda| \cdot |\mu| \\ &= (\|x\| + |\lambda|)(\|y\| + |\mu|) = \|(x, \lambda)\| \cdot \|(y, \mu)\|,\end{aligned}$$

establish (C.1.iv) for the algebra A_u. □

We next establish some technical properties of normed algebras.

(C.4) Theorem. *Let A be a normed algebra. For every x in A the limit* $\lim\limits_{n\to\infty} \|x^n\|^{1/n}$ *exists,*

(i) $\lim\limits_{n\to\infty} \|x^n\|^{1/n} = \inf\{\|x^n\|^{1/n} : n=1, 2, \ldots\}$,

and

(ii) $\lim\limits_{n\to\infty} \|x^n\|^{1/n} \leqq \|x\|$.

Proof. Let $\alpha = \inf\{\|x^n\|^{1/n} : n=1, 2, \ldots\}$; obviously $\alpha \leqq \underline{\lim}_{n\to\infty} \|x^n\|^{1/n}$. Consider $\varepsilon > 0$ and choose a positive integer m for which $\|x^m\|^{1/m} < \alpha + \varepsilon$. For every positive integer n, we write $n = a_n m + b_n$ where $0 \leqq b_n < m$. Then $\|x^n\| \leqq \|x^m\|^{a_n} \|x\|^{b_n}$, and therefore

$$\|x^n\|^{1/n} \leqq \|x^m\|^{a_n/n} \|x\|^{b_n/n} \leqq (\alpha+\varepsilon)^{m a_n/n} \|x\|^{b_n/n}.$$

Since $\lim\limits_{n\to\infty} \frac{m a_n}{n} = 1$ and $\lim\limits_{n\to\infty} \frac{b_n}{n} = 0$, we infer that $\overline{\lim}_{n\to\infty} \|x^n\|^{1/n} \leqq \alpha + \varepsilon$. It follows that $\alpha = \lim\limits_{n\to\infty} \|x^n\|^{1/n}$. Relation (ii) is obvious, since $\|x^n\| \leqq \|x\|^n$ for all n. □

For algebras without unit, the notion of quasi-inverse is important.

(C.5) Definition. Let A be an algebra and x an element of A. An element y in A is called a *left quasi-inverse* [*right quasi-inverse*] *for* x if $x + y - yx = 0$ [$x + y - xy = 0$]. If y is both a left and a right quasi-inverse for x, then y is called a *quasi-inverse for* x.[1]

Suppose that A has a unit u. Then x has a quasi-inverse y if and only if $x - u$ has a [multiplicative] inverse $y - u$. The element $u - x$ has an inverse z if and only if x has a quasi-inverse $u - z$. Thus quasi-inverses are substitutes for inverses in algebras without unit.

Let A be an arbitrary algebra. If x in A has a left and a right quasi-inverse, these elements are equal. Therefore quasi-inverses are unique.

(C.6) Theorem. *Let A be a Banach algebra and let x be in A. If* $\lim\limits_{n\to\infty} \|x^n\|^{1/n} < 1$, *then x has a quasi-inverse y and*

(i) $y = -\sum\limits_{k=1}^{\infty} x^k$.

Proof. From the hypothesis $\lim\limits_{n\to\infty} \|x^n\|^{1/n} < 1$, it follows that the series $\sum\limits_{k=1}^{\infty} \|x^k\|$ converges. If $y_n = -\sum\limits_{k=1}^{n} x^k$ for $n = 1, 2, \ldots$, then $\{y_n\}_{n=1}^{\infty}$ is a Cauchy sequence in A, since $\|y_n - y_m\| = \left\|\sum\limits_{k=m+1}^{n} x^k\right\| \leqq \sum\limits_{k=m+1}^{n} \|x^k\|$ for $n > m$. Let y in A be the limit of the sequence $\{y_n\}_{n=1}^{\infty}$. Clearly, we have

$$x y_n = y_n x = x + y_{n+1}, \tag{1}$$

[1] At least three different terms have been employed for what we call quasi-inverses; we follow RICKART [1], p. 16, but not LOOMIS [2], p. 64 or HILLE and PHILLIPS [1], pp. 680–681.

for each n. Taking limits in (1) as $n \to \infty$, we obtain $xy = yx = x + y$; that is, $x + y - xy = x + y - yx = 0$. □

(C.7) Corollary. *Let A be a Banach algebra with unit u. If x is in A and $\lim_{n\to\infty} \|(u-x)^n\|^{1/n} < 1$, then x^{-1} exists and*

(i) $x^{-1} = u + \sum_{k=1}^{\infty} (u-x)^k$.

(C.8) Corollary. *Let A be a Banach algebra and x an element of A such that $\|x\| < 1$. Then x has a quasi-inverse y and*

(i) $\dfrac{\|x\|}{1+\|x\|} \leqq \|y\| \leqq \dfrac{\|x\|}{1-\|x\|}$.

Proof. By (C.6) and (C.4.ii), x has a quasi-inverse y. Since $x + y - xy = 0$, we have $\|y\| = \|xy - x\| \leqq \|x\| \cdot \|y\| + \|x\|$ and $\|x\| = \|y - xy\| \leqq \|y\| + \|x\| \cdot \|y\|$. □

(C.9) Theorem. *Let A be a Banach algebra. If x and y are in A, if y has a quasi-inverse y', and if $\|x\| < \dfrac{1}{1+\|y'\|}$, then*

(i) *$x + y$ has a quasi-inverse z,*

and

(ii) $\|z - y'\| \leqq \dfrac{\|x\| (\|y'\|+1)^2}{1-(\|y'\|+1)\|x\|}$.

The set $E = \{y \in A : y$ has a quasi-inverse$\}$ is open in A. Let y' denote the quasi-inverse of y; then the mapping $y \to y'$ is a homeomorphism of E onto E.

Proof. Since $\|x - y'x\| \leqq \|x\| (1 + \|y'\|) < 1$, $x - y'x$ has a quasi-inverse v (C.8). Then $-vy' + y' + v$ is a *left* quasi-inverse for $x + y$ since

$$(x+y) + (-vy' + y' + v) - (-vy' + y' + v)(x+y)$$
$$= (y + y' - y'y) - v(y + y' - y'y) + [(x - y'x) + v - v(x - y'x)] = 0.$$

Similarly, $x - xy'$ has a quasi-inverse w and $-y'w + y' + w$ is a *right* quasi-inverse of $x + y$. Consequently $z = -vy' + y' + v$ is the quasi-inverse of $x + y$. This proves (i).

Using (C.8.i), we obtain

$$\|z - y'\| = \|-vy' + y' + v - y'\| = \|-vy' + v\| \leqq \|v\| \cdot (\|y'\| + 1)$$
$$\leqq \frac{\|x - y'x\|}{1 - \|x - y'x\|} (\|y'\| + 1) \leqq \frac{\|x\| (\|y'\|+1)^2}{1 - (\|y'\|+1)\|x\|}.$$

This proves (ii).

Suppose now that y is in E, and consider any $v \in A$ for which $\|v - y\| < \dfrac{1}{1+\|y'\|}$. Then by (i), the element $v - y + y = v$ has a quasi-inverse. That is, the set E is open in A. Plainly the mapping $y \to y'$

is one-to-one and onto and is its own inverse. If $y \in E$ and $\|v-y\| < \frac{1}{1+\|y'\|}$, then (ii) implies that

$$\|v'-y'\| \leq \frac{\|v-y\|(\|y'\|+1)^2}{1-(\|y'\|+1)\|v-y\|}. \tag{1}$$

If $\|v-y\|$ is sufficiently small, the right side of (1) is arbitrarily small. This implies that the mapping $y \to y'$ is continuous. □

(C.10) Corollary. *Let A be a Banach algebra with unit u. The set $E_0 = \{y \in A : y^{-1}$ exists$\}$ is open and the mapping $y \to y^{-1}$ is a homeomorphism of E_0 onto E_0.*

Proof. This follows from (C.9) and the identity $y^{-1} = u - (u-y)'$, which is valid for $y \in E_0$, where $'$ is as in (C.9). □

(C.11) Theorem [GEL'FAND-MAZUR]. *Let A be a Banach algebra with unit u in which every nonzero element has an inverse. Then $A = \{\lambda u : \lambda \in K\}$, and hence A can be identified with K.*

Proof. Let x be an element in A and assume that $x - \lambda u \neq 0$ for all $\lambda \in K$. Then $(x-\lambda u)^{-1}$ exists for all $\lambda \in K$. Let L be any bounded linear functional on A such that $L(x^{-1}) = 1$, and let $g(\lambda) = L\left((x-\lambda u)^{-1}\right)$ for $\lambda \in K$. Then $g(0) = 1$; and we will show that g is an entire function. It is easy to verify that $(x-\lambda u)^{-1}(x-\lambda_0 u)^{-1} = (x-\lambda_0 u)^{-1}(x-\lambda u)^{-1}$ for all $\lambda, \lambda_0 \in K$. Using this, we find that

$$\begin{aligned}(x-\lambda u)^{-1} - (x-\lambda_0 u)^{-1} &= (x-\lambda u)^{-1}(x-\lambda_0 u)^{-1}\left[(x-\lambda_0 u) - (x-\lambda u)\right] \\ &= (x-\lambda u)^{-1}(x-\lambda_0 u)^{-1}(\lambda-\lambda_0),\end{aligned}$$

and therefore

$$\left.\begin{aligned}\frac{g(\lambda)-g(\lambda_0)}{\lambda-\lambda_0} &= \frac{1}{\lambda-\lambda_0} L\left((x-\lambda u)^{-1} - (x-\lambda_0 u)^{-1}\right) \\ &= L\left((x-\lambda u)^{-1}(x-\lambda_0 u)^{-1}\right).\end{aligned}\right\} \tag{1}$$

The mapping $\lambda \to x - \lambda u$ is plainly continuous. By (C.10), the mapping $\lambda \to (x-\lambda u)^{-1}$ is also continuous, as is the mapping $\lambda \to (x-\lambda u)^{-1}(x-\lambda_0 u)^{-1}$. Since L is bounded, it follows that

$$\lim_{\lambda \to \lambda_0} L\left((x-\lambda u)^{-1}(x-\lambda_0 u)^{-1}\right) = L\left((x-\lambda_0 u)^{-2}\right). \tag{2}$$

Relations (1) and (2) show that g has a derivative at λ_0. Thus g is an entire function.

For $\lambda \neq 0$, we have $(x-\lambda u)^{-1} = \lambda^{-1}\left(\frac{x}{\lambda} - u\right)^{-1}$. Using continuity of the mapping $y \to y^{-1}$, we obtain $\lim\limits_{|\lambda| \to \infty} \left(\frac{x}{\lambda} - u\right)^{-1} = -u^{-1}$ and hence

$$\lim_{|\lambda| \to \infty} (x-\lambda u)^{-1} = \lim_{|\lambda| \to \infty} -\lambda^{-1} u^{-1} = 0;$$

these limits are taken in the norm topology of A. Since L is bounded, we infer that

$$\lim_{|\lambda|\to\infty} g(\lambda) = \lim_{|\lambda|\to\infty} L\left((x-\lambda u)^{-1}\right) = 0. \tag{3}$$

Accordingly, g is bounded and by LIOUVILLE's theorem, g is a constant. We infer also from (3) that $g(\lambda)=0$ for all $\lambda\in K$. Since $g(0)=1$, we have a contradiction. □

(C.12) Definition. Let A be an algebra over K. A *multiplicative linear functional* τ *on* A is a *nonzero* linear functional on A that is multiplicative: $\tau(xy)=\tau(x)\,\tau(y)$ for all $x, y\in A$. In other words, τ is an algebra homomorphism of A onto K.[1]

The following theorem applies in particular to multiplicative linear functionals, which play a fundamental rôle in the theory of commutative Banach algebras.

(C.13) Theorem. *Let A be an algebra over a field F, and let τ be an algebra homomorphism of A onto a simple algebra B over F that is different from $\{0\}$. Then $\tau^{-1}(0)$ is a maximal proper two-sided ideal in A. If M is a maximal proper two-sided ideal in A, then A/M is simple.*

The proof is easy and is omitted.

The study of multiplicative linear functionals on commutative Banach algebras can be reduced to the study of such functionals on commutative Banach algebras with unit because of the following theorem.

(C.14) Theorem. *Let A be a Banach algebra and let A_u be as in* (C.3); *A is identified with $\{(x, 0): x\in A\}$. Every multiplicative linear functional τ on A admits a unique extension τ_u on A_u that is a multiplicative linear functional on A_u. Every multiplicative linear functional on A_u is a multiplicative linear functional on A when restricted to A, except for the functional τ_∞ defined by $\tau_\infty((x, \lambda))=\lambda$.*

Proof. Let τ be a multiplicative linear functional on A and suppose that τ_u is a multiplicative linear functional on A_u that extends τ. Then $\tau_u((x, 0))=\tau(x)$ is defined for all $x\in A$ and $\tau_u((0, 1))=1$ since $(0, 1)$ is the unit of A_u. It follows that $\tau_u((x, \lambda))=\tau(x)+\lambda$ for all $(x, \lambda)\in A_u$. Hence there is at most one extension of τ over A_u that is a multiplicative linear functional. Further, it is easy to see that τ_u as defined above is a multiplicative linear functional on A_u.

The remainder of the theorem is likewise easily proved; we omit the details. □

(C.15) Definition. Let A be any algebra. A left ideal I in A is said to be a *regular left ideal* if there is an element v in A such that

$$(x+I)\,(v+I)=xv+I=x+I$$

[1] It is arbitrary but convenient to require that multiplicative linear functionals be nonzero; see for example (C.17).

for all $x \in A$. We say that v is a *right unit relative to I*. *Regular right ideals* are defined analogously. A two-sided ideal I in A is *regular* if A/I has a unit $v+I$:

$$vx+I=xv+I=x+I$$

for all $x \in A$.

Obviously every left, right, and two-sided ideal in an algebra with unit is regular.

(C.16) Theorem. *Let A be an algebra and I a proper regular left ideal in A. Then I is contained in a maximal proper left ideal which is also regular. If A is a Banach algebra and I is a maximal proper regular left ideal in A, then I is closed. Similar assertions hold for right and two-sided ideals.*

Proof. Let v be a right unit relative to I; then plainly v is not in I. Let $\mathscr{J}$ denote the family of all left ideals J such that $I \subset J$ and $v \notin J$. If $\mathscr{L}$ is a subfamily of $\mathscr{J}$ linearly ordered by inclusion, then $\cup\{J : J \in \mathscr{L}\}$ also belongs to $\mathscr{J}$. Thus ZORN's lemma applies, and $\mathscr{J}$ contains a maximal element J_0. The ideal J_0 is a maximal proper left ideal containing I, and it is regular with relative right unit v.

Now let A be a Banach algebra and let I be a maximal proper regular left ideal in A with relative right unit v. It is simple to verify that I^- is a left ideal in A; we omit the argument. If v is a right unit relative to I, then clearly v is a right unit relative to I^-; that is, I^- is a regular left ideal. Assume that $I \subsetneqq I^-$; then $I^- = A$ since I is maximal. Thus there is a z in A such that $\|z\| < 1$ and $v-z$ belongs to I. By (C.8), z has a quasi-inverse z': $z+z'-z'z=0$. Now we have

$$v=(v-z)-z'(v-z)+z'v-z'.$$

Obviously $v-z$ and $z'(v-z)$ belong to I; $z'v-z'$ also belongs to I since $z'+I=z'v+I$. Therefore $v \in I$, and it follows that $I=A$, contrary to hypothesis.

The statements involving right and two-sided ideals are proved similarly. □

Theorem (C.13) implies that the kernel of a multiplicative linear functional on a commutative Banach algebra A is a maximal proper regular ideal in A. We now prove the converse of this statement.

(C.17) Theorem. *Let A be a commutative Banach algebra. Then every maximal proper regular ideal M in A is the kernel of a [unique] multiplicative linear functional τ on A.*

Proof. By (C.16), M is closed. By (C.2), A/M is a Banach algebra. Since M is maximal, A/M is simple (C.13). Since M is regular, A/M has a unit $v+M$. Since A is commutative, A/M is also commutative. From all this, it is easy to see that every nonzero element of A/M has an inverse. Theorem (C.11) implies that $A/M=\{\lambda v+M:\lambda\in K\}$. For each x in A, there is a unique complex number $\tau(x)$ such that $x\in\tau(x)v+M$. It is easy to check that τ is a multiplicative linear functional on A, that $M=\tau^{-1}(0)$, and that τ is uniquely determined by these two requirements. □

We next define the spectrum of an element in an algebra, and proceed to relate this notion with multiplicative linear functionals (C.20).

(C.18) Definitions. Let A be an algebra with unit u and let x be an element of A. The *spectrum of* x is the set of all λ in K for which $(x-\lambda u)^{-1}$ does not exist. Plainly, a nonzero λ in K belongs to the spectrum of x if and only if $\left(\frac{x}{\lambda}-u\right)^{-1}$ does not exist. This motivates our next definition.

Let A be an algebra without unit. The *spectrum* of an element x in A is defined to be the set of all nonzero λ in K for which x/λ does not have a quasi-inverse, together with 0.

The virtue of having 0 in the spectrum of all x in A whenever A fails to have a unit is seen in the following theorem.

(C.19) Theorem. *Let A be an algebra without unit, and let A_u be as in* (C.3). *If $x\in A$, then the spectrum of x as an element in A coincides exactly with the spectrum of x as an element in A_u.*

Proof. This is immediate from the definitions. Note that an element $(x, 0)$ in A_u has no inverse in A_u. □

(C.20) Theorem. *Let A be a commutative Banach algebra and let x be an element of A. If A has a unit, then λ belongs to the spectrum of x if and only if there is a multiplicative linear functional τ on A such that $\tau(x)=\lambda$. If A has no unit and $\lambda\neq 0$, then λ belongs to the spectrum of x if and only if there is a multiplicative linear functional τ such that $\tau(x)=\lambda$.*

Proof. Suppose that A has a unit u, and consider $x\in A$ and $\lambda\in K$. If λ belongs to the spectrum of x, then $(x-\lambda u)^{-1}$ does not exist. Hence $\{(x-\lambda u)z:z\in A\}$ is a proper ideal containing $x-\lambda u$. By (C.16), there is a maximal proper ideal M in A such that $\{(x-\lambda u)z:z\in A\}\subset M$. By (C.17), M is the kernel of some multiplicative linear functional τ. Since $\tau(x-\lambda u)=0$, we have $\tau(x)=\lambda\tau(u)=\lambda$. On the other hand, if $(x-\lambda u)^{-1}$ exists, then for every multiplicative linear functional τ, we have $\tau(x-\lambda u)\neq 0$, since otherwise $\tau(u)=\tau((x-\lambda u)(x-\lambda u)^{-1})=0$. That is, we have $\tau(x)\neq\lambda$.

The second statement of the theorem is proved from the first statement together with (C.14) and (C.19). □

(C.21) Theorem. *Let A be a Banach algebra. Every multiplicative linear functional τ on A is bounded, and $\|\tau\| \leqq 1$. If A has a unit u and $\|u\| = 1$, then $\|\tau\| = 1$.*

Proof. Consider x in A and assume that $\tau(x) = \alpha$ where $|\alpha| > \|x\|$. Since $\|x/\alpha\| < 1$, x/α has a quasi-inverse y (C.8). Therefore we have $\frac{x}{\alpha} + y - \frac{xy}{\alpha} = 0$, $\tau\left(\frac{x}{\alpha}\right) + \tau(y) - \tau\left(\frac{x}{\alpha}\right)\tau(y) = 0$, and hence $1 + \tau(y) - \tau(y) = 0$. This obvious contradiction shows that $|\tau(x)| \leqq \|x\|$ and hence that $\|\tau\| \leqq 1$.

If A has a unit u and $\|u\| = 1$, then $\|\tau\| \geqq 1$ since $\tau(u) = 1$. □

(C.22) Theorem. *Let A be a commutative Banach algebra. The spectrum of every element x in A is compact and nonvoid.*

Proof. In view of (C.19) and (C.3), we may suppose that A has a unit u. If λ belongs to the spectrum of x, then $u - \frac{x}{\lambda}$ has no inverse in A. Hence by (C.7), we have

$$\left\|\frac{x}{\lambda}\right\| \geqq \lim_{n\to\infty} \left\|\left(u - \left(u - \frac{x}{\lambda}\right)\right)^n\right\|^{1/n} \geqq 1.$$

Thus $|\lambda| \leqq \|x\|$ and the spectrum of x is a subset of $\{\lambda \in K : |\lambda| \leqq \|x\|\}$. Since the set $\{y \in A : y^{-1} \text{ exists}\}$ is open (C.10) and the mapping $\lambda \to (x - \lambda u)$ is continuous, we see that the set $\{\lambda \in K : (x - \lambda u)^{-1} \text{ exists}\}$ is open. Thus the spectrum of x is closed and bounded, *i.e.*, compact. The argument used in proving (C.11) shows that $(x - \lambda u)^{-1}$ fails to exist for some complex number λ. That is, x has nonvoid spectrum. □

Let X be a locally compact Hausdorff space. The algebra $\mathfrak{C}_0(X)$ [all operations are pointwise], with the norm $\|f\|_u = \max\{|f(x)| : x \in X\}$ is obviously a commutative Banach algebra. Its structure is particularly simple: and a great deal of effort is expended in harmonic analysis in trying to do for $\boldsymbol{M}(G)$ and its subalgebras things that are obvious for algebras $\mathfrak{C}_0(X)$. The algebra $\mathfrak{C}_0(X)$ has a unit if and only if X is compact; the spectrum of $f \in \mathfrak{C}_0(X)$ is the set $f(X)$ [plus $\{0\}$ if X is noncompact]; every multiplicative linear functional on $\mathfrak{C}_0(X)$ has the form $f \to f(p)$ for some $p \in X$ (C.29); all closed ideals of $\mathfrak{C}_0(X)$ are easily identified (C.30). It turns out that a reasonably large class of commutative Banach algebras are algebraically [but not necessarily normwise] isomorphic with subalgebras of certain algebras $\mathfrak{C}_0(X)$. To establish this fact, we first make a definition.

(C.23) Definition. Let A be a commutative Banach algebra. Let X denote the set of all multiplicative linear functionals τ on A; X is called the *structure space for A*.[1] For every $x \in A$, the *Fourier transform* $\hat{x}$ is defined by

(i) $\hat{x}(\tau) = \tau(x)$

for all $\tau \in X$. Observe that $\hat{x}$ is a function on X. The *Gel'fand topology for X* is the weakest topology on X under which all of the functions $\hat{x}$ are continuous. Unless the contrary is specified, X has this topology. The family of functions $\hat{x}$ on X is denoted by $\hat{A}$.

For the remainder of Appendix C, if a normed algebra has a unit u, it will be tacitly assumed that $\|u\| = 1$.

(C.24) Theorem [Gel'fand]. *Let A be a commutative Banach algebra and let $x \in A$. Then*

(i) $\|\hat{x}\|_u = \sup\{|\tau(x)| : \tau \in X\} = \lim\limits_{n\to\infty} \|x^n\|^{1/n}$.

Proof. The first equality in (i) is merely the definition of $\|\hat{x}\|_u$. In what follows, $\|\hat{x}\|$ means $\|\hat{x}\|_u$. Let A_u be as in (C.3) and let X_u denote its structure space. Since $\tau_\infty((x, 0)) = 0$ for all $x \in A$, we find that $\sup\{|\tau(x)| : \tau \in X\} = \sup\{|\tau((x, 0))| : \tau \in X_u\}$ for $x \in A$ [see (C.14)]. Hence it suffices to prove the present theorem for the case in which A has a unit u. Plainly, we may also suppose that $x \neq 0$.

Using (C.21), we have

$$|\tau(x)|^n = |\tau(x^n)| \leqq \|\tau\| \cdot \|x^n\| = \|x^n\|$$

or all $\tau \in X$ and $n = 1, 2, \ldots$. Thus $|\tau(x)| \leqq \lim\limits_{n\to\infty} \|x^n\|^{1/n}$ and

$$\|\hat{x}\| \leqq \lim_{n\to\infty} \|x^n\|^{1/n}. \tag{1}$$

Consider first a complex number λ such that $|\lambda| < \frac{1}{\|x\|}$. Then $\|\lambda x\| < 1$; hence by (C.7), $(u - \lambda x)^{-1}$ exists and

$$(u - \lambda x)^{-1} = u + \sum_{k=1}^{\infty} (\lambda x)^k; \tag{2}$$

this series converges in the norm topology of A. If λ is in K and $0 < |\lambda| < \frac{1}{\|\hat{x}\|}$ $\left[\text{not } \frac{1}{\|x\|}!\right]$, then $\left|\frac{1}{\lambda}\right| > |\tau(x)|$ for all $\tau \in X$, so that by (C.20), $\left(x - \frac{1}{\lambda} u\right)^{-1}$ and $(u - \lambda x)^{-1}$ exist. Let L be an element in A^*, the conjugate space of the normed linear space A. For λ in K and $|\lambda| < \frac{1}{\|\hat{x}\|}$, let $g(\lambda) = L((u - \lambda x)^{-1})$. By the argument used in the proof of (C.11), it can be shown that g is analytic in the open disk $\left\{\lambda \in K : |\lambda| < \frac{1}{\|\hat{x}\|}\right\}$; we omit the details. The continuity of L and the identity (2) imply that

$$g(\lambda) = L((u - \lambda x)^{-1}) = L(u) + \sum_{k=1}^{\infty} \lambda^k L(x^k)$$

[1] Many writers call X the *maximal ideal space for A*; this is reasonable in view of Theorem (C.17) and the remark preceding it.

for all λ such that $|\lambda| < \frac{1}{\|x\|} \leq \frac{1}{\|\hat{x}\|}$. Thus the series $L(u) + \sum_{k=1}^{\infty} \lambda^k L(x^k)$ is the Taylor series for g. Since g is analytic in $\left\{\lambda \in K : |\lambda| < \frac{1}{\|\hat{x}\|}\right\}$, it follows from the elementary theory of analytic functions that the series $\sum_{k=1}^{\infty} \lambda^k L(x^k)$ converges absolutely not only for $|\lambda| < \frac{1}{\|x\|}$ but also for $|\lambda| < \frac{1}{\|\hat{x}\|}$. The point of all this is that

$$\lim_{k \to \infty} |L(\lambda^k x^k)| = 0 \tag{3}$$

for all λ such that $|\lambda| < \frac{1}{\|\hat{x}\|}$ and for all $L \in A^*$.

We now use the BANACH-STEINHAUS theorem (B.20). Choose any $\lambda \in K$ such that $|\lambda| < \frac{1}{\|\hat{x}\|}$. For each $n = 1, 2, \ldots$, and $L \in A^*$, let

$$T_n(L) = L(\lambda^n x^n).$$

Clearly T_n is the functional $(\lambda^n x^n)'$ on A^* where y' is as defined in (B.23). As a result, we have

$$\|T_n\| = \|\lambda^n x^n\| \tag{4}$$

for all n. By (3), we have $\sup\{\|T_n(L)\| : n = 1, 2, \ldots\} = \sup\{|L(\lambda^n x^n)| : n = 1, 2, \ldots\} < \infty$ for each $L \in A^*$. Hence, by (B.20) there is a $\beta > 0$ such that $\|T_n\| \leq \beta$ for all n. Using (4), we see that $\|\lambda^n x^n\| \leq \beta$, and hence that $\|x^n\|^{1/n} \leq \frac{\beta^{1/n}}{|\lambda|}$, for all n. Therefore $\lim_{n \to \infty} \|x^n\|^{1/n} \leq \frac{1}{|\lambda|}$ whenever $|\lambda| < \frac{1}{\|\hat{x}\|}$. It follows that $\lim_{n \to \infty} \|x^n\|^{1/n} \leq \|\hat{x}\|$; this together with (1) proves (i). □

(C.25) Theorem. *Let A be a commutative Banach algebra with unit u. Under the Gel'fand topology, X is a compact Hausdorff space. The mapping $x \to \hat{x}$ is an algebraic homomorphism of A onto a subalgebra $\hat{A}$ of $\mathfrak{C}(X)$, and $\|\hat{x}\|_u \leq \|x\|$ for $x \in A$. The algebra $\hat{A}$ separates points of X and contains all constant functions. The mapping $x \to \hat{x}$ is an isomorphism if and only if $\lim_{n \to \infty} \|x^n\|^{1/n} > 0$ for all $x \neq 0$.*

Proof. It is easy to see that the following sets form an open basis for the Gel'fand topology of X:

$$U_{F,\varepsilon}(\tau_0) = \{\tau \in X : |\hat{x}(\tau) - \hat{x}(\tau_0)| < \varepsilon \text{ for } x \in F\},$$

where F is a finite subset of A, $\tau_0 \in X$, and ε is a positive real number. It then follows that X can be regarded as a subspace of the product space $Y = \mathsf{P}_{x \in A}\{\lambda \in K : |\lambda| \leq \|x\|\}$; in fact, the element τ in X is identified with the element $(\tau(x)) = (\hat{x}(\tau))$ in Y. To see that X is closed in Y, consider (t_x) in X^-. It is easy to verify that $x \to t_x$ is linear and multiplicative [compare with the proof of (B.25)]. Also, we have $t_u = 1$ since $\tau(u) = 1$ for all $\tau \in X$. Thus $x \to t_x$ is not the zero functional and hence (t_x) belongs to X. We have shown that X is homeomorphic with a closed subspace of the compact space Y, and is therefore itself compact.

It is routine to verify that the mapping $x \to \hat{x}$ is an algebraic homomorphism, that $\|\hat{x}\|_u \leq \|x\|$ for $x \in A$, that $\hat{A}$ separates points of X, and that $\hat{A}$ contains the constant functions. The mapping $x \to \hat{x}$ is an isomorphism of A into $\mathfrak{C}(X)$ if and only if $\hat{x} \neq 0$ whenever $x \neq 0$; *i.e.*, if and only if $\|\hat{x}\|_u > 0$ whenever $x \neq 0$. By (C.24), the last condition is equivalent to the relation $\lim_{n\to\infty} \|x^n\|^{1/n} > 0$ whenever $x \neq 0$. □

(C.26) Theorem. *Let A be a commutative Banach algebra and let A_u be as in* (C.3). *Let X and X_u be the structure spaces for A and A_u, respectively. Then $X_u = X \cup \{\tau_\infty\}$* (C.14), *$X$ is a locally compact Hausdorff space, X_u is the one-point compactification of X, and $\hat{A} \subset \mathfrak{C}_0(X)$.*

Proof. As in (C.25), a generic neighborhood of τ in X is given by

$$U_{F,\varepsilon}(\tau) = \{\tau' \in X : |\tau'(x) - \tau(x)| < \varepsilon \text{ for } x \in F\},$$

where $F \subset A$ is finite and $\varepsilon > 0$. For τ in X, let τ_u denote its extension to A_u. Suppose that $F_u = \{(x, 0) \in A_u : x \in F\}$. If $|\tau(x)| < \varepsilon$ for all $x \in F$, then

$$U_{F_u,\varepsilon}(\tau_u) = \{\tau'_u \in X_u : \tau' \in U_{F,\varepsilon}(\tau)\} \cup \{\tau_\infty\};$$

otherwise, we have

$$U_{F_u,\varepsilon}(\tau_u) = \{\tau'_u \in X_u : \tau' \in U_{F,\varepsilon}(\tau)\}.$$

Thus neighborhoods in X are relatively open as subsets of X_u. A like argument shows that subsets of X that are relatively open in X_u are also open in the Gel'fand topology of X. In other words, the Gel'fand topology of X is the relative topology of X as a subset of X_u. Since $\{\tau_\infty\}$ is closed, X is open in the compact Hausdorff space X_u and hence is locally compact.

Consider any element x in A and $\varepsilon > 0$. Let $F_u = \{(x, 0)\}$; then we have

$$\begin{aligned} U_{F_u,\varepsilon}(\tau_\infty) &= \{\tau_\infty\} \cup \{\tau_u \in X_u : |\tau_u((x,0)) - \tau_\infty((x,0))| < \varepsilon\} \\ &= \{\tau_\infty\} \cup \{\tau_u \in X_u : |\tau(x)| < \varepsilon\}. \end{aligned}$$

This means that $|\hat{x}|$ is arbitrarily small in neighborhoods of τ_∞. Equivalently, $|\hat{x}|$ is arbitrarily small outside of compact sets in X, and so $\hat{x} \in \mathfrak{C}_0(X)$. □

(C.27) We now take up certain Banach $\sim$-algebras: for the definition of a $\sim$-algebra, see (21.6). We recall that in a *normed* $\sim$-algebra A, the identity

(i) $\|x\| = \|x^{\sim}\|$

for all $x \in A$ is postulated. We shall also need to consider an additional property, which may or may not be satisfied:

(ii) $\|xx^{\sim}\| = \|x\|^2$.

The following theorem is of great importance.

(C.28) Theorem [GEL'FAND-NAĬMARK]. *Let A be a commutative Banach $\sim$-algebra satisfying* (C.27.ii), *and let X be the structure space of A. Then the mapping $x \to \hat{x}$ is a norm-preserving algebra isomorphism of A onto $\mathfrak{C}_0(X)$. Furthermore, we have $\widehat{x^{\sim}} = \bar{\hat{x}}$ for all $x \in A$.*

Proof. This simplification of the proof given in the first edition of this book was kindly shown to us by FRED THOELE.

(I) We first prove: if x is in A and $x = x^{\sim}$, then $\tau(x)$ is a real number for all $\tau \in X$. Let t be an arbitrary real number and consider any element y in A such that $\|y\| \leqq 1$. Then we have

$$|\tau(y)|^2 |\tau(x) + it|^2 = |\tau(xy + ity)|^2 \leqq \|\tau\|^2 \|xy + ity\|^2$$

[note (C.21)] and

$$\begin{aligned} \|xy + ity\|^2 &= \|(xy + ity)\,(xy + ity)^{\sim}\| = \|(xy + ity)\,(xy^{\sim} - ity^{\sim})\| \\ &= \|x^2 yy^{\sim} + t^2 yy^{\sim}\| \leqq \|yy^{\sim}\|\,(\|x\|^2 + t^2) \\ &= \|y\|^2 (\|x\|^2 + t^2) \leqq \|x\|^2 + t^2, \end{aligned}$$

and therefore

$$|\tau(y)|^2\, |\tau(x) + it|^2 \leqq \|\tau\|^2 (\|x\|^2 + t^2). \tag{1}$$

The supremum of the left side of (1) as y runs over all y in A such that $\|y\| \leqq 1$ is exactly $\|\tau\|^2 |\tau(x) + it|^2$ and hence we have

$$|\tau(x) + it|^2 \leqq \|x\|^2 + t^2.$$

Now write $\tau(x) = a + ib$. We have

$$|\tau(x) + it|^2 = |a + ib + it|^2 = a^2 + b^2 + 2bt + t^2,$$

$$a^2 + b^2 + 2bt + t^2 \leqq \|x\|^2 + t^2,$$

and

$$a^2 + b^2 + 2bt \leqq \|x\|^2. \tag{2}$$

Since (2) holds for all real numbers t, the number b must be 0.

(II) For every x in A, we can write

$$x = \frac{x + x^{\sim}}{2} + i\,\frac{x - x^{\sim}}{2i} = x_1 + i x_2,$$

where $x_1 = x_1^{\sim}$ and $x_2 = x_2^{\sim}$. Then $\tau(x) = \tau(x_1) + i\tau(x_2)$ and $\tau(x^{\sim}) = \tau((x_1 + ix_2)^{\sim}) = \tau(x_1 - ix_2) = \tau(x_1) - i\tau(x_2) = \overline{\tau(x)}$. In other words, $\widehat{x^{\sim}} = \bar{\hat{x}}$ for all $x \in A$. Thus $\hat{A}$ is a subalgebra of $\mathfrak{C}_0(X)$ that separates points of X and is closed under complex conjugation. The STONE-WEIERSTRASS theorem [see (d) of the footnote on p. 151] implies that $\hat{A}$ is dense in the uniform topology of $\mathfrak{C}_0(X)$.

Next we show that $\|\hat{x}\|_u = \|x\|$ for $x \in A$; it is evident from (C.21) that $\|\hat{x}\|_u \leqq \|x\|$. Suppose first that $x = x^{\sim}$. Then $\|x^2\| = \|xx^{\sim}\| = \|x\|^2$ and $\|x^2\|^{\frac{1}{2}} = \|x\|$. Similarly, we have $\|x^4\|^{\frac{1}{2}} = \|x^2\|$ and hence $\|x^4\|^{\frac{1}{4}} = \|x^2\|^{\frac{1}{2}} = \|x\|$. By finite induction, we have $\|x^{2^n}\|^{1/2^n} = \|x\|$ for all n. By (C.24), we have

$$\|\hat{x}\|_u = \lim_{n\to\infty} \|x^n\|^{1/n} = \lim_{n\to\infty} \|x^{2^n}\|^{1/2^n} = \|x\|.$$

Consider an arbitrary x in A; plainly $(xx^{\sim})^{\sim} = xx^{\sim}$. Therefore

$$\|x\|^2 = \|xx^{\sim}\| = \|\widehat{xx^{\sim}}\|_u = \|\hat{x}\,\widehat{x^{\sim}}\|_u = \|\hat{x}\,\bar{\hat{x}}\|_u = \|\,|\hat{x}|^2\|_u = \|\hat{x}\|_u^2.$$

Since the mapping $x \to \hat{x}$ preserves norms and A is complete, $\hat{A}$ is complete. Consequently, $\hat{A}$ is closed in $\mathfrak{C}_0(X)$ and so $\hat{A} = \mathfrak{C}_0(X)$. □

Theorem (C.28) characterizes very simply in terms of the behavior of norm and adjoint the Banach algebras $\mathfrak{C}_0(X)$ for locally compact Hausdorff spaces X. We complete our analysis of these algebras by identifying all of the closed ideals and multiplicative linear functionals in them.

(C.29) Discussion. Let Y be a locally compact Hausdorff space. For $p \in Y$ and $f \in \mathfrak{C}_0(Y)$, let $\tau_p(f) = f(p)$. It is obvious that τ_p is a multiplicative linear functional on $\mathfrak{C}_0(Y)$. Theorems (C.13) and (C.16) imply that $\tau_p^{-1}(0) = \{f \in \mathfrak{C}_0(Y) : f(p) = 0\}$ is a maximal proper regular closed ideal in $\mathfrak{C}_0(Y)$. We will show that these are the only maximal proper closed ideals in $\mathfrak{C}_0(Y)$ and that the structure space of $\mathfrak{C}_0(Y)$ is $\{\tau_p : p \in Y\}$.

(C.30) Theorem. *Let Y be a locally compact Hausdorff space and let $\mathfrak{J}$ be a proper closed ideal in $\mathfrak{C}_0(Y)$. There exists a unique closed subset E of Y such that*

(i) $\mathfrak{J} = \{f \in \mathfrak{C}_0(Y) : f(E) = 0\}$.

Also every set as in (i) *is a closed ideal of* $\mathfrak{C}_0(Y)$.

Proof. Let $E = \{p \in Y : f(p) = 0$ for all $f \in \mathfrak{J}\}$ and $\mathfrak{M}_E = \{f \in \mathfrak{C}_0(Y) : f(E) = 0\}$; plainly E is closed in Y and $\mathfrak{J} \subset \mathfrak{M}_E$.

Let F be a compact subset of $Y \cap E'$. For each $x \in F$, there is a function $f_x \in \mathfrak{J}$ such that $f_x(x) \neq 0$. Then $h_x = \bar{f}_x f_x$ belongs to $\mathfrak{J}$ and $h_x(x) > 0$. For each $x \in F$, there is a neighborhood U_x of x such that $h_x(y) > 0$ for all $y \in U_x$. There are points $x_1, \ldots, x_n$ in F such that $\bigcup_{k=1}^{n} U_{x_k} \supset F$. Let $h = \sum_{k=1}^{n} h_{x_k}$; then h belongs to $\mathfrak{J}$, and $h(y) > 0$ for all $y \in F$.

Let $\mathfrak{N}$ consist of those functions φ in $\mathfrak{C}_{00}(Y)$ such that $\{y \in Y : \varphi(y) \neq 0\}^- \cap E = \varnothing$. Consider any φ in $\mathfrak{N}$ and let $F = \{y \in Y : \varphi(y) \neq 0\}^-$. The previous paragraph shows that there is a function h in $\mathfrak{J}$ such that $h(y) > 0$ for all $y \in F$. We define $g(y) = \frac{\varphi(y)}{h(y)}$ for $y \in F$ and $g(y) = 0$ for $y \notin F$. It is easy to see that g belongs to $\mathfrak{C}_{00}(Y)$ and hence that $g \cdot h = \varphi$ belongs to $\mathfrak{J}$. We have shown that $\mathfrak{N} \subset \mathfrak{J}$. A very routine argument shows that $\mathfrak{N}$ is uniformly dense in $\mathfrak{M}_E$.[1] Since $\mathfrak{J}$ and $\mathfrak{M}_E$ are closed, it follows that $\mathfrak{J} = \mathfrak{M}_E$.

The uniqueness of E is trivial to prove, as is the fact that every $\mathfrak{M}_E$ is a closed ideal in $\mathfrak{C}_0(Y)$. □

(C.31) Corollary. *Let Y be a locally compact Hausdorff space. For each point p in Y, let $\mathfrak{M}_p = \{f \in \mathfrak{C}_0(Y) : f(p) = 0\}$. Then every $\mathfrak{M}_p$ is a maximal proper closed ideal in the Banach algebra $\mathfrak{C}_0(Y)$, and there are no other maximal proper closed ideals in $\mathfrak{C}_0(Y)$. Every proper closed ideal $\mathfrak{J}$ is the intersection of maximal proper ideals.*

The last statement follows from the identity $\mathfrak{M}_E = \bigcap_{p \in E} \mathfrak{M}_p$.

(C.32) Corollary. *Let Y be a locally compact Hausdorff space. Then the structure space X of $\mathfrak{C}_0(Y)$ is identical with $\{\tau_p : p \in Y\}$ and the mapping $p \to \tau_p$ is a homeomorphism of Y onto X.*

Proof. We have already seen that $\{\tau_p : p \in Y\} \subset X$. Since the only maximal closed ideals in $\mathfrak{C}_0(Y)$ are the ideals $\mathfrak{M}_p = \tau_p^{-1}(0)$, it follows that $\{\tau_p : p \in Y\} = X$ (C.17).

For a finite subset $\mathfrak{F}$ of $\mathfrak{C}_0(Y)$, we have

$$\begin{aligned} U_{\mathfrak{F},\varepsilon}(\tau_q) &= \{\tau_p \in X : |\tau_p(f) - \tau_q(f)| < \varepsilon \text{ for } f \in \mathfrak{F}\} \\ &= \{\tau_p \in X : |f(p) - f(q)| < \varepsilon \text{ for } f \in \mathfrak{F}\}. \end{aligned}$$

It follows that $U_{\mathfrak{F},\varepsilon}(\tau_q)$ is the image of an open set under the correspondence $p \to \tau_p$. If U is an arbitrary open neighborhood of q in Y and f in $\mathfrak{C}_0(Y)$ is chosen so that $f(q) = 1$ and $f(U') = 0$, then $\{p \in Y : \tau_p \in U_{\{f\},1}(\tau_q)\} \subset U$. Consequently, the mapping $p \to \tau_p$ is a homeomorphism. □

[1] If $\varphi \in \mathfrak{M}_E$ and $\varepsilon > 0$, then $F_0 = \{y \in Y : |\varphi(y)| \geqq \varepsilon\}$ is compact and disjoint from E. Let f be a function in $\mathfrak{C}_{00}^+(Y)$ such that $f(F_0) = 1$, $f(Y) \subset [0, 1]$, and $\{y \in Y : f(y) \neq 0\}^- \cap E = \varnothing$. Then $f \cdot \varphi \in \mathfrak{N}$ and $\|f \cdot \varphi - \varphi\|_u < 2\varepsilon$.

Theorems (C.28), (C.34), and (11.37) can be applied in a striking way to give a proof of the spectral theorem for [bounded] Hermitian operators on a Hilbert space (C.42). Throughout (C.33)—(C.43), H will denote an arbitrary Hilbert space. Ancillary notation and terminology are as in (B.53) and (B.56); I will denote the identity operator on H.

(C.33) Lemma. *Let H be a Hilbert space and A a normal operator in $\mathfrak{B}(H)$. Then A^{-1} exists in $\mathfrak{B}(H)$ if and only if there is a positive real number α such that $\|A\xi\| \geqq \alpha \|\xi\|$ for all $\xi \in H$.*

Proof. If A^{-1} exists, take $\alpha = \|A^{-1}\|^{-1}$. Suppose conversely that our condition holds. Since $\langle A\xi, A\xi\rangle = \langle A^{\sim}A\xi, \xi\rangle = \langle AA^{\sim}\xi, \xi\rangle = \langle A^{\sim}\xi, A^{\sim}\xi\rangle$, we have $\|A\xi\| = \|A^{\sim}\xi\| \geqq \alpha\|\xi\|$ for all $\xi \in H$. It follows that A and $A^{\sim}$ are one-to-one. Let ζ in H be such that $\langle A\xi, \zeta\rangle = 0$ for all $\xi \in H$. Then $\langle \xi, A^{\sim}\zeta\rangle = 0$ for all ξ, so that $A^{\sim}\zeta = 0$, and hence $\zeta = 0$. Therefore $A(H)$ is a dense subspace of H. If $A\xi_n \to \eta$, then $\|\xi_n - \xi_m\| \leqq \alpha^{-1}\|A\xi_n - A\xi_m\|$, so that $\{\xi_n\}_{n=1}^{\infty}$ is a Cauchy sequence in H. Hence $\xi_n \to \zeta$ for some $\zeta \in H$, and by continuity of A, $A\zeta = \eta$. We have thus proved that $A(H) = H$. Since A is one-to-one, its inverse A^{-1} exists. It is obvious that A^{-1} is linear and bounded, with norm not exceeding α^{-1}. □

(C.34) Theorem. *Let A be an Hermitian operator in the algebra $\mathfrak{B}(H)$. Then the spectrum of A is real. If A is positive-definite, then the spectrum of A is also nonnegative.*

Proof. If $\lambda \in K$ and λ is not real, then we have

$$\begin{aligned} |\lambda - \bar{\lambda}| \cdot \|\xi\|^2 &= |\langle A\xi - \lambda\xi, \xi\rangle - \langle \xi, A\xi - \lambda\xi\rangle| \\ &\leqq \|A\xi - \lambda\xi\| \cdot \|\xi\| + \|\xi\| \cdot \|A\xi - \lambda\xi\| = 2\|A\xi - \lambda\xi\| \cdot \|\xi\| \end{aligned}$$

for all $\xi \in H$. By (C.33), $(A - \lambda I)^{-1}$ exists and λ does not belong to the spectrum of A.

If A is positive-definite and λ is real and negative, then we have

$$\begin{aligned} \|(A - \lambda I)\xi\|^2 &= \langle A\xi - \lambda\xi, A\xi - \lambda\xi\rangle \\ &= \langle A\xi, A\xi\rangle - 2\lambda\langle A\xi, \xi\rangle + \lambda^2\langle \xi, \xi\rangle \geqq \lambda^2\|\xi\|^2. \end{aligned}$$

Once again (C.33) shows that $(A - \lambda I)^{-1}$ exists. □

(C.35) Theorem. *Let A be a positive-definite operator on a Hilbert space H. Then there is a unique positive-definite operator B on H such that $B^2 = A$. An operator T in $\mathfrak{B}(H)$ commutes with A if and only if it commutes with B.*

Proof. (I) Let $\mathfrak{A}$ be the smallest norm closed subalgebra of $\mathfrak{B}(H)$ that contains A, I, and all operators $(A - \lambda I)^{-1}$ that are in $\mathfrak{B}(H)$, for complex numbers λ. It is easy to show that $\mathfrak{A}$ is a commutative Banach $\sim$-algebra with unit satisfying the hypotheses of (C.28). Every τ in the structure space X of $\mathfrak{A}$ is completely determined by the number $\tau(A)$. By (C.20) and (C.34), every number $\tau(A)$ is real and nonnegative. Hence we can and do identify X with a certain compact subset of $[0, \infty[$ (C.22). We write $\hat{A}(x) = x$ for all $x \in X$. By (C.28), $\mathfrak{A}$ can be identified with $\mathfrak{C}(X)$.[1]

Let $\psi(x) = x^{\frac{1}{2}}$ for all $x \in X$, and let $\{\psi_n\}_{n=1}^{\infty}$ be a sequence of polynomials in x with *real* coefficients such that $\lim_{n\to\infty} \max\{|\psi_n(x) - \psi(x)| : x \in X\} = 0$. By (C.28), there are operators $C, C_1, C_2, \ldots$ in $\mathfrak{A}$ such that $\psi = \hat{C}$ and $\psi_n = \hat{C}_n$. Each C_n is a polynomial in the operator A with real coefficients and hence is Hermitian. Therefore C is Hermitian, since it is the limit in the norm topology of the Hermitian operators C_n. Let $B = C^2$; as the square of an Hermitian operator, B is positive-definite. It is also obvious that $B^2 = A$.

The operator B is obviously the limit of polynomials in A. Hence every operator in $\mathfrak{B}(H)$ that commutes with A also commutes with B.

(II) We now prove the uniqueness of B. Suppose that $B_0^2 = A$ and that B_0 is a positive-definite operator. By (I), there are positive-definite operators C and C_0 such that $C^2 = B$ and $C_0^2 = B_0$. Since $B_0 A = B_0^3 = A B_0$, B_0 also commutes with B. Let ξ be any element in H and $\eta = (B - B_0)\xi$. Then

$$\begin{aligned}\|C\eta\|^2 + \|C_0\eta\|^2 &= \langle C^2\eta, \eta\rangle + \langle C_0^2\eta, \eta\rangle = \langle (B + B_0)\eta, \eta\rangle \\ &= \langle (B + B_0)(B - B_0)\xi, \eta\rangle = \langle (B^2 - B_0^2)\xi, \eta\rangle = \langle 0, \eta\rangle = 0.\end{aligned}$$

Therefore $C\eta = C_0\eta = 0$, $B\eta = C^2\eta = 0$, and $B_0\eta = 0$. Finally, we have

$$\|(B - B_0)\xi\|^2 = \langle (B - B_0)^2\xi, \xi\rangle = \langle (B - B_0)\eta, \xi\rangle = \langle 0, \xi\rangle = 0,$$

so that $B\xi = B_0\xi$. □

The spectral theorem, as we shall see, is really a theorem about $\sim$-representations of algebras $\mathfrak{C}_0(X)$ by bounded operators on H. We use some properties of $\sim$-representations established in §21 to study these representations. In the two following theorems, we find all $\sim$-representations of an algebra $\mathfrak{C}_0(X)$, in terms of certain measures on X.

---

[1] The STONE-WEIERSTRASS theorem [see footnote to p. 151] implies that if $\lambda \in K$ and λ is not in the spectrum of A, then the function $x \to (x - \lambda)^{-1}$ on X is the uniform limit on X of polynomials in x. Transferring this assertion back to the algebra $\mathfrak{A}$, we see that every operator $(A - \lambda I)^{-1}$ in $\mathfrak{B}(H)$ is the norm limit of polynomials in the operator A.

(C.36) Theorem. *Let X be a locally compact Hausdorff space, and let T be a cyclic $\sim$-representation of $\mathfrak{C}_0(X)$ by operators on a Hilbert space H. Then there are a nonnegative measure ι on X as in* §11 *and a linear isometry W of H onto $\mathfrak{L}_2(X, \iota)$ such that*

(i) $WT_fW^{-1}(\xi) = f\xi$ *for all* $\xi \in \mathfrak{L}_2(X, \iota)$ *and* $f \in \mathfrak{C}_0(X)$.

That is, T is equivalent to multiplication by f in $\mathfrak{L}_2(X, \iota)$.

Proof. Let ζ be a cyclic vector in H of norm 1. For $f \in \mathfrak{C}_0(X)$, let $p(f) = \langle T_f\zeta, \zeta\rangle$. The functional p is obviously linear and positive. Since p arises from a $\sim$-representation, (21.18.i) and (21.18.ii) hold for p. Since $\mathfrak{C}_0(X)$ is a Banach $\sim$-algebra, p is bounded (21.20); *i.e.*, $|p(f)| \leqq \alpha \|f\|_u$ for all $f \in \mathfrak{C}_0(X)$. If $f \in \mathfrak{C}_0^+(X)$, there is a $g \in \mathfrak{C}_0^+(X)$ such that $g^2 = g^\sim g = f$. Hence we have $p(f) = p(g^\sim g) \geqq 0$. This means that p is nonnegative in the sense of (11.4) and so there is a measure ι as constructed in §11 for which

$$p(f) = \langle T_f\zeta, \zeta\rangle = \int_X f\, d\iota \quad \text{for all} \quad f \in \mathfrak{C}_{00}(X). \tag{1}$$

For every compact subset F of X, it is easy to see that $\iota(F) \leqq \alpha$; hence we have $\iota(X) \leqq \alpha < \infty$. Therefore the mapping $f \to \int_X f\, d\iota$ is continuous in the norm topology of $\mathfrak{C}_0(X)$; as p is also continuous in this topology, equality (1) must persist throughout $\mathfrak{C}_0(X)$. For $f \in \mathfrak{C}_0(X)$, let $W(T_f\zeta)$ be the function $f \in \mathfrak{C}_0(X) \subset \mathfrak{L}_2(X, \iota)$. It is easy to see that W is a linear isometry of $\{T_f\zeta : f \in \mathfrak{C}_0(X)\}$ [which is by hypothesis a dense linear subspace of H] onto $\mathfrak{C}_0(X)$, which we regard as a subspace of $\mathfrak{L}_2(X, \iota)$. Since $\mathfrak{C}_0(X)$ is dense in $\mathfrak{L}_2(X, \iota)$ (12.10), (B.59) implies that W can be extended to a linear isometry carrying H onto $\mathfrak{L}_2(X, \iota)$. For every $g \in \mathfrak{C}_0(X)$, we have

$$W^{-1}g = T_g\zeta;$$

hence

$$T_fW^{-1}g = T_fT_g\zeta = T_{fg}\zeta \quad \text{for} \quad f \in \mathfrak{C}_0(X);$$

finally

$$WT_fW^{-1}g = fg.$$

Since $\mathfrak{C}_0(X)$ is dense in $\mathfrak{L}_2(X, \iota)$, this proves (i). □

(C.37) Theorem. *Let X be a locally compact Hausdorff space and let T be any $\sim$-representation of $\mathfrak{C}_0(X)$ by operators on a Hilbert space H such that for every nonzero ξ in H, there is an $f \in \mathfrak{C}_0(X)$ for which $T_f\xi \neq 0$. Then T is equivalent to a representation S of the following sort. There is a locally compact Hausdorff space Y which is the union of pairwise disjoint homeomorphs $\tau_\gamma(X)$ of X and in which a subset is open if and only if its intersection with each $\tau_\gamma(X)$ is open. There is a nonnegative, possibly in-*

finite, measure ι *on* Y *as constructed in* §11. *For* $f \in \mathfrak{C}_0(X)$, *let* ϱf *be the function in* $\mathfrak{C}(Y)$ *such that* $\varrho f(\tau_\gamma(x)) = f(x)$. *Then* S_f *is the operator on* $\mathfrak{L}_2(Y, \iota)$ *such that* $S_f \xi = (\varrho f) \cdot \xi$ *for all* $\xi \in \mathfrak{L}_2(Y, \iota)$.

Proof. Let N and $\{H_\gamma\}_{\gamma \in \Gamma}$ be as in (21.13), for the representation T of the algebra $\mathfrak{C}_0(X)$. Our hypothesis implies that $N = \{0\}$. For each $\gamma \in \Gamma$ and $f \in \mathfrak{C}_0(X)$, let $T_{f,\gamma}$ denote $T_f | H_\gamma$, and let T_γ be the representation $f \to T_{f,\gamma}$ of $\mathfrak{C}_0(X)$ by operators on H_γ. Since T_γ is cyclic, (C.36) shows that there is a finite, nonnegative measure ι_γ as in §11 on X such that T_γ is equivalent to multiplication in $\mathfrak{L}_2(X, \iota_\gamma)$: let W_γ be the linear isometry of H_γ onto $\mathfrak{L}_2(X, \iota_\gamma)$ such that (C.36.i) holds.

Since $N = \{0\}$, H is the Hilbert space direct sum $\bigoplus_{\gamma \in \Gamma} H_\gamma$. We form a corresponding concrete realization of the direct sum $\bigoplus_{\gamma \in \Gamma} \mathfrak{L}_2(X, \iota_\gamma)$ as follows. For each $\gamma \in \Gamma$ we consider a homeomorphism τ_γ of X onto a space $\tau_\gamma(X)$ such that the spaces $\tau_\gamma(X)$ are pairwise disjoint. Write Y for $\bigcup_{\gamma \in \Gamma} \tau_\gamma(X)$, and let $A \subset Y$ be open if and only if $A \cap \tau_\gamma(X)$ is open for all $\gamma \in \Gamma$. It is trivial that Y is a locally compact Hausdorff space in which each $\tau_\gamma(X)$ is open and closed. For $\psi \in \mathfrak{C}_{00}(Y)$, let ψ_γ be $\psi | \tau_\gamma(X)$, and let $I(\psi) = \sum_{\gamma \in \Gamma} \int_X (\psi_\gamma \circ \tau_\gamma)\, d\iota_\gamma$. It is easy to see that I is a nonnegative linear functional on $\mathfrak{C}_{00}(Y)$ in the sense of §11, and that the corresponding measure ι on Y is in an obvious sense the sum of the measures ι_γ. For ξ in $\mathfrak{L}_2(Y, \iota)$, let ξ_γ be the function equal to ξ on $\tau_\gamma(X)$ and 0 elsewhere on Y. The space of all ξ_γ's is identifiable *qua* Hilbert space with $\mathfrak{L}_2(X, \iota_\gamma)$, and we can [again in an obvious sense] look at $\mathfrak{L}_2(Y, \iota)$ as the Hilbert space direct sum of the spaces $\mathfrak{L}_2(X, \iota_\gamma)$. Each element η in the original Hilbert space H has a unique representation $\sum_{\gamma \in \Gamma} \eta_\gamma$, where each η_γ is in H_γ. Now define $W(\eta)$ as $\sum_{\gamma \in \Gamma} W_\gamma(\eta_\gamma)$, where we regard $\mathfrak{L}_2(X, \iota_\gamma)$ as imbedded as a direct summand in $\mathfrak{L}_2(Y, \iota)$. Plainly W is a linear isometry of H onto $\mathfrak{L}_2(Y, \iota)$. For $\xi \in \mathfrak{L}_2(Y, \iota)$, we have

$$W^{-1}(\xi) = W^{-1}\Big(\sum_{\gamma \in \Gamma} \xi_\gamma\Big) = \sum_{\gamma \in \Gamma} W_\gamma^{-1}(\xi_\gamma)$$

so that

$$T_f W^{-1}(\xi) = \sum_{\gamma \in \Gamma} T_{f,\gamma} W_\gamma^{-1}(\xi_\gamma).$$

Applying W to this identity, we obtain

$$W T_f W^{-1}(\xi) = \sum_{\gamma \in \Gamma} W_\gamma T_{f,\gamma} W_\gamma^{-1}(\xi_\gamma).$$

Since $W_\gamma T_{f,\gamma} W_\gamma^{-1}(\xi_\gamma) = (f \circ \tau_\gamma^{-1}) \cdot \xi_\gamma$ for $\gamma \in \Gamma$, we see that

$$W T_f W^{-1}(\xi) = \sum_{\gamma \in \Gamma} (f \circ \tau_\gamma^{-1}) \cdot \xi_\gamma = (\varrho f) \cdot \xi,$$

which in turn implies the present theorem. $\square$

(C.38) Note. We can deal with the case $N \neq \{0\}$, but in a totally uninteresting way. A measure μ can plainly be imposed upon X or the union of a number of replicas of X, say Y, so that N is identifiable with $\mathfrak{L}_2(Y, \mu)$. Then T restricted to N is plainly equivalent with multiplication in $\mathfrak{L}_2(Y, \mu)$ by the function 0.

(C.39) Discussion. The concrete realization of the representation $f \to T_f$ described in Theorem (C.37) can be used to write the operators T_f as "integrals" of certain projections. All notation in the present discussion is as in (C.37). Since H and $\mathfrak{L}_2(Y, \iota)$ are connected by a linear isometry under which T_f and multiplication by ϱf are equivalent, we lose no generality by considering only $\mathfrak{L}_2(Y, \iota)$ and multiplication by ϱf: an abstract formulation appears in (C.40) *infra*.

For a subset A of X, let ϱA be the subset $\bigcup_{\gamma \in \Gamma} \tau_\gamma(A)$. Let $\mathscr{S}$ be the family of all subsets of Y having the form ϱE for *Borel* subsets E of X. For every $\varrho E \in \mathscr{S}$, the characteristic function $\xi_{\varrho E}$ is a bounded Borel measurable function on Y, so that $\xi_{\varrho E} \cdot \zeta$ is in $\mathfrak{L}_2(Y, \iota)$ for all $\zeta \in \mathfrak{L}_2(Y, \iota)$. The mapping

$$\zeta \to \xi_{\varrho E} \cdot \zeta = M_E(\zeta) \tag{1}$$

is plainly an idempotent Hermitian operator on $\mathfrak{L}_2(Y, \iota)$; that is, M_E is a projection, and it is projection onto the subspace of all $\zeta \in \mathfrak{L}_2(Y, \iota)$ that vanish off of ϱE.[1] The following identities are immediate consequences of (1):

$$M_{E_1 \cup \cdots \cup E_n \cup \cdots} = \sum_{n=1}^{\infty} M_{E_n} \tag{2}$$

if $E_1, \ldots, E_n, \ldots$ are pairwise disjoint Borel sets in X;

$$M_{E_1} M_{E_2} \cdots M_{E_k} = M_{E_1 \cap \cdots \cap E_k} \tag{3}$$

for any Borel sets $E_1, \ldots, E_k$ in X;

$$M_X = I, \tag{4}$$

where I is the identity operator;

$$M_\varnothing = 0, \tag{5}$$

where 0 is the zero operator.

Now consider any $f \in \mathfrak{C}_0(X)$ and any $\varepsilon > 0$. There is plainly a partition $E_1, \ldots, E_m$ of X into Borel sets such that $|f(x) - f(x')| < \varepsilon$ for x, x' in E_j. Let α_j be any complex number such that $|f(x) - \alpha_j| < \varepsilon$ for all $x \in E_j$.

[1] We can obviously define the mapping $\zeta \to \xi_A \cdot \zeta$ for every ι-measurable subset A of Y. For our present purpose, we wish to consider only sets ϱE.

Then for every $\zeta \in \mathfrak{L}_2(Y, \iota)$, we have

$$\int_Y \left|\varrho f \cdot \zeta - \left(\sum_{j=1}^m \alpha_j \xi_{\varrho E_j}\right)\zeta\right|^2 d\iota$$

$$\leqq \left[\sup\left\{\left|\varrho f(y) - \sum_{j=1}^m \alpha_j \xi_{\varrho E_j}(y)\right| : y \in Y\right\}\right]^2 \int_Y |\zeta|^2 d\iota;$$

equivalently,

$$\left\|S_f\zeta - \left(\sum_{j=1}^m \alpha_j M_{E_j}\right)\zeta\right\|_2 \leqq \varepsilon \|\zeta\|_2. \tag{6}$$

The inequality (6) is an abstract formulation of the spectral theorem. That is, there is a mapping $E \to M_E$ of the Borel subsets of X into the family of all projections on $\mathfrak{L}_2(Y, \iota)$ satisfying (2), (3), (4), and (5) such that S_f is arbitrarily close to linear combinations of the projections M_E, in the sense of (6).

The projections M_E have another important property. Let B be *any* bounded operator on $\mathfrak{L}_2(Y, \iota)$ that commutes with all of the operators S_f. Then B commutes with all M_E. To prove this, let ζ and η be any elements of $\mathfrak{L}_2(Y, \iota)$. Since $\langle S_f B\zeta, \eta\rangle = \langle BS_f\zeta, \eta\rangle = \langle S_f\zeta, B^\sim\eta\rangle$, we have

$$\int_Y (\varrho f)(B\zeta)\bar{\eta}\, d\iota = \int_Y (\varrho f)\zeta\overline{(B^\sim\eta)}\, d\iota. \tag{7}$$

We wish to prove that

$$\int_{\varrho E} (B\zeta)\bar{\eta}\, d\iota = \int_{\varrho E} \zeta\overline{(B^\sim\eta)}\, d\iota \tag{8}$$

for all Borel sets $E \subset X$. This will show that $\langle M_E B\zeta, \eta\rangle = \langle M_E\zeta, B^\sim\eta\rangle$ for all ζ and η, and hence that $M_E B = BM_E$. Let $\varphi = (B\zeta)\bar{\eta} - \zeta\overline{(B^\sim\eta)}$. Then φ belongs to $\mathfrak{L}_1(Y, \iota)$,

$$\int_Y (\varrho f)\varphi\, d\iota = 0 \tag{7'}$$

for all $f \in \mathfrak{C}_0(X)$, and we wish to prove that

$$\int_{\varrho E} \varphi\, d\iota = 0 \tag{8'}$$

for all Borel sets E in X. Assume that $\int_{\varrho E} \varphi\, d\iota = \alpha \neq 0$ for some Borel set $E \subset X$. The function φ is 0 outside of the union of a countable subfamily of the sets $\tau_y(X)$: for convenience, we write $\{\tau_n(X)\}_{n=1}^\infty$ for this family. Since $\sum_{n=1}^\infty \int_{\tau_n(X)} |\varphi|\, d\iota < \infty$, there is a positive integer m for which $\sum_{n=m+1}^\infty \int_{\tau_n(X)} |\varphi|\, d\iota < \frac{|\alpha|}{2}$. For each integer $n = 1, \ldots, m$, there are a compact set $F_n \subset E$ and an open set $U_n \supset E$ such that $\int_{\tau_n(U_n \cap F_n')} |\varphi|\, d\iota < \frac{|\alpha|}{2m}$. Let $U = \bigcap_{n=1}^m U_n$ and $F = \bigcup_{n=1}^m F_n$, and choose a g in $\mathfrak{C}_{00}(X)$ satisfying $g(F) = 1$,

$g(U')=0$, and $g(X)\subset[0,1]$. Plainly, we have

$$\left.\begin{aligned}\left|\int\limits_{\bigcup\limits_{n=m+1}^{\infty}\tau_n(X)}(\varrho g-\xi_{\varrho E})\,\varphi\,d\iota\right| &\leqq \sum_{n=m+1}^{\infty}\int\limits_{\tau_n(X)}|\varrho g-\xi_{\varrho E}|\cdot|\varphi|\,d\iota\\ &\leqq \sum_{n=m+1}^{\infty}\int\limits_{\tau_n(X)}|\varphi|\,d\iota<\frac{|\alpha|}{2},\end{aligned}\right\}\tag{9}$$

and for $n=1,\ldots,m$, we have

$$\left|\int\limits_{\tau_n(X)}(\varrho g-\xi_{\varrho E})\,\varphi\,d\iota\right|\leqq\int\limits_{\tau_n(U\cap F')}|\varphi|\,d\iota\leqq\int\limits_{\tau_n(U_n\cap F_n')}|\varphi|\,d\iota<\frac{|\alpha|}{2m}\tag{10}$$

Relations (9) and (10) imply that

$$\left|\int\limits_{Y}(\varrho g-\xi_{\varrho E})\,\varphi\,d\iota\right|<|\alpha|.$$

Since $\int\limits_{\varrho E}\varphi\,d\iota=\alpha$, it follows that $\int\limits_{Y}(\varrho g)\,\varphi\,d\iota\neq 0$, contrary to (7′).

(C.40) Theorem. *Let X be a locally compact Hausdorff space, and let $f\to T_f$ be any $\sim$-representation of $\mathfrak{C}_0(X)$ by operators on a Hilbert space H. Then there is a mapping $E\to P_E$ of the Borel sets $E\subset X$ onto projections in H with the following properties:*

(i) $P_{E_1\cup E_2\cup\cdots\cup E_n\cup\cdots}=\sum\limits_{n=1}^{\infty}P_{E_n}$ *if the sets $E_1, E_2,\ldots,E_n,\ldots$ are pairwise disjoint;*

(ii) $P_{E_1\cap E_2}=P_{E_1}P_{E_2}$;

(iii) *the projections P_E commute with all bounded operators B that commute with all operators T_f.*

Furthermore, if $\{E_1,\ldots,E_m\}$ is any partition of X into Borel sets such that $|f(x)-f(x')|<\varepsilon$ for $x, x'\in E_j$ and $|f(x)-\alpha_j|<\varepsilon$ for $x\in E_j$ ($j=1, 2\ldots,m$), then

(iv) $\left\|T_f\zeta-\sum\limits_{j=1}^{m}\alpha_j P_{E_j}\zeta\right\|\leqq\varepsilon\|\zeta\|$

for all $\zeta\in H$. Finally, if ζ,η are any elements of H, the mapping $E\to\langle P_E\zeta,\eta\rangle=\mu_{\zeta,\eta}(E)$ is a complex measure on X in the sense of §14, and

(v) $\langle T_f\zeta,\eta\rangle=\int\limits_X f\,d\mu_{\zeta,\eta}$

for all $f\in\mathfrak{C}_0(X)$.

(C.41) Theorem. *Let $\mathfrak{A}$ be any commutative algebra of bounded operators on a Hilbert space H such that $A^{\sim}\in\mathfrak{A}$ whenever $A\in\mathfrak{A}$ and such that $\mathfrak{A}$ is closed in the norm topology for the algebra of all bounded operators on H. Let X be the structure space of $\mathfrak{A}$ and $\hat{A}$ the Fourier transform of $A\in\mathfrak{A}$; thus $\{\hat{A}:A\in\mathfrak{A}\}=\mathfrak{C}_0(X)$. Then all of the assertions of (C.40) hold with T_f replaced everywhere by A and the mapping $f\to T_f$ replaced by the identity mapping $A\to A$. The relation (C.40.iv) becomes the following.*

Let $E_1, \ldots, E_m$ *be any partition of* X *into Borel sets such that* $|\hat{A}(x) - \hat{A}(x')| < \varepsilon$ *for* $x, x' \in E_j$, *and* $|\hat{A}(x) - \alpha_j| < \varepsilon$ *for* $x \in E_j$ $(j = 1, 2, \ldots, m)$. *Then*

(iv) $\left\| A\zeta - \sum_{j=1}^{m} \alpha_j P_{E_j}\zeta \right\| \leq \varepsilon \|\zeta\|$

for all $\zeta \in H$. *The relation* (C.40.v) *becomes*

(v) $\langle A\zeta, \eta \rangle = \int_X \hat{A}\, d\mu_{\zeta,\eta}$.

Having come this far, we may as well state the classical form of the spectral theorem for *bounded* Hermitian operators.

(C.42) Theorem. *Let* B *be an Hermitian operator on a Hilbert space* H. *There is a one-parameter family of projections* $\{P_t : t \in R\}$ *on* H *with the following properties:*

(i) $P_t P_u = P_u$ *if* $u \leq t$;

(ii) *every* P_t *commutes with every operator that commutes with* B;

(iii) *for every* $\zeta \in H$, $\lim_{\varepsilon \downarrow 0} \|P_{t-\varepsilon}\zeta - P_t\zeta\| = 0$;

(iv) *there are real numbers* a *and* b, $a < b$, *such that* $P_t = 0$ *for* $t \leq a$ *and* $P_t = I$ *for* $t \geq b$;

(v) *for every* $\zeta \in H$ *and every subdivision* $a = t_0 < t_1 < \cdots < t_m = b$ *such that* $\max\{|t_j - t_{j-1}| : j = 1, \ldots, m\} < \varepsilon$, *we have*

$$\left\| B\zeta - \sum_{j=1}^{m} t_{j-1}(P_{t_j} - P_{t_{j-1}})\zeta \right\| < \varepsilon \|\zeta\|;$$

(vi) *for* $\zeta, \eta \in H$, *we have* $\langle B\zeta, \eta \rangle = \int_a^b t\, d\langle P_t\zeta, \eta \rangle$, *the integral being an ordinary Riemann-Stieltjes integral.*

Proof. Let $\mathfrak{A}$ be the smallest norm closed subalgebra of $\mathfrak{B}(H)$ containing the operators B and I. Arguing as in the proof of (C.35), we see that the structure space X of $\mathfrak{A}$ can be identified with a compact subset of R, where we have $\hat{B}(x) = x$ for all $x \in X$. Choose real numbers a and b such that $a \leq \min X$ and $b > \max X$. Let $P_t = P_{[a,t[\cap X}$, where P_E is the projection operator defined in (C.39) and (C.40). All assertions of the present theorem now follow from (C.41). □

(C.43) Note. We could also give the classical spectral theorem for a bounded normal operator on H; the assertion and proof are obvious from what has already been done.

Bibliography

ABE, MAKOTO: [1] Über Automorphismen der lokal-kompakten abelschen Gruppen. Proc. Imp. Acad. Tokyo **16**, 59–62 (1940).

ALEKSANDROV, A. D. [Александров, А. Д.]: [1] On groups with an invariant measure [О группах с инвариантной мерой]. Dokl. Akad. Nauk SSSR, N. S. **34**, 5–9 (1942).

ALEKSANDROV, P. S. [Александров, П. С.]: [1] Combinatorial topology [Комбинаторная топология]. OGIZ, Gostehizdat, Moscow, 1947. English translation, Graylock Press, Rochester, New York, 1956.

—, and H. HOPF: [1] Topologie I. Grundlehren der Mathematischen Wissenschaften, Bd. XLV. Berlin: Springer 1935.

ALEXANDER, JAMES W.: [1] On the characters of discrete abelian groups. Ann. of Math. (2) **35**, 389–395 (1934).

—, and LEON W. COHEN: [1] A classification of the homology groups of compact spaces. Ann. of Math. (2) **33**, 538–566 (1932).

—, and LEO ZIPPIN: [1] Discrete abelian groups and their character groups. Ann. of Math. (2) **36**, 71–85 (1935).

ANZAI, HIROTADA, and SHIZUO KAKUTANI: [1] Bohr compactifications of a locally compact abelian group. I, II. Proc. Imp. Acad. Tokyo **19**, 476–480, 533–539 (1943).

APOSTOL, TOM M.: [1] Mathematical analysis. Reading, Mass.: Addison-Wesley Publ. Co. Inc. 1957.

ARENS, RICHARD F.: [1] Topologies for homeomorphism groups. Amer. J. Math. **68**, 593–610 (1946).

ARHANGEL'SKIĬ, A. [Архангельский, А.]: [1] On the identity of the dimension ind G and dim G for locally bicompact groups [О совпадении размерности ind G и dim G для локально бикомпактных групп]. Dokl. Akad. Nauk SSSR, N. S. **132**, 980–981 (1960).

AUBERT, K. E.: [1] Convex ideals in ordered group algebras and the uniqueness of the Haar measure. Math. Scand. **6**, 181–188 (1958).

BAER, REINHOLD: [1] Zur Topologie der Gruppen. J. Reine Angew. Math. **160**, 208–226 (1929).

—, and FRIEDRICH LEVI: [1] Stetige Funktionen in topologischen Räumen. Math. Z. **34**, 110–130 (1932).

BALCERZYK, S.: [1] On algebraically compact groups of I. Kaplansky. Fund. Math. **44**, 91–93 (1957).

BEURLING, ARNE: [1] Sur les intégrales de Fourier absolument convergentes et leur application à une transformation fonctionnelle. Nionde Skandinaviska Matematikerkongr., Helsingfors, 1938, 345–366. Helsingfors: Mercator 1939.

BIRKHOFF, GARRETT: [1] A note on topological groups. Compositio Math. **3**, 427–430 (1936).

— [2] Lattice theory. Amer. Math. Soc. Colloquium Publications, Vol. XXV, 2nd Edition. Amer. Math. Soc., New York, N. Y., 1948.

BOCHNER, SALOMON: [1] Vorlesungen über Fouriersche Integrale. Leipzig: Akademische Verlagsgesellschaft 1932. Reprinted by Chelsea Publ. Co., New York, 1948.

BOURBAKI, NICOLAS: [1] Éléments de mathématique. III. Première partie. Les structures fondamentales de l'analyse. Livre III. Topologie générale. Chap. III.

Groupes topologiques (Théorie élémentaire). Actualités Sci. et Ind. 916. Paris: Hermann & Cie. 1942.

BOURBAKI, NICOLAS: [2] Éléments de mathématique. XIII. Première partie. Les structures fondamentales de l'analyse. Livre VI. Intégration. Chap. I–IV. Actualités Sci. et Ind. 1175. Paris: Hermann & Cie. 1952.

— [3] Éléments de mathématique. XXI. I. Les structures fondamentales de l'analyse. Livre VI. Intégration. Chap. V. Intégration des mesures. Actualités Sci. et Ind. 1244. Paris: Hermann & Cie. 1956.

— [4] Éléments de mathématique. Première partie. Livre VI. Intégration. Chap. VI. Intégration vectorielle. Actualités Sci. et Ind. 1281. Paris: Hermann & Cie. 1959.

BRACONNIER, JEAN: [1] Sur les groupes topologiques localement compacts. J. Math. Pures Appl., N.S. **27**, 1–85 (1948).

BUCK, R. CREIGHTON: [1] Generalized group algebras. Proc. Nat. Acad. Sci. U.S.A. **36**, 747–749 (1950).

— [2] Operator algebras and dual spaces. Proc. Amer. Math. Soc. **3**, 681–687 (1952).

CALDERÓN, ALBERTO P.: [1] A general ergodic theorem. Ann. of Math. (2) **58**, 182–191 (1953).

CARTAN, ÉLIE: [1] La Théorie des Groupes Finis et Continus et l'Analysis Situs. Mémorial des Sci. Math., Fasc. 42. Paris: Gauthier-Villars & Cie. 1930.

CARTAN, HENRI: [1] Sur la mesure de Haar. C.R. Acad. Sci. Paris **211**, 759–762 (1940).

— [2] Sur les fondements de la théorie du potentiel. Bull. Soc. Math. France **69**, 71–96 (1941).

—, and JEAN DIEUDONNÉ: [1] Notes de tératopologie. — III. Rev. Sci. **77**, 413–414 (1939).

—, and ROGER GODEMENT: [1] Théorie de la dualité et analyse harmonique dans les groupes abéliens localement compacts. Ann. Sci. École Norm. Sup. (3) **64**, 79–99 (1947).

CHEVALLEY, CLAUDE: [1] Theory of Lie groups. I. Princeton, N.J.: Princeton University Press 1946.

COHEN, LEON W.: See ALEXANDER, JAMES W.

COHEN, PAUL J.: [1] On a conjecture of Littlewood and idempotent measures. Amer. J. Math. **82**, 191–212 (1960).

COTLAR, MISCHA, and R. RICABARRA: [1] On the existence of characters in topological groups. Amer. J. Math. **76**, 375–388 (1954).

DANIELL, P. J.: [1] A general form of integral. Ann. of Math. (2) **19**, 279–294 (1917–1918).

DANTZIG, D. VAN: [1] Über topologisch homogene Kontinua. Fund. Math. **15**, 102–125 (1930).

— [2] Studien over topologische Algebra. Dissertation, Amsterdam 1931.

— [3] Zur topologischen Algebra. I. Komplettierungstheorie. Math. Ann. **107**, 587–626 (1932).

— [4] Zur topologischen Algebra. II. Abstrakte b_ν-adische Ringe. Compositio Math. **2**, 201–223 (1935).

— [5] Nombres universels ou ν!-adiques avec une introduction sur l'algèbre topologique. Ann. Sci. École Norm. Sup. (3) **53**, 275–307 (1936).

— [6] Zur topologischen Algebra. III. Brouwersche und Cantorsche Gruppen. Compositio Math. **3**, 408–426 (1936).

DAUBLEBSKY VON STERNECK, R.: [1] Ableitung zahlentheoretischer Relationen mit Hilfe eines mehrdimensionalen Systemes von Gitterpunkten. Monatsh. Math. **5**, 255–266 (1894).

Davis, Harry F.: [1] A note on Haar measure. Proc. Amer. Math. Soc. **6**, 318–321 (1955).

Day, Mahlon M.: [1] The spaces L^p with $0<p<1$. Bull. Amer. Math. Soc. **46**, 816–823 (1940).

— [2] Ergodic theorems for Abelian semi-groups. Trans. Amer. Math. Soc. **51**, 399–412 (1942).

— [3] Amenable groups. Preliminary report. Bull. Amer. Math. Soc. **56**, 46 (1950).

— [4] Means for the bounded functions and ergodicity of the bounded representations of semi-groups. Trans. Amer. Math. Soc. **69**, 276–291 (1950).

— [5] Amenable semigroups. Illinois J. Math. **1**, 509–544 (1957).

— [6] Normed linear spaces. Ergebnisse Math., neue Folge, H. 21. Berlin-Göttingen-Heidelberg: Springer 1958.

Devinatz, A., and A. E. Nussbaum: [1] Real characters of certain semi-groups with applications. Duke Math. J. **28**, 221–237 (1961).

Dieudonné, Jean: [1] Sur les fonctions continues numériques definies dans un produit de deux espaces compacts. C. R. Acad. Sci. Paris **205**, 593–595 (1937).

— [2] Notes de tératopologie. — I. Rev. Sci. **77**, 39–40 (1939).

— See Cartan, Henri.

Dirichlet, P. G. L.: [1] Sur la convergence des séries trigonométriques qui servent à représenter une fonction arbitraire entre des limites données. J. Reine Angew. Math. **4**, 157–169 (1829).

Dixmier, Jacques: [1] Les moyennes invariantes dans les semi-groupes et leurs applications. Acta Sci. Math. Szeged **12**. Leopoldo Fejér et Frederico Riesz LXX annos natis dedicatus, Pars A, 213–227 (1950).

— [2] Quelques propriétés des groupes abéliens localement compacts. Bull. Sci. Math. **81**, 38–48 (1957).

Dunford, Nelson, and Jacob T. Schwartz: [1] Linear operators. Part I: General theory. New York, N. Y.: Interscience Publishers, Inc. 1958.

Eckmann, Beno: [1] Über monothetische Gruppen. Comment. Math. Helv. **16**, 249–263 (1943–1944).

Edwards, Robert E.: [1] A theory of Radon measures on locally compact spaces. Acta Math. **89**, 133–164 (1953).

Erdős, Paul: [1] The dimension of the rational points in Hilbert space. Ann. of Math. (2) **41**, 734–736 (1940).

Følner, Erling: [1] Generalization of a theorem of Bogolioùboff to topologica. Abelian groups with an appendix on Banach mean values in non-Abelian groups. Math. Scand. **2**, 5–18 (1954).

— [2] Note on a generalization of a theorem of Bogolioùboff. Math. Scand. **2**, 224–226 (1954).

— [3] On groups with full Banach mean value. Math. Scand. **3**, 243–254 (1955).

— [4] Note on groups with and without full Banach mean value. Math. Scand. **5**, 5–11 (1957).

Freudenthal, Hans: [1] Einige Sätze über topologische Gruppen. Ann. of Math. (2) **37**, 46–56 (1936).

— [2] Topologische Gruppen mit genügend vielen fastperiodischen Funktionen. Ann. of Math. (2) **37**, 57–77 (1936).

— [3] Entwicklungen von Räumen und ihren Gruppen. Compositio Math. **4**, 145–234 (1937).

— [4] Die Haarschen Orthogonalsysteme von Gruppencharakteren im Lichte der Pontrjaginschen Dualitätstheorie. Compositio Math. **5**, 354–356 (1938).

FROBENIUS, G.: [1] Über Gruppencharaktere. Sitzungsber. Preuss. Akad. Wiss., phys.-math. Kl. **1896**, 985—1021.

FUCHS, L.: [1] Abelian groups. Budapest: Publishing House of the Hungarian Academy of Sciences 1958. London: Pergamon Press 1960.

— [2] On character groups of discrete Abelian groups. Acta Math. Acad. Sci. Hungar. **10**, 133—140 (1959).

GELBAUM, B., G.K. KALISCH and J.M.H. OLMSTED: [1] On the embedding of topological semigroups and integral domains. Proc. Amer. Math. Soc. **2**, 807—821 (1951).

GEL'FAND, I. M. [Гельфанд, И. М.]: [1] Über absolut konvergente trigonometrische Reihen und Integrale. Mat. Sb., N.S. **9** (**51**), 51—66 (1941).

—, and M.A. NAĬMARK [M. A. Наймарк]: [1] On the imbedding of normed rings into the ring of operators in Hilbert space [О включении нормированного кольца в кольцо операторов в гилбертовом пространстве]. Mat. Sb., N.S. **12** (**54**), 197—217 (1943).

— [2] Normed rings with involution and their representations [Нормированные кольца с инволюцией и их представления]. Izv. Akad. Nauk SSSR, Ser. Mat. **12**, 445—480 (1948).

—, and D. A. RAĬKOV [Д. А. Райков]: [1] On the theory of characters of commutative topological groups [К теории характеров коммутативных топологических групп]. Dokl. Akad. Nauk SSSR, N.S. **28**, 195—198 (1940).

— [2] Irreducible unitary representations of locally bicompact groups [Неприводимые унитарные представления локально бикомпактных групп]. Mat. Sb., N.S. **13** (**55**), 301—316 (1943).

— [3] Irreducible unitary representations of locally bicompact groups [Неприводимые унитарные представления локально бикомпактных групп]. Dokl. Akad. Nauk SSSR, N.S. **42**, 199—201 (1944).

GLICKSBERG, IRVING: [1] The representation of functionals by integrals. Duke Math. J. **19**, 253—261 (1952).

GODEMENT, ROGER: [1] Les fonctions de type positif et la théorie des groupes. Trans. Amer. Math. Soc. **63**, 1—84 (1948).

— [2] Sur la théorie des représentations unitaires. Ann. of Math. (2) **53**, 68—124 (1951).

— See CARTAN, HENRI.

GRAEV, M.I. [Граев, М. И.]: [1] On the theory of complete direct products of groups [К теории польных прямых произведений групп]. Mat. Sb., N.S. **17** (**59**), 85—104 (1945).

— [2] Free topological groups [Свободные топологические группы]. Izv. Akad. Nauk SSSR, Ser. Mat. **12**, 279—324 (1948).

— [3] On free products of topological groups [О свободных произведениях топологических групп]. Izv. Akad. Nauk SSSR, Ser. Mat. **14**, 343—354 (1950).

— [4] Theory of topological groups. I. Norms and metrics on groups. Complete groups. Free topological groups. [Теория топологических групп. I. Нормы и метрики на группах. Польные группы. Свободные топологические группы.] Uspehi Mat. Nauk, N.S. **5**, vyp. 2 (36), 3—56 (1950).

GUREVIČ, ANNA [Гуревич, Анна]: [1] Unitary representation in Hilbert space of a compact topological group [Унитарное представление в гильбертовом пространстве компактной топологической группы]. Mat. Sb., N.S. **13** (**55**), 79—86 (1943).

HAAR, ALFRÉD: [1] Über unendliche kommutative Gruppen. Math. Z. **33**, 129—159 (1931). Also in Gesammelte Arbeiten, pp. 486—516. Budapest: Akadémiai Kiadó 1959.

HAAR, ALFRÉD: [2] Über die Gruppencharaktere gewisser unendlichen Gruppen. Acta Litt. Sci. Szeged **5**, 172—186 (1932). Also in Gesammelte Arbeiten, pp. 562—576. Budapest: Akadémiai Kiadó 1959.

— [3] Der Maßbegriff in der Theorie der kontinuierlichen Gruppen. Ann. of Math. (2) **34**, 147—169 (1933). Also in Gesammelte Arbeiten, pp. 600—622. Budapest: Akadémiai Kiadó 1959.

HALL, MARSHALL, JR.: [1] A topology for free groups and related groups. Ann. of Math. (2) **52**, 127—139 (1950).

HALMOS, PAUL R.: [1] Comment on the real line. Bull. Amer. Math. Soc. **50**, 877—878 (1944).

— [2] Measure theory. New York, N.Y.: D. Van Nostrand Co. 1950.

— [3] Introduction to Hilbert space and the theory of spectral multiplicity. New York, N.Y.: Chelsea Publ. Co. 1951.

—, and HANS SAMELSON: [1] On monothetic groups. Proc. Nat. Acad. Sci. U.S.A. **28**, 254—258 (1942).

—, and HERBERT E. VAUGHAN: [1] The marriage problem. Amer. J. Math. **72**, 214—215 (1950).

HAMEL, GEORG: [1] Eine Basis aller Zahlen und die unstetigen Lösungen der Funktionalgleichung: $f(x+y)=f(x)+f(y)$. Math. Ann. **60**, 459—462 (1905).

HARDY, G.H., J.E. LITTLEWOOD and G. PÓLYA: [1] Inequalities. Cambridge: Cambridge University Press 1934.

HARRISON, D.K.: [1] Infinite abelian groups and homological methods. Ann. of Math. (2) **69**, 366—391 (1959).

HARTMAN, S., and A. HULANICKI: [1] Les sous-groupes purs et leurs duals. Fund. Math. **45**, 71—77 (1957).

—, and JAN MYCIELSKI: [1] On the imbedding of topological groups into connected topological groups. Colloq. Math. **5**, 167—169 (1958).

—, and C. RYLL-NARDZEWSKI: [1] Zur Theorie der lokal-kompakten abelschen Gruppen. Colloq. Math. **4**, 157—188 (1957).

HAUSDORFF, FELIX: [1] Mengenlehre, 3. Aufl. Göschens Lehrbücherei, Bd. 7. Berlin u. Leipzig: W. de Gruyter 1935. Reprinted by Dover Publications, New York, N.Y., 1944.

HELSON, HENRY: [1] On the ideal structure of group algebras. Ark. Mat. **2**, 83—86 (1952).

HENSEL, KURT: [1] Zahlentheorie. Berlin u. Leipzig: G. J. Göschensche Verlagshandlung 1913.

HEWITT, EDWIN: [1] Integral representation of certain linear functionals. Ark. Mat. **2**, 269—282 (1952).

— [2] Linear functionals on almost periodic functions. Trans. Amer. Math. Soc. **74**, 303—322 (1953).

— [3] Remarks on the inversion of Fourier-Stieltjes transforms. Ann. of Math. (2) **57**, 458—474 (1953).

— [4] Compact monothetic semigroups. Duke Math. J. **23**, 447—457 (1956).

— [5] A survey of abstract harmonic analysis. Some aspects of analysis and probability, pp. 105—168. Surveys in Applied Mathematics, Vol. 4. New York, N.Y.: John Wiley and Sons, Inc. 1958.

—, and KARL R. STROMBERG: [1] A remark on Fourier-Stieltjes transforms. An. Acad. Brasil. Ci. **34**, 175—180 (1962).

— [2] Real and abstract analysis. Graduate Texts in Mathematics, Vol. 25, 2nd printing, corrected. Berlin, Heidelberg, New York: Springer 1969.

HEWITT, EDWIN, and HERBERT S. ZUCKERMAN: [1] A group-theoretic method in approximation theory. Ann. of Math. (2) **52**, 557—567 (1950).
— [2] On convolution algebras. Proc. Intern. Cong. Mathematicians Cambridge, Mass., 1950. Amer. Math. Soc., Providence, R. I., 1952. Vol. I, p. 455.
— [3] Finite dimensional convolution algebras. Acta Math. **93**, 67—119 (1955).
— [4] The l_1-algebra of a commutative semigroup. Trans. Amer. Math. Soc. **83**, 70—97 (1956).
— [5] The irreducible representations of a semigroup related to the symmetric group. Illinois J. Math. **1**, 188—213 (1957).
HILLE, EINAR, and RALPH S. PHILLIPS: [1] Functional analysis and semi-groups. Amer. Math. Soc. Colloquium Publications, Vol. XXXI. Revised Edition. Amer. Math. Soc., Providence, R. I., 1957.
HOFMANN, KARL HEINRICH: [1] Eine Bemerkung über die zentralen Untergruppen in zusammenhängenden topologischen Gruppen. Arch. Math. **9**, 33—38 (1958).
HOPF, H.: See ALEKSANDROV, P. S.
HULANICKI, A.: [1] On locally compact topological groups of power of continuum. Fund. Math. **44**, 156—158 (1957).
— [2] Algebraic characterization of abelian divisible groups which admit compact topologies. Fund. Math. **44**, 192—197 (1957).
— [3] Algebraic structure of compact abelian groups. Bull. Acad. Polon. Sci., Sér. Sci. Math. Astr. Phys. **6**, 71—73 (1958).
— [4] On the topological structure of 0-dimensional topological groups. Fund. Math. **46**, 317—320 (1959).
— See HARTMAN, S.
HURWITZ, ADOLF: [1] Über die Erzeugung der Invarianten durch Integration. Nachr. k. Ges. Göttingen, math.-phys. Kl. **1897**, 71—90. Also in Mathematische Werke, Bd. II, pp. 546—564. Basel: Birkhäuser 1933.
ISIWATA, TAKESI: [1] Duality of topological groups. Sci. Rep. Tokyo Kyoiku Daigaku, Sect. A. **5**, 82—87 (1955).
IVANOVSKIĬ, L. N. [Ивановский, Л. Н.]: [1] On an hypothesis of P. S. Aleksandrov [Об одной гипотезе П. С. Александрова]. Dokl. Akad. Nauk SSSR, N. S. **123**, 785—786 (1958).
IWAMURA, TURANE: See YOSIDA, KÔSAKU.
IWASAWA, KENKICHI: [1] Einige Sätze über freie Gruppen. Proc. Imp. Acad. Tokyo **19**, 272—274 (1943).
— [2] On some types of topological groups. Ann. of Math. (2) **50**, 507—558 (1949).
JONES, F. BURTON: [1] On the first countability axiom for locally compact Hausdorff spaces. Colloq. Math. **7**, 33—34 (1959).
KAKUTANI, SHIZUO: [1] Über die Metrisation der topologischen Gruppen. Proc. Imp. Acad. Tokyo **12**, 82—84 (1936).
— [2] On the uniqueness of Haar's measure. Proc. Imp. Acad. Tokyo **14**, 27—31 (1938).
— [3] Concrete representation of abstract (M)-spaces (A characterization of the space of continuous functions). Ann. of Math. (2) **42**, 994—1024 (1941).
— [4] On cardinal numbers related with a compact abelian group. Proc. Imp. Acad. Tokyo **19**, 366—372 (1943).
— [5] Free topological groups and infinite direct product topological groups. Proc. Imp. Acad. Tokyo **20**, 595—598 (1944).
—, and KUNIHIKO KODAIRA: [1] Über das Haarsche Maß in der lokal bikompakten Gruppe. Proc. Imp. Acad. Tokyo **20**, 444—450 (1944).

KAKUTANI, SHIZUO, and JOHN C. OXTOBY: [1] Construction of a non-separable invariant extension of the Lebesgue measure space. Ann. of Math. (2) **52**, 580—590 (1950).

— See ANZAI, HIROTADA.

KALISCH, G. K.: See GELBAUM, B.

KAMPEN, E. R. VAN: [1] Locally bicompact Abelian groups and their character groups. Ann. of Math. (2) **36**, 448—463 (1935).

KAPLAN, SAMUEL: [1] Extensions of the Pontrjagin duality. I. Infinite products. Duke Math. J. **15**, 649—658 (1948).

— [2] Extensions of the Pontrjagin duality. II. Direct and inverse sequences. Duke Math. J. **17**, 419—435 (1950).

KAPLANSKY, IRVING: [1] Infinite abelian groups. Ann Arbor, Michigan: University of Michigan Press 1954.

KASUGA, TAKASHI: [1] On the isomorphism of topological groups. Proc. Japan Acad. **29**, 435—438 (1953).

KAWADA, YUKIYOSI: [1] Über den Mittelwert der meßbaren fastperiodischen Funktionen auf einer Gruppe. Proc. Imp. Acad. Tokyo **19**, 264—266 (1943).

KEINER, HORST: [1] Verallgemeinerte fastperiodische Funktionen auf Halbgruppen. Arch. Math. **8**, 129—134 (1957).

KELLEY, JOHN L.: [1] Note on a theorem of Kreĭn and Mil'man. J. Osaka Inst. of Sci. and Tech. **3**, 1—2 (1951).

— [2] General topology. New York, N. Y.: D. Van Nostrand Co. 1955.

KEMPERMAN, J. H. B.: [1] On small sumsets in an abelian group. Acta Math. **103**, 63—88 (1960).

KERTÉSZ, A., and T. SZELE: [1] On the existence of non-discrete topologies in infinite abelian groups. Publ. Math. Debrecen **3**, 187—189 (1953).

KESTEN, HARRY: [1] Full Banach mean values on countable groups. Math. Scand. **7**, 146—156 (1959).

KNESER, MARTIN: [1] Summenmengen in lokalkompakten abelschen Gruppen. Math. Z. **66**, 88—110 (1956).

KODAIRA, KUNIHIKO: See KAKUTANI, SHIZUO.

KOOSIS, PAUL: [1] An irreducible unitary representation of a compact group is finite dimensional. Proc. Amer. Math. Soc. **8**, 712—715 (1957).

KÖTHE, GOTTFRIED: [1] Topologische lineare Räume I. Grundlehren der Mathematischen Wissenschaften, Bd. 107. Berlin-Göttingen-Heidelberg: Springer 1960.

KREĬN, M. G. [Крейн, М. Г.]: [1] Integral equations on the half-line with kernel depending on the difference of the arguments [Интегральные уравнения на полупрямой с ядром, зависящим от разности аргументов]. Uspehi Mat. Nauk **13**, vyp. 5 (83), 3—120 (1958).

KRISTENSEN, LEIF: [1] Invariant metrics in coset spaces. Math. Scand. **6**, 33—36 (1958).

KRULL, WOLFGANG: [1] Über separable, insbesondere kompakte separable Gruppen. J. Reine Angew. Math. **184**, 19—48 (1942).

KUROŠ, A. G. [Курош, А. Г.]: [1] Theory of groups, 2nd Edition. [Теория групп, издание второе]. Gostehizdat, Moscow, 1953.

KUZ'MINOV, V. [Кузьминов, В.]: [1] On an hypothesis of P. S. Aleksandrov in the theory of topological groups [О гипотезе П. С. Александрова в теории топологических групп]. Dokl. Akad. Nauk SSSR, N. S. **125**, 727—729 (1959).

LEBESGUE, HENRI: [1] Leçons sur l'intégration et la recherche des fonctions primitives. Paris: Gauthier-Villars & Cie. 1904.

LEHMER, D.H.: [1] Arithmetic of double series. Trans. Amer. Math. Soc. **33**, 945—957 (1931).
LEJA, F.: [1] Sur la notion du groupe abstrait topologique. Fund. Math. **9**, 37—44 (1927).
LEPTIN, HORST: [1] Über eine Klasse linear kompakter abelscher Gruppen. I, II. Abh. Math. Sem. Univ. Hamburg **19**; 23—40 (1954); 221—243 (1955).
— [2] Abelsche Gruppen mit kompakten Charaktergruppen und Dualitätstheorie gewisser linear topologischer abelscher Gruppen. Abh. Math. Sem. Univ. Hamburg **19**, 244—263 (1955).
— [3] Zur Dualitätstheorie projektiver Limites abelscher Gruppen. Abh. Math. Sem. Univ. Hamburg **19**, 264—268 (1955).
— [4] Bemerkung zu einem Satz von S. Kaplan. Arch. Math. **6**, 139—144 (1955).
LEVI, FRIEDRICH: See BAER, REINHOLD.
LITTLEWOOD, J.E.: See HARDY, G.H.
LOOMIS, LYNN H.: [1] Haar measure in uniform structures. Duke Math. J. **16**, 193—208 (1949).
— [2] An introduction to abstract harmonic analysis. Princeton, N. J.: D. Van Nostrand Co. 1953.
— [3] Linear functionals and content. Amer. J. Math. **76**, 168—182 (1954).
LORENTZ, G. G.: [1] A contribution to the theory of divergent sequences. Acta Math. **80**, 167—190 (1948).
ŁOŚ, J.: [1] Abelian groups that are direct summands of every abelian group which contains them as a pure subgroup. Bull. Acad. Polon. Sci., Sér. Sci. Math. Astr. Phys. **4**, 73 (1956).
— [2] Abelian groups that are direct summands of every abelian group which contains them as pure subgroups. Fund. Math. **44**, 84—90 (1957).
LUTHAR, INDAR S.: [1] Uniqueness of the invariant mean on an Abelian semigroup. Illinois J. Math. **3**, 28—44 (1959).
LYUBARSKIĬ, G. YA. [Любарский, Г. Я.]: [1] On the integration in the mean of almost periodic functions on topological groups [Об интегрировании в среднем почти периодических функций на топологических группах]. Uspehi Mat. Nauk, N.S. **3**, vyp. 3 (25), 195—201 (1948).
MAAK, WILHELM: [1] Fastperiodische Funktionen. Grundlehren der Mathematischen Wissenschaften, Bd. LXI. Berlin-Göttingen-Heidelberg: Springer 1950.
— [2] Integralmittelwerte von Funktionen auf Gruppen und Halbgruppen. J. Reine Angew. Math. **190**, 34—48 (1952).
MACBEATH, A.M.: [1] On the measure of product sets in a topological group. J. London Math. Soc. **35**, 403—407 (1960).
MACKEY, GEORGE W.: [1] A remark on locally compact Abelian groups. Bull. Amer. Math. Soc. **52**, 940—944 (1946).
— [2] The Laplace transform for locally compact Abelian groups. Proc. Nat. Acad. Sci. U.S.A. **34**, 156—162 (1948).
— [3] Borel structures in groups and their duals. Trans. Amer. Math. Soc. **85**, 134—165 (1957).
MARKOV, A.A. [Марков, А. А.]: [1] Über endlich-dimensionale Vektorräume. Ann. of Math. (2) **36**, 464—506 (1935).
— [2] On free topological groups [О свободных топологических группах]. Dokl. Akad. Nauk SSSR, N.S. **31**, 299—301 (1941).
— [3] On the existence of periodic connected topological groups [О существовании периодических связных топологических групп]. Izv. Akad. Nauk SSSR, Ser. Mat. **8**, 225—232 (1944).

MARKOV, A. A. [Марков, А. А.]: [4] On free topological groups [О свободных топологических группах]. Izv. Akad. Nauk SSSR, Ser. Mat. **9**, 3—64 (1945).
MONTGOMERY, DEANE, and LEO ZIPPIN: [1] Topological transformation groups. Interscience Tracts in Pure and Applied Mathematics 1. New York, N.Y.: Interscience Publishers, Inc. 1955.
MUNN, W. D.: [1] On semigroup algebras. Proc. Cambridge Philos. Soc. **51**, 1—15 (1955).
MUNROE, M. E.: [1] Introduction to measure and integration. Cambridge, Mass.: Addison-Wesley Publ. Co. 1953.
MYCIELSKI, JAN: See HARTMAN, S.
NACHBIN, LEOPOLDO: [1] On the finite dimensionality of every irreducible representation of a compact group. Proc. Amer. Math. Soc. **12**, 11—12 (1961).
NAĬMARK, M. A. [Наймарк, М. А.]: [1] Normed rings [Нормированные кольца]. Gostehizdat, Moscow, 1956. Revised German translation: Normierte Algebren. Berlin: VEB Deutscher Verlag der Wiss. 1959. English translation: Normed rings. Groningen: P. Noordhoff N.V. 1959.
— See GEL'FAND, I. M.
NAKAYAMA, TADASI: [1] Note on free topological groups. Proc. Imp. Acad. Tokyo **19**, 471—475 (1943).
NEUMANN, JOHN VON: [1] Zur Prüferschen Theorie der idealen Zahlen. Acta Litt. Sci. Szeged **2**, 193—227 (1926).
— [2] Zur allgemeinen Theorie des Maßes. Fund. Math. **13**, 73—116 (1929).
— [3] Die Einführung analytischer Parameter in topologischen Gruppen. Ann. of Math. (2) **34**, 170—190 (1933).
— [4] Almost periodic functions in a group. I. Trans. Amer. Math. Soc. **36**, 445—492 (1934).
— [5] Zum Haarschen Maß in topologischen Gruppen. Compositio Math. **1**, 106—114 (1934).
— [6] The uniqueness of Haar's measure. Mat. Sb., N. S. **1** (**43**), 721—734 (1936).
—, and EUGENE P. WIGNER: [1] Minimally almost periodic groups. Ann. of Math. (2) **41**, 746—750 (1940).
NIKODÝM, OTTON M.: [1] Sur une généralisation des intégrales de M. J. Radon. Fund. Math. **15**, 131—179 (1930).
NUMAKURA, KATSUMI: [1] On bicompact semigroups. Math. J. Okayama Univ. **1**, 99—108 (1952).
NUSSBAUM, A. E.: See DEVINATZ, A.
OLMSTED, J. M. H.: See GELBAUM, B.
OXTOBY, JOHN C.: [1] Invariant measures in groups which are not locally compact. Trans. Amer. Math. Soc. **60**, 215—237 (1946).
— See KAKUTANI, SHIZUO.
PALEY, R. E. A. C.: See WIENER, NORBERT.
PASYNKOV, B. [Пасынков, Б.]: [1] On the coincidence of different definitions of dimension for locally bicompact groups [О совпадении различных определений размерности для локально бикомпактных групп]. Dokl. Akad. Nauk SSSR, N. S. **132**, 1035—1037 (1960).
PÉRÈS, JOSEPH: See VOLTERRA, VITO.
PETER, F.: See WEYL, HERMANN.
PHILLIPS, RALPH S.: See HILLE, EINAR.
PITT, H. R.: See WIENER, NORBERT.
PÓLYA, G.: See HARDY, G. H.
PONTRYAGIN, L. S. [Понтрягин, Л. С.]: [1] Der allgemeine Dualitätssatz für abgeschlossene Mengen. Verhandlungen des Internat. Math.-Kongr. Zürich, 1932. Zürich u. Leipzig: Orell Füssli, Bd. II, pp. 195—197.

PONTRYAGIN, L. S. [Понтрягин, Л. С.]: [2] Sur les groupes topologiques compacts et le cinquième problème de M. Hilbert. C. R. Acad. Sci. Paris **198**, 238—240 (1934).

— [3] Sur les groupes abéliens continus. C. R. Acad. Sci. Paris **198**, 328—330 (1934).

— [4] The theory of topological commutative groups. Ann. of Math. (2) **35**, 361—388 (1934).

— [5] Linear representations of compact topological groups [Линейные представления компактных топологических групп]. Mat. Sb., N.S. **1** (**43**), 267—271 (1936).

— [6] Continuous groups [Непрерывные группы]. Moscow, 1938. English translation: Topological groups. Princeton, N. J.: Princeton University Press 1939.

— [7] Continuous groups, 2nd Edition [Непрерывные группы, издание второе]. Gostehizdat, Moscow, 1954. German translation: Topologische Gruppen, I, II. Leipzig: B.G. Teubner 1957 and 1958.

PRESTON, GERALD C.: [1] On locally compact totally disconnected Abelian groups and their character groups. Pacific J. Math. **6**, 121—134 (1956).

PRÜFER, HEINZ: [1] Neue Begründung der algebraischen Zahlentheorie. Math. Ann. **94**, 198—243 (1925).

RADON, JOHANN: [1] Theorie und Anwendungen der absolut additiven Mengenfunktionen. Sitzungsber. math.-naturwiss. Kl. Akad. Wiss. Wien **122**, Abt. IIa (1913), 1295—1438.

RAĬKOV, D. A. [Райков, Д. А.]: [1] Generalized duality theorem for commutative groups with an invariant measure [Обобщенный закон двойственности для коммутативных групп с инвариантной мерой]. Dokl. Akad. Nauk SSSR, N.S. **30**, 589—591 (1941).

— [2] A new proof of the uniqueness of Haar's measure [Новое доказательство единственности меры Хаара]. Dokl. Akad. Nauk SSSR, N.S. **34**, 211—212 (1942).

— [3] Harmonic analysis on commutative groups with Haar measure and the theory of characters [Гармонический анализ на коммутативных группах и теория характеров]. Trudy Mat. Inst. Steklov **14** (1945). German translation in Sowjetische Arbeiten zur Funktionalanalysis, pp. 11—87. Berlin: Kultur u. Fortschritt 1954.

— [4] On the theory of normed rings with involution [К теории нормированных колец с инволюцией]. Dokl. Akad. Nauk SSSR, N.S. **54**, 387—390 (1946).

— See GEL'FAND, I.M.

RAIMI, RALPH A.: [1] On a theorem of E. Følner. Math. Scand. **6**, 47—49 (1958).

RICABARRA, R.: See COTLAR, MISCHA.

RICKART, CHARLES E.: [1] General theory of Banach algebras. Princeton, N. J.: D. Van Nostrand Co. 1960

RIESZ, FRÉDÉRIC: [1] Sur les opérations fonctionnelles linéaires. C. R. Acad. Sci. Paris **149**, 974—977 (1909). Also in Gesammelte Arbeiten Bd. I, pp. 400—402. Budapest: Akadémiai Kiadó 1960.

— [2] Untersuchungen über Systeme integrierbarer Funktionen. Math. Ann. **69**, 449—497 (1910). Also in Gesammelte Arbeiten, Bd. I, pp. 441—489. Budapest: Akadémiai Kiadó 1960.

— [3] Les systèmes d'équations linéaires à une infinité d'inconnues. Paris: Gauthier-Villars & Cie. 1913. Also in Gesammelte Arbeiten, Bd. II, pp. 829—1016. Budapest: Akadémiai Kiadó 1960.

RISS, JEAN: [1] Représentations continues des groupes topologiques abéliens dans le groupe additif des nombres réels. C. R. Acad. Sci. Paris **224**, 987—988 (1947).

RISS, JEAN: [2] Sur les représentations réelles des groupes topologiques abéliens. C. R. Acad. Sci. Paris **224**, 1095—1097 (1947).
— [3] Éléments de calcul différentiel et théorie des distributions sur les groupes abéliens localement compacts. Acta Math. **89**, 45—105 (1953).
ROBERTSON, WENDY: [1] Thesis, Cambridge University 1954.
ROBISON, GERSON B.: [1] Invariant integrals over a class of Banach spaces. Pacific J. Math. **4**, 123—150 (1954).
ROSEN, WILLIAM G.: [1] On invariant means over compact semigroups. Proc. Amer. Math. Soc. **7**, 1076—1082 (1956).
RUDIN, WALTER: [1] Independent perfect sets in groups. Michigan Math. J. **5**, 159—161 (1958).
— [2] Measure algebras on abelian groups. Bull. Amer. Math. Soc. **65**, 227—247 (1959).
RYLL-NARDZEWSKI, C.: See HARTMAN, S.
SAKS, STANISŁAW: [1] Theory of the integral, 2nd Edition. Monografie Matematyczne, Tom VII. Warszawa-Lwów 1937.
SAMELSON, HANS: See HALMOS, PAUL R.
SCHÖNEBORN, HEINZ: [1] Über Linearformenmoduln unendlichen Ranges. I. Primäre, kompakte Linearformenmoduln. II. Nichtarchimedisch perfekt bewertete, operatorreduzierte Linearformenmoduln. J. Reine Angew. Math. **189**, 168—185, 193—203 (1952).
— [2] Über gewisse Topologien in Abelschen Gruppen. I, II. Math. Z. **59**, 455—473 (1954); **60**, 17—30 (1954).
— [3] Über den Zusammenhang zwischen Dualitäts- und Vollständigkeitseigenschaften bei gewissen topologischen Abelschen Gruppen. Math. Z. **65**, 429—441 (1956).
SCHREIER, OTTO: [1] Abstrakte kontinuierliche Gruppen. Abh. Math. Sem. Univ. Hamburg **4**, 15—32 (1925).
SCHUR, ISSAI: [1] Neue Anwendungen der Integralrechnung auf Probleme der Invariantentheorie. I, II, III. Sitzungsber. Preuss. Akad. Wiss., phys.-math. Kl. **1924**, 189—208, 297—321, 346—355.
SCHWARTZ, JACOB T.: See DUNFORD, NELSON.
SCHWARZ, ŠTEFAN: [1] Theory of characters of commutative semigroups [Теория характеров коммутативных полугрупп]. Czechoslovak Math. J. **4** (**79**), 219—247 (1954).
— [2] On the existence of invariant measures on certain types of bicompact semigroups [О существовании инвариантных мер на некоторых типах бикомпактных полугрупп]. Czechoslovak Math. J. **7** (**82**), 165—182 (1957).
SEGAL, IRVING E.: [1] The group ring of a locally compact group. I. Proc. Nat. Acad. Sci. U.S.A. **27**, 348—352 (1941).
— [2] Irreducible representations of operator algebras. Bull. Amer. Math. Soc. **53**, 73—88 (1947).
— [3] The group algebra of a locally compact group. Trans. Amer. Math. Soc. **61**, 69—105 (1947).
SHIGA, KÔJI: [1] Representations of a compact group on a Banach space. J. Math. Soc. Japan **7**, 224—248 (1955).
SILVERMAN, ROBERT J.: [1] Invariant linear functions. Trans. Amer. Math. Soc. **81**, 411—424 (1956).
— [2] Means on semigroups and the Hahn-Banach extension property. Trans. Amer. Math. Soc. **83**, 222—237 (1956).
— [3] Invariant means and cones with vector interiors. Trans. Amer. Math. Soc. **88**, 75—79 (1958).

SMITH, MARIANNE FREUNDLICH: [1] The Pontrjagin duality theorem in linear spaces. Ann. of Math. (2) **56**, 248—253 (1952).

ŠREĬDER, YU. A. [Шрейдер, Ю. А.]: [1] Construction of maximal ideals in rings of completely additive measures [Строение максимальных идеалов в кольцах вполне аддитивных мер]. Dokl. Akad. Nauk SSSR, N. S. **63**, 359—361 (1948).

— [2] Construction of maximal ideals in rings of measures with convolution [Строение максимальных идеалов в кольцах мер со сверткой]. Mat. Sb., N. S. **27** (**69**), 297—318 (1950).

STEINHAUS, HUGO: [1] Additive und stetige Funktionaloperationen. Math. Z. **5**, 186—221 (1919).

STONE, A. H.: [1] Paracompactness and product spaces. Bull. Amer. Math. Soc. **54**, 977—982 (1948).

STROMBERG, KARL R.: See HEWITT, EDWIN.

STRUBLE, RAIMOND A.: [1] Almost periodic functions on locally compact groups. Proc. Nat. Acad. Sci. U.S.A. **39**, 122—126 (1953).

ŚWIERCZKOWSKI, S.: [1] Measures equivalent to the Haar measure. Proc. Glasgow Math. Assoc. **4**, 157—162 (1960).

— [2] Jacobians for measures in coset spaces. Proc. Glasgow Math. Assoc. **4**, 208—212 (1960).

SZELE, T.: See KERTÉSZ, A.

SZ.-NAGY, BÉLA: [1] Sur la mesure invariante dans des groupes topologiques. C. R. Acad. Sci. Paris **202**, 1248—1250 (1936).

— [2] Zur Theorie der Charaktere Abelscher Gruppen. Math. Ann. **114**, 373—384 (1937).

TARSKI, ALFRED: [1] Ideale in vollständigen Mengenkörpern. I. Fund. Math. **32**, 45—63 (1939).

TÔYAMA, HIRAKU: [1] On Haar measure of some groups. Proc. Japan Acad. **24**, 13—16 (1948).

URBANIK, K.: [1] On the isomorphism of Haar measures. Fund. Math. **46**, 277—284 (1959).

VAUGHAN, HERBERT E.: See HALMOS, PAUL R.

VILENKIN, N. YA. [Виленкин, Н. Я.]: [1] Direct decompositions of topological groups. I, II, III, Corrections [Прямые разложения топологических групп. I, II, III, Исправления]. Mat. Sb., N. S. **19** (**61**), 85—154, 311—340 (1946); **22** (**64**), 191—192 (1948); **29** (**71**), 519—528 (1951).

— [2] On the theory of general topological groups [К теории общих топологических групп]. Dokl. Akad. Nauk SSSR, N. S. **58**, 1573—1575 (1947).

— [3] On the theory of weakly separable groups [К теории слабо сепарабельных групп]. Mat. Sb., N. S. **22** (**64**), 135—177 (1948).

— [4] Fibered abelian topological groups and their character theory [Волокнистые абелевые топологические группы и их теория характеров]. Mat. Sb., N. S. **24** (**66**), 189—226 (1949).

— [5] On the classification of separable and coseparable topological abelian groups [О классификации сепарабельных и косепарабельных топологических абелевых групп]. Mat. Sb., N. S. **27** (**69**), 85—102 (1950).

— [6] On the theory of general noncommutative topological groups [К теории общих некоммутативных топологических групп]. Dokl. Akad. Nauk SSSR, N. S. **71**, 1013—1015 (1950).

— [7] Theory of topological groups. II. Direct products. Direct sums of groups of rank 1. Locally bicompact abelian groups. Fibered and weakly separable groups [Теория топологических групп. II. Прямые произведения. Прямые суммы групп ранга 1. Локально бикомпактные абелевы группы. Волокнистые и слабо сепарабельные группы]. Uspehi Mat. Nauk, N. S. **5**, vyp. 4 (38), 19—74 (1950).

Vilenkin, N.Ya. [Виленкин, Н.Я.]: [8] Direct and inverse spectra of topological groups and their character theories [Прямые и обратные спектры топологических групп и их теория характеров]. Izv. Akad. Nauk SSSR, Ser. Mat. **15**, 503—532 (1951).

— [9] Direct operations on topological groups [Прямые операции над топологическими группами]. Mat. Sb., N. S. **29** (**71**), 371—402 (1951).

— [10] On the classification of zero-dimensional locally compact periodic abelian groups without elements of finite order [К классификации нульмерных локально компактных периодических абелевых групп без элементов конечного порядка]. Mat. Sb., N. S. **28** (**70**), 503—536 (1951).

— [11] On the existence of locally compact groups with given Ulm factors [О существовании локально компактных групп с заданными ульмовскими факторами]. Mat. Sb., N. S. **29** (**71**), 13—30 (1951).

— [12] On the isomorphism of locally compact zero-dimensional abelian groups with isomorphic factors [Об изоморфизме локально компактных нульмерных абелевых групп с изоморфными факторами]. Mat. Sb., N. S. **29** (**71**), 31—62 (1951).

— [13] Theory of characters of topological abelian groups with a given boundedness [Теория характеров топологических абелевых групп с заданной ограниченностью]. Izv. Akad. Nauk SSSR, Ser. Mat. **15**, 439—462 (1951).

— [14] Generalized normal divisors of topological groups and their applications to combinatorial topology [Обобщенные нормальные делители топологических групп и их приложения к комбинаторной топологии]. Trudy Moskov. Mat. Obšč. **3**, 15—88 (1954).

— [15] On the classification of zero-dimensional locally compact abelian groups with an everywhere dense set of elements of finite order [К классификации нульмерных локально компактных абелевых групп со всюду плотным множеством элементов конечного порядка]. Mat. Sb., N. S. **34** (**76**), 55—80 (1954).

— [16] On a class of locally compact zero-dimensional topological groups [Об одном классе локально компактных нульмерных топологических групп]. Mat. Sb., N. S. **40** (**82**), 479—496 (1956).

— [17] On the dyadicity of the group space of bicompact commutative groups [О диадичности группого пространства бикомпактных групп]. Uspehi Mat. Nauk, N. S. **13**, vyp. 6 (84), 79—80 (1958).

Volterra, Vito, and Joseph Pérès: [1] Leçons sur la composition et les fonctions permutables. Paris: Gauthier-Villars & Cie. 1924.

Waerden, Bartel L. van der: [1] Algebra. Teil I. 4. Aufl. Grundlehren der Mathematischen Wissenschaften, Bd. XXXIII. Berlin-Göttingen-Heidelberg: Springer 1955.

Wallace, A. D.: [1] The structure of topological semigroups. Bull. Amer. Math. Soc. **61**, 95—112 (1955).

Weierstrass, Karl: [1] Über die analytische Darstellbarkeit sogenannter willkürlicher Funktionen reeller Argumente. Sitzungsber. Preuss. Akad. Wiss. **1885**, 633—640, 789—906. Also in Math. Werke, Bd. III, pp. 1—37. Berlin: Mayer u. Müller 1903.

Weil, André: [1] La mesure invariante dans les espaces de groupes et les espaces homogènes. Enseignement Math. **35**, 241 (1936).

WEIL, ANDRÉ: [2] Sur les groupes topologiques et les groupes mesurés. C. R. Acad. Sci. Paris **202**, 1147—1149 (1936).

— [3] Sur les espaces à structure uniforme et sur la topologie générale. Actualités Sci. et Ind. 551. Paris: Hermann & Cie. 1938.

— [4] L'intégration dans les groupes topologiques et ses applications. Actualités Sci. et Ind. 869, 1145. Paris: Hermann & Cie. 1941 and 1951.

WENDEL, JAMES G.: [1] Left centralizers and isomorphisms of group algebras. Pacific J. Math. **2**, 251—261 (1952).

WEYL, HERMANN: [1] Theorie der Darstellung kontinuierlicher halb-einfacher Gruppen durch lineare Transformationen. I, II, III, Nachtrag. Math. Z. **23**, 271—309 (1925); **24**, 328—376, 377—395, 789—791 (1926). Also in Selecta Hermann Weyl, pp. 262—366. Basel: Birkhäuser 1956.

— [2] Integralgleichungen und fastperiodische Funktionen. Math. Ann. **97**, 338—356 (1927). Also in Selecta Hermann Weyl, pp. 367—386. Basel: Birkhäuser 1956.

— [3] The classical groups, their invariants and representations. Princeton, N. J.: Princeton University Press 1939.

—, and F. PETER: [1] Die Vollständigkeit der primitiven Darstellungen einer geschlossenen kontinuierlichen Gruppe. Math. Ann. **97**, 737—755 (1927). Also in Selecta Hermann Weyl, pp. 387—404. Basel: Birkhäuser 1956.

WIENER, NORBERT, and R. E. A. C. PALEY: [1] Analytic properties of the characters of infinite Abelian groups. Verhandlungen des Internat. Math.-Kongr. Zürich, 1932. Zürich u. Leipzig: Orell Füssli, Bd. II, p. 95.

— [2] Characters of Abelian groups. Proc. Nat. Acad. Sci. U.S.A. **19**, 253—257 (1933).

—, and H. R. PITT: [1] On absolutely convergent Fourier-Stieltjes transforms. Duke Math. J. **4**, 420—436 (1938).

WIGNER, EUGENE P.: [1] On unitary representations of the inhomogeneous Lorentz group. Ann. of Math. (2) **40**, 149—204 (1939).

— See NEUMANN, JOHN VON.

YOSIDA, KÔSAKU, and TURANE IWAMURA: [1] Equivalence of two topologies of abelian groups. Proc. Imp. Acad. Tokyo **20**, 451—453 (1944).

YOUNG, W. H.: [1] On classes of summable functions and their Fourier series. Proc. Roy. Soc. London, Ser. A **87**, 225—229 (1912).

ZIPPIN, LEO: See ALEXANDER, JAMES W., and MONTGOMERY, DEANE.

ZUCKERMAN, HERBERT S.: See HEWITT, EDWIN.

ZYGMUND, ANTONI: [1] Trigonometric series, 2nd Edition. 2 Vols. Cambridge: Cambridge University Press 1959.

Index of symbols

Step (I) of the proof of Theorem (22.10), page 341. (I) In order to apply (22.7) in this proof, we need the following fact. Let h be a real- or complex-valued λ-measurable function on G. Then there is a λ-measurable function h' on G such that $h = h'$ locally λ-almost everywhere, and $h'\xi_B$ is Borel measurable for all σ-compact sets $B \subset G$.

We may suppose that h is nonnegative. By (5.7), G contains an open and closed σ-compact subgroup H. Let $\{x_\alpha H\}_{\alpha\in A}$ denote the distinct cosets of H, and let $h_\alpha = h\xi_{x_\alpha H}$. For each fixed $\alpha \in A$, we have $x_\alpha H = \bigcup_{n=1}^{\infty} F_n$, where $\{F_n\}_{n=1}^{\infty}$ is an increasing sequence of compact sets. Each function $\min(h_\alpha, n)\xi_{F_n}$ belongs to $\mathfrak{L}_1(G)$ and hence by (11.41), $\min(h_\alpha, n)\xi_{F_n} = g_n$ λ-almost everywhere for some Borel measurable function g_n. It follows that $\lim_{n\to\infty} g_n(x)$ exists and is equal to $h_\alpha(x)$ λ-almost everywhere. Let $h'_\alpha = \left(\lim_{n\to\infty} g_n\right)\xi_{x_\alpha H}$ for $\alpha \in A$.

Now for $x \in G$, we define $h'(x) = h'_\alpha(x)$, where $x \in x_\alpha H$. If B is σ-compact, there exist at most countably many distinct elements $\alpha_1, \alpha_2, \ldots, \alpha_n, \ldots$ in A such that $B \subset \bigcup_{n=1}^{\infty} x_{\alpha_n} H$. From the foregoing, $h' \cdot \sum_{n=1}^{\infty} \xi_{x_{\alpha_n} H}$ is Borel measurable and therefore $h'\xi_B$ is Borel measurable.

Index of authors and terms

Grundlehren der mathematischen Wissenschaften
Continued from page ii

81. Schneider: Einführung in die transzendenten Zahlen
82. Specht: Gruppentheorie
84. Conforto: Abelsche Funktionen und algebraische Geometrie
86. Richter: Wahrscheinlichkeitstheorie
88. Müller: Grundprobleme der mathematischen Theorie elektromagnetischer Schwingungen
89. Pfluger: Theorie der Riemannschen Flächen
90. Oberhettinger: Tabellen zur Fourier-Transformation
91. Prachar: Primzahlverteilung
93. Hadwiger: Vorlesungen über Inhalt, Oberfläche und Isoperimetrie
94. Funk: Variationsrechnung und ihre Anwendung in Physik und Technik
95. Maeda: Kontinuierliche Geometrien
97. Greub: Linear Algebra
98. Saxer: Versicherungsmathematik. 2. Teil
99. Cassels: An Introduction to the Geometry of Numbers
100. Koppenfels/Stallmann: Praxis der konformen Abbildung
101. Rund: The Differential Geometry of Finsler Spaces
103. Schütte: Beweistheorie
104. Chung: Markov Chains with Stationary Transition Probabilities
105. Rinow: Die innere Geometrie der metrischen Räume
106. Scholz/Hasenjaeger: Grundzüge der mathematischen Logik
107. Köthe: Topologische lineare Räume I
108. Dynkin: Die Grundlagen der Theorie der Markoffschen Prozesse
110. Dinghas: Vorlesungen über Funktionentheorie
111. Lions: Equations différentielles opérationnelles et problèmes aux limites
112. Morgenstern/Szabó: Vorlesungen über theoretische Mechanik
113. Meschkowski: Hilbertsche Räume mit Kernfunktion
114. MacLane: Homology
115. Hewitt/Ross: Abstract Harmonic Analysis. Vol. 1: Structure of Topological Groups, Integration Theory, Group Representations, Second Edition
116. Hörmander: Linear Partial Differential Operators
117. O'Meara: Introduction to Quadratic Forms
118. Schäike: Einführung in die Theorie der speziellen Funktionen der mathematischen Physik
119. Harris: The Theory of Branching Processes
120. Collatz: Funktionalanalysis und numerische Mathematik
121. Dynkin: Markov Processes
122. Dynkin: Markov Processes
123. Yosida: Functional Analysis
124. Morgenstern: Einführung in die Wahrscheinlichkeitsrechaung und mathematische Statistik
125. Ito/McKean: Diffusion Processes and Their Sample Paths
126. Lehto/Virtanen: Quasikonforme Abbildungen
127. Hermes: Enumerability, Decidability, Computability
128. Braun/Koecher: Jordan-Algebren
129. Nikodym: The Mathematical Apparatus for Quantum Theories
130. Morrey: Multiple Integrals in the Calculus of Variations
131. Hirzebruch: Topological Methods in Algebraic Geometry
132. Kato: Perturbation Theory for Linear Operators
133. Haupt/Künneth: Geometrische Ordnungen

Grundlehren der mathematischen Wissenschaften

134. Huppert: Endliche Gruppen I
135. Handbook for Automatic Computation. Vol. 1/Part a: Rutishauser: Description of ALGOL 60
136. Greub: Multilinear Algebra
137. Handbook for Automatic Computation. Vol. 1/Part b: Grau/Hill/Langmaack: Translation of ALGOL 60
138. Hahn: Stability of Motion
139. Mathematische Hilísmittel des Ingenieurs. Herausgeber: Sauer/Szabó. 1. Teil
140. Mathematische Hilísmittel des Ingenieurs. Herausgeber: Sauer/Szabó. 2. Teil
141. Mathematische Hilísmittel des Ingenieurs. Herausgeber: Sauer/Szabó. 3. Teil
142. Mathematische Hilísmittel des Ingenieurs. Herausgeber: Sauer/Szabó. 4. Teil
143. Schur/Grunsky: Vorlesungen über Invariantentheorie
144. Weil: Basic Number Theory
145. Butzer/Berens: Semi-Groups of Operators and Approximation
146. Treves: Locally Convex Spaces and Linear Partial Differential Equations
147. Lamotke: Semisimpliziale algebraische Topologie
148. Chandrasekharan: Introduction to Analytic Number Theory
149. Sario/Oikawa: Capacity Functions
150. Iosifescu/Theodorescu: Random Processes and Learning
151. Mandl: Analytical Treatment of One-Dimensional Markov Processes
152. Hewitt/Ross: Abstract Harmonic Analysis. Vol. 2: Structure and Analysis for Compact Groups. Analysis on Locally Compact Abellan Groups
153. Federer: Geometric Measure Theory
154. Singer: Bases in Banach Spaces I
155. Müller: Foundations of the Mathematical Theory of Electromagnetic Waves
156. van der Waerden: Mathematical Statistics
157. Prohorov/Rozanov: Probability Theory
158. Constantinescu/Comea: Potential Theory on Harmonic Spaces
159. Köthe: Topological Vector Spaces I
160. Agrest/Maksimov: Theory of Incomplete Cylindrical Functions and Their Applications. In preparation
161. Bhatia/Szegö: Stability Theory of Dynamical Systems
162. Nevanlinna: Analytic Functions
163. Stoer/Witzgall: Convexity and Optimization in Finite Dimensions I
164. Sario/Nakai: Classification Theory of Riemann Surfaces
165. Mitrinovic: Analytic Inequalities
166. Grothendieck/Dieudonné: Eléments de Géometrie Algébrique 1. En préparation
167. Chandrasekharan: Arithmetical Functions
168. Palamodov: Linear Differential Operators with Constant Coefficients
169. Rademacher: Topics in Analytic Number Theory. In preparation
170. Lions: Optimal Control Systems Governed by Partial Differential Equations
171. Singer: Best Approximation in Normed Linear Spaces by Elements of Linear Subspaces
172. Bühlmann: Mathematical Methods in Risk Theory. In preparation
173. Maeda/Maeda: Theory of Symmetric Lattices
174. Stiefel/Scheifele: Linear and Regular Celestial Mechanics. Perturbed Two-body Motion—Numerical Methods—Canonical Theory
175. Larsen: An Introduction to the Theory of Multipliers
176. Grauert/Remmert: Analytische Stellenalgebren
177. Flügge: Practical Quantum Mechanics I

Grundlehren der mathematischen Wissenschaften

178. Flügge: Practical Quantum Mechanics II
179. Giraud: Cohomologie non abélienne
180. Landkof: Foundations of Modern Potential Theory
181. Lions/Magenes: Non-Homogeneous Boundary Value Problems and Applications I
182. Lions/Magenes: Non-Homogeneous Boundary Value Problems and Applications II
183. Lions/Magenes: Non-Homogeneous Boundary Value Problems and Applications III
184. Rosenblatt: Markov Processes, Structure and Asymptotic Behavior
185. Rubinowicz: Sommerfeldsche Polynommethode
186. Handbook for Automatic Computation. Vol. 2. Wilkinson/Reinsch: Linear Algebra
187. Siegel/Moser: Lectures on Celestial Mechanics
188. Warner: Harmonic Analysis on Semi-Simple Lie Groups I
189. Warner: Harmonic Analysis on Semi-Simple Lie Groups II
190. Faith: Algebra: Rings, Modules, and Categories I
191. Faith: Algebra II: Ring Theory
192. Mal'cev: Algebraic Systems
193. Pólya/Szegö: Problems and Theorems in Analysis I
194. Igusa: Theta Functions
195. Berberian: Baer *-Rings
196. Athreya/Ney: Branching Processes
197. Benz: Vorlesungen über Geometric der Algebren
198. Gaal: Linear Analysis and Representation Theory
199. Nitsche: Vorlesungen über Minimalflächen
200. Doid: Lectures on Algebraic Topology
201. Beck: Continuous Flows in the Plane
202. Schmetterer: Introduction to Mathematical Statistics
203. Schoeneberg: Elliptic Modular Functions
204. Popov: Hyperstability of Control Systems
205. Nikol'skiĭ: Approximation of Functions of Several Variables and Imbedding Theorems
206. André: Homologie des Algèbres Commutatives
207. Donoghue: Monotone Matrix Functions and Analytic Continuation
208. Lacey: The Isometric Theory of Classical Banach Spaces
209. Ringel: Map Color Theorem
210. Gihman/Skorohod: The Theory of Stochastic Processes I
211. Comfort/Negrepontis: The Theory of Ultrafilters
212. Switzer: Algebraic Topology—Homotopy and Homology
213. Shafarevich: Basic Algebraic Geometry
214. van der Waerden: Group Theory and Quantum Mechanics
215. Schaefer: Banach Lattices and Positive Operators
216. Pólya/Szegö: Problems and Theorems in Analysis II
217. Stenström: Rings of Quotients
218. Gihman/Skorohod: The Theory of Stochastic Processes II
219. Duvant/Lions: Inequalities in Mechanics and Physics
220. Kirikov: Elements of the Theory of Representations
221. Mumford: Algebraic Geometry I: Complex Projective Varieties
222. Lang: Introduction to Modular Forms
223. Bergh/Löfström: Interpolation Spaces. An Introduction
224. Gilbarg/Trudinger: Elliptic Partial Differential Equations of Second Order
225. Schütte: Proof Theory

Grundlehren der mathematischen Wissenschaften

226. Karoubi: K-Theory. An Introduction
227. Grauert/Remmert: Theorie der Steinschen Räume
228. Segal/Kunze: Integrals and Operators
229. Hasse: Number Theory
230. Klingenberg: Lectures on Closed Geodesics
231. Lang: Elliptic Curves: Diophantine Analysis
232. Gihman/Skorohod: The Theory of Stochastic Processes III
233. Stroock/Varadhan: Multidimensional Diffusion Processes
234. Aigner: Combinatorial Theory
235. Dynkin/Yushkevich: Controlled Markov Processes
236. Grauert/Remmert: Theory of Stein Spaces
237. Köthe: Topological Vector Spaces II
238. Graham/McGehee: Essays in Commutative Harmonic Analysis
239. Elliott: Probabilistic Number Theory I
240. Elliott: Probabilistic Number Theory II
241. Rudin: Function Theory in the Unit Ball of C^R
242. Huppert/Blackburn: Finite Groups II
243. Huppert/Blackburn: Finite Groups III
244. Kubert/Lang: Modular Units
245. Cornfeld/Fomin/Sinai: Ergodic Theory
246. Naimark/Stern: Theory of Group Representations
247. Suzuki: Group Theory I
248. Suzuki: Group Theory II
249. Chung: Lectures from Markov Processes to Brownian Motion
250. Arnold: Geometrical Methods in the Theory of Ordinary Differential Equations
251. Chow/Hale: Methods of Bifurcation Theory
252. Aubin: Nonlinear Analysis on Manifolds. Monge-Ampère Equations
253. Dwork: Lectures on ρ-adic Differential Equations·
254. Freitag: Siegelsche Modulfunktionen
255. Lang: Complex Multiplication
256. Hörmander: The Analysis of Linear Partial Differential Operators I
257. Hörmander: The Analysis of Linear Partial Differential Operators II
258. Smoller: Shock Waves and Reaction-Diffusion Equations
259. Duren: Univalent Functions
260. Freidlin/Wentzell: Random Perturbations of Dynamical Systems
261. Bosch/Güntzer/Remmert: Non Archimedian Analysis—A System Approach to Rigid Analytic Geometry
262. Doob: Classical Potential Theory and Its Probabilistic Counterpart
263. Krasnosel'skiĭ/Zabreĭko: Geometrical Methods of Nonlinear Analysis
264. Aubin/Cellina: Differential Inclusions
265. Grauert/Remmert: Coherent Analytic Sheaves
266. de Rham: Differentiable Manifolds
267. Arbarello/Cornalba/Griffiths/Harris: Geometry of Algebraic Curves, Vol. I
268. Arbarello/Cornalba/Griffiths/Harris: Geometry of Algebraic Curves, Vol. II
269. Schapira: Microdifferential Systems in the Complex Domain
270. Scharlau: Quadratic and Hermitian Forms
271. Ellis: Entropy, Large Deviations, and Statistical Mechanics

Grundlehren der mathematischen Wissenschaften

272. Elliott: Arithmetic Functions and Integer Products
273. Nikol'skiĭ: Treatise on the Shift Operator
274. Hörmander: The Analysis of Linear Partial Differential Operators III
275. Hörmander: The Analysis of Linear Partial Differential Operators IV
276. Liggett: Interacting Particle Systems
277. Fulton/Lang: Riemann-Roch Algebra
278. Barr/Wells: Toposes, Triples and Theories
279. Bishop/Bridges: Constructive Analysis
280. Neukirch: Class Field Theory
281. Chandrasekharan: Elliptic Functions
282. Lelong/Gruman: Entire Functions of Several Complex Variables
283. Kodaira: Complex Manifolds and Deformation of Complex Structures
284. Finn: Equilibrium Capillary Surfaces
285. Burago/Zalgaller: Geometric Inequalities
286. Andrianov: Quadratic Forms and Hecke Operators
287. Maskit: Kleinian Groups
288. Jacod/Shiryaev: Limit Theorems for Stochastic Processes
289. Manin: Gauge Field Theory and Complex Geometry
290. Conway/Sloane: Sphere Packings, Lattices and Groups
291. Hahn/O'Meara: The Classical Groups and K-Theory
292. Kashiwara/Schapira: Sheaves on Manifolds
293. Revuz/Yor: Continuous Martingales and Brownian Motion
294. Knus: Quadratic and Hermitian Forms over Rings
295. Dierkes/Hildebrandt/Küster/Wohlrab: Minimal Surfaces I
296. Dierkes/Hildebrandt/Küster/Wohlrab: Minimal Surfaces II
297. Pastur/Figotin: Spectra of Random and Almost-Periodic Operators
298. Berline/Getzler/Vergne: Heat Kernels and Dirac Operators
299. Pommerenke: Boundary Behaviour of Conformal Maps
300. Orlik/Terao: Arrangements of Hyperplanes
301. Loday: Cyclic Homology
302. Lange/Birkenhake: Complex Abelian Varieties
303. DeVore/Lorentz: Constructive Approximation
304. Lorentz/v. Golitschek/Makovoz: Constructive Approximation. Advanced Problems
305. Hiriart-Urruty/Lemaréchal: Convex Analysis and Minimization Algorithms I. Fundamentals
306. Hiriart-Urruty/Lemaréchal: Convex Analysis and Minimization Algorithms II. Advanced Theory and Bundle Methods
307. Schwarz: Quantum Field Theory and Topology